Spon's Architects' and Builders' Price Book

2022

WGC

Ebook Set up and Use Instructions

Use of the ebook is subject to the single user licence at the back of this book.

Electronic access to your price book is now provided as an ebook on the VitalSource® Bookshelf platform. You can access it online or offline on your PC/Mac, smartphone or tablet. You can browse and search the content across all the books you've got, make notes and highlights and share these notes with other users.

Setting up

1. Create a VitalSource Bookshelf account at https://online.vitalsource.com/user/new (or log into your existing account if you already have one).

2. Retrieve the code by scratching off the security-protected label inside the front cover of this book. Log in to Bookshelf and click the **Redeem** menu at the top right of the screen. and Enter the code in the **Redeem code** box and press **Redeem**. Once the code has been redeemed your Spon's Price Book will download and appear in your **library**. N.B. the code in the scratch-off panel can only be used once, and has to be redeemed before end December 2022.

When you have created a Bookshelf account and redeemed the code you will be able to access the ebook online or offline on your smartphone, tablet or PC/Mac. Your notes and highlights will automatically stay in sync no matter where you make them.

Use ONLINE

1. Log in to your Bookshelf account at https://online.vitalsource.com).
2. Double-click on the title in your **library** to open the ebook.

Use OFFLINE

Download BookShelf to your PC, Mac, iOS device, Android device or Kindle Fire, and log in to your Bookshelf account to access your ebook, as follows:

On your PC/Mac
Go to https://support.vitalsource.com/hc/en-us/ and follow the instructions to download the free VitalSource Bookshelf app to your PC or Mac. Double-click the VitalSource Bookshelf icon that appears on your desktop and log into your Bookshelf account. Select **All Titles** from the menu on the left – you should see your price book on your Bookshelf. If your Price Book does not appear, select **Update Booklist** from the **Account** menu. Double-click the price book to open it.

On your iPhone/iPod Touch/iPad
Download the free VitalSource Bookshelf App available via the iTunes App Store. Open the Bookshelf app and log into your Bookshelf account. Select **All Titles** – you should see your price book on your Bookshelf. Select the price book to open it. You can find more information at https://support.vitalsource.com/hc/en-us/categories/200134217-Bookshelf-for-iOS

On your Android™ smartphone or tablet
Download the free VitalSource Bookshelf App available via Google Play. Open the Bookshelf app and log into your Bookshelf account. You should see your price book on your Bookshelf. Select the price book to open it. You can find more information at https://support.vitalsource.com/hc/en-us/categories/200139976-Bookshelf-for-Android-and-Kindle-Fire

On your Kindle Fire
Download the free VitalSource Bookshelf App from Amazon. Open the Bookshelf app and log into your Bookshelf account. Select All Titles – you should see your price book on your Bookshelf. Select the price book to open it. You can find more information at https://support.vitalsource.com/hc/en-us/categories/200139976-Bookshelf-for-Android-and-Kindle-Fire

Support

If you have any questions about downloading Bookshelf, creating your account, or accessing and using your ebook edition, please visit https://support.vitalsource.com/

Free Updates

with three easy steps…

1. Register today on www.pricebooks.co.uk/updates

2. We'll alert you by email when new updates are posted on our website

3. Then go to www.pricebooks.co.uk/updates
   and download the update.

All four Spon Price Books – *Architects' and Builders'*, *Civil Engineering and Highway Works*, *External Works and Landscape* and *Mechanical and Electrical Services* – are supported by an updating service. Two or three updates are loaded on our website during the year, typically in November, February and May. Each gives details of changes in prices of materials, wage rates and other significant items, with regional price level adjustments for Northern Ireland, Scotland and Wales and regions of England. The updates terminate with the publication of the next annual edition.

As a purchaser of a Spon Price Book you are entitled to this updating service for this 2022 edition – free of charge. Simply register via the website www.pricebooks.co.uk/updates and we will send you an email when each update becomes available.

If you haven't got internet access or if you've some questions about the updates please write to us at Spon Price Book Updates, Spon Press Marketing Department, 4 Park Square, Milton Park, Abingdon, Oxfordshire, OX14 4RN.

Find out more about Spon books
Visit www.pricebooks.co.uk for more details.

Spon's Architects' and Builders' Price Book

Edited by

AƎCOM

2022

One hundred and forty-seventh edition

First edition 1873
One hundred and forty-seventh edition published 2022
by CRC Press
2 Park Square, Milton Park, Abingdon, Oxon, OX14 4RN

and by CRC Press
Taylor & Francis, 6000 Broken Sound Parkway, NW, Suite 300, Boca Raton, FL 33487

CRC Press is an imprint of the Taylor & Francis Group, an informa business

British Library Cataloguing in Publication Data
A catalogue record for this book is available from the British Library

ISBN: 978-1-032-05216-8
Ebook: 978-1-003-19660-0

ISSN: 0306-3046

Typeset in Arial by Taylor & Francis Books
Printed and bound by CPI Group (UK) Ltd, Croydon, CR0 4YY

Contents

PART 4: PRICES FOR MEASURED WORKS

PART 5: FEES FOR PROFESSIONAL SERVICES

PART 6: DAYWORK AND PRIME COST

PART 7: USEFUL ADDRESSES FOR FURTHER INFORMATION

PART 8: TABLES AND MEMORANDA

Preface to the One Hundred and Forty-Seventh edition

Aside from a boom in infrastructure, the first quarter of 2021 has not seen a continuation of the late 2020 rally in output. But while sentiment is high for a recovery, new orders are not fully mirroring this optimism, and a supply crunch is hitting the materials chain.

Output and New Orders

Since rallying through the second half of 2020, construction output has not continued its upward momentum into 2021, according to the latest data from the Office for National Statistics (ONS). After returning to somewhere near 90% of its pre-pandemic levels, new work output is plotting a sideways track in the early months of 2021. All work construction output grew by 1.6% between January and February 2021. Following data revisions, 2020 annual construction growth was revised down 1.5% points to mark a drop of 14%. This is now the largest decline in yearly growth since annual records began in 1997. Looking ahead, new orders data in Q4 2020 slipped again after posting a modest improvement over the 12 months to Q3 2020.

Infrastructure is clearly the strongest construction sub-sector, surpassing its pre-pandemic peak. All other sectors appear to be struggling to return to pre-pandemic levels of output though. Notably, the private housing and commercial sub-sectors are some way off their early 2020 output levels, especially since they are the two largest segments, by monetary value, of the UK construction sector. The most recent data indicates that the output gap for these sub-sectors compared to late 2019 and early 2020 did not close significantly over 2021 Q1.

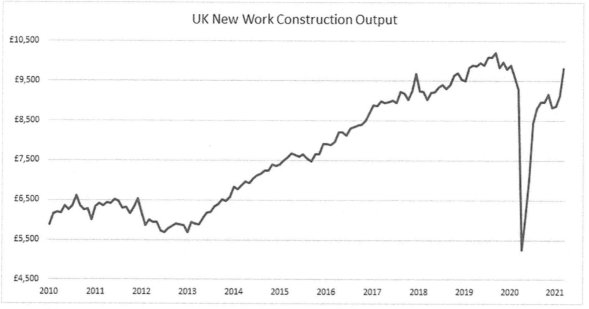

UK New Work Construction Output

Source ONS

A range of factors are all combining to propel input cost inflation. Higher international logistics costs, returning demand from industry workload, and higher global metals prices are some of the ingredients to quicker building cost inflation. Some EU materials exporters to the UK are adjusting to Brexit. EU markets have been a significant source of supply for a wide range of construction materials and prices.

Some EU exporters – initially smaller or specialist firms are proportionally harder – are now deciding not to supply the UK at all, because the additional red tape involved makes it not commercially viable. If this trend replicates across the whole EU to UK supply chain, visible disruption and higher prices are likely until other sources of supply begin to increase in volume and step in to address these supply issues. Among other exporters continuing to supply the UK, significant price increases are being applied to cover the permanent Brexit non-tariff barriers and additional administrative processes.

At the time of compiling this year's book it should be highlighted that many material suppliers are extending delivery periods and fixing prices with very short acceptance periods. Readers are recommended to register for the two Updates which will highlight changes to market rates for materials which will inevitably follow.

Materials

As site activity picks up, a combination of restart demand, lockdown constraints and low supply inventories will create above-average cost inflation in 2021 Q2 and Q3. Elevated input cost inflation will endure until supplier and distributor channels return to more typical output and operating levels.

Rapid lockdowns across the world have left distribution and logistics chains fragmented or disordered. Until these channels return to some form of normal sequencing, supply constraints will endure with impacts to the procurement of both domestically sourced and imported construction materials. Impaired project delivery is a very likely risk too.

Labour

Aggregate wage levels are nominally rising as lockdown ends, the economy reopens, and more construction activity picks up. ONS data on regular pay for construction employees shows that weekly wages are back to their pre-pandemic levels. But other aggregate measures provide a more nuanced picture where wage levels are not quite at parity with those seen immediately prior to the pandemic.

Book labour rates are based on the CIJC Working Rule Agreement. Early in 2020 representatives met and agreed not to engage in any pay negotiations for 2020/21. Discussions have recently started, and the authors have assumed a 2.75% increase for this year's book.

Book Price Level

The price level of Spon's A&B 2022 has been indexed at **132** (2015 = 100). Readers of Spon's A&B are reminded that Spon's is the only known price book in which key rates are checked against current tender prices. Users should note that this number is based on book prices arrived at by pricing our usual tender price models with prices taken from the book. It does not reflect a particular date which would appear in our normal published indices.

Profits and Overheads

The 2022 edition includes a 3% allowance main contractor's overheads and profit.

Preliminaries

There are continuing signs that main contractor preliminaries costs are levelling off, having been adjusted from the one-way downward flow of recent years. Indirect costs have steadily increased for the last few years. Whilst this is not across the board and depends to a large degree on the project parameters, this is a likely response to industry output changes, lingering uncertainty for the medium-term and the need to retain a competitive edge. We have set our example provision for preliminaries this year at 12%, but the range can be over 20% for some specific projects and procurement routes.

NOTE:

- Preliminaries are **not included** within the main *Prices for Measured Works* or in the *Approximate Estimating Rates* sections of the book.
- Preliminaries are **included** in the rates within *Building Prices per Functional Unit* and *Building Prices per Square Metre* sections.

Prices included within this edition do not include for VAT, professional fees etc. which must be added if appropriate.

PARTS OF THIS BOOK

Part 1: General

This section contains advice on various construction specialism's; capital allowances; legislation; taxes; insurances; building cost and tender price indices and regional price variations.

Part 2: Rates of Wages

Book labour rates are based on the CIJC Working Rule Agreement. Early in 2020 representatives met and agreed not to engage in any pay negotiations for 2020/21. Discussions have recently started and the authors have assumed a 2.75% increase for this year's book.

Part 3: Approximate Estimating

This section contains distinct areas:

- *Building Prices per Functional Unit; Building Prices per Square Metre and Building Cost Models.* It should be noted that these sections all *include site preliminaries.* The only occasion this happens within the book.
- *Approximate Estimating Rates* shows typical composite built-up rates organized by building elements. Please note these rates *do not include for any site preliminaries.*

There is also a section where we show typical preliminaries build up for a project valued at approximately £4,000,000. This is intended for guidance only and should not be used as part of any tender submission.

Part 4: Prices for Measured Works

These sections contain Prices for Measured Work organized using the NRM2 Work Sections for building works.

NOTE: All prices in Part 4 exclude the main contractor's preliminaries costs.

Part 5: Fees for Professional Services

This section contains guidance on fee levels for professional services; Quantity Surveyors; Architects and Consulting Engineers. Readers should always obtain fee proposals for their project prior to commencement as there are many factors that influence fee submissions.

NOTE: Professional fees are not included in any rates in the book.

Part 6: Daywork and Prime Cost

This section contains Daywork and Prime Cost allowances issued by the Royal Institute of Chartered Surveyors.

Part 7: Useful Addresses

A list of useful trade associations, professional bodies contact details.

Part 8: Tables and Memoranda

This section contains general formulae, weights and quantities of materials, other design criteria and useful memoranda associated with each trade.

While every effort is made to ensure the accuracy of the information given in this publication, neither the Editors nor Publishers in any way accept liability for loss of any kind resulting from the use of such information.

AECOM Ltd
Aldgate Tower
2 Leman Street
London
E1 8FA

Architect's Legal Handbook:
The Law for Architects
10th edition

Edited By Anthony Speaight QC and Matthew Thorne

The *Architect's Legal Handbook* is the most widely used reference on the law for practicing architects and the established textbook on law for architectural students.

Since the last edition of this book in 2010, the legal landscape in which architecture is practised has changed significantly: the long-standing procurement model with an architect as contract administrator has been challenged by the growing popularity of design and build contracts, contract notices in place of certificates, and novation of architect's duties.

The tenth edition features all the latest developments in the law which affect an architect's work, as well as providing comprehensive coverage of relevant UK law topics. Key highlights of this edition include:
*   an overview of the legal environment, including contract, tort, and land law;
*   analysis of the statutory framework, including planning law, health and safety, construction legislation, and building regulations in the post-Grenfell legal landscape;
*   procurement and the major industry construction contract forms;
*   building dispute resolution, including litigation, arbitration, adjudication, and mediation;
*   key fields for the architect in practice, including architects' registration and professional conduct, contracts with clients and collateral warranties, liability in negligence, and insurance;
*   entirely new chapters on various standard form contracts, architects' responsibility for the work of others, disciplinary proceedings, and data protection;
*   tables of cases, legislation, statutes, and statutory instruments give a full overview of references cited in the text.

The *Architect's Legal Handbook* is the essential legal reference work for all architects and students of architecture.

January 2021: 434 pp
ISBN: 9780367233686

To Order
Tel:+44 (0) 1235 400524
Email: tandf@bookpoint.co.uk

For a complete listing of all our titles visit:
www.tandf.co.uk

Special Acknowledgements

Acodrain
Tel: 01462 816666
buildingdrainage@aco.co.uk
www.aco.co.uk
Drainage channels

Altro Floors
Tel: 0870 6065432/01462 707600
info@altro.co.uk
www.altro.co.uk
Floor coverings

ASSA ABLOY

ASSA ABLOY Entrance Systems Ltd
Tel: 0333 006 3443
sales.uk.entrance@assaabloy.com
www.assaabloyentrance.co.uk
Entrance systems

Building Innovation
Tel: 01926 888808
info@building-innovation.co.uk
www.building-innovation.co.uk
Tapered insulation

GRADUS

Gradus
Tel: 01625 428 922
www.gradus.com
Floor coverings and accessories

Halfen Ltd
Tel: 01582 470 300
brick@halfen.co.uk
www.halfen.co.uk
Brick accessories – channels and special products

Kingspan Insulated Panels
Tel: 01358 716100
info@kingspanpanels.com
www.kingspanels.co.uk
Insulated wall and roof panels

Parker & Highland Joinery Ltd
Tel: 01903 756 283
sales@parker-joinery.com
www.parker-joinery.com
Purpose-made joinery

Profile 22
Tel: 01952 290910
mail@profile22.co.uk
www.profile22.co.uk
uPVC windows

Severfield Ltd
Tel: 01845 577896
www.severfield.com
Structural steel

Acknowledgements

Acodrain
Tel: 01462 816666
buildingdrainage@aco.co.uk
www.aco.co.uk
Drainage channels

Allgood plc
Tel: 020 7387 9951
info@allgood.co.uk
www.allgood.co.uk
Ironmongery

Altro Floors
Tel: 0870 6065432 / 01462 707600
info@altro.co.uk
www.altro.co.uk
Floor coverings

Alumasc Exterior Building Products Ltd
Tel: 01744 648400
info@alumasc-exteriors.co.uk
www.alumasc-exteriors.co.uk
Aluminium rainwater goods

Alumasc Interior Building Products Ltd
Tel: 01952 580590
sales@pendock.co.uk
www.pendock.co.uk
Skirting trunking & casings

Amwell Systems Ltd
Tel: 01763 276200
sales@amwell-systems.com
www.amwell-systems.com
Toilet cubicles

Andrews Marble Tiles
Tel: 0113 262 4751
sales@andrews-tiles.co.uk
www.andrews-tiles.co.uk
Floor and wall tiles

Armitage Shanks Group
Tel: 01543 490253
arm-idealinfo@idealstandard.com
www.armitage-shanks.co.uk
Sanitary fittings

Armstrong Floor Products UK Ltd
Tel: 01642 768660
commercial_uk@armstrong.com
www.armstrong-flooring.co.uk
Flooring products

ASSA ABLOY Entrance Systems Ltd
Tel: 0333 006 3443
sales.uk.entrance@assaabloy.com
www.assaabloyentrance.co.uk
Entrance systems

ASSA ABLOY Ltd
Tel: 01902 364 648
info@assaabloy.com
www.assaabloy.co.uk
Ironmongery

Bison Concrete Products Ltd
Tel: 01283 495000
concrete@bison.co.uk
www.bison.co.uk
Precast concrete floors

Bolton Gate Co
Tel: 01204 871000
sales@boltongate.co.uk
www.boltongate.co.uk
Insulated shutter doors

Bradstone Structural Sales
Tel: 01285 646 884
bradstone-structural@aggregate.com
www.bradstone-structural.com
Walling, dressings and other products

British Gypsum Ltd
Tel: 01509 817 200
bgtechnical.enquiries@bpb.com
www.british-gypsum.com
Plasterboard and plaster products

Building Innovation
Tel: 01926 888808
info@building-innovation.co.uk
www.building-innovation.co.uk
Tapered insulation

Burlington Slate Ltd
Tel: 01229 889 681
sales@burlingtonstone.co.uk
www.burlingtonstone.co.uk
Westmoreland slating

Buttles
Tel: 01727 834 242
beeweb@buttle.co.uk
www.buttles.co.uk
Timber products

Catnic Ltd
Tel: 02920 337900
www.catnic.com
Steel lintels

Cavity Trays Ltd
Tel: 01935 474769
sales@cavitytrays.co.uk
www.cavitytrays.com
Cavity trays, closers and associated products

Cementation Foundations Skansa Ltd
Tel: 01923 423100
cementation.foundations@skanska.co.uk
www.skanska.co.uk
Piling systems

Colour Centre
www.colourcentre.com
Paints, varnishes and stains

Concrete Canvas Ltd (UK)
Tel: 0845 680 1908
info@concretecanvas.co.uk
www.concretecanvas.co.uk
Concrete canvas

Cox Building Products
sales@coxspan.co.uk
www.coxbp.com
Rooflights

Custom Metal Fabrications
Tel: 020 8844 0940
dgibbs@cmf.co.uk
www.cmf.co.uk
Metalwork and balustrading

Decra Roof Systems (UK) Ltd
Tel: 01293 545 058
sales@decra.co.uk
www.decra.co.uk
Roofing system and accessories

Dow Construction Products
Tel: 020 8917 5050
styrofoam-uk@dow.com
www.styrofoameurope.com
Insulation products

Dreadnought Clay Roof Tiles
Tel: 01384 77405
sales@dreadnought-tiles.co.uk
www.dreadnought-tiles.co.uk
Clay roof tiling

Dufaylite Developments Ltd
Tel: 01480 215000
enquiries@dufaylite.com
www.dufaylite.com
Clayboard void former

Envirodoor Ltd
Tel: 01482 659375
sales@envirodoor.com
www.envirodoor.com
Sliding and folding doors

Expamet Building Products
Tel: 01429 866611
sales@expamet.net
www.expamet.co.uk
Builder's metalwork products

Forbo Flooring UK Ltd
Tel: 01773 744 121
www.forbo-flooring.co.uk
Mats and matwells

Forterra Building Products Ltd
Tel: 01604 707600
www.forterra.co.uk
Facing bricks; aircrete blocks; aggregate blocks

Forticrete Ltd
Tel: 01525 244 900
masonry@forticrete.com;
roofing@forticrete.com
nwblocksales@forticrete.com
www.forticrete.co.uk
Blocks and roof tiles

F P McCann Ltd
www.fpmccann.co.uk
Precast concrete goods

Garador Ltd
Tel: 01935 443700
www.garador.co.uk
Garage doors

GCP Applied Technologies
Tel: 01753 490000
SalesOrders.gcp-IRL@gcpat.com
https://gcpat.uk/en-gb
Expansion joint fillers and waterbars

Gradus
Tel: 01625 428 922
imail@gradus.com
www.gradus.com
Flooring and accessories

GRP Tanks
Tel: 0871 200 2082
www.grptanks.net
GRP water tanks

H+H Celcon UK Ltd
Tel: 01732 886333
info@hhcelcon.co.uk
www.hhcelcon.co.uk
Concrete blocks

Halfen Ltd
Tel: 01582 470 300
brick@halfen.co.uk
www.halfen.co.uk
Brick accessories – channels and special products

Hathaway Roofing Ltd
Tel: 01388 605 636
www.hathaway-roofing.co.uk
Sheet wall and roof claddings

Hillaldam Coburn Ltd
Tel: 020 8545 6680
sales@hillaldam.co.uk
www.coburn.co.uk
Sliding and folding door gear

HSS Hire Shops
Tel: 020 8260 3100
www.hss.co.uk
Tool hire

Hudevad
Tel: 02476 881200
sales@hudevad.co.uk
www.hudevad.co.uk
Radiators

Hunter Plastics Ltd
info@hunterplastics.co.uk
www.hunterplastics.co.uks
Plastic rainwater goods

Ibstock Building Products
Tel: 01530 261999
marketing@ibstock.co.uk
www.ibstock.co.uk
Facing bricks; tilebricks

Icopal Ltd
Tel: 0843 224 7400
info.uksales@icopal.com
www.icopal.co.uk
Damp-proof products

James Latham
Tel: 01454 315 421
marketing@lathams.co.uk
www.lathamtimber.co.uk
Hardwood and panel products

Jeld-Wen UK Ltd
Tel: 0870 126 0000
marketing@jeld-wen.co.uk
www.jeld-wen.co.uk
Doors and windows

John Brash and Co Ltd
Tel: 01427 613858
info@johnbrash.co.uk
www.johnbrash.co.uk
Roofing shingles

John Guest Speedfit Ltd
Tel: 01895 449 233
www.speedfit.co.uk
Speedfit product range

Junkers Ltd
Tel: 01376 534 700
sales@junkers.co.uk
www.junkers.co.uk
Hardwood flooring

Kalzip Ltd
Tel: 01942 295500
kalzip-uk@corusgroup.com
www.kalzip.com
Kalzip roofing

KB Rebar Ltd
Tel: 0161 790 8635
www.kbrebar.co.uk
Reinforcement bar and mesh

Kingspan Access Floors
Tel: 01482 781 710
enquiries@kingspanaccessfloors.co.uk
www.kingspanaccessfloors.co.uk
Raised access floors

Kingspan Environmental
Tel: 01296 633000
pollutiongb@kingspanenv.com
www.kingspanenv.com
Interceptors and septic tanks

Kingspan Insulated Panels
Tel: 01358 716100
info@kingspanpanels.com
www.kingspanels.co.uk
Insulated wall and roof panels

Kingspan Insulation Ltd
Tel: 0870 850 8555
info.uk@insulation.kingspan.com
www.insulation.kingspan.com
Insulation products

Kingspan Structural Products
Tel: 01944 712 000
sales@kingspanstructural.com
www.kingspan.com
Multibeam purlins

Knauf Insulation Ltd
Tel: 01744 766 666
sales@knaufinsulation.com
www.knaufinsulation.co.uk
Insulation products

Landpro Ltd
Tel: 01252 795030
info@landpro.co.uk
www.landpro.co.uk
Landscaping consultants

Lignacite
Tel: 01992 464 441
info@lignacite.co.uk
www.lignacite.co.uk
Concrete blocks

Maccaferri
Tel: 01865 770 555
marketing@maccaferri.co.uk
www.maccaferri.co.uk
Gabions

Magrini Ltd
Tel: 01543 375311
sales@magrini.co.uk
www.magrini.co.uk
Baby equipment

Manhole Covers Ltd
Tel: 01296 668850
sales@manholecovers
www.manholecovers.com
Manhole covers

Marshalls CPM
Tel: 01179 814500
sales@cpm-group.co.uk
www.cpm-group.co.uk
Concrete pipes etc.

Marshalls Mono Ltd (drainage)
Tel: 01422 312000
drainage@marshalls.co.uk
www.marshalls.co.uk
Drainage channels

Marshalls Mono Ltd
Tel: 01422 312000
marshallspaving@marshalls.co.uk
www.marshalls.co.uk
Pavings

Metsec Lattice Beams Ltd
technical@metseclatticebeams.com
www.metseclatticebeams.com
Lattice beams

Monier Ltd
Tel: 01293 618418
roofing.redland@monier.com
www.redland.co.uk
Redland roof tiles

NDM Metal Roofing & Cladding Ltd
Tel: 020 8991 7310
enquiries@ndmltd.com
www.ndmltd.com
Metal cladding and roofing

Parker & Highland Joinery Ltd
Tel: 01903 756 283
sales@parker-joinery.com
www.parker-joinery.com
Purpose-made Joinery

Plumb Centre
Tel: 0870 1622 557
www.plumbcentre.co.uk
Cylinders & general plumbing

Polyflor Ltd
Tel: 0161 767 1111
www.polyflor.com
Polyfloor contract flooring

Polypipe Terrain
Tel: 01622 717811
commercialenquiries@polypipe.com
www.terraindrainage.com
Drainage goods

Premdor Ltd
Tel: 01793 708200
enquiries@premdor.com
www.premdor.co.uk
Doors and windows

Premier Loft Ladders
Tel: 0845 9000 195
sales@premierloftladders.com
www.premierloftladders.com
Loft ladders

Pressalit Care plc
Tel: 0844 880 6950
www.pressalitcare.com
Bathroom equipment

Profile 22
mail@profile22.co.uk
www.profile22.co.uk
uPVC windows

Promat UK
Tel: 01344 381 301
promat@promat.co.uk
www.promat.co.uk
Fireproofing materials

Protim Solignum Ltd
Tel: 01628 486644
info@osmose.co.uk
www.osmose.co.uk
Paints and timber treatment

Radius Systems Ltd
Tel: 01773 811 112
www.radius-systems.co.uk
MDPE pipes and fittings

Rawlplug Ltd
Tel: 0141 638 225
info@rawlplug.co.uk
www.rawlplug.co.uk
Anchoring and fixing systems

Richard Potter Timber Merchants
Tel: 01270 625791
richardpotter@fortimber.demon.co.uk
www.fortimber.demon.co.uk
Carcasssing softwood

Rockwool Ltd
Tel: 01656 862 261
info@rockwool.co.uk
www.rockwool.co.uk
Pipe and other insulation products

Ryton's Building Products
Tel: 01536 511874
admin@rytons.com
www.vents.co.uk
Roof ventilation products

Safeguard Europe Ltd
Tel: 01403 212004
www.safeguardeurope.com
Damp-proofing and waterproofing

Saint Gobain PAM UK
Tel: 0115 930 5000
sales.uk.pam@saint-gobain.com
www.saint-gobain-pam.co.uk/
Cast iron soil, water and rainwater pipes and fittings

Sandtoft Roof Tiles Ltd
Tel: 01427 871200
info@sandtoft.co.uk
www.sandtoft.co.uk
Clay roof tiling

Schiedel Rite-Vent
Tel: 0191 416 1150
sales@schiedel.co.uk
www.isokern.co.uk
Flue pipes and gas blocks

Screeduct Ltd
Tel: 01789 459 211
sales@screeduct.com
www.screeduct.com
Trunking systems and conduits

Severfield Limited
Tel: 01845 577896
www.severfield.com
Structural steel

Sheet Piling UK Ltd
Tel: 01772 794 141
enquiries@sheetpilinguk.com
www.sheetpilinguk.com
Sheet piling

Sheffield Insulation
Tel: 020 8477 9500
barking@sheffins.co.uk
www.SheffieldInsulation.co.uk
Insulation products

Siderise Insulation Ltd
Tel: 01656 730833
sales@siderise.com
www.siderise.com
Fire barriers

Slate UK David Wallace International Ltd
Tel: 015395 59289
www.slate.uk.com
Spanish roof slates

Stainless UK Ltd
Tel: 0114 244 1333
sales@stainless-uk.co.uk
www.stainless-uk.co.uk
Stainless steel rebar

Sterling Hydraulics (Huntley & Sparks) Ltd
Tel: 01460 722 22
Rigifix column guards

Stirling Lloyd Polychem Ltd
Tel: 01565 633 111
www.stirlinglloyd.com
Integritank products

Stressline Ltd
Tel: 0870 7503167
sales@stressline.ltd.uk
www.stressline.ltd.uk
Concrete lintels

Swish Building Products
Tel: 01827 317200
marketing@swishbp.co.uk
www.swishbp.co.uk
Swish Celuka

Szerelmey Ltd
Tel: 020 7735 9995
info@szerelmey.com
www.szerelmey.com
Stonework

Tarkett-Marley Floors Ltd
Tel: 01622 854000
uksales@tarkett.com
www.tarkett-floors.com
Sheet and tile flooring

Tarmac Ltd
Tel: 020 8555 2415
enquiries@tarmac.co.uk
www.tarmac.co.uk
Ready-mixed concrete

Tarmac Mortar and Screeds
Tel: 08701 116 116
mortar@tarmac.co.uk
www.tarmac.co.uk
Readymix screeds and mortar

Tarmac Topblock Ltd
Tel: 01902 754 131
enquiries@tarmac.co.uk
www.topblock.co.uk
Concrete blocks

TATA Steel Ltd
Tel: 01724 405 060
www.tatasteeleurope.com
Steel

Timbmet
Tel: 01865 860351
marketing@timbmet.com
www.timbmet.com
Hardwood

Travis Perkins Trading Company
Tel: 01604 752484
www.travisperkins.co.uk
Builders merchant

VA Hutchison Flooring Ltd
Tel: 01243 841 175
www.hutchisonflooring.co.uk
Hardwood flooring

Vandex
Tel: 01403 210204
info@safeguardeurope.com
www.vandex.com
Vandex super and premix products

Velfac Ltd
Tel: 01223 897 100
post@velfac.co.uk
www.velfac.co.uk
Composite windows

Velux Company Ltd
Tel: 0870 166 7676
enquires@velux.co.uk
www.velux.co.uk
Velux roof windows & flashings

Visqueen Building Products
Tel: 01685 840 672
enquiries@visqueenbuilding.co.uk
www.visqueenbuilding.co.uk
Visqueen products

Wavin Plastics Ltd
Tel: 01249 766600
info@wavin.co.uk
www.wavin.co.uk
uPVC drainage goods

Web Dynamics Ltd
Tel: 01204 695 666
www.webdynamics.co.uk
Breather membranes

Welco
Tel: 0121 4219000
enquiries@welconstruct.co.uk
www.welconstruct.co.uk
Lockers and shelving systems

Welsh Slate Ltd
Tel: 01248 600 656
enquiries@welshslate.com
www.welshslate.com
Natural Welsh slates

Wilde Contracts Ltd
Tel: 0161 624 6824
www.rogerwilde.com
Glass blocks

Yeoman Aggregates Ltd
Tel: 020 8896 6800
debra.ward@yeoman-aggregates.co.uk
www.yeoman-aggregates.co.uk
Hardcore, gravels, sand

Yorkshire Copper Tube Ltd
Tel: 0151 546 2700
info@yct.com
www.yorkshirecopper.com
Copper tube

How to use this Book

First time users of *Spon's Architects' and Builders' Price Book* (*Spon's A & B*) and others who may not be familiar with the way in which prices are compiled may find it helpful to read this section before starting to calculate the costs of building work. The level of information on a scheme and availability of detailed specifications will determine which section of the book and which level of prices users should refer to.

We have rebased our **TENDER PRICE INDEX to 2015 = 100** (from 1976 = 100) Users wishing to reference to or from the previous set of indices with a base date of 1976 = 100 need to apply a factor of **5.123**. Multiply the TPI 2015 series below by 5.123 to give the equivalent in the 1976 series.

Prices in the book do not necessarily and are not intended to represent the lowest possible prices achievable but are intended as a guide to expected price levels for the items described. AECOM cost a series of building models using current Spon's rates to calculate a book Tender Price Index (TPI).

For this edition of the book TPI has been calculated at 132 (2015 = 100).

Rates in the book include overhead and recovery margins but do not include main contractors' preliminaries: except for two sections:

- *Building Prices per Functional Unit* and
- *Building Prices per Square Metre* which are both within *Part 3 Approximating Estimating*.

New Rules of Measurement (NRM)

The NRM suite covers the life cycle of cost management and means that, at any point in a building's life, quantity surveyors will have a set of rules for measuring and capturing cost data. In addition, the BCIS *Standard Form of Cost Analysis* (SFCA) 4th edition has been updated so that it is aligned with the NRM suite. This book is intended for use with NRM1 and NRM2.

The three volumes of the NRM are as follows.

- NRM1 – Order of cost estimating and cost planning for capital building works.
- NRM2 – Detailed measurement for building works.
- NRM3 – Order of cost estimating and cost planning for building maintenance works.

APPROXIMATE ESTIMATING

For preliminary estimates/indicative costs before drawings are prepared or very little information is available, users are advised to refer to the average overall *Building Prices per Functional Unit* and multiply this by the proposed number of units to be contained within the building (i.e. number of bedrooms etc.) or *Building Prices per Square Metre* rates and multiply this by the gross internal floor area of the building (the sum of all floor areas measured within external walls) to arrive at an overall initial *Order of Cost Estimate*. These rates include preliminaries but make no allowance for the cost of external works, VAT, or fees for provisional services.

Where preliminary drawings are available, one should be able to measure approximate quantities for all the major components of a building and multiply these by individual rates contained in the *Building Cost Models* or *Approximate Estimating Rates* sections to produce an *Elemental Cost Plan*. This should produce a more accurate estimate of cost than simply using overall prices per square metre. Labour and other incidental associated items, although normally measured separately within Bills of Quantities, are deemed included within approximate estimating rates. These rates do not include preliminaries or fees for professional services.

MEASURED WORKS

For more detailed estimates or documents such as Bills of Quantities (Quantities of supplied and fixed components in a building, measured from drawings), use rates from *Prices for Measured Works*. Depending upon the overall value of the contract readers may want to adjust the value and we have added a simple chart which shows typical adjustments that could be applied as shown later in this chapter. Items within the Measured Works sections are made up of many components: the cost of the material or product; any additional materials needed to carry out the work; plant required; and the labour involved in unloading and fixing, etc.

Measured Works Rates

These components are usually broken down into:

Labour

This figure covers the cost of the operation and is calculated on the gang wage rate (skilled or unskilled) and the time needed for the job. A full explanation and build-up is provided. Large regular or continuous areas of work are more economical to install than smaller complex areas.

Plant

Plant covers the use of machinery ranging from excavators and dumpers to static plant and includes running costs such as fuel, water supply, electricity and waste disposal.

For this year's book we are showing Plant costs separately. Previously it has been included in the Materials column. We believe this will make the rates more transparent and understandable to users who did not always realize that Plant costs were included. Small hand-held plant costs are not included.

Materials

Material prices include the cost of any ancillary materials, nails, screws, waste, etc., which may be needed in association with the main material product/s. If the material being priced varies from a standard measured rate, then identify the difference between the original PC price and the material price and add this to your alternative material price before adding to the labour cost to produce a new overall total rate. Alternative material prices, where given, are largely based upon list prices, before the deduction of quantity discounts etc., and therefore require discount adjustment before they can be substituted in place of PC figures given for Measured Work items.

Prime Cost

Commonly known as the PC; Prime Cost is the actual price of the material such as bricks, blocks, tiles or paint, as sold by suppliers. Prime Cost may be given as *per square metre, per 100 bags* or *each* according to the way the supplier sells the product. Unless otherwise stated, prices in *Spon's A & B* are deemed to be delivered to site (in which case transport costs will be included) and take account of trade and quantity discounts. Part loads generally cost more than whole loads but, unless otherwise stated, Prime Cost figures are based on average prices for full loads delivered to a hypothetical site in Acton, West London. Actual prices for live projects will vary depending on the contractor, supplier, the distance from the supplier to the site, the accessibility of the site, whether the whole quantity ordered is to be supplied in one delivery or at specified dates and market conditions prevailing at the time. Prime Cost figures for commonly used alternative materials are supplied in listed form at the beginning of some work sections.

Where a PC rate is entered alongside an item rate then the cost allowed for that item is in the overall material cost. For instance, bricks need mortar; paving needs sand bedding, so the PC cost is simply for the cost of bricks or paving, thus allowing the user to simply substitute an alternative product cost if desired.

Example:

Item *(NB example data only)*	PC £	Labour £	Material £	Unit	Total Rate £
Half brick wall in common bricks	380.00	18.05	27.14	m²	**45.19**
Replace with Forterra Brick; Brecken Grey @	480.00				

Calculation: (PC) £480.00 – £380.00 = £100 + 3% (OHP) = £103.00
add £1.75/m² to materials rate of £27.14 = £28.89

Item *(NB example data only)*	PC £	Labour £	Material £	Unit	Total Rate £
Therefore, Brecken Grey bricks =	480.00	18.05	28.89	m²	**46.94**

Unit

The Unit is generally based upon measurement guidelines laid out in the New Rules of Measurement – Detailed measurement for building works (NRM2).

Total Rate

Prices in the Total Rate column generally include for the supply and fix of items, unless otherwise described.

Overheads and Profit

The general overheads of the Main Contractor's business – the head office overheads and any profit sought on capital and turnover employed, is usually covered under a general item of overheads and profit which is applied either to all measured rates as a percentage, or alternatively added to the tender summary or included within Preliminaries for site specific overhead costs.

Within this edition we are including an allowance of 3% for overheads and profit on built-up labour rates and material prices.

Preliminaries

Site specific Main Contractor's overheads on a contract, such as insurances, site accommodation, security, temporary roads and the statutory health and welfare of the labour force, are not directly assignable to individual items so they are generally added as a percentage or calculated allowances after all building component items have been costed and totalled. Preliminaries will vary from project to project according to the type of construction, difficulties of the site, labour shortage, or involvement with other contractors, etc. The overall addition for a scheme should be adjusted to allow for these factors. For this edition we have shown a calculated typical Preliminaries cost example of approximately 12%.

Sub/Specialist-Contractor's Costs

For the purpose of this book, these are deemed to include all the above costs, and assume a 2.5% main contractor's discount.

With the exclusion of main contractor's preliminaries, the above items combine to form item rates in the Prices for Measured Works sections. It should be appreciated that a variation in any one item in any group will affect the final measured work price. Any cost variation must be weighed against the total cost of the contract, and a small variation in Prime Cost where the items are ordered in thousands may have more effect on the total cost than a large variation on a few items, while a change in design which introduces the need to use, for example earth moving equipment, which must be brought to the site for that one task, will cause a dramatic rise in the contract cost. Similarly, a small saving on multiple items could provide a useful reserve to cover unforeseen extras.

COST PLANNING

Order of Cost Estimate

The purpose of an Order of Cost Estimate is to establish if the proposed building project is affordable and, if affordable, to establish a realistic cost limit for the project. The cost limit is the maximum expenditure that the employer is prepared to make in relation to the completed building project, which will be managed by the project team.

An Order of Cost Estimate is produced as an intrinsic part of RIBA Work Stages A: Appraisal and B: Design Brief or OGC Gateways 1 (Business Justification) and 2 (Delivery Strategy).

There are comprehensive guidelines within the NRM documentation and readers are recommended to read the relevant sections of the NRM where more detailed explanations and examples can be found.

At this early stage, for the estimate to be representative of the proposed design solution, the key variables that a designer needs to have developed to an appropriate degree of certainty are:

- The floor areas upon which the estimate is based
- Proposed elevations
- The implied level of specification

Rates will need to be updated to current estimate base date by the amount of inflation occurring from the base date of the cost data to the current estimate base date. The percentage addition can be calculated using published indices (i.e. tender price indices [TPI]).

Example 1:

New secondary school

Note: example data only

Gross Internal Floor Area (GIFA) = 15,000 m^2

Cost plan prepared with a TPI = 124

Start on site TPI, say = 128.4 (actual index to be selected from current published TPI)

Location: North West adjustment, say = 0.89

From *Building Prices per Square Metre*

Assume rate of say £1,800 per m^2

		Cost (£)
School rate, say	£1,800 /m^2 × 15,000 m^2 =	27,000,000
Adjust for inflation to start date	(128.4/124) say +3.5%	945,000
	subtotal	27,945,000
Adjust for location	−10%, say	−2,795,000
	subtotal	25,150,000
Allow for contingencies	say 10%	250,000
Total Order of Cost Estimate		**25,400,000**

Main contractor's preliminaries, overheads and profit need not be added to the cost of building works as they are included within the Spon's building prices per square metre rates, but you will need to add on professional fees and other enabling works costs such as site clearance, demolition, external works, car parking, bringing services to site etc.

Elemental Cost Planning

Elemental cost plans are produced as an intrinsic part of RIBA Work Stages C: Concept, D: Design Development, E: Technical Design and F: Production Information; or when the OGC Gateway Process is used, Gateways 3A (Design Brief and Concept Approval) and 3B (Detailed Design Approval).

Cost Models can be used to quickly extract £/m² of GIFA:

Example 2:

Health Centre

Note: example data only

Gross Internal Floor Area = 1,000 m²

Cost plan prepared with TPI = 124

Start on site TPI, say = 128.4 (actual index to be selected from current published TPI)

Location: South West (adjustment, say = 0.90)

Element	Rate (£/m²)	Cost (£)
Substructures	106.53	106,530
Frame and upper floors	160	160,000
Roof	24	24,000
And so on for each element to give a total of	**1,300.56**	**1,300,560**
Contractors preliminaries, overheads and profit	say 15%	195,000
	subtotal	1,495,560
Adjust for inflation to start date	(128.4/124) say +3.5%	52,000
	subtotal	1,495,560
Adjust for location	say −10%	−155,000
	subtotal	1,392,560
Allow for contingencies	say 5%	70,000
	Total Elemental Cost Plan	**1,462,560**

Other allowances such as consultants' fees, design fee, VAT, risk allowance, client costs, fixed price adjustment may need to be added to each of the examples above.

Formal Cost Planning Stages

The NRM schedules several formal cost planning stages, which are comparable with the RIBA Design and Pre-Construction Work Stages and OGC Gateways 3A (Design Brief and Concept Approval) and 3B (Detailed Design Approval) for a building project. The employer is required to 'approve' the cost plan on completion of each RIBA Work Stage before authorizing commencement of the next RIBA Work Stage.

Formal Cost Plan	RIBA Work Stage
1	1] Concept design
2	2] Spatial coordination
3	3] Technical design

Formal Cost Plan 1 is prepared at a point where the scope of work is fully defined, and key criteria are specified but no detailed design has begun. Formal Cost Plan 1 will provide the frame of reference for Formal Cost Plan 2. Likewise, Formal Cost Plan 2 will provide the frame of reference for Formal Cost Plan 3. Neither Formal Cost Plans 2 nor 3 involve the preparation of a completely new Elemental Cost Plan; they are progressions of the previous cost plans, which are developed through the cost checking of price significant components and cost targets as more design information and further information about the site becomes available.

Cost plans can be developed from Elemental Cost Plans using both Approximate Estimating Rates and/or Prices for Measured Works depending upon the level of information available.

The cost targets within each formal cost plan approved by the employer will be used as the baseline for future cost comparisons. Each subsequent cost plan will require reconciliation with the preceding cost plan and explanations relating to changes made. In view of this, it is essential that records of any transfers made to or from the risk allowances and any adjustments made to cost targets are maintained, so that explanations concerning changes can be provided to both the employer and the project team.

Adjustment According to Contract Sum Cost

The construction costs for a project will depend on the size, type of building, standard of finish required, and location, the economic climate of the construction industry, i.e. if there is a shortage of construction work available firms will reduce their tender in order to try and attract work. If the opposite is the case and there is a lot of work available, firms will increase their tenders, as they will not be too keen to obtain the contract which will stretch their resources unless it is worth their while financially. In a recession, construction firms can literally buy work in order to keep their workforce and to ensure some cash flow.

Building Costs can vary between builders/developers. This can be due to the size or purchasing abilities of a company or the discount that it receives from suppliers.

Contract Sum	% adjustment
£5,000,000	−1%
£3,500,000	0%
£2,750,000	1%
£2,000,000	2%
£1,250,000	3%
£1,000,000	4%

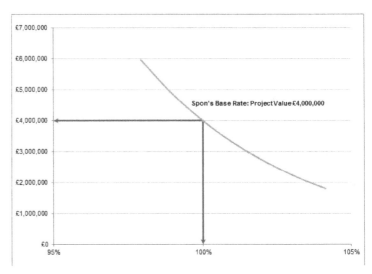

Spon's Architects' & Builders' Price Book is targeted at new build projects with a value range of approximately £3,000,000 – £5,000,000.

Users should not simply apply percentage adjustments to any project regardless of size. We recommend project values between £2,000,000 and £5,000,000 for rates found in *Spon's A & B* could be adjusted according to the above table. This is given only as an indication and users should always remember that there are many factors that affect the overall project costs.

These numbers are only intended as a guide and are not explicit and should be applied to the total project value only, not individual rates.

SPON'S 2022 PRICEBOOKS from AECOM

Spon's Architects' and Builders' Price Book 2022

Editor: AECOM

New items this year include:
- a London fringe office cost model
- a Higher Education Refurbishment cost model
- Pecafil® permanent formwork
and an expanded range of cast iron rainwater products

Hbk & VitalSource® ebook 868pp approx.
978-0-367-51402-0 £175
VitalSource® ebook
978-1-003-05367-5 £175
(inc. sales tax where appropriate)

Spon's Civil Engineering and Highway Works Price Book 2022

Editor: AECOM

This year gives more items on shafts, tunnelling, drainage and water proofing – covering some brand new materials and methods. Notes have been added to tunnelling, viaducts, D-walls and piling under the output section. The book partially reflects costs and new ways of working resulting from the Covid-19 pandemic.

Hbk & VitalSource® ebook 704 pp approx.
978-0-367-51403-7 £195
VitalSource® ebook
978-1-003-05368-2 £195
(inc. sales tax where appropriate)

Spon's External Works and Landscape Price Book 2022

Editor: AECOM in association with LandPro Ltd Landscape Surveyors

Now with a revised and updated street furniture section.
Plus several new items:
- Kinley systems: Metal edgings and systems for landscapes and podiums
- New cost evaluations of water features
- Stainless steel landscape channel drainage

Hbk & VitalSource® ebook 680pp approx. 978-0-367-51404-4 £165
VitalSource® ebook
978-1-003-05369-9 £165
(inc. sales tax where appropriate)

Spon's Mechanical and Electrical Services Price Book 2022

Editor: AECOM

An overhaul of the uninterruptible power supply section, and revised costs for air source heat pumps.
Plus new items, including:
HDPE pipe for above ground drainage systems; fire protection mist systems; and electric vehicle chargers

Hbk & VitalSource® ebook 864pp approx.
978-0-367-51405-1 £175
VitalSource® ebook
978-1-003-05370-5 £175
(inc. sales tax where appropriate)

Receive our VitalSource® ebook free when you order any hard copy Spon 2022 Price Book

Visit www.pricebooks.co.uk

To order:
Tel: 01235 400524
Email: book.orders@tandf.co.uk
A complete listing of all our books is on www.routledge.com

CRC Press
Taylor & Francis Group

Breathing new life into urban spaces

Copr Bay is a vibrant new neighbourhood for Swansea, part of the city's ambitious regeneration programme.

Our cost management and procurement teams are working behind the scenes to bring this vision to life, so that the people of Swansea can enjoy world-class amenities for generations to come.

 **aecom.com**

SWANSEA ARENA

AECOM

opr Bay, Swansea – Phase One
eveloper: Swansea Council
age courtesy of: AFL Architects

Delivering a better world

Architect's Legal Pocket Book, 3rd edition

Matthew Cousins

A little book that's big on information, the *Architect's Legal Pocket Book* is the definitive reference guide on legal issues for architects and architectural students. This handy pocket guide covers key legal principles which will help you to quickly understand the law and where to go for further information.

Now in its third edition, this bestselling book has been fully updated throughout to provide you with the most current information available. Subjects include contract administration, building legislation, planning, listed buildings, contract law, negligence, liability and dispute resolution. This edition also contains new cases, legislation, contract terms and certificates including the RIBA contract administration certificates, inspection duties, practical completion, the Hackitt review, the Report of the Independent Inquiry into the Construction of Edinburgh Schools and practical issues facing architects.

Illustrated with clear diagrams and featuring key cases, this is a comprehensive guide to current law for architects and an invaluable source of information. It is a book no architect should be without.

September 2019: 384pp
ISBN: 9781138506695

To Order
Tel: +44 (0) 1235 400524
Email: book.orders@tandf.co.uk

For a complete listing of all our titles visit:
www.tandf.co.uk

Taylor & Francis
Taylor & Francis Group

PART 1

General

This part contains the following sections:

Architect's Pocket Book, 5th edition

Jonathan Hetreed, Ann Ross
and Charlotte Baden-Powell

This handy pocket book brings together a wealth of useful information that architects need on a daily basis – on-site or in the studio. It provides clear guidance and invaluable detail on a wide range of issues, from planning policy through environmental design to complying with Building Regulations, from structural and services matters to materials characteristics and detailing. This fifth edition includes the updating of regulations, standards and sources across a wide range of topics.

Compact and easy to use, the *Architect's Pocket Book* has sold well over 90,000 copies to the nation's architects, architecture students, designers and construction professionals who do not have an architectural background but need to understand the basics, fast.

This is the famous little blue book that you can't afford to be without.

April 2017: 412 pp
ISBN: 9781138643994

To Order
Tel:+44 (0) 1235 400524
Email: tandf@bookpoint.co.uk

For a complete listing of all our titles visit:
www.tandf.co.uk

Brexit and the UK Construction Sector

The EU–UK Trade and Cooperation Agreement (TCS) came into effect at 11pm on 31 December 2020. The principle of the TCA is to lower or remove tariffs on traded goods between the two geographies. However, as a result of the TCA and choices made during the negotiation, a raft of non-tariff barriers (NTBs) also apply in a material and significant way to trade between the UK and the European Union. NTBs introduce significant administrative overheads and burdens that will almost certainly add to the costs and impact the running of UK construction, not least additional time risks.

The UK points-based immigration scheme will affect labour supply to the construction industry, while at the same time emigration trends from the UK amplify prevailing workforce capacity constraints.

All of these dramatic changes will take some time to settle down as the industry begins to fully understand the raft of new rules and regulations which are affecting nearly every facet of project delivery in the UK and throughout the EU.

Tariffs

Tariffs can be viewed as a tax by one country on the imports of goods and services from another country. They can also be seen as protections for domestic producers. Tariffs are applied either as fixed percentages irrespective of the type or value of good imported, or they are levied as a percentage of the value of the imported good. A consequence of tariffs can be to raise the prices of goods or services in the market where they are imported, with consumers ultimately bearing the cost.

Non-Trade Barriers

Non-tariff barriers (NTBs) are trade restrictions from policies implemented by a country's government. They are technical and operational restrictions, which take a variety of forms: regulations, checks, inspections, quotas, subsidies, rules of origin, prohibitions, and controls for example. Ultimately, they add friction, red tape, time and cost to trade between countries, and will act as inflationary drivers for the UK construction market.

Data suggest that NTBs introduce larger cost and time impediments and/or additions to trade than direct tariffs. Broadly, the UK government assessed NTBs in a UK/EU FTA scenario – as a percent of trade value – as 5–11% for goods and 3–14% for services (UK Government, 'EU Exit: long-term economic analysis, November 2018'). Rest of the World NTBs are at the lower end of these respective ranges. At the higher end of these ranges, NTBs are likely to introduce greater cost impacts on traded goods and services across borders than tariffs otherwise would – especially where time delays and their impacts factor in to overall cost.

Further Information

- Brexit and the UK Construction Sector   https://www.gov.uk/transition
- Research and Development Tax Relief   https://www.gov.uk/trade-tariff

Value Chain Analysis

Value chain analysis (VCA) is a means of evaluating each of the activities in a company's value chain to understand where improvement opportunities exist. A value chain analysis allows each step in the manufacturing or service process to be assessed and whether it increases or reduces value from a product or service. Typically, increasing the performance of one of the secondary activities can benefit at least one of the primary activities. VCA aims to enable or maintain competitive advantage through, for example:

- Cost reduction, by making each activity in the value chain more efficient and, therefore, less expensive
- Product differentiation, by investing more time and resources into activities like research and development, design, or marketing that can help your product stand out from the crowd

AECOM assessed the indicative impacts of Brexit on a typical value chain for the UK construction sector and its contracting supply chain firms:

INTRINSIC ISSUES	STRUCTURAL ISSUES
Business product; scope of service	Scale; sourcing, supply economics
Core service/product offer largely unchanged but shifts in strategy to focus on sectors where demand exists over the medium-term. Adjustment necessary for changes to or divergence in regulatory requirements and understanding the location and nature of new trade markets	Primary impacts to sourcing strategies as suppliers to subcontractors and the wider supply chain review operations. Reduced availability of supply into UK markets, particularly from the EU, will result in procurement risks. Compounded by significant customs and port delays, and additional regulations for road freight permits required in EU countries
IMPACTS: COST ◑ TIME ◔	**IMPACTS: COST ◕ TIME ◕**
SYSTEMIC ISSUES	**PERFORMANCE ISSUES**
Business infrastructure; processes; structure	Productivity; labour; materials
Administrative and legal issues bring proportionally higher cost burden for long tail of small supply chain firms. Management capacity redirected to immediate operational issues. Inability to submit tenders to principal contractors in allotted time. Some mitigation to general 'no deal' risks from lower overall workload	EU labour supply decreases, reducing overall industry capacity, adding inflationary cost pressures. EU nationals comprise approximately 11% of the construction workforce generally, but almost 30% in London. Supply chain productivity impacted. Significant reductions in sterling result in higher imported materials and component costs
IMPACTS: COST ◑ TIME ◕	**IMPACTS: COST ● TIME ◕**

With risks continuing to increase, business planning and preparations should, amongst other things, include:

- Understand and appraise your current situation
- Assess connections, risks and opportunities
- Analyse any exposure to Brexit – qualitatively and quantitatively
- Implement risk management and mitigation procedures

Research & Development (R&D) Tax Relief

Introduction

Research and Development (R&D) tax relief, or credit, is a Corporation Tax benefit that can either reduce a company's tax bill or, in some circumstances, provide a cash sum. It is based on the company's expenditure on R&D. R&D tax relief/credits are aimed at encouraging expenditure by companies on innovative activities that advance the overall knowledge or capability in the fields of science and technology. The tax credits consist of additional tax relief beyond the amount of the expenditure incurred on qualifying R&D activities and are given against Corporation Tax. In some situations, where the company is making losses, the tax credits can be surrendered for a cash payment.

For there to be R&D for the purpose of the tax relief, a company must be undertaking activity as part of a specific project that seeks to make an advance in science or technology. It is necessary to be able to state what the intended advance is, and to show how, through the resolution of scientific or technological uncertainty, the activity seeks to achieve this.

HMRC has confirmed that qualifying activities include those typically undertaken in the construction and property sector, such as engineering design and developing new construction techniques.

Definition of R&D for Tax Purposes

The primary legislation is contained in Part 13 of the Corporation Tax Act 2009 (CTA 2009) and covers two separate schemes:

- R&D expenditure credits (s.1040 A), commonly referred to as RDEC and is available to large companies.
- Additional tax relief for small or medium sized enterprises (s.1044), commonly referred to as the SME scheme.

R&D is defined in section 1138 of the Corporation Tax Act 2010 as 'activities that fall to be treated as research and development in accordance with generally accepted accounting practice'. This is subject to regulations made by HM Treasury on the definition of R&D, which refer to the detailed guidelines published by the Secretary of State for Business, Energy and Industrial Strategy (BEIS). The BEIS guidelines define R&D for the purpose of obtaining relief as follows:

- R&D for tax purposes takes place when a project seeks to achieve an advance in science or technology. The activities which directly contribute to achieving this advance in science or technology through the resolution of scientific or technological uncertainty are R&D. Certain qualifying indirect activities related to the project are also R&D. Activities other than qualifying indirect activities which do not directly contribute to the resolution of the project's scientific or technological uncertainty are not R&D.

Guidance is provided by HM Revenue & Customs (HMRC) on what constitutes R&D for the purposes of the tax relief and can be found within the Corporate Intangibles R&D Manual at CIRD81900. The work that qualifies for R&D tax relief must be part of a specific project to make an advance in science or technology. It cannot be an advance within arts, humanities, a social science such as economics or a theoretical field such as pure maths. The project must relate to the company's trade.

Advance in Science or Technology

An advance in science or technology means an advance in overall knowledge or capability in a field of science or technology (not a company's own state of knowledge or capability alone). This includes the adaptation of knowledge or capability from another field of science or technology in order to make such an advance where this adaptation was not readily deducible. A process, material, device, product, service, or source of knowledge does not become an advance in science or technology simply because science or technology is used in its creation.

A project which seeks to either,

- Extend overall knowledge or capability in a field of science or technology; or

- Create a process, material, device, product, or service which incorporates or represents an increase in overall knowledge or capability in a field of science or technology; or
- Make an appreciable improvement (for example, changing or adapting the scientific or technological characteristics of something to the point where it is 'better' than the original and beyond minor or routine upgrading) to an existing process, material, device, product, or service through scientific or technological changes; or
- Use science or technology to duplicate the effect of an existing process, material, device, product, or service in a new or appreciably improved way (for example, a product which has exactly the same performance characteristics as existing models, but is built in a fundamentally different manner) will therefore be R&D.

Even if the advance in science or technology sought by a project is not achieved or not fully realized, R&D still takes place and the project could qualify for relief. Moreover, if a particular advance in science or technology has already been made or attempted but details are not readily available (for example, if it is a trade secret), work to achieve such an advance can still be an advance in science or technology.

Scientific or Technological Uncertainty

Scientific or technological uncertainty exists when knowledge of whether something is scientifically possible or technologically feasible, or how to achieve it in practice, is not readily available or deducible by a competent professional (an individual who is knowledgeable about the relevant scientific and technological principles involved, aware of the current state of knowledge and has accumulated experience) working in the field. This includes system uncertainty, or a scientific or technological uncertainty that results from the complexity of a system rather than uncertainty about how its individual components behave. Scientific or technological uncertainty will often arise from turning something that has already been established as scientifically feasible into a cost-effective, reliable, and reproducible process, material, device, product, or service.

Uncertainties that can readily be resolved by a competent professional working in the field are not scientific or technological uncertainties. Similarly, improvements, optimizations and fine-tuning which do not materially affect the underlying science or technology do not constitute work to resolve scientific or technological uncertainty.

Qualifying Indirect Activity

These are activities which form part of a project but do not directly contribute to the resolution of the scientific or technological uncertainty.

This includes the following:

- Scientific and technical information services, insofar as they are conducted for the purpose of R&D support (such as the preparation of the original report of R&D findings);
- Indirect supporting activities such as maintenance, security, administration and clerical activities, and finance and personnel activities, insofar as undertaken for R&D;
- Ancillary activities essential to the undertaking of R&D (for example, taking on and paying staff, leasing laboratories, and maintaining R&D equipment including computers used for R&D purposes);
- Training required to directly support an R&D project;
- Research by students and researchers carried out at universities;
- Research (including related data collection) to devise new scientific or technological testing, survey, or sampling methods, where this research is not R&D in its own right; and
- Feasibility studies to inform the strategic direction of a specific R&D activity.

Start and End of R&D for Tax Purposes

R&D begins when work to resolve the scientific or technological uncertainty starts and ends when that uncertainty is resolved or work to resolve it ceases. For example, R&D could end once a working prototype or a material, device, product, or process is ready for testing, production or if the project does not continue. However, R&D can take place even after production starts. For example, if any new problems arise involving scientific or technological uncertainty after the product has been put into production or into use, then the R&D process may start again.

Types of R&D Tax Relief

There are different types of R&D relief, depending on the size of a company and if the project has been subcontracted to the company or not.

The two principal tax reliefs available to companies undertaking R&D in the UK are as follows:

- SME R&D Tax Relief.
- Research and Development Expenditure Credit (RDEC).

SME R&D Tax Relief

An SME for R&D tax relief purposes is a company or organization with less than 500 staff and either of the following:

- Annual turnover not exceeding €100 million.
- Balance sheet not exceeding €86 million.

A company may need to take into account its own data, a proportion of a partner enterprise's data or the data of a linked enterprise. A company may not be considered an SME if it is part of a larger enterprise that would fail these tests if taken as a whole.

SME R&D Tax relief allows companies to:

- Deduct an extra 130% of their qualifying costs from their yearly profit, as well as the normal 100% deduction, to make a total 230% deduction. For profit making SMEs, R&D tax credits reduce the Corporation Tax bill and businesses can receive up to £24.70 for every £100 spent on qualifying R&D activities.
- Claim a tax credit if the company is loss making, worth up to 14.5% of the surrenderable loss. Loss-making SMEs can receive up to £33.35 for every £100 spent on qualifying R&D activities.

A company can only claim R&D tax relief as an SME if the company is a going concern and not in administration or liquidation when making the claim.

In some circumstances, an SME company may be allowed to claim only the RDEC in respect of particular expenditure (for example, for work subcontracted to it or that is otherwise subsidized due to a grant or subsidy). The SME scheme is a notifiable State Aid, and a company can't get the SME Tax relief if it is receiving any other notifiable State Aids for the same R&D project. If an SME company has received a grant which is notifiable State Aid, for an R&D project, it cannot claim R&D tax relief under the SME scheme, but eligible expenditure will qualify under the RDEC scheme.

The UK government has recently announced that SME R&D tax credit claims will be subject to a cap in order to deter abuse of the scheme. This measure limits the amount of payable R&D tax credit which an SME can claim to £20,000 plus 300% of its total Pay-As-You-Earn (PAYE) and National Insurance contributions (NICs) liability for the period. The measure will have effect for accounting periods beginning on or after 1 April 2021.

The measure amends CTA 2009 Part 13 Chapter 2. It adds a new condition at section 1058 (which sets the amount of payable tax credit) limiting the amount that can be claimed to the level of the new cap, which is defined as £20,000 plus three times the company's relevant expenditure on workers.

A company is however exempt from the cap if:

- Its employees are creating, preparing to create or managing Intellectual Property (IP); and
- It does not spend more than 15% of its qualifying R&D expenditure on subcontracting R&D to, or the provision of externally provided workers by, connected persons.

Research and Development Expenditure Credit (RDEC)

Large companies (a company which does not meet the SME test for R&D tax relief purposes) can claim RDEC for working on R&D projects. It can also be claimed by SMEs who have been subcontracted to do R&D work by a large company or who have received a grant or subsidy for their R&D project.

The RDEC is an 'above the line' tax credit equal to 13% of qualifying R&D costs. RDEC is a taxable credit. This means that large companies receive up to £10.53 for every £100 spent on qualifying R&D activities. For profit making companies, the credit discharges the Corporation Tax liability that the company would have to pay. Companies with no Corporation Tax liability will benefit from the RDEC either through a cash payment or a reduction of tax or other duties due. Companies in groups can also surrender the RDEC against another group companies' Corporation Tax liability.

Categories of Qualifying Expenditure

- Staffing costs – This includes salaries, wages, Class 1 NICs, pension fund contributions and certain reimbursed expenses for staff directly and actively engaged in the R&D project. Support staff costs, for example administrative or clerical staff, do not qualify, except when they relate to qualifying indirect activities. These can be activities like maintenance, clerical, administrative and security work.
- Externally provided workers – These are the staff costs paid to an external agency for staff who are directly and actively engaged in the R&D project. These are not employees and subcontractors. Relief is usually given on 65% of the payments made to the staff provider. Special rules apply if the company and staff provider are connected or elect to be connected.
- Consumable items
  - o Cost of items that are directly employed and consumed in qualifying R&D projects. These include materials and the proportion of water, fuel and power of any kind consumed in the R&D process.
  - o From 1 April 2015, the costs of materials incorporated in products that are sold are not eligible for relief.
  - o Where a prototype is created to test the R&D being undertaken, the design, construction and testing costs will normally be qualifying expenses. However, if the prototype is also built with a view to selling the prototype itself (for example, the construction of a bespoke machine), HMRC considers that to be production and outside the R&D scheme, even if R&D was undertaken to create the prototype. In this case, the R&D expenditure and production costs will need to be split. For example, the construction costs and materials consumed would not be qualifying expenses, but design, modelling and testing costs could still qualify for R&D tax relief.
- Software – A company may claim for the cost of software that is directly employed in the R&D activity. Where software is only partly employed in direct R&D, an appropriate apportionment should be made.
- Subcontracted expenditure
  - o SME Scheme: A company can generally claim for 65% of the payments made to unconnected parties. The subcontracted work may be further subcontracted to any third party. Special rules apply where the parties are connected or elect to be connected. Under the SME Scheme, the subcontractor does not need to be a UK resident and there is no requirement for the subcontracted R&D to be performed in the UK.
  - o RDEC Scheme: R&D expenditure subcontracted to other persons is generally not allowable unless it is directly undertaken by a charity, higher education institute, scientific research organization, health service body, or by an individual or a partnership of individuals.

Qualifying expenditure must be allowable as a deduction in calculating the profits of the trade. Capital expenditure is therefore excluded; it may however qualify for R&D allowances.

Costs that Cannot be Claimed for R&D Tax Relief

Companies cannot receive R&D tax relief for the following:

- The production and distribution of goods and services.
- Capital expenditure under either the SME tax relief or RDEC scheme. However, a 100% Research and Development Allowance may be due on capital assets, such as plant, machinery and buildings used for R&D activity.
- The cost of land.
- Cost of patents and trademarks.
- Rent or rates.

Examples of Qualifying R&D Activities in the Construction Industry

Design

- Integrating new or improved technology into buildings.

- Redesigning to improve performance, reliability, quality, safety and life-cycle costs.
- Designing prefabricated components for off-site assembly of large and complex systems.
- Developing innovative sustainable solutions to achieve net zero carbon.
- Developing advanced Buildings Information Management (BIM) capabilities.

On-site

- Solving technical problems that arise during or after construction.
- Developing new techniques to remove contaminants and remediate difficult ground conditions.
- Complex diversions of utilities such as gas, water and electricity.
- Developing new, more efficient construction techniques.
- Developing bespoke temporary solutions on challenging sites.

Research and Development Allowances

Research and Development Allowance (RDA) gives relief for capital expenditure incurred on R&D by a business. RDA is only due if the research and development expenditure is related to the trade being carried on or about to be carried on. Qualifying RDA expenditure is capital expenditure that a business incurs for providing facilities for carrying out R&D, for example, a laboratory in which to carry on R&D. RDA is only available to traders and is not available to a person carrying on a profession or vocation. The rate of RDA is 100% of the qualifying expenditure. However, if a disposal value is brought into account on expenditure in a chargeable period in which an RDA is made, the RDA is 100% of the expenditure less the disposal value. RDA is made in the chargeable period in which the qualifying expenditure was incurred unless it was incurred before the trade began, in which case RDA is made in the chargeable period in which trade began.

How to Claim R&D Tax Relief from HMRC

A company can make a claim for R&D tax relief up to two years after the end of the accounting period it relates to. To make a claim, a company must enter their total qualifying expenditure on the full Company Tax Return form, CT600, to HMRC. If claiming under the SME scheme, a company enters the enhanced expenditure figure into the CT600, and the payable tax credit figure if applicable. If claiming under the RDEC scheme, a company must enter both the total qualifying expenditure and RDEC figures into the CT600. After submitting their full CT600 form, a claimant company can use the HMRC online service to support their R&D tax relief claim and send any additional details, as required.

The Architecture of Natural Cooling, 2nd edition

Brian Ford *et al.*

Overheating in buildings is commonplace. This book describes how we can keep cool without conventional air-conditioning: improving comfort and productivity while reducing energy costs and carbon emissions. It provides architects, engineers and policy makers with a 'how-to' guide to the application of natural cooling in new and existing buildings. It demonstrates, through reference to numerous examples, that natural cooling is viable in most climates around the world.

This completely revised and expanded second edition includes:
- An overview of natural cooling past and present.
- Guidance on the principles and strategies that can be adopted.
- A review of the applicability of different strategies.
- Explanation of simplified tools for performance assessment.
- A review of components and controls.
- A detailed evaluation of case studies from the USA, Europe, India and China.

This book is not just for the technical specialist, as it also provides a general grounding in how to avoid or minimise air-conditioning. Importantly, it demonstrates that understanding our environment, rather than fighting it, will help us to live sustainably in our rapidly warming world..

November 2019: 280 pp
ISBN: 9781138629073

To Order
Tel: +44 (0) 1235 400524
Email: book.orders@tandf.co.uk

For a complete listing of all our titles visit:
www.tandf.co.uk

Taylor & Francis
Taylor & Francis Group

Capital Allowances

Introduction

Capital Allowances provide tax relief by prescribing a statutory rate of depreciation for tax purposes in place of that used for accounting purposes. They are utilized by government to provide an incentive to invest in capital equipment, including assets within commercial property, by allowing the majority of taxpayers a deduction from taxable profits for certain types of capital expenditure, thereby reducing or deferring tax liabilities.

The capital allowances most commonly applicable to real estate are those given for capital expenditure on existing commercial buildings in disadvantaged areas, and plant and machinery in all buildings other than residential dwellings, except for common areas. Relief for certain expenditure on industrial buildings and hotels was withdrawn from April 2011, although the ability to claim plant and machinery remains.

Enterprise Zone Allowances are also available for capital expenditure within designated areas only where there is a focus on high value manufacturing. Enhanced rates of allowances are also available on certain types of energy and water saving plant and machinery assets.

The Act

The primary legislation is contained in the Capital Allowances Act 2001. Major changes to the system were introduced in 2008, 2014 and 2018 affecting the treatment of tax relief to include plant and machinery allowances and the newly introduced Structures and Building Allowances (SBAs).

Plant and Machinery

Various legislative changes and case law precedents in recent years have introduced major changes to the availability of Capital Allowances on property expenditure. The Capital Allowances Act 2001 excludes expenditure on the provision of a building from qualifying for plant and machinery, with prescribed exceptions.

List A in Section 21 of the 2001 Act sets out those assets treated as parts of buildings:

- Walls, floors, ceilings, doors, gates, shutters, windows and stairs.
- Mains services, and systems, for water, electricity and gas.
- Waste disposal systems.
- Sewerage and drainage systems.
- Shafts, or other structures, in which lifts, hoists, escalators and moving walkways are installed.
- Fire safety systems.

Similarly, List B in Section 22 identifies excluded structures and other assets.

Both sections are, however, subject to Section 23. This section sets out expenditure, which, although being part of a building, may still be expenditure on the provision of Plant and Machinery.

Sections 21 and 22 do not affect the question whether expenditure on any item in List C is expenditure on the provision of Plant or Machinery.

List C in Section 23 is reproduced below:

1. Machinery (including devices for providing motive power) not within any other item in this list.
2. Gas and sewerage systems provided mainly –
    a.   to meet the particular requirements of the qualifying activity, or
    b.   to serve particular plant or machinery used for the purposes of the qualifying activity.
3. Omitted.
4. Manufacturing or processing equipment; storage equipment (including cold rooms); display equipment; and counters, checkouts and similar equipment.
5. Cookers, washing machines, dishwashers, refrigerators and similar equipment; washbasins, sinks, baths, showers, sanitaryware and similar equipment; and furniture and furnishings.
6. Hoists.
7. Sound insulation provided mainly to meet the particular requirements of the qualifying activity.
8. Computer, telecommunication and surveillance systems (including their wiring or other links).
9. Refrigeration or cooling equipment.
10. Fire alarm systems; sprinkler and other equipment for extinguishing or containing fires.
11. Burglar alarm systems.
12. Strong rooms in bank or building society premises; safes.
13. Partition walls, where moveable and intended to be moved in the course of the qualifying activity.
14. Decorative assets provided for the enjoyment of the public in hotel, restaurant or similar trades.
15. Advertising hoardings; signs, displays and similar assets.
16. Swimming pools (including diving boards, slides & structures on which such boards or slides are mounted).
17. Any glasshouse constructed so that the required environment (namely, air, heat, light, irrigation and temperature) for the growing of plants is provided automatically by means of devices forming an integral part of its structure.
18. Cold stores.
19. Caravans provided mainly for holiday lettings.
20. Buildings provided for testing aircraft engines run within the buildings.
21. Moveable buildings intended to be moved in the course of the qualifying activity.
22. The alteration of land for the purpose only of installing Plant or Machinery.
23. The provision of dry docks.
24. The provision of any jetty or similar structure provided mainly to carry Plant or Machinery.
25. The provision of pipelines or underground ducts or tunnels with a primary purpose of carrying utility conduits.
26. The provision of towers to support floodlights.
27. The provision of –
    a.   any reservoir incorporated into a water treatment works, or
    b.   any service reservoir of treated water for supply within any housing estate or other particular locality.
28. The provision of –
    a.   silos provided for temporary storage, or
    b.   storage tanks.
29. The provision of slurry pits or silage clamps.
30. The provision of fish tanks or fish ponds.
31. The provision of rails, sleepers and ballast for a railway or tramway.
32. The provision of structures and other assets for providing the setting for any ride at an amusement park or exhibition.
33. The provision of fixed zoo cages.

Main Pool Plant and Machinery

Capital Allowances on main pool plant and machinery are currently given in the form of writing down allowances at the current rate of 18% per annum on a reducing balance basis. For every £100 of qualifying expenditure £18 is claimable in year 1, £14.76 in year 2 and so on until either all of the allowances have been claimed, or the asset is sold.

Integral Features

The category of qualifying expenditure on 'integral features' was introduced with effect from April 2008. The following items are integral features:

- An electrical system (including a lighting system)
- A cold water system
- A space or water heating system, a powered system of ventilation, air cooling or air purification, and any floor or ceiling comprised in such a system
- A lift, an escalator or a moving walkway
- External solar shading

Integral Features attract a writing down allowance of 8% per annum which reduces to 6% per annum from April 2019, on a reducing balance basis.

The Integral Features legislation introduced certain assets which were not previously allowable as plant and machinery. These assets include general power and lighting, thermal insulation to existing buildings, external solar shading and cold-water systems.

Thermal Insulation

From April 2008, expenditure incurred on the installation of thermal insulation to existing buildings qualifies as Integral Feature plant and machinery which is available on a reducing balance basis at 8% per annum.

Long-Life Assets

A writing down allowance of 8% per annum is available on long-life assets.

A long-life asset is defined as plant and machinery that has an expected useful economic life of at least 25 years. The useful economic life is taken as the period from of first use until it is likely to cease to be used as a fixed asset of any business. It is important to note that this is likely to be a shorter period than an asset's physical life.

Plant and machinery provided for use in a building used wholly, or mainly, as a dwelling house, showroom, hotel, office, retail shop, or similar premises, or for purposes ancillary to such use, cannot be classified as a long-life asset.

In contrast certain plant and machinery assets in buildings such as factories, cinemas, hospitals, etc. could potentially be treated as long-life assets.

Case Law

The fact that an item appears in List C does not automatically mean that it will qualify for capital allowances. It only means that it may potentially qualify.

Guidance about what can qualify as plant is found in case law dating back to 1887. The case of *Wimpy International Ltd and Associated Restaurants Ltd v Warland* in the late 1980s is one of the most important case law references for determining what can qualify as plant.

The Judge in that case applied three tests when considering whether, or not, an item is plant.

1. Is the item stock in trade? If the answer is yes, then the item is not plant.
2. Is the item used for carrying on the business? In order to pass the business use test the item must be employed in carrying on the business; it is not enough for the asset to be simply used in the business. For example, product display lighting in a retail store may be plant but general lighting in a warehouse would fail the test. *(Please note, this case law relates to the pre-Integral Feature Legislation, which introduced lighting as an eligible asset.)*
3. Is the item the business premises, or part of the business premises? An item cannot be plant if it fails the premises test, i.e. if the business use is the premises itself, or part of the premises, or a place in which the business is conducted. The meaning of 'part of the premises' in this context should not be confused with real property law. HMRC's

internal manuals suggest there are four general factors to be considered, each of which is a question of fact and degree:

- Does the item appear visually to retain a separate identity?
- With what degree of permanence has it been attached to the building?
- To what extent is the structure complete without it?
- To what extent is it intended to be permanent, or alternatively, is it likely to be replaced within a short period?

Certain assets will qualify as plant in most cases. However, many others need to be considered on a case-by-case basis. For example, decorative assets in a hotel restaurant may be plant, but similar assets in an office reception area may be ineligible.

Refurbishment Schemes

Building refurbishment projects will typically be a mixture of capital costs and revenue expenses, unless the works are so extensive that they are more appropriately classified as a redevelopment. A straightforward repair or a 'like for like' replacement of part of an asset would be a revenue expense, meaning that the entire amount can be deducted from taxable profits in the same year.

Where capital expenditure is incurred which is incidental to the installation of plant or machinery then Section 25 of the Capital Allowances Act 2001 allows it to be treated as part of the expenditure on the qualifying item. Incidental expenditure will often include parts of the building that would be otherwise disallowed, as shown in the Lists reproduced above. For example, the cost of forming a lift shaft inside an existing building would be deemed to be part of the expenditure on the provision of the new lift.

The extent of the application of section 25 was reviewed for the first time by the Special Commissioners in December 2007 and by the First Tier Tribunal (Tax Chamber) in December 2009, in the case of JD Wetherspoon. The key areas of expenditure considered were overheads and preliminaries where it was held that such costs could be allocated on a pro-rata basis; decorative timber panelling which was found to be part of the premises and so ineligible for allowances; toilet lighting which was considered to provide an attractive ambience and qualified for allowances; and incidental building alterations of which enclosing walls to toilets and kitchens and floor finishes did not qualify except for tiled splash backs, bespoke toilet cubicles and drainage did qualify, along with the related sanitary fittings and kitchen equipment.

The Enhanced Capital Allowances Scheme

The scheme is one of a series of measures introduced to ensure that the UK meets its target for reducing greenhouse gases under the Kyoto Protocol. 100% first year allowances are available on products included on the Energy Technology List published on the website at www.eca.gov.uk and other technologies supported by the scheme. All businesses will be able to claim the enhanced allowances, but only investments in new and unused Machinery and Plant can qualify.

There are currently 18 technologies with multiple sub-technologies currently covered by the scheme:

- Air-to-air energy recovery
- Automatic monitoring and targeting (AMT)
- Boiler equipment
- Combined heat and power (CHP)
- Compressed air equipment
- Heat pumps
- Heating ventilation and air conditioning (HVAC) equipment
- High speed hand air dryers
- Illuminated signs
- Lighting
- Motors and drives
- Pipework insulation
- Portable energy monitoring equipment
- Radiant and warm air heaters
- Refrigeration equipment
- Solar thermal systems and collectors
- Uninterruptible power supplies
- Waste heat to electricity conversion equipment

Finance Act 2003 introduced a new category of environmentally beneficial plant and machinery qualifying for 100% first-year allowances. The Water Technology List includes 14 technologies:

- Cleaning in place equipment
- Efficient showers
- Efficient taps
- Efficient toilets
- Efficient washing machines
- Flow controllers
- Greywater recovery and reuse equipment
- Leakage detection equipment
- Meters and monitoring equipment
- Rainwater harvesting equipment
- Small scale slurry and sludge dewatering equipment
- Vehicle wash water reclaim units
- Water efficient industrial cleaning equipment
- Water management equipment for mechanical seals

Buildings and structures and long-life assets, as defined above, cannot qualify under the scheme. However, following the introduction of the integral features rules, lighting in any non-residential building may potentially qualify for enhanced capital allowances, if it meets the relevant criteria. The same would apply to residential common areas.

A limited payable ECA tax credit equal to 19% of the loss surrendered was also introduced for UK companies in April 2008.

From April 2012, expenditure on plant and machinery for which tariff payments are received under the renewable energy schemes introduced by the Department of Energy and Climate Change (Feed-in Tariffs or Renewable Heat Incentives), will not be entitled to enhanced capital allowances.

The 2018 Budget announced that from April 2020, this accelerated ECA benefit will no longer be available.

Structure and Building Allowances

On 29 October 2018 the government published a Technical Note containing details of a new Structures and Buildings Allowance (SBA) for expenditure incurred on non-residential structures and buildings, following an announcement in the October 2018 budget.

Initially, the SBA was available at a flat rate of 2% per annum over a 50-year period, but this increased to 3% per annum over a 33-year period following the March 2020 Budget. This relates to expenditure incurred on new commercial structures and buildings, including costs for new conversions, or renovations, incurred on or after 29 October 2018, but only where all the contracts for the physical construction works were entered into on, or after 29 October 2018.

The SBA extends to landlord capital contribution expenditure.

The relief will be available when the building, or structure, first comes into use and will be available for both UK and overseas assets, provided the business is within the charge to UK tax. Excluded from this relief is expenditure on dwelling houses.

The claimant must have an interest in the land on which the structure, or building is constructed.

The SBA expenditure will not qualify for the annual investment allowance (AIA).

The relief will cease to be available if the building, or structure, is brought into residential use, or if it is demolished. However, SBA continues to be available for periods of temporary disuse.

If the building, or structure is sold, the new owner takes over the remainder of the residue of the SBA over the remaining 50-year period.

Structures and buildings include offices, retail and wholesale premises, walls, bridges, tunnels, factories and warehouses. Capital expenditure on renovations, or conversions, of existing commercial structures, or buildings, will also qualify. Other qualifying uses include mines, transport undertakings, investment management businesses, and professions, or vocations, but UK, or EEA, furnished holiday lettings businesses will not qualify for SBA.

The costs of construction will include only the net direct costs related to physically constructing the building, or structure, after any discounts, refunds, or other adjustments. This will include demolition costs, or any land alterations necessary for construction and direct costs required to bring the building, or structure, into existence.

SBA qualifying expenditure will also be applicable to purchases of second-hand buildings, or structures. The basis of claim will be dependent upon the vendor's holding structure and whether the property has been used, or unused, at the time of acquisition.

Where a building, or structure, is acquired and subsequently altered, or renovated, this will trigger a new 50-year SBA qualification period, as will any further new streams of qualifying expenditure.

It is important to note that expenditure on plant and machinery does not qualify for SBA. At present, however, the draft legislation excludes the facility for a purchaser to reclassify expenditure on plant and machinery, which has been included within a prior claim for SBA.

Expenditure on qualifying land remediation will also not qualify for SBA.

Annual Investment Allowance

The Annual Investment Allowance (AIA) is available to all businesses of any size and allows a deduction for the whole AIA of qualifying expenditure on plant and machinery, including integral features and long-life assets. The AIA rates have fluctuated over the years:

- 1 April 2014 to 31 December 2015 – £500,000
- 1 January 2016 to 31 December 2018 – £200,000
- Currently, the AIA rate from 1 January 2019 to 31 December 2020 is £1,000,000.

For accounting periods less, or greater than, 12 months, or if claiming in periods where the rates have changed, time apportionment rules will apply to calculate hybrid rates applicable to the period of claim.

Enterprise Zones

The creation of 11 Enterprise Zones was announced in the 2011 Budget. Additional zones have since been added bringing the number to 24 in total. Originally introduced in the early 1980s as a stimulus to commercial development and investment, these zones had become virtually non-existent.

Enterprise zones benefit from a number of reliefs, including a 100% first year allowance for new and unused non-leased plant and machinery assets, where there is a focus on high-value manufacturing.

Business Premises Renovation Allowance

The Business Premises Renovation Allowance (BPRA) was first announced in December 2003. The idea behind this scheme was to bring long-term vacant properties back into productive use by providing 100% capital allowances for the cost of renovating and converting unused premises in disadvantaged areas. The legislation was included in Finance Act 2005 and was finally implemented on 11 April 2007 following EU state aid approval.

The scheme will apply to properties within the areas specified in the Assisted Areas Order 2007 and Northern Ireland.

BPRA is available to both individuals and companies who own, or lease, business property which has been unused for 12 months, or more. Allowances will be available to a person who incurs qualifying capital expenditure on the renovation of business premises.

An announcement to extend the scheme by a further five years to 2017 was made within the 2011 Budget, along with a further 11 new designated Enterprise Zones.

Legislation was introduced in Finance Bill 2014 to clarify the scope of expenditure qualifying for relief to actual costs of construction and building work and for certain specified activities, such as architectural and surveying services. The changes came into effect for qualifying expenditure incurred on, or after, 1 April 2014 for businesses within the charge to corporation tax and 6 April 2014 for businesses within the charge to income tax.

Other Capital Allowances

Other types of allowances include those available for capital expenditure on Mineral Extraction, Research and Development, Know-How, Patents, Dredging and Assured Tenancy.

International Tax Depreciation

The UK is not the only tax regime offering investors, owners and occupiers valuable incentives to invest in plant and machinery and environmentally friendly equipment. Ireland, Australia, Malaysia and Singapore also have capital allowances regimes that are broadly similar to the UK and provide comparable levels of tax relief to businesses.

Many other overseas countries have tax depreciation regimes based on accounting treatment, instead of Capital Allowances. Some use a systematic basis over the useful life of the asset and others have prescribed methods spreading the cost over a statutory period, not always equating to the asset's useful life. Some regimes have prescribed statutory rates, whilst others have rates which have become acceptable to the tax authorities through practice.

Brickwork for Apprentices, 6th edition

By J.C. Hodge, Revised by Malcolm Thorpe

Expanded and edited by Malcolm Thorpe, a former CITB Advisor, this latest edition incorporates all the latest industry-based requirements and technical developments in the construction industry. A new feature is the e-resource facility that includes multiple choice questions and answers, short oral questions and answers, and practical competency checks with marking schemes. These are all matched to current programmes, providing students with essential practice and revision for exam preparation.

A classic text, *Brickwork for Apprentices* has been the established reference on brickwork for generations of bricklayers. Continuously in print since 1944, John Hodge's classic text has now been revised in its sixth edition and brought fully in line with the latest Building Regulations and requirements for City & Guilds courses. This is an essential text for qualified bricklayers and other professionals working in construction, as well as students new to the industry and wishing to embark on a career in bricklaying.

December 2020: 416 pp
ISBN: 9780367624347

To Order
Tel:+44 (0) 1235 400524
Email: tandf@bookpoint.co.uk

For a complete listing of all our titles visit:
www.tandf.co.uk

Taylor & Francis
Taylor & Francis Group

Value Added Tax

Introduction

Value Added Tax (VAT) is a tax on the consumption of goods and services. The UK introduced a domestic VAT regime when it joined the European Community in 1973. The principal source of European law in relation to VAT is Council Directive 2006/112/EC, a recast of Directive 77/388/EEC, which is currently restated and consolidated in the UK through the VAT Act 1994 and various Statutory Instruments, as amended by subsequent Finance Acts.

VAT Notice 708: Buildings and construction (December 2020) provides HMRC's interpretation of the VAT law in connection with construction works, however, the UK VAT legislation should always be referred to in conjunction with the publication. Recent VAT tribunals and court decisions since the date of this publication will affect the application of the VAT law in certain instances. The Notice is available on HM Revenue & Customs website at www.hmrc.gov.uk.

The Scope of VAT

VAT is payable on:

- Supplies of goods and services made in the UK;
- By a taxable person;
- In the course or furtherance of business; and
- Which are not specifically exempted or zero-rated.

Rates of VAT

There are three rates of VAT:

- A standard rate, currently 20% since January 2011;
- A reduced rate, currently 5%; and
- A zero rate of 0%.

Additionally, some supplies are exempt from VAT and others are considered outside the scope of VAT.

Recovery of VAT

When a taxpayer makes taxable supplies, he must account for VAT, known as output VAT, at the appropriate rate of either 20%, 5% or 0%. Any VAT due then has to be declared and submitted on a VAT submission to HM Revenue & Customs and will normally be charged to the taxpayer's customers.

As a VAT registered person, the taxpayer is entitled to reclaim from HM Revenue & Customs, commonly referred to as input VAT, the VAT incurred on their purchases and expenses directly related to its business activities in respect of standard-rated, reduced-rated and zero-rated supplies. A taxable person cannot, however, reclaim VAT that relates to any non-business activities (but see below) or, depending on the amount of exempt supplies they made, input VAT may be restricted or not recoverable.

At predetermined intervals the taxpayer will pay to HM Revenue & Customs the excess of VAT collected over the VAT they can reclaim. However, if the VAT reclaimed is more than the VAT collected, the taxpayer, who will be in a net repayment position, can reclaim the difference from HM Revenue & Customs.

Example

X Ltd constructs a block of flats. It sells long leases to buyers for a premium. X Ltd has constructed a new building designed as a dwelling and will have granted a long lease. This first sale of a long lease is a VAT zero-rated supply. This means any

VAT incurred in connection with the development, which X Ltd will have properly paid (e.g. payments for consultants and certain preliminary services) will be recoverable. For reasons detailed below, the contractor employed by X Ltd will have charged VAT on his construction services at the zero rate of VAT.

Use for Business and Non-Business Activities

Where a supply relates partly to business use and partly to non-business use, then the basic rule is that it must be apportioned on a fair and reasonable basis so that only the business element is potentially recoverable. In some cases, VAT on land, buildings and certain construction services, purchased for both business and non-business use, could be recovered in full by applying what is known as 'Lennartz' accounting, to reclaim VAT relating to the non-business use and account for VAT on the non business use over a maximum period of 10 years. Following an ECJ case restricting the scope of this approach, its application to immovable property was removed completely in January 2011 by HMRC (business brief 53/10), when UK VAT law was amended to comply with EU Directive 2009/162/EU.

Taxable Persons

A taxable person is an individual, firm, company, etc., who is required to be registered for VAT. A person who makes taxable supplies above certain turnover limits is compulsorily required to be VAT registered. The current registration limit (since 1 April 2017), known as the VAT threshold, is £85,000. If the threshold is exceeded in any 12 month rolling period, or there is an expectation that the value of the taxable supplies in a single 30 day period, or goods are received into the UK from the EU worth more than the £85,000, then registration for UK VAT is compulsory.

A person who makes taxable supplies below the limit is still entitled to be registered on a voluntary basis if they wish, for example, in order to recover input VAT incurred in relation to those taxable supplies, however output VAT will then become due on the sales and must be accounted for.

In addition, a person who is not registered for VAT in the UK but acquires goods from another EC member state or make distance sales in the UK above certain value limits, may be required to register for VAT in the UK.

VAT Exempt Supplies

Where a supply is exempt from VAT this means that no output VAT is payable – but equally the person making the exempt supply cannot normally recover any of the input VAT on their own costs relating to that exempt supply.

Generally commercial property transactions such as leasing of land and buildings are exempt unless a landlord chooses to standard-rate its interest in the property by a applying for an option to tax. This means that VAT is added to rental income and also that VAT incurred, on, say, an expensive refurbishment, is recoverable.

Supplies outside the scope of VAT

Supplies are outside the scope of VAT if they are:

- Made by someone who is not a taxable person;
- Made outside the UK; or
- Not made in the course or furtherance of business.

In course or furtherance of business

VAT must be accounted for on all taxable supplies made in the course or furtherance of business, with the corresponding recovery of VAT on expenditure incurred.

If a taxpayer also carries out non-business activities, then VAT incurred in relation to such supplies is generally not recoverable.

In VAT terms, business means any activity continuously performed which is mainly concerned with making supplies for a consideration. This includes:

- Anyone carrying on a trade, vocation or profession;
- The provision of membership benefits by clubs, associations and similar bodies in return for a subscription or other consideration; and
- Admission to premises for a charge.

It may also include the activities of other bodies including charities and non-profit making organizations.

Examples of non-business activities are:

- Providing free services or information.
- Maintaining some museums or particular historic sites.
- Publishing religious or political views.

Construction Services

In general, the provision of construction services by a contractor will be VAT standard rated at 20%, however, there are a number of exceptions for construction services provided in relation to certain relevant residential properties and charitable buildings.

The supply of building materials is VAT standard rated at 20%, however, where these materials are supplied and installed as part of the construction services, the VAT liability of those materials follows that of the construction services supplied.

Zero-rated construction services

The following construction services are VAT zero-rated, including the supply of related building materials.

The construction of new dwellings

The supply of services in the course of the construction of a new building designed for use as a dwelling or number of dwellings is zero-rated, other than the services of an architect, surveyor or any other person acting as a consultant or in a supervisory capacity.

The following basic conditions must ALL be satisfied in order for the works to qualify for zero-rating:

1. A qualifying building has been, is being, or will be constructed;
2. Services are made 'in the course of the construction' of that building;
3. Where necessary, you hold a valid certificate; and
4. Your services are not specifically excluded from zero-rating.

The construction of a new building for 'relevant residential or charitable' use

The supply of services in the course of the construction of a building designed for use as a relevant residential Purpose (RRP), or relevant charitable purpose (RCP), is zero-rated, other than the services of an architect, surveyor or any other person acting as a consultant or in a supervisory capacity.

A 'relevant residential' use building means:

1. A home or other institution providing residential accommodation for children;
2. A home or other institution providing residential accommodation with personal care for persons in need of personal care by reason of old age, disablement, past or present dependence on alcohol or drugs or past or present mental disorder;
3. A hospice;
4. Residential accommodation for students or school pupils;
5. Residential accommodation for members of any of the armed forces;

6.   A monastery, nunnery, or similar establishment; or
7.   An institution which is the sole or main residence of at least 90% of its residents.

A 'relevant residential' purpose building does not include use as a hospital, a prison, or similar institution, or as a hotel, inn, or similar establishment.

A 'relevant charitable' purpose means use by a charity in either, or both of the following ways:

1.   Otherwise than in the course or furtherance of a business; or
2.   As a village hall, or similarly in providing social or recreational facilities for a local community.

Non-qualifying use, which is not expected to exceed 10% of the time the building is normally available for use, can be ignored. The calculation of business use can be based on time, floor area, or head count, subject to approval being acquired from HM Revenue & Customs.

The construction services can only be zero-rated if a certificate is given by the end user to the contractor carrying out the works, confirming that the building is to be used for a qualifying purpose, i.e. for a 'relevant residential or charitable' purpose. It follows that such services can only be zero-rated when supplied to the end user and, unlike supplies relating to dwellings, supplies by subcontractors cannot be zero-rated.

The construction of an annex used for a 'relevant charitable' purpose

Construction services provided in the course of construction of an annexe for use entirely, or partly for a 'relevant charitable' purpose, can be zero-rated.

In order to qualify, the annexe must:

1.   Be capable of functioning independently from the existing building;
2.   Have its own main entrance; and
3.   Be covered by a qualifying use certificate.

The conversion of a non-residential building into dwellings, or the conversion of a building from non-residential use to 'relevant residential' use, where the supply is to a 'relevant' housing association

The supply to a 'relevant' housing association in the course of conversion of a non-residential building, or non-residential part of a building, into:

1.   A new eligible dwelling designed as a dwelling, or number of dwellings; or
2.   A building, or part of a building, for use solely for a relevant residential purpose.

Any services related to the conversion, other than the services of an architect, surveyor or any person acting as a consultant or in a supervisory capacity, are zero-rated.

A 'relevant' housing association is defined as:

1.   A private registered provider of social housing;
2.   A registered social landlord within the meaning of Part I of the Housing Act 1996 (Welsh registered social landlords);
3.   A registered social landlord within the meaning of the Housing (Scotland) Act 2001 (Scottish registered social landlords); or
4.   A registered housing association within the meaning of Part II of the Housing (Northern Ireland) Order 1992 (Northern Irish registered housing associations).

If the building is to be used for a 'relevant residential' purpose, the housing association should issue a qualifying use certificate to the contractor completing the works. Subcontractors' services that are not made directly to a relevant housing association are standard-rated.

The development of a residential caravan park

The supply in the course of the construction of any civil engineering work 'necessary for' the development of a permanent park for residential caravans, and any services related to the construction, other than the services of an architect, surveyor or any person acting as a consultant or in a supervisory capacity, are zero-rated when a new permanent park is being developed, the civil engineering works are necessary for the development of the park and the services are not specifically excluded from zero-rating. This includes access roads, paths, drainage, sewerage and the installation of mains water, power and gas supplies.

Certain building alterations for 'disabled' persons

Certain goods and services supplied to a 'disabled' person, or a charity making these items and services available to 'disabled' persons, can be zero-rated. The recipient of these goods or services needs to give the supplier an appropriate written declaration that they are entitled to benefit from zero rating.

The following services (amongst others) are zero-rated:

1. The installation of specialist lifts and hoists and their repair and maintenance;
2. The construction of ramps, widening doorways or passageways including any preparatory work and making good work;
3. The provision, extension and adaptation of a bathroom, washroom or lavatory; and
4. Emergency alarm call systems.

Sale of Reconstructed Buildings

A protected building is not to be regarded as substantially reconstructed unless, when the reconstruction is completed, the reconstructed building incorporates no more of the original building than the external walls, together with other external features of architectural or historical interest.

DIY Builders and Converters

Private individuals who decide to construct their own home are able to reclaim VAT they pay on goods they use to construct their home by use of a special refund mechanism made by way of an application to HM Revenue & Customs. This also applies to services provided in the conversion of an existing non-residential building to form a new dwelling.

The scheme is meant to ensure that private individuals do not suffer the burden of VAT if they decide to construct their own home.

Charities may also qualify for a refund on the purchase of materials incorporated into a building used for non-business purposes where they provide their own free labour for the construction of a 'relevant charitable' use building.

Reduced-rated Construction Services

The following construction services are subject to the reduced rate of VAT of 5%, including the supply of related building materials.

Conversion – changing the number of dwellings

In order to qualify for the 5% rate, there must be a different number of 'single household dwellings' within a building than there were before commencement of the conversion works. A 'single household dwelling' is defined as a dwelling that is designed for occupation by a single household.

These conversions can be from 'relevant residential' purpose buildings, non-residential buildings and houses in multiple occupation.

A house in multiple occupation conversion

This relates to construction services provided in the course of converting a 'single household dwelling', a number of 'single household dwellings', a non-residential building or a 'relevant residential' purpose building into a house for multiple occupation, such as bed sit accommodation.

A special residential conversion

A special residential conversion involves the conversion of a 'single household dwelling', a house in multiple occupation, or a non-residential building into a 'relevant residential' purpose building. such as student accommodation or a care home.

Renovation of derelict dwellings

The provision of renovation services in connection with a dwelling or 'relevant residential' purpose building that has been empty for two or more years prior to the date of commencement of construction works can be carried out at a reduced rate of VAT of 5%.

Installation of energy saving materials

The supply and installation of certain energy saving materials including insulation, draught stripping, central heating, hot water controls and solar panels in a residential building, or a building used for a relevant charitable purpose.

Buildings that are used by charities for non-business purposes, and/or as village halls, are not eligible for the reduced rate for the supply of energy saving materials.

Grant-funded installation of heating equipment or connection of a gas supply

The grant-funded supply and installation of heating appliances, connection of a mains gas supply, supply, installation, maintenance and repair of central heating systems, and supply and installation of renewable source heating systems, to qualifying persons. A qualifying person is someone aged 60 or over or is in receipt of various specified benefits.

Grant-funded installation of security goods

The grant-funded supply and installation of security goods to a qualifying person.

Housing alterations for the elderly

Certain home adaptations that support the needs of elderly people are reduced rated.

Building Contracts

Design and build contracts

If a contractor provides a design and build service relating to works to which the reduced or zero rate of VAT is applicable, then any design costs incurred by the contractor will follow the VAT liability of the principal supply of construction services.

Management contracts

A management contractor acts as a main contractor for VAT purposes and the VAT liability of his services will follow that of the construction services provided. If the management contractor only provides advice without engaging trade contractors, his services will be VAT standard rated.

Construction Management and Project Management

The project manager or construction manager is appointed by the client to plan, manage and coordinate a construction project. This will involve establishing competitive bids for all the elements of the work and the appointment of trade contractors. The trade contractors are engaged directly by the client for their services. The VAT liability of the trade contractors will be determined by the nature of the construction services they provide, and the building being constructed.

The fees of the construction manager or project manager will be VAT standard rated. If the construction manager also provides some construction services, these works may be zero or reduced rated if the works qualify.

Liquidated and Ascertained Damages

Liquidated damages are outside of the scope of VAT as compensation. The employer should not reduce the VAT amount due on a payment under a building contract on account of a deduction of damages. In contrast, an agreed reduction in the contract price will reduce the VAT amount.

Similarly, in certain circumstances, HM Revenue & Customs may agree that a claim by a contractor under a JCT, or other form of contract, is also compensation payment and outside the scope of VAT.

Reverse Charge for Building and Construction Services

A VAT reverse charge was introduced on 1 March 2021 for certain building and construction services. This measure has been introduced to combat missing trader fraud in the construction industry, removing the opportunity for fraudsters to charge VAT and then go missing, before paying it over to the HMRC.

For certain supplies of construction services, the customer will now be liable to account to HMRC for the VAT in respect of those purchases rather than the supplier. This reverse charge will apply through the supply chain where payments are required to be reported through the Construction Industry Scheme (CIS) up to the point where the customer receiving the supply is no longer a business that makes supplies of specified services (end users).

The introduction of a reverse charge does not change the liability of the supply of the specified services: it is the way in which the VAT on those supplies is accounted for. Rather than the supplier charging and accounting for the VAT, the recipient of those supplies accounts for the VAT its return instead of paying the VAT amount to its supplier. It will be able to reclaim that VAT amount as input tax, subject to the normal rules. The supplier will need to issue a VAT invoice that indicates the supplies are subject to the reverse charge.

The types of construction services covered by the reverse charge are based on the definition of 'construction operations' used in CIS under section 74 of the Finance Act 2004.

As well as excluding supplies of specified services to end users, the reverse charge does not capture supplies of specified services where the supplier and customer are connected in a particular way, and for supplies between landlords and tenants.

Building Regulations Pocket Book

Ray Tricker and Samantha Alford

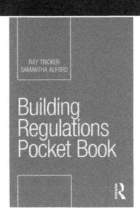

This handy guide provides you with all the information you need to comply with the UK Building Regulations and Approved Documents. On site, in the van, in the office, wherever you are, this is the book you'll refer to time and time again to double check the regulations on your current job. The *Building Regulations Pocket Book* is the must have reliable and portable guide to compliance with the Building Regulations.

Part 1 provides an overview of the Building Act

Part 2 offers a handy guide to the dos and don'ts of gaining the Local Council's approval for Planning Permission and Building Regulations Approval

Part 3 presents an overview of the requirements of the Approved Documents associated with the Building Regulations

Part 4 is an easy to read explanation of the essential requirements of the Building Regulations that any architect, builder or DIYer needs to know to keep their work safe and compliant on both domestic or non-domestic jobs

This book is essential reading for all building contractors and sub-contractors, site engineers, building engineers, building control officers, building surveyors, architects, construction site managers and DIYers. Homeowners will also find it useful to understand what they are responsible for when they have work done on their home (ignorance of the regulations is no defence when it comes to compliance!).

February 2018: 456 pp
ISBN: 9780815368380

To Order
Tel:+44 (0) 1235 400524
Email: tandf@bookpoint.co.uk

For a complete listing of all our titles visit:
www.tandf.co.uk

Taylor & Francis
Taylor & Francis Group

The Aggregates Levy

The Aggregates Levy came into operation on 1 April 2002 in the UK, except for Northern Ireland where it was phased in over five years from 2003.

It was introduced to ensure that the external costs associated with the exploitation of aggregates are reflected in the price of aggregate, and to encourage the use of recycled aggregate. There continues to be strong evidence that the levy is achieving its environmental objectives, with sales of primary aggregate down and production of recycled aggregate up. The Government expects that the rates of the levy will at least keep pace with inflation over time, although it accepts that the levy is still bedding in.

The rate of the levy will continue to be £2.00 per tonne from 1 April 2021 and is levied on anyone considered to be responsible for commercially exploiting 'virgin' aggregates in the UK and should naturally be passed by price increase to the ultimate user.

All materials falling within the definition of 'Aggregates' are subject to the levy unless specifically exempted.

It does not apply to clay, soil, vegetable or other organic matter.

The intention is that it will:

- Encourage the use of alternative materials that would otherwise be disposed of to landfill sites
- Promote development of new recycling processes, such as using waste tyres and glass
- Promote greater efficiency in the use of virgin aggregates
- Reduce noise and vibration, dust and other emissions to air, visual intrusion, loss of amenity and damage to wildlife habitats

Definitions

'Aggregates' means any rock, gravel or sand which is extracted or dredged in the UK for aggregates use. It includes whatever substances are for the time being incorporated in it or naturally occur mixed with it.

'Exploitation' is defined as involving any one or a combination of any of the following:

- Being removed from its original site, a connected site which is registered under the same name as the originating site or a site where it had been intended to apply an exempt process to it, but this process was not applied
- Becoming subject to a contract or other agreement to supply to any person
- Being used for construction purposes
- Being mixed with any material or substance other than water, except in permitted circumstances

The definition of 'aggregate being used for construction purposes' is when it is:

- Used as material or support in the construction or improvement of any structure
- Mixed with anything as part of a process of producing mortar, concrete, tarmacadam, coated roadstone or any similar construction material

Incidence

It is a tax on primary aggregates production – i.e. 'virgin' aggregates won from a source and used in a location within the UK territorial boundaries (land or sea). The tax is not levied on aggregates which are exported or on aggregates imported from outside the UK territorial boundaries.

It is levied at the point of sale.

Exemption from Tax

An 'aggregate' is exempt from the levy if it is:

- Material which has previously been used for construction purposes
- Aggregate that has already been subject to a charge to the Aggregates Levy
- Aggregate which was previously removed from its originating site before the start date of the levy
- Aggregate which is moved between sites under the same Aggregates Levy Registration
- Aggregate which is removed to a registered site to have an exempt process applied to it
- Aggregate which is removed to any premises where china clay or ball clay will be extracted from the aggregate
- Aggregate which is being returned to the land from which it was won provided that it is not mixed with any material other than water
- Aggregate won from a farm land or forest where used on that farm or forest
- Rock which has not been subjected to an industrial crushing process
- Aggregate won by being removed from the ground on the site of any building or proposed building in the course of excavations carried out in connection with the modification or erection of the building and exclusively for the purpose of laying foundations or of laying any pipe or cable
- Aggregate won by being removed from the bed of any river, canal or watercourse or channel in or approach to any port or harbour (natural or artificial), in the course of carrying out any dredging exclusively for the purpose of creating, restoring, improving or maintaining that body of water
- Aggregate won by being removed from the ground along the line of any highway or proposed highway in the course of excavations for improving, maintaining or constructing the highway otherwise than purely to extract the aggregate
- Drill cuttings from petroleum operations on land and on the seabed
- Aggregate resulting from works carried out in exercise of powers under the New Road and Street Works Act 1991, the Roads (Northern Ireland) Order 1993 or the Street Works (Northern Ireland) Order 1995
- Aggregate removed for the purpose of cutting of rock to produce dimension stone, or the production of lime or cement from limestone
- Aggregate arising as a waste material during the processing of the following industrial minerals:
  - anhydrite
  - ball clay
  - barites
  - calcite
  - china clay
  - clay, coal, lignite and slate
  - feldspar
  - flint
  - fluorspar
  - fuller's earth
  - gems and semi-precious stones
  - gypsum
  - any metal or the ore of any metal
  - muscovite
  - perlite
  - potash
  - pumice
  - rock phosphates
  - sodium chloride
  - talc
  - vermiculite
  - spoil from the separation of the above industrial minerals from other rock after extraction
  - material that is mainly but not wholly the spoil, waste or other by-product of any industrial combustion process or the smelting or refining of metal

Anything that consists 'wholly or mainly' of the following is exempt from the levy (note that 'wholly' is defined as 100% but 'mainly' as more than 50%, thus exempting any contained aggregates amounting to less than 50% of the original volumes):

- Clay, soil, vegetable or other organic matter
- Drill cuttings from oil exploration in UK waters
- Material arising from utility works, if carried out under the New Roads and Street Works Act 1991

However, when ground that is more than half clay is mixed with any substance (for example, cement or lime) for the purpose of creating a firm base for construction, the clay becomes liable to Aggregates Levy because it has been mixed with another substance for the purpose of construction.

Anything that consists completely of the following substances is exempt from the levy:

- Spoil, waste or other by-products from any industrial combustion process or the smelting or refining of metal – for example, industrial slag, pulverized fuel ash and used foundry sand. If the material consists completely of these substances at the time it is produced it is exempt from the levy, regardless of any subsequent mixing
- Aggregate necessarily arising from the footprint of any building for the purpose of laying its foundations, pipes or cables. It must be lawfully extracted within the terms of any planning consent
- Aggregate necessarily arising from navigation dredging
- Aggregate necessarily arising from the ground in the course of excavations to improve, maintain or construct a highway or a proposed highway
- Aggregate necessarily arising from the ground in the course of excavations to improve, maintain or construct a railway, monorail or tramway

Relief from the levy either in the form of credit or repayment is obtainable where:

- It is subsequently exported from the UK in the form of aggregate
- It is used in an exempt process
- Where it is used in a prescribed industrial or agricultural process
- It is waste aggregate disposed of by dumping or otherwise, e.g. sent to landfill or returned to the originating site

An exemption for aggregate obtained as a by-product of railway, tramway and monorail improvement, maintenance and construction was introduced in 2007.

Discounts

Water which is added to the aggregate after the aggregate has been won (washing, dust dampening, etc.) may be discounted from the tax calculations. There are two accepted options by which the added water content can be calculated.

The first is to use HMRC's standard added water percentage discounts listed below:

- Washed sand 7%
- Washed gravel 3.5%
- Washed rock/aggregate 4%

Alternatively a more exact percentage can be agreed for dust dampening of aggregates.

Whichever option is adopted, it must be agreed in writing in advance with HMRC.

Impact

The British Aggregates Association suggested that the additional cost imposed by quarries is more likely to be in the order of £3.40 per tonne on mainstream products, applying an above average rate on these in order that by-products and low grade waste products can be held at competitive rates, as well as making some allowance for administration and increased finance charges.

With many gravel aggregates costing in the region of £20.00 per tonne, there is a significant impact on construction costs.

Avoidance

An alternative to using new aggregates in filling operations is to crush and screen rubble which may become available during the process of demolition and site clearance as well as removal of obstacles during the excavation processes.

Example: Assuming that the material would be suitable for fill material under buildings or roads, a simple cost comparison would be as follows (note that for the purpose of the exercise, the material is taken to be 1.80 tonnes per m³ and the total quantity involved less than 1,000 m³):

Disposing of site material:	£/m³	£/tonne
Cost of removing materials from site	30.92	17.18
Importing fill material:		
Cost of 'new' aggregates delivered to site	34.31	19.06
Addition for Aggregates Tax	3.60	2.00
Total cost of disposing waste aggregate and importing fill materials	68.83	38.24
Crushing site materials:		
Transportation of material from excavations or demolition to stockpiles	0.89	0.50
Transportation of material from temporary stockpiles to the crushing plant	2.39	1.33
Establishing plant and equipment on site; removing on completion	2.39	1.33
Maintain and operate plant	10.75	5.97
Crushing hard materials on site	15.53	8.62
Screening material on site	2.39	1.33
Total cost of crushing site materials ready for reuse	34.34	19.08

From the above it can be seen that potentially there is a great benefit in crushing site materials for filling rather than importing fill materials.

Setting the cost of crushing against the import price would produce a saving of £6.78 per m³. If the site materials were otherwise intended to be removed from the site, then the cost benefit increases by the saved disposal cost to £33.41 per m³.

Even if there is no call for any or all of the crushed material on site, it ought to be regarded as a useful asset and either sold on in crushed form or else sold with the prospects of crushing elsewhere.

Specimen Unit Rates	Unit³	£
Establishing plant and equipment on site; removing on completion		
crushing plant	trip	1,450.00
screening plant	trip	750.00
Maintain and operate plant		
crushing plant	week	9,000.00
screening plant	week	2,200.00
Transportation of material from excavations or demolition places to temporary stockpiles	m³	2.00
Transportation of material from temporary stockpiles to the crushing plant	m³	2.00
Breaking up material on site using impact breakers		
mass concrete	m³	50.00
reinforced concrete	m³	71.00
brickwork	m³	36.00
Crushing material on site		
mass concrete not exceeding 1000 m³	m³	16.00
mass concrete 1000–5000 m³	m³	15.00
mass concrete over 5000 m³	m³	14.00
reinforced concrete not exceeding 1000 m³	m³	19.00
reinforced concrete 1000–5000 m³	m³	17.00
reinforced concrete over 5000 m³	m³	16.00
brickwork not exceeding 1000 m³	m³	15.00
brickwork 1000–5000 m³	m³	14.00
brickwork over 5000 m³	m³	13.00
Screening material on site	m³	3.00

More detailed information can be found on the HMRC website (www.hmrc.gov.uk) in Notice AGL1 Aggregates Levy published 1 April 2014 (updated 9 February 2017).

Building Surveyor's Pocket Book

Melanie Smith and Christopher Gorse

Building Surveyor's Pocket Book is an accessible encyclopaedia of matters vital to building surveyors. Well-illustrated with diagrams, pictures, tables, and graphs, it covers all essential elements of building pathology, building performance, and building construction terminology in a simple, accessible way for the practitioner and student. This Pocket Book provides a practical and portable reference text, working as a first-stop publication for those wishing to refresh their knowledge or in need of guidance on surveying practice. Working through fundamental principles in key practice areas, the book is not overly bound by the regulation and legislation of one region, and the principles can be applied internationally. This book is ideal reading for individual surveyors, practitioners, and students in building surveying, facilities management, refurbishment, maintenance, renovation, and services management. It is also of use for those interested in building forensics, building performance, pathology, and anyone studying for their RICS APC. Many other professions in architecture, contracting, engineering, and safety will also find the book of use when undertaking similar practice.

May 2021: 390pp.
ISBN: 9781138307919

To Order
+44 (0) 1235 400524
tandf@bookpoint.co.uk

For a complete listing of all our titles visit:
www.tandf.co.uk

Taylor & Francis
Taylor & Francis Group

Land Remediation

The purpose of this section is to review the general background of ground contamination, the cost implications of current legislation and to consider the various remedial measures and to present helpful guidance on the cost of Land Remediation.

It must be emphasized that the cost advice given is an average and that costs can vary considerably from contract to contract depending on individual Contractors, site conditions, type and extent of contamination, methods of working and various other factors as diverse as difficulty of site access and distance from approved tips.

We have structured this Unit Cost section to cover as many aspects of Land Remediation works as possible.

The introduction of the Landfill Directive in July 2004 has had a considerable impact on the cost of Remediation works in general and particularly on the practice of Dig and Dump. The number of Landfill sites licensed to accept Hazardous Waste has drastically reduced and inevitably this has led to increased costs.

Market forces will determine future increases in cost resulting from the introduction of the Landfill Directive and the cost guidance given within this section will require review in light of these factors.

Statutory Framework

In July 1999 new contaminated land provisions, contained in Part IIa of the Environmental Protection Act 1990 were introduced. Primary objectives of the measures included a legal definition of Contaminated Land and a framework for identifying liability, underpinned by a 'polluter pays' principle meaning that remediation should be paid for by the party (or parties) responsible for the contamination. A secondary, and indirect, objective of Part IIA is to provide the legislative context for remediation carried out as part of development activity which is controlled through the planning system. This is the domain where other related objectives, such as encouraging the recycling of brownfield land, are relevant.

Under the Act action to remediate land is required only where there are unacceptable actual or 'significant possibility of significant harm' to health, controlled waters or the environment. Only Local Authorities have the power to determine a site as Contaminated Land and enforce remediation. Sites that have been polluted from previous land use may not need remediating until the land use is changed; this is referred to as 'land affected by contamination'. This is a risk-based assessment on the site specifics in the context of future end uses. As part of planning controls, the aim is to ensure that a site is incapable of meeting the legal definition of Contaminated Land post-development activity. In addition, it may be necessary to take action only where there are appropriate, cost-effective remediation processes that take the use of the site into account.

The Environment Act 1995 amended the Environment Protection Act 1990 by introducing a new regime designed to deal with the remediation of sites which have been seriously contaminated by historic activities. The regime became operational on 1 April 2000. Local authorities and/or the Environment Agency regulate seriously contaminated sites which are known as 'special sites'. The risks involved in the purchase of potentially contaminated sites are high, particularly considering that a transaction can result in the transfer of liability for historic contamination from the vendor to the purchaser.

The contaminated land provisions of the Environmental Protection Act 1990 are only one element of a series of statutory measures dealing with pollution and land remediation that have been and are to be introduced. Others include:

* Groundwater regulations, including pollution prevention measures
* An integrated prevention and control regime for pollution
* Sections of the Water Resources Act 1991, which deals with works notices for site controls, restoration and clean up

April 2012 saw the first revision of the accompanying Part IIa Statutory Guidance. This has introduced a new categorization scheme for assessing sites under Part IIA. Category 1 is land which definitely is Contaminated Land and Category 4 is for land which definitely is not Contaminated Land. This is intended to assist prioritization of sites which pose the greatest risk.

Still included in the statutory guidance are matters of inspection, definition, remediation, apportionment of liabilities and recovery of costs of remediation. The measures are to be applied in accordance with the following criteria:

- The planning system
- The standard of remediation should relate to the present use
- The costs of remediation should be reasonable in relation to the seriousness of the potential harm
- The proposals should be practical in relation to the availability of remediation technology, impact of site constraints and the effectiveness of the proposed clean-up method

Liability for the costs of remediation rests with either the party that 'caused or knowingly permitted' contamination, or with the current owners or occupiers of the land.

Apportionment of liability, where shared, is determined by the local authority. Although owners or occupiers become liable only if the polluter cannot be identified, the liability for contamination is commonly passed on when land is sold.

If neither the polluter nor owner can be found, the clean-up is funded from public resources.

The ability to forecast the extent and cost of remedial measures is essential for both parties, so that they can be accurately reflected in the price of the land.

At the end of March 2012, the National Planning Policy Framework replaced relevant planning guidance relating to remediation, most significantly PPS 23 Planning and Pollution Control. This has been replaced by a need to investigate and assess land contamination, which must be carried out by a competent person.

The EU Landfill Directive

The Landfill (England and Wales) Regulations 2002 came into force on 15 June 2002 followed by Amendments in 2004 and 2005. These new regulations implement the Landfill Directive (Council Directive 1999/31/EC), which aims to prevent, or to reduce as far as possible, the negative environmental effects of landfill. These regulations have had a major impact on waste regulation and the waste management industry in the UK.

The Scottish Executive implements the Landfill Directive through the Landfill (Scotland) Regulations 2003 and subsequent 2013 Amendments, and the Northern Ireland Assembly implements the Directive through the Landfill Regulations (Northern Ireland) 2003, the Waste and Emissions Trading Act 2003 and the Landfill Allowances Scheme (Northern Ireland) Regulations 2004 (as amended).

In summary, the Directive requires that:

- Sites are to be classified into one of three categories: hazardous, non-hazardous or inert, according to the type of waste they will receive
- Higher engineering and operating standards will be followed
- Biodegradable waste will be progressively diverted away from landfills
- Certain hazardous and other wastes, including liquids, explosive waste and tyres will be prohibited from landfills
- Pretreatment of wastes prior to landfilling will become a requirement

On 15 July 2004 the co-disposal of hazardous and non-hazardous waste in the same landfill site ended and in July 2005 new waste acceptance criteria (WAC) were introduced which also prevents the disposal of materials contaminated by coal tar.

The effect of this Directive has been to dramatically reduce the hazardous disposal capacity post July 2004, resulting in a SIGNIFICANT increase in remediating costs. This has significantly increased travelling distance and cost for disposal to landfill. The increase in operating expenses incurred by the landfill operators has also resulted in higher tipping costs.

However, there is now a growing number of opportunities to dispose of hazardous waste to other facilities such as soil treatment centres, often associated with registered landfills potentially eliminating landfill tax. Equally, improvements in on-site treatment technologies are helping to reduce the costs of disposal by reducing the hazardous properties of materials going off site.

All hazardous materials designated for disposal off site are subject to WAC tests. Samples of these materials are taken from site to laboratories in order to classify the nature of the contaminants. These tests, which cost approximately £200 each, have resulted in increased costs for site investigations and as the results may take up to 3 weeks this can have a detrimental effect on programme.

As from 1 July 2008 the WAC derogations which have allowed oil contaminated wastes to be disposed in landfills with other inert substances were withdrawn. As a result the cost of disposing oil contaminated solids has increased.

There has been a marked slowdown in brownfield development in the UK with higher remediation costs, longer clean-up programmes and a lack of viable treatment options for some wastes.

The UK Government established the Hazardous Waste Forum in December 2002 to bring together key stakeholders to advise on the way forward on the management of hazardous waste.

Effect on Disposal Costs

Although most landfills are reluctant to commit to future tipping prices, tipping costs have generally stabilized. However, there are significant geographical variances, with landfill tip costs in the North of England typically being less than their counterparts in the Southern regions.

For most projects to remain viable there is an increasing need to treat soil in situ by bioremediation, soil washing or other alternative long-term remediation measures. Waste untreatable on site such as coal tar remains a problem. Development costs and programmes need to reflect this change in methodology.

Types of Hazardous Waste

- Sludges, acids and contaminated wastes from the oil and gas industry
- Acids and toxic chemicals from chemical and electronics industries
- Pesticides from the agrochemical industry
- Solvents, dyes and sludges from leather and textile industries
- Hazardous compounds from metal industries
- Oil, oil filters and brake fluids from vehicles and machines
- Mercury-contaminated waste from crematoria
- Explosives from old ammunition, fireworks and airbags
- Lead, nickel, cadmium and mercury from batteries
- Asbestos from the building industry
- Amalgam from dentists
- Veterinary medicines

[Source: Sepa]

Foam insulation materials containing ODP (Ozone Depletant Potential) are also considered as hazardous waste under the EC Regulation 2037/2000.

Land Remediation Techniques

There are two principal approaches to remediation – dealing with the contamination in situ or ex situ. The selection of the approach will be influenced by factors such as: initial and long term cost, timeframe for remediation, types of contamination present, depth and distribution of contamination, the existing and planned topography, adjacent land uses, patterns of surface drainage, the location of existing on-site services, depth of excavation necessary for foundations and below-ground services, environmental impact and safety, interaction with geotechnical performance, prospects for future changes in land use and long-term monitoring and maintenance of in situ treatment.

On most sites, contamination can be restricted to the top couple of metres, although gasholder foundations for example can go down 10 to 15 metres. Underground structures can interfere with the normal water regime and trap water pockets.

There could be a problem if contaminants get into fissures in bedrock.

In situ techniques

A range of in situ techniques is available for dealing with contaminants, including:

- Clean cover – a layer of clean soil is used to segregate contamination from receptor. This technique is best suited to sites with widely dispersed contamination. Costs will vary according to the need for barrier layers to prevent migration of the contaminant.
- On-site encapsulation – the physical containment of contaminants using barriers such as slurry trench cut-off walls. The cost of on-site encapsulation varies in relation to the type and extent of barriers required, the costs of which range from £50/m² to more than £175/m².

There are also in situ techniques for treating more specific contaminants, including:

- Bioremediation – for removal of oily, organic contaminants through natural digestion by microorganisms. Most bioremediation is ex situ, i.e. it is dug out and then treated on site in bio-piles. The process can be slow, taking up to three years depending upon the scale of the problem, but is particularly effective for the long-term improvement of a site, prior to a change of use.
- Phytoremediation – the use of plants that mitigate the environmental problem without the need to excavate the contaminant material and dispose of it elsewhere. Phytoremediation consists of mitigating pollutant concentrations in contaminated soils, water or air, with plants able to contain, degrade or eliminate metals, pesticides, solvents, explosives, crude oil and its derivatives and various other contaminants from the media that contain them.
- Vacuum extraction – involving the extraction of volatile organic compounds (e.g. benzene) from soil and groundwater by vacuum.
- Thermal treatment – the incineration of contaminated soils on site. Thermal processes use heat to increase the volatility to burn, decompose, destroy or melt the contaminants. Cleaning soil with thermal methods may take only a few months to several years.
- Stabilization – cement or lime, is used to physically or chemically bind oily or metal contaminants to prevent leaching or migration. Stabilization can be used in both in situ and ex situ conditions.
- Aeration – if the ground contamination is highly volatile, e.g. fuel oils, then the ground can be ploughed and rotovated to allow the substance to vaporize.
- Air sparging – the injection of contaminant-free air into the subsurface enabling a phase transfer of hydrocarbons from a dissolved state to a vapour phase.
- Chemical oxidization – the injection of reactive chemical oxidants directly into the soil for the rapid destruction of contaminants.
- Pumping – to remove liquid contaminants from boreholes or excavations. Contaminated water can be pumped into holding tanks and allowed to settle; testing may well prove it to be suitable for discharging into the foul sewer subject to payment of a discharge fee to the local authority. It may be necessary to process the water through an approved water treatment system to render it suitable for discharge.

Ex situ techniques

Removal for landfill disposal has, historically, been the most common and cost-effective approach to remediation in the UK, providing a broad-spectrum solution by dealing with all contaminants. As mentioned above, the implementation of the Landfill Directive has resulted in other techniques becoming more competitive for the disposal of hazardous waste.

If used in combination with material-handling techniques such as soil washing, the volume of material disposed at landfill sites can be significantly reduced. The disadvantages of these techniques include the fact that the contamination is not destroyed, there are risks of pollution during excavation and transfer; road haulage may also cause a local nuisance. Ex situ techniques include:

- Soil washing – involving the separation of a contaminated soil fraction or oily residue through a washing process. This also involves the excavation of the material for washing ex situ. The dewatered contaminant still requires disposal to landfill. In order to be cost effective, 70–90% of soil mass needs to be recovered. It will involve constructing a hard area for the washing, intercepting the now-contaminated water and taking it away in tankers.
- Thermal treatment – the incineration of contaminated soils ex situ. The uncontaminated soil residue can be recycled. By-products of incineration can create air pollution and exhaust air treatment may be necessary.

Soil treatment centres are now beginning to be established. These use a combination of treatment technologies to maximize the potential recovery of soils and aggregates and render them suitable for disposal to the landfill. The technologies include:

- Physicochemical treatment – a method which uses the difference in grain size and density of the materials to separate the different fractions by means of screens, hydrocyclones and upstream classification.
- Bioremediation – the aerobic biodegradation of contaminants by naturally occurring microorganisms placed into stockpiles/windrows.
- Stabilization/solidification – a cement or chemical stabilization unit capable of immobilizing persistent leachable components.

Cost Considerations

Cost drivers

Cost drivers relate to the selected remediation technique, site conditions and the size and location of a project.

The wide variation of indicative costs of land remediation techniques shown below is largely because of differing site conditions.

Indicative costs of land remediation techniques for 2021
(excluding general items, testing, landfill tax and backfilling)

Remediation technique	Unit	Rate (£/unit)
Removal – non-hazardous	disposed material (m³)	40–100
Removal – hazardous Note: excluding any pretreatment of material	disposed material (m³)	75–200
Clean cover	surface area of site (m²)	20–45
On-site encapsulation	encapsulated material (m³)	30–95
Bioremediation (in situ)	treated material (m³)	15–45
Bioremediation (ex situ)	treated material (m³)	20–45
Chemical oxidation	treated material (m³)	30–80
Stabilization/solidification	treated material (m³)	20–65
Vacuum extraction	treated material (m³)	25–75
Soil washing	treated material (m³)	40–95
Thermal treatment	treated material (m³)	100–400

Many other on-site techniques deal with the removal of the contaminant from the soil particles and not the wholesale treatment of bulk volumes. Costs for these alternative techniques are very much Engineer designed and site specific.

Factors that need to be considered include:

- waste classification of the material
- underground obstructions, pockets of contamination and live services
- ground water flows and the requirement for barriers to prevent the migration of contaminants
- health and safety requirements and environmental protection measures
- location, ownership and land use of adjoining sites
- distance from landfill tips, capacity of the tip to accept contaminated materials, and transport restrictions
- the cost of diesel fuel, currently approximately £1.29 per litre (at April 2021 prices)

Other project related variables include size, access to disposal sites and tipping charges; the interaction of these factors can have a substantial impact on overall unit rates.

The tables below set out the costs of remediation using dig-and-dump methods for different sizes of project, differentiated by the disposal of non-hazardous and hazardous material. Variation in site establishment and disposal cost accounts for 60–70% of the range in cost.

Variation in the costs of land remediation by removal: Non-hazardous Waste

Item	Disposal Volume (less than 3000 m³) (£/m³)	Disposal Volume (3000–10,000 m³) (£/m³)	Disposal Volume (more than 10,000 m³) (£/m³)
General items and site organization costs	55–90	25–40	7–20
Site investigation and testing	5–12	2–7	2–6
Excavation and backfill	18–35	12–25	10–20
Disposal costs (including tipping charges but not landfill tax)	20–35	20–35	20–35
Haulage	15–35	15–35	15–35
Total (£/m³)	**113–207**	**74–142**	**54–116**
Allowance for site abnormals	*0–10 +*	*0–15 +*	*0–10 +*

Variation in the costs of land remediation by removal: Hazardous Waste

Item	Disposal Volume (less than 3000 m³) (£/m³)	Disposal Volume (3000–10,000 m³) (£/m³)	Disposal Volume (more than 10,000 m³) (£/m³)
General items and site organization costs	55–90	25–40	7–20
Site investigation and testing	10–18	5–12	5–12
Excavation and backfill	18–35	12–25	10–20
Disposal costs (including tipping charges but not landfill tax)	80–170	80–170	80–170
Haulage	25–120	25–120	25–120
Total (£/m³)	**188–433**	**147–367**	**127–342**
Allowance for site abnormals	*0–10 +*	*0–15 +*	*0–10 +*

The strict health and safety requirements of remediation can push up the overall costs of site organization to as much as 50% of the overall project cost (see the above tables). A high proportion of these costs are fixed and, as a result, the unit costs of site organization increase disproportionally on smaller projects.

Haulage costs are largely determined by the distances to a licensed tip. Current average haulage rates, based on a return journey, range from £1.95 to £4.50 per mile. Short journeys to tips, which involve proportionally longer standing times, typically incur higher mileage rates, up to £9.00 per mile.

A further source of cost variation relates to tipping charges. The table below summarizes typical tipping charges for 2021, exclusive of landfill tax:

Typical 2021 tipping charges (excluding landfill tax)

Waste classification	Charges (£/tonne)
Non-hazardous wastes	15–45
Hazardous wastes	35–90
Contaminated liquid	40–75
Contaminated sludge	125–400

Tipping charges fluctuate in relation to the grades of material a tip can accept at any point in time. This fluctuation is a further source of cost risk. Furthermore, tipping charges in the North of England are generally less than other areas of the country.

Prices at licensed tips can vary by as much as 50%. In addition, landfill tips generally charge a tip administration fee of approximately £25 per load, equivalent to £1.25 per tonne. This charge does not apply to non-hazardous wastes.

Landfill Tax, increased on 1 April 2021 to £96.70 per tonne for active waste, is also payable. Exemptions are no longer available for the disposal of historically contaminated material (refer also to 'Landfill Tax' section).

Tax Relief for Remediation of Contaminated Land

The Finance Act 2001 included provisions that allow companies (but not individuals or partnerships) to claim tax relief on capital and revenue expenditure on the 'remediation of contaminated land' in the United Kingdom. The relief is available for expenditure incurred on or after 11 May 2001. From April 2020, non-resident companies are within the scope of UK Corporation Tax and can also claim relief.

From 1 April 2009 there was an increase in the scope of costs that qualify for Land Remediation Relief where they are incurred on long-term derelict land. The list includes costs that the Treasury believe to be primarily responsible for causing dereliction, such as additional costs for removing building foundations and machine bases. However, while there is provision for the list to be extended, the additional condition for the site to have remained derelict since 1998 is likely to render this relief redundant in all but a handful of cases. The other positive change is the fact that Japanese Knotweed removal and treatment (on site only) will now qualify for the relief under the existing legislation, thereby allowing companies to make retrospective claims.

A company is able to claim an additional 50% deduction for 'qualifying land remediation expenditure' allowed as a deduction in computing taxable profits and may elect for the same treatment to be applied to qualifying capital expenditure.

With Landfill Tax Exemption (LTE) now phased out, Land Remediation Relief (LRR) for contaminated and derelict land is the Government's primary tool to create incentives for brownfield development. LRR is available to companies engaged in land remediation that are not responsible for the original contamination.

Over 7 million tonnes of waste each year were being exempted from Landfill Tax in England alone, so this change could have a major impact on the remediation industry. The modified LRR scheme, which provides Corporation Tax relief on any costs incurred on qualifying land remediation expenditure, is in the long run designed to yield benefits roughly equal to those lost through the withdrawal of LTE, although there is some doubt about this stated equity in reality.

However, with much remediation undertaken by polluters or public authorities, who cannot benefit from tax relief benefits, the change could result in a net withdrawal of Treasury support to a vital sector. Lobbying and consultation continues to ensure the Treasury maintains its support for remediation.

While there are no financial penalties for not carrying out remediation, a steep escalator affected the rate of landfill tax for waste material other than inert or inactive wastes, which has increased to £94.15 per tonne from 1 April 2020. This means that for schemes where there is no alternative to dig and dump and no pre-existing LTE, the cost of remediation has risen to prohibitive levels should contaminated material be disposed off site to licensed landfills.

Looking forward, tax-relief benefits under LRR could provide a significant cash contribution to remediation. Careful planning is the key to ensure that maximum benefits are realized, with actions taken at the points of purchase, formation of JV arrangements, procurement of the works and formulation of the Final Account (including apportionment of risk premium) all influencing the final value of the claim agreed with HM Revenue & Customs.

The Relief

Qualifying expenditure may be deducted at 150% of the actual amount expended in computing profits for the year in which it is incurred.

For example, a property trading company may buy contaminated land for redevelopment and incurs £250,000 on qualifying land remediation expenditure that is an allowable for tax purposes. It can claim an additional deduction of £125,000, making a total deduction of £375,000. Similarly, a company incurring qualifying capital expenditure on a fixed asset of the business is able to claim the same deduction provided it makes the relevant election within 2 years.

What is Remediation?

Land remediation is defined as the doing of works including preparatory activities such as condition surveys, to the land in question, any controlled waters affected by the land, or adjoining or adjacent land for the purpose of:

- preventing or minimizing, or remedying or mitigating the effects of, any relevant harm, or any pollution of controlled waters, by reason of which the land is in a contaminated state

Definitions

Contaminated land is defined as land that, because of substances on or under it, is in such a condition that relevant harm is or has the significant possibility of relevant harm being caused to:

- the health of living organisms
- ecological systems
- quality of controlled waters
- property

Relevant harm is defined as meaning:

- death of living organisms or significant injury or damage to living organisms,
- significant pollution of controlled waters,
- a significant adverse impact on the ecosystem, or
- structural or other significant damage to buildings or other structures or interference with buildings or other structures that significantly compromises their use.

Land includes buildings on the land, and expenditure on asbestos removal is expected to qualify for this tax relief. It should be noted that the definition is not the same as that used in the Environmental Protection Act Part IIA.

Sites with a nuclear license are specifically excluded.

Conditions

To be entitled to claim LRR, the general conditions for all sites, which must all be met, are:

- must be a company (non-resident companies can also claim from April 2020)
- must be land in the United Kingdom
- must acquire an interest in the land
- must not be the polluter or have a relevant connection to the polluter
- must not be in receipt of a subsidy
- must not also qualify for Capital Allowances (particular to capital expenditure only)

Additional conditions introduced since 1 April 2009:

- the interest in land must be major – freehold or leasehold longer than 7 years
- must not be obligated to carry out remediation under a statutory notice

Additional conditions particular to derelict land:

- must not be in or have been in productive use at any time since at least 1 April 1998
- must not be able to be in productive use without the removal of buildings or other structures

In order for expenditure to become qualifying, it must relate to substances present at the point of acquisition.

Furthermore, it must be demonstrated that the expenditure would not have been incurred had those substances not been present.

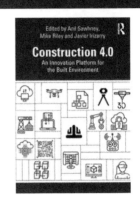

Construction 4.0:
An Innovation Platform for
the Built Environment

Edited By Anil Sawhney, Michael Riley, Javier Irizarry

Modelled on the concept of Industry 4.0, the idea of Construction 4.0 is based on a confluence of trends and technologies that promise to reshape the way built environment assets are designed, constructed, and operated.

With the pervasive use of Building Information Modelling (BIM), lean principles, digital technologies, and offsite construction, the industry is at the cusp of this transformation. The critical challenge is the fragmented state of teaching, research, and professional practice in the built environment sector. This handbook aims to overcome this fragmentation by describing Construction 4.0 in the context of its current state, emerging trends and technologies, and the people and process issues that surround the coming transformation. Construction 4.0 is a framework that is a confluence and convergence of the following broad themes discussed in this book:

- Industrial production (prefabrication, 3D printing and assembly, offsite manufacture)
- Cyber-physical systems (actuators, sensors, IoT, robots, cobots, drones)
- Digital and computing technologies (BIM, video and laser scanning, AI and cloud computing, big data and data analytics, reality capture, Blockchain, simulation, augmented reality, data standards and interoperability, and vertical and horizontal integration)

The aim of this handbook is to describe the Construction 4.0 framework and consequently highlight the resultant processes and practices that allow us to plan, design, deliver, and operate built environment assets more effectively and efficiently by focusing on the physical-to-digital transformation and then digital-to-physical transformation. This book is essential reading for all built environment and AEC stakeholders who need to get to grips with the technological transformations currently shaping their industry, research, and teaching.

February 2020: 526 pp
ISBN: 9780367027308

To Order
Tel:+44 (0) 1235 400524
Email: tandf@bookpoint.co.uk

For a complete listing of all our titles visit:
www.tandf.co.uk

The Landfill Tax

The Tax

The Landfill Tax came into operation on 1 October 1996. It is levied on operators of licensed landfill sites in England, Wales and Northern Ireland at the following rates with effect from 1 April 2021:

- Inactive or inert wastes [Lower Rate] £3.10 per tonne   Included are soil, stones, brick, plain and reinforced concrete, plaster and glass – lower rate

- All other taxable wastes [Standard Rate] £96.70 per tonne   Included are timber, paint and other organic wastes generally found in demolition work and builders skips – standard rate

The 2019 increase in the standard and lower rates of Landfill Tax are in line with the RPI, rounded to the nearest 5 pence and compared with previous rates as follows:

- Lower Rate   From 1 April 2019   £2.90/tonne
- Lower Rate   From 1 April 2020   £3.00/tonne
- Standard Rate   From 1 April 2019   £91.35/tonne
- Standard Rate   From 1 April 2020   £94.15/tonne
- Standard Rate   From 1 April 2021   £96.70/tonne

There has been no change in the loss on ignition threshold for fines which is a major eligibility factor for lower rate Landfill Tax since its introduction in April 2015. The threshold remains at 10%. However, Excise Notice LFT1: a general guide to Landfill Tax was updated on 9 April 2020 in response to the current COVID–19 pandemic to include the following temporary change to the loss of ignition retesting condition requiring a retest to be carried out within 21 days of the disposal of the material (Section 6.9):

> 'If you're unable to get the retest processed within this time frame due to disruptions caused by coronavirus, you can carry this out as soon as is reasonably possible and you should keep evidence to demonstrate this'

Please note that this change is temporary and Excise Notice LFT1 will be re-issued when the change ends. Nothing has been issued to date.

The Landfill Tax (Qualifying Material) Order 2011 came into force on 1 April 2011 and set out the qualifying criteria that are eligible for the lower rate of Landfill Tax. A waste will be lower rated for Landfill Tax only if it is listed as a qualifying material in the Landfill Tax (Qualifying Material) Order 2011 and is disposed of at an authorized waste site.

The principal qualifying materials are:

- Naturally occurring rocks and subsoils
- Ceramic or concrete materials
- Processed or prepared minerals
- Furnace slags
- Ash arising from wood or waste combustion and from the burning of coal and petroleum coke (including when burnt with biomass
- Low activity inorganic compounds
- Calcium sulphate
- Calcium hydroxide and brine

From 1 April 2018, Landfill Tax is due on disposals of material at unauthorized sites in England and Northern Ireland. This also applies to disposals made prior to 1 April 2018 which are still on the site on 1 April 2018. Currently disposals on a site operated as a 'relevant regulated facility' which (as defined under the Environmental Permitting (England and Wales) Regulations 2016) includes sites which are 'a waste operation, such as a transfer station or treatment facility' are not considered to be taxable disposals. The Landfill Tax (Disposals of Material) Order 2018 provides more clarity on what constitutes a 'taxable disposal'. In particular it clarifies which elements of a landfill cell are not taxable under the requirements of Part 3 of the Finance Act 1996, listing them as:

a)   material that forms a layer which performs the function of drainage at the base of a landfill cell;
b)   material that forms the impermeable layer that delimits the landfill cell;
c)   a pipe, pump or associated infrastructure inserted into a landfill cell for the purposes of the extraction or control of surplus liquid or gas from or within that cell;
d)   material used for restoration of a landfill cell that only contains inert material; or
e)   material placed in an information area within the meaning given by regulation 16 A(1) of the Landfill Tax Regulations 1996.

Calculating the Weight of Waste

There are two options:

- If licensed sites have a weighbridge, tax will be levied on the actual weight of waste.
- If licensed sites do not have a weighbridge, tax will be calculated by one of the three Specified Methods described in Excise Notice LFT1: a general guide to Landfill Tax (updated on 9 April 2020 and available via the HMRC website).

Water

If water greater than or equal to 25% of the material by weight has been added to the material to facilitate disposal, for the extraction of minerals or in the course of an industrial process, a landfill operator may apply to discount the water content of the material before tax is levied.

Effect on Prices

The tax is paid by landfill site operators only. Tipping charges reflect this additional cost.

As an example, *Spon's A & B* rates for mechanical disposal will be affected as follows:

- Inactive waste   *Spon's A & B 2022* net rate   £28.24/m³
                     Tax, 1.9 tonne per m³ (unbulked) @ £3.10   £5.89/m³
                     Spon's gross rate including tax   £34.13/m³

- Active waste   Active waste will normally be disposed of by skip and will probably be mixed with inactive waste. The tax levied will depend on the weight of materials in the skip which can vary significantly.

Exemptions

The following disposals are exempt from Landfill Tax subject to meeting certain conditions:

- dredgings which arise from the maintenance of inland waterways and harbours
- naturally occurring materials arising from mining or quarrying operations
- pet cemeteries
- inert waste used to restore landfill sites and to fill working and old quarries where a planning condition or obligation is in existence
- waste from visiting NATO forces

Devolution of Landfill Tax to Scotland from 1 April 2015

The Scotland Act 2012 provided for Landfill Tax to be devolved to Scotland. From 1 April 2015, operators of landfill sites in Scotland are no longer liable to pay UK Landfill Tax for waste disposed at their Scottish sites. Instead, they are liable to register and account for Scottish Landfill Tax (SLfT).

Operators of landfill sites only in Scotland were deregistered from UK Landfill Tax with effect from 31 March 2015.

The Scottish Government announced rates of Landfill Tax from 1 April 2021 to be as follows:

- lower Rate – £3.10/tonne
- standard Rate – £96.70/tonne

Both rates mirror those applicable in England and Northern Ireland.

Devolution of Landfill Tax to Wales from April 2018

The Wales Act 2014 provides for Landfill Tax to be devolved to Wales. This took effect from 1 April 2018. Operators of landfill sites in Wales are liable to register and account for Landfill Disposals Tax (LDT); information is available on the Welsh Revenue Authority website, but in summary there are 3 rates of LDT:

- a lower rate of £3.10/tonne
- a standard rate of £96.70/tonne
- an unauthorized disposals rate of £145.05/tonne (set at 150% of the standard rate)

For further information contact the HMRC VAT and Excise Helpline, Tel: 0300 200 3700.

The Devolution of Landfill Tax (Wales) (Consequential Transitional and Savings Provisions) Order 2018 provides for the closure of the Landfill Communities Fund (LCF) in Wales and the arrangements for a two-year transitional period during which environmental bodies may continue to spend funds on projects in Wales.

The LCF was replaced by the Landfill Disposals Tax (LDT) Communities Scheme which is distributed under contract for the period 1 April 2018 to 31 March 2022 by the Wales Council for Voluntary Action (WCVA). More information is available on the WCVA website.

Construction Contracts: Questions & Answers, 4th edition

David Chappell

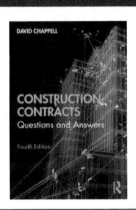

This fully revised and updated edition of *Construction Contracts: Questions and Answers* includes 300 questions and incorporates 42 new judicial decisions, the JCT 2016 updates and the RIBA Building Contracts and Professional Services Contracts 2018 updates.

Construction professionals of all kinds frequently need legal advice that is straightforward as well as authoritative and legally rigorous. Building on the success of previous editions, David Chappell continues to provide answers to real-world FAQs from his experience as consultant and Specialist Advisor to the RIBA. Questions range in content from extensions of time, liquidated damages and loss and/or expense to issues of practical completion, defects, valuation, certificates and payment, architects' instructions, adjudication and fees.

Every question included has been asked of David Chappell during his career and his answers are authoritative but written as briefly and simply as possible. Legal language is avoided but legal cases are given to enable anyone interested to read more deeply into the reasoning behind the answers. This is not only a useful reference for architects, project managers, quantity surveyors and lawyers, but also a useful student resource to stimulate interesting discussions about real-world construction contract issues.

November 2020: 302 pp
ISBN: 9780367532086

To Order
Tel:+44 (0) 1235 400524
Email: tandf@bookpoint.co.uk

For a complete listing of all our titles visit:
www.tandf.co.uk

Taylor & Francis
Taylor & Francis Group

Property Insurance

The problem of adequately covering by insurance the loss and damage caused to buildings by fire and other perils has been highlighted in recent years by the increasing rate of inflation.

Traditionally the insured value must be sufficient to cover the actual cost of reinstating the building. This means that in addition to assessing the current value an estimate has also to be made of the cost increases likely to occur during the period of the policy and of rebuilding which, for a moderate size building, could amount to a total of three years. Obviously, such an estimate is difficult to make with any degree of accuracy; if it is too low the insured may be penalized under the terms of the policy and if too high will result in the payment of unnecessary premiums.

There are variations on the traditional method of insuring which aim to reduce the effects of overestimating and details of these are available from the appropriate offices. For the convenience of readers who may wish to make use of the information contained in this publication in calculating insurance cover required the following may be of interest.

1 Present Cost

The current rebuilding costs may be ascertained in several ways:

- where the actual building cost is known this may be updated by reference to tender price changes
- by reference to average published prices per square metre of floor area. In this case it is important to understand clearly the method of measurement used to calculate the total floor area on which the rates have been based
- by professional valuation
- by comparison with the known cost of another recently built similar building

Whichever of these methods is adopted regard must be paid to any special conditions that may apply, e.g. a confined site; complexity of design; any demolition works and site clearance that may be required.

2 Allowances for Inflation

The Present Cost when established will usually, under the conditions of the policy, be the rebuilding cost on the first day of the policy period. To this must be added a sum to cover future price increases over the period of the policy. For this purpose, using the historical indices as a base and taking account of the likely change in future building costs and the tender climate, an anticipated rebuild cost can be arrived at.

3 Fees

To the total of 1 and 2 above must be added an allowance for any design and other consultancy fees.

4 Value Added Tax (VAT)

To the total of 1 to 3 above must be added Value Added Tax.

Example:

An assessment for insurance cover is required from December 2020 for a property which cost £500,000 when initially built in 2010. This calculation uses AECOM Tender Price Indices and Building Cost Indices which can be found elsewhere in this book. There are various methods for determining insurance cover needed. This is simply an example of one method and users would need to allow for any items that would apply to their own project. Indices would need to be amended/updated to reflect the most recent published data which is in the free Spon's *Updates* and in the *Market Forecast* articles, published quarterly in *Building* magazine

A) Update original build costs (2010Q4) to present day costs (2022Q4), assuming that is the end of the insurance policy

Known cost to build at end of 2010			£500,000
Calculation for present day build cost at end of policy:			
Forecast tender index 2010Q4 =	73		
Forecast tender index 2022Q4 =	128		
Increase in tender index (128/73)	74%		£370,900
Estimated present build cost (2022Q4)			**£870,900**

B) Calculate allowance from end of policy to start rebuild: 2023Q4

Assuming total building damage is suffered on the last day of the policy: 31 December 2022, and planning, documentation and general project set up would require a period of 12 months before rebuilding could commence in 2023Q4, then a further allowance must be made to cover any inflation for that period. Again, use the tender price index.

Calculate allowance for tender inflation to start of rebuild.

Estimated build cost at end of policy (2022Q4)			£870,900
Inflation calculation to tender date:			
Forecast tender index 2022Q4	128		
Forecast tender index 2023Q4	131		
Increase in tender index over period (131/128)	2.4%		£21,200
Estimated rebuild cost (2023Q4)			**£892,100**

C) Cost Inflation during construction

Assuming rebuild would take one year, allowance must be made for the increase in costs met under a building contract you now need to add on any other costs such as fees, demolition, VAT as applicable etc.

Estimated build cost at end of policy (2022Q4)			£892,100
Inflation calculation to tender date:			
Forecast cost index 2022Q4	129.2		
Forecast cost index 2023Q4	133.1		
Increase input costs over period (133.1/129.2 × 50%)	1.5%		£13,500
Estimated rebuild cost			**£905,600**

D) Summary

Estimated cost of reinstatement		£905,600
Demolitions and site clearance		£50,000
Add professional fees	20%	£191,120
		£1,146,720
Add for VAT	20%	£229,344
		£1,376,064
Total insurance cover required	**say**	**£1,380,000**

Please note that this is just a simple example of one method to calculate the cost of rebuilding for insurance purposes. Numbers and indices are made up for the purpose of showing the calculation steps.

Construction Project Manager's Pocket Book, 2nd edition

Duncan Cartlidge

The second edition of the *Construction Project Manager's Pocket Book* maintains its coverage of a broad range of project management skills, from technical expertise to leadership, negotiation, team building and communication. However, this new edition has been updated to include:

- revisions to the CDM regulations,
- changes to the standard forms of contract and other documentation used by the project manager,
- the impact of BIM and emerging technologies,
- implications of Brexit on EU public procurement,
- other new procurement trends, and
- ethics and the project manager.

Construction project management activities are tackled in the order they occur on real projects, with reference made to the RIBA Plan of Work throughout. This is the ideal concise reference which no project manager, construction manager, architect or quantity surveyor should be without.

May 2020: 286pp
ISBN: 9780367435936

To Order
Tel: +44 (0) 1235 400524
Email: book.orders@tandf.co.uk

For a complete listing of all our titles visit:
www.tandf.co.uk

Building Costs Indices; Tender Price Indices and Location Factors

AUTHORS NOTE

We have rebased our indices to 2015 = 100 (from 1976 = 100)

For this edition of the book TPI has been calculated at 132 (2015 = 100).

To avoid confusion, it is essential that the terms, building costs and tender prices, are clearly defined and understood.

- **Building Costs Indices** are the costs incurred by the contractor in the course of his business, the principal ones being those for labour and materials, i.e. cost to contractor
- **Tender Price Indices** represents the price for which the contractor offers to do the project for, i.e. cost to client

Readers are reminded that the following adjustments for time and location should be applied to the whole project and not just to individual elements or materials.

Building Cost Indices

Building costs are the costs incurred by the builder in the course of business, i.e. wages, material prices, plant costs, rates, rents, overheads and taxes. AECOM's Building Cost Index measures movement of basic labour and materials costs to the builder. It is a composite index.

This table reflects the fluctuations since 2015 (base date 2015 = 100) in costs to the contractor. No allowance has been made for changes in productivity or for rates or hours worked which may occur in particular conditions and localities.

Year	First quarter	Second quarter	Third quarter	Fourth quarter	Annual average	% change year-on-year
2015	99.5	99.3	100.9	100.3	100.0	1.1%
2016	100.1	100.6	102.5	102.9	101.5	1.5%
2017	104.0	104.3	106.3	107.1	105.4	3.9%
2018	108.4	109.6	111.4	112.5	110.5	4.8%
2019	112.8	114.2	115.2	114.4	114.2	3.4%
2020	113.7	110.5	113.4	115.8	113.4	−0.7%
2021	117.3 (P)	118.6 (F)	119.8	121.1	119.2	5.2%
2022	122.3	123.4	124.6	125.7	124.0	4.0%
2023	126.7	127.7	128.6	129.6	128.2	3.3%

Note: (P) Provisional, (F) = Forecast thereafter

Users wishing to reference to or from the previous set of indices which were in use in earlier editions of the book, which had a base date of 1976 = 100 need to apply a factor to convert. Multiply the BCI series above by **8.7608** to give the equivalent in the 1976 series.

Tender Price Indices

The extract tables below show the changes in building tender prices since 2010 (base date 2015 = 100). They include building costs but also consider market factors; the tendering climate; constructor sentiment and its assessment of prevailing market conditions amongst other things. They therefore include allowances for profits, risks and other oncosts such as preliminaries. The constructor must also anticipate cost changes during the lifetime of the contract. This means that, for example, in busier times tender prices may increase at a greater rate than building costs, whilst in a downturn the opposite may apply. The same concepts apply to trade contractors, and there is often a compounding of these trends up through the supply chain. AECOM's Tender Price Index is derived from analysis of project data and tender returns on AECOM's UK projects.

Tender prices are similar to building costs but also take into account market considerations such as the availability of labour and materials, and the prevailing economic situation. This means that in boom periods, when there is a surfeit of building work to be done, tender prices may increase at a greater rate than building costs, whilst in a period when work is scarce, tender prices may actually fall when building costs are rising.

Spon's 2022 has been calculated to give a book index of 132.

Year	First quarter	Second quarter	Third quarter	Fourth quarter	Annual average	% change year-on-year
2015	96.1	98.6	101.5	103.8	100.0	9.2%
2016	105.8	107.7	108.7	109.8	108.0	8.0%
2017	110.9	111.3	112.2	112.6	111.7	3.5%
2018	113.2	113.6	115.4	117.3	114.9	2.8%
2019	117.9	118.2	119.3	119.7	118.8	3.4%
2020	120.5	121.0	119.1	119.1	119.9	1.0%
2021	120.0 (P)	121.0 (F)	121.8	122.6	121.4	1.2%
2022	123.2	123.9	124.9	125.9	124.5	2.6%
2023	126.7	127.7	128.8	130.0	128.3	3.1%

Note: (P) = Provisional, (F) = Forecast thereafter

Readers can be kept abreast of tender price movements in the free Spon's *Updates* and also in the *Market Forecast* articles, published quarterly in *Building* magazine.

Users wishing to reference to or from the previous set of indices used in older versions of the book, which had a base date of 1976 = 100 need to apply a factor to convert. Multiply the TPI series above by **5.123** to give the equivalent in the 1976 series.

Using Tender Price Indices to adjust Spon's Book Rates

Example 1:

New secondary school

Note: example data only

Gross Internal Floor Area (GIFA) = 15,000 m^2

Cost plan prepared with a TPI = 124

Start on site TPI, say = 128.4 (actual index to be selected from current published TPI)

Location: North West adjustment, say = 0.89

From *Building Prices per Square Metre*

Assume rate of say £1,800 per m^2

		Cost (£)
School rate, say	£1,800 /m^2 × 15,000 m^2 =	27,000,000
Adjust for inflation to start date	(128.4/124) say +3.5%	945,000
	subtotal	27,945,000
Adjust for location	−10%, say	−2,795,000
	subtotal	25,150,000
Allow for contingencies	say 10%	250,000
Total Order of Cost Estimate		**25,400,000**

Main contractor's preliminaries, overheads and profit need not be added to the cost of building works as they are included within the Spon's building prices per square metre rates, but you will need to add on professional fees and other enabling works costs such as site clearance, demolition, external works, car parking, bringing services to site etc.

Elemental Cost Planning

Elemental cost plans are produced as an intrinsic part of RIBA Work Stages C: Concept, D: Design Development, E: Technical Design and F: Production Information; or when the OGC Gateway Process is used, Gateways 3 A (Design Brief and Concept Approval) and 3B (Detailed Design Approval).

Cost Models can be used to quickly extract £/m² of GIFA:

Example 2:

Health Centre

Note: example data only

Gross Internal Floor Area = 1,000 m²

Cost plan prepared with TPI = 124

Start on site TPI, say = 128.4 (actual index to be selected from current published TPI)

Location: South West (adjustment, say = 0.90)

Element	Rate (£/m²)	Cost (£)
Substructures	106.53	106,530
Frame and Upper Floors	160	160,000
Roof	24	24,000
And so on for each element to give a total of	**1,300.56**	**1,300,560**
Contractors preliminaries, overheads and profit	say 15%	195,000
	subtotal	1,495,560
Adjust for inflation to start date	(128.4/124) say +3.5%	52,000
	subtotal	1,495,560
Adjust for location	say −10%	−155,000
	subtotal	1,392,560
Allow for contingencies	say 5%	70,000
Total Elemental Cost Plan		**1,462,560**

Other allowances such as consultants' fees, design fee, VAT, risk allowance, client costs, fixed price adjustment may need to be added to each of the examples above.

AECOM Tender Price, Building Cost and Consumer Price Indices Chart

This chart shows the relative movement of the indices since 2010. This makes it much easier to see the relative movement between the indices.

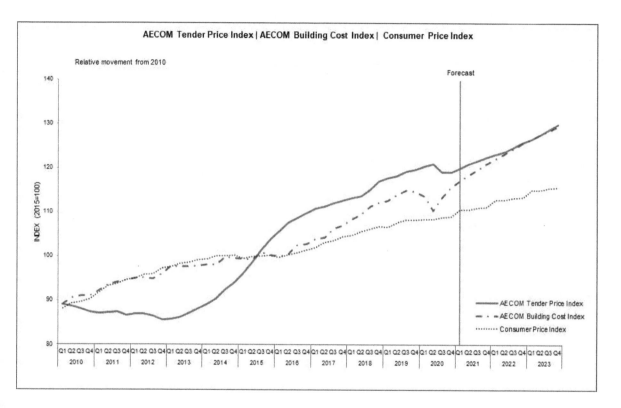

Regional Variations

As well as being aware of inflationary trends when preparing an estimate, it is also important to establish the appropriate price level for the project location.

Prices throughout this book are at a price level index of 132 for Outer London (location factor = 1.00). Regional variations for certain inner London boroughs can be up to 10% higher, while prices in the Yorkshire and Humberside can be as much as 10% lower. Broad regional adjustment factors used to assist with the preparation of initial estimates are shown in the table below and map on the next page.

Over time, price differentials can change depending on regional workloads and local hot spots. Spon's *Updates* together with the *Market Forecast* features in *Building* magazine will keep readers informed of the latest regional developments and changes as they occur.

The regional variations shown in the following table are based on our forecast of price differentials in each of the economic planning regions at the time the book was published. Historically the adjustments do not vary little year-on-year; however, there can be wider variations within any region, particularly between urban and rural locations. The following table shows suggested adjustments required to cost plans or estimates prepared using the *Spon's A&B 2021*. Please note: these adjustments need to be applied to the total cost plan value, not to individual material or element items.

Region	Adjustment to Measured Works section
Outer London (Spon's 2022)	1.00
Inner London	1.05
South East	0.94
South West	0.90
East of England	0.91
East Midlands	0.87
West Midlands	0.85
North East	0.83
North West	0.87
Yorkshire and Humberside	0.86
Wales	0.86
Scotland	0.87
Northern Ireland	0.73

Examples:

Adjust Location:

To adjust a cost plan derived using *Spon's A & B Price book*. Location is East of England (0.91 in table)

Cost Plan = £3,200,000 × 0.91 = £2,912,000 (*adjusted*)

Spons Architects' and Builders' Price Book
2022 Edition
(Tender Index 132)

Scotland
0.87

Northern
Ireland
0.73

North East
0.83

Yorkshire and
Humberside
0.86

North West
0.87

East Midlands
0.87

West Midlands
0.85

East of England
0.91

Wales
0.86

South East
(excluding London) 0.94

South West 0.90

Inner London 1.05
Outer London 1.00

CPD in the Built Environment

Greg Watt and Norman Watts

The aim of this book is to provide a single source of information to support continuing professional development (CPD) in the built environment sector.

The book offers a comprehensive introduction to the concept of CPD and provides robust guidance on the methods and benefits of identifying, planning, monitoring, actioning, and recording CPD activities. It brings together theories, standards, professional and industry requirements, and contemporary arguments around individual personal and professional development. Practical techniques and real-life best practice examples outlined from within and outside of the industry empower the reader to take control of their own built environment-related development, whilst also providing information on how to develop fellow staff members. The contents covered in this book align with the requirements of numerous professional bodies, such as the Royal Institution of Chartered Surveyors (RICS), the Institution of Civil Engineers (ICE), and the Chartered Institute of Builders (CIOB).

The chapters are supported by case studies, templates, practical advice, and guidance. The book is designed to help all current and future built environment professionals manage their own CPD as well as managing the CPD of others. This includes helping undergraduate and postgraduate students complete CPD requirements for modules as part of a wide range of built environment university degree courses and current built environment professionals of all levels and disciplines who wish to enhance their careers through personal and professional development, whether due to professional body requirements or by taking control of identifying and achieving their own educational needs.

April 2021: 182 pp
ISBN: 9780367372156

To Order
Tel: +44 (0) 1235 400524
Email: tandf@bookpoint.co.uk

For a complete listing of all our titles visit:
www.tandf.co.uk

Taylor & Francis
Taylor & Francis Group

PART 2

Rates of Wages and Labour

This part contains the following sections:

Estimator's Pocket Book, 2nd edition

Duncan Cartlidge

The *Estimator's Pocket Book, Second Edition* is a concise and practical reference covering the main pricing approaches, as well as useful information such as how to process sub-contractor quotations, tender settlement and adjudication. It is fully up to date with NRM2 throughout, features a look ahead to NRM3 and describes the implications of BIM for estimators.
It includes instructions on how to handle:

*   the NRM order of cost estimate;
*   unit-rate pricing for different trades;
*   pro-rata pricing and dayworks;
*   builders' quantities;
*   approximate quantities.

Worked examples show how each of these techniques should be carried out in clear, easy-to-follow steps. This is the indispensable estimating reference for all quantity surveyors, cost managers, project managers and anybody else with estimating responsibilities. Particular attention is given to NRM2, but the overall focus is on the core estimating skills needed in practice. Updates to this edition include a greater reference to BIM, an update on the current state of the construction industry as well as up-to-date wage rates, legislative changes and guidance notes.

February 2019: 292pp
ISBN: 9781138366701

To Order
Tel: +44 (0) 1235 400524
Email: book.orders@tandf.co.uk

For a complete listing of all our titles visit:
www.tandf.co.uk

Taylor & Francis
Taylor & Francis Group

Rates of Wages

CIJC Basic Rates of Pay

BUILDING INDUSTRY – ENGLAND, WALES AND SCOTLAND

At the time of compiling this year's books no announcements had been made regarding any rates increases construction for 2020/2021 and beyond. Current agreement dated June 2018 and applicable up to June 2020 is still current.

The authors have therefore assumed an increase of 2.75% to CIJC and BATJIC rates, effective from June 2021.

Readers are reminded to register for Update document which will be produced towards the end of the year where details will be published.

Rates of pay assumed from June 2021	Rate per 39-hour week (£)	Rate per hour (£)
Craft Rate	507.72	13.02
Skill Rate 1	483.27	12.39
Skill Rate 2	465.64	11.94
Skill Rate 3	435.59	11.17
Skill Rate 4	411.54	10.55
General Operative	381.89	9.79

Apprentices/Trainees

The Construction Apprenticeship Scheme (CAS) is aimed mainly at school leavers entering the construction industry for the first time but is open to all ages and operates throughout England and Wales.

CAS is owned by the construction industry and administered by CITB, the Construction Industry Training Board.

For further information telephone CAS helpdesk – 0844 875 2274.

Rates of pay assumed from June 2021	Rate per 39-hour week (£)	Rate per hour (£)
Year 1	228.41	5.86
Year 2	272.49	6.99
Year 3 without NVQ2	317.77	8.15
Year 3 with NVQ2	406.74	10.43
Year 3 with NVQ3	507.72	13.02
On Completion of Apprenticeship with NVQ2	507.72	13.02

Note: If an apprentice is 22 years and over, and in his/her second year of training, then the National Minimum Wage applies.

BUILDING INDUSTRY – ENGLAND, WALES, AND SCOTLAND

BUILDING AND ALLIED TRADES JOINT INDUSTRIAL COUNCIL

The Building and Allied Trades Joint Industrial Council (BATJIC), the partnership between the Federation of Master Builders (FMB) and the Transport and General Workers Union (TGWU).

Rates of pay assumed from June 2021	Rate per 39-hour week (£)	Rate per hour (£)
Craft Operative (NQV3)	527.75	13.53
Craft Operative (NQV2)	454.82	11.66
Adult General Operative	403.53	10.35

For the latest wage/condition's information, go to www.fmb.org.uk.

BUILDING INDUSTRY – ENGLAND, WALES, AND SCOTLAND

PLUMBING AND MECHANICAL ENGINEERING SERVICES INDUSTRY

Authorized rates of wages agreed by the Joint Industry Board for the Plumbing and Mechanical Engineering Services Industry in England and Wales.

The authors have assumed an increase of 3% effective from January 2021. Readers are reminded to register for Updates which will be produced towards the end of the year where details will be published, if an agreement is made.

The Joint Industry Board for Plumbing and Mechanical Engineering Services in England and Wales

Lovell House
Sandpiper Court
Eaton Socon
St Neots
Huntingdon
Cambridgeshire
PE19 8EP
Tel: 01480 476 925
Fax: 01480 403 081
Email: info@jib-pmes.org.uk

Rates of pay assumed from January 2021	Rate per 37.5-hour week (£)	Rate per hour (£)
Operatives		
Technical Plumber and Gas Service Technician	684.05	18.24
Advanced Plumber and Gas Service Engineer	615.68	16.42
Trained Plumber and Gas Service Fitter	528.78	14.10
Apprentices		
1st year of Training	255.70	6.82
2nd year of Training	293.94	7.84
3rd year of Training	331.40	8.84
3rd year of Training with NVQ Level 2	402.47	10.73
4th year of Training	407.49	10.87
4th year of Training with NVQ Level 3	463.11	12.35
4th year of Training with NVQ Level 4	511.40	13.64
Adult Trainees		
1st 6 months of Employment	412.52	11.00
2nd 6 months of Employment	442.26	11.79
3rd 6 months of Employment	460.41	12.28

BUILDING INDUSTRY – ENGLAND, WALES, AND SCOTLAND

The Joint Industry Board for the Plumbing Industry in Scotland and Northern Ireland

Bellevue House
22 Hopetoun Street
Edinburgh
EH7 4GH
Tel: 0131 556 0600
Email: info@snijib.org

Effective from July 2021	Rate per 37.5-hour week (£)	Rate per hour (£)
Operatives Plumbers & Gas Service Operatives		
Plumber and Gas Service Fitter	490.50	13.08
Advanced Plumber and Gas Service Engineer	558.00	14.88
Technician Plumber and Gas Service Technician	617.63	16.47
Plumbing Labourer	437.25	11.66
Apprentice Plumbers and Fitters		
1st year Apprentice	160.50	4.28
2nd year Apprentice	212.25	5.66
3rd year Apprentice	257.25	6.86
4th year Apprentice	331.50	8.84

Labour Rate Calculations

The format of this section is so arranged that, in the case of work normally undertaken by the Main Contractor, the constituent parts of the total rate are shown enabling the reader to make such adjustments as may be required. Similar details have also been given for work normally sublet although it has not been possible to provide this in all instances.

At the time of compiling this year's books no announcements had been made regarding any rates increases for 2021/2022 and beyond.

The authors have therefore assumed an increase of 2.75% from June 2021. Readers are reminded to register for Update document which will be produced towards the end of the year where details will be published if they are available.

As explained in the Preface, there is a facility available to readers, which enables a comparison to be made between the level of prices in this section and current tenders by means of a tender index.

To adjust prices for other regions and times, the reader is recommended to refer to the explanations and examples on how to apply these tender indices.

There follow explanations and definitions of the basis of costs in the Prices for Measured Works section under the following headings:

- Overhead charges and profit
- Labour hours and Labour £ column
- Plant £ column
- Material £ column
- Total rate £ column

Overhead and Profit Charges

Rates checked against winning tenders include overhead charges and profit at current levels.

Labour Hours and Labour £ columns

Labour rates are based upon typical gang costs divided by the number of primary working operatives for the trade concerned, and for general building work include an allowance for trade supervision (see below). Labour hours multiplied by Labour rate with the appropriate addition for overhead charges and profit gives Labour £. In some instances, due to variations in gangs used, Labour rate figures have not been indicated, but can be calculated by dividing Labour £ by Labour hours.

Building Craft Operatives and Labourers

Effective from June 2021 our assumed calculated minimum weekly earnings for craft operatives and general operatives' rates are £507.72 and £381.89 respectively; to these rates have been added allowances for the items below in accordance with the recommended procedure of the Chartered Institute of Building in its Code of Estimating Practice. The resultant hourly rates on which the Prices for Measured Works have generally been based are £17.26 for craft operative and £12.84 for general operatives.

We continue to use CIJC Working Rule Agreement rates to help ensure consistency between editions. Actual wage rates do vary, particularly between trades and regions. It is relatively straightforward for users to apply their own rates.

Calculations need to consider the following:

- Lost time
- Construction Industry Training Board Levy
- Holidays with pay
- Accidental injury, retirement and death benefits scheme
- Sick pay
- National Insurance
- Severance pay and sundry costs
- Employer's liability and third-party insurance

The table which follows illustrates how the all-in hourly rates have been calculated based upon CIJC Working Rule Agreement. Productive time has been based on a total of 1727 hours (44.3 weeks) worked per year.

		Craft Operatives		General Operatives	
		£	£	£	£
Wages at standard basic rate					
Productive time	44.1 weeks	507.72	22,390.39	381.89	16,841.39
Lost time allowance, say	0.9 weeks	507.72	456.95	381.89	343.70
			22,847.34		17,185.09
Holiday pay & Public holiday pay	6.0 weeks	507.72	3,046.31	381.89	2,291.35
Employer's contribution to:					
CITB Levy (say 0.50% of payroll)	0.50 year		129.47		97.38
EasyBuild Stakeholder Pension	52 weeks	7.50	390.00	7.50	390.00
National Insurance (average weekly payment)	52 weeks	43.16	2,252.94	26.17	1,367.36
			28,666.05		21,331.18
Severance pay and sundry costs, say	1.5%		429.99		319.97
			29,096.05		21,651.15
Employer's Liability and Third-Party Insurance, say	2%		581.92		433.02
Total cost per annum			**29,677.97**		**22,084.17**
Total cost per week			**672.97**		**500.77**
Total cost per hour			**17.26**		**12.84**

Similar calculations for other CIJC labour classifications result in the following cost per hour:

Based on assumed basic rates of pay from June 2021	Rate per hour (£)
Craft Operative	17.26
Skill Rate 1	16.40
Skill Rate 2	15.78
Skill Rate 3	14.72
Skill Rate 4	13.88
General Operative	12.84
Rates for additional responsibilities:	
Foreman	18.30
Ganger	13.88

Notes:

1. Absence due to sickness has been assumed to be for periods not exceeding 3 days for which no payment is due
2. All NI payments rates applicable from April 2021
3. Public holiday pay includes for 8 public holiday days
4. This table is an example only, showing a typical breakdown

The labour rates used in the Measured Work sections have been based on the following gang calculations which generally include an allowance for supervision by a foreman or ganger. Alternative labour rates are given showing the effect of various degrees of bonus.

Typical Gang Build-up	Rate £/hour	Unit rate £/hour	Productive Hours	Labour rates with bonus payment £/hour			
				Normal	+10%	+20%	+30%
Groundwork Gang							
1 Ganger	1.00	13.88					
6 Labourers	6.00	12.84					
		90.92	6.50 =	13.99	15.48	16.97	18.47
Concreting Gang							
1 Foreman	1.00	18.30					
4 Skilled labourers	4.00	13.88					
		73.82	4.50 =	16.40	18.16	19.93	21.70
Steel Fixing Gang							
1 Foreman	1.00	18.30					
4 Steel fixers	4.00	17.26					
		87.32	4.50 =	19.40	21.46	23.53	25.60
Formwork Gang							
1 Foreman	1.00	18.30					
10 Carpenters	10.00	17.26					
1 Labourer	1.00	12.84					
		203.69	10.50 =	19.40	21.43	23.46	25.49
Bricklaying/Blockwork Gang							
1 Foreman	1.00	18.30					
6 Bricklayers	6.00	17.26					
1 Labourer	1.00	12.84					
		134.67	6.50 =	20.72	22.90	25.08	27.27
Carpentry/Joinery Gang							
1 Foreman	1.00	18.30					
5 Carpenters	5.00	17.26					
1 Labourer	1.00	12.84					
		117.41	5.50 =	21.35	23.60	25.86	28.12
Painter, Slater, etc.							
1 Craft operative	1.00	17.26					
		17.26	1.00 =	17.26	19.04	20.82	22.60
1 and 1 gang							
1 Craft operative	1.00	17.26					
1 Skilled labourer	1.00	13.88					
		31.14	1.00 =	31.14	34.36	37.59	40.81
2 and 1 Gang							
2 Craft operatives	2.00	17.26					
1 Skilled labourer	1.00	13.88					
		48.39	2.00 =	24.20	26.70	29.20	31.71
Small Labouring Gang (making good)							
1 Foreman	1.00	18.30					
4 Skilled labourers	4.00	13.88					
		73.82	4.50 =	16.40	18.16	19.93	21.70
Drain Laying Gang/Clayware							
2 Skilled labourers	2.00	13.88					
		27.76	2.00 =	13.88	15.33	16.77	18.21

Plumbing Operatives

From January 2021 the hourly earnings for technical and trained plumbers are assumed to be £18.24 and £14.10 respectively; to these rates have been added allowances like those added for building operatives (see below). The resultant average hourly rate on which the Prices for Measured Works have been based is £26.46. The items referred to above for which allowance has been made are:

- Tool allowance
- Plumbers' welding supplement
- Holidays with pay
- Pension and welfare stamp
- National Insurance
- Severance pay and sundry costs
- Employer's liability and third-party insurance

No allowance has been made for supervision as we have assumed the use of a team of technical or trained plumbers who are able to undertake such relatively straightforward plumbing works, e.g. on housing schemes, without supervision.

The table which follows shows how the average hourly rate referred to above has been calculated. Productive time has been based on a total of 1,672.5 hours worked per year. NB Additional days holiday included.

Assumed from January 2021		Technical Plumber		Trained Plumber	
Wages at standard basic rate		Rate £	Total £	Rate £	Total £
Productive time	1672.5 hrs	18.24	30,508.57	14.10	23,583.42
Overtime (paid at basic rate)	336 hrs	18.24	6,129.08	14.10	4,737.84
Plumbers welding supplement	2009 hrs	0.50	1,004.25		
			37,641.90		28,321.26
Employer's contribution to:					
Holiday credit/welfare					
Holiday pay for 32 days	32 days	136.81	4,377.91	105.76	3,384.17
Additional holiday pay	60 credits	2.25	135.00	2.25	135.00
Pension, say	7.5%	71.59	3,161.61	47.58	2,388.03
National Insurance	52 weeks		3,722.90		2,474.11
			49,039.32		36,702.57
Severance pay and sundry costs, say	1.5%		735.59		550.54
Liability and third-party insurance, say	2.0%		995.50		745.06
Total cost per annum			**50,770.41**		**37,998.17**
Total cost per hour			**30.36**		**22.72**
Average all-in rate per hour			**26.54**		

Notes:

1. First 37.5 hours paid at normal rate
2. 1.5 overtime rate after 37.5 hrs up to 8 pm, after which overtime rate is at double time
3. National Insurance is payable on top-up funding as this is payable by the employer
4. Absence due to sickness has been assumed to be for periods not exceeding 3 days for which no payment is due. The entitlement is payable from and including the fourth day of illness onwards
5. This table is an example only showing a typical breakdown

LABOUR CATEGORIES

Schedule 1 to the WRA lists specified work establishing entitlement to the Skilled Operative Pay Rate 4, 3, 2, 1 or the Craft Rate as follows:

CIJC WORKING RULE AGREEMENT FOR THE CONSTRUCTION INDUSTRY – SCHEDULE 1 **Specified Work Establishing Entitlement to the Skilled Operative Pay Rate 4, 3, 2, 1 or Craft Rate**	
BAR BENDERS AND REINFORCEMENT FIXERS	
• Bender and fixer of Concrete Reinforcement capable of reading and understanding drawings and bending schedules and able to set out work	Craft Rate
CONCRETE	
• Concrete Leveller or Vibrator Operator	4
• Screeder and Concrete Surface Finisher working from datum such as road-form, edge beam or wire	4
• Operative required to use trowel or float (hand or powered) to produce high quality finished concrete	4
DRILLING AND BLASTING	
• Drills, rotary or percussive: mobile rigs, operator of	3
• Operative attending drill rig	4
• Shotfirer, operative in control of and responsible for explosives including placing, connecting and detonating charges	3
• Operatives attending on shotfirer, including stemming	4
DRYLINERS	
• Operatives undergoing approved training in drylining	4
• Operatives who can produce a certificate of training achievement indicating satisfactory completion of at least one unit of approved drylining training	3
• Dryliners who have successfully completed their training in drylining fixing and finishing	Craft Rate
FORMWORK CARPENTERS	
• 1st year trainee	4
• 2nd year trainee	3
• Formwork carpenters	Craft Rate
GANGERS AND TRADE CHARGEHANDS (Higher grade payments may be made at the employer's discretion)	4
GAS NETWORK OPERATIONS Operatives who have successfully completed approved training to the standard of:	
• GNO Trainee	4
• GNO Assistant	3
• Team Leader – Services	2
• Team Leader – Mains	1
• Team Leader – Mains and Services	1
LINESMEN – ERECTORS	
• 1st Grade	2
Skilled in all works associated with the assembly, erection, maintenance and dismantling of Overhead Lines Transmission Lines on steel towers, concrete or wood poles, including all overhead lines construction elements.	
• 2nd Grade	3
As above but lesser degree of skill – or competent and fully skilled to carry out some of the elements of construction listed above.	
• Linesmen-erector's mate	4
Semi-skilled in works specified above and a general helper	
MASON PAVIORS	
• Operative assisting a Mason Pavior undertaking kerblaying, block and sett paving, flag laying, in natural stone and precast products	4

LABOUR CATEGORIES

• Operative engaged in stone pitching or dry stone walling	3

MECHANICS

• **Maintenance Mechanic** capable of carrying out field service duties, maintenance activities and minor repairs	2
• **Plant Mechanic** capable of carrying out major repairs and overhauls including welding work, operating metal turning lathe or similar machine and using electronic diagnostic equipment	1
• **Maintenance/Plant Mechanics' Mate** on site or in depot	4
• **Tyre Fitter,** heavy equipment tyres	2

MECHANICAL PLANT DRIVERS AND OPERATORS
Backhoe Loaders (with rear excavator bucket and front shovel and additional equipment such as blades, hydraulic hammers and patch planers)

• Backhoe, up to and including 50 kW net engine power; driver of	4
• Backhoe, over 50 kW up to and including 100 kW net engine power; driver of	3
• Backhoe, over 100 kW net engine power; driver of	2
• **Compressors and Generators** Air compressor or generators over 10 kW; operator of	4

Concrete Mixers

• Operative responsible for operating a concrete mixer or mortar pan up to and including 400 litres drum capacity	4
• Operative responsible for operating a concrete mixer over 400 litres and up to and including 1,500 litres drum capacity	3
• Operative responsible for operating a concrete mixer over 1,500 litres drum capacity	2
• Operative responsible for operating a mobile self-loading and batching concrete mixer up to 2,500 litres drum capacity	2
• Operative responsible for operating a mechanical drag-shovel	4

Concrete Placing Equipment

• Trailer mounted or static concrete pumps: self-propelled concrete placers: concrete placing booms; operator of	3
• Self-propelled mobile concrete pump, with or without boom, mounted on lorry or lorry chassis; driver/operator of	2

Cranes/Mobile Cranes
Self-propelled mobile crane on road wheels, rough terrain wheels or caterpillar tracks including lorry mounted:

• Maximum lifting capacity at minimum radius, up to and including 5 tonnes; driver of	4
• Maximum lifting capacity at minimum radius, over 5 tonnes and up to and including 10 tonnes; driver of	3
• Maximum lifting capacity at minimum radius, over 10 tonnes	Craft Rate

Tower Cranes (including static or travelling: standard trolley or luffing jib)
Up to and including 2 tonnes max. Lifting capacity at min. radius; driver of

Up to and including 2 tonnes max. Lifting capacity at min. radius; driver of	4
• Over 2 tonnes up to and including 10 tonnes max. Lifting capacity at min. radius; driver of	3
• Over 10 tonnes up to and including 20 tonnes max. Lifting capacity at minimum radius; driver of	2
• Over 20 tonnes max. Lifting capacity at minimum radius; driver of	1

Miscellaneous Cranes and Hoists

• Overhead bridge crane or gantry crane up to and including 10 tonnes capacity; driver of	3
• Overhead bridge crane or gantry crane over 10 tonnes up to and including 20 tonnes capacity; driver of	2
• Power driven hoist or jib crane; operator of	4
• **Slinger/Signaller** appointed to attend Crane or hoist to be responsible for fastening or slinging loads and generally to direct lifting operations	4

Dozers

• Crawler dozer with standard operating weight up to and including 10 tonnes; driver of	3
• Crawler dozer with standard operating weight over 10 tonnes and up to and including 50 tonnes; driver of	2
• Crawler dozer with standard operating weight over 50 tonnes; driver of	1

LABOUR CATEGORIES

Dumpers and Dump Trucks	
• Up to and including 10 tonnes rated payload; driver of	4
• Over 10 tonnes and up to and including 20 tonnes rated payload; driver of	3
• Over 20 tonnes and up to and including 50 tonnes rated payload; driver of	2
• Over 50 tonnes and up to and including 100 tonnes rated payload; driver of	1
• Over 100 tonnes rated payload; driver of	Craft Rate
Excavators (360° slewing)	
• Excavators with standard operating weight up to and including 10 tonnes; driver of	3
• Excavator with standard operating weight over 10 tonnes and up to and including 50 tonnes; driver of	2
• Excavator with standard operating weight over 50 tonnes; driver of	1
• **Banksman** appointed to attend excavator or responsible for positioning vehicles during loading or tipping	4
Fork-Lifts Trucks and Telehandlers	
• Smooth or rough terrain fork lift trucks (including side loaders) and telehandlers up to and including 3 tonnes lift capacity; driver of	4
• Over 3 tonnes lift capacity; driver of	3
Motor Graders: driver of	2
Motorized Scrapers: driver of	2
Motor Vehicles (Road Licensed Vehicles) Driver and Vehicle Licensing Agency (DVLA)	
• Vehicles requiring a driving licence of category C1; driver of (Goods vehicle with maximum authorized mass (mam) exceeding 3.5 tonnes but not exceeding 7.5 tonnes and including such a vehicle drawing a trailer with a mam not over 750 kg)	4
• Vehicles requiring a driving licence of category C; driver of (Goods vehicle with a maximum authorized mass (mam) exceeding 3.5 tonnes and including such a vehicle drawing a trailer with mam not over 750 kg)	2
• Vehicles requiring a driving licence of category C plus E; driver of (Combination of a vehicle in category C and a trailer with maximum authorized mass over 750 kg)	1
Power Driven Tools	
• Operatives using power-driven tools such as breakers, percussive drills, picks and spades, rammers and tamping machines	4
Power Rollers	
• Roller, up to and including 4 tonnes operating weight; driver of	4
• Roller, over 4 tonnes operating weight and upwards; driver of	3
Pumps, Power-driven pump(s); attendant of	4
Shovel Loaders, (Wheeled or tracked, including skid steer)	
• Up to and including 2 cubic metre shovel capacity; driver of	4
• Over 2 cubic metre and up to and including 5 cubic metre shovel capacity; driver of	3
• Over 5 cubic metre shovel capacity; driver of	2
Tractors (Wheeled or Tracked)	
• Tractor, when used to tow trailer and/or with mounted compressor, up to and including 100 kW rated engine power; driver of	4
• Tractor, ditto, over 100 KW up to and including 250 kW rated engine power; driver of	3
• Tractor, ditto, over 250 kW rated engine power; driver of	2
Trenchers (Type wheel, chain or saw)	
• Trenching Machine, up to and including 50 kW gross engine power; driver of	4
• Trenching Machine, over 50 kW and up to and including 100 kW gross engine power; driver of	3
• Trenching Machine, over 100 kW gross engine power; driver of	2
Winches, Power driven winch; driver of	4
PILING	
• General Skilled Piling Operative	4
• Piling Chargehand/Ganger	3
• Pile Tripod Frame Winch Driver	3

LABOUR CATEGORIES

Winches, Power driven winch; driver of	4
• CFA or Rotary or Driven Mobile Piling Rig Driver	2
• Concrete Pump Operator	3
PIPE JOINTERS	
• Jointers, pipes up to and including 300 mm dia.	4
• Jointers, pipes over 300 mm dia. and up to 900 mm dia.	3
• Jointers, pipes over 900 mm dia.	2
• **except** in HDPE mains when experienced in butt fusion and/or electrofusion jointing operations	2
PIPELAYERS	
• Operative preparing the bed and laying pipes up to and including 300 mm dia.	4
• Operative preparing the bed and laying pipes over 300 mm dia. and up to and including 900 mm dia.	3
• Operative preparing the bed and laying pipes over 900 mm dia.	2
PRESTRESSING CONCRETE	
• Operative in control of and responsible for hydraulic jacks and other tensioning devices engaged in post-tensioning and/or pre-tensioning concrete elements	3
ROAD SURFACING WORK (includes rolled asphalt, dense bitumen macadam and surface dressings)	
Operatives employed in this category of work to be paid as follows:	
• Chipper	4
• Gritter Operator	4
• Raker	3
• Paver Operator	3
• Leveller on Paver	3
• Road Planer Operator	3
• Road Roller Driver, 4 Tonne and upwards	3
• Spray Bar Operator	4
TIMBERMAN	
• Timberman, installing timber supports	3
• Highly skilled timber man working on complex supports using timbers of size 250 mm by 125 mm and above	2
• Operative attending	4
WELDERS	
• Grade 4 (Fabrication Assistant)	
• Welder able to tack weld using SMAW or MIG welding processes in accordance with verbal instructions and including mechanical preparation such as cutting and grinding	3
• Grade 3 (Basic Skill Level)	
• Welder able to weld carbon and stainless steel using at least one of the following processes SMAW, GTAW, GMAW for plate-plate fillet welding in all major welding positions, including mechanical preparation and complying with fabrication drawings	2
• Grade 2 (Intermediate Skill Level)	
• Welder able to weld carbon and stainless steel using manual SMAW, GTAW, semi-automatic MIG or MAG, and FCAW welding processes including mechanical preparation, and complying with welding procedures, specifications and fabrication drawings.	1
• Grade 1 (Highest Skill Level)	
• Welder able to weld carbon and stainless steel using manual SMAW, GTAW, semi-automatic GMAW or MIG or MAG, and FCA w elding processes in all modes and directions in accordance with BSEN287–1 and/or 287–2 Aluminium Fabrications including mechanical preparation and complying with welding procedures, specifications and fabrication drawings.	Craft Rate
YOUNG WORKERS	
• Operatives below 18 years of age will receive payment 60% of the General Operative Basic Rate.	
• At 18 years of age or over the payment is 100% of the relevant rate.	

PART 3

Approximate Estimating

This part contains the following sections:

Facilities Management Models, Methods and Tools: Research Results for Practice

Per Anker Jensen

This book presents research tested models, methods and tools that can make the work of the facilities manager more robust and sustainable, help long-term strategic planning and support students and practitioners in FM to improve the way they approach and deal with challenges in practice.

The 34 models, methods and tools are presented in relation to five typical challenges for facilities managers:

- Strategy development
- Organisational design
- Space planning
- Building projects
- Optimisation

The chapters are short and concise, presenting a central illustration of one model, method or tool with explanatory text and short, exemplary case studies. Each chapter includes references to further reading, and the book includes a keyword index. Essential reading for all involved in the management of built assets, this book bridges the gap between robust academic research and practical industry tools. It can also be used as a handy student reference..

June 2019: 334pp
ISBN: 9780367028725

To Order
Tel: +44 (0) 1235 400524
Email: book.orders@tandf.co.uk

For a complete listing of all our titles visit:
www.tandf.co.uk

Building Prices per Functional Unit

Prices given under this heading are given as a typical range of prices for new work. These should not be interpreted as a minimum or maximum range.

Prices includes for Preliminaries, and Overheads and Profit recovery.

Unless otherwise stated, prices do not allow for external works, furniture or special equipment.

On certain types of buildings there exists a close relationship between its cost and the number of functional units that it accommodates. During the very early stages of a project an approximate estimate can be quickly derived by multiplying the proposed unit of accommodation (i.e. hotel bedrooms, car parking spaces etc.) by an appropriate rate.

The following indicative unit areas and costs have been derived from historical data and construction standards. It should be emphasized that the prices must be treated with reserve, as they represent the average of prices from our records and cannot provide more than a rough guide to the cost of a building. There are limitations when using this method of estimating, for example, the functional areas and costs of football stadia, hotels, restaurants etc. are strongly influenced by the extent of front and back of house facilities housed within it, and these areas can vary considerably from scheme to scheme.

The areas may also be used as a 'rule of thumb' in order to check on economy of designs. Where we have chosen not to show indicative areas, this is because either ranges are extensive or such figures may be misleading.

Costs have been expressed within a range, although this is not to suggest that figures outside this range will not be encountered, but simply that the calibre of such a type of building can itself vary significantly.

For assistance with the compilation of an *Order of Cost Estimate*, or of an *Elemental Cost Plan*, the reader is directed to the *Building Prices per Square Metre*, *Approximate Estimating Rates* and *Cost Models* sections of this book.

As elsewhere in this edition, prices do not include Value Added Tax or fees for professional services.

BUILDING PRICES PER FUNCTIONAL UNIT

Building type	Function	Range £		
Utilities, Civil Engineering facilities				
Car parking				
surface car parking; 20 to 22 m²/car	car	2300.00	to	3200.00
surface car parking; landscaped 20 to 22 m²/car	car	3000.00	to	4100.00
multi-storey car parks flat slab; 24 to 28 m²/car	car	19000.00	to	26000.00
underground car parks partially underground under buildings; natural ventilation; 27 to 30 m² / car	car	23000.00	to	32000.00
underground car parks completely underground with landscaped roof; mechanical ventilation; 28 to 32 m²/car	car	40500.00	to	56000.00
underground car parks completely underground under buildings; mechanical ventilation; 28 to 32 m²/car	car	33500.00	to	46500.00
Offices				
business park; medium quality, non air-conditioned	person	14500.00	to	20000.00
low rise; air-conditioned; high quality speculative; 4–8 storeys	person	18000.00	to	25000.00
medium rise; air-conditioned; high quality speculative; 8–20 storeys	person	27500.00	to	38000.00
Health and welfare facilities				
district hospitals; 65 to 85 m²/bed	bed	205000.00	to	280000.00
hospital teaching centres; 120+ m²/bed	bed	225000.00	to	310000.00
private hospitals; 75 to 100 m²/bed	bed	240000.00	to	330000.00
residential homes; 40 to 60 m²/bedroom	bed	78000.00	to	110000.00
nursing homes; 40 to 60 m²/bedroom	bed	100000.00	to	140000.00
Recreational facilities				
Theatres				
large – over 500 seats	seat	32000.00	to	44000.00
studio/workshop – less than 500 seats	seat	16000.00	to	22500.00
refurbishment	seat	13000.00	to	18000.00
Sports halls				
indoor bowls halls	rink	190000.00	to	270000.00
squash courts	court	110000.00	to	150000.00
Education facilities				
Schools				
nursery schools; 3 to 5 m²/pupil	pupil	8600.00	to	12000.00
secondary/middle schools; 6 to 10 m²/pupil	pupil	21500.00	to	30000.00
special schools; 18 to 20 m²/pupil	pupil	29000.00	to	40000.00
Universities				
arts buildings	student	17000.00	to	24000.00
computer buildings	student	23000.00	to	32000.00
science buildings	student	23000.00	to	32000.00
laboratories	student	45000.00	to	63000.00
Residential facilities				
Housing				
detached; four bedroom; 100 to 140 m² GIFA	house	195000.00	to	270000.00
semi-detached; three bedroom; 70 to 90 m² GIFA	house	120000.00	to	160000.00
terraced; two bedroom; 55 to 65 m² GIFA	house	75000.00	to	105000.00
low rise flats; two bedroom; 55 to 65 m² GIFA	flat	90000.00	to	125000.00
medium rise flats; two bedroom; 55 to 65 m² GIFA	flat	110000.00	to	150000.00

BUILDING PRICES PER FUNCTIONAL UNIT

Building type	Function	Range £		
Hotels				
luxury; city centre with conference and wet leisure facilities; 70 to 120 m²/bedroom	bed	490000.00	to	680000.00
business; town centre with conference and wet leisure facilities; 70 to 100 m²/bedroom	bed	300000.00	to	420000.00
mid-range; town centre with conference and leisure facilities; 50 to 60 m²/bedroom	bed	140000.00	to	190000.00
budget; city centre with dining and bar facilities; 35 to 45 m²/bedroom	bed	92000.00	to	130000.00
budget; roadside excluding dining facilities; 28 to 35 m²/bedroom	bed	62000.00	to	86000.00
Student residences				
large budget schemes; 200+ units with en-suite accommodation; 18 to 20 m²/bedroom	bed	31500.00	to	44000.00
smaller schemes (40–100 units) with mid range specifications; some with en-suite accommodation; 19 to 24 m²/bedroom	bed	42000.00	to	59000.00
smaller high quality courtyard schemes, college scheme; 24 to 28 m²/bedroom	bed	68000.00	to	94000.00

Fire Safety Design for Tall Buildings

By Feng Fu

Fire Safety Design for Tall Buildings provides structural engineers, architects, and students with a systematic introduction to fire safety design for tall buildings based on current analysis methods, design guidelines, and codes. It covers almost all aspects of fire safety design that an engineer or an architect might encounter—such as performance-based design and the basic principles of fire development and heat transfer.

It also sets out an effective way of preventing the progressive collapse of a building in fire, and it demonstrates 3D modeling techniques to perform structural fire analysis with examples that replicate real fire incidents such as the Twin Towers and WTC7. This helps readers to understand the design of structures and analyze their behavior in fire.

February 2021: 250 pp
ISBN: 9780367444525

To Order
Tel:+44 (0) 1235 400524
Email: tandf@bookpoint.co.uk

For a complete listing of all our titles visit:
www.tandf.co.uk

Building Prices per Square Metre

Prices given under this heading are given as a typical range of prices for new work for a range of typical building types.

Prices in this include for Preliminaries and Overheads and Profit recovery.

Unless otherwise stated, prices do not allow for external works, furniture or special equipment.

Prices are based upon the total gross internal area (GIA) of all storeys measured in accordance with the 6th edition of the RICS Code of Measuring Practice, September 2007.

As in previous editions it is emphasized that the prices must be treated with a degree of reserve as they represent the average of prices from historical records and should be used to provide only an approximate order of cost for a building.

Costs have been expressed within a range, although this is not to suggest that figures outside this range will not be encountered, but simply that the calibre of such a type of building can itself vary significantly.

Prices can vary greatly between regions and even between individual sites. Professional interpretation will always need to be applied in the final analysis.

For assistance with the compilation of an *Order of Cost Estimate*, or of an *Elemental Cost Plan*, the reader is directed to the *Cost Models* and *Approximate Estimating Rates* sections in the book.

As elsewhere in this edition, prices do not include Value Added Tax or fees for professional services.

AUTHORS NOTE

RICS Property Measurement

Overview

The Royal Institution of Chartered Surveyors (RICS) has launched 'RICS Property Measurement', which in part, replaces the Code of Measuring Practice for office buildings. This came into effect on 1 January 2016.

The RICS Property Measurement incorporates the International Property Measurement Standards (IPMS) for offices, whilst the Code of Measuring Practice 6th edition remains in effect for all other building types.

IPMS was formed in 2013 at the World Bank, with the objective to provide global standardization for measurement of the built environment. It now comprises more than 70 organizations including RICS. Each member organization is responsible for setting domestic measurement standards. Having successfully launched IPMS office buildings, they are currently working on IPMS residential buildings, followed by industrial, retail and mixed use.

Timetable

RICS Property Measurement was launched in May 2015. The use of IPMS office buildings becomes mandatory for RICS Members from 1 January 2016, unless clients instruct otherwise or local laws specify the use of a different standard.

Proposed changes from Code of Measuring Practice 6th edition: Office Buildings

The first key change is the use of new vocabulary. Gross External Area (GEA), Gross Internal Area (GIA) or Net Internal Area (NIA) are no longer used. This terminology has been replaced with IPMS 1, IPMS 2 and IPMS 3. Whilst these are not direct replacements they can be closely related to GEA, GIA and NIA respectively.

BUILDING PRICES PER SQUARE METRE

The second key change is the concept that whilst IPMS 1, IPMS 2 and IPMS 3 are more inclusive in measurement than GEA, GIA and NIA, these additional measures need to be stated separately. To ensure transparency, marked up plans should accompany measurement reports, which should be detailed to include the breakdown of the measure, this should provide benefit to the end user.

The third key change is the concept of limited use area. Whilst these areas are included in the measurement, they are then stated separately. They include areas with limited height (less than 1.5 metres), areas with limited natural light, internal structural walls/columns, area difference between internal dominant face and the wall/floor junction.

IPMS 1

The IPMS 1 measure includes the area of the building incorporating the external walls. The key differences between IPMS 1 and GEA are the inclusion of balconies under IPMS 1. However covered galleries and balconies are stated separately.

IPMS 2

IPMS 2 is measured to the internal dominant face. The internal dominant face is defined as the 'surface comprising more than 50% of the surface area for each vertical section forming an internal perimeter'. So for example, for an office building with unitized cladding, this would mean the glass surface.

Additional areas which would not be included in a GIA measure but is included under IPMS 2 are the balconies and covered canopies. These would then be stated separately.

IPMS 3

The most significant changes appear in the comparison between IPMS 3 and NIA. IPMS 3 include area taken by columns, piers/external columns and window reveals. All of these are excluded under NIA. Additionally space taken by balconies, covered galleries, roof terraces are all included but stated separately. Lift lobbies and area with headroom of less than 1.5 metres are also included but stated separately as limited use area.

It will take the wider market some time to wake up to this change, especially with the impact on end user market for valuation purposes, as it will have a direct impact on their comparable databases. There will be a period where the existing GEA, GIA and NIA will remain in favour and these are still being used in this book.

BUILDING PRICES PER SQUARE METRE

Item	Unit	Range £		
UNICLASS D1 UTILITIES, CIVIL ENGINEERING FACILITIES				
Car parking				
surface car parking	m²	96.00	to	120.00
surface car parking; landscaped	m²	120.00	to	160.00
Multi-storey car parks				
flat slab	m²	650.00	to	820.00
Underground car parks				
partially underground under buildings; naturally ventilated	m²	730.00	to	930.00
completely underground under buildings with mechanical ventilation	m²	1000.00	to	1275.00
completely underground with landscaped roof and mechanical ventilation	m²	1200.00	to	1550.00
Transport facilities				
railway stations	m²	4200.00	to	5300.00
bus and coach stations	m²	2950.00	to	3750.00
bus garages	m²	1100.00	to	1425.00
petrol stations	m²	3650.00	to	4650.00
Vehicle showrooms with workshops, garages etc.				
up to 2,000 m²	m²	1425.00	to	1825.00
over 2,000 m²	m²	1225.00	to	1575.00
Vehicle showrooms without workshops, garages etc.				
up to 2,000 m²	m²	1425.00	to	1825.00
Vehicle repair and maintenance buildings				
up to 500 m²	m²	2125.00	to	2700.00
over 500 m² up to 2000 m²	m²	1425.00	to	1825.00
car wash buildings	m²	1225.00	to	1575.00
Airport facilities (excluding aprons)				
airport terminals	m²	3750.00	to	4800.00
airport piers/satellites	m²	3950.00	to	5100.00
Airport campus facilities				
cargo handling bases	m²	950.00	to	1225.00
distribution centres	m²	530.00	to	680.00
hangars (type C and D facilities)	m²	1650.00	to	2075.00
TV, radio and video studios	m²	2700.00	to	3500.00
Telephone exchanges	m²	1425.00	to	1825.00
Telephone engineering centres	m²	970.00	to	1250.00
Branch post offices	m²	1425.00	to	1825.00
Postal delivery offices/sorting offices	m²	1425.00	to	1825.00
Electricity substations	m²	2050.00	to	2600.00
Garages, domestic	m²	880.00	to	1125.00
UNICLASS D2 INDUSTRIAL FACILITIES				
B1 Light industrial/offices buildings				
economical shell, and core with heating only	m²	870.00	to	1100.00
medium shell and core with heating and ventilation	m²	1225.00	to	1575.00
high quality shell and core with air conditioning	m²	1925.00	to	2450.00
developers Category A fit-out	m²	700.00	to	900.00
tenants Category B fit-out	m²	465.00	to	590.00
Agricultural storage	m²	740.00	to	940.00

BUILDING PRICES PER SQUARE METRE

Item	Unit	Range £		
UNICLASS D2 INDUSTRIAL FACILITIES – cont				
Factories				
for letting (incoming services only)	m^2	750.00	to	960.00
for letting (including lighting, power and heating)	m^2	960.00	to	1225.00
nursery units (including lighting, power and heating)	m^2	960.00	to	1225.00
workshops	m^2	890.00	to	1125.00
maintenance/motor transport workshops	m^2	1150.00	to	1475.00
owner occupation for light industrial use	m^2	1225.00	to	1575.00
owner occupation for heavy industrial use	m^2	2050.00	to	2600.00
Factory/office buildings high technology production				
for letting (shell and core only)	m^2	810.00	to	1025.00
for owner occupation (controlled environment fully finished)	m^2	1075.00	to	1375.00
High technology laboratory workshop centres, air-conditioned	m^2	3500.00	to	4450.00
Warehouse and distribution centres				
high bay (10–15 m high) for owner occupation (no heating) up to 10,000 m^2	m^2	390.00	to	500.00
high bay (10–15 m high) for owner occupation (no heating) over 10,000 m^2 up to 20,000 m^2	m^2	295.00	to	375.00
high bay (16–24 m high) for owner occupation (no heating) over 10,000 m^2 up to 20,000 m^2	m^2	430.00	to	550.00
high bay (16–24 m high) for owner occupation (no heating) over 20,000 m^2	m^2	355.00	to	455.00
Fit-out cold stores, refrigerated stores inside warehouse	m^2	780.00	to	1000.00
Industrial buildings				
Shell with heating to office areas only				
500–1,000 m^2	m^2	640.00	to	810.00
1,000–2,000 m^2	m^2	540.00	to	690.00
greater than 2,000 m^2	m^2	590.00	to	750.00
Unit including services to production area				
500–1,000 m^2	m^2	880.00	to	1125.00
1,000–2,000 m^2	m^2	800.00	to	1025.00
greater than 2,000 m^2	m^2	800.00	to	1025.00
UNICLASS D3 ADMINISTRATIVE, COMMERCIAL, PROTECTIVE SERVICES FACILITIES				
Embassies	m^2	2600.00	to	3300.00
County courts	m^2	3150.00	to	4000.00
Magistrates courts	m^2	1700.00	to	2175.00
Civic offices				
non air-conditioned	m^2	1700.00	to	2175.00
fully air-conditioned	m^2	2050.00	to	2600.00
Probation/registrar offices	m^2	1275.00	to	1650.00
Offices for owner occupation				
low rise, medium quality; air-conditioned	m^2	1850.00	to	2350.00
medium rise, medium quality; air-conditioned	m^2	2325.00	to	3000.00
Offices – City and West End				
To developers Cat A finish only				
high quality, speculative 8–20 storeys, air-conditioned	m^2	2600.00	to	3300.00
high rise, air-conditioned, iconic speculative towers	m^2	3950.00	to	5100.00

BUILDING PRICES PER SQUARE METRE

Item	Unit	Range £		
Business park offices				
Generally up to 3 storeys only				
functional non air-conditioned less than 2,000 m^2	m^2	1500.00	to	1925.00
functional non air-conditioned more than 2,000 m^2	m^2	1575.00	to	2000.00
medium quality non air-conditioned less than 2,000 m^2	m^2	1650.00	to	2075.00
medium quality non air-conditioned more than 2,000 m^2	m^2	1750.00	to	2225.00
medium quality air-conditioned less than 2,000 m^2	m^2	1925.00	to	2450.00
medium quality air-conditioned more than 2,000 m^2	m^2	2050.00	to	2600.00
good quality – naturally ventilated to meet BCO specification (exposed soffits, solar shading) less than 2,000 m^2	m^2	1850.00	to	2350.00
good quality – naturally ventilated to meet BCO specification (exposed soffits, solar shading) more than 2,000 m^2	m^2	1925.00	to	2450.00
high quality air-conditioned less than 2,000 m^2	m^2	2325.00	to	3000.00
high quality air-conditioned more than 2,000 m^2	m^2	2375.00	to	3050.00
Large trading floors in medium sized offices	m^2	3750.00	to	4800.00
Two storey ancillary office accommodation to warehouses/factories	m^2	1500.00	to	1925.00
Fitting out offices (NIA – Net Internal Area)				
City and West End				
basic fitting out including carpets, decorations, partitions and services	m^2	680.00	to	870.00
good quality fitting out including carpets, decorations, partitions and services	m^2	750.00	to	960.00
high quality fitting out including carpets, decorations, partitions, ceilings, furniture, air conditioning and electrical services	m^2	820.00	to	1050.00
reception areas	m^2	3400.00	to	4400.00
conference suites	m^2	3300.00	to	4200.00
meeting areas	m^2	4100.00	to	5200.00
back of house/storage	m^2	750.00	to	960.00
sub-equipment room	m^2	2175.00	to	2800.00
kitchen	m^2	4800.00	to	6100.00
Out-of-town (South East England)				
basic fitting out including carpets, decorations, partitions and services	m^2	550.00	to	700.00
good quality fitting out including carpets, decorations, partitions and services	m^2	620.00	to	790.00
high quality fitting out including carpets, decorations, partitions, ceilings, furniture, air conditioning and electrical services	m^2	750.00	to	960.00
reception areas	m^2	2050.00	to	2600.00
conference suites	m^2	2800.00	to	3600.00
meeting areas	m^2	2700.00	to	3500.00
back of house/storage	m^2	620.00	to	790.00
sub-equipment room	m^2	2175.00	to	2800.00
kitchen	m^2	3150.00	to	4000.00
Office refurbishment (including developers finish – (Gross Internal Floor Area – GIFA; Central London)				
minor refurbishment	m^2	640.00	to	820.00
medium refurbishment	m^2	1375.00	to	1750.00
major refurbishment	m^2	2125.00	to	2700.00
Banks and building societies				
banks; local	m^2	1975.00	to	2500.00
banks; city centre/head office	m^2	2800.00	to	3600.00
building society branches	m^2	1925.00	to	2450.00
building society; refurbishment	m^2	1225.00	to	1575.00

BUILDING PRICES PER SQUARE METRE

Item	Unit	Range £		
UNICLASS D3 ADMINISTRATIVE, COMMERCIAL, PROTECTIVE SERVICES FACILITIES – cont				
Shop shells				
small	m²	830.00	to	1050.00
large, including department stores and supermarkets	m²	760.00	to	970.00
Fitting out shell for small shop (including shop fittings)				
simple store	m²	790.00	to	1000.00
designer/fashion store	m²	1575.00	to	2000.00
Fitting out shell for department store or supermarket	m²	2250.00	to	2900.00
Retail warehouses				
shell	m²	570.00	to	730.00
fitting out, including all display and refrigeration units, check outs and IT systems	m²	340.00	to	440.00
Hypermarkets/Supermarkets				
shell	m²	690.00	to	880.00
supermarket fit-out	m²	1375.00	to	1750.00
hypermarket fit-out	m²	960.00	to	1225.00
Shopping centres				
Malls, including fit-out				
comfort cooled	m²	4400.00	to	5600.00
air-conditioned	m²	5600.00	to	7200.00
food court	m²	4850.00	to	6200.00
factory outlet centre – enclosed	m²	3950.00	to	5100.00
factory outlet centre – open	m²	790.00	to	1000.00
anchor tenants; capped off services	m²	1275.00	to	1650.00
medium/small units; capped off services	m²	1225.00	to	1575.00
centre management	m²	2800.00	to	3600.00
enclosed surface level service yard	m²	1975.00	to	2500.00
landlords back of house and service corridors	m²	1975.00	to	2500.00
Refurbishment				
mall; limited scope	m²	1425.00	to	1825.00
mall; comprehensive	m²	2050.00	to	2600.00
Rescue/aid facilities				
ambulance stations	m²	1225.00	to	1575.00
ambulance control centres	m²	1925.00	to	2450.00
fire stations	m²	2050.00	to	2600.00
Police stations	m²	1975.00	to	2500.00
Prisons	m²	2450.00	to	3150.00
UNICLASS D4 MEDICAL, HEALTH AND WELFARE FACILITIES				
District hospitals	m²	2275.00	to	2900.00
refurbishment	m²	1650.00	to	2075.00
Hospices	m²	1850.00	to	2350.00
Private hospitals	m²	2275.00	to	2900.00
Pharmacies	m²	2450.00	to	3150.00
Hospital laboratories	m²	2700.00	to	3500.00
Ward blocks	m²	2050.00	to	2600.00
refurbishment	m²	1225.00	to	1575.00
Geriatric units	m²	2050.00	to	2600.00
Psychiatric units	m²	1925.00	to	2450.00

BUILDING PRICES PER SQUARE METRE

Item	Unit	Range £		
Psychogeriatric units	m²	1850.00	to	2350.00
Maternity units	m²	2050.00	to	2600.00
Operating theatres	m²	2700.00	to	3500.00
Outpatients/casualty units	m²	2050.00	to	2600.00
Hospital teaching centres	m²	1775.00	to	2275.00
Health centres	m²	1650.00	to	2075.00
Welfare centres	m²	1775.00	to	2275.00
Day centres	m²	1650.00	to	2075.00
Group practice surgeries	m²	1850.00	to	2350.00
Homes for the mentally handicapped	m²	1500.00	to	1925.00
Homes for the physically handicapped	m²	1700.00	to	2175.00
Geriatric day hospital	m²	1925.00	to	2450.00
Accommodation for the elderly				
residential homes	m²	1225.00	to	1575.00
nursing homes	m²	1650.00	to	2075.00
Homes for the aged	m²	1925.00	to	2450.00
refurbishment	m²	1025.00	to	1300.00
Observation and assessment units	m²	1375.00	to	1750.00
Primary Health Care				
doctors surgery – basic	m²	1425.00	to	1825.00
doctors surgery/medical centre	m²	1700.00	to	2175.00
Hospitals				
diagnostic and treatment centres	m²	3500.00	to	4450.00
acute services hospitals	m²	3350.00	to	4300.00
radiotherapy and ocology units	m²	3500.00	to	4450.00
community hospitals	m²	2800.00	to	3600.00
trauma unit	m²	2600.00	to	3300.00
UNICLASS D5 RECREATIONAL FACILITIES				
Public houses	m²	2050.00	to	2600.00
Dining blocks and canteens	m²	1650.00	to	2075.00
Restaurants	m²	1850.00	to	2350.00
Community centres	m²	1775.00	to	2275.00
General purpose halls	m²	1500.00	to	1925.00
Visitors' centres	m²	2450.00	to	3150.00
Youth clubs	m²	2050.00	to	2600.00
Arts and drama centres	m²	2175.00	to	2800.00
Theatres, including seating and stage equipment				
large – over 500 seats	m²	4650.00	to	5900.00
studio/workshop – less than 500 seats	m²	3100.00	to	3900.00
refurbishment	m²	2450.00	to	3150.00
Concert halls, including seats and stage equipment	m²	3500.00	to	4450.00
Cinema				
shell	m²	1075.00	to	1375.00
multiplex; shell only	m²	2175.00	to	2800.00
fitting out, including all equipment, air-conditioned	m²	1375.00	to	1750.00
Exhibition centres	m²	2050.00	to	2600.00

BUILDING PRICES PER SQUARE METRE

Item	Unit	Range £		
UNICLASS D5 RECREATIONAL FACILITIES – cont				
Swimming pools				
international standard	m²	4650.00	to	5900.00
local authority standard	m²	3100.00	to	3900.00
school standard	m²	1425.00	to	1825.00
leisure pools, including wave making equipment	m²	3600.00	to	4600.00
Ice rinks	m²	1650.00	to	2075.00
Rifle ranges	m²	1375.00	to	1750.00
Leisure centres				
dry	m²	1925.00	to	2450.00
extension to hotels; shell and fit-out, including small pool	m²	2700.00	to	3500.00
wet and dry	m²	2900.00	to	3700.00
Sports halls including changing rooms	m²	1275.00	to	1650.00
School gymnasiums	m²	1225.00	to	1575.00
Squash courts	m²	1275.00	to	1650.00
Indoor bowls halls	m²	920.00	to	1175.00
Bowls pavilions	m²	1125.00	to	1450.00
Health and fitness clubs	m²	1925.00	to	2450.00
Sports pavilions	m²	1375.00	to	1750.00
changing only	m²	1575.00	to	2000.00
social and changing	m²	1575.00	to	2000.00
Clubhouses	m²	1275.00	to	1650.00
UNICLASS D6 RELIGIOUS FACILITIES				
Temples, mosques, synagogues	m²	1775.00	to	2275.00
Churches	m²	1650.00	to	2075.00
Mission halls, meeting houses	m²	1775.00	to	2275.00
Convents	m²	1925.00	to	2450.00
Crematoria	m²	2175.00	to	2800.00
Mortuaries	m²	3000.00	to	3850.00
UNICLASS D7 EDUCATION, SCIENTIFIC AND INFORMATION FACILITIES				
Nursery schools	m²	1650.00	to	2075.00
Primary/junior schools	m²	1700.00	to	2175.00
Secondary/middle schools	m²	2050.00	to	2600.00
Secondary schools and further education colleges				
classrooms	m²	1375.00	to	1750.00
laboratories	m²	1500.00	to	1925.00
craft design and technology	m²	1500.00	to	1925.00
music	m²	1700.00	to	2175.00
Extensions to schools				
classrooms	m²	1650.00	to	2075.00
laboratories	m²	1700.00	to	2175.00
Sixth form colleges	m²	1650.00	to	2075.00
Special schools	m²	1375.00	to	1750.00
Training colleges	m²	1375.00	to	1750.00
Management training centres	m²	1775.00	to	2275.00

BUILDING PRICES PER SQUARE METRE

Item	Unit	Range £		
Universities				
arts buildings	m^2	1500.00	to	1925.00
science buildings	m^2	1850.00	to	2350.00
College/University libraries	m^2	1500.00	to	1925.00
Laboratories and offices, low level servicing	m^2	2900.00	to	3700.00
Computer buildings	m^2	2450.00	to	3150.00
Museums and art galleries				
national standard museum	m^2	6600.00	to	8500.00
national standard independent specialist museum, excluding fit-out	m^2	4650.00	to	5900.00
regional, including full air conditioning	m^2	3950.00	to	5100.00
local, including full air conditioning	m^2	2950.00	to	3750.00
conversion of existing warehouse to regional standard museum	m^2	1850.00	to	2350.00
conversion of existing warehouse to local standard museum	m^2	1575.00	to	2000.00
Galleries				
international standard art gallery	m^2	4000.00	to	5200.00
national standard art gallery	m^2	3100.00	to	3900.00
independent commercial art gallery	m^2	1700.00	to	2175.00
Arts and drama centre	m^2	1650.00	to	2075.00
Learning resource centre				
economical	m^2	1500.00	to	1925.00
high quality	m^2	1925.00	to	2450.00
Libraries				
regional; 5,000 m^2	m^2	3400.00	to	4400.00
national; 15,000 m^2	m^2	4450.00	to	5700.00
international; 20,000 m^2	m^2	5800.00	to	7400.00
Conference centres	m^2	2375.00	to	3050.00
UNICLASS D8 RESIDENTIAL FACILITIES				
Local Authority and Housing Association schemes				
Housing Asociation Developments (Code for Sustainable Homes Level 3)				
Bungalows				
semi-detached	m^2	1325.00	to	1675.00
terraced	m^2	1250.00	to	1575.00
Two storey housing				
detached	m^2	1325.00	to	1675.00
semi-detached	m^2	1250.00	to	1575.00
terraced	m^2	1100.00	to	1400.00
Three storey housing				
semi-detached	m^2	1325.00	to	1675.00
terraced	m^2	1100.00	to	1400.00
Apartments/flats				
low rise	m^2	1325.00	to	1675.00
medium rise	m^2	1625.00	to	2075.00
Sheltered housing with wardens accommodation	m^2	1250.00	to	1575.00
Private Developments				
single detached houses	m^2	1775.00	to	2250.00
two and three storey houses	m^2	1825.00	to	2350.00

BUILDING PRICES PER SQUARE METRE

Item	Unit	Range £		
UNICLASS D8 RESIDENTIAL FACILITIES – cont				
Private Developments – cont				
Apartments/flats generally				
standard quality; 3–5 storeys	m^2	2200.00	to	2800.00
warehouse/office conversion to apartments	m^2	2425.00	to	3100.00
high quality apartments in residential tower – Inner London	m^2	2900.00	to	3750.00
Hotels (including fittings, furniture and equipment)				
luxury city centre with conference and wet leisure facilities	m^2	3900.00	to	5000.00
business town centre with conference and wet leisure facilities	m^2	2850.00	to	3650.00
mid-range with conference and leisure facilities	m^2	2225.00	to	2800.00
budget city centre with dining and bar facilities	m^2	1950.00	to	2475.00
budget roadside excluding dining facilities	m^2	1675.00	to	2175.00
Hotel accommodation facilities (excluding fittings, furniture and equipment)				
bedroom areas	m^2	1175.00	to	1475.00
front of house and reception	m^2	1475.00	to	1925.00
restaurant areas	m^2	1750.00	to	2250.00
bar areas	m^2	1575.00	to	1975.00
function rooms/conference facilities	m^2	1475.00	to	1925.00
Students residences				
large budget schemes with en-suite accommodation	m^2	1475.00	to	1925.00
smaller schemes (40–100 units) with mid range specifications	m^2	1825.00	to	2325.00
smaller high quality courtyard schemes, college style	m^2	2475.00	to	3150.00

Building Cost Models

The authors have been publishing cost models in *Building* magazine for many years. Although the scope and coverage of the cost models has expanded considerably, the objectives have always remained constant. They are:

- To provide detailed elemental cost information derived from a generic building that can be applied to other projects
- To provide a commentary on cost drivers and other design and specification issues
- To compare suitable procurement routes that secure the clients objectives

Models are based on actual projects and are not compiled from current book rates. Users should remember that rates are always intended purely as a guide and are not a substitute for prices received for actual projects, with specific location, specifications, conditions etc. which can all affect the cost.

For this edition we have published updated elemental cost data for several building types. Locations do vary for each model, so please note the location and location factor for any adjustments that may need to be made. The TPI for each individual model reflects the location of that model, which may not be the same as the general book TPI which has a location factor of 1.00. This is important if users are adjusting model costs to a different location or time.

A summary of UK location factors can be found in the *Building Costs Indices, Tender Price Indices and Location Factors* section of this book, together with examples showing how to apply those factors.

Readers should note that these models are based on actual projects and are intended as an indication to build costs only. There are many factors which need to be considered when putting together an early cost plan or estimate and all need to be considered when arriving at a Cost Plan for any particular project. Rates and totals are rounded.

Readers may refer to the *Approximate Estimating Rates* section of this book to make adjustments to the models for alternative specifications within any of the elements.

As elsewhere in this edition prices do not include Value Added Tax or professional services fees.

RETAIL DISTRIBUTION UNIT

This cost model is based on a single-storey, rectangular-shaped warehouse new building (12 m height to haunch) with a two-storey office, with gross internal floor areas of 29,640 m^2 and 1,680 m^2 respectively. The development is on a brownfield site, which has previously been cleared. Any land remediation has been undertaken as part of a separate contract. The building has a reinforced in situ concrete ground-bearing slab, a steel portal frame with reinforced concrete pads, and an aluminium built-up cladding system to roof and walls. External works consist of access and circulation road, car parking, service yard, pedestrian areas, landscaping areas and gatehouse.

Warehouse: Gross internal floor area	29,640 m^2
Office Shell: Gross internal floor area	1,680 m^2
Model location is based on West Midland location	Model TPI = 112; LF = 0.85

This cost model is copyright of AECOM Ltd

Retail Distribution Unit	Quantity	Unit	Rate (£)	Total (£)	Cost (£/m²)
Substructure				**2,106,700**	**71.08**
Pad foundations; excavations; concrete; reinforcement; holding down bolts; averaged size 2.50 × 2.50 × 0.75 m	45	nr	1,450		
Reinforced column casings below slab finished level; including formwork	45	nr	110		
Reinforced precast concrete perimeter upstand/ground beam below cladding; 425 × 300 mm wide	608	m	140		
Reinforced concrete perimeter retaining wall 1400 mm high	120	m	420		
Foundations to dock levellers: 2975 mm × 325 mm deep reinforced concrete	24	m	1,100		
Strip foundations to retaining walls; concrete; formwork; reinforcement	120	m	300		
Backfilling retaining walls	203	m^3	33		
Imported fill to make up levels – 250 mm thick	7,410	m^3	28		
Ground slab 200 mm thick minimum. Slab to have approved sealant/surface hardener, FM2 special finish	29,640	m^2	48		
150 mm type 1 under slab	4,446	m^2	42		
Design calculations; laser level; floor survey; sacrificial sealant to joints (maybe included with slab quote)		item	6,000		
Inspections by a specialist flooring consultant		item	8,900		
Frame				**1,636,200**	**55.20**
Structural steel portal frame 12 m; hit and miss valley columns, painted, including cold rolled wall and roof sections; service loading 0.25kn/m^2	29,640	m^2	54		
Site treatment/decoration	29,640	m^2	1		
Roof				**2,085,100**	**70.35**
LPCB approved trapezoidal roof panels, standard colour	29,640	m^2	42		
GRP triple skin roof lights to 15% of the ambient	4,446	m^2	70		
Valley gutters; membrane lined system, including stop end and the like	741	m^2	140		

RETAIL DISTRIBUTION UNIT

Retail Distribution Unit	Quantity	Unit	Rate (£)	Total (£)	Cost (£/m²)
Eaves gutter abutment; steel outer skin, insulation board, membrane lined system, including stop ends and the like	494	m	120		
Ridge/hip flashing	988	m	24		
Access hatches, including flashings	2	nr	3,000		
Access ladder to roof from warehouse floor	2	nr	12,000		
Allow for safe travel across roof by means of safety ropes; Top-fix fall arrest cable system, includes user equipment	734	m	42		
Allow access equipment (hoisting, distribution and power)		item	107,500		
Penetrations and air testing		item	32,000		
Syphonic drainage HDPE pipework; including primary and secondary system and external concrete splash pads	1,235	m	110		
Hydrabrake at first manhole		item	6,000		
External Walls, Windows and Doors				**939,300**	**31.69**
Micro-rib composite panels LPC Grade B approved; 1000 mm cover width panels, standard colour; including vertical top hat, joint flashing and the like; laid horizontally	3,910	m²	95		
Extra over; 1-hour fire-rated	3,910	m²	5		
Trapezoidal composite panels LPC Grade B approved; standard colour, including end laps, flashing and the like; laid vertically	4,928	m²	53		
Profiled cladding to rear of parapet; laid vertically	704	m²	60		
Sill closure cill flashing	704	m	36		
Parapet/coping flashing detail 750 mm girth	704	m	51		
Single personnel fire escape doors with internal panic bars	16	nr	1,025		
Louvres in external wall cladding to plant rooms	200	m²	360		
Allowance for access equipment	1	nr	90,000		
Inspections by a specialist cladding consultant		item	6,000		
Dock Leveller Installations				**294,000**	**9.92**
Insulated sectional overhead type dock door (2.75 m × 3 m) finished in a standard colour range; including two vision panels 1500 mm above FFL	24	nr	1,700		
Dock shelters heavy duty scissor retracting frame type with 650 mm wide curtain and 1200 mm deep top curtain in high frequency welded PVC	24	nr	1,350		
Electrohydraulic dock levellers with 480, lip and platform size 2500 mm long × 2000 mm wide; dynamic capacity 6000 kg gross	24	nr	2,800		

RETAIL DISTRIBUTION UNIT

Retail Distribution Unit	Quantity	Unit	Rate (£)	Total (£)	Cost (£/m²)
2 nr internal bollards 138 mm × 11000 mm high to protect door track	24	nr	180		
Heavy duty rubber dock buffers; 175 mm thick × 250 mm wide 4450 mm deep	24	nr	390		
Internal dock loading lights	24	nr	260		
External red/green traffic lights	24	nr	170		
Precast concrete pits to levellers	24	nr	3,300		
External escape stairs – (allowance only)	14	nr	3,000		
Bollards to external stairs 225 mm x 1500 mm	28	nr	300		
Level Access Doors				**4,900**	**0.17**
Insulated sectional overhead type level access doors finished in a colour from a standard range with emergency hand chain; including two vision panels; Size 4000 mm × 5000 mm high	1	nr	4,150		
2 nr internal bollards 225 mm × 1100 mm to protect door track	1	pair	300		
2 nr external bollards 225 mm × 1500 mm to protect door jambs	1	pair	360		
Preliminaries and Contingency				**1,413,200**	**47.68**
Overheads and profit, site establishment and supervision @		15%			
Contingency @		5%			
Construction cost (Warehouse shell only, rate based on GIFA)				**8,479,400**	**286.09**

Office Shell and Fit-Out	Quantity	Unit	Rate (£)	Total (£)	Cost (£/m²)
Substructure				**113,300**	**67.44**
Imported fill (type 1) making up levels; 250 thick	210	m³	36		
Pad foundations; excavations; concrete; reinforcement; holding down bolts; average size 1.75 × 1.75 × 0.75 m	24	nr	900		
Reinforced column casings below slab finished level; including formwork	24	m	115		
Reinforced precast concrete perimeter upstand/ ground beam below cladding; 425 × 300 mm wide	86	m	90		
Lift pit	2	n	3,600		
Ground floor slab reinforced concrete 150 mm thick; 250 mm type 1 granular fill material	840	m²	36		
Reinforced concrete perimeter retaining wall 1500 mm high	86	m	420		
Frame, Upper Floors and Stairs				**454,400**	**270.48**
Fabricated steel; painted, including cold rolled sections	1,680	m²	85		
Site treatment/decoration	1,680	m²	1		
Board fire protection to compartment wall		item	30,000		

RETAIL DISTRIBUTION UNIT

Retail Distribution Unit	Quantity	Unit	Rate (£)	Total (£)	Cost (£/m²)
1 hr intumescent fire protection to columns and beams		item	42,000		
Office bridges structure supports		item	36,000		
Composite concrete floor on structural metal decking	830	m²	60		
Accommodation staircase to serve upper office levels; including finishes, balustrades and hand railing	2	nr	36,000		
Accommodation staircase; ground to plant deck including finishes and handrails	2	nr	18,000		
Precast concrete/metal stairs, ground to second floor plant deck; including finishes, balustrading and hand railing	1	nr	24,000		
Fire stopping to perimeter of offices	116	m	30		
Handrail to plant deck	116	m	140		
Roof				**61,300**	**36.49**
Either Kingspan composite KS1000RW LPC Grade B approved trapezoidal roof panels or decked/insulated single ply membrane system	840	m²	36		
Eaves gutter abutment; 3 mm thick post-galvanized steel outer, 45 mm thick insulation board, 0.7 mm inner skin; including stop ends	116	m	150		
Access hatches, 900 x 900 mm, including	2	nr	2,400		
Allow for safe travel across roof by means of safety ropes; Latchways Mansafe	116	m	55		
Allow access equipment (hoisting, distribution and power)		item	2,400		
External Walls, Windows and Doors				**269,800**	**160.60**
Micro-rib composite panels LPC grade B approved, laid horizontally	1,022	m²	70		
Sill closure drip flashing	116	m	30		
Profiled cladding to rear of parapet; vertically laid (height 1.00 m)	116	m²	36		
Parapet/coping flashing detail 750 mm girth	116	m	70		
Powder coated, thermally broken curtain walling by Schuco/Kawneer	100	m²	420		
Automatic entrance doors to match curtain walling (Only required if no lobby)	1	nr	3,000		
Powder coated, thermally broken double-glazed windows; glazing to be anti-sun outer pane with low-e coating	270	m²	420		
Glass entrance canopy supported on powder coated ground bearing structure; minimum size to be 15 m²	1	nr	24,000		
Internal Partitions and Doors				**571,200**	**340.00**
Office fit-out ground & upper floors (including wall, floor, ceiling finishes, doors etc.)	1,680	m²	340		

RETAIL DISTRIBUTION UNIT

Retail Distribution Unit	Quantity	Unit	Rate (£)	Total (£)	Cost (£/m²)
Mechanical Installations				368,500	219.35
Heating/Cooling	1,680	m²	140		
Ventilation	1,680	m²	60		
Rainwater harvesting tank; 20,000 litres		item	18,000		
Solar Thermal		item	14,500		
Electrical Installations				438,700	261.13
Power to doors	18	nr	900		
LV installations	1,680	m²	24		
Sub-mains cabling	1,680	m²	10		
Containment installations	1,680	m²	12		
Lighting installations	1,680	m²	60		
Small power installations	1,680	m²	30		
Earthing and bonding installations	1,680	m²	1		
Gatehouse		item	18,000		
Testing and commissioning		3%			
Subcontractor preliminaries		12%			
Lift installation; 8 persons electrohydraulic	2	nr	48,000		
Allowance for builder's work in connection with services		5%			
Preliminaries and Contingency				451,500	268.75
Overheads and profit, site establishment and supervision @		15%			
Contingency @		5%			
Construction cost (Office shell and fit-out only, rate based on Office GIA)				2,728,700	1,624.24
External Works	**Quantity**	**Unit**	**Rate (£)**	**Total (£)**	**Cost (£/m²)**
Site Works				2,922,400	93.31
Site clearance including removal of trees, shrubs etc.	80,098	m²	0.50		
Disposal of excavated material off site	3,231	m³	21		
Excavate to reduce levels and stockpile on site	16,020	m²	4		
Filling to make to levels with excavated material	8,010	m³	3		
Level and compact surfaces of excavation or filling	80,098	m²	0.50		
Geotextile membrane	80,098	m²	0.60		
Disposal of excavated material to landscaped areas	4,779	m²	3		
SW drainage to building	30,480	m²	10		
SW drainage to HGV yard and road	16,562	m²	10		
SW drainage to car parking	8,135	m²	10		
Paving	8,163	m²	10		
FW drainage to foul drain runs	387	m	180		

RETAIL DISTRIBUTION UNIT

Retail Distribution Unit	Quantity	Unit	Rate (£)	Total (£)	Cost (£/m²)
FW drainage to manholes	10	nr	1,450		
Petrol interceptors	3	nr	6,600		
Connection to main sewer	1	nr	1,800		
Heavy duty access road and service yard; 175 mm brushed concrete on and including 150 type 1 subbase with 2800 g polythene membrane; sealed construction joints	16,562	m²	42		
Trief kerbs on and including concrete foundations/haunching	182	m	115		
Standard kerbs	2,157	m	18		
Road marking/linings, signage		item	12,000		
Car parking/light duty road; comprising 30 mm dense bitumen wearing course; 80 mm dense bitumen base course; 150 mm type 1 subbase	8,135	m²	27		
Standard kerbs	532	m	18		
Road marking/linings, signage		item	6,000		
Car parking/light duty road; comprising 30 mm dense bitumen wearing course; 80 mm dense bitumen base course; 150 mm type 1 subbase	8,135	m²	27		
Standard kerbs	532	m	18		
Road marking/linings, signage		item	6,000		
Public footpaths to be 600 mm × 600 mm × 50 mm precast concrete paving slabs to provide a minimum 1200 wide path; on and including 50 mm sand bed on 150 mm type 1 granular material	8,163	m²	21		
Perimeter fencing; 2.4 m high Paladin fence with concrete posts	1,142	m	85		
Entrance barriers	1	nr	24,000		
Entrance gates	1	nr	30,000		
Grassed/planted areas	13,655	m²	12		
Tree planting allowance	51	nr	600		
12 months maintenance to begin from completion of planting		item	12,000		
Perimeter fencing; 2.4 m high Paladin fence with concrete posts	1,142	m	85		
Entrance barriers	1	nr	24,000		
Entrance gates	1	nr	30,000		
External Services				**404,400**	**12.91**
External lighting		item	120,000		
Gas; including housing		item	48,000		
Electricity; including hosing		item	60,000		
Water; including hosing		item	42,000		

RETAIL DISTRIBUTION UNIT

Retail Distribution Unit	Quantity	Unit	Rate (£)	Total (£)	Cost (£/m²)
Telecoms		item	2,400		
External hydrant main and hydrants		item	90,000		
Service trenches to electricity, gas, water, hydrant mains distribution	1,000	m	42		
Ancillary Buildings				**98,400**	**3.14**
Prefabricated gatehouse 20 m² (215 sq ft)		item	85,000		
Allowance for foundations, base, drainage and services required		item	5,050		
Vehicle barriers and BWIC (1 nr barrier/lane)	2	nr	4,150		
Preliminaries and Contingency				**685,100**	**21.87**
Overheads; profit, site establishment and site supervision		15%			
Contingency		5%			
Construction cost (External works only, rate based on GIFA)				**4,110,300**	**131.23**
Total distribution centre construction cost (rate based on GIFA)				**15,318,400**	**489.09**

LONDON FRINGE OFFICE

This cost model is based on a new-build seven-storey commercial office building in a fringe London location (as defined above). The parameters are set around a gross internal area (GIA) of 250,000 ft², including a single-storey basement, with minimal site constraints. Overall net to gross efficiency is assumed at 77% of the overall GIA, with an average wall-to-floor ratio of 0.40. The building's HVAC system uses four-pipe fan coils combined with air-handling units and traditional gas boilers and chillers. The external walls are predominantly unitized curtain walling with retail shop frontages at ground floor.

Gross internal floor area	23,225 m²
Model location is Inner London	Model TPI = 139; LF = 1.05

This cost model is copyright of AECOM Ltd

London Fringe Office	Quantity	Unit	Rate (£)	Total (£)	Cost (£/m²)
Substructure				8,030,700	345.78
Allowance for temporary works		item	250,000		
Foundations – rotary bored piles (900–1,200 mm dia.; average 20 m depth), including piling mat	2,500	m²	1,300		
Secant piled basement retaining wall	4,950	m²	350		
Basement excavation and disposal of inert excavated material	13,500	m²	90		
Allowance for dewatering excavations		item	150,000		
Allowance for cavity drainage system to basement, including sumps	275	m	575		
Allowance for below-ground drainage generally	2,500	m²	90		
Reinforced concrete basement slab, 300 mm thick, including waterproofing etc.	2,500	m²	290		
Ground-floor slab – lightweight reinforced concrete on 140 mm profiled metal decking:	2,500	m²	130		
Frame				6,645,400	286.13
Structural steel frame, based on 70 kg/m² of GIA on a 9 m × 9 m column grid:	1,626	tonne	3,050		
Allowance for secondary steelwork, based on extra 5 kg/m² of GIA	116	tonne	3,050		
Fire protection to steel frame (generally 90-minute intumescent paint)	1,742	m²	650		
Allowance for other structures (e.g. lift motor rooms)		item	150,000		
Allowance for expansion joints and other sundries		item	50,000		
Upper Floors and Stairs				3,925,600	169.02
Lightweight reinforced concrete on 140 mm profiled metal decking	17,500	m²	130		
Duragrid/GRP riser protection system	465	m²	530		
Allowance for upstands, plinths, bund walls etc.		item	175,000		
Allowance for notional soft spots (one per upper floor slab)		item	35,000		
Allowance for slab edge detailing	2,012	m	95		

LONDON FRINGE OFFICE

London Fringe Office	Quantity	Unit	Rate (£)	Total (£)	Cost (£/m²)
Precast concrete stairs and landings, including finishes, painted handrails and balustrades	36	flights	23,000		
Feature stairs to reception area		item	75,000		
Allowance for ladders/stepovers/walkways etc.		item	100,000		
Roof				**1,298,200**	**55.90**
Lightweight reinforced concrete on 160 mm profiled metal decking	2,500	m²	140		
RC perimeter upstand; 300 × 300 mm	150	m	160		
Proprietary roof covering, insulation, paving slabs etc.	2,500	m²	190		
Extra over for green roof (assumed 25% of roof area)	625	m²	95		
Allowance for enhanced finishes to terrace areas		item	150,000		
Balustrading to terrace areas	95	m²	1,400		
Secondary steelwork for photovoltaics and louvred plant screen	534	m	200		
External Wall, Windows and Doors				**10,942,300**	**471.14**
Unitized curtain walling system	7,277	m²	1,200		
Retail unit shopfront glazing at ground level	488	m²	1,750		
Lift overrun and core cladding at roof level	525	m²	575		
Notional allowance for blockwork walling		item	20,000		
Louvered plant screening	280	m²	975		
Visual mock-ups and samples		item	75,000		
Building maintenance unit cleaning equipment and track		item	300,000		
Main entrance revolving doors	2	nr	92,500		
Allowance for single doors (including doors to accessible roof terrace)	6	nr	17,500		
Allowance for double-pass doors	2	nr	23,000		
Allowance for loading bay door		item	50,000		
Internal Walls and Partitions and Doors				**2,567,100**	**110.53**
Plasterboard internal walls within landlord shell and core areas, incl. access panels	7,742	m²	140		
Blockwork internal walls within basement and plant areas, incl. windposts and lintels	2,323	m²	160		
Timber internal doors, incl. door frames, architraves and ironmongery	235	nr	3,350		
Riser doors generally within landlord shell and core areas	108	nr	1,150		
Allowance for internal doors to basement and plant areas		item	200,000		

LONDON FRINGE OFFICE

London Fringe Office	Quantity	Unit	Rate (£)	Total (£)	Cost (£/m²)
Internal Finishes				3,228,500	139.01
Wall Finishes: Allowance for new finishes to circulation, WC and reception areas	23,226	m²	29		
Floor Finishes: Allowance for new finishes to circulation, WC and reception areas	23,226	m²	75		
Ceiling Finishes: Allowance for new finishes to circulation, WC and reception areas	23,226	m²	35		
Fittings, Furnishers and Equipment				3,407,100	146.70
Reception desk		item	75,000		
Reception furniture		item	25,000		
Tenant directory sign board		item	25,000		
WC cubicles, IPS, vanity units, washroom fixtures and fittings etc. (excl. MEP, sanitary and finishes elsewhere)	150	nr	1,550		
Allowance for wayfinding and statutory signage	23,226	m²	2		
Toilet fittings (WC cubicles, vanity units etc.)	23,226	m²	100		
Allowance for landlord back-of-house areas including refuse, cycle parking, showers etc.	23,226	m²	29		
Mechanical, Electrical and Plumbing Services				13,935,600	600.03
MEP services installations complete including builder's work in connection (four-pipe fan coil unit system)	23,226	m²	600		
Fit-Out				8,646,000	372.27
Category A fit-out works		item	8,646,000		
Preliminaries and Contingencies				17,642,000	759.61
Main contractor's preliminaries @		17.5%			
Main contractor's design and build risk allowance @		1.5%			
Main contractor's overhead and profit @		4.5%			
Risk allowances and design reserve @		5%			
Construction cost (shell and core only; rate based on GIFA)				80,268,500	3,456.12

Building Cost Models

CAR DEALERSHIP

A purpose-built new dealership on an automotive retail park, comprising showroom and workshop of 900 m² and 1,000 m² respectively, with additional office and ancillary space, giving a gross internal floor area of 3,500 m². It has a reinforced in situ concrete ground-bearing slab, a steel frame with piled foundations, reinforced concrete pile caps and ground beams. There is a full height glazed front, aluminium cladding on the side walls and single ply roofing and rooftop car parking. There is full comfort cooling in all areas and the cost of fitting out this and the display lighting is included. External works consist of car parking, forecourt display areas, landscaping.

Gross internal floor area   3,500 m²

Model location is South East England   Model TPI = 124; LF = 0.94

This cost model is copyright of AECOM Ltd

Car Dealership	Quantity	Unit	Rate (£)	Total (£)	Cost (£/m²)
Substructure				**906,600**	**259.03**
Excavation and disposal off site	7,000	m³	18		
CFA piled foundations; excavations; concrete; reinforcement; 600 mm dia.; 15 m deep	277	nr	1,200		
Reinforced concrete pile caps	94	nr	675		
Reinforced concrete ground beams 500 mm × 600 mm	500	m	180		
Reinforced concrete ground slab, hardcore, DPM, ground beams and column bases for steel frame (0.25 W/m² K)	2,200	m²	115		
Lift pit	1	nr	3,550		
Recessed slab for specialist equipment		item	41,000		
Frame, Upper Floors and Stairs				**956,000**	**273.14**
Steel propped portal frame, cold rolled purlins, surface treatments	220	tonne	2,200		
Intumescent paint to give 60-minute fire protection to steelwork:		item	47,000		
Decorations to exposed steelwork		item	12,000		
150 mm reinforced concrete suspended slab	2,900	m²	110		
Feature customer staircase; steel with cement screed infill; including finishes, balustrading and hand railing	1	nr	47,000		
Internal precast concrete staircase to serve upper office levels; including finishes, balustrades and hand railing	2	nr	23,500		
Envelope				**1,318,000**	**376.57**
Engineering brick plinth	30	m²	300		
Composite wall cladding – front of house	600	m²	200		
Composite wall cladding – back of house	1,300	m²	90		
Glazed curtain walling	400	m²	530		
Vehicle access doors in curtain walling	3	nr	12,000		
Revolving access doors	3	nr	47,000		

CAR DEALERSHIP

Car Dealership	Quantity	Unit	Rate (£)	Total (£)	Cost (£/m²)
Entrance feature		item	23,500		
Personnel doors	3	nr	3,550		
Workshop doors	7	nr	9,400		
Roof to dealership; single ply membrane with tapered insulation	780	m²	110		
Reinforced concrete slab to rooftop car park	1,300	m²	120		
Kingspan Styrozone car park system to rooftop parking; including screed, waterproofing, asphalt, etc.	1,300	m²	150		
Steel ramp to rooftop parking, single vehicle width		item	60,000		
Vehicle barrier to rooftop parking	270	m	180		
Syphonic roof drainage	2,080	m²	14		
Escape stairs enclosure	10	m²	825		
Internal Walls and Partitions and Doors				**318,400**	**90.97**
Plasterboard and metal stud partitions; generally, 2 layers of plasterboard each side; various levels of fire and sound insulations	3,000	m²	85		
Glazed partition system	80	m	625		
Workshop lining system (Whiterock)	100	m²	75		
Column casings		item	5,900		
Internal Finishes				**470,200**	**134.34**
Barrier matting	90	m²	360		
Floor tiles	2,200	m²	80		
Tiled skirting	80	m	36		
Carpet	450	m²	36		
Timber skirting	500	m	18		
Timber (MDF) skirting	200	m	12		
Vinyl flooring	320	m²	47		
Floor paint	300	m²	24		
Painting wall and timber	5,000	m²	14		
Wall tiles	300	m²	75		
Suspended ceilings	1,300	m²	36		
Metal framed ceiling and bulkheads	1,200	m²	53		
Fire barriers		item	5,900		
Space Heating and Air Treatment Ventilation				**803,000**	**229.43**
Incoming services and metering		item	21,500		
Ventilation	3,500	m²	90		
Heat recovery system	3,500	m²	90		
Hot and cold-water services		item	50,000		
Gas distribution		item	12,000		

CAR DEALERSHIP

Car Dealership	Quantity	Unit	Rate (£)	Total (£)	Cost (£/m²)
Workshop heating		item	36,000		
Above ground drainage		item	12,000		
Photovoltaic system		item	19,000		
Testing and commissioning		item	22,500		
Electrical and Gas Installations				**756,000**	**216.00**
Mains switchgear and distribution		item	26,000		
Sub-mains cabling		item	11,100		
Lighting and emergency lighting installations	3,500	m²	90		
Small power installations	3,500	m²	90		
Power to doors and workshop equipment		item	31,000		
External lighting		item	41,000		
Earthing and bonding		item	2,350		
Testing and commissioning		item	14,500		
Lift Installation				**60,000**	**17.14**
8 person electrohydraulic serving 3 floors	1	nr	60,000		
General Fixtures, Furnishings and Equipment				**136,400**	**38.97**
Feature car display lighting trusses		item	23,500		
Benches and seating		item	12,000		
Signage		item	5,900		
WC sanitary ware and cubicles		item	65,000		
Kitchen units and coffee station		item	30,000		
Builder's Work in Connection				**3,000**	**0.86**
Forming holes and chases etc. @		item	3,000		
Preliminaries and Contingency				**1,056,800**	**301.94**
Overheads, profit, site establishment and site supervision @		15%			
Contingency @		3%			
Construction cost for shell only (rate based on GIFA)				**6,784,400**	**1,938**
Car Dealership – External Works	**Quantity**	**Unit**	**Rate**	**Total (£)**	**Cost (£/m²)**
Hard and Soft Landscaping				**260,200**	**74.34**
Grassed/planted areas	650	m²	12		
Car parking and access road	3,700	m²	53		
Standard kerbs	610	nr	26		
Road markings		item	2,350		
Paving to pedestrian areas	200	m²	41		
Stainless steel bollards	100	nr	300		

CAR DEALERSHIP

Car Dealership	Quantity	Unit	Rate (£)	Total (£)	Cost (£/m²)
Perimeter Fencing				18,600	5.31
Perimeter fencing 2.4 m high palading welded mesh fence	102	m²	53		
Entrance vehicle barrier		item	8,800		
Entrance gates	3	nr	1,450		
Drainage and External Services				219,100	62.60
Storm water drainage	720	m	95		
Storm water manholes	17	nr	1,450		
Gullies, rainwater outlets		item	4,700		
Connection to main sewer		item	5,900		
Foul water drainage	500	m	95		
Foul water manholes	11	nr	1,450		
Petrol interceptor	1	nr	9,400		
Connection to main sewer		item	3,550		
Utilities (gas, water, electricity)		item	39,000		
Preliminaries and Contingency				91,900	26.26
Contractors preliminaries, overhead and profit		15%			
Contingency @		3%			
Construction cost for external works only (rate based on GIFA)				589,800	168.51
Total construction cost (rate based on GIFA)				7,374,200	2,107

Building Cost Models

HIGHER EDUCATION REFURBISHMENT

This cost model is based on a refurbished teaching and learning building in a London city location. It is a stand-alone structure as part of a campus location, four storeys in height, containing teaching, office space and associated ancillary/welfare facilities and assume a SKA for Higher Education Rating Gold (Ska HE). It is for a 'medium' refurbishment which assume retaining of the existing structure and frame but introduces extensive new internal finishes and furniture, fixtures and equipment with part renewal of M&E, IT and communication installations.

Gross internal floor area	1,500 m²
Model location is Inner London	Model TPI = 139; LF = 1.05

This cost model is copyright of AECOM Ltd

Higher Education Refurbishment	Rate (£/m²) GIFA	Total (£)	Cost (£/m²)
Facilitating Works		40,500	27.00
Internal demolition of existing building; 50% strip of services, demolition of non-structural internal walls	27		
Frame, Upper Floors and Stairs		324,900	216.60
Work to existing frame only – adaption/fire-rating	90		
Work to existing floors where partitions removed, making good, form and trim openings	19		
Replace roof coverings and insulation	75		
Works to existing stairs and stairwells	33		
Envelope		90,000	60.00
Making good and light alterations to existing building envelope	14		
Double glazed windows replacement to 20% of area	46		
Internal Walls and Partitions and Doors		268,500	179.00
Combination of glazed screens to teaching areas, blockwork and acoustic partitions and fire-rated areas	130		
Solid core fire-rated internal doors in hardwood frames; stainless steel ironmongery; lock suiting; DDA fire compliant	49		
Internal Finishes		381,000	254.00
Wall Finishes			
Includes for acoustic panelling to internal finishes to seminar rooms and two lecture theatres and some fire-rated enclosures. Plasterwork to blockwork walls, painting, wall tiling	49		
Floor Finishes			
Combination of full screed and raised access floor replacement; carpet tiles; sheet vinyl; feature tiling; barrier matting to entrance area	110		
Ceiling Finishes			
General acoustic panelling finishes to seminar rooms and two lecture theatres; fire-rated enclosures	95		
Fittings, Furniture and Equipment		300,000	200.00
Lecture theatre seating; joinery; lockers; perch benching. Excludes loose FF&E and bespoke joinery	200		

HIGHER EDUCATION REFURBISHMENT

Higher Education Refurbishment	Rate (£/m²) GIFA	Total (£)	Cost (£/m²)
Services		29,900	19.93
WCs and fittings, including disabled facilities; urinals and fittings; wash hand basins and fittings; cubicle showers; classroom sinks; drinking fountains	12		
Adaptions to waste, soil; vent work; rainwater adaptions (20% of existing)	8		
Space Heating and Air Treatment Ventilation		431,100	287.40
Mechanical installations to comply with landlord and client requirements	210		
Hot and cold-water services; 40% of existing services retained	2		
Space heating via LTHW radiators with all supporting plant and distribution; part natural, part mechanical ventilation; cooling; extract to toilet and kitchen areas; air handling units and limited local cooling; localized extract to plant rooms; 40% of exiting services retained	75		
Electrical Installations		331,500	221.00
Extension of mains and sub-mains distribution; small power generally; electrical supplies to mechanical plant and equipment; lighting; emergency lighting; external lighting. Assumes all lighting replaced	180		
Lightning protection retained and made good where required	41		
Communications Installations		210,000	140.00
Upgrade of existing fire alarm and smoke detection systems; interface with door hold system; disabled refuge communications; induction loop alarm interface; security system; intruder alarm; CCTV; public address; induction loop; disabled WC alarm system; data network mainly through wireless, including containment. 40% services retained	140		
Specialist Installations		49,500	33.00
BMS and security systems installed; 60% replacement; new set points; reprogrammable existing main controller unit	33		
Lift Installation		52,500	35.00
Refurbishment of lifts and making good	35		
Builder's Work in Connection		24,500	16.33
Forming holes and chases etc. @	16		
Preliminaries and Contingency		1,050,000	700.00
Preliminaries	270		
Contractors overheads and profit	100		
Contingency	220		
Design reserve	110		
Construction cost (rate based on GIFA)		**3,583,900**	**2,389**

HIGHER EDUCATION REFURBISHMENT

We have demonstrated a range of costs to cover the variances you would expect between a 'light, medium and heavy refurbishment' option, when considering budgets. Such is the variable nature of building refurbishments, it is expected that any one refurbishment will vary considerably from the next

Full refurbishment   Strip the building back to its primary frame, retain structural floors, provide a new envelope, resurface roof and fully fit-out internally including mechanical and electrical (M&E), IT and communication installations.

Medium refurbishment   Retain the existing structural fabric and envelope of the building and introduce extensive new internal finishes and furniture, fixtures and equipment (FF&E) with part renewal of M&E, IT and communication installations.

Minimal refurbishment   Retain the building in its present form, with limited elements only of new finishes internally including part FF&E.

As a basis for calculating refurbishment allowances, we have assumed a medium type refurbishment, for purposes of the cost model. Our guidance would be to allow a factor of +30% for a full refurbishment and −45% for minimal refurbishment, to be read also in conjunction with other assumptions and exclusions.

SCHOOL REFURBISHMENT

Looking ahead into the next decade, it seems the government's focus is likely to be on investing in refurbishments rather than new-build projects. This summer, the Department for Education announced £560 m in funding for academies, sixth-form colleges and voluntary aided schools across England to transform facilities and improve school buildings. The cash is part of the government's stimulus package, which is hoped to jump-start UK construction activity. Suggested projects include boiler upgrades and classroom refurbishments.

The cost model captures the full cost of a school refurbishment in the North-east of England. The project was procured via a framework made up of pre-approved consultants and contractors with a focus on transparency and collaboration between parties at all times.

Gross internal floor area	3,043 m²
Model location is North East England	Model TPI = 110; LF = 0.83

This cost model is copyright of AECOM Ltd

School Refurbishment	Quantity	Unit	Rate (£)	Total (£)	Cost (£/m²)
Facilitating Works				197,800	65.00
Internal demolition of existing building; strip out existing services	3,043	m²	65		
Substructure				17,100	5.62
Minor substructure additions	3,043	m²	6		
Roof				456,500	150.02
New bitumen hot melt roof: capping sheet; underlay; base layer; 130 mm insulation; vapour control layer; water membrane	3,043	m²	150		
450 mm-deep painted plywood fascia with Icopal aluminium trim	322	m²	*Included above*		
Velux rooflights 1300 mm × 1300 mm	3	nr	*Included above*		
External Walls				126,600	41.60
New precast L-shaped window lintel and jamb; temporary propping; take down brickwork, rebuild; wedge and pin	14	nr	4,400		
External repairs for damp penetration; hacking off 1 m high plaster; re-render in waterproof render	75	m²	200		
Concrete defects repair and replacement		item	50,000		
Windows and External; Doors				222,900	73.25
New uPVC windows double-glazed: 30 db acoustic rating; PPC cill; EPDM membrane; DPM; insulation; teleflex openers to opening lights	445	m²	400		
External double door: aluminium louvred; high-security mesh; frame; ironmongery; 1,100 mm × 1,590 mm high	1	nr	17,000		
PPC cill, EPDM membrane; DPM; insulation; PPC frame; ironmongery	66	m²	380		
Canopy over small hall exit		item	2,700		

Building Cost Models

SCHOOL REFURBISHMENT

School Refurbishment	Quantity	Unit	Rate (£)	Total (£)	Cost (£/m²)
Internal Walls and Partitions				**77,200**	**25.37**
IPS solid grade laminate, TBS Zenith range SGL – 100 m², 18 nr; toilet cubicles, fully framed, solid grade laminate, TBS Futura, 2,000 mm high	36	cubicles	1,900		
Stud partitioning: 70 C studs at 600 mm centres; 15 mm thick impact-resistant plasterboard either side; 25 mm thick insulation	118	m²	50		
Drylining: 2 layers of 15 mm thick impact-resistant plasterboard lining	136	m²	21		
Internal Doors				**129,200**	**42.46**
New internal doors: paint grade; frames and architraves; ironmongery, single doors 105 nr, double doors 17 nr		item	95,000		
New sliding internal doors: paint grade; frames and architraves; ironmongery	3	nr	1,550		
Internal gate: to suit 1,300 mm high wall; ironmongery	1	nr	18,000		
Paint to doorsets: undercoat; 2 coats gloss		item	11,500		
Wall Finishes				**184,800**	**60.73**
Acoustic wall panels; Gyptone boards	243	m²	290		
Plaster skim to existing walls	1,593	m²	9		
Plaster patching	6,402	m²	4		
Dim out vertical blinds: Décor 250 roller blinds; manual	473	m²	60		
Clean down walls following floor sanding prior to painting	3,186	m²	1		
Splashbacks: Altro Whiterock; 2.5 mm thick; adhesive bonding	38	m²	100		
Eggshell emulsion paint: normal humid hydrothermal; 3 coats high-performance eggshell paint		item	37,000		
Floor Finishes				**149,800**	**49.23**
Soft flooring	2,860	m²	32		
Clean existing vinyl floors	179	m²	5		
Repair and reseal Granwood floor	985	m²	45		
MDF skirting; paint; 18 mm × 100 mm high	679	m²	16		
Epoxy floor paint: 1 initial coat, 2 finishing coats	98	m²	19		
Ceiling Finishes				**87,100**	**28.62**
Suspended ceiling: CEP Classcare; suspension system; insulation; 600 mm × 600 mm tiles; all fittings and fixings; cavity fire barriers	1,184	m²	42		
Plasterboard suspended ceiling: 15 mm-thick wallboard lining; skim coat plaster finish; grid system	1,651	m²	14		

SCHOOL REFURBISHMENT

School Refurbishment	Quantity	Unit	Rate (£)	Total (£)	Cost (£/m²)
Clean down ceilings following floor sanding prior to painting	1,947	m²	1		
Ecosure Matt Dulux paint: 1 initial coat of 10% primer, 2 undercoat/finishing coat, to new plasterboard ceilings	1,575	m²	9		
Fittings, Furniture and Equipment				182,700	60.04
Permanent installation		item	150,000		
Storing/decanting existing FF&E and sundry items		item	20,500		
Removing existing FF&E and disposal		item	4,700		
Signage		item	7,500		
ICT				195,000	64.08
Permanent installation		item	195,000		
Services				1,546,700	508.28
Sanitary fittings		item	30,000		
Disposal installations		item	15,000		
Water installations		item	125,000		
Heat source		item	165,000		
Space heating and air conditioning		item	115,000		
Ventilation		item	60,000		
Electrical installations		item	515,000		
Fire and lightning protection		item	100,000		
Communication, security and control systems (including internal CCTV)		item	100,000		
Specialist installations		item	190,000		
ICT infrastructure works		item	37,000		
Builder's work in connection		item	90,000		
Stripping existing M&E; disposal		item	4,700		
External Works				200,200	65.79
Macadam repairs and tree works		item	9,500		
Handrail: 40 mm dia. CHS uprights; two 40 mm dia. CHS rails; raking; fixings	32	m	270		
Door barriers: 40 mm dia. CHS uprights at 900 mm × 1,000 mm high; all fixings etc.	24	nr	525		
Brickwork		item	1,900		
External drainage		item	120,000		
External service		item	47,500		
Abnormal Costs				352,000	115.68
Temporary accommodation		item	325,000		
Asbestos removal		item	27,000		

SCHOOL REFURBISHMENT

School Refurbishment	Quantity	Unit	Rate (£)	Total (£)	Cost (£/m²)
Preliminaries and Contingency				1,865,000	612.88
Preliminaries		item	960,000		
Contractors overheads and profit		item	165,000		
Risk allowance and design reserve		item	110,000		
Fees		item	630,000		
Construction cost (rate based on GIFA)				5,990,600	1,969

PRIMARY CARE HEALTH CENTRE

A three-storey new build primary care health centre for GPs, community nurses, midwifery services, mental health services. The building in the cost study contains a gross floor area of 2,500 m² located in Greater London. Costs include for group 1 and fitting of group 2 furniture, fixtures and equipment, but exclude external works, utilities, design/construction/client contingency, professional fees.

Gross internal floor area	2,500 m²
Model location is Greater London	Model TPI = 132; LF = 1.00

This cost model is copyright of AECOM Ltd

Primary Care Health Centre	Quantity	Unit	Rate (£)	Total (£)	Cost (£/m²)
Substructure				**497,200**	**198.88**
Allowance for site clearance		item	23,500		
Excavate and disposal to formation		item	53,000		
Perimeter footings	115	m³	600		
Pad foundations	220	m³	550		
Ground bearing slab	1,150	m²	150		
Concrete lift pit		item	12,000		
Drainage	1,150	m²	35		
Allowance for soft spots and below ground obstructions		item	5,900		
Frame and Upper Floors				**688,500**	**275.40**
Steel frame, including fittings, brackets and supports		item	365,000		
Allowance for intumescent paint, 60 m fire-rating		item	97,500		
150 mm precast concrete floor	1,350	m²	90		
Allowance for 75 mm structural screed	1,350	m²	60		
Blockwork to lifts and stairwell, 200 mm thick		item	23,500		
Roof				**566,800**	**226.72**
150 mm deep precast concrete floor	1,200	m²	90		
Bitumen felt roof waterproofing system; including underlay, vapour control layer, insulation, flashings, edge trims,	1,200	m²	150		
Double-glazed rooflights and frames, including ironmongery, restrictors and vents	80	m²	1,000		
Rainwater goods	1,200	m²	24		
Mansafe system		item	23,500		
Parapet capping	230	m	115		
Plant screening		item	41,000		
Main entrance canopy		item	30,000		
Roof terrace and balustrade	40	m²	350		
Roof sundries: walkways, access hatches, plant bases		item	35,000		

Building Cost Models

PRIMARY CARE HEALTH CENTRE

Primary Care Health Centre	Quantity	Unit	Rate (£)	Total (£)	Cost (£/m²)
Stairs				70,000	28.00
Precast concrete staircases, 2 nr from ground to second floor, including handrails and finishes		item	70,000		
External walls, windows and doors				794,500	317.80
External insulated render and cavity wall	1,460	m²	260		
Extra for timber cladding	380	m²	210		
Aluminium framed glazed system	220	m²	725		
Aluminium framed double-glazed windows	250	m²	450		
Main entrance aluminium automatic double doors and ironmongery		item	12,000		
Aluminium external double doorsets	4	nr	2,950		
Aluminium louvre double plant room door, FD 60	2	nr	3,500		
Insulated render soffit	80	m²	330		
Building signage		item	5,900		
Internal Walls and Partitions				498,200	199.28
Metal stud internal walls; 2 layers plasterboard to each side	965	m	310		
Metal wall lining to rear of external wall	1,460	m²	100		
Allow for column boxing		item	53,000		
Internal Doors				336,500	134.60
Internal single leaf timber doors	85	nr	1,125		
Internal 1.5 leaf timber doors	45	nr	1,950		
Internal double leaf timber doors	6	nr	2,200		
Additional acoustic and fire protection to walls and floors		item	60,000		
Internal glazing		item	17,500		
IPS panelling		item	30,000		
Glazed balustrade	15	m	775		
Sliding partition	5	m	1,750		
Shutter to reception	1	nr	12,000		
Wall Finishes				115,200	46.08
Emulsion paint to walls	7,250	m²	12		
Coloured Perspex screens to reception		item	12,000		
Splashback tiling to wet areas		item	2,950		
Whiterock finish to wet areas	100	m²	95		
Sundry items, handrails, protection barriers		item	5,900		
Floor Finishes				152,500	61.00
Self-levelling screed	2,500	m²	12		
Vinyl sheet flooring	2,125	m²	45		

PRIMARY CARE HEALTH CENTRE

Primary Care Health Centre	Quantity	Unit	Rate (£)	Total (£)	Cost (£/m²)
Extra for wet areas	100	m²	12		
Carpet to office areas	150	m²	35		
Ceramic tiling to entrance areas	150	m²	90		
Epoxy resin floor paint to plant rooms	75	m²	24		
Barrier matting		item	5,900		
Ceiling Finishes				**146,000**	**58.40**
Mineral fibre suspended ceilings	2,275	m²	41		
Acoustic timber slatted ceiling	150	m²	300		
Bulkheads and access panels to services		item	5,900		
Moisture-resistant ceiling tiles	100	m²	18		
Fittings and Furnishings				**207,900**	**83.16**
Reception area desk		item	17,500		
Secondary welcome points with storage	2	nr	7,600		
General storage, worktops and room cabinetry	2,500	m²	41		
Statutory signage and wayfinding		item	17,500		
Built-in specimen and prescription-off boxing		item	5,900		
Window blinds	470	m²	30		
Curtain tracks		item	8,800		
Mirrors		item	5,900		
Staff lockers		item	2,950		
Sundry items and fitting of Group 2 FF&E items		item	17,500		
Sanitary Fittings and Disposal Installations				**176,000**	**70.40**
WCs including pan, concealed cistern and waste; WHBs and clinical WHBs including taps, waste and trap; disabled WCs including pan, concealed cistern, waste and grab rails; showers including tray, thermostatic valves and head; dirty utility disposal units, cleaners' sinks including bib taps, waste and trap, tea point sink units including taps, waste, trap and boiling/chilled water dispensers,	2,500	m²	47		
Rainwater disposal installation including gullies and downpipes	2,500	m²	6		
Above ground SVP disposal and condensate installation	2,500	m²	18		
Water Installations				**169,700**	**67.88**
Cold water system	2,500	m²	31		
HWS distribution including incoming storage calorifiers, circulating pumpset and connections to sanitaryware,	2,500	m²	35		
Miscellaneous Cat 5 water supplies		item	4,700		

Building Cost Models

PRIMARY CARE HEALTH CENTRE

Primary Care Health Centre	Quantity	Unit	Rate (£)	Total (£)	Cost (£/m²)
Space Heating and Air Treatment				721,000	288.40
Low-nox condensing boilers, primary pumpset, primary distribution and flue		item	30,000		
LTHW distribution including heat emitters/ underfloor heating and over door heaters	2,500	m²	49		
Supply and extract AHUs, DX condenser unit, supply and extract ductwork distribution, grilles and diffusers	2,350	m²	210		
DX cooling systems to serve reception/waiting areas/office/IT hub rooms	2,500	m²	30		
Ventilation Installations				17,500	7.00
Combined extract systems linked to general extract distribution serving WC areas, dirty utility rooms, baby change	2,500	m²	7		
Gas Installations				7,000	2.80
Gas installations, all costs associated with the supply and installation of gas		item	7,000		
Electrical Installations				676,000	270.40
Incoming LV supply, main distribution panel, containment, sub-mains cabling, distribution boards, PV panels, mechanical supplies, earthing and bonding, leak detection to IT hub rooms	2,500	m²	90		
Photovoltaics	50	m²	700		
Luminaires including emergency luminaires with 3 hr battery packs, final circuits, lighting control, medical lighting, enhanced lighting to reception	2,500	m²	115		
Medical lighting		item	41,000		
Small power distribution	2,500	m²	35		
Protective Installations				6,000	2.40
Lightning protection; earthing installations:	2,500	m²	2		
Communications Installations				427,300	170.92
Data outlets, cabling and wireways	2,500	m²	20		
Fire detection and alarm system and disabled refuge	2,500	m²	38		
Intruder detection, CCTV, access control, nurse call, disabled WC alarm systems, intercoms	2,500	m²	60		
BMS controls and occupancy monitoring	2,500	m²	53		
Lifts				140,000	56.00
1 nr MRL stretcher evacuation lift; 1600 kg, 21 person serving 2 levels; 0.63 m/s minimum speed; 1 nr MRL passenger lift; 1,275 kg, 13 person serving 3 levels		item	140,000		

PRIMARY CARE HEALTH CENTRE

Primary Care Health Centre	Quantity	Unit	Rate (£)	Total (£)	Cost (£/m²)
Builder's Work in Connection with Services				70,300	28.12
Builder's work in connection with services		3%			
Preliminaries and Contingency				1,296,800	518.72
Contractors preliminaries, overheads and profit		20%			
Construction cost (rate based on GIFA)				7,780,900	3,112.36

PRIMARY HEALTHCARE CENTRE FIT-OUT

A three-storey new build primary care health centre for GPs, community nurses, midwifery services, mental health services. The building in the cost study contains a gross floor area of 2,500 m² located in Greater London. Costs include for group 1 and fitting of group 2 furniture, fixtures and equipment, but exclude external works, utilities, design/construction/client contingency, professional fees.

Gross internal floor area   1,550 m²

Model location is Greater London   Model TPI = 132; LF = 1.00

This cost model is copyright of AECOM Ltd

Primary Healthcare Centre Fit-Out	Quantity	Unit	Rate (£)	Total (£)	Cost (£/m²)
Internal Walls and Partitions				**386,000**	**249.03**
Metal stud plasterboard partitions: 2 layers plasterboard each side, skim coat plaster finish		item	205,000		
Shaftwall system; metal stud plasterboard partitions: 2 layers plasterboard each side, skim coat plaster finish		item	37,000		
Metal stud wall lining system; 2 layers plasterboard each side, skim coat plaster finish		item	48,000		
Glazed partition systems; including integrated doorsets		item	28,000		
Fixed internal glazing		item	10,500		
IPS panelling to en-suite and WCs		item	57,500		
Internal Doors				**187,900**	**121.23**
Single-leaf timber doorset: laminate faced, with ironmongery		item	16,000		
Single-leaf timber doorset: PVC faced, with ironmongery to en-suite		item	35,000		
Single-leaf timber doorset: wood veneer faced, with ironmongery, including overhead glazed panels		item	60,000		
Double-leaf timber doorset: wood veneer faced, with ironmongery		item	2,250		
Leaf-and-a-half timber doorset: laminate faced, with ironmongery		item	3,150		
Leaf-and-a-half timber doorset: PVC faced, with ironmongery		item	6,500		
Leaf-and-a-half timber doorset: wood veneer faced, with ironmongery, including overhead glazed panels		item	20,000		
Leaf-and-a-half timber doorset: PVC faced, with ironmongery, including overhead glazed panels		item	7,400		
Concealed frame riser cupboard doors, including ironmongery		item	9,600		
Sliding doorset system: internal lobby double door to main entrance; bi-parting with glazed side screens		item	17,000		

PRIMARY HEALTHCARE CENTRE FIT-OUT

Primary Healthcare Centre Fit-Out	Quantity	Unit	Rate (£)	Total (£)	Cost (£/m²)
Allowance for manifestations to doors and screens		item	11,000		
Wall Finishes				149,500	96.45
Two coats emulsion paint to internal partitions		item	32,000		
Wood veneered panel lining system		item	100,000		
PVC decorative wall coverings		item	7,000		
Wall protection		item	3,000		
Corner guards; 1 m high		item	2,000		
Allowance for pattressing		item	5,500		
Floor Finishes				271,900	175.42
Bonded self-levelling cementitious screed to plant room, including scabbling existing slab		item	2,600		
Raised access floor system, including pedestals, 6 mm plywood flooring fixed to raised access system and filled joints		item	200,000		
Vinyl sheet flooring with 100 mm high coved skirting		item	54,000		
Extra over for non-slip vinyl flooring to wet areas		item	290		
Extra over for anti-static vinyl flooring to hub rooms		item	150		
Waterproof resin floor coating to plant rooms		item	3,300		
Barrier matting to entrances		item	11,500		
Ceiling Finishes				134,400	86.71
Plasterboard suspended ceiling		item	9,400		
Decoration to plasterboard suspended ceiling		item	2,700		
Suspended ceiling: ceiling tiles to general areas		item	18,500		
Suspended ceiling: ceiling tiles to treatment areas, acoustic		item	45,000		
Suspended ceiling: washable ceiling tiles, hygienic		item	7,000		
Suspended ceiling: ceiling tiles waiting and corridors, decorative		item	29,000		
Extra over for access panels		item	3,300		
Allowance for bulkheads, incl decorative panelling		item	19,500		
Fittings and Fixtures				162,000	104.52
Main reception desk at building entrance		item	22,000		
FF&E: group 1, supply and fix		item	115,000		
FF&E: group 2, fix only		item	8,500		
Allowance for wayfinding		item	16,500		

PRIMARY HEALTHCARE CENTRE FIT-OUT

Primary Healthcare Centre Fit-Out	Quantity	Unit	Rate (£)	Total (£)	Cost (£/m²)
Sanitary Fittings and Disposal Installations				**70,800**	**45.68**
Wash hand basin, including taps, waste, trap		Item	5,150		
Clinical wash hand basin, including taps, waste, trap WCs: pan, concealed cistern and frame, flush plate		Item	21,500		
WCs: pan, concealed cistern and frame, flush plate		Item	9,400		
Disabled WCs: concealed cistern and frame, flush plate		Item	8,400		
Urinals: concealed cistern and frame, flush plate		Item	550		
Shower, including thermostatic mixer, tray, enclosure, flush plate		Item	2,100		
Disabled/assisted shower, including thermostatic mixer, tray, enclosure, waste, trap		Item	2,550		
Dirty utility disposal unit, sink and slop hopper, including bib taps, waste, trap		Item	9,900		
Cleaners' sink, including bib taps, waste, trap		Item	3,000		
Stainless steel sink, including taps, waste, trap		Item	2,150		
Multi-faith foot wash, including automatic taps, waste and trap		Item	1,650		
Water dispenser: zip-tap type water cooler		Item	4,400		
Disposal Installations				**35,800**	**23.10**
SVP disposal installation to sanitaryware, including equipment		item	27,000		
Condensate drainage to FCUs and chilled beams		item	4,400		
Allowance for leak detection to under-floor SVP distribution		item	4,400		
Water Installations				**191,500**	**123.55**
Connection to incoming mains cold water supply, booster set, UV treatment, duplex water softener, primary distribution, secondary distribution to HWS generation		item	39,000		
Potable water storage tank		item	11,000		
Cold water distribution		item	44,000		
Cat 5 cold water system break tank and booster set; distribution		item	15,000		
Hot water system packaged heat exchanger/ buffer vessels, pumps, primary distribution		item	33,000		
Hot water distribution		item	44,000		
Trace heating of all distribution pipework externally exposed for frost protection		item	5,500		

PRIMARY HEALTHCARE CENTRE FIT-OUT

Primary Healthcare Centre Fit-Out	Quantity	Unit	Rate (£)	Total (£)	Cost (£/m²)
Space Heating and Air Treatment				671,900	433.48
Primary low-temperature hot water distribution from connection to landlord heat exchanger in pressurization unit, pump set and primary distribution circuit		item	22,000		
Very low-temperature hot water distribution including secondary pump set and distribution circuit		item	22,000		
Low-temperature hot water distribution including secondary pump set and distribution circuit		item	44,000		
Heat emitters (radiant panels; warm air curtains)		item	22,000		
Primary chilled water distribution from connection to landlord heat exchanger including pressurization unit, pump set and primary distribution circuit		item	11,000		
Entering chilled water distribution including secondary pump set and distribution circuit		item	11,000		
Chilled water distribution including secondary pump set and distribution circuit		item	44,000		
Mechanical ventilation with heat recovery units	10	nr	9,900		
Active chilled beams	17	nr	1,950		
Fan coil units	25	nr	1,350		
General supply and extract air-handling units and ventilation ductwork distribution including grilles, dampers etc.		item	330,000		
Ventilation Installations				16,500	10.65
Plant room ventilation		item	16,500		
Electrical Installations				443,300	286.00
Incoming low-voltage distribution from base-build CT panel and main low-voltage panel		item	18,500		
Low-voltage distribution including sub-mains cabling, distribution boards, motorized breakers, power factor correction, metering		item	39,000		
700 A static inverter for examination lamps		item	5,500		
Containment		item	33,000		
General earthing and bonding		item	3,300		
Lighting installation including emergency lighting via 3 hr battery packs and final circuits		item	140,000		
Supply and installation of medical lights		item	62,500		
Lighting control system		item	36,000		
Final circuits to examination lights		item	5,500		
Small power distribution		item	100,000		

PRIMARY HEALTHCARE CENTRE FIT-OUT

Primary Healthcare Centre Fit-Out	Quantity	Unit	Rate (£)	Total (£)	Cost (£/m²)
Communications Installations				226,500	146.13
Data installation: containment and wire-ways		item	8,800		
Data installation: cabling and outlets		item	47,000		
Fire detection and alarm system		item	77,500		
Disabled WC alarm system		item	5,500		
Intruder detection and alarm system		item	4,400		
CCTV		item	11,000		
Access control system including head-end		item	44,000		
Nurse call/panic alarm		item	19,500		
Public address system		item	3,300		
Intercom system		item	2,200		
Induction loop		item	3,300		
Specialist Installations				130,000	83.87
Control/BMS		item	130,000		
Builder's Work in Connection with Services				53,600	34.58
Builder's work in connection with services		3%			
Preliminaries and OHP				901,400	581.55
Testing and commissioning of services installations		3%			
M&E subcontractors preliminaries		10%			
Main contractor's preliminaries (management costs, site establishment and site supervision)		14%			
Main contractor's OH&P		5%			
Construction cost (rate based on GIFA)				4,033,000	2,601.95

MENTAL HEALTH FACILITY

The cost model is based on a new build 49 bed adult acute mental health facility in Greater London. The building has a gross internal floor area of 4,258 m² and is aiming to achieve BREEAM Excellent. Professional fees, external works excluded.

Gross internal floor area   4,258 m²

Model location is Greater London   Model TPI = 132; LF = 1.00

This cost model is copyright of AECOM Ltd

Mental Health Facility	Quantity	Unit	Rate (£)	Total (£)	Cost (£/m²)
Substructure				1,136,100	266.82
Allowance for site clearance		item	44,000		
Excavate and disposal to formation level		item	67,500		
Plain strip foundations to ground-floor external walls, including concrete, brickwork, DPC etc.	609	m	330		
Plain strip foundations to ground-floor internal walls, including concrete, brickwork, DPC etc.	554	m	300		
Ground-floor slab; assumed 225 mm deep beam and block floor; structural screed; damp-proof membrane; insulation	3,802	m²	170		
Concrete lift pit and walls; excavate and disposal		item	11,000		
Frame and Upper Floors				615,300	144.50
Allowance for lower ground floor steelwork		item	540,000		
Upper floors and roof deck; 250 mm thick reinforced concrete	4,275	m²	18		
Roof				846,600	198.83
Single-ply membrane roof, insulation, vapour barrier etc.	3,802	m²	170		
Rainwater goods including anti-scale downpipes and connection to drains	3,802	m²	22		
Rooflights; 2.00 m × 1.00 m	24	nr	1,650		
Mansafe system		item	22,000		
Allowance for screening to rooftop plant		item	55,000		
Stairs				22,000	5.17
Main feature staircase incl. balustrades, handrails and half landings		item	22,000		
External Walls, Windows and Doors				1,973,000	463.36
Traditional external wall construction of medium quality facing bricks, cavity, inner blockwork skin and insulation	3,434	m²	280		
Extra over for timber cladding on brickwork	343	m²	170		
Automatic main entrance aluminium double door including all ironmongery double door including all ironmongery	1	nr	16,500		
Automatic FM entrance aluminium double door including all ironmongery	1	nr	11,000		

MENTAL HEALTH FACILITY

Mental Health Facility	Quantity	Unit	Rate (£)	Total (£)	Cost (£/m²)
Automatic rear entrance aluminium double door including all ironmongery	1	nr	11,000		
External aluminium single door including all ironmongery	4	nr	2,750		
External aluminium double door including all ironmongery	14	nr	3,850		
Security windows including all ironmongery	1,030	m²	825		
Internal Walls, Partitions, Windows and Doors				1,223,800	287.41
Plasterboard partitions with ply backing, 4.00 m high	5,448	m²	100		
IPS panelling to en-suite and WCs	63	nr	1,100		
Internal security-glazed screens	272	m²	390		
Single-leaf timber doorset with anti-ligature ironmongery and vistamatic to bedrooms	49	nr	2,650		
Single-leaf timber doorset with anti-ligature ironmongery to en-suite	49	nr	1,550		
Single-leaf timber doorset with anti-ligature ironmongery to other areas	133	nr	1,550		
Double-leaf timber doorset with anti-ligature ironmongery to other areas	15	nr	2,100		
Airlock double doors between wards	3	nr	8,300		
Airlock double door to main entrance	1	nr	11,000		
Access timber doors with anti-ligature ironmongery	22	nr	1,100		
Wall Finishes				383,800	90.14
2 coats emulsion to inner leaf of external wall	3,434	m²	9		
2 coats emulsion to internal partitions	10,896	m²	9		
Extra over for PVC Altro to en-suite/WCs/shower areas	1,500	m²	140		
Sheet-cushioned vinyl to seclusion room and de-escalation room	105	m²	110		
Window boards; 25 mm primed softwood; 1 coat underseal and 2 coats of gloss paint	256	m²	55		
Allowance for enhanced finishes		item	22,000		
Floor Finishes				391,600	91.97
Self-levelling screed and DPM	4,258	m²	22		
Extra over for screed to falls in showers	50	m²	170		
Vinyl sheet flooring with 100 mm high coved skirting	4,258	m²	65		
Extra for non-slip vinyl flooring to en-suite and bathrooms	177	m²	11		
Extra for resin flooring to seclusion room and de-escalation room	35	m²	90		
Extra barrier matting	15	m²	250		
Extra for timber-sprung floor to fitness suite	40	m²	95		

MENTAL HEALTH FACILITY

Mental Health Facility	Quantity	Unit	Rate (£)	Total (£)	Cost (£/m²)
Ceiling Finishes				**295,100**	**69.30**
Taped and jointed plasterboard ceiling with ply backing	4,258	m²	50		
Extra over for moisture-resistant to wet areas	177	m²	11		
Decoration to the above	4,258	m²	11		
Extra over for tamper-proof access panels to services	30	nr	330		
Allowance for bulkheads		item	23,500		
Fittings and Furnishings and Equipment				**433,000**	**101.69**
FF&E; Group 1; supply and fix		item	400,000		
FF&E; Group 2; fix only		item	33,000		
Mechanical Installations				**420,900**	**98.85**
Above-ground drainage to sanitaryware, utility rooms etc.	256	nr	390		
Above-ground drainage to central kitchen		item	3,850		
Incoming water supply, storage, water treatment and plant room pipework and equipment		item	47,000		
Cold water distribution	256	outlets	500		
Hot-water calorifiers including heat exchanger		item	23,500		
Hot-water distribution	174	outlets	625		
Allowance for providing hot and cold-water supply to main kitchen		item	4,400		
Allowance for Cat 5 cold water supply to refuse areas		item	5,500		
Sanitaryware				**270,800**	**63.60**
Clinical anti-ligature wash hand basins	55	nr	825		
Clinical anti-ligature WCs	67	nr	675		
Clinical anti-ligature showers	48	nr	500		
Clinical anti-ligature assisted WCs	14	nr	1,100		
Clinical anti-ligature assisted wash hand basins	14	nr	825		
Clinical anti-ligature assisted showers	10	nr	500		
Patient assisted bathroom complete with WC and wash hand basins	4	nr	15,000		
WCs	9	nr	550		
Wash hand basins	13	nr	440		
Showers	4	nr	280		
Multi-faith foot wash	1	nr	900		
Kitchen sinks	5	nr	550		
Combined sink and slop hoppers	4	nr	4,400		
Cleaner's sinks	4	nr	1,400		
Installation of the above	256	nr	100		

MENTAL HEALTH FACILITY

Mental Health Facility	Quantity	Unit	Rate (£)	Total (£)	Cost (£/m²)
Heat Source, Space Heating, Air Treatment and Ventilation Systems				**1,466,300**	**344.36**
Natural gas supplies to primary plant		item	14,000		
Natural gas to therapy kitchen/servery etc.		item	14,500		
Provisional allowance for providing 3 nr 500 kW gas-fired condensing boilers, including gas train, controls and flue	3	nr	22,000		
Plant room pipework, valves and fittings etc.		item	95,000		
CHP including flue		item	145,000		
Plant room pipework, valves and fittings etc.		item	47,000		
LTHW plant room pipework, valves and fittings (secondary) including equipment etc.		item	47,000		
LTHW heating distribution including insulation		item	145,000		
Heat emitters (radiators, air curtains)		item	12,000		
Underfloor heating		item	165,000		
Chilled water plant and distribution		item	120,000		
Local cooling		item	22,000		
General supply & extract AHUs and ventilation ductwork distribution including grilles, dampers etc.		item	445,000		
Dirty extract fans and associated ductwork, grilles, dampers etc.		item	120,000		
Kitchen extract system		item	8,800		
Electrical Installations				**1,111,000**	**260.92**
Main LV switch panel and incoming supply		item	72,500		
Containment		item	57,500		
LV sub-mains distribution including distribution boards		item	85,000		
Small power (1,250 pts)		item	235,000		
Lighting (combination of non-anti-ligature and anti-ligature)		item	330,000		
Emergency lighting (combination of non-anti-ligature and anti-ligature)		item	120,000		
Lighting control including wiring		item	120,000		
Earthing and bonding		item	23,500		
Supply and installation of 13-person machine room-less passenger lift to 2 floors, DDA compliant		item	67,500		
Protective, Communication and Specialist Installation				**2,781,400**	**653.22**
Lighting protection		item	14,500		
Gas suppression		item	14,500		
Fire alarm including void detection		item	145,000		
CCTV installation (internally only)		item	120,000		
Nurse call alarms including WCs		item	145,000		

MENTAL HEALTH FACILITY

Mental Health Facility	Quantity	Unit	Rate (£)	Total (£)	Cost (£/m²)
Staff attack system		item	190,000		
Data and communication (circa 1,300 points)		item	165,000		
TV/radio cabling (wire-ways only)		item	19,000		
Induction loop		item	4,700		
Intruder alarm		item	14,500		
Access control based on circa 170 nr doors		item	285,000		
Video entry		item	11,000		
Public address/voice alarm		item	11,000		
Containment		item	47,000		
Controls/BMS		item	235,000		
Leak detection		item	235,000		
Testing and commissioning		7.5%			
Mechanical and electrical subcontractor preliminaries		12%			
Builder's work in connection with services		item	120,000		
Preliminaries and Contingency				**2,674,100**	**628.02**
Contractors preliminaries, overheads and profit		20%			
Construction cost (rate based on GIFA)				**16,044,800**	**3,768**

PALLIATIVE CARE UNIT

The cost model is for a three-storey extension to an existing palliative care unit, creating inpatient and office accommodation with associated ancillary support services; both internal and external finishes to a good quality.

Gross internal floor area, including tiers	1,444 m²
Model location is Greater London	Model TPI = 132; LF = 1.00

This cost model is copyright of AECOM Ltd

Palliative Care Unit	Quantity	Unit	Rate (£)	Total (£)	Cost (£/m²)
Substructure				**375,000**	**259.70**
Allowance for site clearance, excavation into gradient site & disposal		item	112,500		
Excavate and dispose of slab and foundations		item	42,000		
450 mm wide × 500 mm deep strip foundation with 140 mm load bearing blockwork to DPC	25	m³	350		
650 mm wide × 500 mm deep strip foundation with 140 mm load bearing blockwork inner skin and 103 mm facing brickwork to DPC	40	m³	350		
175 mm thick reinforced concrete ground bearing slab with 2 layers of A142 mesh	630	m²	140		
Concrete lift pit and walls; excavate and disposal		item	11,200		
500 mm thick reinforced slab to underside of PCC retaining wall	87	m²	280		
Allowance for 75 mm structural screed to ground floor slab	630	m²	42		
Installation of new drainage system, including connection and integration within new substructure works	630	m²	42		
Waterproofing details		item	21,000		
Frame and Upper Floors				**284,300**	**196.88**
Allowance for lower ground floor steelwork		item	11,200		
Upper floors – 150 mm precast concrete hollow core floor units (5.00 kN/m²)	500	m²	100		
Allowance for 75 mm structural screed	500	m²	42		
Padstones – allowance for lower ground floor		item	2,100		
Allowance for timber framing to ground floor, first floor & roof		item	200,000		
Roof				**284,300**	**196.88**
Concrete roof tiles, including roof coverings, ridges, troughs, hips etc., 370 m² @ £80	370	m²	115		
Double glazed rooflights and frames, including ironmongery, restrictors and vents	12	nr	1,400		
Rainwater goods	700	m²	28		
Mansafe system		item	10,500		
Parapet wall detail; aluminium capping	40	m	230		

PALLIATIVE CARE UNIT

Palliative Care Unit	Quantity	Unit	Rate (£)	Total (£)	Cost (£/m²)
Single ply membrane roof, insulation, vapour barrier etc.	340	m²	190		
Allowance for roof access hatch		item	5,600		
Stairs				**35,000**	**24.24**
Main feature staircase including balustrades, handrails and half landings		item	35,000		
External Walls, Windows and Doors				**368,900**	**255.47**
Traditional external wall construction of medium quality facing bricks, cavity, inner blockwork skin and insulation	740	m²	260		
Extra over wall tiles; including battens	190	m²	115		
Extra over render; assumed 2 coats at 15 mm thick	180	m²	70		
Aluminium framed glazed system, including all ironmongery	50	m²	630		
Double glazed windows and frames, including all ironmongery, lintels, restrictors and vents	70	m²	530		
Window boards; 25 mm primed softwood; 1 coat underseal and 2 coats gloss	40	m	70		
Main entrance aluminium double doors including all ironmongery	1	nr	14,000		
Aluminium external doorsets, including all ironmongery, double	4	nr	3,350		
Aluminium external doorsets, including all ironmongery, single leaf and a half	12	nr	2,550		
Aluminium louvre double plant room door and frame, including ironmongery; fire-rated FD60	3	nr	4,200		
Internal Walls, Partitions, Windows and Doors				**546,900**	**378.74**
140 mm thick blockwork partitions/metsec stud	870	m²	85		
Metal stud internal walls; 2 layers of plasterboard to each side	960	m²	120		
Internal glazed screens in softwood framing	60	m²	700		
IPS panelling		item	28,000		
Allowance for bed head trunking		item	63,000		
Extra over for pattressing walls		item	4,200		
Allowance for lintels to door openings		item	7,000		
Allowance for single leaf timber doors	40	nr	1,200		
Allowance for one and a half leaf timber doors	55	nr	2,000		
Allowance for double leaf timber doors	9	nr	2,550		
Allowance for single leaf glazed doors	5	nr	2,250		
Allowance for double leaf glazed doors	2	nr	3,650		
Extra over for fire protection (FD30)		item	14,000		

PALLIATIVE CARE UNIT

Palliative Care Unit	Quantity	Unit	Rate (£)	Total (£)	Cost (£/m²)
Wall Finishes				163,300	113.09
Plaster to blockwork and skim finish to stud walls	1,240	m²	21		
Allowance for splashback to wet areas		item	2,800		
Whiterock to bathrooms	780	m²	95		
100 mm skirting; 1 coat of underseal and 2 coats of high gloss paint		item	2,800		
Allowance for sundry items (corner protection, handrails)		item	14,000		
Allowance to paint all walls with two coats of emulsion paint	3,110	m²	14		
Floor Finishes				88,500	61.29
Self-levelling screed	1,380	m²	14		
Heavy duty carpet, including underlay etc.	368	m²	42		
Altro vinyl sheet flooring with 100 mm high coved skirting	1,000	m	49		
E/O wet areas	110	m²	14		
Epoxy resin floor paint; Watco Dustop or equivalent	12	m²	28		
Barrier matting – generally		item	2,800		
Ceiling Finishes				124,800	86.43
Taped and jointed plasterboard ceiling	1,380	m²	49		
E/O for access panels to services		item	14,000		
E/O for moisture-resistant		item	2,800		
E/O for fire protection		item	21,000		
Decoration to the above	1,380	m²	14		
Fittings and Furnishings				211,600	146.54
Nurse base station, with built-in storage	2	nr	11,200		
Shelving		item	7,000		
Mirrors		item	4,200		
Hand/Grab rail		item	28,000		
Window blinds/curtain		item	7,000		
Worktops, base units and the like		item	45,000		
Ceiling hoist track		item	70,000		
Staff lockers		item	7,000		
Client and statutory signage		item	7,000		
Sundry and fitting of Group 2 items		item	14,000		
Mechanical Installations				568,300	393.56
WC	29	nr	925		
Clinical wash hand basin	32	nr	520		
Wash hand basins	29	nr	590		

PALLIATIVE CARE UNIT

Palliative Care Unit	Quantity	Unit	Rate (£)	Total (£)	Cost (£/m²)
Stainless steel sink kitchen sink with drainer	3	nr	490		
Tea point	3	nr	490		
Shower, with thermostatic mixer	29	nr	1,400		
Sluice	3	nr	4,200		
Installation of the above		item	14,000		
Disposal installations		item	112,500		
Water installations		item	170,000		
Heat source		item	42,000		
Space heating and air conditioning		item	85,000		
Ventilating systems		item	28,000		
Electrical Installations				**860,500**	**595.91**
Electrical mains and sub-mains distributions		item	100,000		
Lighting		item	170,000		
Allowance for external lighting		item	49,000		
Small power		item	63,000		
Fuel installations/systems		item	7,000		
Lift and conveyor installations/systems		item	112,500		
Fire and lightning protection		item	7,000		
Communication, security and control systems		item	240,000		
External services		item	14,000		
Testing and commissioning of building services installations		item	37,000		
Builder's work in connection		item	61,000		
Preliminaries and Contingency				**759,200**	**525.76**
Management costs, site establishment and site supervision. Contractor's preliminaries, overheads and profit @		20%			
Construction cost (rate based on GIFA)				**4,555,200**	**3,155**

MULTIPLEX CINEMA

This model is based on a 12-screen multiplex cinema situated above retail units within a retail-led development in the South East England. The cinema occupies the entire first floor of the block and includes mezzanine accommodation for projection rooms and administration space. The ground floor of the development consists of retail shells and escape routes from the cinema complex. The breakdown separates the costs of the shell and fit-out construction.

Gross internal floor area	18,000 m²
Model location is South East England	Model TPI = 124; LF = 0.94

This cost model is copyright of AECOM Ltd

Multiplex Shell and Core	Quantity	Unit	Rate (£)	Total (£)	Cost(£/m²)
Frame, Upper Floors and Stairs				**2,072,400**	**115.13**
Structural steel frame including fire protection (based on 80 kg/m²)	422	tonne	2,100		
Allowance for secondary steel – bracings, support to signage & spiral stairs etc.		item	87,500		
Acoustic pads to cinema columns and support stubs	169	nr	190		
100 mm thick Bison solid precast concrete plank with concrete topping	4,136	m²	70		
Acoustic floor slab and insulation to ramps, lower seating in auditoria	1,400	m²	200		
Forming ramps/concrete steps into auditoria		item	120,000		
Elastomeric strip to top of steel where precast plank rests		item	23,500		
Composite metal slab	3,090	m²	95		
Form steps for seating		item	60,000		
Roof				**1,288,500**	**71.58**
Structural steel to roof (based on 48 kg/m²)	158	tonne	2,100		
200 mm thick hollowcore PCC planks with structural topping	972	m²	95		
450 mm thick hollowcore PCC planks with structural topping	2,330	m²	140		
Reinforced laminated PVC single layer polymeric warm roof with closed cell rigid insulation tapered to falls	2,354	2.0%	75		
Reinforced laminated PVC single layer polymeric inverted roof system	1,964	m²	120		
Extra over for maintenance walkway		item	12,000		
Allowance for sundry metalworks, cat ladder, access hatches etc.		item	29,000		
Glazed roof canopy	70	m²	1,200		
External Walls				**1,859,000**	**103.28**
Proprietary metal cladding panel	3,690	m²	380		
Profiled aluminium channel	1,019	m	290		
Clay facing brickwork	116	m²	260		

MULTIPLEX CINEMA

Multiplex Shell and Core	Quantity	Unit	Rate (£)	Total (£)	Cost(£/m²)
Aluminium copings	356	m	180		
Aluminium framed hinged double door to cinema entrance	3	nr	8,700		
Aluminium framed pivot door	1	nr	17,500		
Steel single doors	6	nr	2,350		
Steel double doors	2	nr	4,650		
Internal Walls and Doors				716,100	39.78
Blockwork at ground/first floor	1,681	m²	70		
Additional layer of wall to inside face of external walls	3,000	m²	180		
Timber doors; single	40	nr	410		
Steel doors; single	8	nr	2,350		
Steel door; double	8	nr	2,900		
Internal Finishes				105,200	5.84
Top coat sealer to walls	4,145	m²	4		
Dry lining to internal face of external walls	1,762	m²	35		
Allowance for secondary ceiling grid to support services		item	29,000		
Floor finishes to cinema (excluded)			–		
Space Heating and Air Treatment and Ventilation				1,206,500	67.03
Sanitary appliances (see fit-out)			–		
Disposal installations		item	92,500		
Water installations		item	100,000		
Space heating and air conditioning		item	35,000		
Ventilation systems		item	130,000		
Electrical installations		item	535,000		
Gas and other fuel installations		item	35,000		
Lift installations (by tenant)			–		
Escalators (by tenant)			–		
Fire and lightning protection		item	29,000		
Communication, security and control systems		item	130,000		
Specialist installations		item	60,000		
Builder's work in connection with services		item	60,000		
Preliminaries and Contingency				1,811,900	100.66
Management costs; site establishment; site supervision @		25%			
Construction cost (rate based on GIFA)				9,059,600	503.30

MULTIPLEX CINEMA

Multiplex Fit-Out	Quantity	Unit	Rate (£)	Total (£)	Cost (£/m²)
Internal Walls and Doors				**2,055,500**	**114.19**
Glazed screen at ground level	60	m²	925		
Standard 150 mm thick partitions	794	m²	120		
150 mm thick acoustic wall	246	m²	240		
300 mm thick acoustic wall	4,323	m²	260		
550 mm thick acoustic wall	1,665	m²	320		
Wall mounted handrails to perimeter corridor, steps/ stairs	196	m	210		
Balustrades to auditoria and steps in slab to projection level	205	m	210		
Allowance for shaftwall construction for hoist enclosures		item	5,800		
Allowance for free-standing screens to auditoria and WC entrances		item	29,000		
Melamine-faced solid laminate access panelling to WCs etc.	16	nr	600		
Allowance for additional partitions to administration areas		item	5,800		
Acoustic doors; fire-rated; steel-faced with insulated cores; steel goalpost frames; ironmongery	12	nr	4,050		
Security shutters; to concession unit entrances; including allowance for subframe		item	5,800		
Internal Finishes				**832,300**	**46.24**
13 mm two-coat plaster to blockwork	1,681	m²	14		
Feature plaster finish to entrance lobby	100	m²	140		
Two coats emulsion paint; to plaster/plasterboard generally	1,762	m²	9		
Ceramic tiling	200	m²	75		
Ceramic floor tiling including screed	100	m²	85		
Vinyl sheet flooring including screed; to part auditoria	1,000	m²	60		
Carpet; on underlay; including screed; to circulation, part auditoria and office spaces	2,000	m²	70		
Allowance for enhanced foyer finishes		item	29,000		
Mass barrier ceiling to soffit of auditorium	3,600	m²	120		
MF ceiling to cinema corridors	1,040	m²	47		
Allowance for ceiling to back of house	460	m²	47		
Allowance for access hatches and fire barriers		item	23,500		
Fittings and Furnishings				**57,200**	**3.18**
Allowance for additional joinery		item	12,000		
Allowance for internal signage		item	17,500		

MULTIPLEX CINEMA

Multiplex Fit-Out	Quantity	Unit	Rate (£)	Total (£)	Cost (£/m²)
Allowance for fixing tenant's supply-only fixtures and fittings		item	17,500		
WCs, urinals, wash-hand basins	23	nr	410		
Allowance for fittings for disabled WCs	2	nr	350		
Preliminaries and Contingency				**736,200**	**40.90**
Management costs; site establishment; site supervision @		25%			
Construction cost for fit-out only (rate based on GIFA)				**3,681,200**	**204.51**
Total construction cost (rate based on GIFA)				**12,740,800**	**707.81**

APARTMENTS – PRIVATE RENTED SECTOR

The apartments are being built for PRS and are arranged in a single 11-storey building with two cores. Each core has its own apartment entrance lobby. The building also includes a residents' gym (GIA 185 m²), a private screening room and cinema (92 m²) and a residents' lounge and multifunction area (185 m²), as well as a dedicated estate management suite (GIA of 139 m²) and two retail units (GIA 280 m²).

Gross internal floor area	15,365 m²
Model location is Central London (zones 2/3)	Model TPI = 139; LF = 1.05

This cost model is copyright of AECOM Ltd

Apartment Shell and Core	Quantity	Unit	Rate (£)	Total (£)	Cost (£/m²)
Substructure				**1,418,400**	**92.31**
Works to existing site; site clearance and removal of existing vegetation, etc.	1,280	m²	12		
Excavation and disposal of material off site; assumed average 1.5 m reduce level dig	2,200	m³	65		
Allow for 15% contaminated soil	330	m³	130		
Allow for removal of unknown obstructions		item	31,000		
Allowance for piling mat to footprint of building; say 400 mm deep; including excavation and disposal and compacting	1,280	m²	50		
Allowance for mobilization and demobilization of piling rig; include for setting up piling rig at pile positions		item	95,000		
CFA piling; approx. 600 mm dia. piling 35 m deep; cut off top of pile; allow for pile cap; disposal of spoil; etc.	2,625	m	190		
Allowance for sundries: integrity tests, NDT, etc.		item	25,000		
Allowance for 500 mm thick reinforced concrete slab; excavation and disposal; 350 mm thick hardcore bed, 50 mm sand blinding, DPM, in situ concrete slab, reinforcement (150 kg/m³), etc.	1,280	m²	310		
Allowance for under slab drainage	1,280	m²	44		
Allowance for plant bases		item	12,500		
Allowance for lift pits	4	nr	9,300		
Frame, Upper Floors and Stairs				**5,556,600**	**361.64**
Allowance for reinforced concrete frame; comprising columns and core walls to ground floor	1,280	m²	250		
Allowance for reinforced concrete frame; including columns, beams and core walls to upper floors	14,085	m²	160		
Upper floor; post-tensioned concrete slab; 225 mm thick; 100 kg/m³ reinforcement; 6 kg/m² post tensioned strands, formwork, etc.	14,085	m²	180		
Precast reinforced concrete stairs; with powder coated metal balustrading and handrails; ground to 11th floor	22	m²	9,900		
Allowance for sundry platforms to plant rooms		item	9,300		
Access ladders to roof	4	m²	3,100		
Extra over allowance for enhancements to podium slab; transfer structures over retail areas, etc.	1,280	m²	130		

APARTMENTS – PRIVATE RENTED SECTOR

Apartment Shell and Core	Quantity	Unit	Rate (£)	Total (£)	Cost (£/m²)
Allowance for steel support to lifts	44	nr	950		
Roof				620,600	40.39
Roof slab; post-tensioned concrete slab; 225 mm thick slab; 100 kg/m³ reinforcement; 6 kg/m² post-tensioned strands, formwork etc.	1,280	m²	180		
Allowance for lift overruns, AOVs, etc.		item	19,000		
Allowance for membrane roof covering to concrete slab including insulation, waterproofing, drainage, etc.	1,280	m²	160		
Extra over allowance for green roof coverings to podium roof	1,280	nr	130		
External Walls, Windows and Doors				7,509,000	488.71
Structurally glazed wall to principle facade, apartment entrance lobbies and retail	878	m²	1,000		
Rainscreen cladding system: comprising timber rainscreen and brickwork panels on insulation and Metsec/SFS backing; 60% of facade area	3,834	m²	725		
Extra over for material interfaces and feature detailing		item	310,000		
Allowance for double glazed thermally broken aluminium windows; powder coated; 40% of facade area	2,556	m²	700		
Extra over for glazed sliding double doors to retail units, apartment entrance lobbies, gym and estate management suite	6	nr	4,350		
Extra over for glazed double door to resident's lounge	1	nr	2,500		
External louvered doorsets to residential refuse stores, retail, plant area and roof	7	nr	6,200		
Allowance for canopies to entrances	6	nr	12,500		
Allowance for scaffolding to external wall areas generally	7,268	m²	75		
Allowance for single doors to balconies	200	nr	1,550		
Allowance for balconies to apartments; bolt on steel balconies; including balustrades; cold bridging details, etc.	200	nr	3,750		
Internal Walls, Partitions and Doors				1,815,400	118.15
Blockwork walls; 300 mm overall thickness to ground floor; including wind posts	1,300	m²	160		
Blockwork walls; 100 mm overall thickness to ground floor; including wind posts	380	m²	115		
Partitions to residential corridor; 260 mm overall thickness, 15 mm soundblock plasterboard, 12.5 mm WBP plywood lining, vapour control layer, 2 nr 50 mm metal studs, 100 mm air gap, 2 nr 15 mm soundblock plasterboard	2,376	m²	150		

APARTMENTS – PRIVATE RENTED SECTOR

Apartment Shell and Core	Quantity	Unit	Rate (£)	Total (£)	Cost (£/m²)
Partitions between apartments; 260 mm overall thickness, 15 mm soundblock plasterboard, 12.5 mm WBP plywood lining, vapour control layer, 2 nr 50 mm metal studs, 100 mm air gap, 2 nr 15 mm soundblock plasterboard	4,800	m²	150		
Double leaf timber doors, to ground floor storage	2	nr	2,250		
Single leaf, timber doors to ground floor storage and WCs	5	nr	1,075		
Single leaf, timber entrance doors to apartments	200	nr	1,125		
Single leaf, timber doors to amenity areas	72	nr	1,075		
Single leaf, timber doors to risers	96	nr	625		
Allowance for WC and shower cubicles to amenity areas, etc.	8	nr	2,500		
Allowance for temporary protection to doors, etc.		item	95,000		
Wall Finishes				**636,000**	**41.39**
Plasterboard lining to core walls, service cupboards, etc.	3,496	m²	50		
Plasterboard linings to blockwork walls	2,843	m²	50		
Paint to internal partitions; residential corridors, linings and back of house areas	8,895	m²	10		
MDF skirtings with painted finish to lift lobbies, residential corridors, gym, estate management suite	2,905	m²	19		
Enhanced finishes to apartment entrances and lift lobbies; at ground level		item	62,500		
Timber panelling to resident's lounge	150	m²	310		
Acoustic panelling to cinema room	115	m²	190		
Extra over for forming recesses for projection screen and other detailing generally; to cinema room		item	6,200		
Service panelling to WCs and showers	75	m²	380		
Porcelain tiles to WCs and showers	120	m²	95		
Floor Finishes				**357,700**	**23.28**
75 mm levelling screed to corridors and lift lobbies, apartment entrance lobbies, cinema room, residents lounge, estate management suite and gym	3,118	m²	38		
Sealant to plant areas	418	m²	19		
Power float finish to retail areas	279	m²	4		
Carpet finish to residential corridors and lift lobbies	2,282	m²	50		
Luxury carpet to cinema room	88	m²	95		
Stone floor finish to apartment entrance lobbies	186	m²	310		
Timber batten floor to resident's lounge and estate management suite	300	m²	44		
Engineered timber flooring to gym area	162	m²	160		

APARTMENTS – PRIVATE RENTED SECTOR

Apartment Shell and Core	Quantity	Unit	Rate (£)	Total (£)	Cost (£/m²)
Porcelain tiles to WCs and showers	53	m²	130		
Epoxy resin to storage areas, refuse areas, etc.	46	m²	25		
Allowance for drainage channels including stainless steel grilles to changing facilities to gym	10	m	310		
Ceiling Finishes				**211,000**	**13.73**
Painted plasterboard ceiling to residential corridors, cinema room, residents lounge and back of house areas	2,561	m²	47		
Acoustic rated demountable suspended ceiling, including paint finish to estate management suite and gym	293	m²	75		
Allowance for specialist acoustic ceiling to cinema room	88	m²	95		
Enhanced finishes to reception and lift lobby areas at ground level		item	31,000		
Painted moisture-resistant plasterboard ceiling to WCs and showers	53	m²	56		
Sealer to exposed ceiling soffit to retail units, plant and storage areas	743	m²	19		
Allowance for additional bulkheads and level changes generally		item	12,500		
Furniture and Fittings				**374,200**	**24.35**
Statutory signage		item	38,000		
Mail boxes	200	nr	160		
Allowance for mirrors to gym		item	12,500		
Allowance for vanity units to changing rooms		item	9,300		
Allowance for cubicles, showers and fittings to gym changing areas	6	nr	3,100		
Allowance for fittings to changing rooms and WCs areas; hand driers, soap dispensers, towel racks, etc.		item	6,200		
Allowance for lockers to changing rooms	20	nr	625		
Allowance for benches to changing rooms		item	3,100		
Allowance for FF&E to cinema room; including projection screen and equipment; joinery cabinetry and millwork, etc.		item	25,000		
Allowance for cinema room seating	12	nr	1,900		
Allowance for back of house storage for estate management suite and maintenance facilities		item	25,000		
Allowance for FF&E to gym areas; vanity units, mirrors, etc.		item	19,000		
Allowance for servery joinery and equipment to residents lounge		item	6,200		
Study desks and panelling to residents lounge	10	nr	1,250		

APARTMENTS – PRIVATE RENTED SECTOR

Apartment Shell and Core	Quantity	Unit	Rate (£)	Total (£)	Cost (£/m²)
Residents lounge bookcases	5	m	1,900		
Sundry joinery to residents lounge; window seating, bars, etc.		item	25,000		
Timber screens and panelling to residents lounge		item	12,500		
Fireplace and wall feature to residents lounge		item	9,300		
Allowance for general storage and joinery to estate management suite		item	19,000		
Sundries; notice boards, signage, etc.		item	6,200		
Reception desks to apartment entrance areas and estate management suite		item	50,000		
Exclusions: meeting room furniture, desks & seats, gym equipment					
Sanitary Fittings				**35,300**	**2.30**
Cleaners sinks	3	nr	560		
Sanitaryware to estate management suite	4	nr	825		
Sanitaryware including showers to Gym	12	nr	975		
Sanitaryware to cinema room	2	nr	825		
Sanitaryware to work residents lounge	8	nr	825		
Extra over disabled provision		item	6,200		
Associated subcontractor preliminaries		item	4,150		
Disposal Installations				**383,600**	**24.97**
Rainwater disposal from roof outlets connection to underground drainage	15,365	m²	4		
Foul disposal to amenity areas and apartment complete with associated vent stacks, connection to underground drainage	15,365	m²	17		
Drainage from plant rooms, bin stores and the like via floor gullies		item	17,500		
Capped connection to retail		item	3,100		
Associated subcontractor preliminaries		item	48,000		
Water Installations				**439,700**	**28.62**
Cold water installation incoming main, storage tank, water treatment, multi-booster pump set, distribution to apartments including tenant sets complete with water meter	15,365	m²	20		
Cold water distribution to sanitaryware including estate management suite, cleaners cupboards and gym		item	23,000		
Cat 5 water points & plant to wash down points and to roof BMU		item	19,000		
Hot water installation to sanitaryware including estate management suite, cleaners cupboards and gym via a plate heat exchanger, including pumps, pipe work etc.)		item	23,000		

APARTMENTS – PRIVATE RENTED SECTOR

Apartment Shell and Core	Quantity	Unit	Rate (£)	Total (£)	Cost (£/m²)
Capped connections to retail		item	6,200		
Associated subcontractor preliminaries		item	55,000		
Heat Source, Space Heating and Air Treatment				**854,300**	**55.60**
Gas fired boiler installation	3	nr	33,000		
Flue to atmosphere		item	97,500		
Plant room pipework, pumps sets, controls etc.		item	48,000		
Associated subcontractor preliminaries		item	34,000		
Distribution to apartments, valves, etc.; terminating in HIUs (HIU's part of fit-out)	15,365	m²	19		
Landlords heating via radiators to stair cores and back of house areas		item	14,500		
Underfloor heating to reception areas		item	8,700		
Air treatment to reception, gym, office areas; VRF installation and mini AHU		item	185,000		
Associated subcontractor preliminaries		item	72,500		
Ventilation Installations				**330,900**	**21.54**
Plant room ventilation	418	m²	115		
Stairwell make up via AoV	2	nr	9,300		
Smoke clearance to corridors	24	vents	9,300		
Associated subcontractor preliminaries (lot @ £33,150)		item	41,000		
Electrical and Gas Installations				**1,350,700**	**87.91**
Primary distribution board, landlords power and lighting boards, cabling and containment, sub-metering, ryefield installation to apartments	15,365	m²	30		
Life safety standby generator and flue		item	140,000		
Life safety cabling & equipment to firefighting lifts, smoke extract and sprinkler installations	15,365	m²	3		
Small power to landlords areas and circulation space		item	9,900		
Power to mechanical installations	15,365	m²	3		
Lighting to landlord areas and circulation areas including lighting control and emergency lighting	2,978	m²	75		
Back of house lighting	15,365	m²	1		
Power and lighting to office and gym		item	117,500		
Feature lighting internal/external		item	50,000		
Earthing and bonding	15,365	m²	2		
Capped connections to retail		item	6,200		
Associated subcontractor preliminaries		item	165,000		
Gas installation to boilers	15,365	m²	2		
Associated subcontractor preliminaries		item	4,100		

Building Cost Models

APARTMENTS – PRIVATE RENTED SECTOR

Apartment Shell and Core	Quantity	Unit	Rate (£)	Total (£)	Cost (£/m²)
Protective, Communications and Special Installations				971,800	63.25
Sprinkler distribution to apartments, monitored floor valves	15,365	m²	12		
Dry risers	24	outlets	2,500		
Lightning & surge protection	15,365	m²	1		
Associated subcontractor preliminaries		item	39,000		
Fire alarm to landlord areas, interlink to apartments	15,365	m²	8		
Disabled refuge and WC alarm installation		item	82,500		
Containment for data/telephone installations	15,365	m²	4		
Satellite farm and aerial installation to roof		item	25,000		
Telephone/TV/Satellite to apartments	15,365	m²	9		
CCTV and access control to ground floor areas and plant rooms (12 nr CCTV cameras, 6 nr access control points)		item	77,500		
Data outlets to landlord areas and Wi-Fi installation to residents lounge		item	31,000		
Door entry and sub-door entry installation		item	38,000		
Associated subcontractor preliminaries		item	82,500		
Special Installations and Lifts				1,285,600	83.67
Remote metering to apartments and billing system for LTHW heating only	200	nr	380		
BMS	15,365	m²	11		
Cinema room installation		item	125,000		
Associated subcontractor preliminaries		item	53,000		
13 person, 1.6 m/s	2	nr	165,000		
17 person, 1.6 m/s	2	nr	190,000		
Enhanced finishes to lift car and 2 sets of drapes	2	nr	31,000		
Associated subcontractor preliminaries		item	87,500		
Builder's Work				515,000	33.52
Builder's work in connection with services installation		item	515,000		
Preliminaries and Contingency				7,708,000	501.66
Management costs; site establishment; site supervision		25%			
Contingency		5%			
Construction cost shell and core (rate based on GIFA)				32,373,800	2,107

APARTMENTS – PRIVATE RENTED SECTOR

Residential Fit-Out	Quantity	Unit	Rate (£)	Total (£)	Cost (£/m²)
Internal Walls and Partitions and Doors				1,329,300	86.51
Plasterboard stud partitions within apartments	8,700	m²	65		
EO allowance for boxing out and acoustic treatment to SVP's including crossovers		item	62,500		
Internal apartment doors: flush painted, solid core, single leaf doorset including ironmongery	475	nr	950		
Utility cupboard doors; flush paint, solid core, double leaf doorset; including ironmongery	200	nr	1,250		
Wall Finishes				978,400	63.68
Painted plasterboard lining to internal face of external wall, core walls, external wall of bathroom pod, etc.)	8,500	m²	50		
Paint finish to all internal partitions	17,400	m²	10		
Plywood behind kitchen walls	1,800	m²	19		
Allowance for glass splashbacks to kitchens generally	500	m²	250		
Allowance for mastic	200	nr	310		
MDF skirting with paint finish	8,633	m²	19		
Floor Finishes				1,369,800	89.15
Acoustic timber batten subfloor to general areas	10,072	m²	44		
Carpet finishes to bedrooms; approximately 30% of net residential area	3,022	m²	50		
Timber flooring to all other areas	7,050	m²	110		
Ceiling Finishes				608,400	39.60
Painted plasterboard ceiling	10,072	m²	47		
Allowance for blind boxes/recess	1,400	m	65		
Access hatches	200	nr	220		
Furniture and Fittings				5,840,000	380.08
Allowance for kitchen including sink, taps and white goods (oven, hob, fridge freezer, washer/dryer, extract hood); studio	15	item	8,700		
Allowance for kitchen including sink, taps and white goods (oven, hob, fridge freezer, washer/dryer, extract hood)	80	nr	9,900		
Allowance for kitchen including sink, tap and white goods (oven, hob, fridge freezer, washer/dryer, extract hood); 2 bed	105	nr	11,200		
Allowance for enhanced warranty provision	200	nr	310		
Roller blinds	2,556	m²	50		
Wardrobes to bedrooms generally	654	m	1,000		
Bathroom pod incorporating; sanitary ware, finishes and fittings	305	nr	9,900		

APARTMENTS – PRIVATE RENTED SECTOR

Residential Fit-Out	Quantity	Unit	Rate (£)	Total (£)	Cost (£/m²)
Sanitary Fittings and Disposal Installations				**128,000**	**8.33**
Soil and waste to white goods, connection to bathroom pods	200	nr	560		
Associated subcontractor preliminaries		item	16,000		
Water Installations				**264,000**	**17.18**
Hot and cold-water pipework to white goods, connection to bathroom pods	200	m²	1,125		
Associated subcontractor preliminaries		item	39,000		
Heat Source, Space Heating and Air Treatment				**1,034,500**	**67.33**
Twin plate heat exchanger for heat and hot water generation with heat meter	200	nr	1,800		
LTHW underfloor heating	11,585	m²	47		
Associated subcontractor preliminaries		item	130,000		
Ventilation Installations				**537,500**	**34.98**
Whole house ventilation with heat recovery and boost function	200	nr	2,350		
Associated subcontractor preliminaries		item	67,500		
Electrical Installation				**1,362,000**	**88.64**
Tenant's distribution board, meter	200	nr	500		
Small power distribution points, connection to pods	200	nr	1,600		
Lighting, wiring, installation and switches, connections to pods	200	nr	3,650		
Earthing and bonding	200	nr	210		
Associated subcontractor preliminaries		item	170,000		
Protective, Communications and Special Installations				**756,000**	**49.20**
Sprinkler protection	200	nr	1,350		
Fire alarm	200	nr	330		
Data/voice/TV outlets, wiring	200	nr	775		
Entry system	200	nr	625		
Local control to heating	200	nr	450		
Associated subcontractor preliminaries		item	50,000		
Builder's Work				**205,000**	**13.34**
Builder's work in connection with services installation		item	205,000		
Preliminaries and Contingency				**4,504,000**	**293.13**
Management costs; site establishment; site supervision		25%			
Contingency		5%			
Construction cost fit-out (rate based on GIFA)				**18,916,900**	**1,231**

APARTMENTS – PRIVATE RENTED SECTOR

External Works and Utilities	Quantity	Unit	Rate (£)	Total (£)	Cost (£/m²)
External Works				**310,000**	**20.18**
Allowance for external works		item	310,000		
Utilities				**575,000**	**37.42**
Water infrastructure charges	200	nr	440		
Plus connection charge		item	31,000		
Foul infrastructure charges	200	nr	440		
Plus connection charge		item	50,000		
Electrical connection – no reinforcement	1	nr	155,000		
Fibre installation		item	125,000		
Gas		item	38,000		
Construction cost external works (rate based on GIFA)				**885,000**	**57.60**
Total building cost; shell and core, residential fit-out and external works				**52,175,700**	**3,396**

Grenfell and Construction Industry Reform: A Guide for the Construction Professional

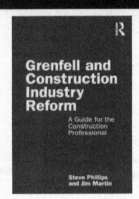

Steve Phillips and Jim Martin

In the wake of the tragic events of the fire at Grenfell Tower, the inquiry into the fire and the independent Hackitt Review revealed deep-rooted and unpalatable truths about the current state of the UK construction industry. Dame Judith Hackitt was scathing in her assessment of the construction industry denouncing it as "an industry that has not reflected and learned for itself, nor looked to other sectors" and defining the key issues as ignorance, indifference, lack of clarity on roles and responsibilities and inadequate regulatory oversight and enforcement tools.

There is an urgent need to change practices and behaviours to prevent a similar tragedy from reoccurring. This book sets out the changes required, why they are required, how they are to be achieved and the progress towards them to date.

Implementation of these major safety reforms will move the construction industry from the conditions that allowed the fire at Grenfell Tower to occur, to a system where construction professionals take greater responsibility for the safety of residents in their buildings. This book provides an overview of how the movement towards implementing a new building safety regime has unfolded over the last three years and details what still needs to be done if residents are to feel safe and be safe in their own homes.

September 2021: 120 pp
ISBN: 9780367552855

To Order
Tel:+44 (0) 1235 400524
Email: tandf@bookpoint.co.uk

For a complete listing of all our titles visit:
www.tandf.co.uk

Preliminaries Build-up Example

The number of items priced in the preliminaries section of tenders and the way they are priced vary considerably between contractors. Some contractors, by modifying their percentage factor for overheads and profit, attempt to cover the costs of preliminary items in their Prices for Measured Works. However, the cost of Preliminaries will vary widely according to job size and complexity, site location, accessibility, degree of mechanization practicable, position of the contractor's head office and relationships with local labour/domestic subcontractors. It is therefore usually far safer to price preliminary items separately on their merits according to the project.

It is not possible for the quantity surveyor/cost manager to quantify the main contractor's preliminaries and it is left for the contractor to interpret the tender documentation and ascertain his resources and method of working they require to complete the works.

This preliminaries bill can therefore be a relatively simple pricing schedule of headings under which the contractor prices their preliminaries items. Allowances should be determined using predicted time durations and quantities for resources.

An example in pricing preliminaries follows, and this assumes the form of contract used is a JCT Standard and the value, including preliminaries, is approximately £4,550,000. The contract is estimated to take 52 weeks to complete and the value is built up as follows:

	£
Contractors' Labour, Plant and Materials value	3,300,000
Subcontracts and Provisional Sums	750,000
Preliminaries, say	£500,000
Project value	**£4,550,000**

At the end of the section the data are summarized to give a total value of preliminaries for the project example.

Note: The term 'Not priced', where used throughout this section means either that the cost implication is negligible or that it is usually included elsewhere in the tender.

Pricing Schedule

Employer's requirements

Site accommodation – not priced
Site records – not priced
Site records – photographs, progress reporting etc. – not priced
Completion and post completion requirements – testing and commissioning, handover plan etc. – not priced
Cleaning – the cost would normally represent an allowance for final clearing of the works on completion with the residue for cleaning throughout the contract period. This amount being reduced with the introduction of waste management plans
Completion and post completion requirements

Defects after completion	Based on 0.2% of the contract sum	£10,000

Main contractor's cost items

Management and staff – main contractor's project specific management and staff

Management and staff	Based on 6% of the tender (excluding preliminaries)	£255,000

Site establishment – main contractor's and common user temporary site accommodation

Site accommodation	Based on 0.4% of the tender (excluding preliminaries)	£17,000

Temporary services

Lighting and power	Based on 1% of the tender (excluding preliminaries)	£43,000
Water for the works (provided from mains)	Based on 0.1% of the tender (excluding preliminaries)	£5,000
Temporary telephones etc.	Based on 0.1% of the tender (excluding preliminaries)	£5,000

Security – allow for staff and security equipment

Security	Based on 0.1% of the tender (excluding preliminaries)	£5,000
Temporary hoarding, fencing etc.	2.4 m high fencing of OSB boarding one side; pair of gates	£15,000

Safety and environmental protection: – compliance with all welfare facilities, first aid etc.

Safety, health and welfare	Based on 0.2% of the tender (excluding preliminaries)	£9,000

Control and protection – allowance for setting out, protection of the works, sampling

Control and protection of the works	Based on 0.1% of the tender (excluding preliminaries)	£5,000
Drying the works	Based on 0.01% of the tender (excluding preliminaries)	£1,000

Mechanical plant – common user mechanical plant and equipment

Plant/Transport	Based on 2% of the tender (excluding preliminaries)	£95,000
Small plant and tools	Based on 0.2% of the tender (excluding preliminaries)	£10,000

Temporary works – common user access scaffolding

Temporary roads and walkways	Say 250 × 3.5 m hardcore/recycled concrete	£10,000
Access scaffolding	Based on 0.75% of the tender (excluding preliminaries)	£32,000
Removing rubbish and cleaning	Based on 0.15% of the tender (excluding preliminaries)	£8,000

Fees and charges – any miscellaneous fees charges (rates etc.) – not priced
Site services – temporary works not specific to an element

Traffic regulations	Based on 0.05% of the tender (excluding preliminaries)	£3,000
Additional temporary items	Based on 0.25% of the tender (excluding preliminaries)	£12,000

Insurances, bonds, guarantees and warranties

Insurance of the works	Based on 0.15% of project value	£9,000

If at the Contractor's risk, the insurance cover must be enough to include the full cost of reinstatement, all increases in cost, professional fees and any consequential costs such as demolition. The average provision for fire risk is 0.15% of the value of the work after adding for increased costs and professional fees.

Note: Insurance premiums are liable to considerable variation, depending on the contractor, the nature of the work and the market in which the insurance is placed.

Summary of Preliminaries Costs Example (NB round to nearest £1000)	
Defects after completion	£10,000
Management and staff	£255,000
Site accommodation	£17,000
Lighting and power for the works	£43,000
Water for the works	£5,000
Temporary telephones	£5,000
Security	£5,000
Hoardings, fans, fencing, etc.	£15,000
Safety, health and welfare	£9,000
Protection of the works	£5,000
Drying the works	£1,000
Large plant/transport	£95,000
Small plant and tools	£10,000
Temporary roads and walkways	£10,000
Access scaffolding	£32,000
Removing rubbish and cleaning	£8,000
Traffic regulations	£3,000
Additional temporary works	£12,000
Insurances	£9,000
TOTAL PRELIMINARIES	**£549,000**

It is emphasized that the above is an example only of the way in which Preliminaries may be priced and it is essential that for any particular contract or project the items set out in Preliminaries should be assessed on their respective values. The value of the Preliminaries items in recent tenders received by the editors varies from a 8% to 14% addition to all other costs. The above example represents approximately a 12% addition to the value of measured work.

A Handbook of Sustainable Building Design and Engineering, 2nd edition

Dejan Mumovic *et al.*

The second edition of this authoritative textbook equips students with the tools they will need to tackle the challenges of sustainable building design and engineering. The book looks at how to design, engineer and monitor energy efficient buildings, how to adapt buildings to climate change, and how to make buildings healthy, comfortable and secure. New material for this edition includes sections on environmental masterplanning, renewable technologies, retrofitting, passive house design, thermal comfort and indoor air quality. With chapters and case studies from a range of international, interdisciplinary authors, the book is essential reading for students and professionals in building engineering, environmental design, construction and architecture.

October 2018: 604pp
ISBN: 9781138215474

To Order
Tel: +44 (0) 1235 400524
Email: book.orders@tandf.co.uk

For a complete listing of all our titles visit:
www.tandf.co.uk

Taylor & Francis
Taylor & Francis Group

Approximate Estimating Rates

Estimating by means of priced approximate quantities is always more accurate than by using overall building prices per square metre. Prices given in this section, which is arranged in elemental order, are derived from *Prices from Measured Works* section, but also include for all the incidental items and labours which are normally measured separately in Bills of Quantities. They have been established with a tender price level as indicated on the book cover. They include overheads and profit but *do not include for main contractor's preliminaries*, for which an example build-up is given in an earlier section.

Whilst every effort is made to ensure the accuracy of these figures, they have been prepared for approximate estimating purposes and guidance only and on no account should they be used for the preparation of tenders.

Unless otherwise described units denoted as m² refer to appropriate unit areas (rather than gross internal floor areas).

As elsewhere in this edition prices do not include Value Added Tax or fees for professional services.

Approximate Estimating Rates are exclusive of any main contractor's preliminaries.

Cost plans can be developed from Elemental Cost Plans using both *Approximate Estimating Rates* and/or *Prices for Measured Works* depending upon the level of information available.

The cost targets within each formal cost plan approved by the employer will be used as the baseline for future cost comparisons. Each subsequent cost plan will require reconciliation with the preceding cost plan and explanations relating to changes made. In view of this, it is essential that records of any transfers made to or from the risk allowances and any adjustments made to cost targets are maintained, so that explanations concerning changes can be provided to both the employer and the project team

1.1 SUBSTRUCTURE

Item	Unit	Range £		

1.1 SUBSTRUCTURE

1.1.1 Standard Foundations

Foundations and floor slab composite
Strip or trenchfill foundations with masonry up to 150 mm above floor level only; blinded hardcore bed; slab insulation; reinforced ground bearing slab 200 mm thick. To suit residential and small commercial developments with good ground bearing capacity

Item	Unit	Range £		
shallow foundations up to 1.00 m deep	m^2	210.00	to	250.00
shallow foundations up to 1.50 m deep	m^2	245.00	to	300.00

Foundations in poor ground; minipiles; typically 300 mm dia.; 15 m long; 175 mm thick reinforced concrete slab; for single storey commercial type development

Item	Unit	Range £		
minipiles to building columns only; 1 per column	m^2	160.00	to	200.00
minipiles to columns and perimeter ground floor beam; piles at 2.00 m centres to ground beams	m^2	320.00	to	390.00
minipiles to entire building at 2.00 m × 2.00 m grid	m^2	350.00	to	425.00

Raft foundations

Item	Unit	Range £		
simple reinforced concrete raft on poorer ground for development up to two storey high	m^2	175.00	to	210.00

Trench fill foundations
Machine excavation, disposal, plain in situ concrete 20.00 N/mm^2 (1:2:4) trench fill, 300 mm high cavity masonry in cement mortar (1:3), pitch polymer damp-proof course
With 3 courses of common brick outer skin up to DPC level (PC bricks @ £380/1000)

Item	Unit	Range £		
600 mm × 1000 mm deep	m	130.00	to	160.00
600 mm × 1500 mm deep	m	180.00	to	220.00

With 3 courses of facing bricks brick outer skin up to DPC level (PC bricks @ £500/1000)

Item	Unit	Range £		
600 mm × 1000 mm deep	m	135.00	to	160.00
600 mm × 1500 mm deep	m	185.00	to	220.00

With 3 courses of facing bricks brick outer skin up to DPC level (PC bricks @ £700/1000)

Item	Unit	Range £		
600 mm × 1000 mm deep	m	140.00	to	170.00
600 mm × 1500 mm deep	m	190.00	to	225.00

1.1 SUBSTRUCTURE

Item	Unit	Range £		

Strip foundations

Excavate trench 600 mm wide, partial backfill, partial disposal, earthwork support (risk item), compact base of trench, plain in situ concrete 20.00 N/mm² (1:2:4), 250 mm thick, 3 courses of common brick up to DPC level, cavity brickwork/blockwork in cement mortar (1:3), pitch polymer damp-proof course machine excavation

With 3 courses of common brick outer skin up to DPC level (PC bricks @ £380/ 1000)

	Unit	Range £ low	to	Range £ high
600 mm × 1000 mm deep	m	140.00	to	170.00
600 mm × 1500 mm deep	m	190.00	to	230.00

With 3 courses of facing brick outer skin up to DPC level (PC bricks @ £500/ 1000)

600 mm × 1000 mm deep	m	150.00	to	180.00
600 mm × 1500 mm deep	m	200.00	to	240.00

With 3 courses of facing brick outer skin up to DPC level (PC bricks @ £700/ 1000)

600 mm × 1000 mm deep	m	150.00	to	185.00
600 mm × 1500 mm deep	m	200.00	to	245.00

Column bases

Excavate pit in firm ground by machine, partial backfill, partial disposal, support, compact base of pit; in situ concrete 25.00 N/mm²; reinforced at 50 kg/m³

rates per base for base size

600 mm × 600 mm × 300 mm; 1000 mm deep pit	nr	450.00	to	550.00
900 mm × 900 mm × 450 mm; 1250 mm deep pit	nr	550.00	to	670.00
1500 mm × 1500 mm × 600 mm; 1500 mm deep pit	nr	920.00	to	1100.00
2700 mm × 2700 mm × 1000 mm; 1500 mm deep pit	nr	2700.00	to	3200.00

rates per m³ for base size

600 mm × 600 mm × 300 mm; 1000 mm deep pit	m³	4050.00	to	4900.00
900 mm × 900 mm × 450 mm; 1250 mm deep pit	m³	1525.00	to	1850.00
1500 mm × 1500 mm × 600 mm; 1500 mm deep pit	m³	700.00	to	840.00
2700 mm × 2700 mm × 1000 mm; 1500 mm deep pit	m³	380.00	to	460.00
extra for reinforcement at 75 kg/m³ concrete, base size	m³	89.00	to	110.00
extra for reinforcement at 100 kg/m³ concrete, base size	m³	110.00	to	130.00

1.1.2 Specialist Foundations

Piling

Enabling works

excavate to form piling mat; supply and lay imported hardcore – recycled brick and similar to form piling mat	m³	11.60	to	14.00
provision of plant (1 nr rig); including bringing to and removing from site; maintenance, erection and dismantling at each pile position	item	20000.00	to	24000.00

Supply and install concrete Continuous Flight Auger (CFA) piles; cart away inactive spoil; measure total length of pile

450 mm dia. reinforced concrete CFA piles	m	55.00	to	66.00
600 mm dia. reinforced concrete CFA piles	m	87.00	to	105.00
750 mm dia. reinforced concrete CFA piles	m	135.00	to	165.00
900 mm dia. reinforced concrete CFA piles	m	195.00	to	240.00

1.1 SUBSTRUCTURE

Item	Unit	Range £		
1.1 SUBSTRUCTURE – cont				
Piling – cont				
Pile testing				
using tension piles as reaction; typical loading test 1000 kN to 2000 kN	item	8200.00	to	9900.00
integrity testing; minimum 20 per visit	nr	16.30	to	19.70
Secant wall piling, 750 mm dia. piles including site mobilization and demobilization; Measure total area of piling (total length × total depth)				
supply and install secant wall piling	m²	280.00	to	335.00
Steel sheet piling; Measure total area of piling (total length × total depth)				
interlocking steel sheet piling to excavation perimeter; Corus LX or similar; extraction on completion	m²	220.00	to	265.00
Pile caps				
Excavate pit in firm ground by machine, partial backfill, partial disposal, support, compaction, cut off pile and prepare reinforcement, formwork; reinforced in situ concrete cap 25 N/mm²; reinforcement at 50 kg/m³; rates per cap for cap size				
900 mm × 900 mm × 1000 mm; 1–2 piles	nr	800.00	to	970.00
2100 mm × 2100 mm × 1000 mm; 2–3 piles	nr	1825.00	to	2175.00
2700 mm × 2700 mm × 1500 mm; 3–5 piles	nr	3650.00	to	4400.00
Excavate pit in firm ground by machine, partial backfill, partial disposal, support, compaction, cut off pile and prepare reinforcement, formwork; reinforced in situ concrete cap 25 N/mm²; reinforcement at 50 kg/m³; rates per m³ for typical cap size				
up to 900 mm × 900 mm × 1000 mm; 1–2 piles	m³	960.00	to	1150.00
up to 2100 mm × 2100 mm × 1000 mm; 2–3 piles	m³	420.00	to	510.00
up to 2700 mm × 2700 mm × 1500 mm; 3–5 piles	m³	340.00	to	410.00
extra for reinforcement at 75 kg/m³ concrete	m³	87.00	to	105.00
extra for reinforcement at 100 kg/m³ concrete	m³	110.00	to	130.00
extra for alternative strength concrete C30 (30.00 N/mm²)	m³	1.45	to	1.75
extra for alternative strength concrete C40 (40.00 N/mm²)	m³	7.30	to	8.80
Concrete ground beams				
Reinforced in situ concrete ground beams; bar reinforcement; formwork				
300 mm × 300 mm, reinforcement at 180 kg/m³	m	56.00	to	68.00
450 mm × 450 mm, reinforcement at 200 kg/m³	m	110.00	to	130.00
450 mm × 600 mm, reinforcement at 270 kg/m³	m	155.00	to	190.00
Precast concrete ground beams; square ends and dowelled. Supplied and installed in lengths to span between stanchion bases; concrete strength C50; maximum UDL of 4.5 kN/m; in situ work at beam ends not included				
350 mm × 425 mm	m	93.00	to	110.00
350 mm × 875 mm	m	220.00	to	270.00
350 mm × 1075 mm	m	300.00	to	365.00
450 mm × 400 mm	m	110.00	to	130.00
Concrete lift pits				
Excavate and disposal; reinforced concrete floor and walls; bitumen tanking as necessary				
1.65 m × 1.81 m × 1.60 m deep pit – 8 person lift/630 kg	nr	2900.00	to	3500.00
1.80 m × 2.50 m × 1.60 m deep pit –13 person lift/1000 kg	nr	3600.00	to	4300.00
3.00 m × 2.50 m × 1.60 m deep pit – 21 person lift/1700 kg	nr	4200.00	to	5100.00

1.1 SUBSTRUCTURE

Item	Unit	Range £		

Underpinning

In stages not exceeding 1500 mm long from one side of existing wall and foundation, excavate preliminary trench by machine and underpinning pit by hand, partial backfill, partial disposal, earthwork support (open boarded), cutting away projecting foundations, prepare underside of existing, compact base of pit, plain in situ concrete 20.00 N/mm² (1:2:4), formwork, brickwork in cement mortar (1:3), pitch polymer damp-proof course, wedge and pin to underside of existing with slates. Commencing at 1.00 m below ground level with common bricks, depth of underpinning

Item	Unit	Range £		
900 mm high, one brick wall	m²	410.00	to	495.00
1500 mm high, one brick wall	m²	590.00	to	710.00
extra for commencing 2.00 m below ground level	m²	80.00	to	97.00
extra for commencing 3.00 m below ground level	m²	155.00	to	190.00
extra for commencing 4.00 m below ground level	m²	220.00	to	265.00

1.1.3 Lowest Floor Construction

Ground floor construction

Mechanical excavation to reduce levels, disposal, level and compact, hardcore bed blinded with sand, 1200 gauge polythene damp-proof membrane, in situ concrete 20.00 N/mm² (1:2:4)

Item	Unit	Range £		
150 mm thick concrete slab with 1 layer of A195 fabric reinforcement	m²	110.00	to	130.00
200 mm thick concrete slab with 1 layer of A252 fabric reinforcement	m²	115.00	to	140.00
250 mm thick concrete slab with 1 layer of A393 fabric reinforcement	m²	120.00	to	140.00
extra for every additional 50 mm thick concrete	m²	8.05	to	9.80
extra for alternative strength concrete C30 (30.00 N/mm²)	m³	1.45	to	1.75
extra for alternative strength concrete C40 (40.00 N/mm²)	m³	7.30	to	8.80
extra for reinforcement at 25 kg/m³ concrete	m³	33.00	to	39.50
extra for reinforcement at 50 kg/m³ concrete	m³	55.00	to	66.00
extra for reinforcement at 75 kg/m³ concrete	m³	87.00	to	105.00
extra for reinforcement at 100 kg/m³ concrete	m³	110.00	to	130.00

Warehouse ground floor

Steel fibre reinforced floor slab placed using large pour construction techniques providing a finish floor flatness complying with FM2 special +/- 15 mm from datum. Note: Excavation, subbase and damp-proof membrane not included

Item	Unit	Range £		
nominal 200 mm thick in situ concrete floor slab, concrete grade C40, reinforced with steel fibres, surface power floated and cured with a spray application of curing and hardening agent	m²	33.00	to	40.00

Suspended ground floor

Beam and block flooring

Item	Unit	Range £		
suspended floor with 150 mm deep precast concrete beams and infill blocks	m²	69.00	to	83.00
suspended floor with 225 mm deep precast concrete beams and infill blocks	m²	110.00	to	135.00

2.1 FRAME

Item	Unit	Range £		

1.1 SUBSTRUCTURE – cont

Board or slab insulation
Kingspan Thermafloor TF70 (Thermal conductivity 0.022 W/mK) rigid urethane
floor insulation for solid concrete and suspended ground floors

50 mm thick	m^2	13.30	to	16.10
75 mm thick	m^2	17.90	to	21.50
100 mm thick	m^2	21.00	to	26.00
125 mm thick	m^2	28.00	to	34.00
150 mm thick	m^2	33.00	to	39.50

Ravatherm XPS X 500 SL (Thermal conductivity 0.033 W/mK) extruded
polystyrene foam or other equal and approved

50 mm thick	m^2	18.60	to	22.50
75 mm thick	m^2	22.50	to	27.00
100 mm thick	m^2	27.00	to	32.50
160 mm thick	m^2	44.00	to	53.00

1.1.5 Basements
Reinforced concrete basement floors; Note: excluding bulk excavation costs
and hardcore fill

200 mm thick waterproof concrete with 2 × layers of A252 mesh reinforcement	m^2	79.00	to	95.00

Reinforced concrete basement walls; Note: excluding bulk excavation costs
and hardcore fill

200 mm thick waterproof reinforced concrete	m^2	390.00	to	475.00

2.1 Frame

2.1.1 Frame and Floor Composite

Concrete frame; flat slab reinforced concrete floors up to 250 mm thick
Suspended slab; no coverings or finishes

up to 6 storeys	m^2	160.00	to	190.00
7-12 storeys	m^2	170.00	to	210.00
13-18 storeys	m^2	200.00	to	240.00

Steel frame; composite beam and slab floors
Suspended slab; permanent steel shuttering with 130 mm thick concrete; no
coverings or finishes

up to 6 storeys	m^2	160.00	to	200.00
7-12 storeys	m^2	190.00	to	230.00
13-18 storeys	m^2	200.00	to	245.00

Concrete frame

Reinforced in situ concrete columns, bar reinforcement, formwork.
Generally all formwork assumes four uses
Reinforcement rate 180 kg/m³; column size

225 mm × 225 mm	m	94.00	to	110.00
300 mm × 600 mm	m	205.00	to	250.00
450 mm × 900 mm	m	370.00	to	450.00

2.1 FRAME

Item	Unit	Range £		
Reinforcement rate 240 kg/m³; column size				
225 mm × 225 mm	m	100.00	to	125.00
300 mm × 600 mm	m	225.00	to	270.00
450 mm × 900 mm	m	410.00	to	500.00
In situ concrete casing to steel column, formwork; column size				
225 mm × 225 mm	m	84.00	to	100.00
300 mm × 600 mm	m	195.00	to	235.00
450 mm × 900 mm	m	330.00	to	400.00
Reinforced in situ concrete beams, bar reinforcement, formwork.				
Generally all formwork assumes four uses				
Reinforcement rate 200 kg/m³; beam size				
225 mm × 450 mm	m	120.00	to	145.00
300 mm × 600 mm	m	195.00	to	235.00
450 mm × 600 mm	m	250.00	to	300.00
600 mm × 600 mm	m	300.00	to	360.00
Reinforcement rate 240 kg/m³; beam size				
225 mm × 450 mm	m	130.00	to	160.00
300 mm × 600 mm	m	205.00	to	245.00
450 mm × 600 mm	m	260.00	to	310.00
600 mm × 600 mm	m	320.00	to	385.00
In situ concrete casing to steel beams, formwork; beam size				
225 mm × 450 mm	m	110.00	to	130.00
300 mm × 600 mm	m	165.00	to	200.00
450 mm × 600 mm	m	220.00	to	265.00
600 mm × 600 mm	m	260.00	to	315.00
Other floor and frame constructions				
reinforced concrete cantilevered balcony; up to 1.50 m wide × 1.20 deep	nr	2900.00	to	3500.00
reinforced concrete cantilevered walkways; up to 1.00 m wide	m²	200.00	to	245.00
reinforced concrete walkways and supporting frame; up to 1.00 m wide × 2.50 m high	m²	265.00	to	320.00
Steel frame				
Fabricated steelwork erected on site with bolted connections, primed				
smaller sections n.e. 40 kg/m	tonne	2350.00	to	2800.00
universal beams; grade S275	tonne	2225.00	to	2700.00
universal beams; grade S355	tonne	2300.00	to	2800.00
universal columns; grade S275	tonne	2225.00	to	2700.00
universal columns; grade S355	tonne	2300.00	to	2800.00
composite columns	tonne	2175.00	to	2600.00
hollow section circular	tonne	2500.00	to	3000.00
hollow section square or rectangular	tonne	2500.00	to	3000.00
cellular beams (FABSEC)	tonne	2500.00	to	3000.00
lattice beams	tonne	3050.00	to	3700.00
roof trusses	tonne	3200.00	to	3900.00
Steel finishes				
grit blast and one coat zinc chromate primer	m²	9.00	to	10.90
touch up primer and one coat of two pack epoxy zinc phosphate primer	m²	5.60	to	6.80
blast cleaning	m²	3.00	to	3.60
galvanizing	m²	20.00	to	24.00
galvanizing	tonne	330.00	to	400.00

2.1 FRAME

Item	Unit	Range £		
2.1 Frame – cont				
Steel frame – cont				
Other floor and frame constructions				
steel space deck on steel frame, unprotected	m²	58.00	to	70.00
18.00 m high bay warehouse; steel propped portal frame, cold rolled purlin sections, primed surface treatment only, excluding decorations and protection	m²	120.00	to	150.00
columns and beams to mansard roof, 60 min fire protection	m²	190.00	to	235.00
feature columns and beams to glazed atrium roof unprotected	m²	200.00	to	245.00
Fire protection to steelwork				
Sprayed mineral fibre; gross surface area				
60 minute protection	m²	18.00	to	22.00
90 minute protection	m²	32.00	to	39.00
Sprayed vermiculite cement; gross surface area				
60 minute protection	m²	20.50	to	25.00
90 minute protection	m²	30.00	to	36.50
Supply and fit fire-resistant boarding to steel columns and beams; noggins, brackets and angles, intumescent paste. Beamclad or similar; measure board area				
30 minute protection for concealed applications	m²	28.00	to	34.00
60 minute protection for concealed applications	m²	61.00	to	74.00
30 minute protection left exposed for decoration	m²	31.00	to	37.00
60 minute protection left exposed for decoration	m²	73.00	to	88.00
Intumescent fire protection coating/decoration to exposed steelwork; Gross surface area (m²) or per tonne; On site application, spray applied				
30 minute protection per m²	m²	12.80	to	15.50
30 minute protection per tonne	tonne	325.00	to	395.00
60 minute protection per m²	m²	16.20	to	19.70
60 minute protection per tonne	tonne	410.00	to	500.00
Intumescent fire protection coating/decoration to exposed steelwork; Gross surface area (m²) or per tonne; Off site application, spray applied				
60 minute protection per m²	m²	25.00	to	31.00
60 minute protection per tonne	tonne	650.00	to	780.00
90 minute protection per m²	m²	46.00	to	56.00
90 minute protection per tonne	tonne	1175.00	to	1425.00
120 minute protection per m²	m²	79.00	to	95.00
120 minute protection per tonne	tonne	2000.00	to	2425.00

2.3 ROOF

Item	Unit	Range £		
2.2 UPPER FLOORS				
2.2.1 Floors				
Composite steel deck flooring				
Composite steel and concrete upper floors (Note: all floor thicknesses are nominal); A142 mesh reinforcement				
TATA Slimdek SD225 steel decking 1.25 mm thick; 130 mm reinforced concrete topping	m^2	100.00	to	120.00
Re-entrant type steel deck 0.90 mm thick; 150 mm reinforced concrete	m^2	88.00	to	105.00
Re-entrant type steel deck 60 mm deep, 1.20 mm thick; 200 mm reinforced concrete	m^2	94.00	to	115.00
Trapezoidal steel decking 0.90 mm thick; 150 mm reinforced concrete topping	m^2	90.00	to	110.00
Trapezoidal steel decking 1.20 mm thick; 150 mm reinforced concrete topping	m^2	94.00	to	110.00
Composite precast concrete and in situ concrete upper floors				
Omnidec flooring system comprising of 50 mm thick precast concrete deck (as permanent shuttering); 210 mm polystyrene void formers; concrete topping and reinforcement	m^2	120.00	to	150.00
Precast concrete suspended floors; 1200 mm wide suspended slab; 75 mm thick screed; no coverings or finishes				
3.00 m span; 150 mm thick planks; 5.00 kN/m^2 loading	m^2	76.00	to	88.00
6.00 m span; 150 mm thick planks; 5.00 kN/m^2 loading	m^2	78.00	to	90.00
7.50 m span; 200 mm thick planks; 5.00 kN/m^2 loading	m^2	82.00	to	95.00
9.00 m span; 250 mm thick planks; 5.00 kN/m^2 loading	m^2	86.00	to	99.00
12.00 m span; 350 mm thick planks; 5.00 kN/m^2 loading	m^2	90.00	to	105.00
3.00 m span; 150 mm thick planks; 8.50 kN/m^2 loading	m^2	75.00	to	86.00
6.00 m span; 200 mm thick planks; 8.50 kN/m^2 loading	m^2	77.00	to	89.00
7.50 m span; 250 mm thick planks; 8.50 kN/m^2 loading	m^2	86.00	to	99.00
3.00 m span; 150 mm thick planks; 12.50 kN/m^2 loading	m^2	78.00	to	90.00
6.00 m span; 250 mm thick planks; 12.50 kN/m^2 loading	m^2	86.00	to	99.00
Post-tensioned concrete upper floors				
reinforced post-tensioned suspended concrete slab 150–225 mm thick, 40 N/mm^2, reinforcement 60 kg/m^3, formwork	m^2	115.00	to	140.00
Softwood floors				
Joisted floor; plasterboard ceiling; skim; emulsion; t&g chipboard, sheet vinyl flooring and painted softwood skirtings	m^2	75.00	to	97.00
2.3 ROOF				
2.3.1 Roof Structure				
Flat roof decking; structure only				
Softwood				
comprising roof joists; 100 mm × 50 mm wall plates; herringbone strutting; 18 mm thick external quality plywood boarding	m^2	58.00	to	70.00
Metal decking				
galvanized steel roof decking; insulation; three layer felt roofing and chippings; 0.70 mm thick steel decking (U-value = 0.25 W/m^{2}K)	m^2	77.00	to	93.00
aluminium roof decking; three layer felt roofing and chippings; 0.90 mm thick aluminium decking (U-value = 0.25 W/m^{2}K)	m^2	8.70	to	10.60

2.3 ROOF

Item	Unit	Range £		
2.3 ROOF – cont				
Flat roof decking – cont				
Concrete				
precast concrete suspended slab with sand: cement screed over	m²	82.00	to	99.00
reinforced concrete slabs; on steel permanent steel shuttering; 150 mm				
reinforced concrete topping	m²	90.00	to	110.00
Screeds/Decks to receive roof coverings				
50 mm thick cement and sand screed	m²	18.00	to	22.00
75 mm thick lightweight bituminous screed and vapour barrier	m²	65.00	to	78.00
Softwood trussed pitched roofs; structure only				
Timber roof trusses and bracings only; plan area				
comprising 75 mm × 50 mm roof trusses at 600 mm centres (measured on plan)	m²	40.00	to	48.00
comprising 100 mm × 38 mm roof trusses at 600 mm centres (measured on plan)	m²	47.00	to	57.00
Mansard type roof comprising 100 mm × 50 mm roof trusses at 600 mm centres; 70° pitch	m²	48.00	to	58.00
forming small dormers up to 1.50 m²	nr	800.00	to	970.00
2.3.2 Roof Coverings				
Timber trusses with tile coverings				
Timber roof trusses; insulation; roof coverings; PVC rainwater goods; plasterboard; skim and emulsion to ceilings (U-value = 0.25 W/m² K); roof plan area				
concrete interlocking tile coverings	m²	200.00	to	240.00
clay plain tile coverings	m²	210.00	to	250.00
clay pantile coverings	m²	225.00	to	270.00
natural Welsh slate coverings	m²	320.00	to	385.00
reconstructed stone coverings	m²	245.00	to	300.00
Timber dormer roof trusses; insulation; roof coverings; PVC rainwater goods; plasterboard; skim and emulsion to ceilings (U-value = 0.25 W/m² K)				
concrete interlocking tile coverings	m²	240.00	to	290.00
clay pantile coverings	m²	225.00	to	270.00
plain clay tile coverings	m²	240.00	to	290.00
natural slate coverings	m²	420.00	to	510.00
reconstructed stone coverings	m²	285.00	to	345.00
Comparative tiling and slating finishes; including underfelt, battening, eaves courses and ridges				
concrete troughed or bold roll interlocking tiles; sloping	m²	73.00	to	88.00
Tudor clay pantiles; sloping	m²	87.00	to	105.00
blue composition (cement fibre) slates; sloping	m²	115.00	to	140.00
Spanish slates; sloping	m²	175.00	to	210.00
Welsh slates; sloping	m²	285.00	to	345.00
reconstructed stone slates; random slates; sloping	m²	160.00	to	195.00

2.3 ROOF

Item	Unit	Range £		
Steel trussed with metal sheet cladding				
Steel roof trusses and beams; thermal and acoustic insulation (U-value = 0.25 W/m^2 K)				
aluminium profiled composite cladding	m^2	300.00	to	365.00
copper roofing on boarding	m^2	360.00	to	440.00
Sheet roof claddings				
Fibre cement sheet profiled cladding				
Profile 6; single skin; natural grey finish	m^2	26.00	to	31.50
P61 Insulated System; natural grey finish; metal inner lining panel (U-value = 0.25 W/m^2 K)	m^2	49.00	to	60.00
extra for coloured fibre cement sheeting	m^2	1.05	to	1.30
double skin GRP translucent roof sheets	m^2	57.00	to	68.00
triple skin GRP translucent roof sheets	m^2	63.00	to	76.00
Steel PVF2 coated galvanized trapezoidal profile cladding fixed to steel purlins (not included); for roof pitches greater than 4°				
single skin trapezoidal sheeting only	m^2	18.30	to	22.00
composite insulated roofing system; 80 mm overall panel thickness (U-value = 0.25 W/m^2 K)	m^2	45.00	to	54.00
composite insulated roofing system; 115 mm overall panel thickness (U-value = 0.18 W/m^2 K)	m^2	47.00	to	57.00
composite insulated roofing system; 150 mm overall panel thickness (U-value = 0.14 W/m^2 K)	m^2	63.00	to	77.00
For very shallow roof pitches				
standing seam joints composite insulated roofing system; 90 mm overall panel thickness (U-value = 0.25 W/m^2 K)	m^2	73.00	to	88.00
standing seam joints composite insulated roofing system; 110 mm overall panel thickness (U-value = 0.20 W/m^2 K)	m^2	78.00	to	95.00
standing seam joints composite insulated roofing system; 125 mm overall panel thickness (U-value = 0.18 W/m^2 K)	m^2	82.00	to	99.00
Aluminium roofing; standing seam				
Kalzip standard natural aluminium; 0.9 mm thick; 180 mm glassfibre insulation; vapour control layer; liner sheets; (U-value = 0.25 W/m^2 K)	m^2	78.00	to	95.00
Copper roofing				
copper roofing with standing seam joints; insulation breather membrane or vapour barrier (U-value = 0.25 W/m^2 K)	m^2	200.00	to	240.00
Stainless steel				
Terne-coated stainless steel roofing on and including Metmatt underlay (U-value = 0.25 W/m^2 K)	m^2	160.00	to	200.00
Zinc; standing seam				
Terne-coated zinc roofing on and including Metmatt underlay (U-value = 0.25 W/m^2 K)	m^2	120.00	to	145.00
Lead				
Roof coverings in welded seam construction including Geotec underlay fixed to prevent lifting and distortion	m^2	165.00	to	200.00
Flat roofing systems				
Includes insulation and vapour control barrier; excludes decking (U-value = 0.25 W/m^2 K)				
single layer polymer roofing membrane	m^2	78.00	to	94.00
single layer polymer roofing membrane with tapered insulation	m^2	150.00	to	180.00
20 mm thick polymer modified asphalt roofing including felt underlay	m^2	105.00	to	130.00

2.3 ROOF

Item	Unit	Range £		
2.3 ROOF – cont				
Flat roofing systems – cont				
Includes insulation and vapour control barrier – cont				
high performance bitumen felt roofing system	m^2	100.00	to	120.00
high performance polymer modified bitumen membrane	m^2	115.00	to	140.00
Kingspan KS1000TD composite Single Ply roof panels for roof pitches greater than 0.7° (after deflection); 1.5 mm Single Ply External Covering, Internal Coating Bright White Polyester (steel)				
71 mm thick (U-value = 0.25 Wm^{2}K)	m^2	59.00	to	72.00
100 mm thick (U-value = 0.18 Wm^{2}K)	m^2	60.00	to	73.00
120 mm thick (U-value = 0.15 Wm^{2}K)	m^2	69.00	to	84.00
Edges to felt flat roofs; softwood splayed fillet				
280 mm × 25 mm painted softwood fascia; no gutter aluminium edge trim	m	49.50	to	60.00
Edges to flat roofs; code 4 lead drip dresses into gutter; 230 mm × 25 mm painted softwood fascia; buildup				
170 mm uPVC gutter	m	120.00	to	145.00
100 mm aluminium gutter	m	130.00	to	160.00
100 mm cast iron gutter; decorated	m	125.00	to	150.00
2.3.3 Specialist Roof Systems				
Landscaped roofs				
Polyester-based elastomeric bitumen waterproofing and vapour equalization layer, copper lined bitumen membrane root barrier and waterproofing layer, separation and slip layers, protection layer, 50 mm thick drainage board, filter fleece, insulation board, Sedum vegetation blanket				
intensive (high maintenance – may include trees and shrubs require deeper substrate layers, are generally limited to flat roofs	m^2	200.00	to	245.00
extensive (low maintenance – herbs, grasses, mosses and drought tolerant succulents such as Sedum)	m^2	170.00	to	210.00
Ethylene tetrafluoroethylene (ETFE) systems				
Multiple layered ETFE inflated cushions supported by a lightweight aluminium or steel structure	m^2	900.00	to	1075.00
2.3.4 Roof Drainage				
Gutters				
100 mm uPVC gutter	m	33.00	to	40.00
170 mm uPVC gutter	m	56.00	to	68.00
100 mm cast iron gutter; decorated	m	50.00	to	60.00
150 mm cast iron gutter; decorated	m	81.00	to	98.00
Rainwater downpipes pipes; fixed to backgrounds; including offsets and shoes				
68 mm dia. uPVC	m	31.50	to	38.00
110 mm dia. uPVC	m	56.00	to	68.00
75 mm dia. cast iron; decorated	m	72.00	to	88.00
100 mm dia. cats iron; decorated	m	95.00	to	115.00

2.3 ROOF

Item	Unit	Range £		
2.3.5 Rooflights/Patent Glazing and Glazed Roofs				
Rooflights				
individual polycarbonate rooflights; rectangular; fixed light	m²	540.00	to	650.00
individual polycarbonate rooflights; rectangular; manual opening	m²	740.00	to	900.00
individual polycarbonate rooflights; rectangular; electric opening	m²	1125.00	to	1350.00
Velux style rooflights to traditional roof construction (tiles/slates)	m²	590.00	to	720.00
Patent glazing; including flashings, standard aluminium alloy bars; Georgian wired or laminated glazing; fixed lights				
single glazed	m²	210.00	to	250.00
low-e clear toughened and laminated double glazed units	m²	550.00	to	660.00
2.3.6 Roof Features				
Flat roof feature				
roof walkways 600 mm × 600 mm × 50 mm precast concrete slabs on support system; pedestrian access	m²	57.00	to	69.00
solar reflective paint	m²	5.35	to	6.45
limestone chipping finish	m²	5.80	to	7.00
grip tiles in hot bitumen	m²	70.00	to	85.00
2.3.7 Insulation				
Glass fibre roll; Crown Loft Roll 44 (Thermal conductivity 0.044 W/mK) or other equal; laid loose				
100 mm thick	m²	2.90	to	3.50
150 mm thick	m²	3.80	to	4.60
200 mm thick	m²	4.80	to	5.80
Glass fibre quilt; Isover Modular roll (Thermal conductivity 0.043 W/mK) or other equal and approved; laid loose				
100 mm thick	m²	2.90	to	3.50
150 mm thick	m²	3.75	to	4.55
200 mm thick	m²	5.60	to	6.75
Crown Rafter Roll 32 (Thermal conductivity 0.032 W/mK) glass fibre flanged building roll; pinned vertically or to slope between timber framing				
50 mm thick	m²	6.00	to	7.30
75 mm thick	m²	7.85	to	9.50
100 mm thick	m²	9.60	to	11.60
Thermafleece EcoRoll (0.039 W/mK)				
50 mm thick	m²	7.00	to	8.50
75 mm thick	m²	8.90	to	10.80
100 mm thick	m²	11.10	to	13.40
140 mm thick	m²	15.00	to	18.10

Approximate Estimating Rates

2.4 STAIRS AND RAMPS

Item	Unit	Range £		

2.4 STAIRS AND RAMPS

2.4.1 Stair/Ramp Structures

Reinforced concrete construction
Escape staircase; granolithic finish; mild steel balustrades and handrails

Item	Unit	Range £		
3.00 m rise; dogleg	nr	6600.00	to	8000.00
plus or minus for each 300 mm variation in storey height	nr	630.00	to	770.00

Staircase; terrazzo finish; mild steel balustrades and handrails; plastered soffit; balustrades and staircase soffit decorated

Item	Unit			
3.00 m rise; dogleg	nr	9900.00	to	12000.00
plus or minus for each 300 mm variation in storey height	nr	960.00	to	1175.00

Staircase; terrazzo finish; stainless steel balustrades and handrails; plastered and decorated soffit

Item	Unit			
3.00 m rise; dogleg	nr	12000.00	to	14000.00
plus or minus for each 300 mm variation in storey height	nr	1150.00	to	1400.00

Staircase; high quality finishes; stainless steel and glass balustrades; plastered and decorated soffit

Item	Unit			
3.00 m rise; dogleg	nr	1875.00	to	2275.00
plus or minus for each 300 mm variation in storey height	nr	2100.00	to	2550.00

Metal construction
Steel access/fire ladder

Item	Unit			
3.00 m high	nr	760.00	to	920.00
4.00 m high; epoxide finished	nr	1150.00	to	1375.00

Light duty metal staircase; galvanized finish; perforated treads; no risers; balustrades and handrails; decorated

Item	Unit			
3.00 m rise; straight; 900 mm wide	nr	4400.00	to	5300.00
plus or minus for each 300 mm variation in storey height	nr	350.00	to	425.00

Light duty circular metal staircase; galvanized finish; perforated treads; no risers; balustrades and handrails; decorated

Item	Unit			
3.00 m rise; straight; 1548 mm dia.	nr	4850.00	to	5900.00
plus or minus for each 300 mm variation in storey height	nr	415.00	to	500.00

Heavy duty cast iron staircase; perforated treads; no risers; balustrades and hand rails; decorated

Item	Unit			
3.00 m rise; straight	nr	5500.00	to	6600.00
plus or minus for each 300 mm variation in storey height	nr	570.00	to	690.00
3.00 m rise; spiral; 1548 mm dia.	nr	6300.00	to	7700.00
plus or minus for each 300 mm variation in storey height	nr	610.00	to	740.00

Feature metal staircase; galvanized finish perforated treads; no risers; decorated

Item	Unit			
3.00 m rise; spiral balustrades and handrails	nr	7200.00	to	8700.00
3.00 m rise; dogleg; hardwood balustrades and handrails	nr	8400.00	to	10000.00
3.00 m rise; dogleg; stainless steel balustrades and handrails	nr	11000.00	to	13000.00
plus or minus for each 300 mm variation in storey height	nr	710.00	to	860.00
galvanized steel catwalk; nylon coated balustrading 450 mm wide	m	400.00	to	480.00

2.5 EXTERNAL WALLS

Item	Unit	Range £		

Timber construction
Softwood staircase; softwood balustrades and hardwood handrail; plasterboard; skim and emulsion to soffit

2.60 m rise; standard; straight flight	nr	1125.00	to	1375.00
2.60 m rise; standard; top three treads winding	nr	1325.00	to	1575.00
2.60 m rise; standard; dogleg	nr	1450.00	to	1750.00

Oak staircase; balustrades and handrails; plasterboard; skim and emulsion to soffit

2.60 m rise; purpose-made; dogleg	nr	9600.00	to	11500.00
plus or minus for each 300 mm variation in storey height	nr	1275.00	to	1550.00

2.4.2 Stair/Ramp Finishes
Finishes to treads and risers; including nosings etc.

vinyl or rubber	m	62.00	to	75.00
carpet (PC sum £22/m²)	m	96.00	to	115.00

2.4.3 Stair/Ramp Balustrades and Handrails
Wall handrails

softwood handrail and brackets	m	85.00	to	100.00
hardwood handrail and brackets	m	120.00	to	140.00
mild steel handrail and brackets	m	160.00	to	190.00
stainless steel handrail and brackets	m	200.00	to	240.00

Balustrading and handrails

mild steel balustrade and steel or timber handrail	m	310.00	to	375.00
balustrade and handrail with metal infill panels	m	390.00	to	475.00
balustrade and handrail with glass infill panels	m	430.00	to	520.00
stainless steel balustrade and handrail	m	510.00	to	620.00
stainless steel and structural glass balustrade	m	900.00	to	1075.00

2.5 EXTERNAL WALLS

2.5.1 External Enclosing Walls Above Ground Level

Brick/block walling; solid walls
Common brick solid walls; bricks PC £380.00/1000

half brick thick	m²	59.00	to	71.00
one brick thick	m²	110.00	to	130.00
one and a half brick thick	m²	160.00	to	190.00
two brick thick	m²	205.00	to	250.00
extra for fair face per side	m²	2.60	to	3.15

Engineering brick walls; class B; bricks PC £590.00/1000

half brick thick	m²	57.00	to	69.00
one brick thick	m²	105.00	to	130.00
one and a half brick thick	m²	150.00	to	180.00
two brick thick	m²	195.00	to	240.00

Facing brick walls; sand faced facings; bricks PC £500.00/1000

half brick thick; pointed one side	m²	75.00	to	90.00
one brick thick; pointed both sides	m²	140.00	to	170.00

2.5 EXTERNAL WALLS

Item	Unit	Range £		
2.5 EXTERNAL WALLS – cont				
Brick/block walling – cont				
Facing bricks solid walls; handmade facings; bricks PC £700.00/1000				
half brick thick; fair face one side	m²	90.00	to	110.00
one brick thick; fair face both sides	m²	175.00	to	210.00
add or deduct for each variation of £10.00/1000 in PC value				
half brick thick	m²	0.75	to	0.90
one brick thick	m²	1.50	to	1.80
one and a half brick thick	m²	2.20	to	2.70
two brick thick	m²	3.00	to	3.60
Aerated lightweight block walls				
100 mm thick	m²	38.00	to	46.50
140 mm thick	m²	52.00	to	62.00
215 mm thick	m²	99.00	to	120.00
Dense aggregate block walls				
100 mm thick	m²	42.50	to	51.00
140 mm thick	m²	52.00	to	63.00
Coloured dense aggregate masonry block walls; Lignacite or similar				
100 mm thick single skin, weathered face, self coloured	m²	88.00	to	105.00
100 mm thick single skin, polished face	m²	135.00	to	165.00
Brick/block walling; cavity walls				
Cavity wall; facing brick outer skin; insulation; with plaster on standard weight block inner skin; emulsion (U-value = 0.30 W/m² K)				
machine-made facings; PC £550.00/1000	m²	140.00	to	175.00
machine-made facings; PC £700.00/1000	m²	160.00	to	190.00
self-finished masonry block, weathered finish outer skin and paint grade block inner skins; fair faced both sides	m²	140.00	to	170.00
self-finished masonry block, polished finish outer skin and paint grade block inner skins; fair faced both sides	m²	190.00	to	230.00
Cavity wall; rendered block outer skin; insulation; with plaster on standard weight block inner skin; emulsion (U-value = 0.30 W/m² K)				
block outer skin; insulation; lightweight block inner skin outer block rendered	m²	175.00	to	210.00
Cavity wall; facing brick outer skin; insulation; plasterboard on stud inner skin; emulsion (U-value = 0.30 W/m² K)				
machine-made facings; PC £550.00/1000	m²	110.00	to	135.00
machine-made facings; PC £700.00/1000	m²	125.00	to	150.00
Reinforced concrete walling				
In situ reinforced concrete 25.00 N/mm²; 15 kg/m² reinforcement; formwork both sides				
150 mm thick	m²	145.00	to	175.00
225 mm thick	m²	160.00	to	200.00
300 mm thick	m²	180.00	to	220.00
Panelled walling				
Precast concrete panels; including insulation; lining and fixings generally 7.5 m × 0.15 thick × storey height (U-value = 0.30 W/m² K)				
standard panels	m²	270.00	to	325.00
standard panels; exposed aggregate finish	m²	290.00	to	350.00
reconstructed stone faced panels	m²	51.00	to	62.00

2.5 EXTERNAL WALLS

Item	Unit	Range £		
brick clad panels (PC £350.00/1000 for bricks)	m²	43.00	to	52.00
natural stone faced panels (Portland Stone or similar)	m²	580.00	to	700.00
marble or granite faced panels	m²	750.00	to	910.00
Kingspan TEK cladding panel; 142 mm thick (15 mm thick OSB board and 112 mm thick rigid eurathane core) fixed to frame; metal edge trim flashings	m²	385.00	to	465.00
Sheet claddings				
Cement fibre profiled cladding				
Profile 6; single skin; natural grey finish	m²	27.00	to	33.00
P61 Insulated System; natural grey finish; metal inner lining panel (U-value = 0.30 W/m² K)	m²	50.00	to	60.00
P61 Insulated System; natural grey finish; 100 mm lightweight concrete block (U-value = 0.30 W/m² K)	m²	99.00	to	120.00
P61 Insulated System; natural grey finish; 12 mm self-finished plasterboard lining (U-value = 0.30 W/m² K)	m²	68.00	to	83.00
extra for coloured fibre cement sheeting	m²	0.75	to	0.90
Metal profiled cladding				
Standard trapezoidal profile				
composite insulated roofing system; 60 mm overall panel thickness (U-value = 0.35 W/m² K)	m²	54.00	to	66.00
composite insulated roofing system; 80 mm overall panel thickness (U-value = 0.26 W/m² K)	m²	57.00	to	69.00
composite insulated roofing system; 100 mm overall panel thickness (U-value = 0.20 W/m² K)	m²	61.00	to	73.00
Micro-rib profile				
composite insulated roofing system; micro-rib panel; 60 mm overall panel thickness (U-value = 0.35 W/m² K)	m²	150.00	to	180.00
composite insulated roofing system; micro-rib panel; 80 mm overall panel thickness (U-value = 0.26 W/m² K)	m²	160.00	to	195.00
composite insulated roofing system; micro-rib panel; 100 mm overall panel thickness (U-value = 0.20 W/m² K)	m²	175.00	to	210.00
Flat panel				
composite insulated roofing system; micro-rib panel; 60 mm overall panel thickness (U-value = 0.35 W/m² K)	m²	240.00	to	290.00
composite insulated roofing system; micro-rib panel; 80 mm overall panel thickness (U-value = 0.26 W/m² K)	m²	230.00	to	280.00
composite insulated roofing system; micro-rib panel; 100 mm overall panel thickness (U-value = 0.20 W/m² K)	m²	270.00	to	330.00
Glazed walling				
Curtain walling				
stick curtain walling with double glazed units, aluminium structural framing and spandrel rails. Standard colour powder coated capped	m²	550.00	to	660.00
unitized curtain walling system with double glazed units, aluminium structural framing and spandrel rails. Standard colour powder coated	m²	1125.00	to	1375.00
unitized curtain walling bespoke project specific system with double glazed units, aluminium structural framing and spandrel rails. Standard colour powder coated.	m²	1125.00	to	1375.00
visual mock-ups for project specific curtain walling solutions	item	36500.00	to	44000.00

2.5 EXTERNAL WALLS

Item	Unit	Range £		

2.5 EXTERNAL WALLS – cont

Glazed walling – cont
Other systems

lift surround of double glazed or laminated glass with aluminium or stainless steel framing	m²	1000.00	to	1225.00

Other cladding systems
Tiles (clay/slate/glass/ceramic)

machine-made clay tiles; including battens	m²	35.00	to	42.50
best handmade sand faced tiles; including battens	m²	48.50	to	59.00
concrete plain tiles; including battens	m²	49.00	to	59.00
natural slates; including battens	m²	140.00	to	170.00
20 mm × 20 mm thick mosaic glass or ceramic; in common colours; fixed on prepared surface	m²	125.00	to	150.00

Steel

vitreous enamelled insulated steel sandwich panel system; with insulation board on inner face	m²	210.00	to	255.00
Formalux sandwich panel system; with coloured lining tray; on steel cladding rails	m²	240.00	to	295.00

Comparative external finishes
Comparative concrete wall finishes

wrought formwork one side including rubbing down	m²	6.20	to	7.45
shotblasting to expose aggregate	m²	7.50	to	9.05
bush hammering to expose aggregate	m²	19.30	to	23.00

Comparative in situ finishes

two coats Sandtex Matt cement paint to render	m²	6.80	to	8.30
13 mm thick cement and sand plain face rendering	m²	25.00	to	30.00
ready-mixed self-coloured acrylic resin render on blockwork	m²	66.00	to	80.00
three coat Tyrolean rendering; including backing	m²	47.50	to	58.00
Stotherm Lamella; 6 mm thick work in 2 coats	m²	93.00	to	110.00

2.5.2 Solar/Rain Screen
Rainscreen

25 mm thick tongued and grooved tanalized softwood boarding; including timber battens	m²	49.50	to	60.00
timber shingles, Western Red Cedar, preservative treated in random widths; not including any subframe or battens	m²	64.00	to	77.00
25 mm thick tongued and grooved Western Red Cedar boarding including timber battens	m²	56.00	to	68.00

2.5 EXTERNAL WALLS

Item	Unit	Range £		
25 mm thick Western Red Cedar tongued and grooved wall cladding on and including treated softwood battens on breather membrane, 10 mm Eternit Blueclad board and 50 mm insulation board; the whole fixed to Metsec frame system; including sealing all joints etc.	m²	135.00	to	160.00
Sucupira Preta timber boarding 19 mm thick × 75 mm wide slats with chamfered open joints fixed to 38 mm × 50 mm softwood battens with stainless steel screws on 100 mm Kingspan K2 insulations and 28 mm WBP plywood fixed to 50 mm × 1000 softwood studs; complete with breather membrane, aluminium cills and Sucupira Petra batten at external joints	m²	310.00	to	380.00
Trespa single skin rainscreen cladding, 8 mm thick panel, with secondary support/frame system; adhesive fixed panels; open joints; 100 mm Kingspan K15 insulation; aluminium subframe fixed to masonry/concrete	m²	260.00	to	320.00
Terracotta rainscreen cladding; aluminium support rails; anti-graffiti coating	m²	460.00	to	550.00
Corium brick tiles in metal tray system; standard colour; including all necessary angle trim on main support system; comprising of polythene vapour check; 75 mm × 50 mm timber studs; 50 mm thick rigid insulation with taped joints; 38 mm × 50 mm timber counter battens	m²	340.00	to	410.00
Solar shading				
fixed aluminium Brise Soleil including uni-strut supports; 300 mm deep	m	180.00	to	220.00

2.5.4 Insulation

Item	Unit	Range £		
Crown Dritherm Cavity Slab 37 (Thermal conductivity 0.037 W/mK) glass fibre batt or other equal; as full or partial cavity fill; including cutting and fitting around wall ties and retaining discs				
50 mm thick	m²	4.40	to	5.30
75 mm thick	m²	6.10	to	7.35
100 mm thick	m²	6.95	to	8.40
Crown Dritherm Cavity Slab 34 (Thermal conductivity 0.034 W/mK) glass fibre batt or other equal; as full or partial cavity fill; including cutting and fitting around wall ties and retaining discs				
65 mm thick	m²	4.60	to	5.55
75 mm thick	m²	5.10	to	6.20
85 mm thick	m²	6.55	to	7.90
100 mm thick	m²	6.70	to	8.10
Crown Dritherm Cavity Slab 32 (Thermal conductivity 0.032 W/mK) glass fibre batt or other equal; as full or partial cavity fill; including cutting and fitting around wall ties and retaining discs				
65 mm thick	m²	5.30	to	6.40
75 mm thick	m²	5.95	to	7.20
85 mm thick	m²	6.45	to	7.80
100 mm thick	m²	7.20	to	8.75
Crown Frametherm Roll 40 (Thermal conductivity 0.040 W/mK) glass fibre semi-rigid or rigid batt or other equal; pinned vertically in timber frame construction				
90 mm thick	m²	6.30	to	7.65
140 mm thick	m²	8.45	to	10.20
Kay-Cel (Thermal conductivity 0.033 W/mK) expanded polystyrene board standard grade SD/N or other equal; fixed with adhesive				
25 mm thick	m²	6.60	to	7.95
50 mm thick	m²	9.55	to	11.60
75 mm thick	m²	12.50	to	15.10
100 mm thick	m²	15.20	to	18.40

2.6 WINDOWS AND EXTERNAL DOORS

Item	Unit	Range £		

2.5 EXTERNAL WALLS – cont

2.5.4 Insulation – cont
Kingspan Kooltherm K8 (Thermal conductivity 0.022 W/mK) zero ODP rigid urethene insulation board or other equal; as partial cavity fill; including cutting and fitting around wall ties and retaining discs

40 mm thick	m²	17.80	to	21.50
50 mm thick	m²	17.80	to	21.50
60 mm thick	m²	20.50	to	25.00
75 mm thick	m²	20.50	to	25.00

Kingspan Thermawall TW50 (Thermal conductivity 0.022 W/mK) zero ODP rigid urethene insulation board or other equal and approved; as partial cavity fill; including cutting and fitting around wall ties and retaining discs

25 mm thick	m²	13.60	to	16.40
50 mm thick	m²	16.60	to	20.00
75 mm thick	m²	22.50	to	27.00
100 mm thick	m²	28.00	to	34.00

Thermafleece TF35 high density wool insulating batts (0.035 W/mK); 60% British wool, 30% recycled polyester and 10% polyester binder with a high recycled content

50 mm thick	m²	15.30	to	18.60
75 mm thick	m²	19.70	to	24.00

2.6 WINDOWS AND EXTERNAL DOORS

2.6.1 External Windows

Softwood windows (U-value = 1.6 W/m² K)
Standard windows

painted; double glazed; up to 1.50 m²	m²	475.00	to	580.00
painted; double glazed; over 1.50 m², up to 3.20 m²	m²	355.00	to	430.00

Purpose-made windows

painted; double glazed; up to 1.50 m²	m²	680.00	to	820.00
painted; double glazed; over 1.50 m²	m²	610.00	to	730.00

Hardwood windows (U-value = 1.4 W/m² K)
Standard windows; stained

double glazed	m²	1050.00	to	1275.00

Purpose-made windows; stained

double glazed	m²	1275.00	to	1575.00

Steel windows (U-value = 1.6 W/m² K)
Standard windows

double glazed; powder coated	m²	640.00	to	780.00

Purpose-made windows

double glazed; powder coated	m²	890.00	to	1075.00

2.6 WINDOWS AND EXTERNAL DOORS

Item	Unit	Range £		
uPVC windows				
Windows; standard ironmongery; sills and factory glazed with low-e 24 mm double glazing				
standard with low-e 24 mm double glazing	m^2	230.00	to	280.00
WER A rating	m^2	230.00	to	280.00
WER C rating	m^2	230.00	to	280.00
Secured by Design accreditation	m^2	240.00	to	290.00
extra for colour finish to uPVC	m^2	66.00	to	79.00
Composite aluminium/timber windows; U-value = 1.5 W/m^2 K				
Purpose-made windows; stainless steel ironmongery; Velfac System 200 or similar				
fixed windows up to 1.50 m^2	m^2	300.00	to	360.00
fixed windows over 1.50 m^2 up to 4.00 m^2	m^2	265.00	to	320.00
outward opening pivot windows up to 1.50 m^2	m^2	730.00	to	890.00
outward opening pivot windows over 1.50 m^2 up to 4.00 m^2	m^2	320.00	to	390.00
round porthole window 1.40 m dia.	nr	1125.00	to	1350.00
round porthole window 1.80 m dia.	nr	170.00	to	205.00
round porthole window 2.60 m dia.	nr	3250.00	to	3950.00
purpose-made entrance screens and doors double glazed	m^2	1375.00	to	1650.00
2.6.2 External Doors				
Softwood external doors				
Standard external softwood doors and hardwood frames; doors painted; including ironmongery				
Matchboarded, framed, ledged and braced door, 838 mm × 1981 mm	nr	580.00	to	700.00
flush door; cellular core; plywood faced; 838 mm × 1981 mm	nr	590.00	to	720.00
Heavy duty solid flush door				
single leaf	nr	1275.00	to	1550.00
double leaf	nr	2100.00	to	2550.00
single leaf; emergency fire exit	nr	1750.00	to	2125.00
double leaf; emergency fire exit	nr	2600.00	to	3100.00
Steel external doors				
Standard doors				
single external steel door, including frame, ironmongery, powder coated finish	nr	1025.00	to	1250.00
single external steel security door, including frame, ironmongery, powder coated finish	nr	2100.00	to	2550.00
double external steel security door, including frame, ironmongery, powder coated finish	nr	3100.00	to	3800.00
Overhead doors				
electric operation standard lift, 42 mm thick insulated sandwich panels	m^2	230.00	to	280.00
rapid lift fabric door, external, electric operation	m^2	990.00	to	1175.00

2.6 WINDOWS AND EXTERNAL DOORS

Item	Unit	Range £		
2.6 WINDOWS AND EXTERNAL DOORS – cont				
Steel external doors – cont				
Dock shelters				
curtain mechanical shelter; extruded aluminium frame; spring loaded-frame connected with parallel bracing arms; 2 side curtains, one top curtain; double-layered high quality polyester, coated on both sides, longitudinal strength: approx. 750 N, transverse strength	n	1275.00	to	1475.00
Inflatable mechanical shelter; hot dipped galvanized surface treatment, polyester painted, top bag with polyester fabric panels; side bags with polyester fabric panels, strong blower for straight movement of top and side bags; side section with steel guards; colour from standard range	n	4000.00	to	4650.00
uPVC external doors				
Entrance doors; residential standard; PVC-u frame; brass furniture (spyhole/ security chain/letter plate/draught excluder/multipoint locking)				
overall 900 × 2100 mm half glazed	nr	520.00	to	630.00
overall 900 × 2100 mm half glazed; WER A rated	nr	530.00	to	640.00
overall 900 × 2100 mm half glazed; WER C rated	nr	520.00	to	630.00
overall 900 × 2100 mm half glazed; coloured	nr	590.00	to	720.00
Composite aluminium/timber entrance screens and doors; U-value = 1.5 W/m^2 K				
Purpose-made doors				
glazed single personnel door; stainless steel ironmongery; Velfac System 100 or similar	nr	2150.00	to	2600.00
glazed double personnel door; stainless steel ironmongery; Velfac System 100 or similar	pair	3450.00	to	4200.00
automatic revolving door; 2000 mm dia.; clear laminated glazing; 4 nr wings; glazed curved walls	nr	43000.00	to	52000.00
automatic sliding door; bi-parting	m^2	7600.00	to	9200.00
Stainless steel entrance screens and doors				
Purpose-made screen; double glazed				
with manual doors	m^2	2500.00	to	3050.00
with automatic doors	m^2	2850.00	to	3450.00
Shop fronts, shutters and grilles				
Purpose-made screen				
temporary timber shop fronts	m^2	84.00	to	100.00
hardwood and glass; including high enclosed window beds	m^2	930.00	to	1125.00
flat facade; glass in aluminium framing; manual centre doors only	m^2	800.00	to	970.00
grilles or shutters	m^2	480.00	to	580.00
fire shutters; powers operated	m^2	470.00	to	570.00

2.7 INTERNAL WALLS AND PARTITIONS

Item	Unit	Range £		

Automatic glazed entrance doors

Automatic glazed sliding doors in aluminium; polyester powder coated; linear sliding doors; inner pivoted pocket screens; glazed with safety units in accordance with BS. ASSA ABLOY SL510 P RC2 Anti-Burglar Sliding Door System complete with 100 mm operator providing automated operation of the sliding door leaves

Item	Unit	Range £		
Double bi-parting; opening 2.00 m × 2.30 m	nr	9400.00	to	12000.00

Automatic revolving door, framed. ASSA ABLOY automatic revolving door system providing hands free operation and escape functionality. Modern modular design comprising four collapsible door leaves with electromechanical breakout, framed curved glazed outer walls, radially segmented ceiling and semi-circular dustcovers, complete with operator and imbedded digital control

Item	Unit	Range £		
3 or 4 wing 2100 mm dia., 2200 mm internal height	nr	31000.00	to	39000.00
3 or 4 wing 3000 mm dia., 2200 mm internal height	nr	32500.00	to	41000.00

Automatic revolving door, structural glass. ASSA ABLOY automatic glass revolving door system providing hands free operation and Hi-Tech functionality. Modern modular design comprising four door leaves, curved glazed outer walls, glass ceiling/canopy, complete with in-floor mounted operator and imbedded digital control

Item	Unit	Range £		
3 or 4 wing 1800 mm dia., 2200 mm internal height	nr	45000.00	to	57000.00
3 or 4 wing 3000 mm dia., 2200 mm internal height	nr	49000.00	to	62000.00

2.7 INTERNAL WALLS AND PARTITIONS

2.7.1 Internal Walls and Partitions

Frame and panel partitions

Timber stud partitions

Item	Unit	Range £		
structure only comprising 100 mm × 38 mm softwood studs at 600 mm centres; head and sole plates	m²	27.00	to	33.00
softwood stud comprising 100 mm × 38 mm softwood studs at 600 mm centres; head and sole plates; 12.5 mm thick plasterboard each side; tape and fill joints; emulsion finish	m²	64.00	to	77.00

Metal stud and board partitions; height range from 2.40 m to 3.30 m

Item	Unit	Range £		
73 mm partition; 48 mm studs and channels; one layer of 12.5 mm Gyproc Wallboard each side; joints filled with joint filler and joint tape; emulsion paint finish; softwood skirtings with gloss finish	m²	46.00	to	56.00
102 mm partition; 70 mm studs and channels; one layer of 15 mm Gyproc Wallboard each side; joints filled with joint filler and joint tape; emulsion paint finish; softwood skirtings with gloss finish	m²	49.00	to	60.00
102 mm thick partition; 70 mm steel studs at 600 mm centres generally; 1 layer 15 mm Fireline board each side; joints filled with joint filler and joint tape; emulsion paint finish; softwood skirtings with gloss finish	m²	56.00	to	68.00
130 mm thick partition; 70 mm steel studs at 600 mm centres generally; 2 layers 15 mm Fireline board each side; joints filled with joint filler and joint tape; emulsion paint finish; softwood skirtings with gloss finish	m²	67.00	to	81.00
102 mm thick partition; 70 mm steel studs at 600 mm centres generally; 1 layer 15 mm SoundBloc board each side; joints filled with joint filler and joint tape; emulsion paint finish; softwood skirtings with gloss finish	m²	56.00	to	67.00
130 mm thick partition; 70 mm steel studs at 600 mm centres generally; 2 layers 15 mm SoundBloc board each side; joints filled with joint filler and joint tape; emulsion paint finish; softwood skirtings with gloss finish	m²	74.00	to	90.00

2.7 INTERNAL WALLS AND PARTITIONS

Item	Unit	Range £		
2.7 INTERNAL WALLS AND PARTITIONS – cont				
Frame and panel partitions – cont				
Alternative board finishes				
12 mm plywood boarding	m²	11.40	to	13.80
one coat Thistle board finish 3 mm thick work to walls; plasterboard base	m²	7.30	to	8.85
extra for curved work	%	15.00	to	20.00
Acoustic insulation to partitions				
Mineral fibre quilt; Isover Acoustic Partition Roll (APR 1200) or other equal; pinned vertically to timber or plasterboard				
25 mm thick	m²	2.75	to	3.30
50 mm thick	m²	3.60	to	4.35
Brick partitions				
common brick half brick thick wall; bricks PC £380.00/1000	m²	56.00	to	68.00
Block partitions				
aerated/lightweight block partitions				
100 mm thick	m²	34.50	to	42.00
140 mm thick	m²	47.00	to	57.00
200/215 mm thick	m²	60.00	to	72.00
dense aggregate block walls				
100 mm thick	m²	42.50	to	51.00
140 mm thick	m²	52.00	to	63.00
extra for fair face (rate per side)	m²	1.00	to	1.25
extra for plaster and emulsion (rate per side)	m²	20.50	to	25.00
extra for curved work	%	15.00	to	20.00
Concrete partitions				
Reinforced concrete walls; C25 strength; standard finish; reinforced at 100 kg/m³				
150 mm thick	m²	175.00	to	210.00
225 mm thick	m²	200.00	to	240.00
300 mm thick	m²	220.00	to	270.00
extra for plaster and emulsion (rate per side)	m²	20.50	to	25.00
extra for curved work	%	20.00	to	30.00
Glass				
Hollow glass block walling; cement mortar joints; reinforced with 6 mm dia. stainless steel rods; pointed both sides with mastic or other equal and approved				
240 mm × 240 mm × 80 mm clear blocks	m²	370.00	to	445.00
240 mm × 240 mm × 80 mm craft blocks	m²	370.00	to	445.00
190 mm × 190 mm × 100 mm glass blocks; 30 minute fire-rated	m²	405.00	to	490.00
190 mm × 190 mm × 100 mm glass blocks; 60 minute fire-rated	m²	670.00	to	810.00
2.7.2 Stair/Ramp Balustrades and Handrails				
Wall handrails				
softwood handrail and brackets	m	85.00	to	100.00
hardwood handrail and brackets	m	120.00	to	140.00
mild steel handrail and brackets	m	160.00	to	190.00
stainless steel handrail and brackets	m	200.00	to	240.00

2.7 INTERNAL WALLS AND PARTITIONS

Item	Unit	Range £		
Balustrading and handrails				
mild steel balustrade and steel or timber handrail	m	310.00	to	375.00
balustrade and handrail with metal infill panels	m	390.00	to	475.00
balustrade and handrail with glass infill panels	m	430.00	to	520.00
stainless steel balustrade and handrail	m	510.00	to	620.00
stainless steel and structural glass balustrade	m	900.00	to	1075.00
2.7.3 Moveable Room Dividers				
Demountable/Folding partitions				
Demountable partitioning; aluminium framing; veneer finish doors				
medium quality; 46 mm thick panels factory finish vinyl faced	m²	380.00	to	460.00
high quality; 46 mm thick panels factory finish vinyl faced	m²	560.00	to	680.00
Demountable aluminium/steel partitioning and doors				
high quality; folding	m²	445.00	to	540.00
high quality; sliding	m²	950.00	to	1150.00
Demountable fire partitions				
enamelled steel; half hour	m²	680.00	to	820.00
stainless steel; half hour	m²	1075.00	to	1300.00
soundproof partitions; hardwood doors luxury veneered	m²	310.00	to	370.00
Aluminium internal patent glazing				
single glazed laminated	m²	205.00	to	250.00
double glazed; 1 layer toughened and 1 layer laminated glass	m²	280.00	to	340.00
Stainless steel glazed manual doors and screens				
high quality; to inner lobby of malls	m²	900.00	to	1075.00
Acoustic folding partition; headtrack suspension and bracing; aluminium framed with high density particle board panel with additional acoustic insulation; melamine laminate finish; acoustic seals. Nominal weight approximately 55 kg per m²				
sound reduction 48 db (Rw)	m²	530.00	to	650.00
sound reduction 55 db (Rw)	m²	620.00	to	750.00
2.7.4 Cubicles				
Framed panel cubicles				
Changing and WC cubicles; high pressure laminate faced MDF; proprietary system				
standard quality WC cubicle partition sets; aluminium framing; melamine face chipboard dividing panels and doors; ironmongery; small range (up to 5 cubicles); standard cubicle set; (rate per cubicle)	nr	480.00	to	580.00
medium quality WC cubicle partition sets; stainless steel framing; real wood veneer face chipboard diiding panels and doors; ironmongery; small range (up to 6 cubicles); standard cubicle set; (rate per cubicle)	nr	930.00	to	1125.00
high end specification flush fronted system floor to ceiling 44 mm doors, no visible fixings; real wood veneer or HPL or high Gloss paint and lacquer; cubicle set; 800 mm × 1500 mm × 2400 mm high per cubicle, with satin finished stainless steel ironmongery; 30 mm high pressure laminated (HPL) chipboard divisions and 44 mm solid cored real wood veneered doors and pilasters; small range (up to 6 cubicles); standard cubicle set; (rate per cubicle)	nr	2500.00	to	3000.00

2.8 INTERNAL DOORS

Item	Unit	Range £		

2.7 INTERNAL WALLS AND PARTITIONS – cont

Framed panel cubicles – cont

Item	Unit	Range £		
IPS duct panel systems; melamine finish chipboard; softwood timber subframe 2.70 m high IPS back panelling system; to accommodate wash hand basins or urinals; access hatch; frame support	m²	150.00	to	180.00

2.8 INTERNAL DOORS

2.8.1 Internal Doors
The following rates include for the supply and hang of doors, complete with all frames, architrave, typical medium standard ironmongery set and appropriate finish

Standard doors
Standard doors; cellular core; softwood; softwood architrave; aluminium ironmongery (latch only)

Item	Unit	Range £		
single leaf; moulded panel; gloss paint finish	nr	325.00	to	390.00
single leaf; Sapele veneered finish	nr	360.00	to	435.00

Purpose-made doors
Softwood panelled; softwood lining; softwood architrave; aluminium ironmongery (latch only); brass or stainless ironmongery (latch only); painting and polishing

Item	Unit	Range £		
single leaf; four panels; mouldings	nr	485.00	to	590.00
double leaf; four panels; mouldings	nr	950.00	to	1150.00

Hardwood panelled; hardwood lining; hardwood architrave; aluminium ironmongery (latch only); brass or stainless ironmongery (latch only); painting and polishing

Item	Unit	Range £		
single leaf; four panelled doors; mouldings	nr	1025.00	to	1225.00
double leaf; four panelled doors; mouldings	nr	1950.00	to	2375.00

Fire doors
Standard fire doors; cellular core; softwood lining; softwood architrave; aluminuim ironmongery (lockable, self-closure); painting or polishing;

Item	Unit	Range £		
single leaf; Oak veneered; 30 min fire resistance; polished	nr	530.00	to	640.00
double leaf; Oak veneered; 30 min fire resistance; polished	nr	1200.00	to	1475.00
single leaf; Oak veneered; 60 min fire resistance; polished	nr	820.00	to	990.00
double leaf; Oak veneered; 60 min fire resistance; polished	nr	1525.00	to	1825.00

Ironmongery sets
Stainless steel ironmongery; euro locks; push plates; kick plates; signage; closures; standard sets

Item	Unit	Range £		
office door; non-locking; fire-rated	nr	335.00	to	410.00
office/store; lockable; fire-rated	nr	380.00	to	460.00
classroom door; lockable; fire-rated	nr	500.00	to	610.00
maintenance/plant room door; lockable; fire-rated	nr	360.00	to	435.00
standard bathroom door (unisex)	nr	295.00	to	360.00
accessible toilet door	nr	170.00	to	210.00
fire escape door	nr	1850.00	to	2225.00

3.1 WALL FINISHES

Item	Unit	Range £		
Overhead doors				
single skin; manual	m²	170.00	to	200.00
single skin; electric	m²	305.00	to	370.00
electric operation standard lift, 42 mm thick insulated sandwich panels	m²	230.00	to	280.00
rapid lift fabric door, internal, electric operation	m²	990.00	to	1175.00

3.1 WALL FINISHES

3.1.1 Wall Finishes
Wall area (unless otherwise described)

Item	Unit	Range £		
In situ wall finishes				
Comparative finishes				
one mist and two coats emulsion paint	m²	3.55	to	4.30
two coats of lightweight plaster	m²	15.10	to	18.30
two coats of lightweight plaster with emulsion finish	m²	18.60	to	22.50
two coat sand cement render and emulsion finish	m²	26.00	to	32.00
plaster and vinyl wallpaper coverings	m²	25.00	to	30.00
polished plaster system; Armourcoat or similar; 11 mm thick first coat and 2 mm thick finishing coat with a polished finish	m²	125.00	to	150.00
Rigid tile/panel/board finishes				
Timber boarding/panelling; on and including battens; plugged to wall				
12 mm thick softwood boarding	m²	41.50	to	50.00
hardwood panelling; t&g & v-jointed	m²	125.00	to	150.00
Ceramic wall tiles; including backing				
economical quality	m²	47.50	to	57.00
medium to high quality	m²	58.00	to	70.00
Porcelain; including backing				
porcelain mosaic tiling; walls and floors	m²	89.00	to	110.00
Marble				
Roman Travertine marble wall linings 20 mm thick; polished	m²	325.00	to	395.00
Granite				
Dakota mahogany granite cladding 20 mm thick; polished finish; jointed and pointed in coloured mortar	m²	430.00	to	520.00
Dakota mahogany granite cladding 40 mm thick; polished finish; jointed and pointed in coloured mortar	m²	720.00	to	870.00
Comparative woodwork finishes				
Knot; one coat primer; two undercoats; gloss on wood surfaces; number of coats:				
two coats gloss	m²	11.50	to	13.90
three coats gloss	m²	12.70	to	15.30
three coats gloss; small girth n.e. 300 mm	m	8.45	to	10.20
Polyurethane lacquer				
two coats	m²	8.80	to	10.60
three coats	m²	9.65	to	11.70

3.2 FLOOR FINISHES

Item	Unit	Range £		
3.1 WALL FINISHES – cont				
Comparative woodwork finishes – cont				
Flame-retardant paint				
Unitherm or similar; two coats	m²	19.60	to	23.50
Polish				
wax polish; seal	m²	14.60	to	17.70
wax polish; stain and body in	m²	18.30	to	22.00
French polish; stain and body in	m²	28.50	to	34.50
Other wall finishes				
Wall coverings				
decorated paper backed vinyl wall paper	m²	9.95	to	12.00
PVC wall linings – Altro or similar; standard satin finish	m²	64.00	to	77.00
3.2 FLOOR FINISHES				
3.2.1 Floor Finishes				
Floor area (unless otherwise described)				
In situ screed and floor finishes; laid level; over 300 mm wide				
Cement and sand (1:3) screeds; steel trowelled				
50 mm thick	m²	18.00	to	22.00
75 mm thick	m²	23.00	to	28.00
100 mm thick	m²	25.00	to	30.50
Flowing screeds				
Latex screeds; self-colour; 3 mm to 5 mm thick	m²	11.40	to	13.80
Lafarge Gyvlon flowing screed 50 mm thick	m²	12.30	to	14.90
Treads, steps and the like (small areas)	m²	34.50	to	42.00
Granolithic; laid on green concrete				
20 mm thick	m²	26.00	to	31.50
38 mm thick	m²	30.00	to	37.00
Resin floor finish				
Altrotect; 2 coat application nominally 350–500 micron thick	m²	17.60	to	21.00
AltroFlow EP; 3 part solvent-system; up to 3 mm thick	m²	38.00	to	46.00
Atro Screed (TB Screed; Quartz; Multiscreed), 3 mm to 4 mm thick	m²	50.00	to	60.00
Altrocrete heavy duty polyurethane screed, 6 mm to 8 mm screed	m²	55.00	to	67.00
Sheet/board flooring				
Chipboard				
18 mm to 22 mm thick chipboard flooring; t&g joints	m²	18.20	to	22.00
Softwood				
22 mm thick wrought softwood flooring; 150 mm wide; t&g joints	m²	35.50	to	43.00
softwood skirting, gloss paint finish	m	17.50	to	21.00
MDF skirting, gloss paint finish	m	18.70	to	22.50
Hardwood				
wrought hardwood t&g strip flooring; polished; including fillets	m²	115.00	to	140.00
hardwood skirting, stained finish	m	23.50	to	28.50

3.2 FLOOR FINISHES

Item	Unit	Range £		
Sprung floors				
Taraflex Combisport 85 System with Taraflex Sports M Plus; t&g plywood 22 mm thick on softwood battens and crumb rubber cradles	m²	180.00	to	215.00
sprung composition block flooring (sports), court markings, sanding and sealing	m²	115.00	to	140.00
Rigid tile/slab finishes (includes skirtings; excludes screeds)				
Quarry tile flooring	m²	76.00	to	92.00
Glazed ceramic tiled flooring				
standard plain tiles	m²	49.00	to	60.00
anti-slip tiles	m²	54.00	to	65.00
designer tiles	m²	110.00	to	135.00
Terrazzo tile flooring 28 mm thick polished	m²	55.00	to	66.00
York stone 50 mm thick paving	m²	180.00	to	215.00
Slate tiles, smooth; straight cut	m²	76.00	to	92.00
Portland stone paving	m²	290.00	to	350.00
Roman Travertine marble paving; polished	m²	290.00	to	350.00
Granite paving 20 mm thick paving	m²	440.00	to	530.00
Parquet/wood block wrought hardwood block floorings; 25 mm thick; polished; t&g joints	m²	150.00	to	185.00
Flexible tiling; welded sheet or butt joint tiles; adhesive fixing				
vinyl floor tiling; 2.00 mm thick	m²	16.30	to	19.80
vinyl safety flooring; 2.00–2.50 mm thick	m²	51.00	to	62.00
vinyl safety flooring; 3.5 mm thick heavy duty	m²	59.00	to	71.00
linoleum tile flooring; 333 mm × 333 mm × 3.20 mm tiles	m²	42.00	to	51.00
linoleum sheet flooring; 2.00 mm thick	m²	33.00	to	39.50
rubber studded tile flooring; 500 mm × 500 mm × 2.50 mm thick	m²	46.00	to	55.00
Carpet tiles; including underlay, edge grippers				
heavy domestic duty; to floors; PC Sum £40/m²	m²	68.00	to	82.00
heavy domestic duty; to treads and risers; PC Sum £40/m²	m	53.00	to	64.00
heavy contract duty; Forbo Flotex HD; PC Sum £25/m²	m²	49.00	to	60.00
Entrance matting				
door entrance and circulation matting; soil and water removal	m²	62.00	to	75.00
Entrance matting and matwell				
18 mm thick Gradus Topguard barrier matting with aluminium frame	m²	420.00	to	510.00
18 mm thick Gradus Topguard barrier matting with stainless steel frame	m²	360.00	to	435.00
18 mm thick Gradus Topguard barrier matting with polished brass frame	m²	500.00	to	610.00
3.2.2 Raised Access Floors				
Raised access floors: including 600 mm × 600 mm steel encased particle boards on height adjustable pedestals < 300 mm				
light grade duty	m²	51.00	to	61.00
medium grade duty	m²	51.00	to	61.00
heavy grade duty	m²	68.00	to	82.00
battened raft chipboard floor with sound insulation fixed to battens; medium quality carpeting	m²	100.00	to	125.00
Common floor coverings bonded to access floor panels				
anti-static vinyl	m²	32.00	to	38.50
heavy duty fully flexible vinyl tiles	m²	26.50	to	32.00
needle punch carpet	m²	19.40	to	23.50

3.3 CEILING FINISHES

Item	Unit	Range £		
3.3 CEILING FINISHES				
3.3.1 Finishes to Ceiling				
Ceiling area (unless otherwise described)				
In situ finishes				
Decoration only to soffits; one mist and two coats emulsion paint				
to exposed steelwork (surface area)	m²	7.50	to	9.10
to concrete soffits (surface area)	m²	4.20	to	5.05
to plaster/plasterboard	m²	3.80	to	4.60
Plaster to soffits				
plaster skim coat to plasterboard ceilings	m²	7.60	to	9.15
lightweight plaster to concrete	m²	20.00	to	24.00
Plasterboard to soffits				
9 mm Gyproc board and skim coat	m²	31.00	to	37.50
12.50 mm Gyproc board and skim coat	m²	31.50	to	38.00
Other board finishes; with fire-resisting properties; excluding decoration				
12.50 mm thick Gyproc Fireline board	m²	24.00	to	29.00
15 mm thick Gyproc Fireline board	m²	25.00	to	30.00
12 mm thick Supalux	m²	60.00	to	73.00
Specialist plasters; to soffits				
sprayed acoustic plaster; self-finished	m²	86.00	to	105.00
rendering; Tyrolean finish	m²	86.00	to	105.00
Other ceiling finishes				
softwood timber t&g boarding	m	57.00	to	75.00
Comparative ceiling finishes				
Emulsion paint to plaster				
two coats	m²	4.30	to	5.20
one mist and two coats	m²	4.40	to	5.35
Gloss paint to timber or metal				
primer and two coats	m²	8.15	to	9.85
primer and three coats	m²	12.00	to	14.50
Other coatings				
Artex plastic compound one coat; textured to plasterboard	m²	8.40	to	10.20
3.3.3 Demountable Suspended Ceilings				
Armstrong suspended ceiling; assume large rooms over 250 m²				
mineral fibre; basic range; Cortega, Tatra, Academy; exposed grid	m²	22.50	to	29.50
mineral fibre; medium quality; Corline; exposed grid	m²	32.50	to	43.00
mineral fibre; medium quality; Corline; concealed grid	m²	48.50	to	64.00
mineral fibre; medium quality; Corline; silhouette grid	m²	61.00	to	80.00
mineral fibre; Specific; Bioguard; exposed grid	m²	27.00	to	35.50
mineral fibre; Specific; Bioguard; clean room grid	m²	42.50	to	56.00
mineral fibre; Specific; Hygiene, clean room; exposed grid	m²	58.00	to	77.00
metal open cell; Cellio Global White; exposed grid	m²	55.00	to	72.00
metal open cell; Cellio Black; exposed grid	m²	66.00	to	87.00
wood veneer board; Microlook 8; plain; exposed grid	m²	72.00	to	95.00
wood veneer; plain; exposed grid	m²	99.00	to	130.00
wood veneers; perforated; exposed grid	m²	170.00	to	225.00
wood veneers; plain; concealed grid	m²	160.00	to	215.00
wood veneers; perforated; concealed grid	m²	225.00	to	295.00

4.1 FITTINGS, FURNISHINGS AND EQUIPMENT

Item	Unit	Range £		
Other suspended ceilings				
perforated aluminium ceiling tiles 600 mm × 600 mm	m²	37.50	to	45.50
perforated aluminium ceiling tiles 1200 mm × 600 mm	m²	45.00	to	55.00
metal linear strip; Dampa/Luxalon	m²	57.00	to	69.00
metal linear strip micro perforated acoustic ceiling with Rockwool acoustic infill	m²	70.00	to	85.00
metal tray	m²	60.00	to	72.00
egg-crate	m²	92.00	to	110.00
open grid; Formalux/Dimension	m²	120.00	to	145.00
Integrated ceilings				
coffered; with steel surfaces	m²	160.00	to	190.00
acoustic suspended ceilings on anti-vibration mountings	m²	73.00	to	88.00

4.1 FITTINGS, FURNISHINGS AND EQUIPMENT

4.1.1 General Fittings, Furnishings and Equipment

Office furniture and equipment
There is a large quality variation for office furniture and we have assumed a medium level, even so prices will vary between suppliers
Reception desk

straight counter; 3500 mm long; 2 person	nr	2375.00	to	3000.00
curved counter; 3500 mm long; 2 person	nr	5700.00	to	7200.00
curved counter; 3500 mm long; 2 person; real wood veneer finish	nr	11500.00	to	14500.00

Furniture and equipment to general office area; standard off the shelf specification

workstation; 2000 mm long desk; drawer unit; task chair	nr	910.00	to	1150.00

4.1.2 Domestic Kitchen Fittings and Equipment

Residential fittings (volume housing)
Kitchen fittings for residential units (not including white goods). NB quality and quantity of units can vary enormously. Always obtain costs from your preferred supplier. The following assume medium standard units from a large manufacturer and includes all units, worktops, stainless steel sink and taps, not including white goods

one person flat	nr	2350.00	to	2950.00
two person flat/house	nr	2600.00	to	3250.00
three person house	nr	4300.00	to	5400.00
four person house; includes utility area	nr	8000.00	to	10000.00
five person house; includes utility area	nr	11000.00	to	14000.00

Hotel bathroom pods
Fully fitted out, finished and furnished bathroom pods; installed; suitable for business class hotels

standard pod (4.50 m plan area)	nr	5700.00	to	7200.00
accessible pod (4.50 m plan area)	nr	7700.00	to	9800.00

5.1 SANITARY INSTALLATIONS

Item	Unit	Range £		
4.1 FITTINGS, FURNISHINGS AND EQUIPMENT – cont				
4.1.3 Special Purpose Fittings, Furnishings and Equipment				
Window blinds				
Louvre blind				
89 mm louvres, manual chain operation, fixed to masonry	m²	61.00	to	77.00
127 mm louvres, manual chain operation, fixed to masonry	m²	53.00	to	67.00
Roller blind				
fabric blinds, roller type, manual chain operation, fixed to masonry	m²	47.00	to	59.00
solar black out blinds, roller type, manual chain operation, fixed to masonry	m²	81.00	to	100.00
Fire curtains				
Electrically operated automatic 2 hr fire curtains to form a virtually continuous barrier against both fire and smoke; size in excess of 4 m² (minimum size 800 mm)	m²	1725.00	to	2175.00
Dock levellers and shelters				
ASSA ABLOY Entrance Systems				
curtain mechanical shelter; 2 side curtains, one top curtain	nr	1350.00	to	1675.00
teledock leveller; electrohydraulic operation with two lifting cylinders; one lip ram cylinder; movable telescopic lip; Control for door and leveller from one communal control panel, including interlocking of leveller/door	nr	4000.00	to	5000.00
inflatable mechanical shelter; top bag with polyester fabric panels, 1000 mm extension; side bags with polyester fabric panels, 650 mm extension, strong blower for straight movement of top and side bags; side section with steel guards; colour from standard range	nr	4200.00	to	5300.00
Stand-alone Load House; a complete loading unit attached externally to building; modular steel skin/insulated cladding; two wall elements, one roof element	nr	6700.00	to	8500.00
5.1 SANITARY INSTALLATIONS				
5.1.1 Sanitary Appliances				
Please refer to *Spon's Mechanical and Electrical Services Price Book* for a more comprehensive range of rates				
Rates are for gross internal floor area – GIFA unless otherwise described				
shopping malls and the like (not tenant fit-out works)	m²	1.10	to	1.50
office building – multi-storey	m²	9.90	to	13.80
office building – business park	m²	8.70	to	12.10
performing arts building (medium specification)	m²	16.00	to	22.00
sports hall	m²	12.70	to	17.60
hotel	m²	42.50	to	59.00
hospital; private	m²	25.00	to	35.00
school; secondary	m²	12.20	to	16.90
residential; multi-storey tower	m²	0.65	to	0.90
supermarket	m²	1.60	to	2.25

5.6 SPACE HEATING AND AIR CONDITIONING

Item	Unit	Range £		
Comparative sanitary fittings/sundries				
Note: Material prices vary considerably, the following composite rates are				
based on average prices for mid priced fittings				
Individual appliances (including fittings)				
low level WCs; vitreous china pan and cistern; black plastic seat; low				
pressure ball valve; plastic flush pipe; fixing brackets	nr	390.00	to	500.00
bowl type wall urinal; white glazed vitreous china flushing cistern; chromium				
plated flush pipes and spreaders; fixing brackets	nr	450.00	to	580.00
sink; glazed fireclay; chromium plated waste; plug and chain	nr	630.00	to	810.00
sink; stainless steel; chromium plated waste; plug and chain; single drainer;				
double bowl (bowl and half)	nr	375.00	to	480.00
sink; stainless steel; chromium plated waste; plug and chain; double drainer;				
double bowl (bowl and half)	nr	400.00	to	510.00
bath; reinforced acrylic; chromium plated taps; overflow; waste; chain and				
plug; P trap and overflow connections	nr	395.00	to	500.00
bath; enamelled steel; chromium plated taps; overflow; waste; chain and				
plug; P trap and overflow connections	nr	850.00	to	1075.00
shower tray; glazed fireclay; chromium plated waste; riser pipe; rose and				
mixing valve	nr	710.00	to	910.00
5.4 WATER INSTALLATIONS				
5.4.1 Mains Water Supply				
Please refer to *Spon's Mechanical and Electrical Services Price Book* for a				
more comprehensive range of rates				
Hot and cold water installations; mains supply; hot and cold water distribution.				
Gross internal floor area (unless described other wise)				
shopping malls and the like (not tenant fit-out works)	m²	13.60	to	18.90
office building – multi-storey	m²	20.00	to	28.00
office building – business park	m²	14.90	to	21.00
performing arts building (medium specification)	m²	23.50	to	32.50
sports hall	m²	17.80	to	24.50
hotel	m²	50.00	to	70.00
hospital; private	m²	60.00	to	83.00
school; secondary; potable and non potable to labs, art rooms	m²	46.00	to	64.00
residential; multi-storey tower	m²	56.00	to	78.00
supermarket	m²	31.00	to	43.50
distribution centre	m²	2.15	to	3.00
5.6 SPACE HEATING AND AIR CONDITIONING				
5.6.1 Space Heating and Air Conditioning				
Please refer to *Spon's Mechanical and Electrical Services Price Book* for a				
more comprehensive range of rates				
Gross internal floor area (unless described otherwise)				
shopping malls and the like (not tenant fit-out works); LTHW, air conditioning,				
ventilation	m²	90.00	to	120.00
office building – multi-storey; LTHW, ductwork, chilled water, ductwork	m²	125.00	to	165.00
office building – business park; LTHW, ductwork, chilled water, ductwork	m²	145.00	to	190.00

5.8 ELECTRICAL INSTALLATIONS

Item	Unit	Range £		

5.6 SPACE HEATING AND AIR CONDITIONING – cont

5.6.1 Space Heating and Air Conditioning – cont
Gross internal floor area (unless described otherwise) – cont

Item	Unit		Range £	
performing arts building (medium specification); LTHW, ductwork, chilled water, ductwork	m²	250.00	to	330.00
sports hall; warm air to main hall, LTHW radiators to ancillary	m²	44.00	to	58.00
hotel; air conditioning	m²	170.00	to	220.00
hospital; private; LPHW	m²	250.00	to	330.00
school; secondary; LTHW, cooling to ICT server room	m²	140.00	to	180.00
residential; multi-storey tower; LTHW to each apartment	m²	76.00	to	99.00
supermarket	m²	18.90	to	25.00
distribution centre; LTHW to offices, displacement system to warehouse	m²	26.00	to	35.00

5.7 VENTILATION SYSTEMS

5.7.1 Ventilation Systems
Please refer to *Spon's Mechanical and Electrical Services Price Book* for a more comprehensive range of rates
Gross internal area (unless otherwise described)

Item	Unit		Range £	
shopping malls and the like (not tenant fit-out works)	m²	43.50	to	57.00
office building – multi-storey; toilet and kitchen areas only	m²	7.45	to	9.80
office building – business park; toilet and kitchen areas only	m²	8.65	to	11.40
performing arts building (medium specification); toilet and kitchen areas, workshop	m²	17.30	to	23.00
sports hall	m²	16.10	to	21.00
hotel; general toilet extraction, bathrooms, kitchens	m²	50.00	to	66.00
hospital; private; toilet and kitchen areas only	m²	9.30	to	12.20
school; secondary; toilet and kitchen areas, science labs	m²	21.00	to	28.00
residential; multi-storey tower	m²	43.50	to	57.00
supermarket	m²	6.30	to	8.25
distribution centre; smoke extract system	m²	6.30	to	8.25

5.8 ELECTRICAL INSTALLATIONS

5.8.1 Electric Mains and Sub-mains Distrubition
Please refer to *Spon's Mechanical and Electrical Services Price Book* for a more comprehensive range of rates
Including LV distribution, HV distribution, lighting, small power. Gross internal floor area (unless described other wise)

Item	Unit		Range £	
shopping malls and the like (not tenant fit-out works)	m²	135.00	to	190.00
office building – multi-storey	m²	120.00	to	160.00
office building – business park	m²	62.00	to	86.00
performing arts building (medium specification)	m²	220.00	to	310.00
sports hall	m²	68.00	to	95.00
hotel	m²	210.00	to	295.00
hospital; private	m²	250.00	to	350.00
school; secondary	m²	150.00	to	210.00
residential; multi-storey tower	m²	99.00	to	140.00
supermarket	m²	80.00	to	110.00
distribution centre	m²	75.00	to	100.00

5.10 LIFT AND CONVEYOR INSTALLATIONS

Item	Unit	Range £		
Comparative fittings/rates per point				
Consumer control unit; 63–100 amp 230 volt; switched and insulated; RCDB protection. Gross internal floor area (unless described other wise)	nr	360.00	to	450.00
Fittings; excluding lamps or light fittings				
lighting point; PVC cables	nr	54.00	to	68.00
lighting point; PVC cables in screwed conduits	nr	61.00	to	77.00
lighting point; MICC cables	nr	80.00	to	100.00
Switch socket outlet; PVC cables				
single	nr	67.00	to	85.00
double	nr	88.00	to	110.00
Switch socket outlet; PVC cables in screwed conduit				
single	nr	100.00	to	125.00
double	nr	120.00	to	150.00
Switch socket outlet; MICC cables				
single	nr	96.00	to	120.00
double	nr	110.00	to	140.00
Other power outlets				
Immersion heater point (excluding heater)	nr	120.00	to	155.00
Cooker point; including control unit	nr	200.00	to	260.00

5.9 FUEL INSTALLATIONS

5.9.2 Fuel Distribution Systems

Please refer to *Spon's Mechanical and Electrical Services Price Book* for a more comprehensive range of rates

Item	Unit	Range £		
Gas mains service to plant room. Gross internal floor area				
shopping mall/supermarket	m²	3.70	to	4.70
warehouse/distribution centre	m²	1.05	to	1.30
office/hotel	m²	1.75	to	2.20

5.10 LIFT AND CONVEYOR INSTALLATIONS

5.10.1 Lifts and Enclosed Hoists

Please refer to *Spon's Mechanical and Electrical Services Price Book* for a more comprehensive range of rates

Item	Unit	Range £		
Passenger lifts				
Passenger lifts (standard brushed stainless steel finish; 2 panel centre opening doors)				
8-person; 4 stops; 1.0 m/s speed	nr	89000.00	to	110000.00
8-person; 4 stops; 1.6 m/s speed	nr	94000.00	to	110000.00
13-person; 6 stops; 1.0 m/s speed	nr	110000.00	to	135000.00
13-person; 6 stops; 1.6 m/s speed	nr	120000.00	to	145000.00
13-person; 6 stops; 2.0 m/s speed	nr	120000.00	to	150000.00
21-person; 8 stops; 1.0 m/s speed	nr	140000.00	to	170000.00
21-person; 8 stops; 1.6 m/s speed	nr	160000.00	to	190000.00
21-person; 8 stops; 2.0 m/s speed	nr	160000.00	to	195000.00
extra for enhanced finishes; mirror; carpet	nr	4400.00	to	5300.00
extra for lift car LCD TV	nr	9500.00	to	11500.00
extra for intelligent group control; 5 cars; 11 stops	nr	45500.00	to	55000.00

5.11 FIRE AND LIGHTNING PROTECTION

Item	Unit	Range £		

5.10 LIFT AND CONVEYOR INSTALLATIONS – cont

Passenger lifts – cont
Special installations

Item	Unit	Range £		
wall climber lift; 10-person; 0.50 m/s; 10 levels	nr	445000.00	to	540000.00
disabled platform lift single wheelchair; 400 kg; 4 stops; 0.16 m/s	nr	11000.00	to	13000.00
Non-passenger lifts				
Goods lifts; prime coated internal finish				
2000 kg load; 4 stops; 1.0 m/s speed	nr	110000.00	to	135000.00
2000 kg load; 4 stops; 1.6 m/s speed	nr	120000.00	to	150000.00
2500 kg load; 4 stops; 1.0 m/s speed	nr	120000.00	to	145000.00
2000 kg load; 8 stops; 1.0 m/s speed	nr	150000.00	to	185000.00
2000 kg load; 8 stops; 1.6 m/s speed	nr	160000.00	to	190000.00
Other goods lifts				
hoist	nr	31000.00	to	37000.00
kitchen service hoist 50 kg; 2 levels	nr	13000.00	to	16000.00
5.10.2 Escalators				
30° escalator; 0.50 m/s; enamelled steel glass balustrades				
3.50 m rise; 800 mm step width	nr	97000.00	to	120000.00
4.60 m rise; 800 mm step width	nr	105000.00	to	130000.00
5.20 m rise; 800 mm step width	nr	110000.00	to	130000.00
6.00 m rise; 800 mm step width	nr	120000.00	to	145000.00
extra for enhanced finish; enamelled finish; glass balustrade	nr	13000.00	to	15500.00
5.10.6 Dock Levellers and Scissor Lifts				
Electrohydraulic operation with two lifting cylinders; integrated frame to suit various pit applications; Control for door and leveller from one communal control pane				
2500 mm long dock leveller with swing lip	nr	3450.00	to	4200.00
2500 mm long dock leveller with telescope lip	nr	3700.00	to	4450.00
5.11 FIRE AND LIGHTNING PROTECTION				
5.11.1 Fire Fighting Systems				
Please refer to *Spon's Mechanical and Electrical Services Price Book* for a more comprehensive range of rates				
Gross internal floor area (unless described otherwise); Including lightning protection, sprinklers unless stated otherwise				
shopping malls and the like (not tenant fit-out works)	m²	16.60	to	23.00
office building – multi-storey	m²	35.50	to	49.00
office building – business park	m²	5.90	to	8.20
performing arts building (medium specification); lightning protection only	m²	1.80	to	2.45
sports hall; lightning protection only	m²	3.55	to	4.90
hospital; private; lightning protection only	m²	2.30	to	3.25
school; secondary; lightning protection only	m²	2.20	to	3.10
hotel	m²	40.50	to	56.00
residential; multi-storey tower	m²	59.00	to	82.00
supermarket; lightning protection only	m²	1.15	to	1.60
distribution centre	m²	45.50	to	63.00

5.14 BUILDER'S WORK IN CONNECTION WITH SERVICES

Item	Unit	Range £		
5.12 COMMUNICATION AND SECURITY INSTALLATIONS				
5.12.1 Communication, Security and Control Systems				
Please refer to *Spon's Mechanical and Electrical Services Price Book* for a more comprehensive range of rates				
Gross internal floor area (unless described otherwise); Including fire alarm, public address system, security installation unless stated otherwise				
shopping malls and the like (not tenant fit-out works)	m²	48.00	to	64.00
office building – multi-storey	m²	35.00	to	46.00
office building – business park; fire alarm and wireways only	m²	16.90	to	22.00
performing arts building (medium specification)	m²	105.00	to	140.00
sports hall	m²	70.00	to	92.00
hotel	m²	120.00	to	155.00
hospital; private	m²	90.00	to	120.00
school; secondary	m²	67.00	to	89.00
residential; multi-storey tower	m²	67.00	to	89.00
supermarket	m²	21.00	to	27.00
distribution centre	m²	25.00	to	32.50
5.13 SPECIAL INSTALLATIONS				
5.13.3 Specialist Mechanical Installations				
Window cleaning equipment				
twin track	m	195.00	to	260.00
manual trolley/cradle	nr	13500.00	to	18000.00
automatic trolley/cradle	nr	27500.00	to	36500.00
Sauna				
2.20 m × 2.20 m internal Finnish sauna; benching; heater; made of Sauna grade Aspen or similar	nr	10500.00	to	14000.00
Jacuzzi Installation				
2.20 m × 2.20 m × 0.95 m; 5 adults; 99 jets; lights; pump	nr	14000.00	to	19000.00
5.13.4 Specialist Electrical/Electronic Installations				
Solar PV systems				
Residential solar water heating including collectors; dual coil cylinders; pump; controller (NB excludes any grant allowance)	m²	970.00	to	1275.00
Solar power including 4 kWp monocrystalline solar modules (12 × 185 Wp) on roof mounting kit; certified inverter; DC and AC isolation switches and connection to the grid and certification	nr	8500.00	to	11000.00
5.14 BUILDER'S WORK IN CONNECTION WITH SERVICES				
5.14.1 Builder's Work in Connection with Services				
Please refer to *Spon's Mechanical and Electrical Services Price Book* for a more comprehensive range of rates				
Warehouses, sports halls and shopping malls. Gross internal floor area				
main supplies, lighting and power to landlord areas	m²	3.90	to	4.40
central heating and electrical installation	m²	11.80	to	13.40
central heating, electrical and lift installation	m²	13.40	to	15.20
air conditioning, electrical and ventilation installations	m²	30.00	to	34.00

8.2 ROADS, PATHS, PAVINGS AND SURFACINGS

Item	Unit	Range £		
5.14 BUILDER'S WORK IN CONNECTION WITH SERVICES – cont				
5.14.1 Builder's Work in Connection with Services – cont				
Offices and hotels. Gross internal floor area				
main supplies, lighting and power to landlord areas	m²	11.40	to	15.20
central heating and electrical installation	m²	16.60	to	22.00
central heating, electrical and lift installation	m²	19.70	to	26.00
air conditioning, electrical and ventilation installations	m²	36.00	to	48.50
8.2 ROADS, PATHS, PAVINGS AND SURFACINGS				
8.2.1 Roads, Paths and Pavings				
Paved areas				
Gravel paving rolled to falls and chambers paving on subbase; including excavation	m²	18.90	to	24.00
Resin bound paving				
16 mm–24 mm deep of natural gravel	m²	73.00	to	93.00
16 mm–24 mm deep of crushed rock	m²	74.00	to	94.00
16 mm–24 mm deep of marble chips	m²	80.00	to	100.00
Tarmacadam paving				
two layers; limestone or igneous chipping finish paving on subbase; including excavation and type 1 subbase	m²	96.00	to	120.00
Slab paving				
precast concrete paving slabs on subbase; including excavation	m²	68.00	to	85.00
precast concrete tactile paving slabs on subbase; including excavation	m²	130.00	to	160.00
York stone slab paving on subbase; including excavation	m²	170.00	to	220.00
imitation York stone slab paving on subbase; including excavation	m²	97.00	to	120.00
Brick/Block/Setts paving				
brick paviors on subbase; including excavation	m²	89.00	to	110.00
precast concrete block paviors to footways including excavation; 150 mm hardcore subbase with dry sand joints	m²	100.00	to	130.00
granite setts on 100 mm thick concrete subbase; including excavation; pointing mortar joints	m²	190.00	to	240.00
cobblestone paving cobblestones on 100 mm thick concrete subbase; including excavation; pointing mortar joints	m²	190.00	to	245.00
Reinforced grass construction				
Grasscrete or similar on 200 mm type 1 subbase; topsoil spread across units and grass seeded upon completion	m²	81.00	to	100.00
Car parking alternatives				
Surface parking; include drains, kerbs, lighting				
surface level parking	m²	110.00	to	140.00
surface car parking with landscape areas	m²	140.00	to	180.00
Rates per car				
surface level parking	car	2450.00	to	3100.00
surface car parking with landscape areas	car	3100.00	to	3950.00

8.2 ROADS, PATHS, PAVINGS AND SURFACINGS

Item	Unit	Range £		
Other surfacing options				
permeable concrete block paving; 80 mm thick Formpave Aquaflow or equal; maximum gradient 5%; laid on 50 mm sand/gravel; subbase 365 mm thick clean crushed angular non plastic aggregate, geotextiles filter layer to top and bottom of subbases	m²	59.00	to	75.00
All purpose roads				
Tarmacadam or reinforced concrete roads, including all earthworks, drainage, pavements, lighting, signs, fencing and safety barriers				
7.30 m wide two lane carriageway	m	3650.00	to	4600.00
10.00 m wide two lane carriageway	m	4500.00	to	5700.00
dual two lane road 7.30 m wide carriageway	m	7300.00	to	9300.00
dual three lane road 11.00 m wide carriageway	m	11500.00	to	14500.00
Road crossings				
Note: Costs include road markings, beacons, lights, signs, advance danger signs etc.				
Zebra crossing	nr	22000.00	to	27500.00
Pelican crossing	nr	38000.00	to	48000.00
Underpass				
Provision of underpasses to new roads, constructed as part of a road building programme				
Precast concrete pedestrian underpass				
3.00 m wide × 2.50 m high	m	6300.00	to	8000.00
Precast concrete vehicle underpass				
7.00 m wide × 5.00 m high	m	28000.00	to	35500.00
14.00 m wide × 5.00 m high	m	66000.00	to	83000.00
8.2.2 Special Surfacing and Pavings Sports Grounds				
Pitch plateau construction and pitch drainage not included				
Hockey and support soccer training (no studs)				
sand filled polypropelene synthetic artificial pitch to FIH national standard. Notts Pad XC FIH shockpad; sport needlepunched polypropelene sand filled synthetic multi-sport carpet; white lining; 3 m high perimeter fence	m²	33.00	to	43.00
FIFA One Star Football and IRB reg 22 Rugby Union				
Lano Rugby Max 60 mm monofilament laid on shockpad on a dynamic base infill with 32 kg/m² sand and 11 kg/m² SBR rubber – IRB Clause 22 compliant; supply and spread to required infill rates, 2EW sand and 0.5–2 mm SBR Rubber. Sand (32 kg/m²) and rubber (11 kg/m²) spread and brushed into surface; white lining; 3.5 m high perimeter fence	m²	27.00	to	36.00
Tennis and netball court				
lay final layer on existing subbase of 50 mm porous stone followed by 2 coats of porous bitumen macadam. 40 mm binder course; 14 mm aggregate and 20 mm surface course with 6 mm aggregate. Laid to fine level tolerances and laser leveled, +/- 6 mm tolerance over 3 m; white lining; 3 m high perimeter fencing	m²	44.00	to	58.00
Natural grass winter sports				
strip topsoil and level existing surface (cut and fill); laser grade formation and proof roll. Load and spread topsoil, grade and consolidate; cultivate and grade topsoil. Supply and spread 25 mm approved sand and mix into topsoil by power harrow; apply pre-seeding fertilizer and sow winter rye sports grass	m²	9.85	to	13.00

Item	Unit	Range £		

8.3 SOFT LANDSCAPING, PLANTING AND IRRIGATION SYSTEMS

8.3.1 Seeding and Turfing
Please refer to *Spon's External Works and Landscape Price Book* for a more comprehensive range of rates
Surface treatment

Item	Unit	Range £		
spread and lightly consolidate top soil from spoil heap 150 mm thick; by machine	m²	2.40	to	2.90
spread and lightly consolidate top soil from spoil heap 150 mm thick; by hand	m²	8.30	to	10.10
Plant supply, planting, maintenance and 12 months guarantee				
seeded areas	m²	5.30	to	6.45
turfed areas	m²	7.10	to	8.60
8.3.2 External Planting				
Planted areas (per m² of planted area)				
herbaceous plants	m²	6.45	to	7.80
climbing plants	m²	10.20	to	12.30
general planting	m²	24.50	to	30.00
woodland	m²	37.00	to	44.50
shrubbed planting	m²	62.00	to	76.00
dense planting	m²	63.00	to	77.00
shrubbed area including allowance for small trees	m²	81.00	to	98.00
Trees				
light standard bare root tree (PC £10)	nr	69.00	to	84.00
standard root balled tree (PC 25)	nr	82.00	to	99.00
heavy standard root ball tree (PC £40)	nr	150.00	to	185.00
semi mature root balled tree (PC £130)	nr	480.00	to	590.00
Parklands				
Note: Work on parklands will involve different techniques of earth shifting and cultivation. The following rates include for normal surface excavation, they include for the provision of any land drainage.				
parklands, including cultivating ground, applying fertilizer, etc. and seeding with parks type grass	ha	24000.00	to	29000.00
Lakes including excavation average 1.0 m deep, laying 1000 micron sheet with welded joints; spreading top soil evenly on top 200 mm deep				
regular shaped lake	1000 m²	27500.00	to	33500.00
8.4 FENCING, RAILINGS AND WALLS				
8.4.1 Fencing and Rails				
Chain link fencing; plastic coated				
1.20 m high	m	24.00	to	29.00
1.80 m high	m	31.50	to	38.00
Timber fencing				
1.20 m high chestnut pale facing	m	24.00	to	29.00
1.80 m high cross-boarded fencing	m	32.50	to	39.00

8.4 FENCING, RAILINGS AND WALLS

Item	Unit	Range £		
8.4.2 Walls and Screens				
Screen walls; one brick thick; including foundations etc.				
1.80 m high facing brick screen wall	m	330.00	to	400.00
1.80 m high smooth faced masonry block boundary wall	m	210.00	to	250.00
8.4.3 Retaining Walls				
Crib retaining walls				
Permacrib timber crib retaining walls on concrete foundation; granular infill material; measure face area m². Excludes backfill material to the rear of the wall				
wall heights up to 2.0 m	m²	210.00	to	255.00
wall heights up to 3.0 m	m²	220.00	to	270.00
wall heights up to 4.0 m	m²	240.00	to	290.00
wall heights up to 5.0 m	m²	250.00	to	300.00
wall heights up to 6.0 m	m²	260.00	to	315.00
Andacrib concrete crib retaining walls on concrete foundation; granular infill material; measure face area m². Excludes backfill material to the rear of the wall				
wall heights up to 2.0 m	m²	225.00	to	270.00
wall heights up to 3.0 m	m²	240.00	to	290.00
wall heights up to 4.0 m	m²	250.00	to	300.00
wall heights up to 5.0 m	m²	265.00	to	320.00
wall heights up to 6.0 m	m²	280.00	to	335.00
Textomur green faced reinforced soil system; geogrid/geotextile reinforcement and compaction of reinforced soil mass.				
wall heights up to 3.0 m	m²	150.00	to	180.00
wall heights up to 4.0 m	m²	160.00	to	190.00
wall heights up to 5.0 m	m²	165.00	to	200.00
wall heights up to 6.0 m	m²	170.00	to	210.00
Titan/Geolock vertical modular block reinforced soil system; steel ladder. Geogrid reinforcement and soil fill material				
wall heights up to 3.0 m	m²	245.00	to	300.00
wall heights up to 4.0 m	m²	260.00	to	320.00
wall heights up to 5.0 m	m²	275.00	to	330.00
wall heights up to 6.0 m	m²	285.00	to	345.00
8.4.4 Barriers and Guardrails				
Post and rail fencing				
open metal post and rail fencing 1.00 m high	m	170.00	to	210.00
galvanized steel post and rail fencing 2.00 m high	m	200.00	to	240.00
Guard rails				
steel guard rails and vehicle barriers	m	68.00	to	83.00
Bollards and barriers				
parking bollards precast concrete or steel	nr	180.00	to	220.00
vehicle control barrier; manual pole	nr	1025.00	to	1250.00

Approximate Estimating Rates

8.6 EXTERNAL DRAINAGE

Item	Unit	Range £		
8.5 EXTERNAL FIXTURES				
8.5.1 Site/Street Furniture and Equipment				
Please refer to *Spon's External Works and Landscape Price Book* for a more comprehensive range of rates				
Street furniture				
Roadsigns				
reflected traffic signs 0.25 m² area on steel post	nr	140.00	to	170.00
internally illuminated traffic signs; dependent on area	nr	225.00	to	270.00
externally illuminated traffic signs; dependent on area	nr	880.00	to	1050.00
Benches; bolted to ground				
benches – hardwood or precast concrete	nr	1150.00	to	1375.00
Litter bins; bolted to ground				
precast concrete	nr	210.00	to	255.00
hardwood slatted	nr	215.00	to	260.00
cast iron	nr	430.00	to	520.00
large aluminium	nr	650.00	to	780.00
Bus stops including basic shelter	nr	2700.00	to	3250.00
Pillar box	nr	650.00	to	780.00
Galvanized steel cycle stand	nr	53.00	to	65.00
Galvanized steel flag staff	nr	1275.00	to	1550.00
Playground equipment				
Modern swings with flat rubber safety seats: four seats; two bays	nr	1625.00	to	1950.00
Stainless steel slide, 3.40 m long	nr	1875.00	to	2275.00
Climbing frame – igloo type 3.20 m × 3.75 m on plan × 2.00 m high	nr	1250.00	to	1500.00
See-saw comprising timber plank on sealed ball bearings				
3960 mm × 230 mm × 70 mm thick	nr	110.00	to	130.00
Wicksteed Tumbleguard type safety surfacing around play equipment	m²	14.10	to	17.00
Bark particles type safety surfacing 150 mm thick on hardcore bed	nr	14.10	to	17.00
8.6 EXTERNAL DRAINAGE				
8.6.1 Surface Water and Foul Water Drainage				
Overall £/m² of drained area allowances				
site drainage (per m² of paved area)	m²	26.00	to	33.00
building storm water drainage (per m² of gross internal floor area)	m²	21.50	to	27.00
Machine excavation, grade bottom, earthwork support, laying and jointing pipes and accessories, backfill and compact, disposal of surplus soil. uPVC pipes and fittings, lip seal coupling joints				
Up to 1.50 m deep; nominal size				
100 mm dia. pipe	m	58.00	to	74.00
160 mm dia. pipe	m	73.00	to	92.00
Over 1.50 m not exceeding 3.00 m deep; nominal size				
100 mm dia. pipe	m	100.00	to	125.00
160 mm dia. pipe	m	115.00	to	145.00

8.6 EXTERNAL DRAINAGE

Item	Unit	Range £		
Machine excavation, grade bottom, earthwork support, laying and jointing pipes and accessories, backfill and compact, disposal of surplus soil. uPVC Ultra-Rib ribbed pipes and fittings, sealed ring push fit joints				
Up to 1.50 m deep; nominal size				
150 mm dia. pipe	m	60.00	to	75.00
225 mm dia. pipe	m	80.00	to	100.00
300 mm dia. pipe	m	95.00	to	120.00
Over 1.50 m not exceeding 3.00 m deep; nominal size				
150 mm dia. pipe	m	100.00	to	130.00
225 mm dia. pipe	m	130.00	to	160.00
300 mm dia. pipe	m	145.00	to	185.00
Machine excavation, grade bottom, earthwork support, laying and jointing pipes and accessories, backfill and compact, disposal of surplus soil. Vitrified clay pipes and fittings, Hepsleve; shingle bed and surround				
Up to 1.00 m deep; nominal size				
150 mm dia. pipe	m	97.00	to	120.00
225 mm dia. pipe	m	150.00	to	190.00
300 mm dia. pipe	m	280.00	to	360.00
Over 1.50 m not exceeding 3.00 m deep; nominal size				
150 mm dia. pipe	m	140.00	to	175.00
225 mm dia. pipe	m	195.00	to	245.00
300 mm dia. pipe	m	320.00	to	410.00
Machine excavation, grade bottom, earthwork support, laying and jointing pipes and accessories, backfill and compact, disposal of surplus soil. Class M tested concrete centrifugally spun pipes and fittings, flexible joints; concrete bed and surround				
Up to 3.00 m deep; nominal size				
300 mm dia. pipe	m	190.00	to	240.00
525 mm dia. pipe	m	295.00	to	370.00
900 mm dia. pipe	m	530.00	to	660.00
1200 mm dia. pipe	m	760.00	to	960.00
Machine excavation, grade bottom, earthwork support, laying and jointing pipes and accessories, backfill and compact, disposal of surplus soil. Cast iron Timesaver drain pipes and fittings, mechanical coupling joints				
Up to 1.50 m deep; nominal size				
100 mm dia. pipe	m	120.00	to	150.00
150 mm dia. pipe	m	170.00	to	210.00
Over 1.50 m not exceeding 3.00 m deep; nominal size				
100 mm dia. pipe	m	160.00	to	200.00
150 mm dia. pipe	m	210.00	to	265.00
Brick manholes Excavate pit in firm ground, partial backfill, partial disposal, earthwork support, compact base of pit, 150 mm plain in situ concrete 20.00 N/mm² aggregate (1:2:4) base, formwork, one brick wall of engineering bricks in cement mortar (1:3) finished fair face, vitrified clay channels, plain in situ concrete 25.00 N/mm² – 20 mm aggregate (1:2:4) cover and reducing slabs, fabric reinforcement, formwork step irons, medium duty cover and frame; Internal size of manhole				

8.6 EXTERNAL DRAINAGE

Item	Unit	Range £		
8.6 EXTERNAL DRAINAGE – cont				
Brick manholes – cont				
600 mm × 450 mm; cover to invert				
not exceeding 1.00 m deep	nr	455.00	to	550.00
over 1.00 m not exceeding 1.50 m deep	nr	600.00	to	730.00
over 1.50 m not exceeding 2.00 m deep	nr	740.00	to	900.00
900 mm × 600 mm; cover to invert				
not exceeding 1.00 m deep	nr	590.00	to	720.00
over 1.00 m not exceeding 1.50 m	nr	800.00	to	970.00
over 1.50 m not exceeding 2.00 m	nr	1000.00	to	1225.00
900 mm × 900 mm; cover to invert				
not exceeding 1.00 m deep	nr	710.00	to	860.00
over 1.00 m not exceeding 1.50 m	nr	950.00	to	1150.00
over 1.50 m not exceeding 2.00 m	nr	1175.00	to	1425.00
1200 × 1800 mm; cover to invert				
not exceeding 1.00 m deep	nr	1425.00	to	1725.00
over 1.00 m not exceeding 1.50 m deep	nr	1825.00	to	2225.00
over 1.50 m not exceeding 2.00 m deep	nr	2225.00	to	2700.00
Concrete manholes				
Excavate pit in firm ground, disposal, earthwork support, compact base of pit, plain in situ concrete 20.00 N/mm² (1:2:4) base, formwork, reinforced precast concrete chamber and shaft rings, taper pieces and cover slabs bedded jointed and pointed in cement; mortar (1:3) weak mix concrete filling to working space, vitrified clay channels, plain in situ concrete 25.00 N/mm² (1:1:5:3) benchings, step irons, medium duty cover and frame; depth from cover to invert; Internal dia. of manhole				
900 mm dia.; cover to invert				
over 1.00 m not exceeding 1.50 m deep	nr	760.00	to	920.00
1050 mm dia.; cover to invert				
over 1.00 m not exceeding 1.50 m deep	nr	830.00	to	1000.00
over 1.50 m not exceeding 2.00 m deep	nr	950.00	to	1150.00
over 2.00 m not exceeding 3.00 m deep	nr	1175.00	to	1400.00
1500 mm dia.; cover to invert				
over 1.50 m not exceeding 2.00 m deep	nr	1575.00	to	1900.00
over 2.00 m not exceeding 3.00 m deep	nr	2025.00	to	2450.00
over 3.00 m not exceeding 4.00 m deep	nr	2475.00	to	3000.00
1800 mm dia.; cover to invert				
over 2.00 m not exceeding 3.00 m deep	nr	2700.00	to	3300.00
over 3.00 m not exceeding 4.00 m deep	nr	3300.00	to	4000.00
2100 mm dia.; cover to invert				
over 2.00 m not exceeding 3.00 m deep	nr	4150.00	to	5000.00
over 3.00 m not exceeding 4.00 m deep	nr	5100.00	to	6200.00
Polypropylene inspection chambers				
475 mm dia. PPIC inspection chamber including all excavations; earthwork support; cart away surplus spoil; concrete bed and surround; lightweight cover and frame				
600 mm deep	nr	580.00	to	700.00
900 mm deep	nr	730.00	to	890.00

8.7 EXTERNAL SERVICES

Item	Unit	Range £		

8.6.2 Ancillary Drainage Systems

Septic tanks
Excavate for, supply and install Klargester glass fibre septic tank, complete with lockable cover

2800 litre capacity	nr	3800.00	to	3800.00
3800 litre capacity	nr	4700.00	to	4700.00
4600 litre capacity	nr	5500.00	to	5500.00

Urban and landscape drainage
Exacavate for and lay oil separators, complete with lockable cover

excavate for, supply and lay 1000 litre polyethelyne by pass oil separator	nr	3900.00	to	3900.00
excavate for, supply and lay 1000 litre polyethelyne full retention oil separator	nr	4050.00	to	4050.00

8.6.4 Land Drainage
Note: If land drainage is required on a project, the propensity of the land to flood will decide the spacing of the land drains. Costs include for excavation and backfilling of trenches and laying agricultural clay drain pipes with 75 mm dia. lateral runs average 600 mm deep, and 100 mm dia. mains runs average 750 mm deep.

land drainage to parkland with laterals at 30 m centres and main runs at 100 m centres	ha	9600.00	to	11000.00

Rainwater harvesting
FP McCann Ltd Easi-Rain domestic and commercial rainwater harvesting tank system designed to reduce demands on the mains network by providing a sustainable water collection and distribution system ideal for non-potable uses such as toilet flushing, washing and landscape or garden watering. Including excavations

1500 litre tank	nr	2600.00	to	2600.00
7500 litre tank	nr	3550.00	to	3550.00

Soakaways: stormwater management

Soakaway crates; heavyweight (60 tonnes) polypropylene high-void box units; including excavation and backfilling as required. Note final surfacing not included	m³	160.00	to	190.00
Soakaway crates; lightweight (20 tonnes) polypropylene high-void box units; including excavation and backfilling as required. Note final surfacing not included	m³	580.00	to	700.00

8.7 EXTERNAL SERVICES

8.7.1 Water Mains Supply
Water main; all laid in trenches including excavation and backfill with excavated material

up to 50 mm dia. MDPE pipe	m	120.00	to	140.00

8.7 EXTERNAL SERVICES

Item	Unit	Range £		

8.7 EXTERNAL SERVICES – cont

Service connection charges

The privatization of telephone, water, gas and electricity has complicated the assessment of service connection charges. Typically, service connection charges will include the actual cost of the direct connection plus an assessment of distribution costs from the main. The latter cost is difficult to estimate as it depends on the type of scheme and the distance from the mains. In addition, service charges are complicated by discounts that may be offered. For instance, the electricity boards will charge less for housing connections if the house is all electric. However, typical charges for a reasonably sized housing estate might be as follows

    Water and sewerage connections

Item	Unit	Range £		
water connections; water main up to 2 m from property	house	560.00	to	680.00
water infrastructure charges for new properties	house	360.00	to	435.00
waste and water sewerage infrastructure charges for new properties	house	720.00	to	870.00

8.7.2 Electricity Mains Supply

Electric main; all laid in trenches including excavation and backfill with excavated material

    600/1000 volt cables. Two core 25 mm dia. cable including 100 mm dia.

Item	Unit	Range £		
clayware duct	m	150.00	to	180.00

Service connection charges

Item	Unit	Range £		
all electric housing	house	2550.00	to	3100.00
pre-packaged sub-station housing	nr	35500.00	to	43000.00

8.7.5 Gas Mains Supply

Gas main; all laid in trenches including excavation and backfill with excavated material

Item	Unit	Range £		
150 mm dia. gas pipe	m	350.00	to	425.00

Service connection charges

    Gas

Item	Unit	Range £		
gas connection to house	house	330.00	to	400.00
governing station	nr	19000.00	to	23000.00

8.7.6 Telecommunication and other Communication System Connections

Telephone duct; all laid in trenches including excavation and backfill with excavated material

Item	Unit	Range £		
100 mm dia. uPVC duct	m	83.00	to	100.00

BT approved communication boxes, including excavations and bedding, precast reinforced concrete box with cast iron cover (medium grade)

Item	Unit	Range £		
communication box 1200 mm long × 600 mm wide × 895 mm high	nr	430.00	to	500.00
communication box 910 mm long × 890 mm wide × 1000 mm high	nr	500.00	to	580.00
telephone connections charges	house	200.00	to	245.00

8.8 ANCILLARY BUILDINGS AND STRUCTURES

Item	Unit	Range £		
8.7.9 External/Street Lighting Systems				
Lighting				
lighting to pedestrian areas and estate roads on 4.00 m–6.00 m columns with up to 70 W lamps	nr	280.00	to	335.00
lighting to main roads;10.00 m–12.00 m columns with 250 W lamps	nr	610.00	to	730.00
lighting to main roads; 12.00 m–15.00 m columns with 400 W high pressure sodium lighting	nr	760.00	to	920.00
8.8 ANCILLARY BUILDINGS AND STRUCTURES				
8.8.2 Ancillary Buildings and Structures				
Footbridges				
Footbridge of either precast concrete or steel construction up to 6.00 m wide, 6.00 m high including deck, access stairs and ramp, parapets etc.				
5 m span between piers or abutments	m²	4550.00	to	5800.00
20 m span between piers or abutments	m²	4250.00	to	5400.00
Footbridge of timber (stress graded with concrete piers)				
12 m span between piers or abutments	m²	1525.00	to	1925.00
Roadbridges				
Reinforced concrete bridge with precast beams; including all excavation, reinforcement, formwork, concrete, bearings, expansion joints, deck water proofing and finishing's, parapets etc. deck area				
10.00 m span	m²	3500.00	to	4450.00
15.00 m span	m²	3400.00	to	4300.00
Reinforced concrete bridge with prefabricated steel beams; including all excavation, reinforcement, formwork, concrete, bearings, expansion joints, deck water proofing and finishing's, parapets etc. deck area				
20.00 m span	m²	3700.00	to	4700.00
30.00 m span	m²	3500.00	to	4450.00
Multiparking systems/stack parkers				
Fully automatic systems				
integrated robotic parking system using robotic car transporter to store vehicles	car	24000.00	to	31000.00
Semi-automatic systems				
integrated parking system, transverse and vertical positioning; semi-automatic parking achieving 17 spaces in a 6 car width × 3 car height grid	car	13500.00	to	17000.00
Integrated stacker systems				
integrated parking system, vertical positioning only; double width, double height pit stacker achieving 4 spaces with each car stacker	car	7100.00	to	9000.00
integrated parking system, vertical positioning only; triple stacker achieving 3 spaces with each car stacker, generally 1 below ground and 2 above ground	car	14000.00	to	18000.00
integrated parking system, vertical positioning only; triple height double width stacker, achieving 6 spaces with each car stacker, generally 2 below ground and 4 above ground	car	10500.00	to	13500.00

9.2 MAIN CONTRACTOR'S PRELIMINARIES

Item	Unit	Range £		

9.2 MAIN CONTRACTOR'S PRELIMINARIES

9.2.2 Site Establishment
Typical rates for site accommodation; Delivery not included; minimum hire periods generally apply

Item	Unit	Range £		
Office cabins – 24 ft × 9 ft (20 m²)	week	31.50	to	40.00
Office cabins – 32 ft × 10 ft (30 m²)	week	40.00	to	51.00
Fire-rated cabins – 24 ft × 9 ft (20 m²)	week	67.00	to	84.00
Fire-rated cabins – 32 ft × 10 ft (30 m²)	week	91.00	to	115.00
Meeting room – 24 ft × 9 ft (20 m²)	week	31.50	to	40.00
Meeting room – 32 ft × 9 ft (30 m²)	week	40.00	to	51.00
Mess cabins (incl. furniture)	week	37.50	to	47.50
Drying rooms (incl. furniture)	week	42.50	to	54.00
Safestore – 20 ft × 8ft (15 m²)	week	22.00	to	27.50
Container with padlock – 20 ft × 8ft (15 m²)	week	12.10	to	15.30
Toilets – 13 ft × 9 ft	week	48.50	to	61.00
Signs and notices	week	61.00	to	77.00
Fire extinguishers	week	9.70	to	12.30
Kitchen with cooker, fridge, sink, water heater; 32 ft × 10 ft	week	180.00	to	230.00
Mess room with wash basin, water heater, seating; 16 ft × 7 ft	week	110.00	to	140.00
Toilets – fitted to mains; three pan unit	week	195.00	to	245.00
Toilets – fitted to mains; four pan unit	week	230.00	to	290.00
Haulage to and from site; Site offices, Storage sheds, Toilets	item	5400.00	to	6800.00

9.2.7 Mechanical Plant
Power and Lighting; Delivery not included; minimum hire periods generally apply

Item	Unit	Range £		
3 kVa transformer	week	11.70	to	14.80
5 kVa transformer	week	19.50	to	24.50
10 kVa transformer	week	39.00	to	49.50
4 Way distribution box; 110 V	week	10.00	to	12.60
14 m extension cable	week	4.80	to	6.05
25 m extension cable	week	12.60	to	15.90
2 × 500 W floodlights	week	14.70	to	18.60
2 × 500 W floodlights on stands	week	18.30	to	23.00

Generators; Delivery not included; minimum hire periods generally apply

Item	Unit	Range £		
1 kVa 240 V portable (low noise)	week	45.50	to	58.00
2 kVa dual volt, portable petrol	week	32.00	to	40.00
5 kVa silenced	week	89.00	to	110.00
7.5 kVa silenced	week	130.00	to	170.00
15 kVa silenced	week	190.00	to	240.00
40 kVa silenced	week	285.00	to	360.00
60 kVa silenced	week	300.00	to	375.00
108 kVa silenced	week	370.00	to	465.00
165 kVa silenced	week	455.00	to	580.00

Mixers and small plant; Delivery not included; minimum hire periods generall apply

Item	Unit	Range £		
small mixer; 5–3 ½ diesel	week	43.00	to	54.00
mixer; 7–5 diesel	week	54.00	to	68.00
compresser	week	110.00	to	140.00

9.2 MAIN CONTRACTOR'S PRELIMINARIES

Item	Unit	Range £		
small dehumidifers	week	29.00	to	37.00
large dehumidifers	week	78.00	to	98.00
turbo dryer	week	48.50	to	62.00
gas jet air heater; 260,000 Btu	week	58.00	to	74.00
9.2.8 Temporary Works				
Roadways				
formation of temporary roads to building perimeter comprising of geotextile membrane and 300 mm MOT type 1; reduce level; spoil to heap on site within 25 m of excavation	m²	26.00	to	31.00
installation of wheel wash facility and maintenance	nr	3100.00	to	3700.00

How to Become a Chartered Surveyor

Jen Lemen

Thinking about a career in property or construction? Thinking of becoming of Chartered Surveyor? *How to Become a Chartered Surveyor* demystifies the process and provides a clear road map for candidates to follow. The book outlines potential pathways and practice areas within the profession and includes the breadth and depth of surveying, from commercial, residential and project management, to geomatics and quantity surveying. Experienced APC assessor and trainer, Jen Lemen BSc (Hons) FRICS, provides invaluable guidance, covering:

- routes to becoming a Chartered Surveyor, including t-levels, apprenticeships and alternative APC routes such as the Senior Professional, Academic and Specialist assessments
- areas of professional practice
- advice for the AssocRICS, APC (MRICS), FRICS and Registered Valuer assessments, including both written and interview elements
- advice on referrals and appeals
- how to support candidates, including the role of the Counsellor and Supervisor
- opportunities   for further career progression, including further qualifications and setting up in practice   as an RICS regulated firm
- global perspectives
- professional ethics for surveyors

Written in clear, concise and simple terms and providing practical advice throughout, this book will help candidates to decode and understand the RICS guidance, plan their career and be successful in their journey to become a Chartered Surveyor. It will also be of relevance to academic institutions, employers, school leavers, apprentices, senior professionals, APC Counsellors/Supervisors and careers advisors.

August 2021: 168 pp
ISBN: 9780367742195

To Order
Tel:+44 (0) 1235 400524
Email: tandf@bookpoint.co.uk

For a complete listing of all our titles visit:
www.tandf.co.uk

PART 4

Prices for Measured Works

This part contains the following NRM2 work sections:

JCT Contract Administration Pocket Book,
2nd edition

Andy Atkinson

This book is quite simply about contract administration using the JCT contracts. The key features of the new and updated edition continue to be its brevity, readability and relevance to everyday practice. It provides a succinct guide written from the point of view of a construction practitioner, rather than a lawyer, to the traditional form of contract with bills of quantities SBC/Q2016, the design and build form DB2016 and the minor works form MWD2016. The book broadly follows the sequence of producing a building from the initial decision to build through to completion. Chapters cover:

- Procurement and tendering
- Payments, scheduling, progress and claims
- Contract termination and insolvency
- Indemnity and insurance
- Supply chain problems, defects and subcontracting issues
- Quality, dealing with disputes and adjudication
- How to administer contracts for BIM-compliant projects

JCT contracts are administered by a variety of professionals including project managers, architects, engineers, quantity surveyors and construction managers. It is individuals in these groups, whether experienced practitioner or student, who will benefit most from this clear, concise and highly relevant book.

December 2020: 248 pp
ISBN: 9780367632786

To Order
Tel:+44 (0) 1235 400524
Email: tandf@bookpoint.co.uk

For a complete listing of all our titles visit:
www.tandf.co.uk

Taylor & Francis
Taylor & Francis Group

INTRODUCTION

The rates contained in Prices for Measured Works are intended to apply to a project in the Outer London area costing about £4,000,000, and assume that reasonable quantities of all types of work are required. Similarly it has been necessary to assume that the size of the project warrants the subletting of all types of work normally sublet. Adjustments should be made to standard rates for time, location, local conditions, site constraints and all the other factors likely to affect costs of any individual project.

Adjustments may need to be made for different project values. In *How to use this Book* you will find some guidance on adjustments accordance to contract cost.

The distinction between builders' work and work normally sublet is stressed because prices for work which can be sublet may well be inadequate for the contractor who is called upon to carry out relatively small quantities of such work themselves.

Measured Works prices are generally based upon wage rates and known material costs from May 2020. Built up prices and subcontractor rates include an allowance of 3% for overheads and profit.

As elsewhere in this edition, prices do not include Value Added Tax or professional services fees.

Measured Works rates are exclusive of any main contractor's preliminaries.

Prices for Measured Works sections is structured to reflect NRM2.

To aid readers with the transition we have scheduled the SMM7 Work Sections and shown the NRM2 Work Sections alongside:

SMM7 Work Section	NRM2 Work Section
	2 Off-site manufactured materials, components and buildings:
C Demolition/Alteration/Renovation	
	3 Demolitions
	4 Alterations, repairs and conservation
D Groundwork	
D20 Excavation and Filling	5 Excavating and filling
D30 Cast In Place Piling	7 Piling
D32 Steel Piling	
D40 Embedded Retaining Walling	8 Underpinning
D41 Crib Walls/Gabions/Reinforced Earthworks	9 Diaphragm walls and embedded retaining walls
D50 Underpinning	10 Crib walls, gabions and reinforced earth
E In Situ Concrete/Large Precast Concrete	
E10 In Situ Concrete Construction	11 In situ concrete works
E20 Formwork For In Situ Concrete	12 Precast/composite concrete
E30 Reinforcement For In Situ Concrete	13 Precast concrete
E40 Designed Joints In Situ Concrete	
E41 Worked Finishes/Cutting To In Situ Concrete	
E42 Accessories Cast Into In Situ Concrete	
E50 Precast Concrete Large Units	
E60 Precast/Composite Concrete Decking	

SMM7 Work Section	NRM2 Work Section
F Masonry	14 Masonry
F10 Brick/Block Walling	
F11 Glass Block Walling	
F20 Natural Stone Rubble Walling	
F22 Cast Stone Walling/Dressings	
F30 Accessories/Sundry Items For Brick/Block/Stone Walling	
F31 Precast Concrete Sills/Lintels/Coping features	
G Structural/Carcassing Metal/Timber	
G10 Structural Steel Framing	15 Structural metalwork
G12 Isolated Structural Metal Members	16 Carpentry
G20 Carpentry/Timber Framing/First Fixing	
H Cladding/Covering	
H10 Patent Glazing	17 Sheet roof coverings
H11 Curtain Walling	18 Tile and slate roof and wall coverings
H20 Rigid Sheet Cladding	21 Cladding and covering
H30 Fibre Cement Profile Sheet Cladding	
H31 Metal Profiled/Flat Sheet Cladding/Covering/Siding	
H32 Plastic Profiled Sheet Cladding/Covering/Siding	
H41 Glass Reinforced Plastic Panel Cladding Features	
H51 Natural Stone Cladding Features	
H53 Clay Slab/Cladding/Features	
H60 Plain Roof Tiling	
H61 Fibre Cement Slating	
H62 Natural Slating	
H63 Reconstructed Stone Slating/Tiling	
H64 Timber Shingling	
H71 Lead Sheet Coverings/Flashes	
H72 Aluminium Sheet Coverings/Flashings	
H73 Copper Strip Sheet Coverings/Flashings	
H74 Zinc Strip Sheet Coverings/Flashings	
H75 Stainless Steel Sheet Coverings/Flashings	
H76 Fibre Bitumen Thermoplastic Sheet Coverings/Flashings	
H92 Rainscreen Cladding	
J Waterproofing	
J10 Specialist Waterproof Rendering	19 Waterproofing
J20 Mastic Asphalt Tanking/Damp-Proof Membranes	
J21 Mastic Asphalt Roofing/Insulation/Finishes	
J30 Liquid Applied Tanking/Damp-Proof Membranes	

SMM7 Work Section	NRM2 Work Section
J40 Flexible Sheet Tanking/Damp-Proof Membranes	
J41 Built Up Felt Roof Coverings	
J42 Single Layer Plastic Roof Coverings	
J43 Proprietary Roof Decking With Felt Finish	
K Linings/Sheathing/Dry partitioning	
K10 Plasterboard Dry Lining/Partitions/Ceilings	20 Proprietary linings and partitions
K11 Rigid Sheet Flooring/Sheathing/Linings/Casings	30 Suspended ceilings
K13 Rigid Sheet Fine Linings/Panelling	
K14 Glass Reinforced Gypsum Linings/Panelling	
K20 Timber Board Flooring/Sheathing/Linings/Casings	
K30 Demountable Partitions	
K32 Framed Panel Cubicle Partitions	
K33 Concrete/Terrazzo Partitions	
K40 Demountable Suspended Ceilings	
K41 Raised Access Floors	
L Windows/Doors/Stairs	
L10 Windows/Rooflights/Screens/Louvres	23 Windows, screens and lights
L20 Doors/Shutters/Hatches	24 Doors, shutters and hatches
L30 Stairs/Walkways/Balustrades	25 Stairs, walkways and balustrades
L40 General Glazing	27 Glazing
M Surface Finishes	
M10 Cement: Sand/Concrete Screeds/Granolithic Screeds/ Topping	28 Floor, wall, ceiling and roof finishings
M11 Mastic Asphalt Flooring/Floor Underlays	29 Decoration
M12 Trowelled Bitumen/Resin/Rubber Latex	
M20 Plastered/Rendered/Roughcast Coating	
M21 Insulation With Rendered Finish	
M22 Sprayed Mineral Fibre Coatings	
M30 Metal Mesh Lathing/Anchored Reinforcement For Plastered Ceilings	
M31 Fibrous Plaster	
M40 Stone/Concrete/Quarry/Ceramic Tiling	
M41 Terrazzo Tiling/In Situ Terrazzo	
M42 Wood Block/Composition Block/Parquet	
M50 Rubber/Plastic/Cork/Lino/Carpet Tiling/Sheeting	
M51 Edge Fixed Carpeting	
M52 Decorative Papers/Fabrics	
M60 Painting/Clear Finishing	

SMM7 Work Section	NRM2 Work Section
N Furniture/Equipment	
N10/11 General Fixtures/Kitchen Fittings	32 Furniture, fittings and equipment
N13 Sanitary Appliances/Fittings	
N15 Signs/Notices	
P Building Fabric Sundries	
P10 Sundry Insulation/Proofing Work/Fire Stops	22 General joinery
P20 Unframed Isolated Trims/Skirtings/Sundry Items	31 Insulation, fire stopping and fire protection
P21 Ironmongery	
P30 Trenches/Pipeways/Pits For Buried Engineering Services	
P31 Holes/Chases/Covers/Supports For Services	
Q Paving/Planting/Fencing/Site Furniture	
Q10 Kerbs/Edgings/Channels/Paving Access	35 Site work
Q20 Hardcore/Granular/Cement Bound Bases	36 Fencing
Q21 In Situ Concrete Roads/Pavings	37 Soft landscaping
Q22 Coated Macadam/Asphalt Roads/Pavings	
Q23 Gravel/Hoggin/Woodchip Roads/Pavings	
Q25 Slab/Brick/Block/Sett/Cobble Pavings	
Q26 Special Surfacings/Pavings For Sport	
Q30 Seeding/Turfing	
Q31 Planting	
Q40 Fencing	
R Disposal Systems	
R10 Rainwater Pipework/Gutters	33 Drainage above ground
R11 Foul Drainage Above Ground	34 Drainage below ground
R12 Drainage Below Ground	
R13 Land Drainage	
S Piped Supply Systems	
S10/S11 Hot And Cold Water	
S13 Pressurized Water	
T Mechanical Heating/Cooling/Refrigeration Systems	
T10 Gas/Oil Fired Boilers	
T31 Low Temperature Hot Water	
V Electrical systems	
V21/V22 General Lighting and Low Voltage Power	39 Electrical services
W Security Systems	
W20 Lightning Protection	
X Transport Systems	41 Builder's work in connection with services

03 DEMOLITIONS

Item	PC £	Labour hours	Labour £	Plant £	Material £	Unit	Total rate £
3.01 DEMOLITIONS							
NOTE: Demolition rates vary enormously from project to project, depending upon access, type of construction, method of demolition, redundant or recycleable materials etc. Always obtain specific quotations for each project under consideration. The following rates for simple demolition works may be of some assistance for comparative purposes. Generally scaffold and other access equipment is not included in the rates.							
Demolishing structures; by machine; note disposal not included							
Demolishing to ground level; single storey brick outbuilding; timber flat roof; grub up shallow foundations							
single storey outbuilding approximately 50 m³	–	–	–	–	–	m³	93.56
single storey outbuilding approximately 200 m³	–	–	–	–	–	m³	64.16
single storey outbuilding approximately 500 m³	–	–	–	–	–	m³	30.74
Demolishing to ground level; light steel framed and sheet roofed cycle shelter; grub up shallow foundations	–	–	–	–	–	m³	66.84
Concrete framed multi-storey car park	–	–	–	–	–	m³	45.44
Demolishing to ground level; multi-storey masonry building; flat roof; 1–4 storeys; volume	–	–	–	–	–	m³	33.42
Warehouse to slab level; 7 m eaves	–	–	–	–	–	m²	18.71
Warehouse to below slab; 7 m eaves; grub up shallow foundations (pads, ground beams and ground slab)	–	–	–	–	–	m²	30.74
Reinforced concrete frame building; easy access	–	–	–	–	–	m³	53.47
Reinforced concrete frame building; restricted access	–	–	–	–	–	m³	100.25
Concrete encased steel frame building; easy access	–	–	–	–	–	m³	37.43
Concrete encased steel frame building; restricted access	–	–	–	–	–	m³	89.56
Demolishing parts of structures; by hand; cost of skip included							
Breaking up plain concrete bed; load into skip							
100 mm thick	–	0.45	15.50	6.88	–	m²	22.38
150 mm thick	–	0.67	22.94	13.64	–	m²	36.58
200 mm thick	–	0.90	31.01	13.72	–	m²	44.73
300 mm thick	–	1.33	45.89	20.12	–	m²	66.01
Breaking up reinforced concrete bed; load into skip							
100 mm thick	–	0.50	17.36	7.63	–	m²	24.99
150 mm thick	–	0.75	25.74	14.40	–	m²	40.14
200 mm thick	–	1.00	34.42	15.29	–	m²	49.71
300 mm thick	–	1.50	51.78	22.79	–	m²	74.57
Demolishing reinforced concrete column or cutting away concrete casing to steel column; load into skip	–	9.99	344.17	103.93	–	m³	448.10
Demolishing reinforced concrete beam or cutting away concrete casing to steel beam; load into skip	–	11.47	395.34	113.15	–	m³	508.49

Prices for Measured Works

03 DEMOLITIONS

Item	PC £	Labour hours	Labour £	Plant £	Material £	Unit	Total rate £
3.01 DEMOLITIONS – cont							
Demolishing parts of structures – cont							
Demolishing reinforced concrete wall; load into skip							
100 mm thick	–	0.99	34.10	10.32	–	m²	**44.42**
150 mm thick	–	1.50	51.78	15.40	–	m²	**67.18**
225 mm thick	–	2.25	77.52	23.17	–	m²	**100.69**
300 mm thick	–	3.00	103.25	31.07	–	m²	**134.32**
Demolishing reinforced concrete suspended slabs; load into skip							
100 mm thick	–	0.84	28.84	9.61	–	m²	**38.45**
150 mm thick	–	1.25	43.10	14.04	–	m²	**57.14**
225 mm thick	–	1.87	64.50	21.05	–	m²	**85.55**
300 mm thick	–	2.50	86.20	28.44	–	m²	**114.64**
Breaking up small concrete plinth; make good structures; load into skip	–	3.83	132.08	50.77	7.80	m³	**190.65**
Breaking up small precast concrete kerb; make good structures; load into skip	–	0.41	14.27	4.21	7.80	m	**26.28**
Remove precast concrete window sill; set aside for reuse	–	1.33	45.89	–	–	m	**45.89**
Remove brick on-edge-coping; prepare walls for raising; load into skip							
one brick thick	–	0.38	12.36	0.76	–	m	**13.12**
one and a half brick thick	–	0.50	16.48	1.14	–	m	**17.62**
Demolishing brick chimney to 300 mm below roof level; sealing off flues with slates; piecing in treated sawn softwood rafters and making good roof coverings; load into skip (excluding scaffold access)							
680 mm × 680 mm × 900 mm high above roof	–	11.40	322.76	21.79	97.90	nr	**442.45**
add for each additional 300 mm height	–	2.08	58.90	5.81	–	nr	**64.71**
680 mm × 1030 mm × 900 mm high above roof	–	16.86	477.28	23.26	139.52	nr	**640.06**
add for each additional 300 mm height	–	3.11	88.20	8.72	–	nr	**96.92**
1030 mm × 1030 mm × 900 mm high above roof	–	25.52	722.62	36.32	189.44	nr	**948.38**
add for each additional 300 mm height	–	4.71	133.39	11.63	–	nr	**145.02**
Demolishing defective brick chimney to roof level; rebuild using 25% new facing bricks; new lead flashings; core flues; re-set chimney pot; load into skip (excluding scaffold access)							
680 mm × 680 mm × 900 mm high above roof	–	10.73	305.36	21.79	146.96	nr	**474.11**
add for each additional 300 mm height	–	2.08	58.90	5.81	17.39	nr	**82.10**
680 mm × 1030 mm × 900 mm high above roof	–	15.66	443.64	23.26	176.53	nr	**643.43**
add for each additional 300 mm height	–	3.11	88.20	8.72	21.47	nr	**118.39**
1030 mm × 1030 mm × 900 mm high above roof	–	24.02	680.57	36.32	205.06	nr	**921.95**
add for each additional 300 mm height	–	4.71	133.39	11.63	27.56	nr	**172.58**
Demolishing external brick walls; load into skip							
half brick thick	–	0.87	24.47	5.70	–	m²	**30.17**
two half brick thick skins if cavity wall	–	1.46	40.86	11.04	–	m²	**51.90**
one brick thick	–	1.50	42.12	11.04	–	m²	**53.16**
one and a half brick thick	–	2.04	57.25	16.56	–	m²	**73.81**
two brick thick	–	2.62	73.40	22.09	–	m²	**95.49**
extra for plaster, render or pebbledash finish per side	–	0.08	2.27	1.14	–	m²	**3.41**

03 DEMOLITIONS

Item	PC £	Labour hours	Labour £	Plant £	Material £	Unit	Total rate £
Demolish external stone walls; load into skip							
300 mm thick	–	1.00	28.03	11.63	–	m²	**39.66**
400 mm thick	–	1.33	37.33	15.11	–	m²	**52.44**
600 mm thick	–	2.00	56.05	22.81	–	m²	**78.86**
Demolish external stone walls; clean off and set aside for reuse							
300 mm thick	–	1.50	42.04	2.90	–	m²	**44.94**
400 mm thick	–	2.00	56.05	4.36	–	m²	**60.41**
600 mm thick	–	3.00	84.07	5.81	–	m²	**89.88**
Remove fireplace surround and hearth							
breaking up 500 mm wide concrete hearth; load into skip	–	1.50	42.12	3.80	–	m	**45.92**
remove fire surround; tiled interior; load into skip	–	1.54	43.13	8.72	–	nr	**51.85**
cast iron surround; set aside for reuse	–	2.58	72.40	–	–	nr	**72.40**
stone surround; set aside for reuse	–	6.74	188.93	–	–	nr	**188.93**
fill in opening with common bricks; air brick; 2 coat plaster; timber skirting; re-screed floor	–	0.95	30.97	–	79.58	m²	**110.55**
Removing roof timbers complete (NB not coverings); including rafters, purlins, ceiling joists, plates etc.; load into skip; measured flat on plan	–	0.28	8.16	4.48	–	m²	**12.64**
Remove softwood floor structure; load into skip							
joists at ground floor level	–	0.20	5.74	1.52	–	m²	**7.26**
joists at first floor level	–	0.41	11.60	1.52	–	m²	**13.12**
joists at second floor	–	0.58	16.39	1.52	–	m²	**17.91**
individual timber members	–	0.23	6.30	0.03	–	m	**6.33**
Remove boarding; withdraw nails; set aside for reuse							
softwood flooring at ground floor level	–	0.30	8.41	–	–	m²	**8.41**
softwood flooring at first floor level	–	0.51	14.35	–	–	m²	**14.35**
softwood flooring at second floor level	–	0.61	17.15	–	–	m²	**17.15**
fascia/barge board/gutter board second floor level; load into skip	–	0.52	14.44	0.03	–	m	**14.47**
Remove doors and windows; set aside for reuse							
solid single door only (frame left in place)	–	0.33	9.34	1.14	–	nr	**10.48**
solid single door; frame to skip	–	0.55	15.42	1.90	–	nr	**17.32**
solid double door only (frame left in place)	–	0.66	18.49	2.28	–	nr	**20.77**
solid double door; frame to skip	–	1.00	28.00	3.80	–	nr	**31.80**
glazed screen and doors	–	1.00	28.03	5.70	–	m²	**33.73**
casement window frame; set aside frame; glass into skip; up to 1.00 m²	–	0.75	21.02	0.29	–	m²	**21.31**
casement window frame; set aside frame; glass into skip; 1.00 to 2.00 m²	–	1.00	28.03	0.29	–	m²	**28.32**
casement window frame; set aside frame; glass into skip; 2.00 to 3.00 m²	–	1.10	30.83	0.29	–	m²	**31.12**
pair French windows and frame; upto 1200 mm × 1200 mm	–	3.33	93.32	5.70	–	nr	**99.02**
removing double hung sash window and frame; store for reuse	–	2.50	70.00	–	–	nr	**70.00**

03 DEMOLITIONS

Item	PC £	Labour hours	Labour £	Plant £	Material £	Unit	Total rate £
3.01 DEMOLITIONS – cont							
Demolishing parts of structures – cont							
Remove doors and windows; load into skip							
remove solid timber door	–	0.30	8.41	1.90	–	nr	**10.31**
remove door frame	–	0.10	2.81	1.90	–	nr	**4.71**
casement window frame; up to 1.00 m^2	–	0.60	16.81	1.90	–	m^2	**18.71**
casement window frame; 1.00 to 2.00 m^2	–	0.80	22.42	1.90	–	m^2	**24.32**
casement window frame; 2.00 to 3.00 m^2	–	0.90	25.22	1.90	–	m^2	**27.12**
Demolishing internal partitions; load into skip							
half brick thick brickwork	–	0.87	24.47	5.70	–	m^2	**30.17**
one brick thick brickwork	–	1.50	42.12	11.04	–	m^2	**53.16**
one and a half brick thick brickwork	–	2.04	57.25	16.56	–	m^2	**73.81**
75 mm blockwork	–	0.58	16.39	4.19	–	m^2	**20.58**
90 mm blockwork	–	0.62	17.40	4.94	–	m^2	**22.34**
100 mm blockwork	–	0.67	18.66	5.70	–	m^2	**24.36**
115 mm blockwork	–	0.71	19.93	5.70	–	m^2	**25.63**
125 mm blockwork	–	0.75	20.94	6.08	–	m^2	**27.02**
140 mm blockwork	–	0.79	22.21	6.47	–	m^2	**28.68**
150 mm blockwork	–	0.84	23.46	7.22	–	m^2	**30.68**
190 mm blockwork	–	0.98	27.49	9.12	–	m^2	**36.61**
215 mm blockwork	–	1.08	30.26	9.89	–	m^2	**40.15**
255 mm blockwork	–	1.25	35.03	11.78	–	m^2	**46.81**
extra for plaster finish per side	–	0.08	2.27	1.14	–	m^2	**3.41**
breaking up brick plinths	–	3.33	93.32	38.02	–	m^3	**131.34**
stud walls (metal or timber); solid board each side	–	0.38	10.60	3.80	–	m^2	**14.40**
stud walls (metal or timber); glazed panels	–	0.50	14.12	3.80	–	m^2	**17.92**
lightweight steel mesh security screen	–	0.41	11.60	1.88	–	m^2	**13.48**
solid steel demountable partition	–	0.62	17.40	2.66	–	m^2	**20.06**
glazed demountable partition including removal of glass	–	0.84	23.46	3.80	–	m^2	**27.26**
glazed screen including any doors incorporated	–	1.00	28.03	5.70	–	m^2	**33.73**
remove large folding partition and track	–	9.90	277.45	57.84	–	m^2	**335.29**
remove timber skirting, architrave, dado rail; load into skip	–	0.10	2.81	0.03	–	m	**2.84**
Removing timber staircase and balustrades							
single straight flight	–	2.92	81.72	38.02	–	nr	**119.74**
single dogleg flight	–	4.17	116.78	57.02	–	nr	**173.80**
Perimeter hoarding							
security fencing; 2400 mm high painted plywood hoarding	–	–	–	–	–	m	**167.10**

03 DEMOLITIONS

Item	PC £	Labour hours	Labour £	Plant £	Material £	Unit	Total rate £
3.02 TEMPORARY SUPPORT OF STRUCTURES, ROADS AND THE LIKE							
Temporary support							
NOTE: The requirement for shoring and strutting for the formation of large openings are dependent upon a number of factors; for example the weight of the superimposed structure to be supported; the number of windows; number of floors; roof type; whether raking shores are required; depth to load bearing surface; duration support is to be left in place. Prices therefore are best built-up by assessing the use and waste of materials and the labours involved. The follow guide needs to be tested against each particular project.							
Support of structures not to be demolished							
strutting to window openings over proposed new openings	–	0.56	18.31	–	10.26	nr	**28.57**
plates, struts, braces and hardwood wedges in supports to floors and roof of openings	–	1.11	36.30	–	30.38	nr	**66.68**
dead shore and needle using die square timber with sole plates, braces, hardwood wedges and steel dogs	–	27.75	907.43	–	151.56	nr	**1058.99**
set of two raking shores using die square timber with 50 mm thick wall pieces, hardwood wedges and steel dogs, including forming holes for needles and making good	–	33.30	1088.90	–	153.61	nr	**1242.51**
cut holes through one brick wall for die square needle and make good on completion of works, including facings externally and plaster internally	–	5.36	175.27	–	28.72	nr	**203.99**

04 ALTERATIONS, REPAIRS AND CONSERVATION

Item	PC £	Labour hours	Labour £	Plant £	Material £	Unit	Total rate £
4.01 REMOVING							
NOTE: It is highly unlikely for exactly the same composite items of alteration works will be encountered on different projects. The following spot items have been included to allow the reader to build up the composite rates for estimating purposes. Rates include for the removal of debris from site, but do not include for any temporary works, shoring, scaffolding etc. or for any redecorations.							
Fixtures and fittings; remove and load into skip							
Handrails and balustrades							
timber or metal handrail and brackets	–	0.27	7.61	0.38	–	m	**7.99**
timber balustrades	–	0.30	8.42	1.14	–	m	**9.56**
metal balustrades	–	0.45	12.50	1.14	–	m	**13.64**
Kitchen fittings							
wall units (up to 500 mm wide)	–	0.41	11.41	5.70	–	nr	**17.11**
floor units (up to 500 mm wide)	–	0.27	7.61	8.36	–	nr	**15.97**
larder units (up to 500 mm wide)	–	0.36	10.06	19.01	–	nr	**29.07**
Bathroom fittings							
bath panels and bearers	–	0.36	10.06	0.38	–	nr	**10.44**
toilet roll holders or soap dispensers; 2 or 3 screw fixings	–	0.27	8.44	0.03	–	nr	**8.47**
towel rails; 4 or 6 screw fixings	–	0.54	15.23	0.03	–	nr	**15.26**
mirror; up to 600 mm × 600 mm; 4 screw fixings	–	0.58	18.08	0.06	–	nr	**18.14**
pipe casings	–	0.27	7.61	1.52	–	m	**9.13**
Other fixtures and fittings							
shelves, window boards and the like; screw fixings	–	0.30	9.35	0.38	–	m	**9.73**
curtain track, plastic; screw fixing at 600 mm centres	–	0.10	3.11	–	–	m	**3.11**
small nameplates or individual numerals/letters; screw fixings	–	0.25	7.76	–	–	nr	**7.76**
small notice board and frame from walls; up to 600 mm × 900 mm; screw fixings	–	0.50	15.54	–	–	nr	**15.54**
Finishes; removing and load into skip unless stated otherwise							
Breaking up floor screeds							
asphalt paving	–	0.54	15.23	0.88	–	m²	**16.11**
floor screed up to 100 mm thick unreinforced	–	0.33	9.25	0.88	–	m²	**10.13**
granolithic floor screed up to 100 mm thick unreinforced	–	0.58	16.31	0.88	–	m²	**17.19**
Floor finishes							
carpet and underfelt	–	0.11	2.99	1.45	–	m²	**4.44**
carpetgrip edge fixing strip	–	0.01	0.31	–	–	m	**0.31**
vinyl or similar sheet flooring	–	0.09	2.45	1.04	–	m²	**3.49**
vinyl or similar tile flooring with tile remover	–	0.17	4.62	1.21	–	m²	**5.83**
ceramic floor tiles with tile remover	–	0.40	11.21	1.28	–	m²	**12.49**
woodblock flooring; set aside for reuse	–	0.67	18.76	–	–	m²	**18.76**

04 ALTERATIONS, REPAIRS AND CONSERVATION

Item	PC £	Labour hours	Labour £	Plant £	Material £	Unit	Total rate £
Hack off wall finishes with chipping hammer							
plaster to walls	–	0.33	9.25	1.16	–	m²	**10.41**
cement rendering, pebbledash or similar	–	0.25	7.01	0.80	–	m²	**7.81**
ceramic wall tiles	–	0.40	11.21	1.16	–	m²	**12.37**
Strip off wallpaper use steam paper stripper							
1 layer	–	0.10	2.81	0.14	–	m²	**2.95**
2 layers	–	0.14	3.92	0.14	–	m²	**4.06**
3 layers	–	0.17	4.91	0.14	–	m²	**5.05**
Remove wall linings including battening behind							
plain sheet boarding	–	0.27	7.61	1.52	–	m²	**9.13**
matchboarding	–	0.36	10.06	2.28	–	m²	**12.34**
Ceiling finishes							
plasterboard and skim; removes nails or screws	–	0.20	5.60	1.14	–	m²	**6.74**
wood lathe and plaster; removes nails or screws	–	0.22	6.17	1.90	–	m²	**8.07**
suspended ceilings; lay in grid; hangers and tee sections	–	0.45	12.50	1.90	–	m²	**14.40**
plain sheet boarding; including battening	–	0.41	11.41	1.52	–	m²	**12.93**
matchboarding; including battening	–	0.54	15.23	2.28	–	m²	**17.51**
Wall panelling							
remove timber wall panelling; clean off and set aside for reuse	–	0.58	16.31	–	–	m²	**16.31**
Removing coverings load into skip							
Roof coverings							
slates	–	0.45	12.50	0.76	–	m²	**13.26**
slates; set aside for reuse	–	0.54	15.23	–	–	m²	**15.23**
nibbed tiles	–	0.36	10.06	0.76	–	m²	**10.82**
nibbed tiles; set aside for reuse	–	0.45	12.50	–	–	m²	**12.50**
underfelt and nails	–	0.04	1.09	0.38	–	m²	**1.47**
felt or polymer membrane	–	0.22	6.25	0.76	–	m²	**7.01**
profiled metal sheeting; insulation; steel liner	–	0.75	21.07	1.45	–	m²	**22.52**
sheet metal coverings	–	0.45	12.50	0.76	–	m²	**13.26**
remove tiling battens	–	0.07	1.91	0.38	–	m²	**2.29**
Remove roof coverings; select and refix including providing 25% new tiles including nails							
natural slates; Welsh Blue	–	1.08	30.18	0.38	60.88	m²	**91.44**
clay plain tiles	–	0.99	27.73	0.38	8.15	m²	**36.26**
asbestos free artificial blue/black slates	–	0.99	27.73	0.38	9.05	m²	**37.16**
concrete interlocking tiles	–	0.63	17.68	0.38	3.58	m²	**21.64**
removing ridge or hip tile; provide and fit new	–	0.45	12.50	–	12.72	m	**25.22**
4.02 CUTTING OR FORMING OPENINGS							
Cutting openings or recesses; spoil into adjacent skip							
Through reinforced concrete walls; make good; not including any props or supports							
150 mm thick wall	–	5.02	169.20	8.03	12.90	m²	**190.13**
225 mm thick wall	–	6.95	232.01	11.62	17.40	m²	**261.03**
300 mm thick wall	–	8.75	295.91	10.96	21.44	m²	**328.31**

04 ALTERATIONS, REPAIRS AND CONSERVATION

Item	PC £	Labour hours	Labour £	Plant £	Material £	Unit	Total rate £
4.02 CUTTING OR FORMING OPENINGS – cont							
Cutting openings or recesses – cont							
Through reinforced concrete suspended slab; make good; not including any props or supports							
150 mm thick wall	–	3.81	128.78	7.55	19.14	m²	**155.47**
225 mm thick wall	–	5.66	191.00	11.62	17.40	m²	**220.02**
300 mm thick wall	–	7.04	238.28	10.96	21.44	m²	**270.68**
Through brick or block walls or partitions							
half brick thick	–	2.38	66.60	6.47	–	m²	**73.07**
one brick thick	–	3.95	110.64	12.92	–	m²	**123.56**
one and a half brick thick	–	5.52	154.68	19.39	–	m²	**174.07**
two brick thick	–	7.09	198.72	25.85	–	m²	**224.57**
100 mm thick blockwork	–	1.75	48.94	5.70	–	m²	**54.64**
140 mm thick blockwork	–	2.10	58.99	7.98	–	m²	**66.97**
215 mm thick blockwork	–	2.60	72.85	12.17	–	m²	**85.02**
Cut opening through wall for new window (not included); 1200 mm × 1200 mm; quoining up jambs; cut and pin galvanized steel lintel; new soldier course in facing bricks							
one brick thick wall or two half brick thick skins	8.00	7.18	195.54	18.61	187.90	nr	**402.05**
one and a half brick thick wall	8.00	9.44	258.96	27.92	187.90	nr	**474.78**
two brick thick wall	8.00	11.70	322.37	37.22	187.90	nr	**547.49**
Cut opening through wall for new door (not included); 1200 mm × 2100 mm; quoining up jambs; cut and pin galvanized steel lintel; new soldier course in facing bricks							
one brick thick wall or two half brick thick skins	8.00	17.11	500.20	32.57	264.60	nr	**797.37**
one and a half brick thick wall	8.00	22.75	665.82	48.86	303.01	nr	**1017.69**
two brick thick wall	8.00	28.39	832.84	65.14	341.36	nr	**1239.34**
Diamond cutting							
Cutting to create openings in block wall							
up to 1.00 m girth; up to 150 mm thick	–	–	–	–	–	m	**30.06**
up to 1.00 m girth; 200 mm thick	–	–	–	–	–	m	**35.87**
up to 1.00 m girth; 250 mm thick	–	–	–	–	–	m	**41.66**
up to 1.00 m girth; 300 mm thick	–	–	–	–	–	m	**47.46**
3 m–3.5 m girth; up to 150 mm thick	–	–	–	–	–	m	**135.26**
3 m–3.5 m girth; 200 mm thick	–	–	–	–	–	m	**161.35**
3 m–3.5 m girth; 250 mm thick	–	–	–	–	–	m	**187.48**
3 m–3.5 m girth; 300 mm thick	–	–	–	–	–	m	**213.55**
9 m–10 m girth; up to 150 mm thick	–	–	–	–	–	m	**448.94**
9 m–10 m girth; 200 mm thick	–	–	–	–	–	m	**598.60**
9 m–10 m girth; 250 mm thick	–	–	–	–	–	m	**748.22**
9 m–10 m girth; 300 mm thick	–	–	–	–	–	m	**897.88**
Cutting to create openings in brick wall							
up to 1.00 m girth; up to 150 mm thick	–	–	–	–	–	m	**37.57**
up to 1.00 m girth; 200 mm thick	–	–	–	–	–	m	**44.83**
up to 1.00 m girth; 250 mm thick	–	–	–	–	–	m	**52.07**

04 ALTERATIONS, REPAIRS AND CONSERVATION

Item	PC £	Labour hours	Labour £	Plant £	Material £	Unit	Total rate £
up to 1.00 m girth; 300 mm thick	–	–	–	–	–	m	59.30
3 m–3.5 m girth; up to 150 mm thick	–	–	–	–	–	m	270.58
3 m–3.5 m girth; 200 mm thick	–	–	–	–	–	m	322.73
3 m–3.5 m girth; 250 mm thick	–	–	–	–	–	m	374.98
3 m–3.5 m girth; 300 mm thick	–	–	–	–	–	m	427.10
9 m–10 m girth; up to 150 mm thick	–	–	–	–	–	m	897.88
9 m–10 m girth; 200 mm thick	–	–	–	–	–	m	1197.19
9 m–10 m girth; 250 mm thick	–	–	–	–	–	m	1496.48
9 m–10 m girth; 300 mm thick	–	–	–	–	–	m²	1795.78
Cutting to create openings in concrete wall							
up to 1.00 m girth; up to 150 mm thick	–	–	–	–	–	m	150.32
up to 1.00 m girth; 200 mm thick	–	–	–	–	–	m	179.28
up to 1.00 m girth; 250 mm thick	–	–	–	–	–	m	208.32
up to 1.00 m girth; 300 mm thick	–	–	–	–	–	m	237.28
3 m–3.5 m girth; up to 150 mm thick	–	–	–	–	–	m	1352.87
3 m–3.5 m girth; 200 mm thick	–	–	–	–	–	m	1613.62
3 m–3.5 m girth; 250 mm thick	–	–	–	–	–	m	1874.76
3 m–3.5 m girth; 300 mm thick	–	–	–	–	–	m	2135.54
9 m–10 m girth; up to 150 mm thick	–	–	–	–	–	m	4489.44
9 m–10 m girth; 200 mm thick	–	–	–	–	–	m	5985.91
9 m–10 m girth; 250 mm thick	–	–	–	–	–	m	7482.38
9 m–10 m girth; 300 mm thick	–	–	–	–	–	m	8978.86
Cutting to create openings in timber wall							
up to 1.00 m girth; up to 150 mm thick	–	–	–	–	–	m	45.08
up to 1.00 m girth; 200 mm thick	–	–	–	–	–	m	53.80
up to 1.00 m girth; 250 mm thick	–	–	–	–	–	m	62.48
up to 1.00 m girth; 300 mm thick	–	–	–	–	–	m	71.20
3 m–3.5 m girth; up to 150 mm thick	–	–	–	–	–	m	405.88
3 m–3.5 m girth; 200 mm thick	–	–	–	–	–	m	484.07
3 m–3.5 m girth; 250 mm thick	–	–	–	–	–	m	562.44
3 m–3.5 m girth; 300 mm thick	–	–	–	–	–	m	640.68
9 m–10 m girth; up to 150 mm thick	–	–	–	–	–	m	1346.83
9 m–10 m girth; 200 mm thick	–	–	–	–	–	m	1795.78
9 m–10 m girth; 250 mm thick	–	–	–	–	–	m	2244.72
9 m–10 m girth; 300 mm thick	–	–	–	–	–	m	2693.66
Quoining up jambs to openings							
Common bricks							
half brick thick wall	–	0.90	29.50	–	9.13	m	38.63
one brick thick wall	–	1.35	44.09	–	18.26	m	62.35
one and a half brick thick wall	–	1.75	57.10	–	27.41	m	84.51
two brick thick wall	–	2.15	70.44	–	36.54	m	106.98
100 mm blockwork wall	–	0.63	20.62	–	13.80	m	34.42
140 mm blockwork wall	–	0.78	25.38	–	18.62	m	44.00
215 mm blockwork wall	–	0.97	31.72	–	32.75	m	64.47
Cutting back projections; spoil into skip							
Brick projection flush with adjacent wall							
225 mm × 112 mm	–	0.27	8.88	0.38	–	m	9.26
225 mm × 225 mm	–	0.45	14.59	0.76	–	m	15.35
337 mm × 112 mm	–	0.63	20.62	1.14	–	m	21.76
450 mm × 225 mm	–	0.81	26.33	1.52	–	m	27.85
chimney breast	–	1.57	51.38	12.17	–	m²	63.55

04 ALTERATIONS, REPAIRS AND CONSERVATION

Item	PC £	Labour hours	Labour £	Plant £	Material £	Unit	Total rate £
4.02 CUTTING OR FORMING OPENINGS – cont							
Temporary screens							
Providing and erecting; maintaining temporary dust proof screens; 50 mm × 75 mm softwood framing and 12 mm thick plywood covering to one side; single layer of polythene sheet to the other; clear away on completion	–	0.90	29.50	1.45	17.36	m²	**48.31**
4.03 FILLING IN OPENINGS							
Fill existing openings							
Fill in small holes							
make good hole where small (up to 12 mm dia.) pipe removed; cement mortar	–	0.19	6.22	–	0.31	nr	**6.53**
Larger holes fill in with common brickwork or blockwork							
half brick thick	–	1.66	54.24	–	34.32	m²	**88.56**
one brick thick	–	2.74	89.45	–	72.86	m²	**162.31**
one and a half brick thick	–	3.77	123.38	–	109.30	m²	**232.68**
two brick thick	–	4.71	154.15	–	145.73	m²	**299.88**
100 mm blockwork	–	0.94	30.77	–	28.79	m²	**59.56**
140 mm blockwork	–	1.13	36.79	–	40.62	m²	**77.41**
215 mm blockwork	–	1.44	46.94	–	61.06	m²	**108.00**
cavity wall; facing bricks; 100 mm cavity fill insulation; 140 mm block	–	4.13	135.12	–	93.72	m²	**228.84**
4.04 REMOVE EXISTING AND REPLACING							
Remove existing material for replacement							
NOTE: Access equipment is not included							
Cutting out decayed, defective or cracked work and replacing with new common bricks; gauged mortar							
small areas; half brick thick walling	–	4.56	149.11	3.64	40.20	m²	**192.95**
small areas; one brick thick walling	–	8.88	290.38	5.81	83.40	m²	**379.59**
small areas; one and a half brick thick walling	–	12.58	411.37	10.18	125.11	m²	**546.66**
small areas; two brick thick walling	–	16.10	526.48	13.08	166.80	m²	**706.36**
individual bricks; half brick thick walling	–	0.28	9.16	–	0.88	nr	**10.04**
Cutting out decayed, defective or cracked work and replacing with new facing bricks; gauged mortar; facing to one side							
small areas; half brick thick walling (PC £ per 1000)	500.00	6.75	220.73	3.64	51.00	m²	**275.37**
small areas; half brick thick walling (PC £ per 1000)	700.00	6.75	220.73	3.64	69.00	m²	**293.37**
individual bricks; half brick thick walling (PC £ per 1000)	500.00	0.42	13.74	–	1.07	m²	**14.81**

04 ALTERATIONS, REPAIRS AND CONSERVATION

Item	PC £	Labour hours	Labour £	Plant £	Material £	Unit	Total rate £
Cutting out decayed, defective or cracked soldier course arch and replace with new							
facing bricks (PC £ per 1000)	500.00	1.80	58.86	0.38	13.80	m²	**73.04**
facing bricks (PC £ per 1000)	700.00	1.80	58.86	0.38	18.60	m²	**77.84**
Cutting out raking cracks in brickwork; stitching in new common bricks							
half brick thick (PC £ per 1000)	380.00	2.96	96.79	3.04	18.38	m²	**118.21**
one brick thick (PC £ per 1000)	380.00	5.41	176.90	5.70	39.78	m²	**222.38**
one and a half brick thick (PC £ per 1000)	380.00	8.09	264.54	8.75	58.18	m²	**331.47**
Cutting out raking cracks in brickwork; stitching in new facing bricks; half brick thick; facing to one side							
facing bricks (PC £ per 1000)	500.00	4.40	143.88	3.04	25.50	m²	**172.42**
facing bricks (PC £ per 1000)	700.00	4.44	145.19	3.04	34.50	m²	**182.73**
Cutting out raking cracks in cavity brickwork; stitching in new common bricks one side; facing bricks the other side; both skins half brick thick; facing to one side							
facing bricks (PC £ per 1000)	500.00	7.59	248.20	6.08	42.18	m²	**296.46**
facing bricks (PC £ per 1000)	700.00	7.59	248.20	6.08	54.60	m²	**308.88**
Cutting away old angle fillets and replacing with new; cement mortar; 50 mm face width	–	0.23	7.52	–	9.00	m	**16.52**
Cutting out ends of joists and plates from walls; making good in common bricks in cement mortar							
175 mm joists; 400 mm centres	–	0.60	19.62	–	15.54	m	**35.16**
225 mm joists; 400 mm centres	–	0.74	24.19	–	17.82	m	**42.01**
Cutting and pinning to existing brickwork ends of joists	–	0.37	12.10	–	–	nr	**12.10**
Remove defective wall and rebuild with new materials							
defective parapet wall; 600 mm high; with two courses of tiles and brick-on-edge coping over; rebuilding with new facing bricks, tiles and coping	–	6.16	201.43	10.64	90.24	m	**302.31**
defective capping stones and haunching; replace stones and re-haunch in cement mortar	–	1.39	45.46	–	18.00	m	**63.46**

4.05 PREPARING EXISTING STRUCTURES FOR CONNECTION

Prepare existing structures for connection

Item	PC £	Labour hours	Labour £	Plant £	Material £	Unit	Total rate £
Making good where intersecting wall has been removed							
half brick thick	–	0.28	9.16	–	3.00	m	**12.16**
one brick thick	–	0.37	12.10	–	6.00	m	**18.10**
100 mm blockwork	–	0.23	7.52	–	3.00	m	**10.52**
150 mm blockwork	–	0.27	8.83	–	3.00	m	**11.83**
215 mm blockwork	–	0.32	10.46	–	6.00	m	**16.46**
225 mm blockwork	–	0.36	11.77	–	6.00	m	**17.77**

04 ALTERATIONS, REPAIRS AND CONSERVATION

Item	PC £	Labour hours	Labour £	Plant £	Material £	Unit	Total rate £
4.06 REPAIRING							
Repairing concrete							
Reinstating plain concrete bed with site mixed in situ concrete; mix 20.00 N, where opening no longer required.							
100 mm thick	–	0.44	14.84	–	18.60	m²	**33.44**
150 mm thick	–	0.72	24.49	–	27.90	m²	**52.39**
Reinstating reinforced concrete bed with site mixed in situ concrete; mix 20.00 N; mesh reinforcement, where opening no longer required							
100 mm thick	–	0.66	22.43	–	24.72	m²	**47.15**
150 mm thick	–	0.91	31.03	–	34.02	m²	**65.05**
Reinstating reinforced concrete suspended floor with site mixed in situ concrete; mix 25.00 N; mesh reinforcement; formwork, where opening no longer required							
150 mm thick	–	2.96	100.06	–	50.62	m²	**150.68**
225 mm thick	–	3.12	105.56	–	65.24	m²	**170.80**
300 mm thick	–	3.60	122.10	–	79.86	m²	**201.96**
Reinstating small hole through concrete suspended slab with site mixed in situ concrete; mix 25.00 N, where opening no longer required							
50 mm dia.	–	0.39	13.44	–	0.08	nr	**13.52**
100 mm dia.	–	0.51	17.57	–	0.30	nr	**17.87**
150 mm dia.	–	0.65	22.39	–	0.67	nr	**23.06**
Clean out minor crack and fill with cement: mortar mixed with bonding agent	–	0.31	8.69	–	1.64	m	**10.33**
Clean out crack to form a 20 mm × 20 mm groove and fill with cement: mortar mixed with bonding agent	–	0.61	17.10	0.14	6.46	m	**23.70**
Repairing metal							
Overhauling and repairing metal casement windows adjust and oil ironmongery; prepare affected parts for redecoration	–	1.39	45.46	–	–	nr	**45.46**
Repairing timber							
Removing or punching projecting nails; refixing timber flooring							
loose boards	–	0.14	4.57	–	–	m²	**4.57**
refix floorboards previously set aside	–	0.74	24.19	–	1.16	m²	**25.35**
Remove damaged softwood flooring and fix new plain edge softwood flooring							
small areas	–	1.06	34.67	–	52.75	m²	**87.42**
individual boards 150 mm wide	–	0.28	9.16	–	5.34	m	**14.50**
Sanding down and resurfacing existing flooring; preparing; body in with wax polish							
softwood	–	0.70	20.33	–	–	m²	**20.33**
hardwood	–	0.85	24.68	–	–	m²	**24.68**

04 ALTERATIONS, REPAIRS AND CONSERVATION

Item	PC £	Labour hours	Labour £	Plant £	Material £	Unit	Total rate £
Fitting existing softwood skirting or architrave to new frames							
75 mm high	–	0.09	2.94	–	0.02	m	**2.96**
150 mm high	–	0.12	3.92	–	0.02	m	**3.94**
225 mm high	–	0.15	4.91	–	0.02	m	**4.93**
Piecing in new 25 mm × 150 mm moulded softwood skirtings to match existing where old has been removed; prepare for decoration	–	0.35	11.17	–	5.59	m	**16.76**
Repair doors							
Easing and adjusting softwood doors; oil ironmongery	–	0.59	19.30	–	–	nr	**19.30**
Remove softwood doors; easing and adjust; oil ironmongery; rehang	–	0.75	24.53	–	–	nr	**24.53**
Remove softwood door; plane 12 mm from bottom edge; rehang	–	1.11	36.30	–	–	nr	**36.30**
Take off existing softwood doorstops; supply and fit new 25 mm × 38 mm doorstop	–	0.10	3.28	–	1.27	m	**4.55**
Cutting out infected or decayed structural members; supply and fix new treated sawn softwood members							
Floors and flat roofs							
50 mm × 125 mm	–	0.37	12.10	–	4.56	m	**16.66**
50 mm × 150 mm	–	0.41	13.40	–	5.47	m	**18.87**
50 mm × 175 mm	–	0.44	14.39	–	5.11	m	**19.50**
Pitched roofs							
38 mm × 100 mm	–	0.33	10.79	–	5.21	m	**16.00**
50 mm × 100 mm	–	0.42	13.74	–	3.65	m	**17.39**
50 mm × 125 mm	–	0.46	15.05	–	4.56	m	**19.61**
50 mm × 150 mm	–	0.51	16.68	–	5.47	m	**22.15**
Kerb bearers							
50 mm × 75 mm	–	0.42	13.74	–	2.20	m	**15.94**
50 mm × 100 mm	–	0.52	17.00	–	2.93	m	**19.93**
75 mm × 100 mm	–	0.63	20.60	–	4.38	m	**24.98**
75 mm × 150 mm	–	0.66	21.58	–	6.58	m	**28.16**
75 mm × 175 mm	–	0.70	22.88	–	7.67	m	**30.55**
4.07 REPOINTING							
Repointing masonry							
Raking out decayed masonry joints and repointing in cement mortar							
brickwork walls generally	–	0.69	22.56	–	3.00	m²	**25.56**
cutting out staggered cracks and repointing to match existing along brick joints	–	0.37	12.10	–	–	m	**12.10**
brickwork and re-wedge horizontal flashing	–	0.23	7.52	–	1.51	m	**9.03**
brickwork and re-wedge stepped flashing	–	0.34	11.12	–	1.51	m	**12.63**
brickwork in chimney stacks	–	1.11	36.30	–	3.00	m²	**39.30**
uncoursed stonework	–	1.11	36.30	–	1.51	m	**37.81**

04 ALTERATIONS, REPAIRS AND CONSERVATION

Item	PC £	Labour hours	Labour £	Plant £	Material £	Unit	Total rate £
4.09 CLEANING SURFACES							
Cleaning surfaces by hand using hand-held manual tools only (brushes, scrapers etc.)							
Masonry/Concrete							
Cleaning surface of moss and lichen from walls	–	0.28	7.85	–	–	m²	**7.85**
Cleaning bricks of mortar, sort and stack for reuse	–	9.25	259.24	–	–	1000	**259.24**
Clean surface of concrete to receive new damp-proof membrane	–	0.14	3.92	–	–	m²	**3.92**
4.10 DECONTAMINATION							
Insecticide or fungicidal treatments							
Treating individual timbers with two coats of proprietary insecticide and fungicide by brush or spray application as appropriate							
general boarding	–	–	–	–	–	m²	**14.08**
structural members	–	–	–	–	–	m²	**12.16**
skirtings, architraves etc.	–	–	–	–	–	m	**12.53**
remove cobwebs, dust and roof insulation; treat exposed timbers with proprietary insecticide and fungicide by spray application	–	–	–	–	–	m²	**23.52**
lifting necessary floorboards; treating flooring with two coats of proprietary insecticide and fungicide by spray application; refix boards	–	–	–	–	–	m²	**22.51**
treating surfaces of adjoining masonry or concrete with two coats of proprietary dry rot by spray application	–	–	–	–	–	m²	**13.81**

05 EXCAVATE AND FILLING

Item	PC £	Labour hours	Labour £	Plant £	Material £	Unit	Total rate £
MATERIAL RATES							
Aggregates; large quantities							
MOT Type 1	–	–	–	–	22.49	tonne	**22.49**
building sand	–	–	–	–	15.34	tonne	**15.34**
20 mm shingle	–	–	–	–	15.85	tonne	**15.85**
20 mm ballast	–	–	–	–	15.85	tonne	**15.85**
Aggregates; bulk bags							
soft building sand	–	–	–	–	35.80	850 kg	**35.80**
coarse sharp sand	–	–	–	–	34.70	850 kg	**34.70**
20 mm ballast	–	–	–	–	34.70	850 kg	**34.70**
10 mm shingle	–	–	–	–	34.70	850 kg	**34.70**
20 mm shingle	–	–	–	–	34.70	850 kg	**34.70**
MOT Type 1	–	–	–	–	36.50	850 kg	**36.50**
5.01 SITE CLEARANCE AND PREPARATION							
Site clearance							
Prices are applicable to excavation in firm soil							
Fell and removing trees							
girth 500–1500 mm	–	18.50	266.53	17.42	–	nr	**283.95**
girth 1500–3000 mm	–	32.50	468.24	17.42	–	nr	**485.66**
girth exceeding 3000 mm	–	46.50	669.94	21.77	–	nr	**691.71**
Removing tree stumps							
girth 500 mm–1.50 m	–	2.00	28.82	22.77	–	nr	**51.59**
girth 1.50–3.00 m	–	3.00	43.22	32.55	–	nr	**75.77**
girth exceeding 3.00 m	–	6.00	86.45	45.65	–	nr	**132.10**
Clearing site vegetation							
bushes, scrub, undergrowth, hedges and trees and tree stumps not exceeding 500 mm girth	–	0.03	0.43	1.27	–	m²	**1.70**
Lifting turf for preservation, by hand							
stacking	–	0.32	4.61	–	–	m²	**4.61**
Topsoil for preservation; to spoil heap less than 50 m from excavations							
average depth 150 mm	–	0.02	0.29	0.91	–	m²	**1.20**
add or deduct for each 25 mm variation in average depth	–	0.01	0.15	0.21	–	m²	**0.36**
5.02 EXCAVATION							
Excavating by machine							
To reduce levels							
maximum depth not exceeding 0.25 m	–	0.03	0.48	0.70	–	m³	**1.18**
maximum depth not exceeding 1.00 m	–	0.13	1.80	0.70	–	m³	**2.50**
maximum depth not exceeding 2.00 m	–	0.04	0.53	0.76	–	m³	**1.29**
maximum depth not exceeding 4.00 m	–	0.04	0.58	0.83	–	m³	**1.41**
Basements and the like; commencing level exceeding 0.25 m below existing ground level							
maximum depth not exceeding 1.00 m	–	0.06	0.87	1.27	–	m³	**2.14**
maximum depth not exceeding 2.00 m	–	0.07	1.01	1.27	–	m³	**2.28**
maximum depth not exceeding 4.00 m	–	0.08	1.15	1.57	–	m³	**2.72**
maximum depth not exceeding 6.00 m	–	0.09	1.30	1.96	–	m³	**3.26**
maximum depth not exceeding 8.00 m	–	0.12	1.73	2.25	–	m³	**3.98**

05 EXCAVATE AND FILLING

Item	PC £	Labour hours	Labour £	Plant £	Material £	Unit	Total rate £
5.02 EXCAVATION – cont							
Excavating by machine – cont							
Pits							
maximum depth not exceeding 0.25 m	–	0.31	4.47	4.59	–	m³	**9.06**
maximum depth not exceeding 1.00 m	–	0.33	4.76	4.59	–	m³	**9.35**
maximum depth not exceeding 2.00 m	–	0.39	5.62	5.18	–	m³	**10.80**
maximum depth not exceeding 4.00 m	–	0.47	6.77	5.86	–	m³	**12.63**
maximum depth not exceeding 6.00 m	–	0.49	7.06	6.16	–	m³	**13.22**
Extra over pit excavating for commencing level exceeding 0.25 m below existing ground level							
1.00 m below	–	0.03	0.43	0.68	–	m³	**1.11**
2.00 m below	–	0.05	0.72	0.98	–	m³	**1.70**
3.00 m below	–	0.06	0.87	1.27	–	m³	**2.14**
4.00 m below	–	0.09	1.30	1.66	–	m³	**2.96**
Trenches; width not exceeding 0.30 m							
maximum depth not exceeding 0.25 m	–	0.26	3.75	3.62	–	m³	**7.37**
maximum depth not exceeding 1.00 m	–	0.28	4.04	3.62	–	m³	**7.66**
maximum depth not exceeding 2.00 m	–	0.33	4.76	4.20	–	m³	**8.96**
maximum depth not exceeding 4.00 m	–	0.40	5.77	5.18	–	m³	**10.95**
maximum depth not exceeding 6.00 m	–	0.46	6.62	6.16	–	m³	**12.78**
Trenches; width exceeding 0.30 m							
maximum depth not exceeding 0.25 m	–	0.23	3.32	3.22	–	m³	**6.54**
maximum depth not exceeding 1.00 m	–	0.25	3.61	3.22	–	m³	**6.83**
maximum depth not exceeding 2.00 m	–	0.30	4.33	3.91	–	m³	**8.24**
maximum depth not exceeding 4.00 m	–	0.35	5.05	4.59	–	m³	**9.64**
maximum depth not exceeding 6.00 m	–	0.43	6.19	5.86	–	m³	**12.05**
Extra over trench excavating for commencing level exceeding 0.25 m below existing ground level							
1.00 m below	–	0.03	0.43	0.68	–	m³	**1.11**
2.00 m below	–	0.05	0.72	0.98	–	m³	**1.70**
3.00 m below	–	0.06	0.87	1.27	–	m³	**2.14**
4.00 m below	–	0.09	1.30	1.66	–	m³	**2.96**
For pile caps and ground beams between piles							
maximum depth not exceeding 0.25 m	–	0.35	5.05	6.25	–	m³	**11.30**
maximum depth not exceeding 1.00 m	–	0.39	5.62	6.16	–	m³	**11.78**
maximum depth not exceeding 2.00 m	–	0.39	5.62	6.84	–	m³	**12.46**
To bench sloping ground to receive filling							
maximum depth not exceeding 0.25 m	–	0.07	1.01	1.66	–	m³	**2.67**
maximum depth not exceeding 1.00 m	–	0.09	1.30	1.66	–	m³	**2.96**
maximum depth not exceeding 2.00 m	–	0.09	1.30	1.96	–	m³	**3.26**
Extra over any types of excavating irrespective of depth							
excavating below ground water level	–	0.13	1.87	2.25	–	m³	**4.12**
next to existing services	–	0.35	5.05	1.27	–	m³	**6.32**
around existing services crossing excavation	–	0.60	8.64	3.62	–	m³	**12.26**
Extra over any types of excavating irrespective of depth for breaking out existing materials							
rock	–	2.95	42.50	20.30	–	m³	**62.80**
concrete	–	2.55	36.74	16.01	–	m³	**52.75**
reinforced concrete	–	3.60	51.87	23.33	–	m³	**75.20**
brickwork, blockwork or stonework	–	1.85	26.66	11.78	–	m³	**38.44**

05 EXCAVATE AND FILLING

Item	PC £	Labour hours	Labour £	Plant £	Material £	Unit	Total rate £
Extra over any types of excavating irrespective of depth for breaking out existing hard pavings, 75 mm thick							
coated macadam or asphalt	–	0.19	2.74	1.01	–	m²	3.75
Extra over any types of excavating irrespective of depth for breaking out existing hard pavings, 150 mm thick							
concrete	–	0.39	5.62	2.43	–	m²	8.05
reinforced concrete	–	0.58	8.35	3.32	–	m²	11.67
coated macadam or asphalt and hardcore	–	0.26	3.75	1.13	–	m²	4.88
Working space allowance to excavations 600 mm wide							
reduce levels, basements and the like	–	0.07	1.01	1.27	–	m²	2.28
pits	–	0.19	2.74	3.62	–	m²	6.36
trenches	–	0.18	2.60	3.22	–	m²	5.82
pile caps and ground beams between piles	–	0.20	2.88	3.62	–	m²	6.50
Extra over excavating for working space for backfilling in with special materials							
hardcore	–	0.13	1.87	1.60	19.99	m²	23.46
sand	–	0.13	1.87	1.60	18.22	m²	21.69
40 mm–20 mm gravel	–	0.13	1.87	1.60	23.11	m²	26.58
plain in situ ready mixed designated concrete C7.5–40 mm aggregate	–	0.93	15.72	1.67	64.39	m²	81.78
Excavating by hand							
Topsoil for preservation							
average depth 150 mm	–	0.23	3.32	–	–	m²	3.32
add or deduct for each 25 mm variation in average depth	–	0.03	0.43	–	–	m²	0.43
To reduce levels							
maximum depth not exceeding 0.25 m	–	1.44	20.74	–	–	m³	20.74
maximum depth not exceeding 1.00 m	–	1.63	23.48	–	–	m³	23.48
maximum depth not exceeding 2.00 m	–	1.80	25.94	–	–	m³	25.94
maximum depth not exceeding 4.00 m	–	1.99	28.68	–	–	m³	28.68
Basements and the like; commencing level exceeding 0.25 m below existing ground level							
maximum depth not exceeding 1.00 m	–	1.90	27.38	–	–	m³	27.38
maximum depth not exceeding 2.00 m	–	2.04	29.39	–	–	m³	29.39
maximum depth not exceeding 4.00 m	–	2.73	39.34	–	–	m³	39.34
maximum depth not exceeding 6.00 m	–	3.33	47.98	–	–	m³	47.98
maximum depth not exceeding 8.00 m	–	4.02	57.92	–	–	m³	57.92
Pits							
maximum depth not exceeding 0.25 m	–	2.13	30.68	–	–	m³	30.68
maximum depth not exceeding 1.00 m	–	2.75	39.62	–	–	m³	39.62
maximum depth not exceeding 2.00 m	–	3.30	47.54	–	–	m³	47.54
maximum depth not exceeding 4.00 m	–	4.18	60.22	–	–	m³	60.22
maximum depth not exceeding 6.00 m	–	5.17	74.49	–	–	m³	74.49
Extra over pit excavating for commencing level exceeding 0.25 m below existing ground level							
1.00 m below	–	0.42	6.05	–	–	m³	6.05
2.00 m below	–	0.88	12.68	–	–	m³	12.68
3.00 m below	–	1.30	18.73	–	–	m³	18.73
4.00 m below	–	1.71	24.64	–	–	m³	24.64

05 EXCAVATE AND FILLING

Item	PC £	Labour hours	Labour £	Plant £	Material £	Unit	Total rate £
5.02 EXCAVATION – cont							
Excavating by hand – cont							
Trenches; width not exceeding 0.30 m							
maximum depth not exceeding 0.25 m	–	1.85	26.66	–	–	m³	**26.66**
maximum depth not exceeding 1.00 m	–	2.76	39.77	–	–	m³	**39.77**
Trenches; width exceeding 0.30 m							
maximum depth not exceeding 0.25 m	–	1.80	25.94	–	–	m³	**25.94**
maximum depth not exceeding 1.00 m	–	2.46	35.44	–	–	m³	**35.44**
maximum depth not exceeding 2.00 m	–	2.88	41.49	–	–	m³	**41.49**
maximum depth not exceeding 4.00 m	–	3.66	52.74	–	–	m³	**52.74**
maximum depth not exceeding 6.00 m	–	4.68	67.42	–	–	m³	**67.42**
Extra over trench excavating for commencing level exceeding 0.25 m below existing ground level							
1.00 m below	–	0.42	6.05	–	–	m³	**6.05**
2.00 m below	–	0.88	12.68	–	–	m³	**12.68**
3.00 m below	–	1.30	18.73	–	–	m³	**18.73**
4.00 m below	–	1.71	24.64	–	–	m³	**24.64**
For pile caps and ground beams between piles							
maximum depth not exceeding 0.25 m	–	2.78	40.06	–	–	m³	**40.06**
maximum depth not exceeding 1.00 m	–	2.96	42.64	–	–	m³	**42.64**
maximum depth not exceeding 2.00 m	–	3.52	50.72	–	–	m³	**50.72**
To bench sloping ground to receive filling							
maximum depth not exceeding 0.25 m	–	1.30	18.73	–	–	m³	**18.73**
maximum depth not exceeding 1.00 m	–	1.48	21.32	–	–	m³	**21.32**
maximum depth not exceeding 2.00 m	–	1.67	24.06	–	–	m³	**24.06**
Extra over any types of excavating irrespective of depth							
excavating below ground water level	–	0.32	4.61	–	–	m³	**4.61**
next existing services	–	0.93	13.40	–	–	m³	**13.40**
around existing services crossing excavation	–	1.85	26.66	–	–	m³	**26.66**
Extra over any types of excavating irrespective of depth for breaking out existing materials							
rock	–	4.63	66.70	18.25	–	m³	**84.95**
concrete	–	4.16	59.94	15.20	–	m³	**75.14**
reinforced concrete	–	5.55	79.96	21.29	–	m³	**101.25**
brickwork, blockwork or stonework	–	2.78	40.06	9.13	–	m³	**49.19**
Extra over any types of excavating irrespective of depth for breaking out existing hard pavings, 60 mm thick							
precast concrete paving slabs	–	0.28	4.04	–	–	m²	**4.04**
Extra over any types of excavating irrespective of depth for breaking out existing hard pavings, 75 mm thick							
coated macadam or asphalt	–	0.37	5.34	1.23	–	m²	**6.57**
Extra over any types of excavating irrespective of depth for breaking out existing hard pavings, 150 mm thick							
concrete	–	0.65	9.36	2.13	–	m²	**11.49**
reinforced concrete	–	0.83	11.96	3.05	–	m²	**15.01**
coated macadam or asphalt and hardcore	–	0.46	6.62	1.52	–	m²	**8.14**

05 EXCAVATE AND FILLING

Item	PC £	Labour hours	Labour £	Plant £	Material £	Unit	Total rate £
Working space allowance to excavations							
reduce levels, basements and the like	–	2.13	30.68	1.87	–	m²	**32.55**
pits	–	2.22	31.98	1.87	–	m²	**33.85**
trenches	–	1.94	27.95	1.87	–	m²	**29.82**
pile caps and ground beams between piles	–	2.31	33.28	1.87	–	m²	**35.15**
Extra over excavation for working space for backfilling with special materials							
hardcore	–	0.74	10.66	1.87	21.42	m²	**33.95**
sand	–	0.74	10.66	1.87	21.05	m²	**33.58**
40 mm–20 mm gravel	–	0.74	10.66	1.87	23.11	m²	**35.64**
plain in situ concrete ready mixed designated concrete; C7.5–40 mm aggregate	–	1.02	17.23	1.87	64.39	m²	**83.49**
5.03 SUPPORT TO EXCAVATIONS							
Earthwork support (average risk prices)							
Maximum depth not exceeding 1.00 m							
distance between opposing faces not exceeding 2.00 m	–	0.10	1.37	–	–	m²	**1.37**
distance between opposing faces 2.00–4.00 m	–	0.10	1.50	–	–	m²	**1.50**
distance between opposing faces exceeding 4.00 m	–	0.11	1.64	–	–	m²	**1.64**
Maximum depth not exceeding 2.00 m							
distance between opposing faces not exceeding 2.00 m	–	0.11	1.64	–	–	m²	**1.64**
distance between opposing faces 2.00–4.00 m	–	0.12	1.78	–	–	m²	**1.78**
distance between opposing faces exceeding 4.00 m	–	0.13	1.92	–	–	m²	**1.92**
Maximum depth not exceeding 4.00 m							
distance between opposing faces not exceeding 2.00 m	–	0.15	2.19	–	–	m²	**2.19**
distance between opposing faces 2.00–4.00 m	–	0.15	2.19	–	–	m²	**2.19**
distance between opposing faces exceeding 4.00 m	–	0.17	2.46	–	–	m²	**2.46**
Maximum depth not exceeding 6.00 m							
distance between opposing faces not exceeding 2.00 m	–	0.17	2.46	–	–	m²	**2.46**
distance between opposing faces 2.00–4.00 m	–	0.18	2.60	–	–	m²	**2.60**
distance between opposing faces exceeding 4.00 m	–	0.21	3.01	–	–	m²	**3.01**
Maximum depth not exceeding 8.00 m							
distance between opposing faces not exceeding 2.00 m	–	0.22	3.15	–	–	m²	**3.15**
distance between opposing faces 2.00–4.00 m	–	0.27	3.83	–	–	m²	**3.83**
distance between opposing faces exceeding 4.00 m	–	0.31	4.52	–	–	m²	**4.52**

05 EXCAVATE AND FILLING

Item	PC £	Labour hours	Labour £	Plant £	Material £	Unit	Total rate £
5.03 SUPPORT TO EXCAVATIONS – cont							
Earthwork support (open boarded)							
Maximum depth not exceeding 1.00 m							
distance between opposing faces not exceeding 2.00 m	–	0.27	3.83	–	–	m^2	**3.83**
distance between opposing faces 2.00–4.00 m	–	0.29	4.24	–	–	m^2	**4.24**
distance between opposing faces exceeding 4.00 m	–	0.33	4.79	–	–	m^2	**4.79**
Maximum depth not exceeding 2.00 m							
distance between opposing faces not exceeding 2.00 m	–	0.33	4.79	–	–	m^2	**4.79**
distance between opposing faces 2.00–4.00 m	–	0.37	5.34	–	–	m^2	**5.34**
distance between opposing faces exceeding 4.00 m	–	0.42	6.03	–	–	m^2	**6.03**
Maximum depth not exceeding 4.00 m							
distance between opposing faces not exceeding 2.00 m	–	0.42	6.03	–	–	m^2	**6.03**
distance between opposing faces 2.00–4.00 m	–	0.47	6.84	–	–	m^2	**6.84**
distance between opposing faces exceeding 4.00 m	–	0.53	7.66	–	–	m^2	**7.66**
Maximum depth not exceeding 6.00 m							
distance between opposing faces not exceeding 2.00 m	–	0.53	7.66	–	–	m^2	**7.66**
distance between opposing faces 2.00–4.00 m	–	0.58	8.35	–	–	m^2	**8.35**
distance between opposing faces exceeding 4.00 m	–	0.67	9.58	–	–	m^2	**9.58**
Maximum depth not exceeding 8.00 m							
distance between opposing faces not exceeding 2.00 m	–	0.70	10.12	–	–	m^2	**10.12**
distance between opposing faces 2.00–4.00 m	–	0.79	11.36	–	–	m^2	**11.36**
distance between opposing faces exceeding 4.00 m	–	0.92	13.28	–	–	m^2	**13.28**
Earthwork support (close boarded)							
Maximum depth not exceeding 1.00 m							
distance between opposing faces not exceeding 2.00 m	–	0.70	10.12	–	–	m^2	**10.12**
distance between opposing faces 2.00–4.00 m	–	0.77	11.08	–	–	m^2	**11.08**
distance between opposing faces exceeding 4.00 m	–	0.85	12.32	–	–	m^2	**12.32**
Maximum depth not exceeding 2.00 m							
distance between opposing faces not exceeding 2.00 m	–	0.88	12.73	–	–	m^2	**12.73**
distance between opposing faces 2.00–4.00 m	–	0.97	13.96	–	–	m^2	**13.96**
distance between opposing faces exceeding 4.00 m	–	1.05	15.19	–	–	m^2	**15.19**

05 EXCAVATE AND FILLING

Item	PC £	Labour hours	Labour £	Plant £	Material £	Unit	Total rate £
Maximum depth not exceeding 4.00 m							
distance between opposing faces not exceeding 2.00 m	–	1.10	15.87	–	–	m²	**15.87**
distance between opposing faces 2.00–4.00 m	–	1.24	17.79	–	–	m²	**17.79**
distance between opposing faces exceeding 4.00 m	–	1.36	19.57	–	–	m²	**19.57**
Maximum depth not exceeding 6.00 m							
distance between opposing faces not exceeding 2.00 m	–	1.37	19.71	–	–	m²	**19.71**
distance between opposing faces 2.00–4.00 m	–	1.49	21.49	–	–	m²	**21.49**
distance between opposing faces exceeding 4.00 m	–	1.67	24.09	–	–	m²	**24.09**
Maximum depth not exceeding 8.00 m							
distance between opposing faces not exceeding 2.00 m	–	1.67	24.09	–	–	m²	**24.09**
distance between opposing faces 2.00–4.00 m	–	1.84	26.55	–	–	m²	**26.55**
distance between opposing faces exceeding 4.00 m	–	2.11	30.39	–	–	m²	**30.39**
Extra over earthwork support for							
curved	–	0.02	0.28	–	–	m²	**0.28**
below ground water level	–	0.27	3.83	–	–	m²	**3.83**
unstable ground	–	0.44	6.29	–	–	m²	**6.29**
next to roadways	–	0.35	5.07	–	–	m²	**5.07**
left in	–	0.57	8.21	–	–	m²	**8.21**
Earthwork support (average risk prices – inside existing buildings)							
Maximum depth not exceeding 1.00 m							
distance between opposing faces not exceeding 2.00 m	–	0.18	2.60	–	–	m²	**2.60**
distance between opposing faces 2.00–4.00 m	–	0.19	2.74	–	–	m²	**2.74**
distance between opposing faces exceeding 4.00 m	–	0.22	3.17	–	–	m²	**3.17**
Maximum depth not exceeding 2.00 m							
distance between opposing faces not exceeding 2.00 m	–	0.22	3.17	–	–	m²	**3.17**
distance between opposing faces 2.00–4.00 m	–	0.24	3.46	–	–	m²	**3.46**
distance between opposing faces exceeding 4.00 m	–	0.32	4.61	–	–	m²	**4.61**
Maximum depth not exceeding 4.00 m							
distance between opposing faces not exceeding 2.00 m	–	0.28	4.04	–	–	m²	**4.04**
distance between opposing faces 2.00–4.00 m	–	0.31	4.47	–	–	m²	**4.47**
distance between opposing faces exceeding 4.00 m	–	0.34	4.90	–	–	m²	**4.90**
Maximum depth not exceeding 6.00 m							
distance between opposing faces not exceeding 2.00 m	–	0.34	4.90	–	–	m²	**4.90**
distance between opposing faces 2.00–4.00 m	–	0.38	5.48	–	–	m²	**5.48**
distance between opposing faces exceeding 4.00 m	–	0.43	6.19	–	–	m²	**6.19**

Prices for Measured Works

05 EXCAVATE AND FILLING

Item	PC £	Labour hours	Labour £	Plant £	Material £	Unit	Total rate £
5.04 DISPOSAL							
Disposal of excavated material off site							
Basic Landfill tax rates (exclusive of site tipping charges)							
inactive waste	–	–	–	–	3.10	tonne	**3.10**
all other taxable wastes	–	–	–	–	96.70	tonne	**96.70**
Excavated material deposited off site							
inactive waste off site; to tip not exceeding 13 km (using lorries); including Landfill Tax	–	0.10	1.40	16.03	21.70	m³	**39.13**
active non-hazardous waste off site; to tip not exceeding 13 km (using lorries); including Landfill Tax	–	0.06	1.04	16.03	274.06	m³	**291.13**
5.05 RETAINING EXCAVATED MATERIAL ON SITE							
Retain excavated material on site							
Excavated material deposited on site using dumpers to transport material							
on site; spreading; average 25 m distance	–	0.20	2.88	0.48	–	m³	**3.36**
on site; depositing in spoil heaps; average 50 m distance	–	–	–	1.18	–	m³	**1.18**
on site; spreading; average 50 m distance	–	0.20	2.88	0.91	–	m³	**3.79**
on site; depositing in spoil heaps; average 100 m distance	–	–	–	2.08	–	m³	**2.08**
on site; spreading; average 100 m distance	–	0.20	2.88	1.39	–	m³	**4.27**
on site; depositing in spoil heaps; average 200 m distance	–	–	–	2.64	–	m³	**2.64**
on site; spreading; average 200 m distance	–	0.20	2.88	1.87	–	m³	**4.75**
5.06 FILLING OBTAINED FROM EXCAVATED MATERIAL							
Filling to make up levels							
By machine							
average thickness over 50 mm not exceeding 500 mm	–	0.22	3.11	2.25	–	m³	**5.36**
average thickness exceeding 500 mm	–	0.18	2.60	1.62	–	m³	**4.22**
By hand							
average thickness over 50 mm not exceeding 500 mm	–	1.25	18.00	4.41	–	m³	**22.41**
average thickness exceeding 500 mm	–	1.02	14.70	3.58	–	m³	**18.28**
Surface treatments							
Surface packing to filling							
To vertical or battered faces	–	0.17	2.45	0.20	–	m²	**2.65**
Compacting							
bottoms of excavations	–	0.04	0.58	0.04	–	m²	**0.62**
Trimming							
sloping surfaces; by hand	–	0.17	2.45	–	–	m²	**2.45**

05 EXCAVATE AND FILLING

Item	PC £	Labour hours	Labour £	Plant £	Material £	Unit	Total rate £
5.07 IMPORTED FILLING							
Basic material prices delivered to site							
Supply only in full loads							
D.O.T. type 1	–	–	–	–	22.61	tonne	**22.61**
D.O.T. type 2	–	–	–	–	22.61	tonne	**22.61**
10 mm single size aggregate	–	–	–	–	21.73	tonne	**21.73**
20 mm single size aggregate	–	–	–	–	21.73	tonne	**21.73**
20 mm all in aggregate	–	–	–	–	24.82	tonne	**24.82**
40 mm all in aggregate	–	–	–	–	24.82	tonne	**24.82**
40 mm scalpings	–	–	–	–	21.73	tonne	**21.73**
hardcore	–	–	–	–	17.72	tonne	**17.72**
soft/building sand	–	–	–	–	15.41	tonne	**15.41**
recycled type 1	–	–	–	–	20.19	tonne	**20.19**
E-blend (50% type 1 and 50% recycled type 1)	–	–	–	–	16.48	tonne	**16.48**
Filling to make up levels; by machine							
Average thickness over 50 mm not exceeding 500 mm							
obtained off site; imported topsoil	20.24	0.22	3.11	1.87	20.85	m³	**25.83**
obtained off site; hardcore	35.49	0.25	3.63	2.16	36.55	m³	**42.34**
obtained off site; granular fill type one	51.90	0.25	3.63	3.76	53.46	m³	**60.85**
obtained off site; granular fill type two	51.90	0.25	3.63	3.76	53.46	m³	**60.85**
obtained off site; sand	35.39	0.25	3.63	3.76	36.45	m³	**43.84**
Average thickness over 500 mm							
obtained off site; imported topsoil	20.24	0.18	2.60	2.10	20.85	m³	**25.55**
obtained off site; hardcore	35.12	0.22	3.11	0.41	36.17	m³	**39.69**
obtained off site; granular fill type one	51.90	0.22	3.11	2.94	53.46	m³	**59.51**
obtained off site; granular fill type two	51.90	0.22	3.11	2.94	53.46	m³	**59.51**
obtained off site; sand	35.39	0.22	3.11	2.94	36.45	m³	**42.50**
Filling to make up levels; by hand							
Average thickness over 50 mm not exceeding 500 mm							
obtained off site; imported topsoil	20.24	1.25	18.00	4.40	20.85	m³	**43.25**
obtained off site; hardcore	35.49	1.39	20.02	4.88	36.55	m³	**61.45**
obtained off site; granular fill type one	51.90	1.54	22.19	5.41	53.46	m³	**81.06**
obtained off site; granular fill type two	51.90	1.54	22.19	5.41	53.46	m³	**81.06**
obtained off site; sand	35.39	1.54	22.19	5.41	36.45	m³	**64.05**
Average thickness over 500 mm							
obtained off site; imported topsoil	20.24	1.02	14.70	3.58	20.85	m³	**39.13**
obtained off site; hardcore	35.49	1.34	19.30	4.72	36.55	m³	**60.57**
obtained off site; granular fill type one	51.90	1.43	20.60	5.06	53.46	m³	**79.12**
obtained off site; granular fill type two	51.90	1.43	20.60	5.06	53.46	m³	**79.12**
obtained off site; sand	35.39	1.43	20.60	5.06	36.45	m³	**62.11**
Surface treatments							
Surface packing to filling							
To vertical or battered faces	–	0.17	2.45	0.20	–	m²	**2.65**

05 EXCAVATE AND FILLING

Item	PC £	Labour hours	Labour £	Plant £	Material £	Unit	Total rate £
5.07 IMPORTED FILLING – cont							
Surface treatments – cont							
Compacting							
bottoms of excavations	–	0.04	0.58	0.04	–	m²	0.62
filling; blinding with sand 50 mm thick	–	0.04	0.58	0.04	1.48	m²	2.10
Trimming							
sloping surfaces; by hand	–	0.17	2.45	–	–	m²	2.45
sloping surfaces; in rock	–	0.93	13.30	4.25	–	m²	17.55
5.08 GEOTEXTILE FABRIC							
Erosion mats							
Filter membrane; one layer; laid on earth to receive granular material; over 500 mm wide							
Terram 500 filter membrane or other equal; one layer; laid on earth. NB mainly agricultural/ landscape use	0.41	0.04	0.58	–	0.45	m²	1.03
Terram 700 filter membrane or other equal and approved; one layer; laid on earth	0.63	0.04	0.58	–	0.70	m²	1.28
Terram 1000; filter membrane or other equal and approved; one layer; laid on earth	0.55	0.04	0.58	–	0.61	m²	1.19
Terram 2000; filter membrane or other equal and approved; one layer; laid on earth	1.24	0.04	0.58	–	1.47	m²	2.05
Terram High Visibility; orange top surface to warn of potential hazardous material underneath	0.74	0.04	0.58	–	0.88	m²	1.46
5.09 MEMBRANES							
Sheeting to prevent moisture loss/gas barrier							
Building paper; lapped joints							
subsoil grade 410; horizontal on foundations	–	0.02	0.34	–	0.85	m²	1.19
standard grade 420; horizontal on slabs	–	0.04	0.68	–	1.28	m²	1.96
Polythene sheeting; lapped joints; horizontal on slabs							
250 microns; 0.25 mm thick	–	0.04	0.68	–	0.47	m²	1.15
Visqueen sheeting or other equal and approved; lapped joints; horizontal on slabs							
250 microns; 0.25 mm thick	–	0.04	0.68	–	0.39	m²	1.07
300 microns; 0.30 mm thick	–	0.05	0.84	–	0.46	m²	1.30
gas-resistant damp-proof membrane taped joints	–	0.05	0.84	–	12.94	m²	13.78
sealing at perimeter of slab	–	0.10	1.69	–	1.48	m	3.17
sealing around penetrations; drain pipes etc.	–	1.00	16.89	–	12.36	nr	29.25

07 PILING

Item	PC £	Labour hours	Labour £	Plant £	Material £	Unit	Total rate £
7.01 INTERLOCKING SHEET PILES							
Arcelor 600 mm wide 'U' shaped steel sheeting piling; or other equal approved; pitched and driven							
Provision of all plant for sheet pile installation; including bringing to and removing from site; maintenance, erection and dismantling; assuming one rig for 1500 m² of piling							
leader rig with vibratory hammer	–	–	–	–	–	item	**4895.10**
conventional rig	–	–	–	–	–	item	**6394.50**
silent vibrationless rig	–	–	–	–	–	item	**7133.18**
Supply only of standard sheet pile sections							
PU12	–	–	–	–	–	m²	**126.79**
PU18–1	–	–	–	–	–	m²	**137.81**
PU22–1	–	–	–	–	–	m²	**157.66**
PU25	–	–	–	–	–	m²	**179.71**
PU32	–	–	–	–	–	m²	**220.50**
Pitching and driving of sheet piles; using the following plant							
leader rig with vibratory hammer	–	–	–	–	–	m²	**27.56**
conventional rig	–	–	–	–	–	m²	**37.48**
silent vibrationless rig	–	–	–	–	–	m²	**58.43**
Provision of all plant for sheet pile extraction; including bringing to and removing from site; maintenance, erection and dismantling; assuming one rig for 1500 m² of piling							
leader rig with vibratory hammer	–	–	–	–	–	item	**5016.38**
conventional rig	–	–	–	–	–	item	**6394.50**
silent vibrationless rig	–	–	–	–	–	item	**6394.50**
Extraction of sheet piles; using the following plant							
leader rig with vibratory hammer	–	–	–	–	–	m²	**19.84**
conventional rig	–	–	–	–	–	m²	**23.81**
silent vibrationless rig	–	–	–	–	–	m²	**34.40**
Credit on extracted piles; recovered in re-usable lengths; for sheet pile sections							
PU12	–	–	–	–	–	m²	**89.96**
PU18–1	–	–	–	–	–	m²	**99.22**
PU22–1	–	–	–	–	–	m²	**112.45**
PU25	–	–	–	–	–	m²	**126.79**
PU32	–	–	–	–	–	m²	**155.45**

07 PILING

Item	PC £	Labour hours	Labour £	Plant £	Material £	Unit	Total rate £
7.02 BORED PILES							
NOTE: The following approximate prices, for the quantities of piling quoted, are for work on clear open sites with reasonable access. They are based on normal concrete mix 35.00 N/mm^2; reinforced and include up to 0.15 m of projecting reinforcement at top of pile. The prices do not allow for removal of spoil.							
Minipile cast-in-place concrete piles							
300 mm nominal dia.							
Establish equipment on site							
Provision of all plant (1 nr rig); including bringing to and removing from site; maintenance, erection and dismantling at each pile position for 50 nr piles	–	–	–	–	–	item	5300.00
Bored piles							
300 mm dia. piles; nominally reinforced; 15 m long	–	–	–	–	–	nr	229.50
set up at each pile position	–	–	–	–	–	nr	24.08
add for additional piles length	–	–	–	–	–	m	15.30
deduct for reduction in pile length	–	–	–	–	–	m	5.47
Delays							
rig standing time	–	–	–	–	–	hour	214.20
Pile tests							
integrity tests (minimum 20 per visit)	–	–	–	–	–	nr	16.32
300 mm nominal dia.; small area with difficult access and limited headroom							
Establish equipment on site							
Provision of all plant (1 nr rig); including bringing to and removing from site; maintenance, erection and dismantling at each pile position for 50 nr piles	–	–	–	–	–	item	2900.00
Bored piles							
300 mm dia. piles; nominally reinforced; 15 m long	–	–	–	–	–	nr	229.50
set up at each pile position	–	–	–	–	–	nr	24.08
add for additional piles length	–	–	–	–	–	m	49.27
deduct for reduction in pile length	–	–	–	–	–	m	19.71
Delays							
rig standing time	–	–	–	–	–	hour	204.00
Pile tests							
integrity tests (minimum 20 per visit)	–	–	–	–	–	nr	16.32
450 mm nominal dia.							
Establish equipment on site							
Provision of all plant (2 nr rigs); including bringing to and removing from site; maintenance, erection and dismantling at each pile position for 100 nr piles	–	–	–	–	–	item	24480.00
Bored piles							
450 mm dia. piles; nominally reinforced; 20 m long	–	–	–	–	–	nr	3570.00
add for additional piles length	–	–	–	–	–	m	163.20
deduct for reduction in pile length	–	–	–	–	–	m	21.83

07 PILING

Item	PC £	Labour hours	Labour £	Plant £	Material £	Unit	Total rate £
Blind bored piles							
450 mm dia.	–	–	–	–	–	m	142.80
Delays							
rig standing time	–	–	–	–	–	hour	204.00
Extra over piling							
breaking through obstructions	–	–	–	–	–	hour	206.04
Pile tests							
working to 600 kN; using tension piles as reaction; first pile	–	–	–	–	–	nr	11750.00
working to 600 kN; using tension piles as reaction; subsequent piles	–	–	–	–	–	nr	11750.00
integrity tests (minimum 20 per visit)	–	–	–	–	–	nr	16.32
Rotary CFA bored cast-in-place concrete piles							
450 mm dia. piles							
Provision of plant							
provision of all plant (1 nr rig); including bringing to and removing from site; maintenance, erection and dismantling at each pile position for 100 nr piles 20 m long	–	–	–	–	–	item	17000.00
Bored piles							
450 mm dia. piles; reinforced; 20 m long	–	–	–	–	–	nr	937.89
add for additional piles length	–	–	–	–	–	m	46.31
deduct for reduction in pile length	–	–	–	–	–	m	18.53
Delays							
rig standing time	–	–	–	–	–	hour	475.95
Pile tests							
working to 600 kN; using tension piles as reaction; first pile	–	–	–	–	–	nr	6650.00
working to 600 kN; using tension piles as reaction; subsequent piles	–	–	–	–	–	nr	6640.00
integrity tests (minimum 20 per visit)	–	–	–	–	–	nr	16.32
600 mm dia. piles							
Provision of plant							
provision of all plant (1 nr rig); including bringing to and removing from site; maintenance, erection and dismantling at each pile position for 100 nr piles 20 m long	–	–	–	–	–	item	17000.00
Bored piles							
600 mm dia. piles; reinforced; 20 m long	–	–	–	–	–	nr	1473.19
add for additional piles length	–	–	–	–	–	m	75.27
deduct for reduction in pile length	–	–	–	–	–	m	32.43
Delays							
rig standing time	–	–	–	–	–	hour	451.57
Pile tests							
working to 1000 kN; using tension piles as reaction; first pile	–	–	–	–	–	nr	8200.00
working to 1000 kN; using tension piles as reaction; subsequent piles	–	–	–	–	–	nr	8200.00
integrity test 600 mm dia. piles; min 20 per visit	–	–	–	–	–	nr	16.32

07 PILING

Item	PC £	Labour hours	Labour £	Plant £	Material £	Unit	Total rate £
7.02 BORED PILES – cont							
750 mm dia. piles							
Provision of plant							
provision of all plant (1 nr rig); including bringing to and removing from site; maintenance, erection and dismantling at each pile position for 100 nr piles 20 m long	–	–	–	–	–	item	20000.00
Bored piles							
750 mm dia. piles; reinforced; 20 m long	–	–	–	–	–	nr	2317.95
add for additional piles length	–	–	–	–	–	m	115.78
deduct for reduction in pile length	–	–	–	–	–	m	46.31
Delays							
rig standing time	–	–	–	–	–	hour	544.20
Pile tests							
working to 1500 kN; using tension piles as reaction; first pile	–	–	–	–	–	nr	10000.00
working to 1500 kN; using tension piles as reaction; subsequent piles	–	–	–	–	–	nr	10000.00
integrity test 600 mm dia. piles; min 20 per visit	–	–	–	–	–	nr	16.32
900 mm dia. piles							
Provision of plant							
provision of all plant (1 nr rig); including bringing to and removing from site; maintenance, erection and dismantling at each pile position for 100 nr piles 20 m long	–	–	–	–	–	item	19800.00
Bored piles							
900 mm dia. piles; reinforced; 20 m long	–	–	–	–	–	nr	3350.00
add for additional piles length	–	–	–	–	–	m	167.89
deduct for reduction in pile length	–	–	–	–	–	m	69.47
Delays							
rig standing time	–	–	–	–	–	hour	544.20
Pile tests							
working to 2000 kN; using tension piles as reaction; first pile	–	–	–	–	–	nr	13000.00
working to 2000 kN; using tension piles as reaction; subsequent piles	–	–	–	–	–	nr	13000.00
integrity test 900 mm dia. piles; min 20 per visit	–	–	–	–	–	nr	16.32
Pile tops							
Cut off and prepare reinforcement to receive reinforced concrete pile cap; not including disposal							
450 mm dia. piles	–	1.00	24.93	–	–	m	24.93
600 mm dia. piles	–	1.20	29.91	–	–	m	29.91
750 mm dia. piles	–	1.50	37.39	–	–	m	37.39
900 mm dia. piles	–	1.80	44.87	–	–	m	44.87

08 UNDERPINNING

Item	PC £	Labour hours	Labour £	Plant £	Material £	Unit	Total rate £
8.01 UNDERPINNING							
Excavating; by machine							
Preliminary trenches							
maximum depth not exceeding 1.00 m	–	0.23	3.32	6.87	–	m³	10.19
maximum depth not exceeding 2.00 m	–	0.28	4.04	8.27	–	m³	12.31
maximum depth not exceeding 4.00 m	–	0.32	4.61	9.68	–	m³	14.29
extra over preliminary trench excavating for breaking out existing hard pavings, 150 mm thick concrete	–	0.65	9.36	2.13	–	m²	11.49
Excavating; by hand							
Preliminary trenches							
maximum depth not exceeding 1.00 m	–	2.68	38.61	–	–	m³	38.61
maximum depth not exceeding 2.00 m	–	3.05	43.94	–	–	m³	43.94
maximum depth not exceeding 4.00 m	–	3.93	56.62	–	–	m³	56.62
extra over preliminary trench excavating for breaking out existing hard pavings, 150 mm thick concrete	–	2.00	28.82	2.94	–	m²	31.76
Underpinning pits; commencing from 1.00 m below existing ground level							
maximum depth not exceeding 0.25 m	–	4.07	58.64	–	–	m³	58.64
maximum depth not exceeding 1.00 m	–	4.44	63.97	–	–	m³	63.97
maximum depth not exceeding 2.00 m	–	5.32	76.64	–	–	m³	76.64
Underpinning pits; commencing from 2.00 m below existing ground level							
maximum depth not exceeding 0.25 m	–	5.00	72.04	–	–	m³	72.04
maximum depth not exceeding 1.00 m	–	5.37	77.36	–	–	m³	77.36
maximum depth not exceeding 2.00 m	–	6.24	89.90	–	–	m³	89.90
Underpinning pits; commencing from 4.00 m below existing ground level							
maximum depth not exceeding 0.25 m	–	5.92	85.29	–	–	m³	85.29
maximum depth not exceeding 1.00 m	–	6.29	90.62	–	–	m³	90.62
maximum depth not exceeding 2.00 m	–	7.17	103.30	–	–	m³	103.30
Extra over any types of excavating irrespective of depth							
excavating below ground water level	–	0.32	4.61	4.44	–	m³	9.05
Cutting away existing projecting foundations							
Concrete							
maximum width 150 mm; maximum depth 150 mm	–	0.15	2.16	0.22	–	m	2.38
maximum width 150 mm; maximum depth 225 mm	–	0.22	3.17	0.33	–	m	3.50
maximum width 150 mm; maximum depth 300 mm	–	0.30	4.33	0.44	–	m	4.77
maximum width 300 mm; maximum depth 300 mm	–	0.58	8.35	0.88	–	m	9.23

08 UNDERPINNING

Item	PC £	Labour hours	Labour £	Plant £	Material £	Unit	Total rate £
8.01 UNDERPINNING – cont							
Cutting away existing projecting foundations – cont							
Masonry							
maximum width one brick thick; maximum depth one course high	–	0.04	0.58	0.08	–	m	**0.66**
maximum width one brick thick; maximum depth two courses high	–	0.13	1.87	0.20	–	m	**2.07**
maximum width one brick thick; maximum depth three courses high	–	0.25	3.61	0.37	–	m	**3.98**
maximum width one brick thick; maximum depth four courses high	–	0.42	6.05	0.62	–	m	**6.67**
Preparing the underside of existing work to receive the pinning up of the new work							
Width of existing work							
380 mm wide	–	0.56	8.06	–	–	m	**8.06**
600 mm wide	–	0.74	10.66	–	–	m	**10.66**
900 mm wide	–	0.93	13.40	–	–	m	**13.40**
1200 mm wide	–	1.11	16.00	–	–	m	**16.00**
Disposal							
Excavated material							
off site; to tip not exceeding 13 km (using lorries); including Landfill Tax based on inactive waste	–	0.75	10.80	16.16	26.35	m³	**53.31**
Filling to excavations; by hand							
Average thickness exceeding 0.25 m							
arising from the excavations	–	0.93	13.40	–	–	m³	**13.40**
Surface treatments							
Compacting							
bottoms of excavations	–	0.04	0.58	0.04	–	m²	**0.62**
Wedging and pinning							
To underside of existing construction with slates in cement mortar (1:3)							
width of wall – half brick thick	–	1.02	22.86	–	14.11	m	**36.97**
width of wall – one brick thick	–	1.20	26.89	–	28.22	m	**55.11**
width of wall – one and a half brick thick	–	1.39	31.16	–	42.33	m	**73.49**

08 UNDERPINNING

Item	PC £	Labour hours	Labour £	Plant £	Material £	Unit	Total rate £
8.02 CONCRETE							
Plain in situ ready mixed designated concrete; C10–40 mm aggregate; poured against faces of excavation							
Underpinning							
thickness not exceeding 150 mm	–	3.42	57.78	–	122.70	m³	**180.48**
thickness 150–450 mm	–	2.87	48.49	–	122.70	m³	**171.19**
thickness exceeding 450 mm	–	2.50	42.24	–	122.70	m³	**164.94**
Plain in situ ready mixed designated concrete; C20–20 mm aggregate; poured against faces of excavation							
Underpinning							
thickness not exceeding 150 mm	–	3.42	57.78	–	126.88	m³	**184.66**
thickness 150–450 mm	–	2.87	48.49	–	126.88	m³	**175.37**
thickness exceeding 450 mm	–	2.50	42.24	–	126.88	m³	**169.12**
extra for working around reinforcement	–	0.28	4.73	–	–	m³	**4.73**
Sawn formwork; sides of foundations in underpinning							
Plain vertical							
height exceeding 1.00 m	–	1.48	29.58	–	–	m²	**29.58**
height not exceeding 250 mm	–	0.51	10.20	–	–	m²	**10.20**
height 250–500 mm	–	0.79	15.79	–	–	m²	**15.79**
height 500 mm–1.00 m	–	1.20	23.98	–	–	m²	**23.98**
8.03 FORMWORK							
Earthwork support to preliminary trenches (open boarded – in 3.00 m lengths)							
Maximum depth not exceeding 1.00 m							
distance between opposing faces not exceeding 2.00 m	–	0.37	5.34	–	–	m²	**5.34**
Maximum depth not exceeding 2.00 m							
distance between opposing faces not exceeding 2.00 m	–	0.46	6.62	–	–	m²	**6.62**
Maximum depth not exceeding 4.00 m							
distance between opposing faces not exceeding 2.00 m	–	0.59	8.50	–	–	m²	**8.50**
Earthwork support to underpinning pits (open boarded – in 3.00 m lengths)							
Maximum depth not exceeding 1.00 m							
distance between opposing faces not exceeding 2.00 m	–	0.41	5.91	–	–	m²	**5.91**
Maximum depth not exceeding 2.00 m							
distance between opposing faces not exceeding 2.00 m	–	0.51	7.34	–	–	m²	**7.34**
Maximum depth not exceeding 4.00 m							
distance between opposing faces not exceeding 2.00 m	–	0.65	9.36	–	–	m²	**9.36**

08 UNDERPINNING

Item	PC £	Labour hours	Labour £	Plant £	Material £	Unit	Total rate £
8.03 FORMWORK – cont							
Earthwork support to preliminary trenches (closed boarded – in 3.00 m lengths)							
Maximum depth not exceeding 1.00 m							
1.00 m deep	–	0.93	13.40	–	–	m²	**13.40**
Maximum depth not exceeding 2.00 m							
distance between opposing faces not exceeding 2.00 m	–	1.16	16.72	–	–	m²	**16.72**
Maximum depth not exceeding 4.00 m							
distance between opposing faces not exceeding 2.00 m	–	1.43	20.60	–	–	m²	**20.60**
Earthwork support to underpinning pits (closed boarded – in 3.00 m lengths)							
Maximum depth not exceeding 1.00 m							
distance between opposing faces not exceeding 2.00 m	–	1.02	14.70	–	–	m²	**14.70**
Maximum depth not exceeding 2.00 m							
distance between opposing faces not exceeding 2.00 m	–	1.28	18.44	–	–	m²	**18.44**
Maximum depth not exceeding 4.00 m							
distance between opposing faces not exceeding 2.00 m	–	1.57	22.62	–	–	m²	**22.62**
Extra over earthwork support for							
left in	–	0.69	9.94	–	–	m²	**9.94**
8.04 REINFORCEMENT							
Reinforcement bar; hot rolled steel bars							
20 mm dia. nominal size							
straight	640.00	22.00	461.79	–	871.44	tonne	**1333.23**
bent	675.00	28.00	587.74	–	908.39	tonne	**1496.13**
16 mm dia. nominal size							
straight	640.00	28.00	587.74	–	895.91	tonne	**1483.65**
bent	675.00	36.00	755.66	–	932.86	tonne	**1688.52**
12 mm dia. nominal size							
straight	640.00	40.00	839.63	–	920.39	tonne	**1760.02**
bent	675.00	50.00	1049.54	–	957.33	tonne	**2006.87**
10 mm dia. nominal size							
straight	640.00	48.00	1007.56	–	948.76	tonne	**1956.32**
bent	675.00	62.00	1301.43	–	985.71	tonne	**2287.14**
8 mm dia. nominal size							
straight	640.00	58.00	1217.46	–	956.75	tonne	**2174.21**
bent	675.00	74.00	1553.31	–	1010.18	tonne	**2563.49**

08 UNDERPINNING

Item	PC £	Labour hours	Labour £	Plant £	Material £	Unit	Total rate £
8.05 BRICKWORK OR BLOCKWORK							
Common bricks; in cement mortar (1:3)							
Walls in underpinning							
one brick thick (PC £ per 1000)	380.00	2.22	49.76	–	59.06	m²	**108.82**
one and a half brick thick	–	3.05	68.36	–	87.77	m²	**156.13**
two brick thick	–	3.79	84.94	–	123.25	m²	**208.19**
Class A engineering bricks; in cement mortar (1:3)							
Walls in underpinning							
one brick thick (PC £ per 1000)	1144.10	2.22	49.76	–	160.59	m²	**210.35**
one and a half brick thick	–	3.05	68.36	–	240.05	m²	**308.41**
two brick thick	–	3.79	84.94	–	326.29	m²	**411.23**
Class B engineering bricks; in cement mortar (1:3)							
Walls in underpinning							
one brick thick (PC £ per 1000)	332.50	2.22	49.76	–	52.75	m²	**102.51**
one and a half brick thick	–	3.05	68.36	–	78.30	m²	**146.66**
two brick thick	–	3.79	84.94	–	110.62	m²	**195.56**
Add or deduct for variation of £10.00/1000 in PC of bricks							
one brick thick	–	–	–	–	1.24	m²	**1.24**
one and a half bricks thick	–	–	–	–	1.85	m²	**1.85**
two bricks thick	–	–	–	–	2.47	m²	**2.47**
8.06 TANKING							
Damp-proof course							
Zedex CPT (Co-Polymer Thermoplastic) damp-proof course or other equal; 200 mm laps; in gauged mortar (1:1:6)							
Horizontal							
width exceeding 225 mm	–	0.23	5.15	–	4.53	m²	**9.68**
width not exceeding 225 mm	–	0.46	10.31	–	4.53	m²	**14.84**
Hyload (pitch polymer) damp-proof course or similar; 150 mm laps; in cement mortar (1:3)							
Horizontal							
width exceeding 225 mm	–	0.23	5.15	–	5.50	m²	**10.65**
width not exceeding 225 mm	–	0.46	10.31	–	5.62	m²	**15.93**
Alumite aluminium cored bitumen gas retardant damp-proof course or other equal and approved; 200 mm laps; in gauged mortar (1:1:6)							
Horizontal							
width exceeding 225 mm	–	0.31	6.95	–	7.65	m²	**14.60**
width not exceeding 225 mm	–	0.60	13.45	–	7.65	m²	**21.10**
Two courses of slates in cement mortar (1:3)							
Horizontal							
width exceeding 225 mm	–	1.39	31.16	–	60.08	m²	**91.24**
width not exceeding 225 mm	–	2.31	51.77	–	61.43	m²	**113.20**

09 DIAPHRAGM WALLS AND EMBEDDED RETAINING WALLS

Item	PC £	Labour hours	Labour £	Plant £	Material £	Unit	Total rate £
9.01 EMBEDDED RETAINING WALLS							
Diaphragm walls; contiguous panel construction; panel lengths not exceeding 5 m							
Plant							
provision of all plant; including bringing to and removing from site; maintenance, erection and dismantling; assuming one rig for 1000 m² of walling	–	–	–	–	–	item	**274050.00**
Excavation for diaphragm wall; excavated material removed from site; Bentonite slurry supplied and disposed of							
600 mm thick walls	–	–	–	–	–	m³	**772.14**
1000 mm thick walls	–	–	–	–	–	m³	**551.53**
Concrete							
ready mixed reinforced in situ concrete; normal Portland cement; C30 in walls	–	–	–	–	–	m³	**165.46**
Reinforcement bar; BS 4449 cold rolled deformed square high yield steel bars; straight or bent							
25 mm–40 mm dia.	–	–	–	–	–	tonne	**1213.35**
20 mm dia.	–	–	–	–	–	tonne	**1213.35**
16 mm dia.	–	–	–	–	–	tonne	**1213.35**
Formwork							
formwork 75 mm thick to form chases	–	–	–	–	–	m²	**99.28**
construct twin guide walls in reinforced concrete; together with reinforcement and formwork along the axis of the diaphragm wall	–	–	–	–	–	m	**1047.90**
Delays							
rig standing	–	–	–	–	–	hour	**2283.75**

10 CRIB WALLS, GABION AND REINFORCED EARTH

Item	PC £	Labour hours	Labour £	Plant £	Material £	Unit	Total rate £
10.01 GABION BASKET WALLS							
Gabion baskets							
Wire mesh gabion baskets; Maccaferri Ltd or other equal; galvanized mesh 80 mm × 100 mm; filling with broken stones 125 mm–200 mm size							
2.00 m × 1.00 m × 0.50 m	28.36	1.00	24.93	13.37	89.42	nr	**127.72**
2.00 m × 1.00 m × 0.50 m; PVC coated	35.87	1.00	24.93	13.37	98.27	nr	**136.57**
2.00 m × 1.00 m × 1.00 m	39.75	2.00	49.85	26.74	160.94	nr	**237.53**
2.00 m × 1.00 m × 1.00 m PVC coated	50.31	2.00	49.85	26.74	172.08	nr	**248.67**
Reno mattress gabion baskets or other equal; Maccaferri Ltd; filling with broken stones 125 mm–200 mm size							
6.00 m × 2.00 m × 0.17 m	126.89	2.00	49.85	20.05	255.31	nr	**325.21**
6.00 m × 2.00 m × 0.23 m	137.35	2.50	62.31	26.74	309.18	nr	**398.23**
6.00 m × 2.00 m × 0.30 m	161.12	3.00	74.78	30.08	384.24	nr	**489.10**

11 IN SITU CONCRETE WORKS

Item	PC £	Labour hours	Labour £	Plant £	Material £	Unit	Total rate £
MATERIAL RATES							
Cement material prices							
Cements							
Ordinary Portland cement	–	–	–	–	138.01	tonne	**138.01**
high alumina cement	–	–	–	–	623.74	tonne	**623.74**
Sulfacrete sulphate-resisting cement	–	–	–	–	168.91	tonne	**168.91**
Ferrocrete rapid hardening cement	–	–	–	–	321.27	tonne	**321.27**
Snowcrete white cement	–	–	–	–	219.69	tonne	**219.69**
Blue Circle general purpose cement	3.29	–	–	–	3.39	25 kg	**3.39**
Blue Circle Mastercrete cement	4.15	–	–	–	4.27	25 kg	**4.27**
Blue Circle sulphate-resisting cement	4.25	–	–	–	4.38	25 kg	**4.38**
Hanson readimix concrete	4.80	–	–	–	4.94	25 kg	**4.94**
Cement admixtures							
Febtone colorant – red, marigold, yellow, brown, black	–	–	–	–	6.13	kg	**6.13**
Febproof waterproof	–	–	–	–	1.44	litre	**1.44**
Febond PVA bonding agent	–	–	–	–	2.49	litre	**2.49**
Febspeed frostproofer and hardener	–	–	–	–	1.65	litre	**1.65**
Basic mixed concrete prices							
NOTE: The following prices are for concrete ready for placing excluding any allowance for waste, discount or overheads and profit. Prices are based upon delivery to site within a 10 mile (16 km) radius of concrete mixing plant, using full loads							
GEN 0 (7.5N)	–	–	–	–	99.83	m^3	**99.83**
GEN 1 (10N)	–	–	–	–	103.77	m^3	**103.77**
GEN 2 (15N)	–	–	–	–	105.77	m^3	**105.77**
GEN 3 (20N)	–	–	–	–	107.30	m^3	**107.30**
RC20/25	–	–	–	–	112.31	m^3	**112.31**
RC25/30	–	–	–	–	113.80	m^3	**113.80**
RC30/37	–	–	–	–	115.31	m^3	**115.31**
RC35/45	–	–	–	–	120.32	m^3	**120.32**
RC40/50	–	–	–	–	121.34	m^3	**121.34**
RC50/60	–	–	–	–	123.35	m^3	**123.35**
FND3	–	–	–	–	120.32	m^3	**120.32**
FND4	–	–	–	–	122.35	m^3	**122.35**
Caltite waterprrof system to RC 35 concrete	–	–	–	–	203.16	m^3	**203.16**
Site Mixed Concrete (on site batching plant)							
mix 7.50 N/mm^2; cement (1:8)	–	–	–	–	108.15	m^3	**108.15**
mix 10.00 N/mm^2; cement (1:8)	–	–	–	–	113.30	m^3	**113.30**
mix 20.00 N/mm^2; cement (1:2:4)	–	–	–	–	127.72	m^3	**127.72**
mix 25.00 N/mm^2; cement to (1:1:5:3)	–	–	–	–	133.90	m^3	**133.90**
Add for							
rapid-hardening cement	–	–	–	–	1.65	kg	**1.65**
polypropylene fibre additive	–	–	–	–	7.39	m^3	**7.39**
air entrained concrete	–	–	–	–	6.49	m^3	**6.49**
water repellent additive	–	–	–	–	1.44	litre	**1.44**
foamed concrete	–	–	–	–	–	%	**25.00**
distance per mile in excess of 10 miles (16 km)	–	–	–	–	0.78	m^3	**0.78**

11 IN SITU CONCRETE WORKS

Item	PC £	Labour hours	Labour £	Plant £	Material £	Unit	Total rate £
11.01 MASS CONCRETE							
SUPPLY AND LAY PRICES							
NOTE: The following concrete material prices include an allowance for shrinkage and waste. PC Sums are concrete supply only prices.							
Plain in situ ready mixed designated concrete; C7.5–40 mm aggregate							
Foundations	96.92	1.02	17.31	–	104.81	m³	**122.12**
Isolated foundations	96.92	1.13	19.00	–	104.81	m³	**123.81**
Plain in situ ready mixed designated concrete; C10–40 mm aggregate							
Foundations	100.75	1.02	17.31	–	108.95	m³	**126.26**
Isolated foundations	100.75	1.13	19.06	–	108.95	m³	**128.01**
Plain in situ ready mixed designated concrete; C10–40 mm aggregate; poured on or against earth or unblinded hardcore							
Foundations	100.75	1.02	17.31	–	111.14	m³	**128.45**
Isolated foundations	100.75	1.13	19.06	–	111.14	m³	**130.20**
Plain in situ ready mixed designated concrete; C20–20 mm aggregate							
Foundations	104.17	0.98	16.56	–	112.66	m³	**129.22**
Isolated foundations	104.17	1.13	19.06	–	112.66	m³	**131.72**
Plain in situ ready mixed; 20N; poured on or against earth or unblinded hardcore							
Foundations	104.17	1.02	17.31	–	114.91	m³	**132.22**
Isolated foundations	104.17	1.13	19.06	–	114.91	m³	**133.97**
11.02 HORIZONTAL WORK							
SUPPLY AND LAY PRICES							
NOTE: The following concrete material prices include an allowance for shrinkage and waste. PC Sums are concrete supply only prices.							
Plain in situ ready mixed designated concrete; C7.5–40 mm aggregate							
Foundations	96.92	1.02	17.31	–	104.81	m³	**122.12**
Isolated foundations	96.92	1.13	19.00	–	104.81	m³	**123.81**
Beds							
thickness not exceeding 300 mm	96.92	1.02	17.31	–	104.81	m³	**122.12**
thickness exceeding 300 mm	96.92	0.93	15.72	–	104.81	m³	**120.53**

11 IN SITU CONCRETE WORKS

Item	PC £	Labour hours	Labour £	Plant £	Material £	Unit	Total rate £
11.02 HORIZONTAL WORK – cont							
Plain in situ ready mixed designated concrete – cont							
Screeded beds; protection to compressible formwork exceeding 600 mm wide							
50 mm thick	96.92	0.10	1.69	–	5.24	m²	**6.93**
75 mm thick	96.92	0.15	2.53	–	7.87	m²	**10.40**
100 mm thick	96.92	0.20	3.38	–	10.49	m²	**13.87**
Plain in situ ready mixed designated concrete; C10–40 mm aggregate							
Foundations	100.75	1.02	17.31	–	108.95	m³	**126.26**
Isolated foundations	100.75	1.13	19.06	–	108.95	m³	**128.01**
Beds							
thickness not exceeding 300 mm	100.75	1.02	17.31	–	108.95	m³	**126.26**
thickness exceeding 300 mm	100.75	0.93	15.63	–	108.95	m³	**124.58**
Plain in situ ready mixed designated concrete; C10–40 mm aggregate; poured on or against earth or unblinded hardcore							
Foundations	100.75	1.02	17.31	–	111.14	m³	**128.45**
Isolated foundations	100.75	1.13	19.06	–	111.14	m³	**130.20**
Beds							
thickness not exceeding 300 mm	100.75	1.02	17.31	–	111.14	m³	**128.45**
thickness exceeding 300 mm	100.75	0.93	15.63	–	111.14	m³	**126.77**
Plain in situ ready mixed designated concrete; C20–20 mm aggregate							
Foundations	104.17	0.98	16.56	–	112.66	m³	**129.22**
Isolated foundations	104.17	1.13	19.06	–	112.66	m³	**131.72**
Beds							
thickness not exceeding 300 mm	104.17	1.02	17.23	–	112.66	m³	**129.89**
thickness exceeding 300 mm	104.17	0.93	15.72	–	112.66	m³	**128.38**
Plain in situ ready mixed; 20N; poured on or against earth or unblinded hardcore							
Foundations	104.17	1.02	17.31	–	114.91	m³	**132.22**
Isolated foundations	104.17	1.13	19.06	–	114.91	m³	**133.97**
Beds							
thickness not exceeding 300 mm	104.17	1.02	17.31	–	114.91	m³	**132.22**
thickness exceeding 300 mm	104.17	0.93	15.72	–	114.91	m³	**130.63**
Reinforced in situ ready mixed designated concrete; 25N							
Foundations	110.49	1.02	17.31	–	125.47	m³	**142.78**
Ground beams	110.49	2.59	43.76	–	119.50	m³	**163.26**
Isolated foundations	110.49	1.13	19.00	–	125.47	m³	**144.47**

11 IN SITU CONCRETE WORKS

Item	PC £	Labour hours	Labour £	Plant £	Material £	Unit	Total rate £
Beds							
thickness not exceeding 300 mm	110.49	1.02	17.23	–	119.50	m³	**136.73**
thickness exceeding 300 mm	110.49	0.93	15.72	–	119.50	m³	**135.22**
Slabs							
thickness not exceeding 300 mm	110.49	1.25	21.13	–	119.50	m³	**140.63**
thickness exceeding 300 mm	110.49	1.15	19.44	–	119.50	m³	**138.94**
Coffered and troughed slabs							
thickness not exceeding 300 mm	110.49	2.96	50.02	–	119.50	m³	**169.52**
thickness exceeding 300 mm	110.49	2.59	43.76	–	119.50	m³	**163.26**
Extra over for sloping work							
not exceeding 15°	–	0.23	3.88	–	–	m³	**3.88**
exceeding 15°	–	–	–	–	–	m³	**–**
Beams							
isolated	110.49	3.70	62.52	–	119.50	m³	**182.02**
isolated deep	110.49	4.07	68.77	–	119.50	m³	**188.27**
attached deep	110.49	3.70	62.52	–	119.50	m³	**182.02**
Beam casings							
isolated	110.49	4.07	68.77	–	119.50	m³	**188.27**
isolated deep	110.49	4.44	75.03	–	119.50	m³	**194.53**
attached deep	110.49	4.07	68.77	–	119.50	m³	**188.27**
Staircases	110.49	5.25	88.70	–	119.50	m³	**208.20**
Upstands	110.49	3.30	55.75	–	119.50	m³	**175.25**
Reinforced in situ ready mixed designated concrete; 30N							
Foundations	111.95	1.02	17.31	–	121.08	m³	**138.39**
Ground beams	111.95	2.59	43.76	–	121.08	m³	**164.84**
Isolated foundations	111.95	1.13	19.10	–	121.08	m³	**140.18**
Beds							
thickness not exceeding 300 mm	111.95	1.02	17.23	–	121.08	m³	**138.31**
thickness exceeding 300 mm	111.95	0.95	16.05	–	121.08	m³	**137.13**
Slabs							
thickness not exceeding 300 mm	111.95	1.25	21.13	–	121.08	m³	**142.21**
thickness exceeding 300 mm	111.95	1.15	19.44	–	121.08	m³	**140.52**
Coffered and troughed slabs							
thickness not exceeding 300 mm	111.95	2.96	50.02	–	121.08	m³	**171.10**
thickness exceeding 300 mm	111.95	2.59	43.76	–	121.08	m³	**164.84**
Extra over for sloping							
not exceeding 15°	–	0.23	3.88	–	–	m³	**3.88**
exceeding 15°	–	0.46	7.78	–	–	m³	**7.78**
Beams							
isolated	111.95	3.70	62.52	–	121.08	m³	**183.60**
isolated deep	111.95	4.07	68.77	–	121.08	m³	**189.85**
attached deep	111.95	3.70	62.52	–	121.08	m³	**183.60**
Beam casings							
isolated	111.95	4.07	68.77	–	121.08	m³	**189.85**
isolated deep	111.95	4.44	75.03	–	121.08	m³	**196.11**
attached deep	111.95	4.07	68.77	–	121.08	m³	**189.85**
Staircases	111.95	5.55	93.77	–	121.08	m³	**214.85**
Upstands	111.95	3.56	60.15	–	121.08	m³	**181.23**

11 IN SITU CONCRETE WORKS

Item	PC £	Labour hours	Labour £	Plant £	Material £	Unit	Total rate £
11.02 HORIZONTAL WORK – cont							
Reinforced in situ ready mixed designated concrete; 35N							
Foundations	116.82	1.02	17.31	–	126.34	m^3	**143.65**
Ground beams	116.82	2.59	43.76	–	126.34	m^3	**170.10**
Isolated foundations	116.82	1.13	19.10	–	126.34	m^3	**145.44**
Beds							
thickness not exceeding 300 mm	116.82	1.02	17.23	–	126.34	m^3	**143.57**
thickness exceeding 300 mm	116.82	0.95	16.05	–	126.34	m^3	**142.39**
Slabs							
thickness not exceeding 300 mm	116.82	1.25	21.13	–	126.34	m^3	**147.47**
thickness exceeding 300 mm	116.82	1.15	19.44	–	126.34	m^3	**145.78**
Coffered and troughed slabs							
thickness not exceeding 300 mm	116.82	2.96	50.02	–	126.34	m^3	**176.36**
thickness exceeding 300 mm	116.82	2.59	43.76	–	126.34	m^3	**170.10**
Extra over for sloping							
not exceeding 15°	–	0.23	3.88	–	–	m^3	**3.88**
exceeding 15°	–	0.46	7.78	–	–	m^3	**7.78**
Beams							
isolated	116.82	3.70	62.52	–	126.34	m^3	**188.86**
isolated deep	116.82	4.07	68.77	–	126.34	m^3	**195.11**
attached deep	116.82	3.70	62.52	–	126.34	m^3	**188.86**
Beam casings							
isolated	116.82	4.07	68.77	–	126.34	m^3	**195.11**
isolated deep	116.82	4.44	75.03	–	126.34	m^3	**201.37**
attached deep	116.82	4.07	68.77	–	126.34	m^3	**195.11**
Staircases	116.82	5.55	93.77	–	126.34	m^3	**220.11**
Upstands	116.82	3.56	60.15	–	126.34	m^3	**186.49**
Reinforced in situ ready mixed designated concrete; 40N							
Foundations	117.81	1.02	17.31	–	127.41	m^3	**144.72**
Isolated foundations	117.81	1.13	19.06	–	127.41	m^3	**146.47**
Ground beams	117.81	2.59	43.76	–	127.41	m^3	**171.17**
Beds							
thickness not exceeding 300 mm	117.81	0.95	16.05	–	127.41	m^3	**143.46**
thickness exceeding 300 mm	117.81	0.93	15.72	–	127.41	m^3	**143.13**
Slabs							
thickness not exceeding 300 mm	117.81	1.20	20.28	–	127.41	m^3	**147.69**
thickness exceeding 300 mm	117.81	1.15	19.44	–	127.41	m^3	**146.85**
Coffered and troughed slabs							
thickness not exceeding 300 mm	117.81	2.96	50.02	–	127.41	m^3	**177.43**
thickness exceeding 300 mm	117.81	2.59	43.76	–	127.41	m^3	**171.17**
Extra over for sloping							
not exceeding 15°	–	0.23	3.88	–	–	m^3	**3.88**
exceeding 15°	–	0.46	7.78	–	–	m^3	**7.78**
Beams							
isolated	117.81	3.70	62.52	–	127.41	m^3	**189.93**
isolated deep	117.81	4.07	68.77	–	127.41	m^3	**196.18**
attached deep	117.81	3.70	62.52	–	127.41	m^3	**189.93**

11 IN SITU CONCRETE WORKS

Item	PC £	Labour hours	Labour £	Plant £	Material £	Unit	Total rate £
Beam casings							
isolated	117.81	4.07	68.77	–	127.41	m³	**196.18**
isolated deep	117.81	4.44	75.03	–	127.41	m³	**202.44**
attached deep	117.81	4.07	68.77	–	127.41	m³	**196.18**
Staircases	117.81	5.55	93.77	–	127.41	m³	**221.18**
Upstands	117.81	3.56	60.15	–	127.41	m³	**187.56**
Proprietary voided Bubbledeck, Cobiax or other equal and approved slab; concrete mix RC35; to achieve design loadings of 5.0 kN/m² live and 3.0 kN/m² dead; with trowelled finish							
Beds							
360 mm overall thickness	–	–	–	–	–	m²	**139.10**
additional concrete up to 600 mm wide at edges where formers omitted at junctions with walls etc.	–	–	–	–	–	m	**56.91**

11.03 VERTICAL WORK

SUPPLY AND LAY PRICES

NOTE: The following concrete material prices include an allowance for shrinkage and waste. PC Sums are concrete supply only prices.

Item	PC £	Labour hours	Labour £	Plant £	Material £	Unit	Total rate £
Plain in situ ready mixed designated concrete; C7.5–40 mm aggregate							
Filling hollow walls							
thickness not exceeding 150 mm	96.92	3.15	53.22	–	104.81	m³	**158.03**
Column casings							
stub columns beneath suspended ground slabs	96.92	4.50	76.03	–	104.81	m³	**180.84**
Plain in situ ready mixed designated concrete; C10–40 mm aggregate							
Filling hollow walls							
thickness not exceeding 300 mm	100.75	3.15	53.22	–	108.95	m³	**162.17**
Plain in situ ready mixed designated concrete; C20–20 mm aggregate							
Filling hollow walls							
thickness not exceeding 300 mm	104.17	3.00	50.69	–	112.66	m³	**163.35**
Reinforced in situ ready mixed designated concrete; 25N							
Walls							
thickness not exceeding 300 mm	110.49	3.15	53.22	–	119.50	m³	**172.72**
thickness exceeding 300 mm	110.49	2.40	40.63	–	119.50	m³	**160.13**
Columns	110.49	4.20	70.97	–	119.50	m³	**190.47**
Column casings	110.49	4.90	82.79	–	119.50	m³	**202.29**
Staircases	110.49	5.25	88.70	–	119.50	m³	**208.20**
Upstands	110.49	3.30	55.75	–	119.50	m³	**175.25**

11 IN SITU CONCRETE WORKS

Item	PC £	Labour hours	Labour £	Plant £	Material £	Unit	Total rate £
11.03 VERTICAL WORK – cont							
Reinforced in situ ready mixed designated concrete; 30N							
Walls							
thickness not exceeding 300 mm	111.95	2.67	45.20	–	121.08	m³	**166.28**
thickness exceeding 300 mm	111.95	2.41	40.67	–	121.08	m³	**161.75**
Columns	111.95	4.44	75.03	–	121.08	m³	**196.11**
Column casings	111.95	4.90	82.79	–	121.08	m³	**203.87**
Staircases	111.95	5.55	93.77	–	121.08	m³	**214.85**
Upstands	111.95	3.56	60.15	–	121.08	m³	**181.23**
Reinforced in situ ready mixed designated concrete; 35N							
Walls							
thickness not exceeding 300 mm	116.82	2.67	45.20	–	126.34	m³	**171.54**
thickness exceeding 300 mm	116.82	2.41	40.67	–	126.34	m³	**167.01**
Columns	116.82	4.44	75.03	–	126.34	m³	**201.37**
Column casings	116.82	4.90	82.79	–	126.34	m³	**209.13**
Staircases	116.82	5.55	93.77	–	126.34	m³	**220.11**
Upstands	116.82	3.56	60.15	–	126.34	m³	**186.49**
Reinforced in situ ready mixed designated concrete; 40N							
Walls							
thickness not exceeding 300 mm	117.81	2.73	46.12	–	127.41	m³	**173.53**
thickness exceeding 300 mm	117.81	2.41	40.72	–	127.41	m³	**168.13**
Columns	117.81	4.44	75.03	–	127.41	m³	**202.44**
Column casings	117.81	4.90	82.79	–	127.41	m³	**210.20**
Staircases	117.81	5.55	93.77	–	127.41	m³	**221.18**
Upstands	117.81	3.56	60.15	–	127.41	m³	**187.56**
11.04 SURFACE FINISHES							
Worked finishes							
Tamping by mechanical means	–	0.02	0.34	0.12	–	m²	**0.46**
Power floating	–	0.16	2.70	0.39	–	m²	**3.09**
Trowelling	–	0.31	5.24	–	–	m²	**5.24**
Hacking							
by mechanical means	–	0.31	5.24	0.45	–	m²	**5.69**
by hand	–	0.65	10.98	–	–	m²	**10.98**
Lightly shot blasting surface of concrete	–	0.37	6.25	–	–	m²	**6.25**
Blasting surface of concrete to produce textured finish	–	0.65	10.98	0.97	–	m²	**11.95**
Sand blasting (blast and vac method)	–	–	–	–	–	m²	**46.48**
Wood float finish	–	0.12	2.03	–	–	m²	**2.03**
Tamped finish							
level or to falls	–	0.06	1.01	–	–	m²	**1.01**
to falls	–	0.09	1.52	–	–	m²	**1.52**
Spade finish	–	0.14	2.37	–	–	m²	**2.37**

11 IN SITU CONCRETE WORKS

Item	PC £	Labour hours	Labour £	Plant £	Material £	Unit	Total rate £
Diamond drilling/cutting							
Cutting holes in block wall							
up to 52 mm dia.; up to 150 mm thick	–	–	–	–	–	nr	18.64
up to 52 mm dia.; 200 mm thick	–	–	–	–	–	nr	23.23
up to 52 mm dia.; 250 mm thick	–	–	–	–	–	nr	27.82
up to 52 mm dia.; 300 mm thick	–	–	–	–	–	nr	32.40
79 mm–107 mm dia.; up to 150 mm thick	–	–	–	–	–	nr	29.24
79 mm–107 mm dia.; 200 mm thick	–	–	–	–	–	nr	34.88
79 mm–107 mm dia.; 250 mm thick	–	–	–	–	–	nr	40.51
79 mm–107 mm dia.; 300 mm thick	–	–	–	–	–	nr	46.14
251 mm–300 mm dia.; n.e. 150 mm thick	–	–	–	–	–	nr	72.13
251 mm–300 mm dia.; 200 mm thick	–	–	–	–	–	nr	91.37
251 mm–300 mm dia.; 250 mm thick	–	–	–	–	–	nr	110.61
251 mm–300 mm dia.; 300 mm thick	–	–	–	–	–	nr	129.82
Cutting holes in brick wall							
up to 52 mm dia.; up to 150 mm thick	–	–	–	–	–	nr	21.28
up to 52 mm dia.; 200 mm thick	–	–	–	–	–	nr	26.53
up to 52 mm dia.; 250 mm thick	–	–	–	–	–	nr	31.80
up to 52 mm dia.; 300 mm thick	–	–	–	–	–	nr	37.06
79 mm–107 mm dia.; up to 150 mm thick	–	–	–	–	–	nr	33.38
79 mm–107 mm dia.; 200 mm thick	–	–	–	–	–	nr	39.84
79 mm–107 mm dia.; 250 mm thick	–	–	–	–	–	nr	46.28
79 mm–107 mm dia.; 300 mm thick	–	–	–	–	–	nr	52.73
251 mm–300 mm dia.; n.e. 150 mm thick	–	–	–	–	–	nr	82.44
251 mm–300 mm dia.; 200 mm thick	–	–	–	–	–	nr	104.42
251 mm–300 mm dia.; 250 mm thick	–	–	–	–	–	nr	126.38
251 mm–300 mm dia.; 300 mm thick	–	–	–	–	–	nr	148.39
Cutting holes in concrete wall							
up to 52 mm dia.; up to 150 mm thick	–	–	–	–	–	nr	26.60
up to 52 mm dia.; 200 mm thick	–	–	–	–	–	nr	33.18
up to 52 mm dia.; 250 mm thick	–	–	–	–	–	nr	39.73
up to 52 mm dia.; 300 mm thick	–	–	–	–	–	nr	46.29
79 mm–107 mm dia.; up to 150 mm thick	–	–	–	–	–	nr	41.75
79 mm–107 mm dia.; 200 mm thick	–	–	–	–	–	nr	49.80
79 mm–107 mm dia.; 250 mm thick	–	–	–	–	–	nr	57.86
79 mm–107 mm dia.; 300 mm thick	–	–	–	–	–	nr	65.92
251 mm–300 mm dia.; n.e. 150 mm thick	–	–	–	–	–	nr	103.04
251 mm–300 mm dia.; 200 mm thick	–	–	–	–	–	nr	130.51
251 mm–300 mm dia.; 250 mm thick	–	–	–	–	–	nr	158.00
251 mm–300 mm dia.; 300 mm thick	–	–	–	–	–	nr	185.44
Cutting holes in timber wall							
up to 52 mm dia.; up to 150 mm thick	–	–	–	–	–	nr	25.26
up to 52 mm dia.; 200 mm thick	–	–	–	–	–	nr	31.50
up to 52 mm dia.; 250 mm thick	–	–	–	–	–	nr	37.76
up to 52 mm dia.; 300 mm thick	–	–	–	–	–	nr	43.98
79 mm–107 mm dia.; up to 150 mm thick	–	–	–	–	–	nr	39.67
79 mm–107 mm dia.; 200 mm thick	–	–	–	–	–	nr	47.32
79 mm–107 mm dia.; 250 mm thick	–	–	–	–	–	nr	54.98
79 mm–107 mm dia.; 300 mm thick	–	–	–	–	–	nr	62.61
251 mm–300 mm dia.; n.e. 150 mm thick	–	–	–	–	–	nr	97.90
251 mm–300 mm dia.; 200 mm thick	–	–	–	–	–	nr	124.00
251 mm–300 mm dia.; 250 mm thick	–	–	–	–	–	nr	150.10
251 mm–300 mm dia.; 300 mm thick	–	–	–	–	–	nr	176.20

11 IN SITU CONCRETE WORKS

Item	PC £	Labour hours	Labour £	Plant £	Material £	Unit	Total rate £
11.04 SURFACE FINISHES – cont							
Diamond drilling/cutting – cont							
Chasing for concealed conduits							
up to 35 mm deep, up to 50 mm wide in block wall	–	–	–	–	–	m	8.65
up to 35 mm deep, up to 50 mm wide in brick wall	–	–	–	–	–	m	9.97
up to 35 mm deep, up to 50 mm wide in concrete wall	–	–	–	–	–	m	33.26
up to 35 mm deep, up to 50 mm wide in concrete floor	–	–	–	–	–	m	39.91
up to 35 mm deep, 75 mm–100 mm wide in block wall	–	–	–	–	–	m	13.51
up to 35 mm deep, 75 mm–100 mm wide in brick wall	–	–	–	–	–	m	15.58
up to 35 mm deep, 75 mm–100 mm wide in concrete wall	–	–	–	–	–	m	51.97
up to 35 mm deep, 75 mm–100 mm wide in concrete floor	–	–	–	–	–	m	62.36
recess 100 mm × 100 mm × 35 mm deep in block wall	–	–	–	–	–	nr	4.63
recess 100 mm × 100 mm × 35 mm deep in brick wall	–	–	–	–	–	nr	5.38
recess 100 mm × 100 mm × 35 mm deep in concrete wall	–	–	–	–	–	nr	21.50
recess 100 mm × 100 mm × 35 mm deep in concrete floor	–	–	–	–	–	nr	17.90
Floor sawing							
diamond floor sawing; 25 mm depth	–	–	–	–	–	m	6.65
Ring sawing							
up to 25 mm deep in block wall	–	–	–	–	–	m	8.65
up to 25 mm deep in brick wall	–	–	–	–	–	m	9.97
up to 25 mm deep in concrete floor	–	–	–	–	–	m	13.31
up to 25 mm deep in concrete wall	–	–	–	–	–	m	15.96
11.05 FORMWORK							
Sides of foundations; 18 mm thick external quality plywood; basic finish							
Plain vertical							
height not exceeding 250 mm	–	0.38	7.55	–	5.20	m	12.75
height not exceeding 250 mm; left in	–	0.38	7.55	–	8.10	m	15.65
height 250–500 mm	–	0.71	14.21	–	12.22	m	26.43
height 250–500 mm; left in	–	0.62	12.41	–	20.94	m	33.35
height 500 mm–1.00 m	–	1.00	19.96	–	16.28	m	36.24
height 500 mm–1.00 m; left in	–	0.95	19.07	–	28.97	m	48.04
height exceeding 1.00 m	–	1.33	26.63	–	15.71	m²	42.34
height exceeding 1.00 m; left in	–	1.17	23.38	–	27.80	m²	51.18

11 IN SITU CONCRETE WORKS

Item	PC £	Labour hours	Labour £	Plant £	Material £	Unit	Total rate £
Sides of foundations; polystyrene sheet formwork; Cordek Claymaster or equal							
50 mm thick; plain vertical							
height not exceeding 250 mm; left in	–	0.09	1.80	–	0.98	m	**2.78**
height 250–500 mm; left in	–	0.14	2.87	–	1.97	m	**4.84**
height 500 mm–1.00 m; left in	–	0.22	4.32	–	3.93	m	**8.25**
height exceeding 1.00 m; left in	–	0.27	5.40	–	3.75	m²	**9.15**
75 mm thick; plain vertical							
height not exceeding 250 mm; left in	–	0.09	1.80	–	1.33	m	**3.13**
height 250–500 mm; left in	–	0.14	2.87	–	2.65	m	**5.52**
height 500 mm–1.00 m; left in	–	0.22	4.32	–	5.55	m	**9.87**
height exceeding 1.00 m; left in	–	0.27	5.40	–	5.30	m²	**10.70**
100 mm thick; plain vertical							
height not exceeding 250 mm; left in	–	0.09	1.80	–	2.16	m	**3.96**
height 250–500 mm; left in	–	0.14	2.87	–	4.32	m	**7.19**
height 500 mm–1.00 m; left in	–	0.22	4.32	–	8.64	m	**12.96**
height exceeding 1.00 m; left in	–	0.27	5.40	–	8.25	m²	**13.65**
150 mm thick; plain vertical							
height not exceeding 250 mm; left in	–	0.09	1.80	–	2.99	m	**4.79**
height 250–500 mm; left in	–	0.14	2.87	–	5.97	m	**8.84**
height 500 mm–1.00 m; left in	–	0.22	4.32	–	11.96	m	**16.28**
height exceeding 1.00 m; left in	–	0.28	5.50	–	11.41	m²	**16.91**
175 mm thick; plain vertical							
height not exceeding 250 mm; left in	–	0.10	2.00	–	3.49	m	**5.49**
height 250–500 mm; left in	–	0.15	3.00	–	6.97	m	**9.97**
height 500 mm–1.00 m; left in	–	0.22	4.40	–	13.95	m	**18.35**
height exceeding 1.00 m; left in	–	0.30	5.99	–	13.31	m²	**19.30**
200 mm thick; plain vertical							
height not exceeding 250 mm; left in	–	0.10	2.00	–	3.99	m	**5.99**
height 250–500 mm; left in	–	0.15	3.00	–	7.97	m	**10.97**
height 500 mm–1.00 m; left in	–	0.22	4.40	–	15.93	m	**20.33**
height exceeding 1.00 m; left in	–	0.30	5.99	–	15.21	m²	**21.20**
250 mm thick; plain vertical							
height not exceeding 250 mm; left in	–	0.15	3.00	–	4.99	m	**7.99**
height 250–500 mm; left in	–	0.17	3.50	–	9.96	m	**13.46**
height 500 mm–1.00 m; left in	–	0.25	5.00	–	19.92	m	**24.92**
height exceeding 1.00 m; left in	–	0.35	6.99	–	19.01	m²	**26.00**
Combined heave pressure relief insulation and compressible board substructure formwork; Cordeck Cellcore CP or other equal and approved; butt joints; securely fixed in place							
Plain horizontal							
200 mm thick; beneath slabs; left in	–	0.54	10.79	–	23.12	m²	**33.91**
250 mm thick; beneath slabs; left in	–	0.58	11.69	–	25.74	m²	**37.43**
300 mm thick; beneath slabs; left in	–	0.63	12.59	–	27.73	m²	**40.32**

11 IN SITU CONCRETE WORKS

Item	PC £	Labour hours	Labour £	Plant £	Material £	Unit	Total rate £
11.05 FORMWORK – cont							
Dufaylite Clayboard void former; butt joints							
Residential (was KN30); compressive strength of							
30 kN/m²; board thickness; typically for residential							
60 mm	–	0.05	0.72	–	22.68	m²	**23.40**
90 mm	–	0.05	0.72	–	25.73	m²	**26.45**
110 mm	–	0.05	0.72	–	27.15	m²	**27.87**
160 mm	–	0.05	0.72	–	30.67	m²	**31.39**
Commercial (was KN90); compressive strength of							
90 kN/m²; board thickness; typically for commercial							
60 mm	–	0.06	0.87	–	26.58	m²	**27.45**
90 mm	–	0.06	0.87	–	29.12	m²	**29.99**
110 mm	–	0.06	0.87	–	31.56	m²	**32.43**
160 mm	–	0.06	0.87	–	33.41	m²	**34.28**
600 mm voidpack pipe							
36 mm dia.	–	0.05	0.72	–	6.82	nr	**7.54**
Sides of ground beams and edges of beds;							
basic finish; 18 mm thick external quality							
plywood; basic finish; four uses							
Plain vertical							
height not exceeding 250 mm	–	0.41	8.27	–	5.14	m	**13.41**
height 250–500 mm	–	0.75	14.92	–	11.58	m	**26.50**
height 500 mm–1.00 m	–	1.04	20.87	–	15.65	m	**36.52**
height exceeding 1.00 m	–	1.38	27.52	–	15.65	m²	**43.17**
Edges of suspended slabs; basic finish; 18 mm							
thick external quality plywood; basic finish; four							
uses							
Plain vertical							
height not exceeding 250 mm	–	0.62	12.41	–	5.25	m	**17.66**
height 250–500 mm	–	0.92	18.34	–	9.49	m	**27.83**
height 500 mm–1.00 m	–	1.46	29.14	–	15.76	m	**44.90**
Sides of upstands; basic finish; 18 mm thick							
external quality plywood; basic finish; four uses							
Plain vertical							
height not exceeding 250 mm	–	0.52	10.43	–	5.37	m	**15.80**
height 250–500 mm	–	0.84	16.73	–	11.80	m	**28.53**
height 500 mm–1.00 m	–	1.46	29.14	–	18.81	m	**47.95**
height exceeding 1.00 m	–	1.67	33.28	–	18.81	m²	**52.09**
Steps in top surfaces; basic finish; 18 mm thick							
external quality plywood; basic finish; four uses							
Plain vertical							
height not exceeding 250 mm	–	0.41	8.27	–	5.43	m	**13.70**
height 250–500 mm	–	0.67	13.31	–	11.86	m	**25.17**

11 IN SITU CONCRETE WORKS

Item	PC £	Labour hours	Labour £	Plant £	Material £	Unit	Total rate £
Steps in soffits; basic finish; 18 mm thick external quality plywood; basic finish; four uses							
Plain vertical							
height not exceeding 250 mm	–	0.46	9.18	–	4.64	m	13.82
height 250–500 mm	–	0.73	14.57	–	8.70	m	23.27
Machine bases and plinths; basic finish; 18 mm thick external quality plywood; basic finish; four uses							
Plain vertical							
height not exceeding 250 mm	–	0.41	8.27	–	5.14	m	13.41
height 250–500 mm	–	0.71	14.21	–	11.58	m	25.79
height 500 mm–1.00 m	–	1.04	20.87	–	15.65	m	36.52
height exceeding 1.00 m	–	1.33	26.63	–	15.65	m²	42.28
Soffits of slabs; basic finish; 18 mm thick external quality plywood; basic finish; four uses							
Slab thickness not exceeding 200 mm							
horizontal; height to soffit not exceeding 1.50 m	–	1.50	30.03	–	13.55	m²	43.58
horizontal; height to soffit 1.50–2.40 m	–	1.46	29.14	–	13.67	m²	42.81
horizontal; height to soffit 2.40–2.70 m	–	1.38	27.52	–	12.34	m²	39.86
horizontal; height to soffit 2.70–3.00 m	–	1.33	26.63	–	10.96	m²	37.59
horizontal; height to soffit 3.00–4.50 m	–	1.41	28.24	–	13.67	m²	41.91
horizontal; height to soffit 4.50–6.00 m	–	1.50	30.03	–	14.23	m²	44.26
Slab thickness 200–300 mm							
horizontal; height to soffit 1.50–3.00 m	–	1.50	30.03	–	18.07	m²	48.10
Slab thickness 300–400 mm							
horizontal; height to soffit 1.50–3.00 m	–	1.54	30.76	–	19.54	m²	50.30
Slab thickness 400–500 mm							
horizontal; height to soffit 1.50–3.00 m	–	1.62	32.37	–	22.13	m²	54.50
Slab thickness 500–600 mm							
horizontal; height to soffit 1.50–3.00 m	–	1.75	34.90	–	22.13	m²	57.03
Extra over soffits of slabs for							
sloping not exceeding 15°	–	0.17	3.42	–	–	m²	3.42
sloping exceeding 15°	–	0.33	6.65	–	–	m²	6.65
Soffits of landings; basic finish; 18 mm thick external quality plywood; basic finish; four uses							
Slab thickness not exceeding 200 mm							
horizontal; height to soffit 1.50–3.00 m	–	1.50	30.03	–	15.83	m²	45.86
Slab thickness 200–300 mm							
horizontal; height to soffit 1.50–3.00 m	–	1.58	31.66	–	19.68	m²	51.34
Slab thickness 300–400 mm							
horizontal; height to soffit 1.50–3.00 m	–	1.62	32.37	–	21.61	m²	53.98
Slab thickness 400–500 mm							
horizontal; height to soffit 1.50–3.00 m	–	1.71	34.18	–	23.54	m²	57.72
Slab thickness 500–600 mm							
horizontal; height to soffit 1.50–3.00 m	–	1.84	36.70	–	23.54	m²	60.24

11 IN SITU CONCRETE WORKS

Item	PC £	Labour hours	Labour £	Plant £	Material £	Unit	Total rate £
11.05 FORMWORK – cont							
Soffits of landings – cont							
Extra over soffits of landings for							
sloping not exceeding 15°	–	0.17	3.42	–	–	m²	**3.42**
sloping exceeding 15°	–	0.33	6.65	–	–	m²	**6.65**
Top formwork; basic finish; 18 mm thick external							
quality plywood; basic finish; four uses							
sloping exceeding 15°	–	1.25	25.00	–	11.92	m²	**36.92**
Walls; basic finish; 18 mm thick external quality							
plywood; basic finish; four uses							
vertical	–	1.50	30.03	–	15.14	m²	**45.17**
vertical; height exceeding 3.00 m above floor							
level	–	1.84	36.70	–	18.92	m²	**55.62**
vertical; interrupted	–	1.75	34.90	–	18.92	m²	**53.82**
vertical; to one side only	–	2.92	58.28	–	22.70	m²	**80.98**
battered	–	2.33	46.59	–	19.43	m²	**66.02**
Beams; basic finish; 18 mm thick external							
quality plywood; basic finish; four uses							
Attached to slabs							
regular shaped; square or rectangular; height to							
soffit 1.50–3.00 m	–	1.84	36.70	–	18.29	m²	**54.99**
regular shaped; square or rectangular; height to							
soffit 3.00–4.50 m	–	1.92	38.32	–	18.63	m²	**56.95**
regular shaped; square or rectangular; height to							
soffit 4.50–6.00 m	–	2.00	39.93	–	18.92	m²	**58.85**
Attached to walls							
regular shaped; square or rectangular; height to							
soffit 1.50–3.00 m	–	1.92	38.32	–	18.29	m²	**56.61**
Isolated							
regular shaped; square or rectangular; height to							
soffit 1.50–3.00 m	–	2.00	39.93	–	18.29	m²	**58.22**
regular shaped; square or rectangular; height to							
soffit 3.00–4.50 m	–	2.08	41.55	–	18.63	m²	**60.18**
regular shaped; square or rectangular; height to							
soffit 4.50–6.00 m	–	2.17	43.35	–	18.92	m²	**62.27**
Extra over beams for							
regular shaped; sloping not exceeding 15°	–	0.25	5.04	–	1.69	m²	**6.73**
regular shaped; sloping exceeding 15°	–	0.50	10.07	–	3.39	m²	**13.46**
Beam casings; basic finish; 18 mm thick							
external quality plywood; basic finish; four uses							
Attached to slabs							
regular shaped; square or rectangular; height to							
soffit 1.50–3.00 m	–	1.92	38.32	–	18.29	m²	**56.61**
regular shaped; square or rectangular; height to							
soffit 3.00–4.50 m	–	2.00	39.93	–	18.63	m²	**58.56**

11 IN SITU CONCRETE WORKS

Item	PC £	Labour hours	Labour £	Plant £	Material £	Unit	Total rate £
Attached to walls							
regular shaped; square or rectangular; height to soffit 1.50–3.00 m	–	2.00	39.93	–	18.29	m²	**58.22**
Isolated							
regular shaped; square or rectangular; height to soffit 1.50–3.00 m	–	2.08	41.55	–	18.29	m²	**59.84**
regular shaped; square or rectangular; height to soffit 3.00–4.50 m	–	2.17	43.35	–	18.63	m²	**61.98**
Extra over beam casings for							
regular shaped; sloping not exceeding 15°	–	0.25	5.04	–	1.69	m²	**6.73**
regular shaped; sloping exceeding 15°	–	0.50	10.07	–	3.39	m²	**13.46**
Columns; basic finish; 18 mm thick external quality plywood; basic finish; four uses							
Attached to walls							
regular shaped; square or rectangular; height to soffit 1.50–3.00 m	–	1.84	36.70	–	15.71	m²	**52.41**
Isolated							
regular shaped; square or rectangular; height to soffit 1.50–3.00 m	–	1.92	38.32	–	15.14	m²	**53.46**
regular shaped; circular; not exceeding 300 mm dia.; height to soffit 1.50–3.00 m	–	3.33	66.55	–	23.93	m²	**90.48**
regular shaped; circular; 300–600 mm dia.; height to soffit 1.50–3.00 m	–	3.12	62.42	–	21.00	m²	**83.42**
regular shaped; circular; 600–900 mm dia.; height to soffit 1.50–3.00 m	–	2.92	58.28	–	21.00	m²	**79.28**
Column casings; basic finish; 18 mm thick external quality plywood; basic finish; four uses							
Attached to walls							
regular shaped; square or rectangular; height to soffit 1.50–3.00 m	–	1.92	38.32	–	15.71	m²	**54.03**
Isolated							
regular shaped; square or rectangular; height to soffit 1.50–3.00 m	–	2.00	39.93	–	15.71	m²	**55.64**
Special shapes and finishes							
Recesses or rebates							
12 × 12 mm	–	0.05	1.08	–	0.46	m	**1.54**
25 × 25 mm	–	0.05	1.08	–	0.96	m	**2.04**
25 × 50 mm	–	0.05	1.08	–	1.22	m	**2.30**
50 × 50 mm	–	0.05	1.08	–	1.18	m	**2.26**
Nibs							
50 × 50 mm	–	0.46	9.18	–	1.71	m	**10.89**
100 × 100 mm	–	0.65	12.95	–	2.52	m	**15.47**
100 × 200 mm	–	0.86	17.26	–	17.46	m	**34.72**
Extra over a basic finish for fine formed finishes							
slabs	–	0.27	5.40	–	–	m²	**5.40**
walls	–	0.27	5.40	–	–	m²	**5.40**
beams	–	0.27	5.40	–	–	m²	**5.40**
columns	–	0.27	5.40	–	–	m²	**5.40**

11 IN SITU CONCRETE WORKS

Item	PC £	Labour hours	Labour £	Plant £	Material £	Unit	Total rate £
11.05 FORMWORK – cont							
Special shapes and finishes – cont							
Add to prices for basic formwork for							
curved radius 6.00 m	–	–	–	–	–	%	**50.00**
curved radius 2.00 m	–	–	–	–	–	%	**100.00**
coating with retardant agent	–	0.01	0.18	–	0.26	m²	**0.44**
Wall kickers; basic finish							
height 150 mm	–	0.41	8.27	–	4.47	m	**12.74**
height 225 mm	–	0.54	10.79	–	5.70	m	**16.49**
Suspended wall kickers; basic finish							
Height 150 mm	–	0.52	10.43	–	3.71	m	**14.14**
Wall ends, soffits and steps in walls; basic finish							
Plain							
width exceeding 1.00 m	–	1.58	31.66	–	18.63	m²	**50.29**
width not exceeding 250 mm	–	0.50	10.07	–	4.35	m	**14.42**
width 250–500 mm	–	0.79	15.83	–	9.89	m	**25.72**
width 500 mm–1.00 m	–	1.25	25.00	–	18.63	m	**43.63**
Openings in walls							
Plain							
width exceeding 1.00 m	–	1.75	34.90	–	18.63	m²	**53.53**
width not exceeding 250 mm	–	0.54	10.79	–	4.35	m	**15.14**
width 250–500 mm	–	0.92	18.34	–	9.89	m	**28.23**
width 500 mm–1.00 m	–	1.41	28.24	–	18.63	m	**46.87**
Stairflights							
Width 1.00 m; 150 mm waist; 150 mm undercut risers							
string, width 300 mm	–	4.17	83.29	–	39.29	m	**122.58**
Width 2.00 m; 200 mm waist; 150 mm undercut risers							
string, width 350 mm	–	7.50	149.83	–	121.89	m	**271.72**
Mortices							
Girth not exceeding 500 mm							
depth not exceeding 250 mm; circular	–	0.13	2.51	–	1.74	nr	**4.25**
Holes							
Girth not exceeding 500 mm							
depth not exceeding 250 mm; circular	–	0.17	3.42	–	2.08	nr	**5.50**
depth 250–500 mm; circular	–	0.25	5.04	–	5.38	nr	**10.42**
Girth 500 mm–1.00 m							
depth not exceeding 250 mm; circular	–	0.21	4.14	–	3.71	nr	**7.85**
depth 250–500 mm; circular	–	0.32	6.29	–	10.09	nr	**16.38**
Girth 1.00–2.00 m							
depth not exceeding 250 mm; circular	–	0.38	7.55	–	10.09	nr	**17.64**
depth 250–500 mm; circular	–	0.56	11.15	–	20.23	nr	**31.38**
Girth 2.00–3.00 m							
depth not exceeding 250 mm; circular	–	0.50	10.07	–	19.58	nr	**29.65**
depth 250–500 mm; circular	–	0.75	14.92	–	124.28	nr	**139.20**

11 IN SITU CONCRETE WORKS

Item	PC £	Labour hours	Labour £	Plant £	Material £	Unit	Total rate £
Pecafil formwork system							
To sides and base of foundations in conjuction with a full rebar cage as support							
not exceeding 500 mm high	–	0.20	4.00	–	18.15	m	**22.15**
exceeding 500 mm high	–	0.15	3.00	–	36.30	m²	**39.30**
Sides of isolated columns							
sides of isolated columns; rectangle or square; regular	–	0.75	14.99	–	36.30	m²	**51.29**
11.06 REINFORCEMENT							
Bars; hot rolled deformed high steel bars; grade 500C							
40 mm dia. nominal size							
straight	–	7.00	146.94	–	798.99	tonne	**945.93**
bent	–	9.00	188.91	–	835.94	tonne	**1024.85**
32 mm dia. nominal size							
straight	–	8.00	167.92	–	805.84	tonne	**973.76**
bent	–	10.00	209.90	–	842.80	tonne	**1052.70**
25 mm dia. nominal size							
straight	–	9.00	188.91	–	813.48	tonne	**1002.39**
bent	–	12.00	251.89	–	850.43	tonne	**1102.32**
20 mm dia. nominal size							
straight	–	11.00	230.90	–	823.46	tonne	**1054.36**
bent	–	14.00	293.87	–	860.41	tonne	**1154.28**
16 mm dia. nominal size							
straight	–	14.00	293.87	–	834.22	tonne	**1128.09**
bent	–	18.00	377.83	–	871.17	tonne	**1249.00**
12 mm dia. nominal size							
straight	–	20.00	419.82	–	851.83	tonne	**1271.65**
bent	–	25.00	524.76	–	888.79	tonne	**1413.55**
10 mm dia. nominal size							
straight	–	24.00	503.77	–	873.36	tonne	**1377.13**
bent	–	31.00	650.71	–	910.30	tonne	**1561.01**
8 mm dia. nominal size							
straight	–	29.00	608.73	–	890.97	tonne	**1499.70**
links	–	37.00	776.65	–	955.35	tonne	**1732.00**
bent	–	37.00	776.65	–	927.93	tonne	**1704.58**
Bars; stainless steel							
40 mm dia. nominal size							
straight	–	7.00	146.94	–	4180.21	tonne	**4327.15**
bent	–	9.00	188.91	–	4344.12	tonne	**4533.03**
32 mm dia. nominal size							
straight	–	17.00	356.84	–	4180.21	tonne	**4537.05**
bent	–	10.00	209.90	–	4344.12	tonne	**4554.02**
25 mm dia. nominal size							
straight	–	9.00	188.91	–	4170.72	tonne	**4359.63**
bent	–	12.00	251.89	–	4334.62	tonne	**4586.51**
20 mm dia. nominal size							
straight	–	11.00	230.90	–	4184.12	tonne	**4415.02**
bent	–	14.00	293.87	–	4348.03	tonne	**4641.90**

11 IN SITU CONCRETE WORKS

Item	PC £	Labour hours	Labour £	Plant £	Material £	Unit	Total rate £
11.06 REINFORCEMENT – cont							
Bars – cont							
16 mm dia. nominal size							
straight	–	14.00	293.87	–	4208.59	tonne	**4502.46**
bent	–	18.00	377.83	–	4372.49	tonne	**4750.32**
12 mm dia. nominal size							
straight	–	20.00	419.82	–	4233.06	tonne	**4652.88**
bent	–	25.00	524.76	–	4396.97	tonne	**4921.73**
10 mm dia. nominal size							
straight	–	24.00	503.77	–	4261.44	tonne	**4765.21**
bent	–	31.00	650.71	–	4425.34	tonne	**5076.05**
8 mm dia. nominal size							
straight	–	29.00	608.73	–	4285.91	tonne	**4894.64**
bent	–	37.00	776.65	–	4449.82	tonne	**5226.47**
Bars; stainless steel; (low nickel alloys or lean duplexes)							
40 mm dia. nominal size							
straight	–	7.00	146.94	–	3327.91	tonne	**3474.85**
bent	–	9.00	188.91	–	3502.74	tonne	**3691.65**
32 mm dia. nominal size							
straight	–	8.00	167.92	–	3330.09	tonne	**3498.01**
bent	–	10.00	209.90	–	3502.74	tonne	**3712.64**
25 mm dia. nominal size							
straight	–	9.00	188.91	–	3318.40	tonne	**3507.31**
bent	–	12.00	251.89	–	3493.23	tonne	**3745.12**
20 mm dia. nominal size							
straight	–	11.00	230.90	–	3331.81	tonne	**3562.71**
bent	–	14.00	293.87	–	3506.65	tonne	**3800.52**
16 mm dia. nominal size							
straight	–	14.00	293.87	–	3356.29	tonne	**3650.16**
bent	–	18.00	377.83	–	3531.12	tonne	**3908.95**
12 mm dia. nominal size							
straight	–	20.00	419.82	–	3380.76	tonne	**3800.58**
bent	–	25.00	524.76	–	3555.59	tonne	**4080.35**
10 mm dia. nominal size							
straight	–	24.00	503.77	–	3409.14	tonne	**3912.91**
bent	–	31.00	650.71	–	3583.97	tonne	**4234.68**
8 mm dia. nominal size							
straight	–	29.00	608.73	–	3441.02	tonne	**4049.75**
bent	–	37.00	776.65	–	3608.43	tonne	**4385.08**
Fabric; in standard 4.8 m × 2.4 m sheets							
Ref D49 (0.77 kg/m^2)							
100 mm minimum laps; bent	–	0.24	5.04	–	2.87	m^2	**7.91**
Ref D98 (1.54 kg/m^2)							
400 mm minimum laps	–	0.12	2.52	–	1.72	m^2	**4.24**
strips in one width; 600 mm width	–	0.15	3.15	–	1.72	m^2	**4.87**
strips in one width; 900 mm width	–	0.14	2.94	–	1.72	m^2	**4.66**
strips in one width; 1200 mm width	–	0.13	2.73	–	1.72	m^2	**4.45**

11 IN SITU CONCRETE WORKS

Item	PC £	Labour hours	Labour £	Plant £	Material £	Unit	Total rate £
Ref A142 (2.22 kg/m²)							
400 mm minimum laps	–	0.12	2.52	–	2.02	m²	**4.54**
strips in one width; 600 mm width	–	0.15	3.15	–	2.02	m²	**5.17**
strips in one width; 900 mm width	–	0.14	2.94	–	2.02	m²	**4.96**
strips in one width; 1200 mm width	–	0.13	2.73	–	2.02	m²	**4.75**
Ref A193 (3.02 kg/m²)							
400 mm minimum laps	–	0.12	2.52	–	2.74	m²	**5.26**
strips in one width; 600 mm width	–	0.15	3.15	–	2.74	m²	**5.89**
strips in one width; 900 mm width	–	0.14	2.94	–	2.74	m²	**5.68**
strips in one width; 1200 mm width	–	0.13	2.73	–	2.74	m²	**5.47**
Ref A252 (3.95 kg/m²)							
400 mm minimum laps	–	0.13	2.73	–	3.53	m²	**6.26**
strips in one width; 600 mm width	–	0.16	3.36	–	3.53	m²	**6.89**
strips in one width; 900 mm width	–	0.15	3.15	–	3.53	m²	**6.68**
strips in one width; 1200 mm width	–	0.14	2.94	–	3.53	m²	**6.47**
Ref A393 (6.16 kg/m²)							
400 mm minimum laps	–	0.15	3.15	–	5.52	m²	**8.67**
strips in one width; 600 mm width	–	0.18	3.78	–	5.52	m²	**9.30**
strips in one width; 900 mm width	–	0.17	3.56	–	5.52	m²	**9.08**
strips in one width; 1200 mm width	–	0.16	3.36	–	5.52	m²	**8.88**
Ref B196 (3.05 kg/m²)							
400 mm minimum laps	–	0.12	2.52	–	3.24	m²	**5.76**
strips in one width; 600 mm width	–	0.15	3.15	–	3.24	m²	**6.39**
strips in one width; 900 mm width	–	0.14	2.94	–	3.24	m²	**6.18**
strips in one width; 1200 mm width	–	0.13	2.73	–	3.24	m²	**5.97**
Ref B283 (3.73 kg/m²)							
400 mm minimum laps	–	0.12	2.52	–	3.47	m²	**5.99**
strips in one width; 600 mm width	–	0.15	3.15	–	3.47	m²	**6.62**
strips in one width; 900 mm width	–	0.14	2.94	–	3.47	m²	**6.41**
strips in one width; 1200 mm width	–	0.13	2.73	–	3.47	m²	**6.20**
Ref B385 (4.53 kg/m²)							
400 mm minimum laps	–	0.13	2.73	–	4.20	m²	**6.93**
strips in one width; 600 mm width	–	0.16	3.36	–	4.20	m²	**7.56**
strips in one width; 900 mm width	–	0.15	3.15	–	4.20	m²	**7.35**
strips in one width; 1200 mm width	–	0.14	2.94	–	4.20	m²	**7.14**
Ref B503 (5.93 kg/m²)							
400 mm minimum laps	–	0.15	3.15	–	5.44	m²	**8.59**
strips in one width; 600 mm width	–	0.18	3.78	–	5.44	m²	**9.22**
strips in one width; 900 mm width	–	0.17	3.56	–	5.44	m²	**9.00**
strips in one width; 1200 mm width	–	0.16	3.36	–	5.44	m²	**8.80**
Ref B785 (8.14 kg/m²)							
400 mm minimum laps	–	0.17	3.56	–	7.47	m²	**11.03**
strips in one width; 600 mm width	–	0.20	4.20	–	7.47	m²	**11.67**
strips in one width; 900 mm width	–	0.19	3.99	–	7.47	m²	**11.46**
strips in one width; 1200 mm width	–	0.18	3.78	–	7.47	m²	**11.25**
Ref B1131 (10.90 kg/m²)							
400 mm minimum laps	–	0.18	3.78	–	10.00	m²	**13.78**
strips in one width; 600 mm width	–	0.24	5.04	–	10.00	m²	**15.04**
strips in one width; 900 mm width	–	0.22	4.61	–	10.00	m²	**14.61**
strips in one width; 1200 mm width	–	0.20	4.20	–	10.00	m²	**14.20**

11 IN SITU CONCRETE WORKS

Item	PC £	Labour hours	Labour £	Plant £	Material £	Unit	Total rate £
11.07 DESIGNED JOINTS							
Formed; Fosroc impregnated fibreboard joint filler or other equal							
Width not exceeding 150 mm							
12.50 mm thick	–	0.14	2.80	–	2.48	m	**5.28**
20 mm thick	–	0.19	3.80	–	3.11	m	**6.91**
25 mm thick	–	0.23	4.59	–	3.68	m	**8.27**
Width 150–300 mm							
12.50 mm thick	–	0.23	4.59	–	4.60	m	**9.19**
20 mm thick	–	0.23	4.59	–	6.09	m	**10.68**
25 mm thick	–	0.23	4.59	–	6.87	m	**11.46**
Width 300–450 mm							
12.50 mm thick	–	0.28	5.59	–	6.63	m	**12.22**
20 mm thick	–	0.28	5.59	–	8.66	m	**14.25**
25 mm thick	–	0.28	5.59	–	9.74	m	**15.33**
Formed; waterproof bonded cork joint filler board							
Width not exceeding 150 mm							
10 mm thick	–	0.14	2.80	–	4.86	m	**7.66**
13 mm thick	–	0.14	2.80	–	4.93	m	**7.73**
19 mm thick	–	0.14	2.80	–	6.43	m	**9.23**
25 mm thick	–	0.14	2.80	–	6.85	m	**9.65**
Width 150–300 mm							
10 mm thick	–	0.19	3.80	–	8.76	m	**12.56**
13 mm thick	–	0.19	3.80	–	8.90	m	**12.70**
19 mm thick	–	0.19	3.80	–	11.89	m	**15.69**
25 mm thick	–	0.19	3.80	–	12.73	m	**16.53**
Width 300–450 mm							
10 mm thick	–	0.23	4.59	–	13.40	m	**17.99**
13 mm thick	–	0.23	4.59	–	13.61	m	**18.20**
19 mm thick	–	0.23	4.59	–	18.09	m	**22.68**
25 mm thick	–	0.23	4.59	–	19.37	m	**23.96**
Sealants; Fosroc Nitoseal MB77 hot poured rubberized bituminous compound or other equal							
Width 10 mm							
25 mm depth	–	0.17	3.40	–	2.56	m	**5.96**
Width 12.50 mm							
25 mm depth	–	0.18	3.59	–	3.15	m	**6.74**
Width 20 mm							
25 mm depth	–	0.19	3.80	–	5.14	m	**8.94**
Width 25 mm							
25 mm depth	–	0.20	4.00	–	6.31	m	**10.31**

11 IN SITU CONCRETE WORKS

Item	PC £	Labour hours	Labour £	Plant £	Material £	Unit	Total rate £
Sealants; Fosroc Thioflex 600 gun grade polysulphide or other equal							
Width 10 mm							
25 mm depth	–	0.05	1.00	–	3.56	m	**4.56**
Width 12.50 mm							
25 mm depth	–	0.06	1.19	–	4.45	m	**5.64**
Width 20 mm							
25 mm depth	–	0.07	1.40	–	7.13	m	**8.53**
Width 25 mm							
25 mm depth	–	0.08	1.60	–	8.91	m	**10.51**
Sealants; two component sealant on primed surface							
Width 10 mm							
25 mm depth	–	0.19	3.21	–	6.66	m	**9.87**
Width 13 mm							
25 mm depth	–	0.19	3.21	–	8.56	m	**11.77**
Width 19 mm							
25 mm depth	–	0.23	3.88	–	12.33	m	**16.21**
Width 25 mm							
25 mm depth	–	0.23	3.88	–	16.11	m	**19.99**
Waterstops; Grace Servicised or other equal							
Hydrophilic strip water stop; lapped joints; cast into concrete							
50 × 20 mm Adcor 500S	9.12	0.30	5.07	–	12.25	m	**17.32**
Servitite Internal 10 mm thick PVC water stop; flat dumbbell type; heat welded joints; cast into concrete							
Servitite 150; 150 mm wide	–	0.23	4.83	–	18.59	m	**23.42**
flat angle	–	0.28	5.88	–	39.83	nr	**45.71**
vertical angle	–	0.28	5.88	–	39.63	nr	**45.51**
flat three way intersection	–	0.37	7.77	–	57.98	nr	**65.75**
vertical three way intersection	–	0.37	7.77	–	65.34	nr	**73.11**
four way intersection	–	0.46	9.65	–	72.17	nr	**81.82**
Servitite 230; 230 mm wide	–	0.23	4.83	–	24.69	m	**29.52**
flat angle	–	0.28	5.88	–	48.25	nr	**54.13**
vertical angle	–	0.28	5.88	–	56.79	nr	**62.67**
flat three way intersection	–	0.37	7.77	–	70.58	nr	**78.35**
vertical three way intersection	–	0.37	7.77	–	120.84	nr	**128.61**
four way intersection	–	0.46	9.65	–	88.95	nr	**98.60**
Servitite AT200; 200 mm wide	–	0.23	4.83	–	27.12	m	**31.95**
flat angle	–	0.28	5.88	–	47.53	nr	**53.41**
vertical angle	–	0.28	5.88	–	51.34	nr	**57.22**
flat three way intersection	–	0.37	7.77	–	81.95	nr	**89.72**
vertical three way intersection	–	0.37	7.77	–	64.97	nr	**72.74**
four way intersection	–	0.46	9.65	–	97.60	nr	**107.25**
Servitite K305; 305 mm wide	–	0.28	5.88	–	39.43	m	**45.31**
flat angle	–	0.32	6.72	–	82.20	nr	**88.92**
vertical angle	–	0.32	6.72	–	88.42	nr	**95.14**
flat three way intersection	–	0.42	8.82	–	117.45	nr	**126.27**
vertical three way intersection	–	0.42	8.82	–	136.94	nr	**145.76**
four way intersection	–	0.51	10.70	–	158.94	nr	**169.64**

11 IN SITU CONCRETE WORKS

Item	PC £	Labour hours	Labour £	Plant £	Material £	Unit	Total rate £
11.07 DESIGNED JOINTS – cont							
Waterstops – cont							
Serviseal External PVC water stop; PVC water stop; centre bulb type; heat welded joints; cast into concrete							
Serviseal 195; 195 mm wide	–	0.23	4.83	–	12.01	m	**16.84**
flat angle	–	0.28	5.88	–	24.97	nr	**30.85**
vertical angle	–	0.28	5.88	–	40.83	nr	**46.71**
flat three way intersection	–	0.37	7.77	–	41.81	nr	**49.58**
four way intersection	–	0.46	9.65	–	61.62	nr	**71.27**
Serviseal 240; 240 mm wide	–	0.23	4.83	–	14.72	m	**19.55**
flat angle	–	0.28	5.88	–	28.79	nr	**34.67**
vertical angle	–	0.28	5.88	–	43.57	nr	**49.45**
flat three way intersection	–	0.37	7.77	–	47.60	nr	**55.37**
four way intersection	–	0.46	9.65	–	69.30	nr	**78.95**
Serviseal AT240; 240 mm wide	–	0.23	4.83	–	31.64	m	**36.47**
flat angle	–	0.28	5.88	–	46.83	nr	**52.71**
vertical angle	–	0.28	5.88	–	44.89	nr	**50.77**
flat three way intersection	–	0.37	7.77	–	71.72	nr	**79.49**
four way intersection	–	0.46	9.65	–	104.51	nr	**114.16**
Serviseal K320; 320 mm wide	–	0.28	5.88	–	19.06	m	**24.94**
flat angle	–	0.32	6.72	–	60.62	nr	**67.34**
vertical angle	–	0.32	6.72	–	32.97	nr	**39.69**
flat three way intersection	–	0.42	8.82	–	89.18	nr	**98.00**
four way intersection	–	0.51	10.70	–	112.21	nr	**122.91**
11.08 ACCESSORIES CAST IN							
Foundation bolt boxes							
Temporary plywood; for group of 4 nr bolts							
75 × 75 × 150 mm	–	0.42	8.39	–	1.81	nr	**10.20**
75 × 75 × 250 mm	–	0.42	8.39	–	2.14	nr	**10.53**
Expanded metal; Expamet Building Products Ltd or other equal and approved							
75 mm dia. × 150 mm long	–	0.28	5.59	–	80.72	nr	**86.31**
75 mm dia. × 300 mm long	–	0.28	5.59	–	90.68	nr	**96.27**
100 mm dia. × 450 mm long	–	0.28	5.59	–	3.02	nr	**8.61**
Foundation bolts and nuts							
Black hexagon							
10 mm dia. × 100 mm long	–	0.23	4.59	–	0.62	nr	**5.21**
12 mm dia. × 120 mm long	–	0.23	4.59	–	0.94	nr	**5.53**
16 mm dia. × 160 mm long	–	0.28	5.59	–	2.60	nr	**8.19**
20 mm dia. × 180 mm long	–	0.28	5.59	–	3.03	nr	**8.62**
Masonry slots							
Stainless steel; dovetail slots; 1.20 mm thick; 18G							
1000 mm long	–	0.25	5.00	–	6.83	m	**11.83**
100 mm long	–	0.07	1.40	–	0.32	nr	**1.72**

11 IN SITU CONCRETE WORKS

Item	PC £	Labour hours	Labour £	Plant £	Material £	Unit	Total rate £
Stainless steel; metal insert slots; Halfen Ltd Ribslot or other equal; 2.50 mm thick; end caps and foam filling							
41 × 41 mm; ref P3270	–	0.37	7.40	–	9.51	m	**16.91**
41 × 41 × 100 mm; ref P3250	–	0.09	1.80	–	1.44	nr	**3.24**
41 × 41 × 150 mm; ref P3251	–	0.09	1.80	–	1.06	nr	**2.86**
Cramps							
Stainless steel; once bent; one end shot fired into concrete; other end fanged and built into brickwork joint							
200 mm girth	–	0.14	3.23	–	1.31	nr	**4.54**
Column guards							
Corner protection; White nylon coated steel; plugging; screwing to concrete; 1.5 mm thick							
75 × 75 × 1200 mm	–	0.74	14.79	–	24.60	nr	**39.39**
Galvanized steel; 3 mm thick							
75 × 75 × 1000 mm	–	0.56	11.20	–	17.45	nr	**28.65**
Galvanized steel; 4.5 mm thick							
75 × 75 × 1000 mm	–	0.56	11.20	–	23.46	nr	**34.66**
Stainless steel; HKW or other equal; Halfen Ltd; 5 mm thick							
50 × 50 × 1200 mm	–	0.93	18.59	–	79.53	nr	**98.12**
50 × 50 × 2000 mm	–	1.11	22.19	–	131.38	nr	**153.57**
Channels							
Stainless steel; Halfen Ltd or other equal							
ref 38/17/HTA channel	–	0.32	6.40	–	45.92	m	**52.32**
ref 41/22/HZA; 80 mm long; including T headed bolts and plate washers	–	0.09	1.80	–	29.69	nr	**31.49**
Channel ties							
Stainless steel; Halfen Ltd or other equal							
ref HTS – B12; 150 mm projection; including insulation retainer	–	0.03	0.67	–	0.58	nr	**1.25**
ref HTS – B12; 200 mm projection; including insulation retainer	–	0.03	0.67	–	0.68	nr	**1.35**
11.09 IN SITU CONCRETE SUNDRIES							
Grouting with cement mortar (1:1)							
Stanchion bases							
10 mm thick	–	0.93	15.72	–	0.19	nr	**15.91**
25 mm thick	–	1.16	19.60	–	0.47	nr	**20.07**
Grouting with epoxy resin							
Stanchion bases							
10 mm thick	–	1.16	19.60	–	4.14	nr	**23.74**
25 mm thick	–	1.39	23.48	–	20.70	nr	**44.18**

Prices for Measured Works

11 IN SITU CONCRETE WORKS

Item	PC £	Labour hours	Labour £	Plant £	Material £	Unit	Total rate £
11.09 IN SITU CONCRETE SUNDRIES – cont							
Grouting with Conbextra GP cementitious grout							
Stanchion bases							
10 mm thick	–	1.16	19.60	–	0.80	nr	**20.40**
25 mm thick	–	1.39	23.48	–	2.04	nr	**25.52**
Grouting with Conbextra HF flowable cementitious grout							
Stanchion bases							
10 mm thick	–	1.16	19.60	–	1.38	nr	**20.98**
25 mm thick	–	1.39	23.48	–	3.54	nr	**27.02**
Sundry in situ concrete work							
Filling; plain in situ concrete; mixed on site							
mortices	–	0.09	1.52	0.23	0.67	nr	**2.42**
holes	–	0.23	3.88	0.59	140.59	m^3	**145.06**
chases not exceeding 300 mm wide or thick	–	0.14	2.37	0.35	1.40	m	**4.12**
chases exceeding 300 mm wide or thick	–	0.19	3.21	0.48	140.59	m^3	**144.28**
extra for reinforcement content over 5%	–	0.51	8.62	–	–	m^3	**8.62**

12 PRECAST/COMPOSITE CONCRETE

Item	PC £	Labour hours	Labour £	Plant £	Material £	Unit	Total rate £
12.01 PRECAST/COMPOSITE CONCRETE WORK							
Prestressed precast concrete structural suspended floors; Bison Hollowcore or other equal; supplied and fixed on hard level bearings, to areas of 500 m² per site visit; top surface screeding and ceiling finishes by others							
Floors to dwellings, offices, car parks, shop retail floors, hospitals, school teaching rooms, staff rooms and the like; superimposed load of 5.00 kN/m²							
floor spans up to 3.00 m; 1200 mm × 150 mm	–	–	–	–	–	m²	60.69
floor spans 3.00 m–6.00 m; 1200 mm × 150 mm	–	–	–	–	–	m²	62.16
floor spans 6.00 m–7.50 m; 1200 mm × 200 mm	–	–	–	–	–	m²	66.43
floor spans 7.50 m–9.50 m; 1200 mm × 250 mm	–	–	–	–	–	m²	70.71
floor spans 9.50 m–12.00 m; 1200 mm × 300 mm	–	–	–	–	–	m²	73.06
floor spans 12.00 m–12.50 m; 1200 mm × 350 mm	–	–	–	–	–	m²	75.41
floor spans 12.50 m–14.00 m; 1200 mm × 400 mm	–	–	–	–	–	m²	83.67
floor spans 14.00 m–15.00 m; 1200 mm × 450 mm	–	–	–	–	–	m²	84.85
Floors to shop stockrooms, light warehousing, schools, churches or similar places of assembly, light factory accommodation, laboratories and the like; superimposed load of 8.50 kN/m²							
floor spans up to 3.00 m; 1200 mm × 150 mm	–	–	–	–	–	m²	58.93
floor spans 3.00 m–6.00 m; 1200 mm × 200 mm	–	–	–	–	–	m²	61.28
floor spans 6.00 m–7.50 m; 1200 mm × 250 mm	–	–	–	–	–	m²	70.71
Floors to heavy warehousing, factories, stores and the like; superimposed load of 12.50 kN/m²							
floor spans up to 3.00 m; 1200 mm × 150 mm	–	–	–	–	–	m²	62.47
floor spans 3.00 m–6.00 m; 1200 mm × 250 mm	–	–	–	–	–	m²	70.71
Prestressed precast concrete staircase, supplied and fixed in conjunction with Bison Hollowcore flooring system or similar; comprising 2 nr 1100 mm wide flights with 7 nr 275 mm treads, 8 nr 185 mm risers and 150 mm waist; 1 nr 2200 mm × 1400 mm × 150 mm half landing and 1 nr top landing							
3.00 m storey height	–	–	–	–	–	nr	2895.39
Prestressed precast concrete beam and block floor; cement and sand grout brushed between beams and blocks; 440 × 215 × 100 mm concrete block infill							
Beam and block flooring at ground level							
beam and block flooring at ground level 150 mm thick beams; up to 3.30 m span	–	0.45	11.22	–	47.83	m²	59.05
beam and block flooring at ground level; 225 mm thick beams; up to 4.20 m span	–	0.45	11.22	–	84.98	m²	96.20
beam and poly block flooring at ground level 150 mm thick beams; up to 3.30 m span (U-value 0.20 W/m² K)	27.35	0.45	11.22	–	28.17	m²	39.39

12 PRECAST/COMPOSITE CONCRETE

Item	PC £	Labour hours	Labour £	Plant £	Material £	Unit	Total rate £
12.01 PRECAST/COMPOSITE CONCRETE WORK – cont							
Prestressed precast concrete beam and block floor – cont							
Beam and block flooring at ground level – cont							
beam and poly block flooring at ground level							
150 mm thick beams; up to 3.30 m span (U-value 0.15 W/m² K)	30.99	0.50	12.46	–	31.92	m²	**44.38**
beam and poly block flooring at ground level							
150 mm thick beams; up to 3.30 m span (U-value 0.12W/m² K)	35.86	0.50	12.46	–	36.94	m²	**49.40**
Composite floor comprising reinforced in situ ready-mixed concrete 30.00 N/mm²; on and including steel deck permanent shutting; complete with A142 anti-crack mesh; NB temporary props may be required, but have not been included in the following rates							
0.9 mm thick re-entrant type deck; 60 mm deep							
150 mm thick suspended slab							
1.50 m–3.00 m high to soffit	–	0.91	16.40	–	61.93	m²	**78.33**
3.00 m–4.50 m high to soffit	–	0.94	16.92	–	61.93	m²	**78.85**
4.50 m–6.00 m high to soffit	–	0.98	17.71	–	61.93	m²	**79.64**
200 mm thick suspended slab							
1.50 m–3.00 m high to soffit	–	0.95	17.08	–	68.12	m²	**85.20**
3.00 m–4.50 m high to soffit	–	0.97	17.60	–	68.12	m²	**85.72**
4.50 m–6.00 m high to soffit	–	1.00	18.05	–	68.12	m²	**86.17**
1.2 mm thick re-entrant type deck							
150 mm thick suspended slab							
1.50 m–3.00 m high to soffit	–	0.91	16.40	–	66.45	m²	**82.85**
3.00 m–4.50 m high to soffit	–	0.94	16.92	–	66.45	m²	**83.37**
4.50 m–6.00 m high to soffit	–	0.98	17.71	–	66.45	m²	**84.16**
200 mm thick suspended slab							
1.50 m–3.00 m high to soffit	–	0.95	17.08	–	72.65	m²	**89.73**
3.00 m–4.50 m high to soffit	–	0.97	17.60	–	72.65	m²	**90.25**
4.50 m–6.00 m high to soffit	–	1.00	18.05	–	72.65	m²	**90.70**
0.9 mm thick trapezoidal deck 60 mm deep							
150 mm thick suspended slab							
1.50 m–3.00 m high to soffit	–	0.91	16.40	–	58.79	m²	**75.19**
3.00 m–4.50 m high to soffit	–	0.94	16.92	–	58.79	m²	**75.71**
4.50 m–6.00 m high to soffit	–	0.98	17.71	–	58.79	m²	**76.50**
200 mm thick suspended slab							
1.50 m–3.00 m high to soffit	–	0.95	17.08	–	64.98	m²	**82.06**
3.00 m–4.50 m high to soffit	–	0.97	17.60	–	64.98	m²	**82.58**
4.50 m–6.00 m high to soffit	–	1.00	18.05	–	64.98	m²	**83.03**

12 PRECAST/COMPOSITE CONCRETE

Item	PC £	Labour hours	Labour £	Plant £	Material £	Unit	Total rate £
0.9 mm thick trapezoidal deck 80 mm deep							
150 mm thick suspended slab							
1.50 m–3.00 m high to soffit	–	0.91	16.40	–	60.30	m²	**76.70**
3.00 m–4.50 m high to soffit	–	0.94	16.92	–	60.30	m²	**77.22**
4.50 m–6.00 m high to soffit	–	0.98	17.71	–	60.30	m²	**78.01**
200 mm thick suspended slab							
1.50 m–3.00 m high to soffit	–	0.95	17.08	–	66.49	m²	**83.57**
3.00 m–4.50 m high to soffit	–	0.97	17.60	–	66.49	m²	**84.09**
4.50 m–6.00 m high to soffit	–	1.00	18.05	–	66.49	m²	**84.54**
1.2 mm thick trapezoidal deck 80 mm deep							
150 mm thick suspended slab							
1.50 m–3.00 m high to soffit	–	0.91	16.40	–	64.41	m²	**80.81**
3.00 m–4.50 m high to soffit	–	0.94	16.92	–	64.41	m²	**81.33**
4.50 m–6.00 m high to soffit	–	0.98	17.71	–	64.41	m²	**82.12**
200 mm thick suspended slab							
1.50 m–3.00 m high to soffit	–	0.95	17.08	–	70.60	m²	**87.68**
3.00 m–4.50 m high to soffit	–	0.97	17.60	–	70.60	m²	**88.20**
4.50 m–6.00 m high to soffit	–	1.00	18.05	–	70.60	m²	**88.65**
Soffits of coffered or troughed slabs; basic finish							
Cordek Correx trough mould or other equal; 300 mm deep; ribs of mould at 600 mm centres and cross ribs at centres of bay; slab thickness 300–400 mm							
horizontal; height to soffit 1.50–3.00 m	–	2.08	41.55	–	23.11	m²	**64.66**
horizontal; height to soffit 3.00–4.50 m	–	2.17	43.35	–	23.40	m²	**66.75**
horizontal; height to soffit 4.50–6.00 m	–	2.25	44.97	–	23.57	m²	**68.54**

13 PRECAST CONCRETE

Item	PC £	Labour hours	Labour £	Plant £	Material £	Unit	Total rate £
13.01 PRECAST CONCRETE GOODS							
Contractor designed precast concrete staircases and landings; including all associated steel supports and fixing in position							
Straight staircases; 280 mm treads; 170 mm undercut risers							
1200 mm wide; 2750 mm rise	–	–	–	–	–	nr	**1896.98**
1200 mm wide; 3750 mm rise	–	–	–	–	–	nr	**2529.31**
Dogleg staircases							
1200 mm wide; one full width half landing; 2750 mm rise	–	–	–	–	–	nr	**2908.69**
1200 mm wide; one full width half landing; 3750 mm rise	–	–	–	–	–	nr	**3793.97**
1200 mm wide 200 mm thick concrete landing support walls	–	–	–	–	–	nr	**885.25**
1800 mm wide; one full width half landing; 2750 mm rise	–	–	–	–	–	nr	**4110.12**
1800 mm wide; one full width half landing; 3750 mm rise	–	–	–	–	–	nr	**5374.78**
1800 mm wide 200 mm thick concrete landing support walls	–	–	–	–	–	nr	**1365.83**
Precast concrete sill, lintels, copings							
Lintels; plate; prestressed bedded							
100 mm × 70 mm × 600 mm long	6.20	0.37	8.29	–	6.43	nr	**14.72**
100 mm × 70 mm × 900 mm long	9.24	0.37	8.29	–	9.57	nr	**17.86**
100 mm × 70 mm × 1100 mm long	11.32	0.37	8.29	–	11.70	nr	**19.99**
100 mm × 70 mm × 1200 mm long	12.33	0.37	8.29	–	12.75	nr	**21.04**
100 mm × 70 mm × 1500 mm long	15.42	0.46	10.31	–	15.97	nr	**26.28**
100 mm × 70 mm × 1800 mm long	18.49	0.46	10.31	–	19.13	nr	**29.44**
100 mm × 70 mm × 2100 mm long	21.58	0.56	12.56	–	22.34	nr	**34.90**
140 mm × 70 mm × 1200 mm long	18.13	0.46	10.31	–	18.76	nr	**29.07**
140 mm × 70 mm × 1500 mm long	22.66	0.56	12.56	–	23.45	nr	**36.01**
Lintels; rectangular; reinforced with mild steel bars; bedded							
100 mm × 145 mm × 900 mm long	4.57	0.56	12.56	–	4.76	nr	**17.32**
100 mm × 145 mm × 1050 mm long	5.34	0.56	12.56	–	5.55	nr	**18.11**
100 mm × 145 mm × 1200 mm long	6.10	0.56	12.56	–	6.33	nr	**18.89**
225 mm × 145 mm × 1200 mm long	23.65	0.74	16.58	–	24.48	nr	**41.06**
225 mm × 225 mm × 1800 mm long	35.38	1.39	31.16	–	36.56	nr	**67.72**
Lintels; boot; reinforced with mild steel bars; bedded							
250 mm × 225 mm × 1200 mm long	26.21	1.11	24.87	–	27.11	nr	**51.98**
275 mm × 225 mm × 1800 mm long	43.24	1.67	37.43	–	44.66	nr	**82.09**
Padstones							
100 mm × 200 mm × 440 mm	22.99	0.28	6.27	–	23.72	nr	**29.99**
150 mm × 215 mm × 440 mm	22.97	0.37	8.29	–	23.74	nr	**32.03**
150 mm × 225 mm × 225 mm	18.05	0.56	12.56	–	18.92	nr	**31.48**

13 PRECAST CONCRETE

Item	PC £	Labour hours	Labour £	Plant £	Material £	Unit	Total rate £
Copings; once weathered; once throated; bedded and pointed							
152 mm × 76 mm	6.87	0.65	14.56	–	7.24	m	**21.80**
178 mm × 64 mm	7.60	0.65	14.56	–	7.99	m	**22.55**
305 mm × 76 mm	12.83	0.74	16.58	–	13.60	m	**30.18**
extra for fair ends	–	–	–	–	5.07	nr	**5.07**
extra for angles	–	–	–	–	5.75	nr	**5.75**
Copings; twice weathered; twice throated; bedded and pointed							
152 mm × 76 mm	6.87	0.65	14.56	–	7.24	m	**21.80**
178 mm × 64 mm	7.56	0.65	14.56	–	7.95	m	**22.51**
305 mm × 76 mm	12.83	0.74	16.58	–	13.60	m	**30.18**
extra for fair ends	–	–	–	–	5.07	nr	**5.07**
extra for angles	–	–	–	–	5.75	nr	**5.75**
Sills; splayed top edge, stooled ends; bedded and pointed							
200 mm × 90 mm	31.87	0.75	16.81	–	32.99	m	**49.80**
200 mm × 90 mm; slip sill	35.23	0.75	16.81	–	36.45	m	**53.26**

14 MASONRY

Item	PC £	Labour hours	Labour £	Plant £	Material £	Unit	Total rate £
MATERIAL RATES							
Basic mortar prices							
Mortar materials only							
cement	–	–	–	–	138.01	tonne	**138.01**
sand	–	–	–	–	15.80	tonne	**15.80**
lime	–	–	–	–	0.43	kg	**0.43**
white cement	–	–	–	–	0.43	kg	**0.43**
mortar plasticizer	–	–	–	–	0.76	litre	**0.76**
Coloured mortar materials (excluding cement)							
light	–	–	–	–	7.13	kg	**7.13**
medium	–	–	–	–	7.13	kg	**7.13**
dark	–	–	–	–	7.13	kg	**7.13**
extra dark	–	–	–	–	7.13	kg	**7.13**
SUPPLY ONLY BRICK PRICES							
Alternative prices below are for supply only bricks.							
NOTE: These costs exclude labour, mortar wastage,							
overheads and profit recovery							
Forterra Brick							
Brown Rustic	–	–	–	–	574.02	1000	**574.02**
Rustic	–	–	–	–	801.13	1000	**801.13**
Brecken Grey	–	–	–	–	787.02	1000	**787.02**
Old English Brindled Red	–	–	–	–	556.82	1000	**556.82**
Chiltern	–	–	–	–	631.29	1000	**631.29**
Claydon Red Multi	–	–	–	–	631.70	1000	**631.70**
Cotswold	–	–	–	–	774.46	1000	**774.46**
Dapple Light	–	–	–	–	642.41	1000	**642.41**
Georgian	–	–	–	–	808.14	1000	**808.14**
Golden Buff	–	–	–	–	715.23	1000	**715.23**
Heather	–	–	–	–	653.54	1000	**653.54**
Hereward Light	–	–	–	–	619.55	1000	**619.55**
Honey Buff	–	–	–	–	636.33	1000	**636.33**
Ironstone	–	–	–	–	630.57	1000	**630.57**
Milton Buff	–	–	–	–	624.49	1000	**624.49**
Regency	–	–	–	–	798.46	1000	**798.46**
Sandfaced	–	–	–	–	695.66	1000	**695.66**
Saxon Gold	–	–	–	–	341.03	1000	**341.03**
Sunset Red	–	–	–	–	705.45	1000	**705.45**
Tudor	–	–	–	–	630.26	1000	**630.26**
Windsor	–	–	–	–	667.03	1000	**667.03**
Selected Regrades	–	–	–	–	646.12	1000	**646.12**
Ibstock Brick							
Aldridge Brown Blend	–	–	–	–	531.27	1000	**531.27**
Aldridge Leicester Anglican Red Rustic	–	–	–	–	531.27	1000	**531.27**
Chailey Stock	–	–	–	–	453.20	1000	**453.20**
Dorking Multi	–	–	–	–	424.67	1000	**424.67**
Funton Second Hard Stock	–	–	–	–	678.77	1000	**678.77**
Leicester Red Stock	–	–	–	–	407.67	1000	**407.67**
Roughdales Red Multi Rustic	–	–	–	–	531.27	1000	**531.27**

14 MASONRY

Item	PC £	Labour hours	Labour £	Plant £	Material £	Unit	Total rate £
Stourbridge Himley Mixed Russet	–	–	–	–	805.46	1000	**805.46**
Stourbridge Kenilworth Multi	–	–	–	–	514.48	1000	**514.48**
Strattford Red Rustic	–	–	–	–	531.27	1000	**531.27**
Swanage Handmade Restoration	–	–	–	–	709.16	1000	**709.16**
Tonbridge Handmade Multi	–	–	–	–	1130.63	1000	**1130.63**
SUPPLY ONLY BLOCK PRICES							
Alternative prices below are for supply only blocks.							
NOTE: These costs exclude labour, mortar wastage,							
overheads and profit recovery							
Celcon Standard aerated blocks; 440 mm × 215 mm							
100 mm Standard; 3.6N	–	–	–	–	14.16	m²	**14.16**
150 mm Standard; 3.6N	–	–	–	–	27.50	m²	**27.50**
215 mm Standard; 3.6N	–	–	–	–	34.93	m²	**34.93**
100 mm Standard; 7N	–	–	–	–	19.90	m²	**19.90**
150 mm Standard; 7N	–	–	–	–	25.95	m²	**25.95**
215 mm Standard; 7N	–	–	–	–	43.00	m²	**43.00**
215 mm × 65 mm × 100 mm coursing block	–	–	–	–	2.48	m	**2.48**
Celcon Solar blocks; 440 mm × 215 mm							
100 mm Solar; 3.6N	–	–	–	–	16.20	m²	**16.20**
150 mm Solar; 3.6N	–	–	–	–	24.30	m²	**24.30**
215 mm Solar; 3.6N	–	–	–	–	34.83	m²	**34.83**
265 mm Solar; 3.6N	–	–	–	–	42.93	m²	**42.93**
Celcon Foundation Blocks							
300 mm × 440 mm × 215 mm; 3.6N	–	–	–	–	11.20	m	**11.20**
Durox Supabloc; Aerated concrete;							
620 mm × 215 mm (7 nr per m²)							
100 mm Supabloc; 3.6N	–	–	–	–	25.60	m²	**25.60**
140 mm Supabloc; 3.6N	–	–	–	–	35.84	m²	**35.84**
200 mm Supabloc; 3.6N	–	–	–	–	51.20	m²	**51.20**
215 mm Supabloc; 3.6N	–	–	–	–	55.04	m²	**55.04**
100 mm Supabloc7; 7.3N	–	–	–	–	27.70	m²	**27.70**
140 mm Supabloc7; 7.3N	–	–	–	–	38.78	m²	**38.78**
200 mm Supabloc7; 7.3N	–	–	–	–	55.40	m²	**55.40**
215 mm Supabloc7; 7.3N	–	–	–	–	59.56	m²	**59.56**
Forterra Fenlite dense block							
100 mm solid; 3.6N	–	–	–	–	12.56	m²	**12.56**
100 mm solid; 7.3N	–	–	–	–	18.34	m²	**18.34**
Lignacite Lignacrete standard blocks;							
440 mm × 215 mm; 7.3N							
100 mm	–	–	–	–	14.66	m²	**14.66**
140 mm	–	–	–	–	20.54	m²	**20.54**
150 mm	–	–	–	–	22.00	m²	**22.00**
190 mm	–	–	–	–	27.88	m²	**27.88**
215 mm	–	–	–	–	31.53	m²	**31.53**
Tarmac Hemelite; 440 mm × 215 mm							
100 mm solid; 3.50 N/mm²	–	–	–	–	17.57	m²	**17.57**
100 mm solid; 7.00 N/mm²	–	–	–	–	15.71	m²	**15.71**
140 mm solid; 7.00 N/mm²	–	–	–	–	23.53	m²	**23.53**
190 mm solid; 7.00 N/mm²	–	–	–	–	31.94	m²	**31.94**
215 mm solid; 7.00 N/mm²	–	–	–	–	36.14	m²	**36.14**

14 MASONRY

Item	PC £	Labour hours	Labour £	Plant £	Material £	Unit	Total rate £
MATERIAL RATES – cont							
SUPPLY ONLY BLOCK PRICES – cont							
Tarmac Toplite standard blocks; 440 mm × 215 mm							
100 mm	–	–	–	–	17.88	m²	**17.88**
125 mm	–	–	–	–	22.35	m²	**22.35**
140 mm	–	–	–	–	25.03	m²	**25.03**
150 mm	–	–	–	–	26.82	m²	**26.82**
215 mm	–	–	–	–	38.44	m²	**38.44**
Plasmor concrete blocks; 440 mm × 215 mm							
100 mm Aglite, standard blocks 7N	–	–	–	–	16.46	m²	**16.46**
140 mm Aglite, standard blocks 7N	–	–	–	–	23.05	m²	**23.05**
190 mm Aglite, standard blocks 7N	–	–	–	–	31.28	m²	**31.28**
215 mm Aglite, standard blocks 7N	–	–	–	–	35.40	m²	**35.40**
100 mm Stranlite, standard blocks 7N	–	–	–	–	15.45	m²	**15.45**
140 mm Stranlite, standard blocks 7N	–	–	–	–	21.63	m²	**21.63**
190 mm Stranlite, standard blocks 7N	–	–	–	–	29.36	m²	**29.36**
215 mm Stranlite, standard blocks 7N	–	–	–	–	33.23	m²	**33.23**
100 mm Stranlite, paint grade blocks 7N	–	–	–	–	30.00	m²	**30.00**
140 mm Stranlite, paint grade blocks 7N	–	–	–	–	29.90	m²	**29.90**
190 mm Stranlite, paint grade blocks 7N	–	–	–	–	40.57	m²	**40.57**
215 mm Stranlite, paint grade blocks 7N	–	–	–	–	45.91	m²	**45.91**
14.01 BRICK WALLING							
SUPPLY AND LAY PRICES							
Common bricks							
Walls; in gauged mortar (1:1:6)							
prime cost for bricks £ per 1000	380.00	–	–	–	380.00	1000	**380.00**
half brick thick	22.80	0.84	18.76	–	29.60	m²	**48.36**
half brick thick; building against other work; concrete	22.80	0.92	20.57	–	32.90	m²	**53.47**
half brick thick; building overhand	22.80	1.04	23.40	–	29.60	m²	**53.00**
half brick thick; curved; 6.00 m radii	22.80	1.08	24.20	–	29.60	m²	**53.80**
half brick thick; curved; 1.50 m radii	26.60	1.41	31.67	–	33.71	m²	**65.38**
one brick thick	45.60	1.41	31.67	–	59.20	m²	**90.87**
one brick thick; curved; 6.00 m radii	49.40	1.84	41.15	–	63.31	m²	**104.46**
one brick thick; curved; 1.50 m radii	49.40	2.29	51.23	–	64.96	m²	**116.19**
one and a half brick thick	68.40	1.92	42.96	–	88.81	m²	**131.77**
one and a half brick thick; battering	68.40	2.21	49.42	–	88.81	m²	**138.23**
two brick thick	91.20	2.33	52.24	–	118.41	m²	**170.65**
two brick thick; battering	91.20	2.75	61.52	–	118.41	m²	**179.93**
337 mm average thick; tapering, one side	68.40	2.41	54.05	–	88.81	m²	**142.86**
450 mm average thick; tapering, one side	91.20	3.12	69.99	–	118.41	m²	**188.40**
337 mm average thick; tapering, both sides	68.40	2.79	62.53	–	88.81	m²	**151.34**
450 mm average thick; tapering, both sides	91.20	3.50	78.47	–	120.06	m²	**198.53**
facework one side, half brick thick	22.80	0.92	20.57	–	29.60	m²	**50.17**
facework one side, one brick thick	45.60	1.50	33.68	–	59.20	m²	**92.88**
facework one side, one and a half brick thick	68.40	2.00	44.77	–	88.81	m²	**133.58**
facework one side, two brick thick	91.20	2.41	54.05	–	118.41	m²	**172.46**
facework both sides, half brick thick	22.80	1.00	22.39	–	29.60	m²	**51.99**

14 MASONRY

Item	PC £	Labour hours	Labour £	Plant £	Material £	Unit	Total rate £
facework both sides, one brick thick	45.60	1.58	35.50	–	59.20	m²	**94.70**
facework both sides, one and a half brick thick	68.40	2.08	46.60	–	88.81	m²	**135.41**
facework both sides, two brick thick	91.20	2.50	56.07	–	118.41	m²	**174.48**
Isolated piers							
one brick thick	45.60	2.36	52.89	–	59.20	m²	**112.09**
two brick thick	91.20	3.70	82.93	–	120.06	m²	**202.99**
three brick thick	136.80	4.67	104.67	–	180.91	m²	**285.58**
Isolated casings							
half brick thick	22.80	1.20	26.89	–	29.60	m²	**56.49**
one brick thick	45.60	2.04	45.72	–	59.20	m²	**104.92**
Chimney stacks							
one brick thick	45.60	2.36	52.89	–	59.20	m²	**112.09**
two brick thick	91.20	3.70	82.93	–	120.06	m²	**202.99**
three brick thick	136.80	4.67	104.67	–	180.91	m²	**285.58**
Projections							
225 mm width; 112 mm depth; vertical	5.07	0.28	6.27	–	6.06	m	**12.33**
225 mm width; 225 mm depth; vertical	10.13	0.56	12.56	–	12.11	m	**24.67**
337 mm width; 225 mm depth; vertical	15.20	0.83	18.60	–	18.17	m	**36.77**
440 mm width; 225 mm depth; vertical	20.27	0.93	20.85	–	24.23	m	**45.08**
Closing cavities							
width of cavity 50 mm, closing with common brickwork half brick thick; vertical	–	0.28	6.27	–	1.51	m	**7.78**
width of cavity 50 mm, closing with common brickwork half brick thick; horizontal	–	0.28	6.27	–	4.46	m	**10.73**
width of cavity 50 mm, closing with common brickwork half brick thick; including damp-proof course; vertical	–	0.37	8.29	–	2.67	m	**10.96**
width of cavity 50 mm, closing with common brickwork half brick thick; including damp-proof course; horizontal	–	0.32	7.17	–	5.60	m	**12.77**
width of cavity 75 mm, closing with common brickwork half brick thick; vertical	–	0.28	6.27	–	2.19	m	**8.46**
width of cavity 75 mm, closing with common brickwork half brick thick; horizontal	–	0.28	6.27	–	6.51	m	**12.78**
width of cavity 75 mm, closing with common brickwork half brick thick; including damp-proof course; vertical	–	0.37	8.29	–	3.35	m	**11.64**
width of cavity 75 mm, closing with common brickwork half brick thick; including damp-proof course; horizontal	–	0.32	7.17	–	7.66	m	**14.83**
Bonding to existing							
half brick thick	–	0.28	6.27	–	1.65	m	**7.92**
one brick thick	–	0.42	9.41	–	3.30	m	**12.71**
one and a half brick thick	–	0.65	14.56	–	4.94	m	**19.50**
two brick thick	–	0.88	19.72	–	6.60	m	**26.32**
Arches							
height on face 102 mm, width of exposed soffit 102 mm, shape of arch – segmental, one ring	2.53	1.57	31.01	–	19.66	m	**50.67**
height on face 102 mm, width of exposed soffit 215 mm, shape of arch – segmental, one ring	5.07	2.04	41.55	–	22.79	m	**64.34**
height on face 102 mm, width of exposed soffit 102 mm, shape of arch – semi-circular, one ring	2.53	1.99	40.43	–	19.66	m	**60.09**

14 MASONRY

Item	PC £	Labour hours	Labour £	Plant £	Material £	Unit	Total rate £
14.01 BRICK WALLING – cont							
Common bricks – cont							
Arches – cont							
height on face 102 mm, width of exposed soffit							
215 mm, shape of arch – semi-circular, one ring	5.07	2.50	51.86	–	22.79	m	**74.65**
height on face 215 mm, width of exposed soffit							
102 mm, shape of arch – segmental, two ring	5.07	1.99	40.43	–	22.80	m	**63.23**
height on face 215 mm, width of exposed soffit							
215 mm, shape of arch – segmental, two ring	10.13	2.45	50.74	–	29.07	m	**79.81**
height on face 215 mm, width of exposed soffit							
102 mm, shape of arch – semi-circular, two ring	5.07	2.68	55.90	–	22.80	m	**78.70**
height on face 215 mm, width of exposed soffit							
215 mm, shape of arch – semi-circular, two ring	10.13	3.05	64.19	–	29.07	m	**93.26**
ADD or DEDUCT to walls for variation of £10.00/							
1000 in prime cost of common bricks							
half brick thick	–	–	–	–	0.62	m²	**0.62**
one brick thick	–	–	–	–	1.24	m²	**1.24**
one and a half brick thick	–	–	–	–	1.85	m²	**1.85**
two brick thick	–	–	–	–	2.47	m²	**2.47**
Class B engineering bricks							
Walls; in cement mortar (1:3)							
prime cost £ per 1000	332.50	–	–	–	332.50	1000	**332.50**
half brick thick	19.95	0.92	20.57	–	26.52	m²	**47.09**
one brick thick	39.90	1.50	33.68	–	53.04	m²	**86.72**
one brick thick; building against other work	39.90	1.79	40.14	–	57.99	m²	**98.13**
one brick thick; curved; 6.00 m radii	39.90	2.00	44.77	–	53.04	m²	**97.81**
one and a half brick thick	59.85	2.00	44.77	–	79.56	m²	**124.33**
one and a half brick thick; building against other							
work	59.85	2.41	54.05	–	79.56	m²	**133.61**
two brick thick	79.80	2.50	56.07	–	106.08	m²	**162.15**
337 mm thick; tapering, one side	59.85	2.58	57.89	–	79.56	m²	**137.45**
450 mm thick; tapering, one side	79.80	3.33	74.63	–	106.08	m²	**180.71**
337 mm thick; tapering, both sides	59.85	3.00	67.17	–	79.56	m²	**146.73**
450 mm thick; tapering, both sides	79.80	3.79	84.92	–	107.73	m²	**192.65**
facework one side, half brick thick	19.95	1.00	22.39	–	26.52	m²	**48.91**
facework one side, one brick thick	39.90	1.58	35.50	–	53.04	m²	**88.54**
facework one side, one and a half brick thick	59.85	2.08	46.60	–	79.56	m²	**126.16**
facework one side, two brick thick	79.80	2.58	57.89	–	106.08	m²	**163.97**
facework both sides, half brick thick	19.95	1.08	24.20	–	26.52	m²	**50.72**
facework both sides, one brick thick	39.90	1.67	37.32	–	53.04	m²	**90.36**
facework both sides, one and a half brick thick	59.85	2.17	48.62	–	79.56	m²	**128.18**
facework both sides, two brick thick	79.80	2.66	59.71	–	106.08	m²	**165.79**
Isolated piers							
one brick thick	39.90	2.59	58.05	–	53.04	m²	**111.09**
two brick thick	79.80	4.07	91.22	–	107.73	m²	**198.95**
three brick thick	119.70	5.00	112.06	–	162.42	m²	**274.48**
Isolated casings							
half brick thick	19.95	1.30	29.14	–	26.52	m²	**55.66**
one brick thick	39.90	2.22	49.76	–	53.04	m²	**102.80**

14 MASONRY

Item	PC £	Labour hours	Labour £	Plant £	Material £	Unit	Total rate £
Projections							
225 mm width; 112 mm depth; vertical	4.43	0.32	7.17	–	5.37	m	**12.54**
225 mm width; 225 mm depth; vertical	8.87	0.60	13.45	–	10.74	m	**24.19**
337 mm width; 225 mm depth; vertical	13.30	0.88	19.72	–	16.11	m	**35.83**
440 mm width; 225 mm depth; vertical	17.73	1.02	22.86	–	21.49	m	**44.35**
Bonding to existing							
half brick thick	–	0.32	7.17	–	1.44	m	**8.61**
one brick thick	–	0.46	10.31	–	2.89	m	**13.20**
one and a half brick thick	–	0.65	14.56	–	4.34	m	**18.90**
two brick thick	–	0.97	21.74	–	5.79	m	**27.53**
ADD or DEDUCT to walls for variation of £10.00/ 1000 in prime cost of bricks							
half brick thick	–	–	–	–	0.62	m²	**0.62**
one brick thick	–	–	–	–	1.24	m²	**1.24**
one and a half brick thick	–	–	–	–	1.85	m²	**1.85**
two brick thick	–	–	–	–	2.47	m²	**2.47**
Class A engineering bricks							
Walls; in cement mortar (1:3)							
bricks prime cost £ per 1000	1144.10	–	–	–	1144.10	1000	**1144.10**
half brick thick	68.65	0.92	20.57	–	79.19	m²	**99.76**
one brick thick	137.29	1.50	33.68	–	158.37	m²	**192.05**
one brick thick; building against other work	137.29	1.79	40.14	–	163.32	m²	**203.46**
one brick thick; curved; 6.00 m radii	137.29	2.00	44.77	–	158.37	m²	**203.14**
one and a half brick thick	205.94	2.00	44.77	–	237.55	m²	**282.32**
one and a half brick thick; building against other work	205.94	2.41	54.05	–	237.55	m²	**291.60**
two brick thick	274.58	2.50	56.07	–	316.74	m²	**372.81**
337 mm thick; tapering, one side	205.94	2.58	57.89	–	237.55	m²	**295.44**
450 mm thick; tapering, one side	274.58	3.33	74.63	–	316.74	m²	**391.37**
337 mm thick; tapering, both sides	205.94	3.00	67.17	–	237.55	m²	**304.72**
450 mm thick; tapering, both sides	274.58	3.79	84.92	–	318.38	m²	**403.30**
facework one side, half brick thick	68.65	1.00	22.39	–	79.19	m²	**101.58**
facework one side, one brick thick	137.29	1.58	35.50	–	158.37	m²	**193.87**
facework one side, one and a half brick thick	205.94	2.08	46.60	–	237.55	m²	**284.15**
facework one side, two brick thick	274.58	2.58	57.89	–	316.74	m²	**374.63**
facework both sides, half brick thick	68.65	1.08	24.20	–	79.19	m²	**103.39**
facework both sides, one brick thick	137.29	1.67	37.32	–	158.37	m²	**195.69**
facework both sides, one and a half brick thick	205.94	2.17	48.62	–	237.55	m²	**286.17**
facework both sides, two brick thick	274.58	2.66	59.71	–	316.74	m²	**376.45**
Isolated piers							
one brick thick	137.29	2.59	58.05	–	158.37	m²	**216.42**
two brick thick	274.58	4.07	91.22	–	318.38	m²	**409.60**
three brick thick	411.88	5.00	112.06	–	478.40	m²	**590.46**
Isolated casings							
half brick thick	68.65	1.30	29.14	–	79.19	m²	**108.33**
one brick thick	137.29	2.22	49.76	–	158.37	m²	**208.13**
Projections							
225 mm width; 112 mm depth; vertical	15.25	0.32	7.17	–	17.08	m	**24.25**
225 mm width; 225 mm depth; vertical	30.51	0.60	13.45	–	34.14	m	**47.59**
337 mm width; 225 mm depth; vertical	45.76	0.88	19.72	–	51.22	m	**70.94**
440 mm width; 225 mm depth; vertical	61.02	1.02	22.86	–	68.30	m	**91.16**

14 MASONRY

Item	PC £	Labour hours	Labour £	Plant £	Material £	Unit	Total rate £
14.01 BRICK WALLING – cont							
Class A engineering bricks – cont							
Bonding to existing							
half brick thick	–	0.32	7.17	–	4.37	m	**11.54**
one brick thick	–	0.46	10.31	–	8.74	m	**19.05**
one and a half brick thick	–	0.65	14.56	–	13.11	m	**27.67**
two brick thick	–	0.97	21.74	–	17.49	m	**39.23**
ADD or DEDUCT to walls for variation of £10.00/ 1000 in prime cost of bricks							
half brick thick	–	–	–	–	0.62	m²	**0.62**
one brick thick	–	–	–	–	1.24	m²	**1.24**
one and a half brick thick	–	–	–	–	1.85	m²	**1.85**
two brick thick	–	–	–	–	2.47	m²	**2.47**
Facing bricks; machine-made facings; in gauged mortar (1:1:6)							
Walls; in gauged mortar (1:1:6)							
bricks prime cost £ per 1000	500.00	–	–	–	500.00	1000	**500.00**
facework one side, half brick thick; stretcher bond	30.00	1.08	24.20	–	37.39	m²	**61.59**
facework one side, half brick thick, flemish bond with snapped headers	40.00	1.25	28.04	–	48.20	m²	**76.24**
facework one side, half brick thick, stretcher bond; building against other work; concrete	30.00	1.17	26.22	–	40.69	m²	**66.91**
facework one side, half brick thick; flemish bond with snapped headers; building against other work; concrete	40.00	1.33	29.85	–	51.50	m²	**81.35**
facework one side, half brick thick, stretcher bond; building overhand	30.00	1.33	29.85	–	37.39	m²	**67.24**
facework one side, half brick thick; flemish bond with snapped headers; building overhand	40.00	1.50	33.68	–	48.20	m²	**81.88**
facework one side, half brick thick; stretcher bond; curved; 6.00 m radii	30.00	1.58	35.50	–	37.39	m²	**72.89**
facework one side, half brick thick; flemish bond with snapped headers; curved; 6.00 m radii	40.00	1.79	40.14	–	48.20	m²	**88.34**
facework one side, half brick thick; stretcher bond; curved; 1.50 m radii	35.00	2.00	44.77	–	42.80	m²	**87.57**
facework one side, half brick thick; flemish bond with snapped headers; curved; 1.50 m radii	40.00	2.33	52.24	–	48.20	m²	**100.44**
facework both sides, one brick thick; two stretcher skins tied together	60.00	1.87	41.95	–	75.94	m²	**117.89**
facework both sides, one brick thick; flemish bond	60.00	1.92	42.96	–	74.78	m²	**117.74**
facework both sides, one brick thick; two stretcher skins tied together; curved; 6.00 m radii	60.00	2.58	57.89	–	75.94	m²	**133.83**
facework both sides, one brick thick; flemish bond; curved; 6.00 m radii	60.00	2.66	59.71	–	74.78	m²	**134.49**
facework both sides, one brick thick; two stretcher skins tied together; curved; 1.50 m radii	60.00	3.20	71.81	–	77.59	m²	**149.40**
facework both sides, one brick thick; flemish bond; curved; 1.50 m radii	60.00	3.33	74.63	–	76.43	m²	**151.06**

14 MASONRY

Item	PC £	Labour hours	Labour £	Plant £	Material £	Unit	Total rate £
Isolated piers							
facework both sides, one brick thick; two stretcher skins tied together	60.00	2.45	54.91	–	76.60	m²	**131.51**
facework both sides, one brick thick; flemish bond	60.00	2.50	56.03	–	76.60	m²	**132.63**
Isolated casings							
facework one side, half brick thick; stretcher bond	30.00	1.85	41.46	–	37.39	m²	**78.85**
facework one side, half brick thick; flemish bond with snapped headers	40.00	2.04	45.72	–	48.20	m²	**93.92**
Projections							
225 mm width; 112 mm depth; stretcher bond; vertical	6.67	0.28	6.27	–	7.79	m	**14.06**
225 mm width; 112 mm depth; flemish bond with snapped headers; vertical	6.67	0.37	8.29	–	7.79	m	**16.08**
225 mm width; 225 mm depth; flemish bond; vertical	13.33	0.60	13.45	–	22.44	m	**35.89**
328 mm width; 112 mm depth; stretcher bond; vertical	10.00	0.56	12.56	–	11.69	m	**24.25**
328 mm width; 112 mm depth; flemish bond with snapped headers; vertical	10.00	0.65	14.56	–	11.69	m	**26.25**
328 mm width; 225 mm depth; flemish bond; vertical	20.00	1.11	24.87	–	23.28	m	**48.15**
440 mm width; 112 mm depth; stretcher bond; vertical	13.33	0.83	18.60	–	15.57	m	**34.17**
440 mm width; 112 mm depth; flemish bond with snapped headers; vertical	13.33	0.88	19.72	–	15.57	m	**35.29**
440 mm width; 225 mm depth; flemish bond; vertical	26.67	1.62	36.31	–	31.15	m	**67.46**
Arches							
height on face 215 mm, width of exposed soffit 102 mm, shape of arch – flat	6.67	0.93	18.99	–	10.28	m	**29.27**
height on face 215 mm, width of exposed soffit 215 mm, shape of arch – flat	13.33	1.39	29.30	–	18.23	m	**47.53**
height on face 215 mm, width of exposed soffit 102 mm, shape of arch – segmental, one ring	6.67	1.76	34.81	–	24.40	m	**59.21**
height on face 215 mm, width of exposed soffit 215 mm, shape of arch segmental, one ring	13.33	2.13	43.11	–	31.60	m	**74.71**
height on face 215 mm, width of exposed soffit 102 mm, shape of arch – semi-circular, one ring	6.67	2.68	55.43	–	24.40	m	**79.83**
height on face 215 mm, width of exposed soffit 215 mm, shape of arch – semi-circular, one ring	13.33	3.61	76.27	–	31.60	m	**107.87**
height on face 215 mm, width of exposed soffit 102 mm, shape of arch – segmental, two ring	6.67	2.17	44.00	–	24.40	m	**68.40**
height on face 215 mm, width of exposed soffit 215 mm, shape of arch – segmental; two ring	13.33	2.82	58.57	–	31.60	m	**90.17**
height on face 215 mm, width of exposed soffit 102 mm, shape of arch – semi-circular, two ring	6.67	3.61	76.27	–	24.40	m	**100.67**
height on face 215 mm, width of exposed soffit 215 mm, shape of arch – semi-circular, two ring	13.33	5.00	107.43	–	31.60	m	**139.03**
Arches; cut voussoirs; prime cost £ per 1000	3450.00	–	–	–	3553.50	1000	**3553.50**
height on face 215 mm, width of exposed soffit 102 mm, shape of arch – segmental, one ring	46.00	1.80	35.71	–	66.94	m	**102.65**
height on face 215 mm, width of exposed soffit 215 mm, shape of arch – segmental, one ring	92.00	2.27	46.24	–	116.68	m	**162.92**
height on face 215 mm, width of exposed soffit 102 mm, shape of arch – semi-circular, one ring	46.00	2.04	41.09	–	66.94	m	**108.03**

14 MASONRY

Item	PC £	Labour hours	Labour £	Plant £	Material £	Unit	Total rate £
14.01 BRICK WALLING – cont							
Facing bricks – cont							
Arches – cont							
height on face 215 mm, width of exposed soffit 215 mm, shape of arch – semi-circular, one ring	92.00	2.59	53.42	–	116.68	m	**170.10**
height on face 320 mm, width of exposed soffit 102 mm, shape of arch – segmental, one and a half ring	92.00	2.41	49.38	–	117.02	m	**166.40**
height on face 320 mm, width of exposed soffit 215 mm, shape of arch – segmental, one and a half ring	184.00	3.15	65.96	–	235.29	m	**301.25**
Arches; bullnosed specials; prime cost £ per 1000	2000.00	–	–	–	2060.00	1000	**2060.00**
height on face 215 mm, width of exposed soffit 102 mm, shape of arch – flat	26.67	0.97	19.89	–	31.91	m	**51.80**
height on face 215 mm, width of exposed soffit 215 mm, shape of arch – flat	54.00	1.43	30.20	–	62.21	m	**92.41**
Bullseye windows; 600 mm dia.; facing bricks							
height on face 215 mm, width of exposed soffit 102 mm, two rings	14.00	4.63	99.14	–	22.84	nr	**121.98**
height on face 215 mm, width of exposed soffit 215 mm, two rings	28.00	6.48	140.59	–	39.67	nr	**180.26**
Bullseye windows; 600 mm; cut voussoirs; prime cost £ per 1000	3450.00	–	–	–	3553.50	1000	**3553.50**
height on face 215 mm, width of exposed soffit 102 mm, one ring	120.75	3.89	82.54	–	138.29	nr	**220.83**
height on face 215 mm, width of exposed soffit 215 mm, one ring	120.75	5.37	115.72	–	270.57	nr	**386.29**
Bullseye windows; 1200 mm dia.; facing bricks							
height on face 215 mm, width of exposed soffit 102 mm, two rings	28.00	7.22	157.18	–	55.42	nr	**212.60**
height on face 215 mm, width of exposed soffit 215 mm, two rings	56.00	10.36	227.56	–	90.65	nr	**318.21**
Bullseye windows; 1200 mm dia.; cut voussoirs	3450.00	–	–	–	3553.50	1000	**3553.50**
height on face 215 mm, width of exposed soffit 102 mm, one ring	207.00	6.11	132.30	–	250.66	nr	**382.96**
height on face 215 mm, width of exposed soffit 215 mm, one ring	414.00	8.70	190.35	–	477.83	nr	**668.18**
ADD or DEDUCT for variation of £10.00 per 1000 in PC of facing bricks in 102 mm high arches with 215 mm soffit	–	–	–	–	0.28	m	**0.28**
Facework sills							
150 mm × 102 mm; headers on edge; pointing top and one side; set weathering; horizontal	6.67	0.51	11.43	–	7.61	m	**19.04**
150 mm × 102 mm; cant headers on edge; pointing top and one side; set weathering; horizontal; prime cost £ per 1000	26.67	0.56	12.56	–	28.73	m	**41.29**
150 mm × 102 mm; bullnosed specials; headers on flat; pointing top and one side; horizontal; prime cost £ per 1000	26.67	0.46	10.31	–	28.73	m	**39.04**

14 MASONRY

Item	PC £	Labour hours	Labour £	Plant £	Material £	Unit	Total rate £
Facework copings							
215 mm × 102 mm; headers on edge; pointing top and both sides; horizontal	6.67	0.42	9.41	–	8.03	m	**17.44**
260 mm × 102 mm; headers on edge; pointing top and both sides; horizontal	10.00	0.65	14.56	–	11.86	m	**26.42**
215 mm × 102 mm; double bullnose specials; headers on edge; pointing top and both sides; horizontal; prime cost £ per 1000	2000.00	0.46	10.31	–	28.29	m	**38.60**
260 mm × 102 mm; single bullnose specials; headers on edge; pointing top and both sides; horizontal; prime cost £ per 1000	2000.00	0.65	14.56	–	55.97	m	**70.53**
ADD or DEDUCT for variation of £10.00 per 1000 in prime cost of facing bricks in copings 215 mm wide, 102 mm high	–	–	–	–	0.14	m	**0.14**
Extra over facing bricks for; facework ornamental bands and the like, plain bands							
flush; horizontal; 225 mm width; entirely of stretchers	–	0.19	4.25	–	0.72	m	**4.97**
Extra over facing brick for; facework quoins							
flush; mean girth 320 mm	–	0.28	6.27	–	0.72	m	**6.99**
Bonding to existing							
facework one side, half brick thick; stretcher bond	–	0.46	10.31	–	2.05	m	**12.36**
facework one side, half brick thick; flemish bond with snapped headers	–	0.46	10.31	–	2.05	m	**12.36**
facework both sides, one brick thick; two stretcher skins tied together	–	0.65	14.56	–	4.10	m	**18.66**
facework both sides, one brick thick; flemish bond	–	0.65	14.56	–	4.10	m	**18.66**
ADD or DEDUCT for variation of £10.00 per 1000 in prime cost of facing bricks; in walls built entirely of facings; in stretcher or flemish bond							
half brick thick	–	–	–	–	0.62	m²	**0.62**
one brick thick	–	–	–	–	1.24	m²	**1.24**
Facing bricks; handmade							
Walls in gauged mortar (1:1:6)							
prime cost for bricks £ per 1000	700.00	–	–	–	700.00	1000	**700.00**
facework one side, half brick thick; stretcher bond	42.00	1.08	24.20	–	50.37	m²	**74.57**
facework one side, half brick thick; flemish bond with snapped headers	56.00	1.25	28.04	–	65.51	m²	**93.55**
facework one side; half brick thick; stretcher bond; building against other work; concrete	42.00	1.17	26.22	–	53.66	m²	**79.88**
facework one side, half brick thick; flemish bond with snapped headers; building against other work; concrete	56.00	1.33	29.85	–	65.92	m²	**95.77**
facework one side, half brick thick; stretcher bond; building overhand	42.00	1.33	29.85	–	50.37	m²	**80.22**
facework one side, half brick thick; flemish bond with snapped headers; building overhand	56.00	1.50	33.68	–	65.51	m²	**99.19**

Prices for Measured Works

14 MASONRY

Item	PC £	Labour hours	Labour £	Plant £	Material £	Unit	Total rate £
14.01 BRICK WALLING – cont							
Facing bricks – cont							
Walls in gauged mortar (1:1:6) – cont							
facework one side, half brick thick; stretcher bond; curved; 6.00 m radii	42.00	1.58	35.50	–	50.37	m²	**85.87**
facework one side, half brick thick; flemish bond with snapped headers; curved; 6.00 m radii	56.00	1.79	40.14	–	65.51	m²	**105.65**
facework one side, half brick thick; stretcher bond; curved 1.50 m radii	42.00	2.00	44.77	–	50.37	m²	**95.14**
facework one side, half brick thick; flemish bond with snapped headers; curved; 1.50 m radii	56.00	2.33	52.24	–	65.51	m²	**117.75**
facework both sides, one brick thick; two stretcher skins tied together	84.00	1.87	41.95	–	101.90	m²	**143.85**
facework both sides, one brick thick; flemish bond	84.00	1.92	42.96	–	100.73	m²	**143.69**
facework both sides; one brick thick; two stretcher skins tied together; curved; 6.00 m radii	84.00	2.58	57.89	–	101.90	m²	**159.79**
facework both sides, one brick thick; flemish bond; curved; 6.00 m radii	84.00	2.66	59.71	–	100.73	m²	**160.44**
facework both sides, one brick thick; two stretcher skins tied together; curved; 1.50 m radii	84.00	3.20	71.81	–	103.55	m²	**175.36**
facework both sides, one brick thick; flemish bond; curved; 1.50 m radii	84.00	3.33	74.63	–	102.38	m²	**177.01**
Isolated piers							
facework both sides, one brick thick; two stretcher skins tied together	84.00	2.45	54.91	–	102.56	m²	**157.47**
facework both sides, one brick thick; flemish bond	84.00	2.50	56.03	–	102.56	m²	**158.59**
Isolated casings							
facework one side, half brick thick; stretcher bond	42.00	1.85	41.46	–	50.37	m²	**91.83**
facework one side, half brick thick; flemish bond with snapped headers	42.00	2.04	45.72	–	50.37	m²	**96.09**
Projections							
225 mm width; 112 mm depth; stretcher bond; vertical	9.33	0.28	6.27	–	10.67	m	**16.94**
225 mm width; 112 mm depth; flemish bond with snapped headers; vertical	9.33	0.37	8.29	–	10.67	m	**18.96**
225 mm width; 225 mm depth; flemish bond; vertical	18.67	0.60	13.45	–	21.34	m	**34.79**
328 mm width; 112 mm depth; stretcher bond; vertical	14.00	0.56	12.56	–	16.03	m	**28.59**
328 mm width; 112 mm depth; flemish bond with snapped headers; vertical	14.00	0.65	14.56	–	16.03	m	**30.59**
328 mm width; 225 mm depth; flemish bond; vertical	28.00	1.11	24.87	–	31.93	m	**56.80**
440 mm width; 112 mm depth; stretcher bond; vertical	18.67	0.83	18.60	–	21.34	m	**39.94**
440 mm width; 112 mm depth; flemish bond with snapped headers; vertical	18.67	0.88	19.72	–	21.34	m	**41.06**
440 mm width; 225 mm depth; flemish bond; vertical	37.33	1.62	36.31	–	42.68	m	**78.99**

14 MASONRY

Item	PC £	Labour hours	Labour £	Plant £	Material £	Unit	Total rate £
Arches							
height on face 215 mm, width of exposed soffit 102 mm, shape of arch – flat	9.33	0.93	18.99	–	13.16	m	**32.15**
height on face 215 mm, width of exposed soffit 215 mm, shape of arch – flat	18.90	1.39	29.30	–	24.25	m	**53.55**
height on face 215 mm, width of exposed soffit 102 mm, shape of arch – segmental, one ring	9.33	1.76	34.81	–	27.28	m	**62.09**
height on face 215 mm, width of exposed soffit 215 mm, shape of arch – segmental, one ring	18.67	2.13	43.11	–	37.37	m	**80.48**
height on face 215 mm, width of exposed soffit 102 mm, shape of arch – semi-circular, one ring	9.33	2.68	55.43	–	27.28	m	**82.71**
height on face 215 mm, width of exposed soffit 215 mm, shape of arch – semi-circular, one ring	18.67	3.61	76.27	–	37.37	m	**113.64**
height on face 215 mm, width of exposed soffit 102 mm, shape of arch – segmental, two ring	9.33	2.17	44.00	–	27.28	m	**71.28**
height on face 215 mm, width of exposed soffit 215 mm, shape of arch – segmental, two ring	18.67	2.82	58.57	–	37.37	m	**95.94**
height on face 215 mm, width of exposed soffit 102 mm, shape of arch – semi-circular, two ring	9.33	3.61	76.27	–	27.28	m	**103.55**
height on face 215 mm, width of exposed soffit 215 mm, shape of arch – semi-circular, two ring	18.67	5.00	107.43	–	37.37	m	**144.80**
Arches; cut voussoirs; prime cost £ per 1000	3450.00	–	–	–	3553.50	1000	**3553.50**
height on face 215 mm, width of exposed soffit 102 mm, shape of arch – segmental, one ring	46.00	1.80	35.71	–	66.94	m	**102.65**
height on face 215 mm, width of exposed soffit 215 mm, shape of arch – segmental, one ring	92.00	2.27	46.24	–	116.68	m	**162.92**
height on face 215 mm, width of exposed soffit 102 mm, shape of arch – semi-circular, one ring	46.00	2.04	41.09	–	66.94	m	**108.03**
height on face 215 mm, width of exposed soffit 215 mm, shape of arch – semi-circular, one ring	92.00	2.59	53.42	–	116.68	m	**170.10**
height on face 320 mm, width of exposed soffit 102 mm, shape of arch – segmental, one and a half ring	92.00	2.41	49.38	–	117.02	m	**166.40**
height on face 320 mm, width of exposed soffit 215 mm, shape of arch – segmental, one and a half ring	184.00	3.15	65.96	–	235.29	m	**301.25**
Arches; bullnosed specials; prime cost £ per 1000	2750.00	–	–	–	2832.50	1000	**2832.50**
height on face 215 mm, width of exposed soffit 102 mm, shape of arch – flat	36.67	0.97	19.89	–	42.72	m	**62.61**
height on face 215 mm, width of exposed soffit 215 mm, shape of arch – flat	74.25	1.43	30.20	–	84.11	m	**114.31**
Bullseye windows; 600 mm dia.							
height on face 215 mm, width of exposed soffit 102 mm, two ring	19.60	4.63	99.14	–	28.89	nr	**128.03**
height on face 215 mm, width of exposed soffit 215 mm, two ring	39.20	6.48	140.59	–	81.44	nr	**222.03**
Bullseye windows; 600 mm dia.; cut voussoirs; prime cost £ per 1000	4000.00	–	–	–	4120.00	1000	**4120.00**
height on face 215 mm, width of exposed soffit 102 mm, one ring	140.00	3.89	82.54	–	159.10	nr	**241.64**
height on face 215 mm, width of exposed soffit 215 mm, one ring	280.00	5.37	115.72	–	312.20	nr	**427.92**

14 MASONRY

Item	PC £	Labour hours	Labour £	Plant £	Material £	Unit	Total rate £
14.01 BRICK WALLING – cont							
Facing bricks – cont							
Bullseye windows; 1200 mm dia.							
height on face 215 mm, width of exposed soffit 102 mm, two ring	39.20	7.22	157.18	–	67.54	nr	**224.72**
height on face 215 mm, width of exposed soffit 215 mm, two ring	160.72	10.36	227.56	–	114.88	nr	**342.44**
Bullseye windows; 1200 mm dia.; cut voussoirs; prime cost £ per 1000	4000.00	–	–	–	4120.00	1000	**4120.00**
height on face 215 mm, width of exposed soffit 102 mm, one ring	240.00	6.11	132.30	–	286.35	nr	**418.65**
height on face 215 mm, width of exposed soffit 215 mm, one ring	480.00	8.70	190.35	–	549.21	nr	**739.56**
ADD or DEDUCT for variation of £10.00 per 1000 in prime cost of facing bricks in 102 mm high arches with 215 mm soffit	–	–	–	–	0.28	m	**0.28**
Facework sills							
150 mm × 102 mm; headers on edge; pointing top and one side; set weathering; horizontal; prime cost £ per m	9.33	0.51	11.43	–	10.67	m	**22.10**
150 mm × 102 mm; cant headers on edge; pointing top and one side; set weathering; prime cost £ per m	36.67	0.56	12.56	–	40.23	m	**52.79**
150 mm × 102 mm; bullnosed specials; headers on edge; pointing top and one side; horizontal; prime cost £ per m	36.67	0.46	10.31	–	40.23	m	**50.54**
Facework copings							
215 mm × 102 mm; headers on edge; pointing top and both sides; horizontal	9.33	0.42	9.41	–	10.92	m	**20.33**
260 mm × 102 mm; headers on edge; pointing top and both sides; horizontal	14.00	0.65	14.56	–	16.18	m	**30.74**
215 mm × 102 mm; double bullnose specials; headers on edge; pointing top and both sides; horizontal; prime cost £ per m	36.67	0.46	10.31	–	40.48	m	**50.79**
260 mm × 102 mm; single bullnose specials; headers on edge; pointing top and both sides; horizontal; prime cost £ per m	73.33	0.65	14.56	–	80.35	m	**94.91**
ADD or DEDUCT for variation of £10.00 per 1000 in prime cost of facing bricks in copings 215 mm wide, 102 mm high	–	–	–	–	0.14	m	**0.14**
Extra over facing bricks for; facework ornamental bands and the like, plain bands							
flush; horizontal; 225 mm width; entirely of stretchers; prime cost £ per m	–	0.19	4.25	–	0.72	m	**4.97**
Extra over facing bricks for; facework quoins							
flush mean girth 320 mm; prime cost £ per m	–	0.28	6.27	–	0.69	m	**6.96**

14 MASONRY

Item	PC £	Labour hours	Labour £	Plant £	Material £	Unit	Total rate £
Bonding ends to existing							
facework one side, half brick thick; stretcher bond	–	0.46	10.31	–	2.05	m	12.36
facework one side, half brick thick; flemish bond with snapped headers	–	0.46	10.31	–	2.05	m	12.36
facework both sides, one brick thick; two stretcher skins tied together	–	0.65	14.56	–	4.10	m	18.66
facework both sides, one brick thick; flemish bond	–	0.65	14.56	–	4.10	m	18.66
ADD or DEDUCT for variation of £10.00/1000 in prime cost of facing bricks; in walls built entirely of facings; in stretcher or flemish bond							
half brick thick	–	–	–	–	0.62	m²	0.62
one brick thick	–	–	–	–	1.24	m²	1.24
Facework steps							
Facework steps; machine cut specials in cement mortar (1:3)							
prime cost £ per 1000 with class A engineering bricks	2000.00	–	–	–	2000.00	1000	2000.00
215 mm × 102 mm Class A engineering bricks; all headers-on-edge; edges set with bullnosed specials; pointing top and one side; set weathering; horizontal; specials	26.67	0.51	11.43	–	45.92	m	57.35
returned ends pointed	–	0.14	3.14	–	0.58	nr	3.72
430 mm × 102 mm; all headers-on-edge; edges set with bullnosed specials; pointing top and one side; set weathering; horizontal; engineering bricks	26.67	0.74	16.58	–	35.49	m	52.07
returned ends pointed	–	0.19	4.25	–	0.58	nr	4.83
Facing bricks; slips							
20 mm thick; in gauged mortar (1:1:6) built up against concrete including flushing up at back (ties not included)							
prime cost brick £ per 1000	1350.00	–	–	–	1350.00	1000	1350.00
walls facework one side, half brick thick; stretcher bond	81.00	1.85	41.46	–	259.41	m²	300.87
edges of suspended slabs; 200 mm wide	–	0.56	12.56	–	18.51	m	31.07
columns; 400 mm wide	–	1.11	24.87	–	37.02	m	61.89
Facing tile bricks; Ibstock Tilebrick or other equal and approved; in gauged mortar (1:1:6)							
Walls							
standard tile brick; facework one side; half brick thick; stretcher bond	–	0.87	19.50	–	98.51	m²	118.01
three quarter tile brick; facework one side; half brick thick; stretcher bond	–	0.87	19.50	–	114.67	m²	134.17
Extra over facing tile bricks for							
fair ends; 79 mm long	–	0.28	6.27	–	42.84	m	49.11
fair ends; 163 mm long	–	0.28	6.27	–	42.84	m	49.11
90° × ½ external return	–	0.28	6.27	–	90.03	m	96.30

14 MASONRY

Item	PC £	Labour hours	Labour £	Plant £	Material £	Unit	Total rate £
14.01 BRICK WALLING – cont							
Facing tile bricks – cont							
Extra over facing tile bricks for – cont							
90° internal return	–	0.28	6.27	–	106.21	m	**112.48**
30° or 45° or 60° external return	–	0.28	6.27	–	90.03	m	**96.30**
30° or 45° or 60° internal return	–	0.28	6.27	–	90.03	m	**96.30**
angled verge	–	0.28	6.27	–	50.26	m	**56.53**
Air bricks							
Air bricks; red terracotta; building into prepared openings							
215 mm × 65 mm	–	0.07	1.57	–	4.45	nr	**6.02**
215 mm × 140 mm	–	0.07	1.57	–	6.61	nr	**8.18**
215 mm × 215 mm	–	0.07	1.57	–	16.49	nr	**18.06**
14.02 BLOCK WALLING							
SUPPLY AND LAY PRICES							
Lightweight aerated concrete blocks; Thermalite Turbo blocks or other equal; in gauged mortar (1:2:9)							
Walls							
100 mm thick	19.20	0.41	9.28	–	22.36	m²	**31.64**
140 mm thick	27.73	0.46	10.29	–	32.23	m²	**42.52**
150 mm thick	29.70	0.46	10.29	–	34.54	m²	**44.83**
190 mm thick	37.63	0.50	11.30	–	43.74	m²	**55.04**
200 mm thick	39.61	0.50	11.30	–	46.05	m²	**57.35**
215 mm thick	42.58	0.50	11.30	–	49.50	m²	**60.80**
300 mm thick	59.40	0.55	12.33	–	69.06	m²	**81.39**
Isolated piers or chimney stacks							
190 mm thick	–	0.83	18.60	–	43.74	m²	**62.34**
215 mm thick	–	0.83	18.60	–	49.50	m²	**68.10**
Isolated casings							
100 mm thick	–	0.51	11.43	–	22.36	m²	**33.79**
140 mm thick	–	0.56	12.56	–	32.23	m²	**44.79**
Extra over for fair face; flush pointing							
walls; one side	–	0.04	0.90	–	–	m²	**0.90**
walls; both sides	–	0.09	2.02	–	–	m²	**2.02**
Closing cavities							
width of cavity 50 mm, closing with lightweight blockwork 100 mm thick; vertical	–	0.23	5.15	–	1.27	m	**6.42**
width of cavity 50 mm, closing with lightweight blockwork 100 mm thick; including damp-proof course; vertical	–	0.28	6.27	–	2.41	m	**8.68**
width of cavity 75 mm, closing with lightweight blockwork 100 mm thick; vertical	–	0.23	5.15	–	1.81	m	**6.96**
width of cavity 75 mm, closing with lightweight blockwork 100 mm thick; including damp-proof course; vertical	–	0.28	6.27	–	2.96	m	**9.23**

14 MASONRY

Item	PC £	Labour hours	Labour £	Plant £	Material £	Unit	Total rate £
Bonding ends to common brickwork							
100 mm thick	–	0.14	3.14	–	2.58	m	**5.72**
140 mm thick	–	0.23	5.15	–	3.74	m	**8.89**
150 mm thick	–	0.23	5.15	–	3.99	m	**9.14**
200 mm thick	–	0.28	6.27	–	5.31	m	**11.58**
215 mm thick	–	0.32	7.17	–	5.73	m	**12.90**
Lightweight aerated concrete blocks; Thermalite Aircrete Shield blocks or other equal; in thin joint mortar							
Walls							
100 mm thick	18.13	0.46	10.31	–	22.03	m²	**32.34**
125 mm thick	19.93	0.48	10.75	–	24.41	m²	**35.16**
140 mm thick	22.32	0.51	11.43	–	27.47	m²	**38.90**
150 mm thick	27.18	0.51	11.43	–	33.05	m²	**44.48**
190 mm thick	34.44	0.56	12.56	–	41.86	m²	**54.42**
200 mm thick	36.25	0.60	13.45	–	44.06	m²	**57.51**
Isolated piers or chimney stacks							
190 mm thick	–	0.60	13.45	–	41.86	m²	**55.31**
Isolated casings							
100 mm thick	–	0.35	7.85	–	22.03	m²	**29.88**
140 mm thick	–	0.38	8.52	–	27.47	m²	**35.99**
Lightweight aerated concrete blocks; Thermalite Aircrete Shield blocks or other equal; in gauged mortar (1:2:9)							
Walls							
100 mm thick	19.49	0.41	9.28	–	21.18	m²	**30.46**
125 mm thick	19.93	0.41	9.19	–	23.60	m²	**32.79**
140 mm thick	24.00	0.46	10.29	–	26.24	m²	**36.53**
150 mm thick	29.22	0.46	10.29	–	31.74	m²	**42.03**
190 mm thick	37.02	0.50	11.30	–	40.21	m²	**51.51**
200 mm thick	38.97	0.50	11.30	–	42.33	m²	**53.63**
Isolated piers or chimney stacks							
190 mm thick	–	0.83	18.60	–	40.21	m²	**58.81**
Isolated casings							
100 mm thick	–	0.51	11.43	–	21.18	m²	**32.61**
140 mm thick	–	0.56	12.56	–	26.24	m²	**38.80**
Extra over for fair face; flush pointing							
walls; one side	–	0.04	0.90	–	–	m²	**0.90**
walls; both sides	–	0.09	2.02	–	–	m²	**2.02**
Closing cavities							
width of cavity 50 mm, closing with lightweight blockwork 100 mm thick; vertical	–	0.23	5.15	–	1.21	m	**6.36**
width of cavity 50 mm, closing with lightweight blockwork 100 mm thick; including damp-proof course; vertical	–	0.28	6.27	–	2.36	m	**8.63**
width of cavity 75 mm, closing with lightweight blockwork 100 mm thick; vertical	–	0.23	5.15	–	1.72	m	**6.87**
width of cavity 75 mm, closing with lightweight blockwork 100 mm thick; including damp-proof course; vertical	–	0.28	6.27	–	2.87	m	**9.14**

14 MASONRY

Item	PC £	Labour hours	Labour £	Plant £	Material £	Unit	Total rate £
14.02 BLOCK WALLING – cont							
Lightweight aerated concrete blocks – cont							
Bonding ends to common brickwork							
100 mm thick	–	0.14	3.14	–	2.44	m	**5.58**
140 mm thick	–	0.23	5.15	–	3.08	m	**8.23**
150 mm thick	–	0.23	5.15	–	3.68	m	**8.83**
190 mm thick	–	0.28	6.27	–	4.65	m	**10.92**
200 mm thick	–	0.28	6.27	–	4.90	m	**11.17**
Smooth face paint grade concrete block; in gauged mortar; flush pointing one side							
Walls							
100 mm thick	14.54	0.50	11.30	–	17.20	m²	**28.50**
140 mm thick	20.34	0.58	13.11	–	24.05	m²	**37.16**
150 mm thick	21.79	0.58	13.11	–	25.77	m²	**38.88**
190 mm thick	27.59	0.67	14.92	–	32.62	m²	**47.54**
200 mm thick	29.05	0.67	14.92	–	34.35	m²	**49.27**
215 mm thick	31.23	0.67	14.92	–	36.94	m²	**51.86**
100 mm thick; Paintgrade Smooth	17.35	0.55	12.35	–	20.31	m²	**32.66**
215 mm thick; Paintgrade Smooth	37.32	0.73	16.27	–	43.67	m²	**59.94**
Isolated piers or chimney stacks							
200 mm thick	–	0.93	20.85	–	34.35	m²	**55.20**
215 mm thick	–	0.93	20.85	–	36.94	m²	**57.79**
Isolated casings							
100 mm thick	–	0.69	15.46	–	17.20	m²	**32.66**
140 mm thick	–	0.74	16.58	–	24.05	m²	**40.63**
Extra over for fair face flush pointing							
walls; one side	–	0.04	0.90	–	–	m²	**0.90**
walls; both sides	–	0.09	2.02	–	–	m²	**2.02**
Bonding ends to common brickwork							
100 mm thick	–	0.23	5.15	–	2.04	m	**7.19**
140 mm thick	–	0.23	5.15	–	2.89	m	**8.04**
150 mm thick	–	0.28	6.27	–	3.07	m	**9.34**
200 mm thick	–	0.32	7.17	–	4.10	m	**11.27**
215 mm thick	–	0.32	7.17	–	4.43	m	**11.60**
Smooth face aerated concrete blocks; Party Wall block; in gauged mortar; flush pointing one side							
Walls							
100 mm thick	10.69	0.56	12.56	–	12.66	m²	**25.22**
215 mm thick	22.98	0.74	16.58	–	27.21	m²	**43.79**
Isolated piers or chimney stacks							
215 mm thick	–	0.93	20.85	–	27.21	m²	**48.06**
Isolated casings							
100 mm thick	–	0.69	15.46	–	12.66	m²	**28.12**
Extra over for fair face flush pointing							
walls; both sides	–	0.04	0.90	–	–	m²	**0.90**
Bonding ends to common brickwork							
100 mm thick	–	0.23	5.15	–	1.44	m	**6.59**
215 mm thick	–	0.32	7.17	–	3.12	m	**10.29**

14 MASONRY

Item	PC £	Labour hours	Labour £	Plant £	Material £	Unit	Total rate £
High strength concrete blocks (7N); Thermalite Hi-Strength 7 blocks or other equal and approved; in cement mortar (1:3)							
Walls							
100 mm thick	16.04	0.41	9.28	–	18.86	m²	**28.14**
140 mm thick	22.44	0.46	10.29	–	26.39	m²	**36.68**
150 mm thick	24.04	0.46	10.29	–	28.26	m²	**38.55**
190 mm thick	30.45	0.50	11.30	–	35.79	m²	**47.09**
200 mm thick	32.06	0.50	11.30	–	37.69	m²	**48.99**
215 mm thick	34.46	0.50	11.30	–	40.51	m²	**51.81**
Isolated piers or chimney stacks							
190 mm thick	–	0.83	18.60	–	35.79	m²	**54.39**
200 mm thick	–	0.83	18.60	–	37.69	m²	**56.29**
215 mm thick	–	0.83	18.60	–	40.51	m²	**59.11**
Isolated casings							
100 mm thick	–	0.51	11.43	–	18.86	m²	**30.29**
140 mm thick	–	0.56	12.56	–	26.39	m²	**38.95**
150 mm thick	–	0.56	12.56	–	28.26	m²	**40.82**
190 mm thick	–	0.69	15.46	–	35.79	m²	**51.25**
200 mm thick	–	0.69	15.46	–	37.69	m²	**53.15**
215 mm thick	–	0.69	15.46	–	40.51	m²	**55.97**
Extra over for flush pointing							
walls; one side	–	0.04	0.90	–	–	m²	**0.90**
walls; both sides	–	0.09	2.02	–	–	m²	**2.02**
Bonding ends to common brickwork							
100 mm thick	–	0.23	5.15	–	2.24	m	**7.39**
140 mm thick	–	0.23	5.15	–	3.15	m	**8.30**
150 mm thick	–	0.28	6.27	–	3.36	m	**9.63**
190 mm thick	–	0.32	7.17	–	4.24	m	**11.41**
200 mm thick	–	0.32	7.17	–	4.48	m	**11.65**
215 mm thick	–	0.32	7.17	–	4.83	m	**12.00**
Dense smooth faced concrete blocks; Lignacite standard and paint grade; in gauged mortar (1:2:9); flush pointing one side							
Walls							
100 mm thick; 3.6N	19.38	0.56	12.50	–	21.06	m²	**33.56**
140 mm thick; 3.6N	27.13	0.65	14.52	–	29.48	m²	**44.00**
100 mm thick; 7N	15.75	0.56	12.50	–	17.33	m²	**29.83**
140 mm thick; 7N	22.08	0.65	14.52	–	24.28	m²	**38.80**
Isolated piers or chimney stacks							
100 mm thick; 7N	–	1.14	25.55	–	18.30	m²	**43.85**
140 mm thick; 7N	–	1.26	28.24	–	25.10	m²	**53.34**
Isolated casings							
100 mm thick; 7N	–	0.78	17.48	–	17.33	m²	**34.81**
140 mm thick; 7N	–	0.90	20.17	–	24.28	m²	**44.45**

14 MASONRY

Item	PC £	Labour hours	Labour £	Plant £	Material £	Unit	Total rate £
14.02 BLOCK WALLING – cont							
Dense aggregate concrete blocks; in cement mortar (1:2:9)							
Walls or partitions or skins of hollow walls							
100 mm thick; solid	18.13	0.62	13.92	–	21.18	m²	**35.10**
140 mm thick; solid	22.32	0.75	16.74	–	26.24	m²	**42.98**
140 mm thick; hollow	22.32	0.67	14.92	–	26.24	m²	**41.16**
190 mm thick; hollow	34.44	0.84	18.76	–	40.21	m²	**58.97**
215 mm thick; hollow	36.25	0.92	20.57	–	42.50	m²	**63.07**
Isolated piers or chimney stacks							
140 mm thick; hollow	–	1.02	22.86	–	26.24	m²	**49.10**
190 mm thick; hollow	–	1.34	30.03	–	40.21	m²	**70.24**
215 mm thick; hollow	–	1.53	34.29	–	42.50	m²	**76.79**
Isolated casings							
100 mm thick; solid	–	0.74	16.58	–	21.18	m²	**37.76**
140 mm thick; solid	–	0.93	20.85	–	26.24	m²	**47.09**
Extra over for fair face; flush pointing							
walls; one side	–	0.09	2.02	–	–	m²	**2.02**
walls; both sides	–	0.14	3.14	–	–	m²	**3.14**
Bonding ends to common brickwork							
100 mm thick solid	–	0.23	5.15	–	2.49	m	**7.64**
140 mm thick solid	–	0.28	6.27	–	3.14	m	**9.41**
140 mm thick hollow	–	0.28	6.27	–	3.14	m	**9.41**
190 mm thick hollow	–	0.32	7.17	–	4.75	m	**11.92**
215 mm thick hollow	–	0.37	8.29	–	5.06	m	**13.35**
Dense aggregate paint grade concrete blocks; (7 N) Forticrete or similar; in cement mortar (1:3)							
Walls							
75 mm thick; solid	15.19	0.50	11.30	–	17.64	m²	**28.94**
100 mm thick; hollow	20.26	0.62	13.92	–	23.54	m²	**37.46**
100 mm thick; solid	20.37	0.62	13.92	–	23.66	m²	**37.58**
140 mm thick; hollow	28.36	0.67	14.92	–	32.94	m²	**47.86**
140 mm thick; solid	28.19	0.75	16.74	–	32.75	m²	**49.49**
190 mm thick; hollow	38.48	0.84	18.76	–	44.69	m²	**63.45**
190 mm thick; solid	38.48	0.92	20.57	–	44.69	m²	**65.26**
215 mm thick; hollow	43.55	0.92	20.57	–	50.57	m²	**71.14**
215 mm thick; solid	43.55	0.94	21.05	–	50.57	m²	**71.62**
Dwarf support wall							
140 mm thick; solid	–	1.16	26.00	–	32.75	m²	**58.75**
190 mm thick; solid	–	1.34	30.03	–	44.69	m²	**74.72**
215 mm thick; solid	–	1.53	34.29	–	50.57	m²	**84.86**
Isolated piers or chimney stacks							
140 mm thick; hollow	–	1.02	22.86	–	32.94	m²	**55.80**
190 mm thick; hollow	–	1.34	30.03	–	44.69	m²	**74.72**
215 mm thick; hollow	–	1.53	34.29	–	50.57	m²	**84.86**
Isolated casings							
75 mm thick; solid	–	0.69	15.46	–	17.64	m²	**33.10**
100 mm thick; solid	–	0.74	16.58	–	23.66	m²	**40.24**
140 mm thick; solid	–	0.93	20.85	–	32.75	m²	**53.60**

14 MASONRY

Item	PC £	Labour hours	Labour £	Plant £	Material £	Unit	Total rate £
Extra over for fair face; flush pointing							
walls; one side	–	0.09	2.02	–	–	m²	**2.02**
walls; both sides	–	0.14	3.14	–	–	m²	**3.14**
Bonding ends to common brickwork							
75 mm thick solid	–	0.14	3.14	–	2.08	m	**5.22**
100 mm thick solid	–	0.23	5.15	–	2.78	m	**7.93**
140 mm thick solid	–	0.28	6.27	–	3.88	m	**10.15**
190 mm thick solid	–	0.32	7.17	–	5.25	m	**12.42**
215 mm thick solid	–	0.37	8.29	–	5.97	m	**14.26**
Architectural finish block walls							
100 mm thick Fairface finish	–	1.50	33.62	–	33.97	m²	**67.59**
100 mm thick Shot-blasted finish	–	1.50	33.62	–	45.26	m²	**78.88**
100 mm thick Splitface finish	–	1.50	33.62	–	44.13	m²	**77.75**
100 mm thick Splitface finish	–	2.00	44.83	–	81.38	m²	**126.21**
High strength concrete blocks (10N); Thermalite Hi-Strength 10 blocks; in cement mortar (1:3)							
Walls							
100 mm thick (NB other thicknesses as a special order item)	–	0.50	11.30	–	20.84	m²	**32.14**
Thermalite Trenchblock 3.6N; with tongued and grooved joints; in cement mortar (1:4)							
Walls							
255 mm thick	–	0.60	13.45	–	49.24	m²	**62.69**
275 mm thick	–	0.65	14.56	–	53.09	m²	**67.65**
305 mm thick	–	0.70	15.69	–	57.86	m²	**73.55**
355 mm thick	–	0.75	16.81	–	68.13	m²	**84.94**
Concrete blocks; Thermalite Trenchblock, 7N; with tongued and grooved joints; in cement mortar (1:4)							
Walls							
255 mm thick	42.39	0.70	15.69	–	49.24	m²	**64.93**
275 mm thick	45.71	0.75	16.81	–	53.09	m²	**69.90**
305 mm thick	49.87	0.80	17.93	–	57.86	m²	**75.79**
355 mm thick	59.01	0.85	19.06	–	68.13	m²	**87.19**

14 MASONRY

Item	PC £	Labour hours	Labour £	Plant £	Material £	Unit	Total rate £
14.02 BLOCK WALLING – cont							
Architectural blockwork							
Lignacite Ltd; 440 × 215 mm face size; solid blocks 17.5N; laid stretcher bond with class 3 Snowstorm mortar with white Portland cement and recessed joints. AUTHORS NOTE: LIGNACITE HAVE WITHDRAWN THEIR FACING MASONRY PRODUCTS. We have left these items in for this year as a guide to price levels only.							
Snowstorm Weathered							
100 mm thick weathered one face	–	1.00	22.41	–	52.05	m²	**74.46**
100 mm thick; weathered one face and one end	–	0.10	2.25	–	25.96	m	**28.21**
217 mm × 100 mm thick; cut half block; weathered one face	–	1.05	23.54	–	71.89	m²	**95.43**
quoins; weathered two external faces; 440 mm × 100 mm × 215 mm with 215 mm external return	–	0.11	2.46	–	105.03	m	**107.49**
100 mm × 440 mm × 65 mm Roman Bricks weathered one face	–	1.00	22.41	–	58.45	m²	**80.86**
Snowstorm Split							
100 mm thick; split one face	–	1.00	22.41	–	54.38	m²	**76.79**
100 mm thick; split one face and weathered one end	–	0.10	2.25	–	26.99	m	**29.24**
217 mm × 100 mm thick; cut half block; split one face	–	1.05	23.54	–	85.53	m²	**109.07**
quoins; split two external faces; 440 mm × 215 mm × 100 mm thick with 215 mm external return	–	0.11	2.46	–	120.52	m	**122.98**
100 mm × 440 mm × 65 mm Roman Bricks; split one face	–	1.00	22.41	–	63.31	m²	**85.72**
Snowstorm Polished							
Note: polished dimensions reduced by 3 mm							
100 mm thick; polished one face	–	1.10	24.66	–	89.70	m²	**114.36**
100 mm thick; polished one face one end	–	0.11	2.46	–	71.83	m	**74.29**
217 mm × 100 mm thick; cut half block; polished one face	–	1.18	26.34	–	108.53	m²	**134.87**
quoins; polished two external faces; 440 mm × 215 mm × 100 mm thick with 215 mm external return	–	0.12	2.69	–	174.55	m	**177.24**
100 mm × 440 mm × 65 mm Roman Bricks; polished one face	–	1.10	24.66	–	109.37	m²	**134.03**
Snowstorm Planished							
Note: planished dimensions reduced by 3 mm							
100 mm thick; planished one face	–	1.10	24.66	–	89.70	m²	**114.36**
100 mm thick; planished one face one end	–	0.11	2.46	–	71.83	m	**74.29**
100 mm thick; cut half block; planished one face	–	1.18	26.34	–	108.53	m²	**134.87**
quoins; planished two external faces; 440 mm × 215 mm × 100 mm thick with 215 mm external return	–	0.12	2.69	–	174.55	m	**177.24**
100 mm × 440 mm × 65 mm Roman Bricks; planished one face	–	1.10	24.66	–	89.08	m²	**113.74**

14 MASONRY

Item	PC £	Labour hours	Labour £	Plant £	Material £	Unit	Total rate £
Snowstorm and Crushed recycled Glass (Polished or Planished). Standard colours							
Note: planished dimensions reduced by 3 mm							
100 mm thick; planished one face	–	1.10	24.66	–	105.01	m²	**129.67**
100 mm thick; planished one face one end	–	0.11	2.46	–	77.74	m	**80.20**
100 mm thick; cut half block; planished one face	–	1.18	26.34	–	124.51	m²	**150.85**
quoins; planished two external faces;							
440 mm × 215 mm × 100 mm thick with 215 mm							
external return	–	0.12	2.69	–	181.39	m	**184.08**
100 mm × 440 mm × 65 mm Roman Bricks;							
planished one face	–	1.10	24.66	–	123.66	m²	**148.32**
Midnight Polished (Note 7.3N strength) 10 mm Black Granite facing bonded to 90 mm thick backing block							
100 mm thick; polished one face	–	1.15	25.77	–	150.39	m²	**176.16**
100 mm thick; polished one face one end	–	0.12	2.69	–	232.75	m	**235.44**
100 mm thick; cut half block; polished one face	–	1.19	26.55	–	213.55	m²	**240.10**
quoins; polished two external faces;							
440 mm × 215 mm × 110 mm thick with 215 mm							
external return	–	0.14	3.14	–	262.93	m	**266.07**
Dense aggregate coloured concrete blocks; Forticrete Bathstone or equal; in coloured gauged mortar (1:1:6); flush pointing one side							
Walls							
100 mm thick hollow	33.11	0.74	16.58	–	36.90	m²	**53.48**
100 mm thick solid	33.11	0.74	16.58	–	36.90	m²	**53.48**
140 mm thick hollow	47.96	0.83	18.60	–	53.40	m²	**72.00**
140 mm thick solid	47.96	0.93	20.85	–	53.40	m²	**74.25**
215 mm thick hollow	54.82	1.16	26.00	–	61.65	m²	**87.65**
Isolated piers or chimney stacks							
140 mm thick solid	–	1.25	28.02	–	53.40	m²	**81.42**
215 mm thick hollow	–	1.57	35.18	–	61.65	m²	**96.83**
Extra over blocks for							
100 mm thick half lintel blocks; ref D14	–	0.23	5.15	–	27.34	m	**32.49**
140 mm thick half lintel blocks; ref H14	–	0.28	6.27	–	48.93	m	**55.20**
140 mm thick quoin blocks; ref H16	–	0.32	7.17	–	41.73	m	**48.90**
140 mm thick cavity closer blocks; ref H17	–	0.32	7.17	–	44.75	m	**51.92**
140 mm thick sill blocks; ref H21	–	0.28	6.27	–	32.87	m	**39.14**
Glazed finish blocks; Forticrete Astra-Glaze or equal; in gauged mortar (1:1:6); joints raked out; gun applied latex grout to joints							
Walls or partitions or skins of hollow walls							
100 mm thick; glazed one side	–	0.93	20.85	–	136.58	m²	**157.43**
extra for glazed square end return	–	0.37	8.29	–	40.27	m	**48.56**
100 mm thick; glazed both sides	–	1.11	24.87	–	173.13	m²	**198.00**
100 mm thick lintel 200 mm high; glazed one side	–	0.83	16.28	–	37.95	m	**54.23**

14 MASONRY

Item	PC £	Labour hours	Labour £	Plant £	Material £	Unit	Total rate £
14.03 GLASS BLOCK WALLING							
NOTE: The following specialist prices for glass block walling; assume standard blocks in panels of at least 50 m²; work in straight walls at ground level; and all necessary ancillary fixing; strengthening; easy access; pointing and expansion materials etc.							
Hollow glass block walling; Pittsburgh Corning sealed Thinline or other equal and approved; in cement mortar joints; reinforced with 6 mm dia. stainless steel rods; pointed both sides with mastic or other equal and approved							
Walls; facework both sides							
190 mm × 190 mm × 80 mm; clear blocks	71.51	6.00	134.48	–	117.12	m²	**251.60**
190 mm × 190 mm × 80 mm; craft blocks, standard colours	108.78	7.00	156.88	–	159.34	m²	**316.22**
Fire-rated walls							
190 mm × 190 mm × 100 mm glass blocks; 30 minute fire-rated	117.56	8.00	179.29	–	169.30	m²	**348.59**
190 mm × 190 mm × 160 mm glass blocks; 60 minute fire-rated	297.30	9.00	201.70	–	372.94	m²	**574.64**
14.04 NATURAL STONE							
Cotswold Guiting limestone or other equal and approved; laid dry							
Uncoursed random rubble walling							
275 mm thick	–	2.07	48.06	–	14.13	m²	**62.19**
350 mm thick	–	2.46	56.83	–	17.97	m²	**74.80**
425 mm thick	–	2.81	64.58	–	21.83	m²	**86.41**
500 mm thick	–	3.15	72.09	–	25.69	m²	**97.78**
Cotswold Guiting limestone or other equal; bedded; jointed and pointed in cement: lime mortar (1:2:9)							
Uncoursed random rubble walling; faced and pointed; both sides							
275 mm thick	–	1.98	45.82	–	24.10	m²	**69.92**
350 mm thick	–	2.18	49.84	–	30.66	m²	**80.50**
425 mm thick	–	2.39	54.12	–	37.23	m²	**91.35**
500 mm thick	–	2.59	58.13	–	43.82	m²	**101.95**
Coursed random rubble walling; rough dressed; faced and pointed one side							
114 mm thick	–	1.48	32.51	–	62.60	m²	**95.11**
150 mm thick	–	1.76	42.58	–	75.60	m²	**118.18**
Fair returns on walling							
114 mm wide	–	0.02	0.45	–	–	m	**0.45**
150 mm wide	–	0.03	0.67	–	–	m	**0.67**
275 mm wide	–	0.06	1.35	–	–	m	**1.35**
350 mm wide	–	0.08	1.79	–	–	m	**1.79**
425 mm wide	–	0.10	2.25	–	–	m	**2.25**
500 mm wide	–	0.12	2.69	–	–	m	**2.69**

14 MASONRY

Item	PC £	Labour hours	Labour £	Plant £	Material £	Unit	Total rate £
Fair raking cutting or circular cutting							
114 mm wide	–	0.20	4.53	–	8.77	m	**13.30**
150 mm wide	–	0.25	5.68	–	10.53	m	**16.21**
Level uncoursed rubble walling for damp-proof courses and the like							
275 mm wide	–	0.19	4.74	–	3.13	m	**7.87**
350 mm wide	–	0.20	4.99	–	3.85	m	**8.84**
425 mm wide	–	0.21	5.23	–	4.74	m	**9.97**
500 mm wide	–	0.22	5.48	–	5.62	m	**11.10**
Copings formed of rough stones; faced and pointed all round							
275 mm × 200 mm (average) high	–	0.56	13.24	–	10.36	m	**23.60**
350 mm × 250 mm (average) high	–	0.75	17.57	–	14.23	m	**31.80**
425 mm × 300 mm (average) high	–	0.97	22.57	–	19.09	m	**41.66**
500 mm × 300 mm (average) high	–	1.23	28.41	–	25.50	m	**53.91**
Sundries – stone walling							
Coating backs of stones with brush applied cold bitumen solution; two coats							
limestone facework	–	0.19	3.21	–	2.74	m²	**5.95**
Cutting grooves in limestone masonry for							
water bars or the like	–	–	–	–	–	m	**13.54**
Mortices in limestone masonry for							
metal dowel	–	–	–	–	–	nr	**2.72**
metal cramp	–	–	–	–	–	nr	**5.41**
14.05 CAST STONE							
Reconstructed walling; Bradstone masonry blocks; standard colours; or other equal; laid to pattern or course recommended; bedded; jointed and pointed in approved coloured cement: lime mortar (1:2:9)							
Walls; facing and pointing one side							
Fyfestone Enviromasonry Rustic	–	1.00	22.41	–	51.29	m²	**73.70**
Fyfestone Enviromasonry Fairfaced	–	1.00	22.41	–	38.95	m²	**61.36**
Fyfestone Enviromasonry Split	–	1.10	24.66	–	62.49	m²	**87.15**
Fyfestone Enviromasonry Textured	–	1.10	24.66	–	67.18	m²	**91.84**
Fyfestone Enviromasonry Polished	–	2.00	44.83	–	166.07	m²	**210.90**
masonry blocks; random uncoursed	–	1.04	23.31	–	91.37	m²	**114.68**
extra for							
returned ends	–	0.37	8.29	–	72.22	m	**80.51**
plain L shaped quoins	–	0.12	2.69	–	89.80	m	**92.49**
traditional walling; coursed squared	–	1.30	29.14	–	91.37	m²	**120.51**
squared coursed rubble	–	1.25	28.02	–	93.68	m²	**121.70**
squared random rubble	–	1.30	29.14	–	93.35	m²	**122.49**
squared and pitched rock faced walling; coursed	–	1.34	30.03	–	93.35	m²	**123.38**
ashlar; 440 × 215 × 100 mm thick	–	1.10	24.66	–	95.34	m²	**120.00**
rough hewn rockfaced walling; random	–	1.39	31.16	–	93.02	m²	**124.18**
extra for							
returned ends	–	0.15	3.36	–	–	m	**3.36**

14 MASONRY

Item	PC £	Labour hours	Labour £	Plant £	Material £	Unit	Total rate £
14.05 CAST STONE – cont							
Reconstructed walling – cont							
Isolated piers or chimney stacks; facing and pointing one side							
Enviromasonry Rustic	–	1.40	31.37	–	51.29	m²	**82.66**
masonry blocks; random uncoursed	–	1.43	32.05	–	91.37	m²	**123.42**
traditional walling; coursed squared	–	1.80	40.35	–	91.37	m²	**131.72**
squared coursed rubble	–	1.76	39.45	–	93.68	m²	**133.13**
squared random rubble	–	1.80	40.35	–	93.35	m²	**133.70**
squared and pitched rock faced walling; coursed	–	1.90	42.58	–	93.35	m²	**135.93**
ashlar; 440 × 215 × 100 mm thick	–	1.54	34.52	–	95.34	m²	**129.86**
rough hewn rockfaced walling; random	–	1.94	43.48	–	93.02	m²	**136.50**
Isolated casings; facing and pointing one side							
Enviromasonry Rustic	–	1.20	26.89	–	51.29	m²	**78.18**
masonry blocks; random uncoursed	–	1.25	28.02	–	91.37	m²	**119.39**
traditional walling; coursed squared	–	1.57	35.18	–	91.37	m²	**126.55**
squared coursed rubble	–	1.53	34.29	–	93.68	m²	**127.97**
squared random rubble	–	1.57	35.18	–	93.35	m²	**128.53**
squared and pitched rock faced walling; coursed	–	1.62	36.31	–	93.35	m²	**129.66**
ashlar; 440 × 215 × 100 mm thick	–	1.32	29.58	–	95.34	m²	**124.92**
rough hewn rockfaced walling; random	–	1.67	37.43	–	93.02	m²	**130.45**
Fair returns 100 mm wide							
Enviromasonry Rustic	–	0.10	2.25	–	–	m²	**2.25**
masonry blocks; random uncoursed	–	0.11	2.46	–	–	m²	**2.46**
traditional walling; coursed squared	–	0.14	3.14	–	–	m²	**3.14**
squared coursed rubble	–	0.13	2.91	–	–	m²	**2.91**
squared random rubble	–	0.14	3.14	–	–	m²	**3.14**
squared and pitched rock faced walling; coursed	–	0.14	3.14	–	–	m²	**3.14**
ashlar; 440 × 215 × 100 mm thick	–	0.14	3.14	–	–	m²	**3.14**
rough hewn rockfaced walling; random	–	0.15	3.36	–	–	m²	**3.36**
Fair raking cutting or circular cutting							
100 mm wide	–	0.17	3.81	–	–	m	**3.81**
Quoin							
ashlar; 440 × 215 × 215 × 100 mm thick	–	0.75	16.81	–	183.46	m	**200.27**
Reconstructed limestone dressings; Bradstone Architectural dressings in weathered Cotswold or North Cerney shades or other equal; bedded, jointed and pointed in approved coloured cement: lime mortar (1:2:9)							
Copings; twice weathered and throated							
610 mm × 140 mm	–	0.37	8.29	–	34.32	m	**42.61**
extra for							
fair end	–	–	–	–	16.69	nr	**16.69**
returned mitred fair end	–	–	–	–	16.69	nr	**16.69**
Copings; once weathered and throated							
610 mm × 140 mm	–	0.37	8.29	–	34.32	m	**42.61**
610 mm × 140 mm	–	0.37	8.29	–	30.93	m	**39.22**
extra for							
fair end	–	–	–	–	16.69	nr	**16.69**
returned mitred fair end	–	–	–	–	16.69	nr	**16.69**

14 MASONRY

Item	PC £	Labour hours	Labour £	Plant £	Material £	Unit	Total rate £
Pier caps; four times weathered and throated							
305 mm × 305 mm	–	0.23	5.15	–	7.92	nr	**13.07**
381 mm × 381 mm	–	0.23	5.15	–	11.42	nr	**16.57**
Splayed corbels							
479 mm × 100 mm × 215 mm	–	0.14	3.14	–	33.92	nr	**37.06**
665 mm × 100 mm × 215 mm	–	0.19	4.25	–	45.28	nr	**49.53**
100 mm × 140 mm lintels; rectangular; reinforced with mild steel bars							
all lengths to 2.07 m	–	0.26	5.83	–	53.82	m	**59.65**
100 mm × 215 mm lintels; rectangular; reinforced with mild steel bars							
all lengths to 2.85 m	–	0.30	6.73	–	55.97	m	**62.70**
Sills to suit standard windows; stooled 100 mm at ends							
150 mm ×140 mm; not exceeding 1.97 m long	–	0.28	6.27	–	70.32	m	**76.59**
197 mm ×140 mm; not exceeding 1.97 m long	–	0.28	6.27	–	78.82	m	**85.09**
Window surround; traditional with label moulding; for single light; sill 146 mm × 133 mm; jambs 146 mm × 146 mm; head 146 mm × 105 mm; including all dowels and anchors							
overall size 508 mm × 1479 mm	–	0.83	18.60	–	245.47	nr	**264.07**
Window surround; traditional with label moulding; three light; for windows 508 mm × 1219 mm; sill 146 mm × 133 mm; jambs 146 mm × 146 mm; head 146 mm × 103 mm; mullions 146 mm × 108 mm; including all dowels and anchors							
overall size 1975 mm × 1479 mm	–	2.17	48.64	–	578.71	nr	**627.35**
Door surround; moulded continuous jambs and head with label moulding; including all dowels and anchors							
door 839 mm × 1981 mm in 102 mm × 64 mm frame	–	1.53	34.29	–	523.68	nr	**557.97**
Portland Whitbed limestone bedded and jointed in cement–lime–mortar (1:2:9); slurrying with weak lime and stone dust mortar; flush pointing and cleaning on completion (cramps etc. not included)							
Facework; one face plain and rubbed; bedded against backing							
50 mm thick stones	–	–	–	–	–	m²	**387.67**
63 mm thick stones	–	–	–	–	–	m²	**441.57**
75 mm thick stones	–	–	–	–	–	m²	**499.39**
100 mm thick stones	–	–	–	–	–	m²	**548.20**
Fair returns on facework							
50 mm wide	–	–	–	–	–	m	**5.41**
63 mm wide	–	–	–	–	–	m	**6.76**
75 mm wide	–	–	–	–	–	m	**9.47**
100 mm wide	–	–	–	–	–	m	**12.18**
Fair raking cutting on facework							
50 mm thick	–	–	–	–	–	m	**24.37**
63 mm thick	–	–	–	–	–	m	**27.07**
75 mm thick	–	–	–	–	–	m	**32.48**
100 mm thick	–	–	–	–	–	m	**35.20**

14 MASONRY

Item	PC £	Labour hours	Labour £	Plant £	Material £	Unit	Total rate £
14.05 CAST STONE – cont							
Portland Whitbed limestone – cont							
Copings; once weathered; and throated; rubbed; set horizontal or raking							
250 mm × 50 mm	–	–	–	–	–	m	189.49
extra for							
external angle	–	–	–	–	–	nr	33.84
internal angle	–	–	–	–	–	nr	33.84
300 mm × 50 mm	–	–	–	–	–	m	200.33
extra for							
external angle	–	–	–	–	–	nr	33.84
internal angle	–	–	–	–	–	nr	40.61
350 mm × 75 mm	–	–	–	–	–	m	223.35
extra for							
external angle	–	–	–	–	–	nr	33.84
internal angle	–	–	–	–	–	nr	43.31
400 mm × 100 mm	–	–	–	–	–	m	268.01
extra for							
external angle	–	–	–	–	–	nr	40.61
internal angle	–	–	–	–	–	nr	56.86
450 mm × 100 mm	–	–	–	–	–	m	324.86
extra for							
external angle	–	–	–	–	–	nr	51.44
internal angle	–	–	–	–	–	nr	70.38
500 mm × 125 mm	–	–	–	–	–	m	494.06
extra for							
external angle	–	–	–	–	–	nr	70.38
internal angle	–	–	–	–	–	nr	87.98
Band courses; plain; rubbed; horizontal							
225 mm × 112 mm	–	–	–	–	–	m	148.89
300 mm × 112 mm	–	–	–	–	–	m	198.97
extra for							
stopped ends	–	–	–	–	–	nr	8.12
external angles	–	–	–	–	–	nr	8.12
Band courses; moulded 100 mm girth on face; rubbed; horizontal							
125 mm × 75 mm	–	–	–	–	–	m	169.20
extra for							
stopped ends	–	–	–	–	–	nr	27.07
external angles	–	–	–	–	–	nr	33.84
internal angles	–	–	–	–	–	nr	67.68
150 mm × 75 mm	–	–	–	–	–	m	196.28
extra for							
stopped ends	–	–	–	–	–	nr	27.07
external angles	–	–	–	–	–	nr	47.39
internal angles	–	–	–	–	–	nr	81.23

14 MASONRY

Item	PC £	Labour hours	Labour £	Plant £	Material £	Unit	Total rate £
200 mm × 100 mm	–	–	–	–	–	m	223.35
extra for							
stopped ends	–	–	–	–	–	nr	27.07
external angles	–	–	–	–	–	nr	67.68
internal angles	–	–	–	–	–	nr	121.82
250 mm × 150 mm	–	–	–	–	–	m	324.86
extra for							
stopped ends	–	–	–	–	–	nr	27.07
external angles	–	–	–	–	–	nr	67.68
internal angles	–	–	–	–	–	nr	135.36
300 mm × 250 mm	–	–	–	–	–	m	527.90
extra for							
stopped ends	–	–	–	–	–	nr	27.07
external angles	–	–	–	–	–	nr	94.75
internal angles	–	–	–	–	–	nr	162.43
Coping apex block; two sunk faces; rubbed							
650 mm × 450 mm × 225 mm	–	–	–	–	–	nr	663.26
Coping kneeler block; three sunk faces; rubbed							
350 mm × 350 mm × 375 mm	–	–	–	–	–	nr	527.90
450 mm × 450 mm × 375 mm	–	–	–	–	–	nr	609.11
Corbel; turned and moulded; rubbed							
225 mm × 225 mm × 375 mm	–	–	–	–	–	nr	433.15
Slab surrounds to openings; one face splayed; rubbed							
75 mm × 100 mm	–	–	–	–	–	m	94.75
75 mm × 200 mm	–	–	–	–	–	m	128.59
100 mm × 100 mm	–	–	–	–	–	m	115.06
125 mm × 100 mm	–	–	–	–	–	m	128.59
125 mm × 150 mm	–	–	–	–	–	m	155.66
175 mm × 175 mm	–	–	–	–	–	m	189.49
225 mm × 175 mm	–	–	–	–	–	m	223.35
300 mm × 175 mm	–	–	–	–	–	m	270.72
300 mm × 225 mm	–	–	–	–	–	m	351.93
Slab surrounds to openings; one face sunk splayed; rubbed							
75 mm × 100 mm	–	–	–	–	–	m	121.82
75 mm × 200 mm	–	–	–	–	–	m	155.66
100 mm × 100 mm	–	–	–	–	–	m	142.13
125 mm × 100 mm	–	–	–	–	–	m	155.66
125 mm × 150 mm	–	–	–	–	–	m	182.73
175 mm × 175 mm	–	–	–	–	–	m	216.56
225 mm × 175 mm	–	–	–	–	–	m	250.42
300 mm × 175 mm	–	–	–	–	–	m	297.79
300 mm × 225 mm	–	–	–	–	–	m	379.00
extra for							
throating	–	–	–	–	–	m	13.54
rebates and grooves	–	–	–	–	–	m	29.78
stooling	–	–	–	–	–	m	51.44

14 MASONRY

Item	PC £	Labour hours	Labour £	Plant £	Material £	Unit	Total rate £
14.05 CAST STONE – cont							
Eurobrick insulated brick cladding systems or other equal and approved; extruded polystyrene foam insulation; brick slips bonded to insulation panels with Eurobrick gun applied adhesive or other equal and approved; pointing with formulated mortar grout							
25 mm insulation to walls							
over 300 mm wide; fixing with proprietary screws and plates to timber	–	1.39	31.52	–	64.73	m²	**96.25**
50 mm insulation to walls							
over 300 mm wide; fixing with proprietary screws and plates; to timber	–	1.39	31.52	–	70.41	m²	**101.93**
14.06 FLUES AND FLUE LININGS							
Flues							
Scheidel Rite-Vent ICS Plus flue system; suitable for domestic multifuel appliances; stainless steel; twin wall; insulated; for use internally or externally							
80 mm pipes; including one locking band (fixing brackets measured separately)	–	0.90	17.97	–	104.58	m	**122.55**
Extra for							
appliance connecter	–	0.80	15.98	–	16.18	nr	**32.16**
30° bend	–	1.80	35.95	–	78.73	nr	**114.68**
45° bend	–	1.80	35.95	–	74.80	nr	**110.75**
135° tee; fully welded	–	2.70	53.92	–	154.83	nr	**208.75**
inspection length	–	0.90	17.97	–	11.19	nr	**29.16**
drain plug and support	–	1.00	19.97	–	72.52	nr	**92.49**
damper	–	0.90	17.97	–	58.97	nr	**76.94**
angled flashing including storm collar	–	1.25	24.97	–	78.12	nr	**103.09**
stub terminal	–	1.00	19.97	–	25.86	nr	**45.83**
tapered terminal	–	1.00	19.97	–	53.88	nr	**73.85**
floor support (2 piece)	–	1.50	29.95	–	42.56	nr	**72.51**
firestop floor support (2 piece)	–	1.50	29.95	–	23.85	nr	**53.80**
wall support (stainless steel)	–	1.00	19.97	–	90.37	nr	**110.34**
wall sleeve	–	1.20	23.97	–	35.86	nr	**59.83**
100 mm pipes; including one locking band (fixing brackets measured separately)	–	1.00	19.97	–	111.38	m	**131.35**
Extra for							
appliance connecter	–	0.90	17.97	–	17.86	nr	**35.83**
30° bend	–	2.00	39.94	–	82.36	nr	**122.30**
45° bend	–	2.00	39.94	–	78.19	nr	**118.13**
135° tee; fully welded	–	3.00	59.92	–	150.33	nr	**210.25**
inspection length	–	1.00	19.97	–	256.36	nr	**276.33**
drain plug and support	–	1.10	21.97	–	75.09	nr	**97.06**
damper	–	1.00	19.97	–	62.09	nr	**82.06**
angled flashing including storm collar	–	1.40	27.95	–	87.05	nr	**115.00**
stub terminal	–	1.10	21.97	–	26.28	nr	**48.25**
tapered terminal	–	1.10	21.97	–	57.52	nr	**79.49**
floor support (2 piece)	–	1.65	32.95	–	48.40	nr	**81.35**

14 MASONRY

Item	PC £	Labour hours	Labour £	Plant £	Material £	Unit	Total rate £
firestop floor support (2 piece)	–	1.65	32.95	–	26.29	nr	**59.24**
wall support (stainless steel)	–	1.10	21.97	–	95.50	nr	**117.47**
wall sleeve	–	1.35	26.97	–	40.73	nr	**67.70**
150 mm pipes; including one locking band (fixing brackets measured separately)	–	1.10	21.97	–	130.03	m	**152.00**
Extra for							
appliance connecter	–	1.00	19.97	–	23.03	nr	**43.00**
30° bend	–	2.20	43.94	–	98.84	nr	**142.78**
45° bend	–	2.20	43.94	–	93.98	nr	**137.92**
135° tee; fully welded	–	3.30	65.90	–	173.60	nr	**239.50**
inspection length	–	1.10	21.97	–	268.85	nr	**290.82**
drain plug and support	–	1.20	23.97	–	95.89	nr	**119.86**
damper	–	1.10	21.97	–	81.15	nr	**103.12**
angled flashing including storm collar	–	1.55	30.95	–	88.32	nr	**119.27**
stub terminal	–	1.20	23.97	–	28.27	nr	**52.24**
tapered terminal	–	1.20	23.97	–	66.11	nr	**90.08**
floor support (2 piece)	–	1.80	35.95	–	48.40	nr	**84.35**
firestop floor support (2 piece)	–	1.80	35.95	–	26.29	nr	**62.24**
wall support (stainless steel)	–	1.20	23.97	–	105.77	nr	**129.74**
wall sleeve	–	1.50	29.95	–	40.73	nr	**70.68**

14.07 FORMING CAVITIES

Forming cavities
In hollow walls

width of cavity 50 mm; polypropylene ties; three wall ties per m²	–	0.05	1.12	–	0.40	m²	**1.52**
width of cavity 50 mm; galvanized steel twisted wall ties; three wall ties per m²	–	0.05	1.12	–	0.80	m²	**1.92**
width of cavity 50 mm; stainless steel butterfly wall ties; three wall ties per m²	–	0.05	1.12	–	0.59	m²	**1.71**
width of cavity 50 mm; stainless steel twisted wall ties; three wall ties per m²	–	0.05	1.12	–	0.73	m²	**1.85**
width of cavity 75 mm; polypropylene ties; three wall ties per m²	–	0.05	1.12	–	0.40	m²	**1.52**
width of cavity 75 mm; galvanized steel twisted wall ties; three wall ties per m²	–	0.05	1.12	–	0.85	m²	**1.97**
width of cavity 75 mm; stainless steel butterfly wall ties; three wall ties per m²	–	0.05	1.12	–	0.82	m²	**1.94**
width of cavity 75 mm; stainless steel twisted wall ties; three wall ties per m²	–	0.05	1.12	–	0.81	m²	**1.93**

Pointing in
Pointing with mastic

wood frames or sills	–	0.09	1.60	–	1.22	m	**2.82**

Pointing with polysulphide sealant

wood frames or sills	–	0.09	1.60	–	3.09	m	**4.69**

14 MASONRY

Item	PC £	Labour hours	Labour £	Plant £	Material £	Unit	Total rate £
14.07 FORMING CAVITIES – cont							
Wedging and pinning							
To underside of existing construction with slates in cement: mortar (1:3)							
width of wall – one brick thick	–	0.74	16.58	–	5.74	m	**22.32**
width of wall – one and a half brick thick	–	0.93	20.85	–	11.47	m	**32.32**
width of wall – two brick thick	–	1.11	24.87	–	17.21	m	**42.08**
Slate and tile sills							
Sills; two courses of machine-made plain roofing tiles							
set weathering; bedded and pointed	–	0.56	12.56	–	5.77	m	**18.33**
Sundries							
Weep holes							
weep vents to perpends; plastic	–	0.02	0.45	–	0.62	nr	**1.07**
Chimney pots; red terracotta; plain roll top or cannon-head; setting and flaunching in cement mortar (1:3)							
185 mm dia. × 450 mm long	52.87	1.67	37.43	–	57.75	nr	**95.18**
185 mm dia. × 600 mm long	81.02	1.85	41.46	–	86.75	nr	**128.21**
185 mm dia. × 900 mm long	153.22	1.85	41.46	–	161.11	nr	**202.57**
Halfen Channels and Brick Support (NB Supply only rates)							
cavity brick tie; Halfen HTS-C12; 225 mm stainless steel A2	–	–	–	–	0.43	nr	**0.43**
cavity brick tie; Halfen HTS-C12; 225 mm pre-galvanized	–	–	–	–	0.18	nr	**0.18**
frame cramp, with slot; Halfen HTS-FS12; 150 mm stainless steel A2	–	–	–	–	0.32	nr	**0.32**
frame cramp, with slot; Halfen HTS-FS12; 150 mm pre-galvanized	–	–	–	–	0.15	nr	**0.15**
head restraint; Halfen CHR-V telescopic concealed tube and HSC−9 tie, in A2/ 304 stainless steel	–	–	–	–	2.66	nr	**2.66**
brick tie channel; Halfen FRS 25/14 2700; supplied in 2700 mm lengths with fixing holes at 110 mm centres; grade 304 stainless steel	–	–	–	–	2.21	nr	**2.21**
channel tie; Halfen FRS−03; 150 mm long safety end to suit 25/14 FRS channel, stainless steel	–	–	–	–	0.43	nr	**0.43**
ribslot brick tie channel; Halfen; 3000 mm long, stainless steel A2, complete with polystyrene filler (vf) and nail holes, to suit HTS ties	–	–	–	–	12.93	nr	**12.93**
channel tie; Halfen HTS-B12; 200 mm long, grade 304 stainless steel safety end to suit 28/15, 25/17HH and ribslot channels	–	–	–	–	0.45	nr	**0.45**
ribslot brick tie channel; Halfen; 3000 mm long, pre-galvanized mild steel, complete with polystyrene filler (vf) and nail holes, to suit HTS ties	–	–	–	–	6.91	nr	**6.91**

14 MASONRY

Item	PC £	Labour hours	Labour £	Plant £	Material £	Unit	Total rate £
channel tie; Halfen HTS-B12; 150 mm long, safety end pre galvanized to suit 28/15, 25/17 HH and ribslot channels	–	–	–	–	0.17	nr	**0.17**
brick support angle swystem; Halfen HK4-U-100 cavity/6 kN incl. fixings stainless steel, A4	–	–	–	–	61.38	m	**61.38**
stone support anchor, grout-in; Halfen UMA-10-1; 150 mm stainless steel	–	–	–	–	1.58	nr	**1.58**
stone restraint anchor, grout-in; Halfen UHA-5-1; 180 mm stainless steel A4	–	–	–	–	0.87	nr	**0.87**
windpost, brickwork; Halfen BW1-1406 × 2500 mm complete with fixings & ties, stainless steel, A4	–	–	–	–	234.18	nr	**234.18**
windpost, cavity; Halfen CW2-3544 × 2500 mm stainless steel complete with fixings & ties	–	–	–	–	152.64	nr	**152.64**
cast-in CE marked channel, cold rolled; Halfen HTA-K-38/17; 200 mm anchor centres, stainless steel A2-70, complete with combi filler (kf) and nail holes supplied in 3000 mm lengths	–	–	–	–	21.90	m	**21.90**
'T' bolt and nut to suit; Halfen HS-38/17 channel; M12 × 50 mm, stainless steel A2-70	–	–	–	–	2.06	nr	**2.06**
cast-in CE marked channel, cold rolled; Halfen HTA-K-38/17; 200 mm anchor centres, hot-dip galvanized (fv) , complete with combi filler (kf) and nail holes supplies in 3000 mm lengths	–	–	–	–	12.99	m	**12.99**
'T' bolt and nut to suit Halfen HS-38/17 channel; M12 × 50 mm, hot-dip galvanized (fv) 4.6 mild steel	–	–	–	–	0.80	nr	**0.80**
cast-in CE marked channel, hot rolled; Halfen HTA-K-40/22; hot-dip galvanized (fv), complete with combi filler (kf) and nail holes supplied in 3000 mm lengths	–	–	–	–	43.81	m	**43.81**
'T' bolt and nut to suit Halfen HS-40/22 channel; M16 × 50 mm, hot-dip galvanized (fv) 4.6 mild steel	–	–	–	–	1.50	nr	**1.50**
cast-in channel, hot rolled; Halfen HTA-52/34; hot-dip galvanized (fv), complete with polystyrene filler (vf) and nail holes supplied in 6070 mm lengths	–	–	–	–	37.84	m	**37.84**
'T' bolt and nut, to suit Halfen HS – 52/34 channel; M16 × 60 mm hot-dip galvanized (fv) 4.6 mild steel	–	–	–	–	2.85	nr	**2.85**
cast-in CE marked channel, hot rolled; Halfen HTA-52/34; stainless steel A4, complete with polystyrene filler (vf) and nail holes supplied in 6070 mm lengths	–	–	–	–	116.78	m	**116.78**
'T' bolt and nut, to suit Halfen HS-52/34 channel; M16 × 60 mm stainless steel A4 (fv) 4.6 mild steel	–	–	–	–	7.57	nr	**7.57**

14 MASONRY

Item	PC £	Labour hours	Labour £	Plant £	Material £	Unit	Total rate £
14.07 FORMING CAVITIES – cont							
Sundries – cont							
Halfen Channels and Brick Support (NB Supply only rates) – cont							
reinforcement, female coupler; Halfen HBM-F16/ M20; 810 mm long	–	–	–	–	3.35	nr	**3.35**
reinforcement, male coupler; Halfen HBM-M16/ M20; 770 mm long	–	–	–	–	4.49	nr	**4.49**
reinforcement continuity system; Halfen Kwikastrip M17Sb12C150; 1200 mm	–	–	–	–	37.13	nr	**37.13**
shear dowel set; Halfen CRET–124 stainless steel A4	–	–	–	–	57.29	nr	**57.29**
shear dowel set; Halfen CRET–10; 20 × 300 stainless steel A4	–	–	–	–	8.36	nr	**8.36**
shear sleeve; Halfen CRET-J-20; 160 stainless steel A4	–	–	–	–	6.90	nr	**6.90**
concrete balcony insulated connection Unit Halfen HIT HP; 80 mm unit for a 200 mm slab for a 1500 mm cantilever balcony (Passivhaus certified)	–	–	–	–	88.87	nr	**88.87**
concrete balcony insulated connection Unit Halfen HIT HP; 80 mm Unit for a 200 mm slab for a 2000 mm cantilever balcony (Passivhaus certified)	–	–	–	–	130.69	nr	**130.69**
steel balcony connection SBC-HBM; M20x4 BZP Cast-in Socket Anchor Assembly, 190 × 110 mm	–	–	–	–	39.73	nr	**39.73**
steel balcony connection; Halfen; SBC-TSS–10; 330 × 200 mm thermal separator pad, 4no. 22 mm holes at 190 × 110 mm	–	–	–	–	94.09	nr	**94.09**
precast lifting spread anchor CE marked; Halfen TPA-FS; 4.0 tonne × 180 mm carbon steel (wb)	–	–	–	–	2.14	nr	**2.14**
precast ring clutch CE marked; Halfen TPA R1; 5.0T (T)	–	–	–	–	119.18	nr	**119.18**
External door and window cavity closures							
Thermabate or equivalent; inclusive of flange clips; jointing strips; wall fixing ties and adhesive tape							
closing cavities; width of cavity 50 mm–60 mm	–	0.14	3.14	–	5.00	m	**8.14**
closing cavities; width of cavity 75 mm–84 mm	–	0.14	3.14	–	4.86	m	**8.00**
closing cavities; width of cavity 90 mm–99 mm	–	0.14	3.14	–	5.31	m	**8.45**
closing cavities; width of cavity 100 mm–110 mm	–	0.14	3.14	–	5.96	m	**9.10**
Kooltherm or equivalent; inclusive of flange clips; jointing strips; wall fixing ties and adhesive tape; 1 hr fire-rating							
closing cavities; width of cavity 50 mm	–	0.14	3.14	–	5.62	m	**8.76**
closing cavities; width of cavity 75 mm	–	0.14	3.14	–	5.85	m	**8.99**
closing cavities; width of cavity 100 mm	–	0.14	3.14	–	6.25	m	**9.39**
closing cavities; width of cavity 125 mm	–	0.14	3.14	–	7.44	m	**10.58**
closing cavities; width of cavity 150 mm	–	0.14	3.14	–	7.78	m	**10.92**

14 MASONRY

Item	PC £	Labour hours	Labour £	Plant £	Material £	Unit	Total rate £
Type WCA cavicloser or simialr; uPVC universal cavity closer, insulator and damp-proof course by Cavity Trays Ltd; built into cavity wall as work proceeds, complete with face closer and ties							
closing cavities; width of cavity up to 100 mm	–	0.07	1.57	–	13.44	m	**15.01**
closing cavities; width of cavity up to 150 mm	–	0.07	1.57	–	15.91	m	**17.48**
Type L durropolyethelene lintel stop ends or other equal; Cavity Trays Ltd; fixing with butyl anchoring strip; building in as the work proceeds							
adjusted to lintel as required	–	0.04	0.90	–	0.78	nr	**1.68**
Type W polypropylene weeps/vents or other equal and approved; Cavity Trays Ltd; built into cavity wall as work proceeds							
100/115 mm × 65 mm × 10 mm including lock fit wedges	–	0.04	0.90	–	0.57	nr	**1.47**
extra; extension duct							
200/225 mm × 65 mm × 10 mm	–	0.07	1.57	–	0.98	nr	**2.55**
Type X polypropylene abutment cavity tray or other equal; Cavity Trays Ltd; built into facing brickwork as the work proceeds; complete with Code 4 flashing; intermediate/catchment tray with short leads (requiring soakers); to suit roof of							
17–20° pitch	–	0.05	1.12	–	6.97	nr	**8.09**
21–25° pitch	–	0.05	1.12	–	6.51	nr	**7.63**
26–45° pitch	–	0.05	1.12	–	6.20	nr	**7.32**
Type X polypropylene abutment cavity tray or other equal; Cavity Trays Ltd; built into facing brickwork as the work proceeds; complete with Code 4 flashing; intermediate/catchment tray with long leads (suitable only for corrugated roof tiles); to suit roof of							
17–20° pitch	–	0.05	1.12	–	9.42	nr	**10.54**
21–25° pitch	–	0.05	1.12	–	8.68	nr	**9.80**
26–45° pitch	–	0.05	1.12	–	8.00	nr	**9.12**
Type X polypropylene abutment cavity tray or other equal; Cavity Trays Ltd; built into facing brickwork as the work proceeds; complete with Code 4 flashing; ridge tray with short/long leads; to suit roof of							
17–20° pitch	–	0.05	1.12	–	19.07	nr	**20.19**
21–25° pitch	–	0.05	1.12	–	14.73	nr	**15.85**
26–45° pitch	–	0.05	1.12	–	13.11	nr	**14.23**
Expamet stainless steel wall starters or other equal and approved; plugged and screwed							
to suit walls 60 mm–75 mm thick	–	0.23	3.29	0.14	2.76	m	**6.19**
to suit walls 100 mm–115 mm thick	–	0.23	3.29	0.14	2.78	m	**6.21**
to suit walls 125 mm–180 mm thick	–	0.37	5.29	0.22	3.75	m	**9.26**
to suit walls 190 mm–260 mm thick	–	0.46	6.57	0.27	4.66	m	**11.50**

14 MASONRY

Item	PC £	Labour hours	Labour £	Plant £	Material £	Unit	Total rate £
14.07 FORMING CAVITIES – cont							
Galvanized steel lintels; Catnic or other equal;							
built into brickwork or blockwork							
70/100 range open back lintel for cavity wall;							
70 mm–85 mm cavity wall							
1200 mm long	–	0.32	7.17	–	119.26	nr	**126.43**
1500 mm long	–	0.37	8.29	–	148.94	nr	**157.23**
1800 mm long	–	0.42	9.41	–	186.42	nr	**195.83**
2100 mm long	–	0.46	10.31	–	213.19	nr	**223.50**
2400 mm long	–	0.56	12.56	–	252.70	nr	**265.26**
2700 mm long	–	0.65	14.56	–	291.63	nr	**306.19**
3000 mm long	–	0.74	16.58	–	414.96	nr	**431.54**
90/100 range open back lintel for cavity wall;							
90 mm–105 mm cavity wall							
1200 mm long	–	0.32	7.17	–	123.91	nr	**131.08**
1500 mm long	–	0.37	8.29	–	154.90	nr	**163.19**
1800 mm long	–	0.42	9.41	–	186.42	nr	**195.83**
2100 mm long	–	0.46	10.31	–	221.71	nr	**232.02**
2400 mm long	–	0.56	12.56	–	267.31	nr	**279.87**
2700 mm long	–	0.65	14.56	–	320.05	nr	**334.61**
3000 mm long	–	0.74	16.58	–	416.35	nr	**432.93**
3300 mm long	–	0.83	18.60	–	471.89	nr	**490.49**
CN92 single lintel; for 75 mm internal walls							
1050 mm long	–	0.28	6.27	–	16.16	nr	**22.43**
1200 mm long	–	0.32	7.17	–	18.27	nr	**25.44**
CN102 single lintel; for 100 mm internal walls							
1050 mm long	–	0.28	6.27	–	20.42	nr	**26.69**
1200 mm long	–	0.32	7.17	–	22.54	nr	**29.71**
CN100 single lintel; for 75 mm internal walls							
1050 mm long	–	0.28	6.27	–	49.78	nr	**56.05**
1200 mm long	–	0.32	7.17	–	61.83	nr	**69.00**
BHD140 single lintel; for 140 mm internal walls							
1050 mm long	–	0.28	6.27	–	108.57	nr	**114.84**
1200 mm long	–	0.32	7.17	–	124.07	nr	**131.24**
1500 mm long	–	0.32	7.17	–	155.07	nr	**162.24**
Precast concrete sill, lintels, copings							
Lintels; plate; prestressed bedded							
100 mm × 70 mm × 600 mm long	6.20	0.37	8.29	–	6.43	nr	**14.72**
100 mm × 70 mm × 900 mm long	9.24	0.37	8.29	–	9.57	nr	**17.86**
100 mm × 70 mm × 1100 mm long	11.32	0.37	8.29	–	11.70	nr	**19.99**
100 mm × 70 mm × 1200 mm long	12.33	0.37	8.29	–	12.75	nr	**21.04**
100 mm × 70 mm × 1500 mm long	15.42	0.46	10.31	–	15.97	nr	**26.28**
100 mm × 70 mm × 1800 mm long	18.49	0.46	10.31	–	19.13	nr	**29.44**
100 mm × 70 mm × 2100 mm long	21.58	0.56	12.56	–	22.34	nr	**34.90**
140 mm × 70 mm × 1200 mm long	18.13	0.46	10.31	–	18.76	nr	**29.07**
140 mm × 70 mm × 1500 mm long	22.66	0.56	12.56	–	23.45	nr	**36.01**

14 MASONRY

Item	PC £	Labour hours	Labour £	Plant £	Material £	Unit	Total rate £
Lintels; rectangular; reinforced with mild steel bars; bedded							
100 mm × 145 mm × 900 mm long	4.57	0.56	12.56	–	4.76	nr	**17.32**
100 mm × 145 mm × 1050 mm long	5.34	0.56	12.56	–	5.55	nr	**18.11**
100 mm × 145 mm × 1200 mm long	6.10	0.56	12.56	–	6.33	nr	**18.89**
225 mm × 145 mm × 1200 mm long	23.65	0.74	16.58	–	24.48	nr	**41.06**
225 mm × 225 mm × 1800 mm long	35.38	1.39	31.16	–	36.56	nr	**67.72**
Lintels; boot; reinforced with mild steel bars; bedded							
250 mm × 225 mm × 1200 mm long	26.21	1.11	24.87	–	27.11	nr	**51.98**
275 mm × 225 mm × 1800 mm long	43.24	1.67	37.43	–	44.66	nr	**82.09**
14.08 DAMP-PROOF COURSES							
Damp-proof courses							
Polythene damp-proof course or other equal; 200 mm laps; in gauged mortar (1:1:6)							
100 mm width exceeding horizontal in brick or block course	–	0.01	0.23	–	0.14	m	**0.37**
150 mm width exceeding horizontal in brick or block course	–	0.01	0.23	–	0.21	m	**0.44**
300 mm width exceeding horizontal in brick or block course	–	0.01	0.23	–	0.41	m	**0.64**
width exceeding 225 mm; horizontal	–	0.23	5.15	–	0.35	m²	**5.50**
width exceeding 225 mm; forming cavity gutters in hollow walls; horizontal	–	0.37	8.29	–	0.35	m²	**8.64**
width not exceeding 225 mm; horizontal	–	0.46	10.31	–	0.35	m²	**10.66**
width not exceeding 225 mm; vertical	–	0.69	15.46	–	0.35	m²	**15.81**
Icopal Polymeric damp-proof course; Xtra-Load or other equal; 200 mm laps; in gauged morter (1:1:6)							
width exceeding 225 mm; horizontal	–	0.23	5.15	–	7.01	m²	**12.16**
width exceeding 225 mm; forming cavity gutters in hollow walls; horizontal	–	0.37	8.29	–	7.01	m²	**15.30**
width not exceeding 225 mm; horizontal	–	0.46	10.31	–	7.01	m²	**17.32**
width not exceeding 225 mm; vertical	–	0.69	15.46	–	7.01	m²	**22.47**
Zedex CPT (Co-Polymer Thermoplastic) damp-proof course or other equal; 200 mm laps; in gauged mortar (1:1:6)							
width exceeding 225 mm; horizontal	–	0.23	5.15	–	4.53	m²	**9.68**
width exceeding 225 mm wide; forming cavity gutters in hollow walls; horizontal	–	0.37	8.29	–	4.53	m²	**12.82**
width not exceeding 225 mm; horizontal	–	0.46	10.31	–	4.53	m²	**14.84**
width not exceeding 225 mm; vertical	–	0.69	15.46	–	4.53	m²	**19.99**
Hyload (pitch polymer) damp-proof course or other equal; 150 mm laps; in gauged mortar (1:1:6)							
width exceeding 225 mm; horizontal	–	0.23	5.15	–	5.12	m²	**10.27**
width exceeding 225 mm; forming cavity gutters in hollow walls; horizontal	–	0.37	8.29	–	5.12	m²	**13.41**
width not exceeding 225 mm; horizontal	–	0.46	10.31	–	5.12	m²	**15.43**
width not exceeding 225 mm; vertical	–	0.69	15.46	–	5.12	m²	**20.58**

14 MASONRY

Item	PC £	Labour hours	Labour £	Plant £	Material £	Unit	Total rate £
14.08 DAMP-PROOF COURSES – cont							
Damp-proof courses – cont							
Nubit bitumen and polyester-based damp-proof course or other equal; 200 mm laps; in gauged mortar (1:1:6)							
width exceeding 225 mm; horizontal	–	0.23	5.15	–	10.21	m²	**15.36**
width exceeding 225 mm wide; forming cavity gutters in hollow walls; horizontal	–	0.37	8.29	–	10.21	m²	**18.50**
width not exceeding 225 mm; horizontal	–	0.46	10.31	–	10.21	m²	**20.52**
width not exceeding 225 mm; vertical	–	0.69	15.46	–	10.21	m²	**25.67**
Permabit bitumen polymer damp-proof course or other equal; 150 mm laps; in gauged mortar (1:1:6)							
width exceeding 225 mm; horizontal	–	0.23	5.15	–	11.02	m²	**16.17**
width exceeding 225 mm; forming cavity gutters in hollow walls; horizontal	–	0.37	8.29	–	11.02	m²	**19.31**
width not exceeding 225 mm; horizontal	–	0.46	10.31	–	11.02	m²	**21.33**
width not exceeding 225 mm; vertical	–	0.69	15.46	–	11.02	m²	**26.48**
Alumite aluminium cored bitumen gas retardant damp-proof course or other equal; 200 mm laps; in gauged mortar (1:1;6)							
width exceeding 225 mm; horizontal	–	0.31	6.95	–	7.65	m²	**14.60**
width exceeding 225 mm; forming cavity gutters in hollow walls; horizontal	–	0.49	10.98	–	7.65	m²	**18.63**
width not exceeding 225 mm; horizontal	–	0.60	13.45	–	7.65	m²	**21.10**
width not exceeding 225 mm; vertical	–	0.83	18.60	–	7.65	m²	**26.25**
Milled lead damp-proof course; BS 1178; 1.80 mm thick (code 4), 175 mm laps; in cement: lime mortar (1:2:9)							
width exceeding 225 mm; horizontal (PC £/kg)	–	1.85	41.46	–	42.32	m²	**83.78**
width not exceeding 225 mm; horizontal	–	2.78	62.30	–	42.32	m²	**104.62**
Two courses slates in cement: mortar (1:3)							
width exceeding 225 mm; horizontal	–	1.39	31.16	–	26.57	m²	**57.73**
width exceeding 225 mm; vertical	–	2.08	46.62	–	26.57	m²	**73.19**
Synthaprufe damp-proof membrane or other equal and approved; three coats brushed on							
width not exceeding 150 mm; vertical	–	0.31	4.43	–	8.22	m²	**12.65**
width 150 mm–225 mm; vertical	–	0.30	4.28	–	8.22	m²	**12.50**
width 225 mm–300 mm; vertical	–	0.28	4.01	–	8.22	m²	**12.23**
width exceeding 300 mm wide; vertical	–	0.26	3.72	–	8.22	m²	**11.94**
14.09 JOINT REINFORCEMENT							
Joint reinforcement							
Brickforce galvanized steel joint reinforcement or other equal							
width 60 mm; ref GBF40W60B25	–	0.05	1.12	–	0.55	m	**1.67**
width 100 mm; ref GBF40W100B25	–	0.07	1.57	–	0.56	m	**2.13**
width 175 mm; ref GBF40W175B25	–	0.10	2.25	–	0.69	m	**2.94**

14 MASONRY

Item	PC £	Labour hours	Labour £	Plant £	Material £	Unit	Total rate £
Brickforce stainless steel joint reinforcement or other equal							
width 60 mm; ref SBF35W60BSC	–	0.05	1.12	–	1.40	m	**2.52**
width 100 mm; ref SBF35W100BSC	–	0.07	1.57	–	1.51	m	**3.08**
width 175 mm; ref SBF35W175BSC	–	0.10	2.25	–	1.77	m	**4.02**
Bricktor galvanized or stainless steel joint reinforcement or similar							
width 160 mm, galvanized steel; ref GBT160CCR	–	0.12	2.69	–	0.75	m	**3.44**
width 160 mm stainless steel; ref SBT160CCR	–	0.13	2.91	–	0.94	m	**3.85**
width 180 mm galvanized steel; ref GBT180CCR	–	0.14	3.14	–	0.94	m	**4.08**
width 180 mm stainless steel; ref SBT180CCR	–	0.14	3.14	–	1.29	m	**4.43**

14.10 FILLETS

Weather fillets
Weather fillets in cement: mortar (1:3)

Item	PC £	Labour hours	Labour £	Plant £	Material £	Unit	Total rate £
50 mm face width	–	0.11	2.46	–	0.11	m	**2.57**
100 mm face width	–	0.19	4.25	–	0.41	m	**4.66**

Angle fillets
Angle fillets in cement: mortar (1:3)

Item	PC £	Labour hours	Labour £	Plant £	Material £	Unit	Total rate £
50 mm face width	–	0.11	2.46	–	0.11	m	**2.57**
100 mm face width	–	0.19	4.25	–	0.41	m	**4.66**

14.11 JOINTS

Joints

Item	PC £	Labour hours	Labour £	Plant £	Material £	Unit	Total rate £
Hacking joints and faces of brickwork or blockwork to form key for plaster	–	0.24	3.43	–	–	m²	**3.43**
Raking out joint in brickwork or blockwork for turned-in edge of flashing							
horizontal	–	0.14	3.14	–	–	m	**3.14**
stepped	–	0.19	4.25	–	–	m	**4.25**
Raking out and enlarging joint in brickwork or blockwork for nib of asphalt							
horizontal	–	0.19	4.25	–	–	m	**4.25**
Cutting grooves in brickwork or blockwork							
for water bars and the like	–	0.23	3.29	1.04	–	m	**4.33**
for nib of asphalt; horizontal	–	0.23	3.29	1.04	–	m	**4.33**
Preparing to receive new walls							
top existing 215 mm wall	–	0.19	4.25	–	–	m	**4.25**
Cleaning and priming both faces; filling with preformed closed cell joint filler and pointing one side with polysulphide sealant; 12 mm deep							
expansion joints; 12 mm wide	–	0.23	4.69	–	5.65	m	**10.34**
expansion joints; 20 mm wide	–	0.28	5.58	–	8.49	m	**14.07**
expansion joints; 25 mm wide	–	0.32	6.24	–	10.34	m	**16.58**

14 MASONRY

Item	PC £	Labour hours	Labour £	Plant £	Material £	Unit	Total rate £
14.11 JOINTS – cont							
Joints – cont							
Fire-resisting horizontal expansion joints; filling with joint filler; fixed with high temperature slip adhesive; between top of wall and soffit							
wall not exceeding 215 mm wide; 10 mm wide joint with 30 mm deep filler (one hour fire seal)	–	0.23	5.15	–	6.07	m	**11.22**
wall not exceeding 215 mm wide; 10 mm wide joint with 30 mm deep filler (two hour fire seal)	–	0.23	5.15	–	6.07	m	**11.22**
wall not exceeding 215 mm wide; 20 mm wide joint with 45 mm deep filler (two hour fire seal)	–	0.28	6.27	–	9.36	m	**15.63**
wall not exceeding 215 mm wide; 30 mm wide joint with 75 mm deep filler (three hour fire seal)	–	0.32	7.17	–	24.09	m	**31.26**
Fire-resisting vertical expansion joints; filling with joint filler; fixed with high temperature slip adhesive; with polysulphide sealant one side; between end of wall and concrete							
wall not exceeding 215 mm wide; 20 mm wide joint with 45 mm deep filler (two hour fire seal)	–	0.37	7.83	–	15.54	m	**23.37**

15 STRUCTURAL METALWORK

Item	PC £	Labour hours	Labour £	Plant £	Material £	Unit	Total rate £
MATERIAL RATES							
BASIC STEEL PRICES – MATERIAL SUPPLY ONLY TATA Steel have recently revised their pricing structure and now publish prices for extras only. The extras cost needs to be added to their steel Base Price. The Base Price given here is a guide price only and subject to normal trading conditions with regards to discounts or plussages. The Base Price assumes a minimum order of 5 tonnes, S355JR steel and standard lengths.							
Universal beams and columns							
Base price for steel	–	–	–	–	800.00	tonne	**800.00**
Extra to be added to base price for (kg/m)							
1016 × 305 mm (222, 249, 272, 314, 349, 393, 438, 487) – Price on application							
914 × 419 mm (343, 388)	–	–	–	–	206.00	tonne	**206.00**
914 × 305 mm (201, 224, 253, 289)	–	–	–	–	175.10	tonne	**175.10**
838 × 292 mm (226)	–	–	–	–	175.10	tonne	**175.10**
838 × 292 mm (176, 194)	–	–	–	–	164.80	tonne	**164.80**
762 × 267 mm (197)	–	–	–	–	164.80	tonne	**164.80**
762 × 267 mm (134, 147, 173)	–	–	–	–	144.20	tonne	**144.20**
686 × 254 mm (152, 170)	–	–	–	–	164.80	tonne	**164.80**
686 × 254 mm (125, 140)	–	–	–	–	144.20	tonne	**144.20**
610 × 305 mm (238)	–	–	–	–	175.10	tonne	**175.10**
610 × 305 mm (149, 179)	–	–	–	–	144.20	tonne	**144.20**
610 × 229 mm (125, 140)	–	–	–	–	10.30	tonne	**10.30**
610 × 229 mm (101, 113)	–	–	–	–	10.30	tonne	**10.30**
610 × 178 mm (82, 92, 100) – Price on application							
533 × 312 mm (150, 182, 219, 272) – Price on application							
533 × 210 mm (138)	–	–	–	–	185.40	tonne	**185.40**
533 × 210 mm (101, 109, 122)	–	–	–	–	10.30	tonne	**10.30**
533 × 210 mm (82, 92)	–	–	–	–	10.30	tonne	**10.30**
533 × 165 mm (66, 74, 85) – Price on application							
457 × 191 mm (161)	–	–	–	–	206.00	tonne	**206.00**
457 × 191 mm (133)	–	–	–	–	185.40	tonne	**185.40**
457 × 191 mm (106)	–	–	–	–	51.50	tonne	**51.50**
457 × 191 mm (74, 82, 89, 98)	–	–	–	–	10.30	tonne	**10.30**
457 × 191 mm (67)	–	–	–	–	10.30	tonne	**10.30**
457 × 152 mm (74, 82)	–	–	–	–	51.50	tonne	**51.50**
457 × 152 mm (67)	–	–	–	–	51.50	tonne	**51.50**
457 × 152 mm (52, 60)	–	–	–	–	10.30	tonne	**10.30**
406 × 178 mm (85)	–	–	–	–	51.50	tonne	**51.50**
406 × 178 mm (74)	–	–	–	–	10.30	tonne	**10.30**
406 × 178 mm (54, 60, 67)	–	–	–	–	10.30	tonne	**10.30**
406 × 140 mm (53)	–	–	–	–	51.50	tonne	**51.50**
406 × 140 mm (46)	–	–	–	–	10.30	tonne	**10.30**
406 × 140 mm (39)	–	–	–	–	10.30	tonne	**10.30**
356 × 171 mm (45, 51, 57, 67)	–	–	–	–	10.30	tonne	**10.30**
305 × 165 mm (40, 46, 54)	–	–	–	–	10.30	tonne	**10.30**
305 × 127 mm (42, 48)	–	–	–	–	10.30	tonne	**10.30**
305 × 127 mm (37)	–	–	–	–	10.30	tonne	**10.30**

15 STRUCTURAL METALWORK

Item	PC £	Labour hours	Labour £	Plant £	Material £	Unit	Total rate £
MATERIAL RATES – cont							
Universal beams and columns – cont							
Extra to be added to base price for (kg/m) – cont							
305 × 102 mm (28, 33)	–	–	–	–	10.30	tonne	**10.30**
305 × 102 mm (25)	–	–	–	–	10.30	tonne	**10.30**
254 × 102 mm (28)	–	–	–	–	20.60	tonne	**20.60**
254 × 102 mm (22, 25)	–	–	–	–	10.30	tonne	**10.30**
254 × 146 (43)	–	–	–	–	10.30	tonne	**10.30**
254 × 146 (31, 37)	–	–	–	–	10.30	tonne	**10.30**
203 × 133 mm (25, 30)	–	–	–	–	10.30	tonne	**10.30**
203 × 102 mm (23)	–	–	–	–	10.30	tonne	**10.30**
178 × 102 mm (19)	–	–	–	–	20.60	tonne	**20.60**
152 × 89 mm (16)	–	–	–	–	20.60	tonne	**20.60**
127 × 76 mm (13)	–	–	–	–	20.60	tonne	**20.60**
Asymmetric Slimflor Beams (ASB) – Price on Application							
Universal columns (kg/m). NB These are to be added to the base steel price. Base price for steel	–	–	–	–	824.00	tonne	**824.00**
356 × 406 mm (634)	–	–	–	–	309.00	tonne	**309.00**
356 × 406 mm (340, 393, 467, 551)	–	–	–	–	206.00	tonne	**206.00**
356 × 406 mm (235, 287)	–	–	–	–	175.10	tonne	**175.10**
356 × 368 mm (177, 202)	–	–	–	–	175.10	tonne	**175.10**
356 × 368 mm (153)	–	–	–	–	164.80	tonne	**164.80**
356 × 368 mm (129)	–	–	–	–	144.20	tonne	**144.20**
305 × 305 mm (240, 283)	–	–	–	–	20.60	tonne	**20.60**
305 × 305 mm (137, 158, 198)	–	–	–	–	10.30	tonne	**10.30**
305 × 305 mm (97, 118)	–	–	–	–	10.30	tonne	**10.30**
254 × 254 mm (132, 167)	–	–	–	–	10.30	tonne	**10.30**
254 × 254 mm (73, 89, 107)	–	–	–	–	10.30	tonne	**10.30**
203 × 203 mm (127)	–	–	–	–	185.40	tonne	**185.40**
203 × 203 mm (100, 113)	–	–	–	–	51.50	tonne	**51.50**
203 × 203 mm (71, 86)	–	–	–	–	10.30	tonne	**10.30**
203 × 203 mm (46, 52, 60)	–	–	–	–	10.30	tonne	**10.30**
152 × 152 mm (44, 51)	–	–	–	–	51.50	tonne	**51.50**
152 × 152 mm (23, 30, 37)	–	–	–	–	10.30	tonne	**10.30**
Channels (kg/m). NB These are to be added to the base steel price. Base price for steel	–	–	–	–	824.00	tonne	**824.00**
430 × 100 mm (64.4)	–	–	–	–	330.89	tonne	**330.89**
380 × 100 mm (54.0)	–	–	–	–	308.46	tonne	**308.46**
300 × 100 mm (45.5)	–	–	–	–	128.99	tonne	**128.99**
300 × 90 mm (41.4)	–	–	–	–	140.21	tonne	**140.21**
260 × 90 mm (34.8)	–	–	–	–	128.99	tonne	**128.99**
260 × 75 mm (27.6)	–	–	–	–	78.52	tonne	**78.52**
230 × 90 mm (32.2)	–	–	–	–	140.21	tonne	**140.21**
230 × 75 mm (25.7)	–	–	–	–	78.52	tonne	**78.52**
200 × 90 mm (29.7)	–	–	–	–	128.99	tonne	**128.99**
200 × 75 mm (23.4)	–	–	–	–	44.87	tonne	**44.87**
180 × 90 mm (26.1)	–	–	–	–	128.99	tonne	**128.99**
180 × 75 mm (20.3)	–	–	–	–	44.87	tonne	**44.87**

15 STRUCTURAL METALWORK

Item	PC £	Labour hours	Labour £	Plant £	Material £	Unit	Total rate £
150 × 90 mm (23.9)	–	–	–	–	128.99	tonne	**128.99**
150 × 75 mm (17.9)	–	–	–	–	22.43	tonne	**22.43**
125 × 65 mm (14.8)	–	–	–	–	22.43	tonne	**22.43**
Equal angles (mm). NB These are to be added to the base steel price. Base price for steel	–	–	–	–	824.00	tonne	**824.00**
200 × 200 mm (16, 18, 20, 24)	–	–	–	–	73.65	tonne	**73.65**
150 × 150 mm (10, 12, 15, 18)	–	–	–	–	73.65	tonne	**73.65**
120 × 120 mm (8, 10, 12, 15)	–	–	–	–	11.33	tonne	**11.33**
Unequal angles (mm). NB These are to be added to the base steel price. Base price for steel	–	–	–	–	824.00	tonne	**824.00**
200 × 150 mm (12, 15, 18)	–	–	–	–	130.29	tonne	**130.29**
200 × 100 mm (10, 12, 15)	–	–	–	–	73.65	tonne	**73.65**
Specification extras							
for Advance275JR steel	–	–	–	–	30.90	tonne	**30.90**
for Advance275JO steel	–	–	–	–	41.20	tonne	**41.20**
for Advance275J2 steel	–	–	–	–	82.40	tonne	**82.40**
for Advance355JO steel	–	–	–	–	10.30	tonne	**10.30**
for Advance355J2 steel	–	–	–	–	41.20	tonne	**41.20**
for Advance355 K2 steel	–	–	–	–	82.40	tonne	**82.40**
Hollow sections							
NOTE: The following prices are for basic quantities of 10 tonnes and over in one size, thickness, length, steel grade and surface finish and include delivery (for delivery to outer London).							
circular hollow sections	–	–	–	–	1133.00	tonne	**1133.00**
rectangular hollow section	–	–	–	–	1133.00	tonne	**1133.00**
square hollow sections	–	–	–	–	1133.00	tonne	**1133.00**
15.01 FRAMED MEMBERS, FRAMING, FABRICATION							
SUPPLY AND FIX PRICES							
Framing, fabrication; weldable steel; Grade S275 or Grade S355; hot rolled structural steel sections; welded fabrication							
Columns							
weight not exceeding 40 kg/m	–	–	–	–	–	tonne	**1853.82**
weight not exceeding 40 kg/m; cellular (Fabsec)	–	–	–	–	–	tonne	**2304.47**
weight not exceeding 40 kg/m; curved	–	–	–	–	–	tonne	**2637.33**
weight not exceeding 40 kg/m; square hollow section	–	–	–	–	–	tonne	**2483.70**
weight not exceeding 40 kg/m; circular hollow section	–	–	–	–	–	tonne	**2483.70**
weight 40–100 kg/m	–	–	–	–	–	tonne	**1746.27**
weight 40–100 kg/m; cellular (Fabsec)	–	–	–	–	–	tonne	**2017.69**
weight 40–100 kg/m; curved	–	–	–	–	–	tonne	**2053.53**
weight 40–100 kg/m; square hollow section	–	–	–	–	–	tonne	**2022.81**
weight 40–100 kg/m; circular hollow section	–	–	–	–	–	tonne	**2022.81**
weight exceeding 100 kg/m	–	–	–	–	–	tonne	**1689.94**

15 STRUCTURAL METALWORK

Item	PC £	Labour hours	Labour £	Plant £	Material £	Unit	Total rate £
15.01 FRAMED MEMBERS, FRAMING, FABRICATION – cont							
Framing, fabrication – cont							
Columns – cont							
weight exceeding 100 kg/m; cellular (Fabsec)	–	–	–	–	–	tonne	1925.51
weight exceeding 100 kg/m; curved	–	–	–	–	–	tonne	1915.26
weight exceeding 100 kg/m; square hollow section	–	–	–	–	–	tonne	3190.41
weight exceeding 100 kg/m; circular hollow section	–	–	–	–	–	tonne	3190.41
Beams							
weight not exceeding 40 kg/m	–	–	–	–	–	tonne	2176.43
weight not exceeding 40 kg/m; cellular (Fabsec)	–	–	–	–	–	tonne	2586.12
weight not exceeding 40 kg/m; curved	–	–	–	–	–	tonne	2709.02
weight not exceeding 40 kg/m; square hollow section	–	–	–	–	–	tonne	2749.99
weight not exceeding 40 kg/m; circular hollow section	–	–	–	–	–	tonne	2749.99
weight 40–100 kg/m	–	–	–	–	–	tonne	1746.27
weight 40–100 kg/m; cellular (Fabsec)	–	–	–	–	–	tonne	2335.19
weight 40–100 kg/m; curved	–	–	–	–	–	tonne	2053.53
weight 40–100 kg/m; square hollow section	–	–	–	–	–	tonne	2570.76
weight 40–100 kg/m; circular hollow section	–	–	–	–	–	tonne	2570.76
weight exceeding 100 kg/m	–	–	–	–	–	tonne	1689.94
weight exceeding 100 kg/m; cellular (Fabsec)	–	–	–	–	–	tonne	1935.75
weight exceeding 100 kg/m; curved	–	–	–	–	–	tonne	1992.08
weight exceeding 100 kg/m; square hollow section	–	–	–	–	–	tonne	3190.41
weight exceeding 100 kg/m; circular hollow section	–	–	–	–	–	tonne	3190.41
Bracings							
weight not exceeding 40 kg/m	–	–	–	–	–	tonne	2176.43
weight not exceeding 40 kg/m; square hollow section	–	–	–	–	–	tonne	2749.99
weight not exceeding 40 kg/m; circular hollow section	–	–	–	–	–	tonne	2749.99
weight 40–100 kg/m	–	–	–	–	–	tonne	1746.27
weight 40–100 kg/m; square hollow section	–	–	–	–	–	tonne	2570.76
weight 40–100 kg/m; circular hollow section	–	–	–	–	–	tonne	2570.76
weight exceeding 100 kg/m	–	–	–	–	–	tonne	1689.94
weight exceeding 100 kg/m; square hollow section	–	–	–	–	–	tonne	3190.41
weight exceeding 100 kg/m; circular hollow section	–	–	–	–	–	tonne	3190.41
Purlins and cladding rails							
weight not exceeding 40 kg/m	–	–	–	–	–	tonne	2176.43
weight not exceeding 40 kg/m; square hollow section	–	–	–	–	–	tonne	2749.99
weight not exceeding 40 kg/m; circular hollow section	–	–	–	–	–	tonne	2749.99
weight 40–100 kg/m	–	–	–	–	–	tonne	1746.27
weight 40–100 kg/m; square hollow section	–	–	–	–	–	tonne	2570.76
weight 40–100 kg/m; circular hollow section	–	–	–	–	–	tonne	2570.76
weight exceeding 100 kg/m	–	–	–	–	–	tonne	1689.94
weight exceeding 100 kg/m; square hollow section	–	–	–	–	–	tonne	3190.41
weight exceeding 100 kg/m; circular hollow section	–	–	–	–	–	tonne	3190.41

15 STRUCTURAL METALWORK

Item	PC £	Labour hours	Labour £	Plant £	Material £	Unit	Total rate £
Grillages							
weight not exceeding 40 kg/m	–	–	–	–	–	tonne	2176.43
weight 40–100 kg/m	–	–	–	–	–	tonne	2176.43
weight exceeding 100 kg/m	–	–	–	–	–	tonne	1746.27
Trestles, towers and built up columns							
straight	–	–	–	–	–	tonne	2749.99
Trusses and built up girders							
straight	–	–	–	–	–	tonne	2749.99
curved	–	–	–	–	–	tonne	292.93
Fittings							
general steel fittings; nuts, bolts, plates etc.	–	–	–	–	–	tonne	2529.79
Framing, erection							
Trial erection	–	–	–	–	–	tonne	238.65
Permanent erection on site	–	–	–	–	–	tonne	288.37
Metsec Lightweight Steel Framing System (SFS); or other equal and approved; as inner leaf to external wall; studs typically at 600 mm centres; including provision for all openings, abutments, junctions and head details etc.							
Inner leaf; with supports and perimeter sections; 12 mm plasterboard internally; 10 mm cement fibre substrate externally (insulation and external cladding measured separately)							
100 mm thick steel walling	–	–	–	–	–	m²	79.75
150 mm thick steel walling	–	–	–	–	–	m²	86.03
200 mm thick steel walling	–	–	–	–	–	m²	92.33
Inner leaf; with 16 mm Pyroc sheething board							
100 mm thick steel walling	–	–	–	–	–	m²	98.63
150 mm thick steel walling	–	–	–	–	–	m²	104.91
200 mm thick steel walling	–	–	–	–	–	m²	111.21
16 mm Pyroc sheething board fixed to slab perimeter							
not exceeding 300 mm	–	–	–	–	–	m	9.79
Inner leaf; with 16 mm Pyroc sheething board and Thermawall TW50 insulation supported by Halfen channels type 28/15 fixed to studs at 450 mm centres.							
100 mm thick steel walling with 50 mm insulation	–	–	–	–	–	m²	114.90
150 mm thick steel walling with 75 mm insulation	–	–	–	–	–	m²	128.02
200 mm thick steel walling with 100 mm insulatiom	–	–	–	–	–	m²	140.40
16 mm Pyroc sheething board and 40 mm Thermawall TW55 insulation fixed to slab perimeter							
not exceeding 300 mm	–	–	–	–	–	m	11.18
Storage costs							
Costs for storing fabricated steelwork							
storage off site	–	–	–	–	18.54	t/week	18.54
storage on extending trailers	–	–	–	–	30.90	t/week	30.90

15 STRUCTURAL METALWORK

Item	PC £	Labour hours	Labour £	Plant £	Material £	Unit	Total rate £
15.02 ISOLATED STRUCTURAL METAL MEMBERS							
Isolated structural member; weldable steel; BS EN 10025: 2004 Grade S275; hot rolled structural steel sections							
Plain member; beams							
weight not exceeding 40 kg/m	–	–	–	–	–	tonne	**1570.00**
weight 40–100 kg/m	–	–	–	–	–	tonne	**1570.00**
weight exceeding 100 kg/m	–	–	–	–	–	tonne	**1570.00**
Metsec open web steel lattice beams or other equal and approved; in single members; raised 3.50 m above ground; ends built in							
Beams; one coat zinc phosphate primer at works							
220 mm deep; to span 6.00 m (11.50 kg/m); ref B22	–	0.19	4.74	–	34.04	m	**38.78**
270 mm deep; to span 7.00 m (11.50 kg/m); ref B27	–	0.19	4.74	–	34.04	m	**38.78**
300 mm deep; to span 8.00 m (12.50 kg/m); ref B30	–	0.23	5.74	–	36.94	m	**42.68**
350 mm deep; to span 9.00 m (14.00 kg/m); ref B35	–	0.23	5.74	–	41.27	m	**47.01**
350 mm deep; to span 10.00 m (20.00 kg/m); ref D35	–	0.28	6.98	–	58.61	m	**65.59**
450 mm deep; to span 11.00 m (21.00 kg/m); ref D45	–	0.32	7.97	–	61.49	m	**69.46**
450 mm deep; to span 12.00 m (32.50 kg/m); ref G45	–	0.46	11.46	–	94.72	m	**106.18**
Beams; galvanized							
220 mm deep; to span 6.00 m (11.50 kg/m); ref B22	–	0.19	4.74	–	38.58	m	**43.32**
270 mm deep; to span 7.00 m (11.50 kg/m); ref B27	–	0.19	4.74	–	38.58	m	**43.32**
300 mm deep; to span 8.00 m (12.50 kg/m); ref B30	–	0.23	5.74	–	41.85	m	**47.59**
350 mm deep; to span 9.00 m (14.00 kg/m); ref B35	–	0.23	5.74	–	46.80	m	**52.54**
350 mm deep; to span 10.00 m (20.00 kg/m); ref D35	–	0.28	6.98	–	66.47	m	**73.45**
450 mm deep; to span 11.00 m (21.00 kg/m); ref D45	–	0.32	7.97	–	69.76	m	**77.73**
450 mm deep; to span 12.00 m (32.50 kg/m); ref G45	–	0.46	11.46	–	107.51	m	**118.97**

15 STRUCTURAL METALWORK

Item	PC £	Labour hours	Labour £	Plant £	Material £	Unit	Total rate £
15.03 COLD ROLLED STEEL PURLINS AND CLADDING RAILS							
Cold formed galvanized steel; Kingspan Multibeam or other equal and approved							
Cold rolled purlins and cladding rails							
175 × 65 × 1.40 mm gauge purlins or rails; fixed to steelwork	–	0.04	0.84	–	9.07	m	9.91
175 × 65 × 1.60 mm gauge purlins or rails; fixed to steelwork	–	0.04	0.84	–	9.65	m	10.49
175 × 65 × 2.00 mm gauge purlins or rails; fixed to steelwork	–	0.04	0.84	–	11.66	m	12.50
205 × 65 × 1.40 mm gauge purlins or rails; fixed to steelwork	–	0.04	0.84	–	10.02	m	10.86
205 × 65 × 1.60 mm gauge purlins or rails; fixed to steelwork	–	0.04	0.84	–	10.92	m	11.76
205 × 65 × 2.00 mm gauge purlins or rails; fixed to steelwork	–	0.04	0.84	–	12.43	m	13.27
Cleats							
weld-on for 175 mm purlin or rail	–	0.10	2.10	–	2.63	nr	4.73
bolt-on for 175 mm purlin or rail; including fixing bolts	–	0.02	0.42	–	5.65	m	6.07
weld-on for 205 mm purlin or rail	–	0.10	2.10	–	2.99	nr	5.09
bolt-on for 205 mm purlin or rail; including fixing bolts	–	0.02	0.42	–	6.09	m	6.51
Tubular ties							
1800 mm long; bolted diagonally across purlins or cladding rails	–	0.02	0.42	–	6.64	m	7.06
15.04 PROFILED METAL DECKING							
Soffits of slabs; galvanized steel permanent re-entrant type shuttering; including safety net							
Slab thickness not exceeding 200 mm							
0.9 mm decking; height to soffit 1.50–3.00 m	34.89	0.27	5.89	–	39.76	m²	45.65
0.9 mm decking; height to soffit 3.00–4.50 m	34.89	0.30	6.54	–	39.76	m²	46.30
1.2 mm decking; height to soffit 3.00–4.50 m	39.18	0.27	5.89	–	44.17	m²	50.06
Edge trim and restraints to decking							
Edge trim 1.2 mm × 300 mm girth	–	0.17	3.59	–	15.54	m	19.13
Edge trim 1.2 mm × 350 mm girth	–	0.17	3.59	–	15.54	m	19.13
Edge trim 1.2 mm × 400 mm girth	–	0.17	3.59	–	15.54	m	19.13
Bearings to decking; connection to steel work with thru-deck welded shear studs							
1995 × 95 mm high studs at 100 mm centres	–	–	–	–	17.40	m	17.40
1995 × 95 mm high studs at 200 mm centres	–	–	–	–	8.69	m	8.69
1995 × 95 mm high studs at 300 mm centres	–	–	–	–	5.79	m	5.79
19120 × 120 mm high studs at 100 mm centres	–	–	–	–	17.40	m	17.40
19120 × 120 mm high studs at 200 mm centres	–	–	–	–	8.69	m	8.69
19120 × 120 mm high studs at 300 mm centres	–	–	–	–	5.79	m	5.79

15 STRUCTURAL METALWORK

Item	PC £	Labour hours	Labour £	Plant £	Material £	Unit	Total rate £
15.04 PROFILED METAL DECKING – cont							
Soffits of slabs; galvanized steel permanent							
trapezoidal type shuttering; including safety net							
Slab thickness not exceeding 200 mm							
0.9 mm decking; height to soffit 1.50–3.00 m	31.91	0.27	5.89	–	36.69	m²	**42.58**
0.9 mm decking; height to soffit 3.00–4.50 m	31.91	0.30	6.54	–	36.69	m²	**43.23**
1.2 mm decking; height to soffit 3.00–4.50 m	37.24	0.27	5.89	–	42.17	m²	**48.06**
15.05 SURFACE TREATMENTS							
Surface preparation							
At works							
blast cleaning	–	–	–	–	–	m²	**2.99**
Surface treatment							
At works							
galvanizing	–	–	–	–	–	tonne	**330.75**
galvanizing	–	–	–	–	–	m²	**20.21**
shotblasting and priming to SA 2.5	–	–	–	–	–	m²	**8.24**
touch up primer and one coat of two pack epoxy zinc phosphate primer	–	–	–	–	–	m²	**5.14**
intumescent paint fire protection (30 minutes); spray applied	–	–	–	–	–	m²	**23.15**
intumescent paint fire protection (60 minutes); spray applied	–	–	–	–	–	m²	**23.15**
intumescent paint fire protection (90 minutes); spray applied	–	–	–	–	–	m²	**41.90**
intumescent paint fire protection (120 minutes); spray applied	–	–	–	–	–	m²	**71.66**
extra over for; separate decorative sealer top coat	–	–	–	–	–	m²	**4.41**
On site							
intumescent paint fire protection (30 minutes); spray applied	–	–	–	–	–	m²	**11.68**
intumescent paint fire protection (30 minutes) to circular columns etc.; spray applied	–	–	–	–	–	m²	**19.57**
intumescent paint fire protection (60 minutes) to UBs etc.; spray applied	–	–	–	–	–	m²	**14.80**
intumescent paint fire protection (60 minutes) to circular columns etc.; spray applied	–	–	–	–	–	m²	**24.82**
extra for separate decorative sealer top coat	–	–	–	–	–	m²	**3.88**

16 CARPENTRY

Item	PC £	Labour hours	Labour £	Plant £	Material £	Unit	Total rate £
16.01 PRIMARY AND STRUCTURAL TIMBERS							
SUPPLY AND FIX PRICES							
Sawn softwood; untreated							
Floor members							
38 mm × 100 mm	–	0.11	2.54	–	2.80	m	**5.34**
38 mm × 150 mm	–	0.13	3.01	–	3.87	m	**6.88**
47 mm × 75 mm	–	0.11	2.54	–	1.79	m	**4.33**
47 mm × 100 mm	–	0.13	3.01	–	2.30	m	**5.31**
47 mm × 125 mm	–	0.13	3.01	–	2.95	m	**5.96**
47 mm × 150 mm	–	0.14	3.23	–	3.39	m	**6.62**
47 mm × 175 mm	–	0.14	3.23	–	4.00	m	**7.23**
47 mm × 200 mm	–	0.15	3.47	–	4.34	m	**7.81**
47 mm × 225 mm	–	0.15	3.47	–	4.97	m	**8.44**
47 mm × 250 mm	–	0.16	3.70	–	5.60	m	**9.30**
75 mm × 125 mm	–	0.15	3.47	–	6.75	m	**10.22**
75 mm × 150 mm	–	0.15	3.47	–	7.32	m	**10.79**
75 mm × 175 mm	–	0.15	3.47	–	8.82	m	**12.29**
75 mm × 200 mm	–	0.16	3.70	–	9.87	m	**13.57**
75 mm × 225 mm	–	0.16	3.70	–	10.76	m	**14.46**
75 mm × 250 mm	–	0.17	3.93	–	16.37	m	**20.30**
100 mm × 150 mm	–	0.20	4.62	–	9.43	m	**14.05**
100 mm × 200 mm	–	0.21	4.86	–	12.56	m	**17.42**
100 mm × 250 mm	–	0.23	5.33	–	15.75	m	**21.08**
100 mm × 300 mm	–	0.25	5.79	–	20.37	m	**26.16**
Wall or partition members							
25 mm × 25 mm	–	0.06	1.39	–	1.09	m	**2.48**
25 mm × 38 mm	–	0.06	1.39	–	1.07	m	**2.46**
25 mm × 75 mm	–	0.08	1.85	–	1.52	m	**3.37**
38 mm × 38 mm	–	0.08	1.85	–	1.43	m	**3.28**
38 mm × 50 mm	–	0.08	1.85	–	1.82	m	**3.67**
38 mm × 75 mm	–	0.11	2.54	–	2.24	m	**4.78**
38 mm × 100 mm	–	0.14	3.23	–	2.80	m	**6.03**
47 mm × 50 mm	–	0.11	2.54	–	1.39	m	**3.93**
47 mm × 75 mm	–	0.14	3.23	–	1.84	m	**5.07**
47 mm × 100 mm	–	0.17	3.93	–	2.35	m	**6.28**
47 mm × 125 mm	–	0.18	4.16	–	3.00	m	**7.16**
75 mm × 75 mm	–	0.17	3.93	–	4.15	m	**8.08**
75 mm × 100 mm	–	0.19	4.40	–	5.62	m	**10.02**
100 mm × 100 mm	–	0.19	4.40	–	6.85	m	**11.25**
Joist strutting; herringbone							
47 mm × 50 mm; depth of joist 150 mm	–	0.46	10.64	–	3.32	m	**13.96**
47 mm × 50 mm; depth of joist 175 mm	–	0.46	10.64	–	3.38	m	**14.02**
47 mm × 50 mm; depth of joist 200 mm	–	0.46	10.64	–	3.44	m	**14.08**
47 mm × 50 mm; depth of joist 225 mm	–	0.46	10.64	–	3.50	m	**14.14**
47 mm × 50 mm; depth of joist 250 mm	–	0.46	10.64	–	3.55	m	**14.19**
Joist strutting; block							
47 mm × 150 mm; depth of joist 150 mm	–	0.28	6.48	–	4.08	m	**10.56**
47 mm × 175 mm; depth of joist 175 mm	–	0.28	6.48	–	4.69	m	**11.17**
47 mm × 200 mm; depth of joist 200 mm	–	0.28	6.48	–	5.03	m	**11.51**
47 mm × 225 mm; depth of joist 225 mm	–	0.28	6.48	–	5.67	m	**12.15**
47 mm × 250 mm; depth of joist 250 mm	–	0.28	6.48	–	6.30	m	**12.78**

16 CARPENTRY

Item	PC £	Labour hours	Labour £	Plant £	Material £	Unit	Total rate £
16.01 PRIMARY AND STRUCTURAL TIMBERS – cont							
Sawn softwood; untreated – cont							
Cleats							
225 mm × 100 mm × 75 mm	–	0.19	4.40	–	1.09	nr	**5.49**
Extra for stress grading to above timbers							
general structural (GS) grade	–	–	–	–	43.32	m³	**43.32**
special structural (SS) grade	–	–	–	–	86.62	m³	**86.62**
Extra for protecting and flameproofing timber with							
Celgard CF protection or other equal and approved							
small sections	–	–	–	–	152.48	m³	**152.48**
large sections	–	–	–	–	147.24	m³	**147.24**
Wrot surfaces							
plain; 50 mm wide	–	0.02	0.46	–	–	m	**0.46**
plain; 100 mm wide	–	0.03	0.69	–	–	m	**0.69**
plain; 150 mm wide	–	0.04	0.93	–	–	m	**0.93**
Sawn softwood; tanalized							
Floor members							
38 mm × 75 mm	–	0.11	2.54	–	2.46	m	**5.00**
38 mm × 100 mm	–	0.11	2.54	–	3.09	m	**5.63**
38 mm × 150 mm	–	0.13	3.01	–	4.31	m	**7.32**
47 mm × 75 mm	–	0.11	2.54	–	2.08	m	**4.62**
47 mm × 100 mm	–	0.13	3.01	–	2.68	m	**5.69**
47 mm × 125 mm	–	0.13	3.01	–	3.43	m	**6.44**
47 mm × 150 mm	–	0.14	3.23	–	3.96	m	**7.19**
47 mm × 175 mm	–	0.14	3.23	–	4.67	m	**7.90**
47 mm × 200 mm	–	0.15	3.47	–	5.10	m	**8.57**
47 mm × 225 mm	–	0.15	3.47	–	5.83	m	**9.30**
47 mm × 250 mm	–	0.16	3.70	–	6.55	m	**10.25**
75 mm × 125 mm	–	0.15	3.47	–	7.46	m	**10.93**
75 mm × 150 mm	–	0.15	3.47	–	8.18	m	**11.65**
75 mm × 175 mm	–	0.15	3.47	–	9.81	m	**13.28**
75 mm × 200 mm	–	0.16	3.70	–	11.00	m	**14.70**
75 mm × 225 mm	–	0.16	3.70	–	12.04	m	**15.74**
75 mm × 250 mm	–	0.17	3.93	–	17.79	m	**21.72**
100 mm × 150 mm	–	0.20	4.62	–	10.57	m	**15.19**
100 mm × 200 mm	–	0.21	4.86	–	14.07	m	**18.93**
100 mm × 250 mm	–	0.23	5.33	–	17.65	m	**22.98**
100 mm × 300 mm	–	0.25	5.79	–	22.65	m	**28.44**
Wall or partition members							
25 mm × 25 mm	–	0.06	1.39	–	1.13	m	**2.52**
25 mm × 38 mm	–	0.06	1.39	–	1.14	m	**2.53**
25 mm × 75 mm	–	0.08	1.85	–	1.67	m	**3.52**
38 mm × 38 mm	–	0.08	1.85	–	1.55	m	**3.40**
38 mm × 50 mm	–	0.08	1.85	–	1.97	m	**3.82**
38 mm × 75 mm	–	0.11	2.54	–	2.46	m	**5.00**
38 mm × 100 mm	–	0.14	3.23	–	3.09	m	**6.32**

16 CARPENTRY

Item	PC £	Labour hours	Labour £	Plant £	Material £	Unit	Total rate £
47 mm × 50 mm	–	0.11	2.54	–	1.58	m	4.12
47 mm × 75 mm	–	0.14	3.23	–	2.13	m	5.36
47 mm × 100 mm	–	0.17	3.93	–	2.73	m	6.66
47 mm × 125 mm	–	0.18	4.16	–	3.48	m	7.64
75 mm × 75 mm	–	0.17	3.93	–	4.57	m	8.50
75 mm × 100 mm	–	0.19	4.40	–	6.19	m	10.59
100 mm × 100 mm	–	0.19	4.40	–	7.60	m	12.00
Roof members; flat							
38 mm × 75 mm	–	0.13	3.01	–	2.46	m	5.47
38 mm × 100 mm	–	0.13	3.01	–	3.09	m	6.10
38 mm × 125 mm	–	0.13	3.01	–	3.68	m	6.69
38 mm × 150 mm	–	0.13	3.01	–	4.31	m	7.32
47 mm × 100 mm	–	0.13	3.01	–	2.68	m	5.69
47 mm × 125 mm	–	0.13	3.01	–	3.43	m	6.44
47 mm × 150 mm	–	0.14	3.23	–	3.96	m	7.19
47 mm × 175 mm	–	0.14	3.23	–	4.67	m	7.90
47 mm × 200 mm	–	0.15	3.47	–	5.10	m	8.57
47 mm × 225 mm	–	0.15	3.47	–	5.83	m	9.30
47 mm × 250 mm	–	0.16	3.70	–	6.55	m	10.25
75 mm × 150 mm	–	0.15	3.47	–	8.18	m	11.65
75 mm × 175 mm	–	0.15	3.47	–	9.81	m	13.28
75 mm × 200 mm	–	0.16	3.70	–	11.00	m	14.70
75 mm × 225 mm	–	0.16	3.70	–	12.04	m	15.74
75 mm × 250 mm	–	0.17	3.93	–	17.79	m	21.72
Roof members; pitched							
25 mm × 100 mm	–	0.11	2.54	–	2.43	m	4.97
25 mm × 125 mm	–	0.11	2.54	–	3.32	m	5.86
25 mm × 150 mm	–	0.14	3.23	–	3.99	m	7.22
25 mm × 175 mm	–	0.16	3.70	–	4.67	m	8.37
25 mm × 200 mm	–	0.17	3.93	–	5.34	m	9.27
38 mm × 100 mm	–	0.14	3.23	–	3.09	m	6.32
38 mm × 125 mm	–	0.14	3.23	–	3.68	m	6.91
38 mm × 150 mm	–	0.14	3.23	–	4.31	m	7.54
38 mm × 175 mm	–	0.16	3.70	–	5.11	m	8.81
38 mm × 200 mm	–	0.17	3.93	–	5.87	m	9.80
47 mm × 50 mm	–	0.11	2.54	–	1.52	m	4.06
47 mm × 75 mm	–	0.14	3.23	–	2.08	m	5.31
47 mm × 100 mm	–	0.17	3.93	–	2.68	m	6.61
47 mm × 125 mm	–	0.17	3.93	–	3.43	m	7.36
47 mm × 150 mm	–	0.19	4.40	–	3.96	m	8.36
47 mm × 175 mm	–	0.19	4.40	–	4.67	m	9.07
47 mm × 200 mm	–	0.19	4.40	–	5.10	m	9.50
47 mm × 225 mm	–	0.19	4.40	–	5.83	m	10.23
75 mm × 100 mm	–	0.23	5.33	–	6.09	m	11.42
75 mm × 125 mm	–	0.23	5.33	–	7.46	m	12.79
75 mm × 150 mm	–	0.23	5.33	–	8.18	m	13.51
100 mm × 150 mm	–	0.28	6.48	–	10.62	m	17.10
100 mm × 175 mm	–	0.28	6.48	–	12.38	m	18.86
100 mm × 200 mm	–	0.28	6.48	–	14.07	m	20.55
100 mm × 225 mm	–	0.31	7.17	–	15.82	m	22.99
100 mm × 250 mm	–	0.31	7.17	–	17.65	m	24.82

16 CARPENTRY

Item	PC £	Labour hours	Labour £	Plant £	Material £	Unit	Total rate £
16.01 PRIMARY AND STRUCTURAL TIMBERS – cont							
Sawn softwood; tanalized – cont							
Plates							
38 mm × 75 mm	–	0.11	2.54	–	2.53	m	**5.07**
38 mm × 100 mm	–	0.14	3.23	–	3.09	m	**6.32**
47 mm × 75 mm	–	0.14	3.23	–	2.08	m	**5.31**
47 mm × 100 mm	–	0.17	3.93	–	2.68	m	**6.61**
75 mm × 100 mm	–	0.19	4.40	–	6.09	m	**10.49**
75 mm × 125 mm	–	0.22	5.09	–	7.41	m	**12.50**
75 mm × 150 mm	–	0.25	5.79	–	8.13	m	**13.92**
Plates; fixing by bolting							
38 mm × 75 mm	–	0.20	4.62	–	2.46	m	**7.08**
38 mm × 100 mm	–	0.23	5.33	–	3.09	m	**8.42**
47 mm × 75 mm	–	0.23	5.33	–	2.08	m	**7.41**
47 mm × 100 mm	–	0.26	6.02	–	2.68	m	**8.70**
75 mm × 100 mm	–	0.29	6.71	–	6.09	m	**12.80**
75 mm × 125 mm	–	0.31	7.17	–	7.41	m	**14.58**
75 mm × 150 mm	–	0.34	7.87	–	8.13	m	**16.00**
Joist strutting; herringbone							
47 mm × 50 mm; depth of joist 150 mm	–	0.46	10.64	–	3.70	m	**14.34**
47 mm × 50 mm; depth of joist 175 mm	–	0.46	10.64	–	3.78	m	**14.42**
47 mm × 50 mm; depth of joist 200 mm	–	0.46	10.64	–	3.84	m	**14.48**
47 mm × 50 mm; depth of joist 225 mm	–	0.46	10.64	–	3.91	m	**14.55**
47 mm × 50 mm; depth of joist 250 mm	–	0.46	10.64	–	3.99	m	**14.63**
Joist strutting; block							
47 mm × 150 mm; depth of joist 150 mm	–	0.28	6.48	–	4.65	m	**11.13**
47 mm × 175 mm; depth of joist 175 mm	–	0.28	6.48	–	5.36	m	**11.84**
47 mm × 200 mm; depth of joist 200 mm	–	0.28	6.48	–	5.79	m	**12.27**
47 mm × 225 mm; depth of joist 225 mm	–	0.28	6.48	–	6.52	m	**13.00**
47 mm × 250 mm; depth of joist 250 mm	–	0.28	6.48	–	7.24	m	**13.72**
Cleats							
225 mm × 100 mm × 75 mm	–	0.19	4.40	–	1.23	nr	**5.63**
Extra for stress grading to above timbers							
general structural (GS) grade	–	–	–	–	43.32	m³	**43.32**
special structural (SS) grade	–	–	–	–	86.62	m³	**86.62**
Extra for protecting and flameproofing timber with							
Celgard CF protection or other equal and approved							
small sections	–	–	–	–	152.48	m³	**152.48**
large sections	–	–	–	–	147.24	m³	**147.24**
Wrot surfaces							
plain; 50 mm wide	–	0.02	0.46	–	–	m	**0.46**
plain; 100 mm wide	–	0.03	0.69	–	–	m	**0.69**
plain; 150 mm wide	–	0.04	0.93	–	–	m	**0.93**

16 CARPENTRY

Item	PC £	Labour hours	Labour £	Plant £	Material £	Unit	Total rate £
16.02 ENGINEERED OR PREFABRICATED ITEMS							
Trussed rafters, stress graded sawn softwood pressure impregnated; raised through two storeys and fixed in position. Roof trusses are always project specific and prices always need to be obtained from a manufacturer for any particular roof design.							
W type truss (Fink); 22.5° pitch; 450 mm eaves overhang							
5.00 m span	–	1.48	34.24	–	66.49	nr	**100.73**
7.60 m span	–	1.62	37.48	–	88.12	nr	**125.60**
10.00 m span	–	1.85	42.80	–	160.21	nr	**203.01**
W type truss (Fink); 30° pitch; 450 mm eaves overhang							
5.00 m span	–	1.48	34.24	–	66.49	nr	**100.73**
7.60 m span	–	1.62	37.48	–	90.52	nr	**128.00**
10.00 m span	–	1.85	42.80	–	157.81	nr	**200.61**
W type truss (Fink); 45° pitch; 450 mm eaves overhang							
4.60 m span	–	1.48	34.24	–	145.79	nr	**180.03**
7.00 m span	–	1.62	37.48	–	282.77	nr	**320.25**
Mono type truss; 17.5° pitch; 450 mm eaves overhang							
3.30 m span	–	1.30	30.08	–	59.27	nr	**89.35**
5.60 m span	–	1.48	34.24	–	80.90	nr	**115.14**
7.00 m span	–	1.71	39.56	–	92.92	nr	**132.48**
Attic type truss; 45° pitch; 450 mm eaves overhang							
5.00 m span	–	2.91	67.32	–	73.69	nr	**141.01**
7.60 m span	–	3.05	70.57	–	109.74	nr	**180.31**
9.00 m span	–	3.24	74.96	–	318.81	nr	**393.77**
Glulam timber beams whitewood; pressure impregnated; adhesive; clean planed finish. NB hoisting into position not included							
Laminated roof beams upto 8 m long							
approximate cubic rate for glulam beams material only	–	–	–	–	1030.00	m³	**1030.00**
66 mm × 225 mm	–	0.20	4.62	–	13.60	m	**18.22**
66 mm × 315 mm	–	0.25	5.79	–	22.56	m	**28.35**
90 mm × 315 mm	–	0.40	9.25	–	29.14	m	**38.39**
90 mm × 405 mm	–	0.50	11.57	–	37.52	m	**49.09**
115 mm × 405 mm	–	0.50	11.57	–	47.97	m	**59.54**
115 mm × 495 mm	–	0.75	17.36	–	58.57	m	**75.93**
115 mm × 630 mm	–	0.90	20.83	–	74.60	m	**95.43**
140 mm × 405 mm	–	0.60	13.88	–	58.41	m	**72.29**
140 mm × 495 mm	–	0.75	17.36	–	71.37	m	**88.73**
140 mm × 630 mm	–	1.00	23.13	–	95.33	m	**118.46**

16 CARPENTRY

Item	PC £	Labour hours	Labour £	Plant £	Material £	Unit	Total rate £
16.03 BOARDING TO FLOORS							
Chipboard boarding and flooring							
Boarding to floors; butt joints							
18 mm thick	4.74	0.28	6.48	–	5.79	m²	**12.27**
Boarding to floors; tongued and grooved joints							
18 mm thick	5.97	0.30	6.94	–	7.15	m²	**14.09**
22 mm thick	6.94	0.32	7.41	–	8.22	m²	**15.63**
Acoustic chipboard flooring							
Boarding to floors; tongued and grooved joints							
chipboard on blue bat bearers	–	–	–	–	–	m²	**26.47**
chipboard on New Era levelling system	–	–	–	–	–	m²	**35.75**
Laminated engineered board flooring; 180 or 240 mm face widths; with 6 mm wear surface down to tongue; pre-finished laquered, oiled or untreated							
Boarding to floors; microbevel or square edge							
Country laquered; on 10 mm Pro Foam	–	–	–	–	–	m²	**77.81**
Rustic laquered; on 10 mm Pro Foam	–	–	–	–	–	m²	**73.92**
Plywood flooring							
Boarding to floors; tongued and grooved joints							
18 mm thick	9.84	0.41	9.49	–	11.43	m²	**20.92**
22 mm thick	11.93	0.45	10.41	–	13.75	m²	**24.16**
Strip boarding; wrought softwood							
Boarding to floors; butt joints							
19 mm × 75 mm boards	–	0.56	12.96	–	23.13	m²	**36.09**
19 mm × 125 mm boards	–	0.51	11.80	–	17.79	m²	**29.59**
22 mm × 150 mm boards	–	0.46	10.64	–	19.84	m²	**30.48**
25 mm × 100 mm boards	–	0.51	11.80	–	21.70	m²	**33.50**
25 mm × 150 mm boards	–	0.46	10.64	–	22.07	m²	**32.71**
Boarding to floors; tongued and grooved joints							
19 mm × 75 mm boards	–	0.65	15.04	–	24.94	m²	**39.98**
19 mm × 125 mm boards	–	0.60	13.88	–	20.05	m²	**33.93**
22 mm × 150 mm boards	–	0.56	12.96	–	20.42	m²	**33.38**
25 mm × 100 mm boards	–	0.60	13.88	–	25.84	m²	**39.72**
25 mm × 150 mm boards	–	0.56	12.96	–	24.18	m²	**37.14**
16.04 BOARDING TO CEILINGS							
Plywood (Eastern European); internal quality							
Lining to ceilings 4 mm thick							
over 300 mm wide	2.53	0.46	10.64	–	3.22	m²	**13.86**
not exceeding 300 mm wide	–	0.30	6.94	–	1.00	m	**7.94**
holes for pipes and the like	–	0.02	0.46	–	–	nr	**0.46**

16 CARPENTRY

Item	PC £	Labour hours	Labour £	Plant £	Material £	Unit	Total rate £
Lining to ceilings 6 mm thick							
over 300 mm wide	3.65	0.49	11.34	–	4.47	m²	**15.81**
not exceeding 300 mm wide	–	0.32	7.41	–	1.37	m	**8.78**
holes for pipes and the like	–	0.02	0.46	–	–	nr	**0.46**
Lining to ceilings 12 mm thick							
over 300 mm wide	6.79	0.56	12.96	–	7.94	m²	**20.90**
not exceeding 300 mm wide	–	0.37	8.56	–	2.41	m	**10.97**
holes for pipes and the like	–	0.03	0.69	–	–	nr	**0.69**
Lining to ceilings 18 mm thick							
over 300 mm wide	9.91	0.60	13.88	–	11.40	m²	**25.28**
not exceeding 300 mm wide	–	0.40	9.25	–	3.45	m	**12.70**
holes for pipes and the like	–	0.03	0.69	–	–	nr	**0.69**
Plywood (Eastern European); external quality							
Lining to ceilings 4 mm thick							
over 300 mm wide	6.26	0.46	10.64	–	7.35	m²	**17.99**
not exceeding 300 mm wide	–	0.30	6.94	–	2.24	m	**9.18**
holes for pipes and the like	–	0.02	0.46	–	–	nr	**0.46**
Lining to ceilings 6.5 mm thick							
over 300 mm wide	7.00	0.49	11.34	–	8.18	m²	**19.52**
not exceeding 300 mm wide	–	0.32	7.41	–	2.48	m	**9.89**
holes for pipes and the like	–	0.02	0.46	–	–	nr	**0.46**
Lining to ceilings 9 mm thick							
over 300 mm wide	9.01	0.53	12.26	–	10.40	m²	**22.66**
not exceeding 300 mm wide	–	0.34	7.87	–	3.15	m	**11.02**
holes for pipes and the like	–	0.03	0.69	–	–	nr	**0.69**
Lining to ceilings 12 mm thick							
over 300 mm wide	11.25	0.56	12.96	–	12.89	m²	**25.85**
not exceeding 300 mm wide	–	0.37	8.56	–	3.89	m	**12.45**
holes for pipes and the like	–	0.03	0.69	–	–	nr	**0.69**
Extra over linings fixed with screws	–	0.10	2.32	–	0.08	m²	**2.40**
Strip boarding; wrought softwood							
Boarding to internal ceilings							
12 mm × 100 mm boards	–	0.93	21.52	–	22.89	m²	**44.41**
16 mm × 100 mm boards	–	0.93	21.52	–	24.72	m²	**46.24**
19 mm × 100 mm boards	–	0.93	21.52	–	28.16	m²	**49.68**
19 mm × 125 mm boards	–	0.88	20.36	–	25.85	m²	**46.21**
19 mm × 125 mm boards; chevron pattern	–	1.30	30.08	–	25.85	m²	**55.93**
25 mm × 125 mm boards	–	0.88	20.36	–	22.19	m²	**42.55**
12 mm × 100 mm boards; knotty pine	–	0.93	21.52	–	13.83	m²	**35.35**
Masterboard or other equal; sanded finish							
Lining to ceilings 6 mm thick							
over 300 mm wide	11.93	0.41	9.49	–	13.53	m²	**23.02**
not exceeding 300 mm wide	–	0.25	5.79	–	4.07	m	**9.86**
holes for pipes and the like	–	0.02	0.46	–	–	nr	**0.46**
Lining to ceilings 9 mm thick							
over 300 mm wide	30.20	0.42	9.71	–	33.75	m²	**43.46**
not exceeding 300 mm wide	–	0.27	6.24	–	10.14	m	**16.38**
holes for pipes and the like	–	0.03	0.69	–	–	nr	**0.69**

16 CARPENTRY

Item	PC £	Labour hours	Labour £	Plant £	Material £	Unit	Total rate £
16.04 BOARDING TO CEILINGS – cont							
Supalux or other equal; sanded finish							
Lining to ceilings 6 mm thick							
over 300 mm wide	23.86	0.41	9.49	–	26.74	m²	**36.23**
not exceeding 300 mm wide	–	0.25	5.79	–	8.03	m	**13.82**
holes for pipes and the like	–	0.03	0.69	–	–	nr	**0.69**
Lining to ceilings 9 mm thick							
over 300 mm wide	31.30	0.42	9.71	–	34.98	m²	**44.69**
not exceeding 300 mm wide	–	0.27	6.24	–	10.51	m	**16.75**
holes for pipes and the like	–	0.03	0.69	–	–	nr	**0.69**
Lining to ceilings 12 mm thick							
over 300 mm wide	40.68	0.49	11.34	–	45.36	m²	**56.70**
not exceeding 300 mm wide	–	0.30	6.94	–	13.62	m	**20.56**
holes for pipes and the like	–	0.04	0.93	–	–	nr	**0.93**
Extra over linings fixed with screws	–	0.10	2.32	–	0.08	m²	**2.40**
16.05 BOARDING TO ROOFS							
Plywood; external quality; 18 mm thick							
Boarding to roofs; butt joints							
flat to falls	15.39	0.37	8.56	–	17.57	m²	**26.13**
sloping	15.39	0.40	9.25	–	17.57	m²	**26.82**
vertical	15.39	0.53	12.26	–	17.57	m²	**29.83**
Plywood; external quality; 12 mm thick							
Boarding to roofs; butt joints							
flat to falls	11.23	0.37	8.56	–	12.97	m²	**21.53**
sloping	11.23	0.40	9.25	–	12.97	m²	**22.22**
vertical	11.23	0.53	12.26	–	12.97	m²	**25.23**
Sawn softwood; untreated							
Boarding to roofs; 150 mm wide boards; butt joints							
19 mm thick; flat; over 600 mm wide	–	0.42	9.71	–	15.27	m²	**24.98**
19 mm thick; flat; not exceeding 600 mm wide	–	0.55	12.72	–	9.27	m	**21.99**
19 mm thick; sloping; over 600 mm wide	–	0.46	10.64	–	15.27	m²	**25.91**
19 mm thick; sloping; not exceeding 600 mm wide	–	0.62	14.35	–	9.27	m	**23.62**
19 mm thick; sloping; laid diagonally; over 600 mm wide	–	0.58	13.42	–	15.27	m²	**28.69**
19 mm thick; sloping; laid diagonally; not exceeding 600 mm wide	–	0.75	17.36	–	9.27	m	**26.63**
25 mm thick; flat; over 600 mm wide	–	0.42	9.71	–	24.98	m²	**34.69**
25 mm thick; flat; not exceeding 600 mm wide	–	0.56	12.96	–	15.09	m	**28.05**
25 mm thick; sloping; over 600 mm wide	–	0.46	10.64	–	24.98	m²	**35.62**
25 mm thick; sloping; not exceeding 600 mm wide	–	0.62	14.35	–	15.09	m	**29.44**
25 mm thick; sloping; laid diagonally; over 600 mm wide	–	0.58	13.42	–	24.98	m²	**38.40**
25 mm thick; sloping; laid diagonally; not exceeding 600 mm wide	–	0.74	17.12	–	15.09	m	**32.21**

16 CARPENTRY

Item	PC £	Labour hours	Labour £	Plant £	Material £	Unit	Total rate £
Boarding to tops or cheeks of dormers; 150 mm wide boards; butt joints							
19 mm thick; laid diagonally; over 600 mm wide	–	0.74	17.12	–	15.27	m²	**32.39**
19 mm thick; laid diagonally; not exceeding 600 mm wide	–	0.92	21.28	–	9.27	m	**30.55**
19 mm thick; laid diagonally; area not exceeding 1.00 m² irrespective of width	–	0.93	21.52	–	14.58	nr	**36.10**
Sawn softwood; tanalized							
Boarding to roofs; 150 wide boards; butt joints							
19 mm thick; flat; over 600 mm wide	–	0.42	9.71	–	16.72	m²	**26.43**
19 mm thick; flat; not exceeding 600 mm wide	–	0.56	12.96	–	10.14	m	**23.10**
19 mm thick; sloping; over 600 mm wide	–	0.46	10.64	–	16.72	m²	**27.36**
19 mm thick; sloping; not exceeding 600 mm wide	–	0.62	14.35	–	10.14	m	**24.49**
19 mm thick; sloping; laid diagonally; over 600 mm wide	–	0.58	13.42	–	16.72	m²	**30.14**
19 mm thick; sloping; laid diagonally; not exceeding 600 mm wide	–	0.74	17.12	–	10.14	m	**27.26**
25 mm thick; flat; over 600 mm wide	–	0.42	9.71	–	26.87	m²	**36.58**
25 mm thick; flat; not exceeding 600 mm wide	–	0.56	12.96	–	15.97	m	**28.93**
25 mm thick; sloping; over 600 mm wide	–	0.46	10.64	–	26.87	m²	**37.51**
25 mm thick; sloping; not exceeding 600 mm wide	–	0.62	14.35	–	15.97	m	**30.32**
25 mm thick; sloping; laid diagonally; over 600 mm wide	–	0.58	13.42	–	26.87	m²	**40.29**
25 mm thick; sloping; laid diagonally; not exceeding 600 mm wide	–	0.74	17.12	–	15.97	m	**33.09**
Boarding to tops or cheeks of dormers; 150 mm wide boards; butt joints							
19 mm thick; laid diagonally; over 600 mm wide	–	0.74	17.12	–	16.72	m²	**33.84**
19 mm thick; laid diagonally; not exceeding 600 mm wide	–	0.92	21.28	–	10.14	m	**31.42**
19 mm thick; laid diagonally; area not exceeding 1.00 m² irrespective of width	–	0.93	21.52	–	16.02	nr	**37.54**
Wrought softwood							
Boarding to roofs; tongued and grooved joints							
19 mm thick; flat to falls	–	0.51	11.80	–	23.68	m²	**35.48**
19 mm thick; sloping	–	0.56	12.96	–	23.68	m²	**36.64**
19 mm thick; sloping; laid diagonally	–	0.72	16.66	–	23.68	m²	**40.34**
25 mm thick; flat to falls	–	0.51	11.80	–	23.83	m²	**35.63**
25 mm thick; sloping	–	0.56	12.96	–	23.83	m²	**36.79**
Boarding to tops or cheeks of dormers; tongued and grooved joints							
19 mm thick; laid diagonally	–	0.93	21.52	–	23.68	m²	**45.20**
Wrought softwood; tanalized							
Boarding to roofs; tongued and grooved joints							
19 mm thick; flat to falls	–	0.51	11.80	–	25.12	m²	**36.92**
19 mm thick; sloping	–	0.56	12.96	–	25.12	m²	**38.08**
19 mm thick; sloping; laid diagonally	–	0.72	16.66	–	25.12	m²	**41.78**
25 mm thick; flat to falls	–	0.51	11.80	–	25.73	m²	**37.53**
25 mm thick; sloping	–	0.56	12.96	–	25.73	m²	**38.69**

16 CARPENTRY

Item	PC £	Labour hours	Labour £	Plant £	Material £	Unit	Total rate £
16.05 BOARDING TO ROOFS – cont							
Wrought softwood; tanalized – cont							
Boarding to tops or cheeks of dormers; tongued and grooved joints							
19 mm thick; laid diagonally	–	0.93	21.52	–	25.12	m²	**46.64**
Masterboard or other equal and approved; 6 mm thick							
Eaves, verge soffit boards, fascia boards and the like							
over 300 mm wide	11.93	0.65	15.04	–	14.28	m²	**29.32**
75 mm wide	–	0.19	4.40	–	1.10	m	**5.50**
150 mm wide	–	0.22	5.09	–	2.14	m	**7.23**
225 mm wide	–	0.26	6.02	–	3.18	m	**9.20**
300 mm wide	–	0.28	6.48	–	4.23	m	**10.71**
Plywood; external quality; 12 mm thick							
Eaves, verge soffit boards, fascia boards and the like							
over 300 mm wide	11.23	0.76	17.58	–	13.50	m²	**31.08**
75 mm wide	–	0.23	5.33	–	1.04	m	**6.37**
150 mm wide	–	0.27	6.24	–	2.03	m	**8.27**
225 mm wide	–	0.31	7.17	–	3.01	m	**10.18**
300 mm wide	–	0.34	7.87	–	4.00	m	**11.87**
Plywood; external quality; 15 mm thick							
Eaves, verge soffit boards, fascia boards and the like							
over 300 mm wide	13.07	0.76	17.58	–	15.53	m²	**33.11**
75 mm wide	–	0.23	5.33	–	1.19	m	**6.52**
150 mm wide	–	0.27	6.24	–	2.33	m	**8.57**
225 mm wide	–	0.31	7.17	–	3.47	m	**10.64**
300 mm wide	–	0.34	7.87	–	4.60	m	**12.47**
Plywood; external quality; 18 mm thick							
Eaves, verge soffit boards, fascia boards and the like							
over 300 mm wide	15.39	0.76	17.58	–	18.10	m²	**35.68**
75 mm wide	–	0.23	5.33	–	1.38	m	**6.71**
150 mm wide	–	0.27	6.24	–	2.72	m	**8.96**
225 mm wide	–	0.31	7.17	–	4.05	m	**11.22**
300 mm wide	–	0.34	7.87	–	5.38	m	**13.25**
Plywood; marine quality; 18 mm thick							
Gutter boards; butt joints							
over 300 mm wide	12.99	0.86	19.90	–	15.44	m²	**35.34**
150 mm wide	–	0.31	7.17	–	2.32	m	**9.49**
225 mm wide	–	0.34	7.87	–	3.50	m	**11.37**
300 mm wide	–	0.38	8.80	–	4.64	m	**13.44**

16 CARPENTRY

Item	PC £	Labour hours	Labour £	Plant £	Material £	Unit	Total rate £
Eaves, verge soffit boards, fascias boards and the like							
over 300 mm wide	12.99	0.76	17.58	–	15.44	m²	**33.02**
75 mm wide	–	0.23	5.33	–	1.18	m	**6.51**
150 mm wide	–	0.27	6.24	–	2.32	m	**8.56**
225 mm wide	–	0.31	7.17	–	3.45	m	**10.62**
300 mm wide	–	0.34	7.87	–	4.58	m	**12.45**
Plywood; marine quality; 25 mm thick							
Gutter boards; butt joints							
over 300 mm wide	18.04	0.93	21.52	–	21.04	m²	**42.56**
150 mm wide	–	0.32	7.41	–	3.16	m	**10.57**
225 mm wide	–	0.37	8.56	–	4.76	m	**13.32**
300 mm wide	–	0.42	9.71	–	6.31	m	**16.02**
Eaves, verge soffit boards, fascia baords and the like							
over 300 mm wide	18.04	0.81	18.74	–	21.04	m²	**39.78**
75 mm wide	–	0.24	5.55	–	1.61	m	**7.16**
150 mm wide	–	0.29	6.71	–	3.16	m	**9.87**
225 mm wide	–	0.29	6.71	–	4.71	m	**11.42**
300 mm wide	–	0.37	8.56	–	6.26	m	**14.82**
Sawn softwood; untreated							
Gutter boards; butt joints							
19 mm thick; 150 mm wide; sloping	–	1.16	26.84	–	16.08	m²	**42.92**
19 mm thick; 75 mm wide	–	0.32	7.41	–	1.25	m	**8.66**
19 mm thick; 150 mm wide	–	0.37	8.56	–	2.35	m	**10.91**
19 mm thick; 225 mm wide	–	0.42	9.71	–	4.28	m	**13.99**
25 mm thick; sloping	–	1.16	26.84	–	25.78	m²	**52.62**
25 mm thick; 75 mm wide	–	0.32	7.41	–	1.58	m	**8.99**
25 mm thick; 150 mm wide	–	0.37	8.56	–	3.80	m	**12.36**
25 mm thick; 225 mm wide	–	0.42	9.71	–	5.99	m	**15.70**
Cesspools with 25 mm thick sides and bottom							
225 mm × 225 mm × 150 mm	–	1.11	25.68	–	5.04	nr	**30.72**
300 mm × 300 mm × 150 mm	–	1.30	30.08	–	6.61	nr	**36.69**
Individual supports; firrings							
50 mm wide × 36 mm average depth	–	0.14	3.23	–	3.32	m	**6.55**
50 mm wide × 50 mm average depth	–	0.14	3.23	–	5.08	m	**8.31**
50 mm wide × 75 mm average depth	–	0.14	3.23	–	6.63	m	**9.86**
Individual supports; bearers							
25 mm × 50 mm	–	0.09	2.08	–	1.39	m	**3.47**
38 mm × 50 mm	–	0.09	2.08	–	1.94	m	**4.02**
50 mm × 50 mm	–	0.09	2.08	–	1.44	m	**3.52**
50 mm × 75 mm	–	0.09	2.08	–	1.90	m	**3.98**
Individual supports; angle fillets							
38 mm × 38 mm	–	0.09	2.08	–	1.33	m	**3.41**
50 mm × 50 mm	–	0.09	2.08	–	1.69	m	**3.77**
75 mm × 75 mm	–	0.11	2.54	–	3.52	m	**6.06**

Prices for Measured Works

16 CARPENTRY

Item	PC £	Labour hours	Labour £	Plant £	Material £	Unit	Total rate £
16.05 BOARDING TO ROOFS – cont							
Sawn softwood; untreated – cont							
Individual supports; tilting fillets							
19 mm × 38 mm	–	0.09	2.08	–	0.78	m	**2.86**
25 mm × 50 mm	–	0.09	2.08	–	1.30	m	**3.38**
38 mm × 75 mm	–	0.09	2.08	–	2.00	m	**4.08**
50 mm × 75 mm	–	0.09	2.08	–	2.59	m	**4.67**
75 mm × 100 mm	–	0.14	3.23	–	4.82	m	**8.05**
Individual supports; grounds or battens							
13 mm × 19 mm	–	0.04	0.93	–	0.64	m	**1.57**
13 mm × 32 mm	–	0.04	0.93	–	0.64	m	**1.57**
25 mm × 50 mm	–	0.04	0.93	–	1.28	m	**2.21**
Individual supports; grounds or battens; plugged and screwed							
13 mm × 19 mm	–	0.14	3.23	–	0.62	m	**3.85**
13 mm × 32 mm	–	0.14	3.23	–	0.62	m	**3.85**
25 mm × 50 mm	–	0.14	3.23	–	1.26	m	**4.49**
Framed supports; open-spaced grounds or battens; at 300 mm centres one way							
25 mm × 50 mm	–	0.14	3.23	–	4.23	m²	**7.46**
25 mm × 50 mm; plugged and screwed	–	0.42	9.71	–	4.20	m²	**13.91**
Framed supports; at 300 mm centres one way and 600 mm centres the other way							
25 mm × 50 mm	–	0.69	15.97	–	6.34	m²	**22.31**
38 mm × 50 mm	–	0.69	15.97	–	9.07	m²	**25.04**
50 mm × 50 mm	–	0.69	15.97	–	6.61	m²	**22.58**
50 mm × 75 mm	–	0.69	15.97	–	8.90	m²	**24.87**
75 mm × 75 mm	–	0.69	15.97	–	20.42	m²	**36.39**
Framed supports; at 300 mm centres one way and 600 mm centres the other way; plugged and screwed							
25 mm × 50 mm	–	1.16	26.84	–	6.49	m²	**33.33**
38 mm × 50 mm	–	1.16	26.84	–	9.22	m²	**36.06**
50 mm × 50 mm	–	1.16	26.84	–	6.76	m²	**33.60**
50 mm × 75 mm	–	1.16	26.84	–	9.05	m²	**35.89**
75 mm × 75 mm	–	1.16	26.84	–	20.57	m²	**47.41**
Framed supports; at 500 mm centres both ways							
25 mm × 50 mm; to bath panels	–	0.83	19.20	–	8.26	m²	**27.46**
Framed supports; as bracketing and cradling around steelwork							
25 mm × 50 mm	–	1.30	30.08	–	8.96	m²	**39.04**
50 mm × 50 mm	–	1.39	32.16	–	9.33	m²	**41.49**
50 mm × 75 mm	–	1.48	34.24	–	12.54	m²	**46.78**
Sawn softwood; tanalized							
Gutter boards; butt joints							
19 mm thick; 150 mm; sloping	–	1.16	26.84	–	17.51	m²	**44.35**
19 mm thick; 75 mm wide	–	0.32	7.41	–	1.35	m	**8.76**
19 mm thick; 150 mm wide	–	0.37	8.56	–	2.56	m	**11.12**
19 mm thick; 225 mm wide	–	0.42	9.71	–	4.61	m	**14.32**
25 mm thick; sloping	–	1.16	26.84	–	27.68	m²	**54.52**

16 CARPENTRY

Item	PC £	Labour hours	Labour £	Plant £	Material £	Unit	Total rate £
25 mm thick; 75 mm wide	–	0.32	7.41	–	1.72	m	9.13
25 mm thick; 150 mm wide	–	0.37	8.56	–	4.09	m	12.65
25 mm thick; 225 mm wide	–	0.42	9.71	–	6.42	m	16.13
Cesspools with 25 mm thick sides and bottom							
225 mm × 225 mm × 150 mm	–	1.11	25.68	–	5.42	nr	31.10
300 mm × 300 mm × 150 mm	–	1.30	30.08	–	7.13	nr	37.21
Individual supports; firrings							
50 mm wide × 36 mm average depth	–	0.14	3.23	–	3.46	m	6.69
50 mm wide × 50 mm average depth	–	0.14	3.23	–	5.27	m	8.50
50 mm wide × 75 mm average depth	–	0.14	3.23	–	6.92	m	10.15
Individual supports; bearers							
25 mm × 50 mm	–	0.09	2.08	–	1.48	m	3.56
38 mm × 50 mm	–	0.09	2.08	–	2.07	m	4.15
50 mm × 50 mm	–	0.09	2.08	–	1.63	m	3.71
50 mm × 75 mm	–	0.09	2.08	–	2.18	m	4.26
Individual supports; angle fillets							
38 mm × 38 mm	–	0.09	2.08	–	1.38	m	3.46
50 mm × 50 mm	–	0.09	2.08	–	1.79	m	3.87
75 mm × 75 mm	–	0.11	2.54	–	3.74	m	6.28
Individual supports; tilting fillets							
19 mm × 38 mm	–	0.09	2.08	–	0.81	m	2.89
25 mm × 50 mm	–	0.09	2.08	–	1.34	m	3.42
38 mm × 75 mm	–	0.09	2.08	–	2.11	m	4.19
50 mm × 75 mm	–	0.09	2.08	–	2.73	m	4.81
75 mm × 100 mm	–	0.14	3.23	–	5.11	m	8.34
Individual supports; grounds or battens							
13 mm × 19 mm	–	0.04	0.93	–	0.65	m	1.58
13 mm × 32 mm	–	0.04	0.93	–	0.66	m	1.59
25 mm × 50 mm	–	0.04	0.93	–	1.38	m	2.31
Individual supports; grounds or battens; plugged and screwed							
13 mm × 19 mm	–	0.14	3.23	–	0.63	m	3.86
13 mm × 32 mm	–	0.14	3.23	–	0.64	m	3.87
25 mm × 50 mm	–	0.14	3.23	–	1.36	m	4.59
Framed supports; open-spaced grounds or battens; at 300 mm centres one way							
25 mm × 50 mm	–	0.14	3.23	–	4.54	m²	7.77
25 mm × 50 mm; plugged and screwed	–	0.42	9.71	–	4.52	m²	14.23
Framed supports; at 300 mm centres one way and 600 mm centres the other way							
25 mm × 50 mm	–	0.69	15.97	–	6.83	m²	22.80
38 mm × 50 mm	–	0.69	15.97	–	9.80	m²	25.77
50 mm × 50 mm	–	0.69	15.97	–	7.56	m²	23.53
50 mm × 75 mm	–	0.69	15.97	–	10.33	m²	26.30
75 mm × 75 mm	–	0.69	15.97	–	22.55	m²	38.52
Framed supports; at 300 mm centres one way and 600 mm centres the other way; plugged and screwed							
25 mm × 50 mm	–	1.16	26.84	–	6.97	m²	33.81
38 mm × 50 mm	–	1.16	26.84	–	9.94	m²	36.78
50 mm × 50 mm	–	1.16	26.84	–	7.70	m²	34.54
50 mm × 75 mm	–	1.16	26.84	–	10.48	m²	37.32
75 mm × 75 mm	–	1.16	26.84	–	22.70	m²	49.54

16 CARPENTRY

Item	PC £	Labour hours	Labour £	Plant £	Material £	Unit	Total rate £
16.05 BOARDING TO ROOFS – cont							
Sawn softwood; tanalized – cont							
Framed supports; at 500 mm centres both ways							
25 mm × 50 mm; to bath panels	–	0.83	19.20	–	8.88	m²	**28.08**
Framed supports; as bracketing and cradling around steelwork							
25 mm × 50 mm	–	1.30	30.08	–	9.63	m²	**39.71**
50 mm × 50 mm	–	1.39	32.16	–	10.65	m²	**42.81**
50 mm × 75 mm	–	1.48	34.24	–	14.53	m²	**48.77**
Wrought softwood							
Gutter boards; tongued and grooved joints							
19 mm thick; 150 mm; sloping	–	1.39	32.16	–	24.48	m²	**56.64**
19 mm thick; 75 mm wide	–	0.37	8.56	–	1.77	m	**10.33**
19 mm thick; 150 mm wide	–	0.42	9.71	–	3.61	m	**13.32**
19 mm thick; 225 mm wide	–	0.46	10.64	–	5.27	m	**15.91**
25 mm thick; sloping	–	1.39	32.16	–	25.16	m²	**57.32**
25 mm thick; 75 mm wide	–	0.37	8.56	–	1.96	m	**10.52**
25 mm thick; 150 mm wide	–	0.42	9.71	–	3.57	m	**13.28**
25 mm thick; 225 mm wide	–	0.46	10.64	–	5.38	m	**16.02**
Eaves, verge soffit boards, fascia boards and the like							
19 mm thick; over 300 mm wide	–	1.15	26.60	–	24.73	m²	**51.33**
19 mm thick; 150 mm wide; once grooved	–	0.19	4.40	–	4.35	m	**8.75**
25 mm thick; 150 mm wide; once grooved	–	0.19	4.40	–	5.30	m	**9.70**
25 mm thick; 175 mm wide; once grooved	–	0.19	4.40	–	5.21	m	**9.61**
32 mm thick; 225 mm wide; once grooved	–	0.23	5.33	–	8.22	m	**13.55**
Wrought softwood; tanalized							
Gutter boards; tongued and grooved joints							
19 mm thick; 150 mm; sloping	–	1.39	32.16	–	25.91	m²	**58.07**
19 mm thick; 75 mm wide	–	0.37	8.56	–	1.88	m	**10.44**
19 mm thick; 150 mm wide	–	0.42	9.71	–	3.82	m	**13.53**
19 mm thick; 225 mm wide	–	0.46	10.64	–	5.60	m	**16.24**
25 mm thick; sloping	–	1.39	32.16	–	27.06	m²	**59.22**
25 mm thick; 75 mm wide	–	0.37	8.56	–	2.10	m	**10.66**
25 mm thick; 150 mm wide	–	0.42	9.71	–	3.86	m	**13.57**
25 mm thick; 225 mm wide	–	0.46	10.64	–	5.80	m	**16.44**
Eaves, verge soffit boards, fascia boards and the like							
19 mm thick; over 300 mm wide	–	1.15	26.60	–	26.17	m²	**52.77**
19 mm thick; 150 mm wide; once grooved	–	0.19	4.40	–	4.57	m	**8.97**
25 mm thick; 150 mm wide; once grooved	–	0.19	4.40	–	5.59	m	**9.99**
25 mm thick; 175 mm wide; once grooved	–	0.20	4.62	–	5.54	m	**10.16**
32 mm thick; 225 mm wide; once grooved	–	0.23	5.33	–	8.76	m	**14.09**

16 CARPENTRY

Item	PC £	Labour hours	Labour £	Plant £	Material £	Unit	Total rate £
16.06 CASINGS							
Chipboard (plain)							
Two-sided 15 mm thick pipe casing; to softwood framing (not included)							
300 mm girth	–	0.56	12.96	–	1.33	m	**14.29**
600 mm girth	–	0.65	15.04	–	2.35	m	**17.39**
Three-sided 15 mm thick pipe casing; to softwood framing (not included)							
450 mm girth	–	1.16	26.84	–	2.00	m	**28.84**
900 mm girth	–	1.39	32.16	–	3.57	m	**35.73**
extra for 400 mm × 400 mm removable access panel; brass cups and screws; additional framing	–	0.93	21.52	–	1.59	nr	**23.11**
Plywood (Eastern European); internal quality							
Two-sided 6 mm thick pipe casings; to softwood framing (not included)							
300 mm girth	–	0.74	17.12	–	1.53	m	**18.65**
600 mm girth	–	0.93	21.52	–	2.75	m	**24.27**
Three-sided 6 mm thick pipe casing; to softwood framing (not included)							
450 mm girth	–	1.06	24.52	–	2.30	m	**26.82**
900 mm girth	–	1.25	28.92	–	4.17	m	**33.09**
Plywood (Eastern European); external quality							
Two-sided 6.5 mm thick pipe casings; to softwood framing (not included)							
300 mm girth	–	0.74	17.12	–	2.65	m	**19.77**
600 mm girth	–	0.93	21.52	–	4.97	m	**26.49**
Three-sided 6.5 mm thick pipe casing; to softwood framing (not included)							
450 mm girth	–	1.06	24.52	–	3.97	m	**28.49**
900 mm girth	–	1.25	28.92	–	7.51	m	**36.43**
Two-sided 12 mm thick pipe casing; to softwood framing (not included)							
300 mm girth	–	0.69	15.97	–	4.06	m	**20.03**
600 mm girth	–	0.83	19.20	–	7.80	m	**27.00**
Three-sided 12 mm thick pipe casing; to softwood framing (not included)							
450 mm girth	–	0.93	21.52	–	6.09	m	**27.61**
900 mm girth	–	1.11	25.68	–	11.74	m	**37.42**
extra for 400 mm × 400 mm removable access panel; brass cups and screws; additional framing	–	1.00	23.13	–	1.59	nr	**24.72**
Preformed white melamine faced plywood casings; Pendock Profiles Ltd or other equal; to softwood battens (not included)							
Skirting trunking profile; plain butt joints in the running length							
45 mm × 150 mm; ref TK150	–	0.11	2.54	–	37.62	m	**40.16**
extra for stop end	–	0.04	0.93	–	23.42	nr	**24.35**
extra for external corner	–	0.09	2.08	–	32.37	nr	**34.45**
extra for internal corner	–	0.09	2.08	–	19.65	nr	**21.73**

Prices for Measured Works

16 CARPENTRY

Item	PC £	Labour hours	Labour £	Plant £	Material £	Unit	Total rate £
16.06 CASINGS – cont							
Preformed white melamine faced plywood							
casings – cont							
Casing profiles							
150 mm × 150 mm; ref MX150/150; 5 mm thick	–	0.11	2.54	–	31.36	m	**33.90**
extra for stop end	–	0.04	0.93	–	8.10	nr	**9.03**
extra for external corner	–	0.09	2.08	–	49.54	nr	**51.62**
extra for internal corner	–	0.09	2.08	–	19.65	nr	**21.73**
16.07 METAL FIXINGS, FIXINGS, FASTENINGS							
AND FITTINGS							
Straps; mild steel; galvanized							
Standard twisted vertical restraint; fixing to softwood							
and brick or blockwork							
27.5 mm × 2.5 mm × 400 mm girth	–	0.23	5.33	–	1.55	nr	**6.88**
27.5 mm × 2.5 mm × 600 mm girth	–	0.24	5.55	–	2.14	nr	**7.69**
27.5 mm × 2.5 mm × 800 mm girth	–	0.25	5.79	–	3.09	nr	**8.88**
27.5 mm × 2.5 mm × 1000 mm girth	–	0.28	6.48	–	4.01	nr	**10.49**
27.5 mm × 2.5 mm × 1200 mm girth	–	–	–	–	–	nr	**–**
Hangers; mild steel; galvanized							
Joist hangers 0.90 mm thick; The Expanded Metal							
Company Ltd Speedy or other equal and approved;							
for fixing to softwood; joist sizes							
50 mm wide; all sizes to 225 mm deep	2.02	0.11	2.54	–	2.27	nr	**4.81**
75 mm wide; all sizes to 225 mm deep	2.11	0.14	3.23	–	2.46	nr	**5.69**
100 mm wide; all sizes to 225 mm deep	2.27	0.17	3.93	–	2.72	nr	**6.65**
Joist hangers 2.50 mm thick; for building in; joist sizes							
50 mm × 150 mm	1.53	0.09	2.07	–	1.84	nr	**3.91**
50 mm × 175 mm	1.51	0.09	2.07	–	1.82	nr	**3.89**
50 mm × 200 mm	1.75	0.11	2.53	–	2.05	nr	**4.58**
50 mm × 225 mm	1.76	0.11	2.53	–	2.16	nr	**4.69**
75 mm × 150 mm	2.66	0.09	2.07	–	3.07	nr	**5.14**
75 mm × 175 mm	2.47	0.09	2.07	–	2.86	nr	**4.93**
75 mm × 200 mm	2.62	0.11	2.53	–	3.09	nr	**5.62**
75 mm × 225 mm	2.82	0.11	2.53	–	3.30	nr	**5.83**
75 mm × 250 mm	2.82	0.13	3.00	–	3.37	nr	**6.37**
100 mm × 200 mm	8.24	0.11	2.53	–	9.17	nr	**11.70**
Metal connectors; mild steel; galvanized							
Round toothed plate; for 10 mm or 12 mm dia. bolts							
38 mm dia.; single sided	–	0.01	0.23	–	0.56	nr	**0.79**
38 mm dia.; double sided	–	0.01	0.23	–	0.62	nr	**0.85**
50 mm dia.; single sided	–	0.01	0.23	–	0.60	nr	**0.83**
50 mm dia.; double sided	–	0.01	0.23	–	0.67	nr	**0.90**
63 mm dia.; single sided	–	0.01	0.23	–	0.88	nr	**1.11**
63 mm dia.; double sided	–	0.01	0.23	–	0.97	nr	**1.20**
75 mm dia.; single sided	–	0.01	0.23	–	1.29	nr	**1.52**
75 mm dia.; double sided	–	0.01	0.23	–	1.34	nr	**1.57**
framing anchor	–	0.14	3.23	–	1.05	nr	**4.28**

16 CARPENTRY

Item	PC £	Labour hours	Labour £	Plant £	Material £	Unit	Total rate £
Bolts; mild steel; galvanized							
Fixing only bolts; 50 mm–200 mm long							
6 mm dia.	–	0.03	0.69	–	–	nr	**0.69**
8 mm dia.	–	0.03	0.69	–	–	nr	**0.69**
10 mm dia.	–	0.04	0.93	–	–	nr	**0.93**
12 mm dia.	–	0.04	0.93	–	–	nr	**0.93**
16 mm dia.	–	0.05	1.15	–	–	nr	**1.15**
20 mm dia.	–	0.05	1.15	–	–	nr	**1.15**
Bolts							
Expanding bolts; Rawlbolt projecting type or other equal and approved; Rawl Fixings; plated; one nut; one washer							
6 mm dia.; ref M6 10P	–	0.09	2.08	–	0.66	nr	**2.74**
6 mm dia.; ref M6 25P	–	0.09	2.08	–	0.77	nr	**2.85**
6 mm dia.; ref M6 60P	–	0.09	2.08	–	0.77	nr	**2.85**
8 mm dia.; ref M8 25P	–	0.09	2.08	–	0.76	nr	**2.84**
8 mm dia.; ref M8 60P	–	0.09	2.08	–	0.77	nr	**2.85**
10 mm dia.; ref M10 15P	–	0.09	2.08	–	1.02	nr	**3.10**
10 mm dia.; ref M10 30P	–	0.09	2.08	–	1.07	nr	**3.15**
10 mm dia.; ref M10 60P	–	0.09	2.08	–	1.04	nr	**3.12**
12 mm dia.; ref M12 15P	–	0.09	2.08	–	1.88	nr	**3.96**
12 mm dia.; ref M12 30P	–	0.10	2.32	–	0.16	nr	**2.48**
12 mm dia.; ref M12 75P	–	0.09	2.08	–	0.22	nr	**2.30**
16 mm dia.; ref M16 35P	–	0.09	2.08	–	4.01	nr	**6.09**
16 mm dia.; ref M16 75P	–	0.09	2.08	–	4.14	nr	**6.22**
Expanding bolts; Rawlbolt loose bolt type or other equal; Rawl Fixings; plated; one bolt; one washer							
6 mm dia.; ref M6 10L	–	0.09	2.08	–	0.71	nr	**2.79**
6 mm dia.; ref M6 25L	–	0.09	2.08	–	0.71	nr	**2.79**
6 mm dia.; ref M6 40L	–	0.09	2.08	–	0.83	nr	**2.91**
8 mm dia.; ref M8 25L	–	0.09	2.08	–	0.92	nr	**3.00**
8 mm dia.; ref M8 40L	–	0.09	2.08	–	0.96	nr	**3.04**
10 mm dia.; ref M10 10L	–	0.09	2.08	–	0.97	nr	**3.05**
10 mm dia.; ref M10 25L	–	0.09	2.08	–	1.08	nr	**3.16**
10 mm dia.; ref M10 50L	–	0.09	2.08	–	1.08	nr	**3.16**
10 mm dia.; ref M10 75L	–	0.09	2.08	–	1.18	nr	**3.26**
12 mm dia.; ref M12 10L	–	0.09	2.08	–	1.87	nr	**3.95**
12 mm dia.; ref M12 25L	–	0.09	2.08	–	1.95	nr	**4.03**
12 mm dia.; ref M12 40L	–	0.09	2.08	–	1.93	nr	**4.01**
12 mm dia.; ref M12 60L	–	0.09	2.08	–	2.44	nr	**4.52**
16 mm dia.; ref M16 30L	–	0.09	2.08	–	4.10	nr	**6.18**
16 mm dia.; ref M16 60L	–	0.09	2.08	–	4.79	nr	**6.87**
Truss clips							
Truss clips; fixing to softwood; joist size							
38 mm wide	0.79	0.14	3.23	–	1.19	nr	**4.42**
50 mm wide	0.75	0.14	3.23	–	1.15	nr	**4.38**
Sole plate angles; mild steel galvanized							
Sole plate angle; fixing to softwood and concrete							
112 mm × 40 mm × 76 mm	0.91	0.19	4.40	–	2.46	nr	**6.86**

16 CARPENTRY

Item	PC £	Labour hours	Labour £	Plant £	Material £	Unit	Total rate £
16.07 METAL FIXINGS, FIXINGS, FASTENINGS AND FITTINGS – cont							
Chemical anchors							
R-CAS Spin-in epoxy acrylate capsules and standard studs or other equal and approved; Rawl Fixings; with nuts and washers; drilling masonry							
capsule ref 60–408; stud ref 60–448	–	0.25	5.79	–	1.03	nr	**6.82**
capsule ref 60–410; stud ref 60–454	–	0.28	6.48	–	1.19	nr	**7.67**
capsule ref 60–412; stud ref 60–460	–	0.31	7.17	–	1.62	nr	**8.79**
capsule ref 60–416; stud ref 60–472	–	0.34	7.87	–	2.01	nr	**9.88**
capsule ref 60–420; stud ref 60–478	–	0.36	8.33	–	2.61	nr	**10.94**
capsule ref 60–424; stud ref 60–484	–	0.40	9.25	–	8.32	nr	**17.57**
R-CAS Spin-in epoxy acrylate capsules and stainless steel studs or other equal and approved; Rawl Fixings; with nuts and washers; drilling masonry							
capsule ref 60–408; stud ref 60–905	–	0.25	5.79	–	2.08	nr	**7.87**
capsule ref 60–410; stud ref 60–910	–	0.28	6.48	–	3.66	nr	**10.14**
capsule ref 60–412; stud ref 60–915	–	0.31	7.17	–	4.66	nr	**11.83**
capsule ref 60–416; stud ref 60–920	–	0.34	7.87	–	9.93	nr	**17.80**
capsule ref 60–420; stud ref 60–925	–	0.36	8.33	–	19.29	nr	**27.62**
capsule ref 60–424; stud ref 60–930	–	0.40	9.25	–	34.35	nr	**43.60**
R-CAS Spin-in epoxy acrylate capsules and standard internal threaded sockets or other equal and approved; Rawl Fixings; drilling masonry							
capsule ref 60–408; socket ref 60–650	–	0.25	5.79	–	1.42	nr	**7.21**
capsule ref 60–410; socket ref 60–656	–	0.28	6.48	–	1.47	nr	**7.95**
capsule ref 60–412; socket ref 60–662	–	0.31	7.17	–	1.57	nr	**8.74**
capsule ref 60–416; socket ref 60–668	–	0.34	7.87	–	2.22	nr	**10.09**
capsule ref 60–420; socket ref 60–674	–	0.36	8.33	–	1.63	nr	**9.96**
capsule ref 60–424; socket ref 60–676	–	0.40	9.25	–	5.12	nr	**14.37**
R-CAS Spin-in epoxy acrylate capsules and stainless steel internal threaded sockets or other equal and approved; Rawl Fixings; drilling masonry							
capsule ref 60–408; socket ref 60–943	–	0.25	5.79	–	2.91	nr	**8.70**
capsule ref 60–410; socket ref 60–945	–	0.28	6.48	–	2.97	nr	**9.45**
capsule ref 60–412; socket ref 60–947	–	0.31	7.17	–	3.70	nr	**10.87**
capsule ref 60–416; socket ref 60–949	–	0.34	7.87	–	5.40	nr	**13.27**
capsule ref 60–420; socket ref 60–951	–	0.36	8.33	–	5.72	nr	**14.05**
capsule ref 60–424; socket ref 60–955	–	0.40	9.25	–	13.74	nr	**22.99**
R-CAS Spin-in epoxy acrylate capsules, perforated sleeves and standard studs or other equal and approved; Rawl Fixings; in low density material; with nuts and washers; drilling masonry							
capsule ref 60–408; sleeve ref 60–538; stud ref 60–448	–	0.25	5.79	–	2.56	nr	**8.35**
capsule ref 60–410; sleeve ref 60–544; stud ref 60–454	–	0.28	6.48	–	2.91	nr	**9.39**
capsule ref 60–412; sleeve ref 60–550; stud ref 60–460	–	0.31	7.17	–	3.40	nr	**10.57**
capsule ref 60–416; sleeve ref 60–562; stud ref 60–472	–	0.34	7.87	–	3.80	nr	**11.67**

16 CARPENTRY

Item	PC £	Labour hours	Labour £	Plant £	Material £	Unit	Total rate £
R-CAS Spin-in epoxy acrylate capsules, perforated sleeves and stainless steel studs or other equal and approved; Rawl Fixings; in low density material; with nuts and washers; drilling masonry							
capsule ref 60–408; sleeve ref 60–538; stud ref 60–905	–	0.25	5.79	–	3.62	nr	**9.41**
capsule ref 60–410; sleeve ref 60–544; stud ref 60–910	–	0.28	6.48	–	5.38	nr	**11.86**
capsule ref 60–412; sleeve ref 60–550; stud ref 60–915	–	0.31	7.17	–	6.44	nr	**13.61**
capsule ref 60–416; sleeve ref 60–562; stud ref 60–920	–	0.34	7.87	–	11.72	nr	**19.59**
R-CAS Spin-in epoxy acrylate capsules, perforated sleeves and standard internal threaded sockets or other equal and approved; The Rawlplug Company; in low density material; with nuts and washers; drilling masonry							
capsule ref 60–408; sleeve ref 60–538; socket ref 60–650	–	0.25	5.79	–	2.96	nr	**8.75**
capsule ref 60–410; sleeve ref 60–544; socket ref 60–656	–	0.28	6.48	–	3.19	nr	**9.67**
capsule ref 60–412; sleeve ref 60–550; socket ref 60–662	–	0.31	7.17	–	3.35	nr	**10.52**
R-CAS Spin-in epoxy acrylate capsules, perforated sleeves and stainless steel internal threaded sockets or other equal and approved; The Rawlplug Company; in low density material; drilling masonry							
capsule ref 60–416; sleeve ref 60–562; socket ref 60–668	–	0.34	7.87	–	4.03	nr	**11.90**
capsule ref 60–408; sleeve ref 60–538; socket ref 60–943	–	0.25	5.79	–	4.45	nr	**10.24**
capsule ref 60–410; sleeve ref 60–544; socket ref 60–945	–	0.28	6.48	–	4.69	nr	**11.17**
capsule ref 60–412; sleeve ref 60–550; socket ref 60–947	–	0.31	7.17	–	5.47	nr	**12.64**
capsule ref 60–416; sleeve ref 60–562; socket ref 60–949	–	0.34	7.87	–	7.19	nr	**15.06**

17 SHEET ROOF COVERINGS

Item	PC £	Labour hours	Labour £	Plant £	Material £	Unit	Total rate £
17.01 BUILT UP FELT ROOF COVERINGS							
NOTE: The following items of felt roofing, unless otherwise described, include for conventional lapping, laying and bonding between layers and to base; and laying flat or to falls, crossfalls or to slopes not exceeding 10° – but exclude any insulation etc.							
Reinforced bitumen membranes							
Three layer coverings							
type S1P1 bitumen glass fibre based felt	–	–	–	–	–	m²	17.50
cover with and bed in hot bitumen 13 mm thick stone chippings	–	–	–	–	–	m²	5.08
two base layers type S2P3 bitumen polyester-based felt; top layer type S4P4 polyester-based mineral surfaced felt; 10 mm stone chipping covering; bitumen bonded	–	–	–	–	–	m²	29.80
cover with and bed in hot bitumen 300 mm × 300 mm × 8 mm g.r.p. tiles	–	–	–	–	–	m²	54.21
Skirtings; three layer; top layer mineral surfaced; dressed over tilting fillet; turned into groove							
not exceeding 200 mm girth	–	–	–	–	–	m	13.04
200 mm–400 mm girth	–	–	–	–	–	m	16.11
Coverings to kerbs; three layer							
400 mm–600 mm girth	–	–	–	–	–	m	20.85
Linings to gutters; three layer							
400 mm–600 mm girth	–	–	–	–	–	m	25.32
Collars around pipes and the like; three layer mineral surface; 150 mm high							
not exceeding 55 mm nominal size	–	–	–	–	–	nr	13.84
55 mm–110 mm nominal size	–	–	–	–	–	nr	13.84
Outlets and dishing to gullies							
300 mm dia.	–	–	–	–	–	nr	15.00
Polyester-based roofing systems							
Andersons high performance polyester-based roofing system or other equal							
two layer coverings; first layer HT 125 underlay; second layer HT 350; fully bonded to wood; fibre or cork base	–	–	–	–	–	m²	24.70
top layer mineral surfaced	–	–	–	–	–	m²	2.10
13 mm thick stone chippings	–	–	–	–	–	m²	5.08
third layer of type 3B as underlay for concrete or screeded base	–	–	–	–	–	m²	6.46
working into outlet pipes and the like	–	–	–	–	–	nr	14.99
Skirtings; two layer; top layer mineral surfaced; dressed over tilting fillet; turned into groove							
not exceeding 200 mm girth	–	–	–	–	–	m	127.27
200 mm–400 mm girth	–	–	–	–	–	m	16.47
Coverings to kerbs; two layer							
400 mm–600 mm girth	–	–	–	–	–	m	21.35
Linings to gutters; three layer							
400 mm–600 mm girth	–	–	–	–	–	m	22.94

17 SHEET ROOF COVERINGS

Item	PC £	Labour hours	Labour £	Plant £	Material £	Unit	Total rate £
Collars around pipes and the like; two layer; 150 mm high							
not exceeding 55 mm nominal size	–	–	–	–	–	nr	14.99
55 mm–110 mm nominal size	–	–	–	–	–	nr	14.99
Ruberoid Challenger SBS high performance roofing or other equal							
two layer coverings; first and second layers Ruberglas 120 GP; fully bonded to wood, fibre or cork base	–	–	–	–	–	m²	16.26
top layer mineral surfaced	–	–	–	–	–	m²	5.75
13 mm thick stone chippings	–	–	–	–	–	m²	5.08
third layer of Rubervent 3G as underlay for concrete or screeded base	–	–	–	–	–	m²	6.44
working into outlet pipes and the like	–	–	–	–	–	nr	14.87
Skirtings; two layer; top layer mineral surfaced; dressed over tilting fillet; turned into groove							
not exceeding 200 mm girth	–	–	–	–	–	m	12.41
200 mm–400 mm girth	–	–	–	–	–	m	16.24
Coverings to kerbs; two layer							
400 mm–600 mm girth	–	–	–	–	–	m	21.05
Linings to gutters; three layer							
400 mm–600 mm girth	–	–	–	–	–	m	22.53
Collars around pipes and the like; two layer, 150 mm high							
not exceeding 55 mm nominal size	–	–	–	–	–	nr	14.87
55 mm–110 mm nominal size	–	–	–	–	–	nr	14.87
Ruberfort HP 350 high performance roofing or other equal							
two layer coverings; first layer Ruberfort HP 180; second layer Ruberfort HP 350; fully bonded; to wood; fibre or cork base	–	–	–	–	–	m²	19.16
top layer mineral surfaced	–	–	–	–	–	m²	7.93
13 mm thick stone chippings	–	–	–	–	–	m²	5.08
third layer of Rubervent 3G; as underlay for concrete or screeded base	–	–	–	–	–	m²	6.44
working into outlet pipes and the like	–	–	–	–	–	nr	15.06
Skirtings; two layer; top layer mineral surface; dressed over tilting fillet; turned into groove							
not exceeding 200 mm girth	–	–	–	–	–	m	12.66
200 mm–400 mm girth	–	–	–	–	–	m	16.55
Coverings to kerbs; two layer							
400 mm–600 mm girth	–	–	–	–	–	m	21.47
Linings to gutters; three layer							
400 mm–600 mm girth	–	–	–	–	–	m	27.74
Collars around pipes and the like; two layer; 150 mm high							
not exceeding 55 mm nominal size	–	–	–	–	–	nr	15.06
55 mm–110 mm nominal size	–	–	–	–	–	nr	15.06

17 SHEET ROOF COVERINGS

Item	PC £	Labour hours	Labour £	Plant £	Material £	Unit	Total rate £
17.01 BUILT UP FELT ROOF COVERINGS – cont							
Polyester-based roofing systems – cont							
Ruberoid Superflex Firebloc high performance roofing or other equal (15 year guarantee specification)							
two layer coverings; first layer Superflex 180; second layer Superflex 250; fully bonded to wood; fibre or cork base	–	–	–	–	–	m²	23.69
top layer mineral surfaced	–	–	–	–	–	m²	5.63
13 mm thick stone chippings	–	–	–	–	–	m²	5.08
third layer of Rubervent 3G as underlay for concrete or screeded base	–	–	–	–	–	m²	6.44
working into outlet pipes and the like	–	–	–	–	–	nr	17.10
Skirtings; two layer; top layer mineral surfaced; dressed over tilting fillet; turned into groove							
not exceeding 200 mm girth	–	–	–	–	–	m	14.80
200 mm–400 mm girth	–	–	–	–	–	m	19.52
Coverings to kerbs; two layer							
400 mm–600 mm girth	–	–	–	–	–	m	26.07
Linings to gutters; three layer							
400 mm–600 mm girth	–	–	–	–	–	m	28.22
Collars around pipes and the like; two layer; 150 mm high							
not exceeding 55 mm nominal size	–	–	–	–	–	nr	17.10
55 mm–110 mm nominal size	–	–	–	–	–	nr	17.10
Ruberoid Ultra Prevent high performance roofing or other equal							
two layer coverings; first layer Ultra prevENt underlay; second layer Ultra prevENt mineral surface cap sheet	–	–	–	–	–	m²	43.60
extra over for							
third layer of Rubervent 3G as underlay for concrete or screeded base	–	–	–	–	–	m²	6.44
working into outlet pipes and the like	–	–	–	–	–	nr	20.68
Skirtings; two layer; dressed over tilting fillet; turned into groove							
not exceeding 200 mm girth	–	–	–	–	–	m	18.52
200 mm–400 mm girth	–	–	–	–	–	m	24.67
Coverings to kerbs; two layer							
400 mm–600 mm girth	–	–	–	–	–	m	34.06
Linings to gutters; three layer							
400 mm–600 mm girth	–	–	–	–	–	m	35.60
Collars around pipes and the like; two layer; 150 mm high							
not exceeding 55 mm nominal size	–	–	–	–	–	nr	20.64
55 mm–110 mm nominal size	–	–	–	–	–	nr	20.64

17 SHEET ROOF COVERINGS

Item	PC £	Labour hours	Labour £	Plant £	Material £	Unit	Total rate £
Accessories							
Eaves trim; extruded aluminium alloy; working felt into trim							
Rubertrim; type FL/G; 65 mm face	–	–	–	–	–	m	**16.69**
extra for external angle	–	–	–	–	–	nr	**16.80**
Roof screed ventilator – aluminium alloy							
Extr-aqua-vent or other equal and approved – set on screed over and including dished sinking and collar	–	–	–	–	–	nr	**52.80**
Insulation board underlays							
Vapour barrier							
reinforced; metal lined	–	–	–	–	–	m²	**15.96**
Rockwool; Duorock flat insulation board							
140 mm thick (0.25 U-value)	–	–	–	–	–	m²	**43.74**
Kingspan Thermaroof TR21 zero OPD urethene insulation board							
50 mm thick	–	–	–	–	–	m²	**26.88**
90 mm thick	–	–	–	–	–	m²	**46.32**
100 mm thick (0.25 U-value)	–	–	–	–	–	m²	**51.45**
Wood fibre boards; impregnated; density 220–350 kg/m³							
12.70 mm thick	–	–	–	–	–	m²	**7.25**
Tapered insulation board underlays							
Tapered insulation £/m² prices can vary dramatically depending upon the factors which determine the scheme layout; these primarily being gutter/outlet locations and the length of fall involved. The following guide assumes a U-value of 0.18 W/m² K as a benchmark.							
As the required insulation value will vary from project to project the required U-value should be determined by calculating the buildings energy consumption at the design stage. Due to tapered insulation scheme prices varying by project, the following prices are indicative. Please contact a specialist for a project specific quotation. U-value must be calculated in accordance with BSENISO 6946:2007 Annex C							
Tapered PIR (Polyisocyanurate) boards; bedded in hot bitumen							
effective thickness achieving 0.18 W/m² K	32.46	–	–	–	–	m²	**68.66**
minimum thickness achieving 0.18 W/m² K	38.61	–	–	–	–	m²	**76.22**
Tapered PIR (Polyisocyanurate) board; mechanically fastened							
effective thickness achieving 0.18 W/m² K	32.46	–	–	–	–	m²	**71.67**
Tapered Aspire; Hybrid EPS/PIR boards; bedded in hot bitumen							
effective thickness achieving 0.18 W/m² K	30.21	–	–	–	–	m²	**68.66**
minimum thickness achieving 0.18 W/m² K	36.37	–	–	–	–	m²	**76.22**

17 SHEET ROOF COVERINGS

Item	PC £	Labour hours	Labour £	Plant £	Material £	Unit	Total rate £
17.01 BUILT UP FELT ROOF COVERINGS – cont							
Tapered insulation board underlays – cont							
Tapered PIR (Polyisocyanurate) board;							
mechanically fastened							
effective thickness achieving 0.18 W/m² K	30.21	–	–	–	–	m²	**71.67**
minimum thickness achieving 0.18 W/m² K	36.37	–	–	–	–	m²	**79.23**
Tapered Rockwool boards; bedded in hot bitumen							
effective thickness achieving 0.18 W/m² K	57.97	–	–	–	–	m²	**101.00**
minimum thickness achieving 0.18 W/m² K	71.63	–	–	–	–	m²	**110.09**
Tapered Rockwool boards; mechanically fastened							
effective thickness achieving 0.18 W/m² K	57.97	–	–	–	–	m²	**104.02**
minimum thickness achieving 0.18 W/m² K	71.63	–	–	–	–	m²	**112.85**
Tapered EPS (Expanded polystyrene) boards;							
bedded in hot bitumen							
effective thickness achieving 0.18 W/m² K	26.53	–	–	–	–	m²	**61.17**
minimum thickness achieving 0.18 W/m² K	32.46	–	–	–	–	m²	**68.66**
Tapered EPS (Expanded polystyrene) boards;							
mechanicaly fastened							
effective thickness achieving 0.18 W/m² K	26.53	–	–	–	–	m²	**64.09**
minimum thickness achieving 0.18 W/m² K	32.46	–	–	–	–	m²	**71.66**
Insulation board overlays							
Dow Roofmate SL extruded polystyrene foam							
boards or other equal and approved; Thermal							
conductivity – 0.028 W/mK							
50 mm thick	–	–	–	–	–	m²	**17.79**
140 mm thick	–	–	–	–	–	m²	**31.06**
160 mm thick	–	–	–	–	–	m²	**33.65**
Dow Roofmate LG extruded polystyrene foam							
boards or other equal and approved; Thermal							
conductivity – 0.028 W/mK							
80 mm thick	–	–	–	–	–	m²	**63.37**
100 mm thick	–	–	–	–	–	m²	**67.91**
120 mm thick	–	–	–	–	–	m²	**72.52**
17.02 FIBRE BITUMEN THERMOPLASTIC SHEET COVERINGS/FLASHINGS							
Glass fibre reinforced bitumen strip slates; Ruberglas 105 or other equal and approved; 1000 mm × 336 mm mineral finish; to external quality plywood boarding (boarding not included)							
Roof coverings	13.03	0.23	5.21	–	14.96	m²	**20.17**
Wall coverings	13.03	0.37	8.38	–	14.96	m²	**23.34**
Extra over coverings for							
double course at eaves; felt soaker	–	0.19	4.31	–	9.99	m	**14.30**
verges; felt soaker	–	0.14	3.17	–	8.29	m	**11.46**
valley slate; cut to shape; felt soaker and cutting both sides	–	0.42	9.52	–	13.03	m	**22.55**

17 SHEET ROOF COVERINGS

Item	PC £	Labour hours	Labour £	Plant £	Material £	Unit	Total rate £
ridge slate; cut to shape	–	0.28	6.34	–	8.29	m	**14.63**
hip slate; cut to shape; felt soaker and cutting both sides	–	0.42	9.52	–	12.95	m	**22.47**
holes for pipes and the like	–	0.48	10.88	–	–	nr	**10.88**
Bostik Findley Flashband Plus sealing strips and flashings or other equal and approved; special grey finish							
Flashings; wedging at top if required; pressure bonded; to walls							
100 mm girth	–	0.23	5.78	–	0.89	m	**6.67**
150 mm girth	–	0.31	7.79	–	1.34	m	**9.13**
225 mm girth	–	0.37	9.30	–	1.87	m	**11.17**
300 mm girth	–	0.42	10.56	–	1.99	m	**12.55**
450 mm girth	–	0.45	11.31	–	3.27	m	**14.58**
600 mm girth	–	0.47	11.94	–	4.08	m	**16.02**
17.03 SINGLE LAYER PLASTIC ROOF COVERINGS							
Kingspan KS1000TD composite single ply roof panels for roof pitches greater than 0.7° (after deflection)							
1.5 mm single ply external covering, internal coating bright white polyester (steel)							
71 mm thick panel; U-value 0.25 W/m^{2}K	–	–	–	–	–	m^2	**51.84**
100 mm thick panel; U-value 0.18 W/m^{2}K	–	–	–	–	–	m^2	**57.34**
120 mm thick panel; U-value 0.15 W/m^{2}K	–	–	–	–	–	m^2	**60.65**
Trocal S PVC roofing or other equal and approved							
Coverings	–	–	–	–	–	m^2	**20.10**
Skirtings; dressed over metal upstands							
not exceeding 200 mm girth	–	–	–	–	–	m	**15.63**
200 mm–400 mm girth	–	–	–	–	–	m	**19.19**
Coverings to kerbs							
400 mm–600 mm girth	–	–	–	–	–	m	**35.15**
Collars around pipes and the like; 150 mm high							
not exceeding 55 mm nominal size	–	–	–	–	–	nr	**10.75**
55 mm–110 mm nominal size	–	–	–	–	–	nr	**10.75**
Trocal metal upstands or other equal and approved							
Sarnafil polymeric waterproofing membrane; cold roof							
Roof coverings							
pitch not exceeding 5°; to metal decking or the like	–	–	–	–	–	m^2	**37.77**
pitch not exceeding 5°; to concrete base or the like; prime concrete with spirrit priming solution	–	–	–	–	–	m^2	**37.77**

17 SHEET ROOF COVERINGS

Item	PC £	Labour hours	Labour £	Plant £	Material £	Unit	Total rate £
17.03 SINGLE LAYER PLASTIC ROOF COVERINGS – cont							
Sarnafil polymeric waterproofing membrane; 1.2 mm thick fleece backed membrane; cold roof							
Roof coverings							
pitch not exceeding 5°; to metal decking or the like	–	–	–	–	–	m²	37.77
pitch not exceeding 5°; to concrete base or the like; prime concrete with spririt priming solution	–	–	–	–	–	m²	37.77
Sarnafil polymeric waterproofing membrane; 120 mm thick Sarnaform G CFC & HCFC free insulation board; vapour control layer; prime concrete with spirit priming solution							
Mechanically fastened system							
Roof coverings							
pitch not exceeding 5°; to metal decking or the like	–	–	–	–	–	m²	57.33
pitch not exceeding 5°; to concrete base or the like	–	–	–	–	–	m²	66.96
Coverings to kerbs; parapet flashing; Sarnatrim 50 mm deep on face 100 mm fixing arm; standard Sarnafil detail 1.1							
not exceeding 200 mm girth	–	–	–	–	–	m	32.15
200 mm–400 mm girth	–	–	–	–	–	m	36.64
400 mm–600 mm girth	–	–	–	–	–	m	41.09
Eaves detail; Sarnatrim drip edge to gutter; standard Sarnafil detail 1.3							
not exceeding 200 mm girth	–	–	–	–	–	m	30.92
Skirtings/Upstands; skirting to brickwork with galvanized steel counter flashing to top edge; standard Sarnafil detail 2.3							
not exceeding 200 mm girth	–	–	–	–	–	m	37.61
200 mm–400 mm girth	–	–	–	–	–	m	45.22
400 mm–600 mm girth	–	–	–	–	–	m	52.93
Skirtings/Upstands; skirting to brickwork with Sarnametal Raglet to chase; standard Sarnafil detail 2.8							
not exceeding 200 mm girth	–	–	–	–	–	m	37.61
200 mm–400 mm girth	–	–	–	–	–	m	45.22
400 mm–600 mm girth	–	–	–	–	–	m	52.93
Collars around pipe standards, and the like							
50 mm dia. × 150 mm high	–	–	–	–	–	nr	48.73
100 mm dia. × 150 mm high	–	–	–	–	–	nr	48.73
Outlets and dishing to gullies							
fix Sarnadrain PVC rainwater outlet; 110 mm dia.; weld membrane to same; fit plastic leafguard	–	–	–	–	–	nr	115.18

17 SHEET ROOF COVERINGS

Item	PC £	Labour hours	Labour £	Plant £	Material £	Unit	Total rate £
Fully adhered system							
Roof coverings							
pitch not exceeding 5°; to metal decking or the like	–	–	–	–	–	m²	63.97
pitch not exceeding 5°; to concrete base or the like	–	–	–	–	–	m²	71.36
Coverings to kerbs; parapet flashing; Sarnatrim 50 mm deep on face 100 mm fixing arm; standard Sarnafil detail 1.1							
not exceeding 200 mm girth	–	–	–	–	–	m	30.61
200 mm–400 mm girth	–	–	–	–	–	m	35.22
400 mm–600 mm girth	–	–	–	–	–	m	39.83
Eaves detail; Sarnametal drip edge to gutter; standard Sarnafil detail 1.3							
not exceeding 200 mm girth	–	–	–	–	–	m	29.32
Skirtings/Upstands; skirting to brickwork with galvanized steel counter flashing to top edge; standard Sarnafil detail 2.3							
not exceeding 200 mm girth	–	–	–	–	–	m	36.21
200 mm–400 mm girth	–	–	–	–	–	m	44.11
400 mm–600 mm girth	–	–	–	–	–	m	49.39
Skirtings/Upstands; skirting to brickwork with Sarnametal Raglet to chase; standard Sarnafil detail 2.8							
not exceeding 200 mm girth	–	–	–	–	–	m	37.53
200 mm–400 mm girth	–	–	–	–	–	m	52.08
400 mm–600 mm girth	–	–	–	–	–	m	52.08
Collars around pipe standards, and the like							
50 mm dia. × 150 mm high	–	–	–	–	–	nr	50.40
100 mm dia. × 150 mm high	–	–	–	–	–	nr	50.40
Outlets and dishing to gullies							
fix Sarnadrain PVC rainwater outlet; 110 mm dia.; weld membrane to same; fit plastic leafguard	–	–	–	–	–	nr	119.15
Options							
extra over for 1.2 mm fleece backed membrane	–	–	–	–	–	m²	6.49
Landscape roofing							
SarnaVert extensive biodiverse roof; sedum blanket; 100 mm growing medium; aquafrain; 1.5 mm thick membrane; 120 mm thick insulation board; vapour control layer							
Pitch not exceeding 5°; to metal decking or the like	–	–	–	–	–	m²	158.31
Pitch not exceeding 5°; to concrete base or the like; prime concrete with spirit priming solution	–	–	–	–	–	m²	164.74
Kerb and eaves; standard Sarnafil details							
Sarnafil kerb; 150 mm above roof level	–	–	–	–	–	m	39.20
Sarnafil eaves detail with gravel stop n.e. 200 mm girth	–	–	–	–	–	m	64.96
Collars around pipes and the like							
50 mm dia. × 150 mm high	–	–	–	–	–	nr	50.40
100 mm dia. × 150 mm high	–	–	–	–	–	nr	50.40
Sarnafil rainwater outlet	–	–	–	–	–	nr	202.75

Prices for Measured Works

17 SHEET ROOF COVERINGS

Item	PC £	Labour hours	Labour £	Plant £	Material £	Unit	Total rate £
17.04 GLASS REINFORCED PLASTIC PANEL							
Glass fibre translucent sheeting grade AB class 3							
Roof cladding; sloping not exceeding 50°; fixing to timber purlins with drive screws; to suit							
Profile 3 or other equal	24.16	0.18	4.08	–	31.32	m²	**35.40**
Profile 6 or other equal	24.16	0.23	5.21	–	31.32	m²	**36.53**
Roof cladding; sloping not exceeding 50°; fixing to timber purlins with hook bolts; to suit							
Profile 3 or other equal	24.16	0.23	5.21	–	32.32	m²	**37.53**
Profile 6 or other equal	24.16	0.28	6.34	–	32.32	m²	**38.66**
Longrib 1000 or other equal	24.16	0.28	6.34	–	32.32	m²	**38.66**
17.05 LEAD SHEET COVERINGS/FLASHINGS							
Milled Lead; BS EN 12588; on and including Geotec underlay							
Roof and dormer coverings							
1.80 mm thick (code 4) roof coverings							
flat (in wood roll construction (Prime Cost £ per kg)	4.17	1.80	45.24	–	51.28	m²	**96.52**
pitched (in wood roll construction)	–	2.00	50.26	–	51.54	m²	**101.80**
pitched (in welded seam construction)	–	1.80	45.24	–	51.28	m²	**96.52**
vertical (in welded seam construction)	–	2.00	50.26	–	48.95	m²	**99.21**
1.80 mm thick (code 4) dormer coverings							
flat (in wood roll construction) (Prime Cost £ per kg)	4.17	1.35	33.93	–	50.70	m²	**84.63**
pitched (in wood roll construction)	–	1.50	37.70	–	50.89	m²	**88.59**
pitched (in welded seam construction)	–	1.35	33.93	–	50.70	m²	**84.63**
vertical (in welded seam construction)	–	3.00	75.40	–	48.95	m²	**124.35**
2.24 mm thick (code 5) roof coverings							
flat (in wood roll construction) (Prime Cost £ per kg)	4.17	1.89	47.50	–	61.81	m²	**109.31**
pitched (in wood roll construction)	–	2.10	52.78	–	62.08	m²	**114.86**
pitched (in welded seam construction)	–	1.89	47.50	–	61.81	m²	**109.31**
vertical (in welded seam construction)	–	2.10	52.78	–	59.37	m²	**112.15**
2.24 mm thick (code 5) dormer coverings							
flat (in wood roll construction) (Prime Cost £ per kg)	4.17	1.42	35.64	–	61.20	m²	**96.84**
pitched (in wood roll construction)	–	1.58	39.60	–	61.41	m²	**101.01**
pitched (in welded seam construction)	–	1.42	35.64	–	61.20	m²	**96.84**
vertical (in welded seam construction)	–	3.15	79.17	–	59.37	m²	**138.54**
2.65 mm thick (code 6) roof coverings							
flat (in wood roll construction) (Prime Cost £ per kg)	4.17	1.98	49.76	–	71.64	m²	**121.40**
pitched (in wood roll construction)	–	2.20	55.29	–	71.92	m²	**127.21**
pitched (in welded seam construction)	–	1.98	49.76	–	71.64	m²	**121.40**
vertical (in welded seam construction)	–	2.20	55.29	–	69.08	m²	**124.37**
2.65 mm thick (code 6) dormer coverings							
flat (in wood roll construction) (Prime Cost £ per kg)	4.17	1.49	37.35	–	71.00	m²	**108.35**
pitched (in wood roll construction)	–	1.65	41.47	–	71.21	m²	**112.68**
pitched (in welded seam construction)	–	1.49	37.35	–	71.00	m²	**108.35**
vertical (in welded seam construction)	–	3.30	82.94	–	69.08	m²	**152.02**

17 SHEET ROOF COVERINGS

Item	PC £	Labour hours	Labour £	Plant £	Material £	Unit	Total rate £
3.15 mm thick (code 7) roof coverings							
flat (in wood roll construction) (Prime Cost £ per kg)	4.17	2.12	53.18	–	83.66	m²	**136.84**
pitched (in wood roll construction)	–	2.35	59.06	–	83.96	m²	**143.02**
pitched (in welded seam construction)	–	2.12	53.18	–	83.66	m²	**136.84**
vertical (in welded seam construction)	–	2.35	59.06	–	80.92	m²	**139.98**
3.15 mm thick (code 7) dormer coverings							
flat (in wood roll construction) (Prime Cost £ per kg)	4.17	1.59	39.86	–	82.97	m²	**122.83**
pitched (in wood roll construction)	–	1.76	44.28	–	83.20	m²	**127.48**
pitched (in welded seam construction)	–	1.59	39.86	–	82.97	m²	**122.83**
vertical (in welded seam construction)	–	3.53	88.61	–	80.92	m²	**169.53**
3.55 mm thick (code 8) roof coverings							
flat (in wood roll construction) (Prime Cost £ per kg)	4.17	2.30	57.70	–	93.37	m²	**151.07**
pitched (in wood roll construction)	–	2.55	64.09	–	93.70	m²	**157.79**
pitched (in welded seam construction)	–	2.30	57.70	–	93.37	m²	**151.07**
vertical (in welded seam construction)	–	2.55	64.09	–	90.40	m²	**154.49**
3.55 mm thick (code 8) dormer coverings							
flat (in wood roll construction) (Prime Cost £ per kg)	4.17	1.72	43.28	–	92.63	m²	**135.91**
pitched (in wood roll construction)	–	1.91	48.05	–	92.87	m²	**140.92**
pitched (in welded seam construction)	–	1.72	43.28	–	92.63	m²	**135.91**
vertical (in welded seam construction)	–	3.83	96.15	–	90.40	m²	**186.55**
Sundries							
patination oil to finished work surfaces	–	0.05	1.26	–	0.25	m²	**1.51**
chalk slurry to underside of panels	–	0.67	16.74	–	2.27	m²	**19.01**
provision of 45 × 45 mm wood rolls at 600 mm centres (per m)	–	0.20	5.03	–	1.07	m	**6.10**
dressing over glazing bars and glass	–	0.50	12.57	–	0.69	m	**13.26**
soldered nail head	–	0.02	0.40	–	0.05	nr	**0.45**
1.32 mm thick (code 3) lead flashings, etc.							
Soakers							
200 × 200 mm	–	0.03	0.75	–	1.15	nr	**1.90**
300 × 300 mm	–	0.03	0.75	–	2.62	nr	**3.37**
1.80 mm thick (code 4) lead flashings, etc.							
Flashings; wedging into grooves							
150 mm girth	–	0.50	12.57	–	6.64	m	**19.21**
200 mm girth	–	0.50	12.57	–	8.85	m	**21.42**
240 mm girth	–	0.50	12.57	–	10.62	m	**23.19**
300 mm girth	–	0.50	12.57	–	13.27	m	**25.84**
Stepped flashings; wedging into grooves							
180 mm girth	–	1.00	25.13	–	7.96	m	**33.09**
270 mm girth	–	1.00	25.13	–	11.95	m	**37.08**
Linings to sloping gutters							
390 mm girth	–	0.80	20.11	–	17.25	m	**37.36**
450 mm girth	–	0.90	22.62	–	19.91	m	**42.53**
600 mm girth	–	1.10	27.65	–	26.54	m	**54.19**
Cappings to hips or ridges							
450 mm girth	–	1.00	25.13	–	19.91	m	**45.04**
600 mm girth	–	1.20	30.16	–	26.54	m	**56.70**
Saddle flashings; at intersections of hips and ridges; dressing and bossing							
450 × 450 mm	–	1.00	25.13	–	11.54	nr	**36.67**
600 × 200 mm	–	1.00	25.13	–	18.51	nr	**43.64**

17 SHEET ROOF COVERINGS

Item	PC £	Labour hours	Labour £	Plant £	Material £	Unit	Total rate £
17.05 LEAD SHEET COVERINGS/FLASHINGS – cont							
1.80 mm thick (code 4) lead flashings – cont							
Slates; with 150 mm high collar							
450 × 450 mm; to suit 50 mm dia. pipe	–	1.50	37.70	–	13.88	nr	**51.58**
450 × 450 mm; to suit 100 mm dia. pipe	–	1.50	37.70	–	14.91	nr	**52.61**
450 × 450 mm; to suit 150 mm dia. pipe	–	1.50	37.70	–	15.96	nr	**53.66**
2.24 mm thick (code 5) lead flashings, etc.							
Flashings; wedging into grooves							
150 mm girth	–	0.50	12.57	–	8.16	m	**20.73**
200 mm girth	–	0.50	12.57	–	10.87	m	**23.44**
240 mm girth	–	0.50	12.57	–	13.05	m	**25.62**
300 mm girth	–	0.50	12.57	–	16.31	m	**28.88**
Stepped flashings; wedging into grooves							
180 mm girth	–	1.00	25.13	–	9.78	m	**34.91**
270 mm girth	–	1.00	25.13	–	14.68	m	**39.81**
Linings to sloping gutters							
390 mm girth	–	0.80	20.11	–	21.20	m	**41.31**
450 mm girth	–	0.90	22.62	–	24.46	m	**47.08**
600 mm girth	–	1.10	27.65	–	32.62	m	**60.27**
Cappings to hips or ridges							
450 mm girth	–	1.00	25.13	–	24.46	m	**49.59**
600 mm girth	–	1.20	30.16	–	32.62	m	**62.78**
Saddle flashings; at intersections of hips and ridges; dressing and bossing							
450 × 450 mm	–	1.00	25.13	–	12.30	nr	**37.43**
600 × 200 mm	–	1.00	25.13	–	20.86	nr	**45.99**
Slates; with 150 mm high collar							
450 × 450 mm; to suit 50 mm dia. pipe	–	1.50	37.70	–	14.23	nr	**51.93**
450 × 450 mm; to suit 100 mm dia. pipe	–	1.50	37.70	–	15.50	nr	**53.20**
450 × 450 mm; to suit 150 mm dia. pipe	–	1.50	37.70	–	16.79	nr	**54.49**
17.06 ALUMINIUM SHEET COVERINGS/ FLASHINGS							
Aluminium roofing; commercial grade; on and including Geotec underlay							
The following rates are based upon nett 'deck' or 'wall' areas							
Roof, dormer and wall coverings							
0.7 mm thick roof coverings; mill finish							
flat (in wood roll construction) (Prime Cost rate £ per kg)	6.56	2.00	50.26	–	25.58	m²	**75.84**
eaves detail ED1	–	0.40	10.05	–	3.21	m	**13.26**
abutment upstands at perimeters	–	0.66	16.58	–	1.28	m	**17.86**
pitched over 3° (in standing seam construction)	–	1.50	37.70	–	21.35	m²	**59.05**
vertical (in angled or flat seam construction)	–	1.60	40.21	–	21.35	m²	**61.56**

17 SHEET ROOF COVERINGS

Item	PC £	Labour hours	Labour £	Plant £	Material £	Unit	Total rate £
0.7 mm thick dormer coverings; mill finish							
flat (in wood roll construction)	–	3.00	75.40	–	25.19	m²	100.59
eaves detail ED1	–	0.40	10.05	–	3.21	m	13.26
pitched over 3° (in standing seam construction)	–	2.50	62.83	–	21.35	m²	84.18
vertical (in angled or flat seam construction)	–	2.70	67.86	–	21.35	m²	89.21
0.7 mm thick roof coverings; Pvf2 finish							
flat (in wood roll construction) (Prime Cost rate £ per kg)	8.33	2.00	50.26	–	30.11	m²	80.37
eaves detail ED1	–	0.40	10.05	–	4.00	m	14.05
abutment upstands at perimeters	–	0.66	16.58	–	1.60	m	18.18
pitched over 3° (in standing seam construction)	–	1.50	37.70	–	26.11	m²	63.81
vertical (in angled or flat seam construction)	–	1.60	40.21	–	26.11	m²	66.32
0.7 mm thick dormer coverings; Pvf2 finish							
flat (in wood roll construction)	–	3.00	75.40	–	30.11	m²	105.51
eaves detail ED1	–	0.40	10.05	–	4.00	m	14.05
pitched over 3° (in standing seam construction)	–	2.50	62.83	–	26.11	m²	88.94
vertical (in angled or flat seam construction)	–	2.70	67.86	–	26.11	m²	93.97
0.7 mm thick aluminium flashings, etc.							
Flashings; wedging into grooves; mill finish							
150 mm girth (PC per kg)	6.56	0.50	12.57	–	1.93	m	14.50
240 mm girth	–	0.50	12.57	–	3.08	m	15.65
300 mm girth	–	0.50	12.57	–	3.86	m	16.43
Stepped flashings; wedging into grooves; mill finish							
180 mm girth	–	1.00	25.13	–	2.31	m	27.44
270 mm girth	–	1.00	25.13	–	3.47	m	28.60
Flashings; wedging into grooves; Pvf2 finish							
150 mm girth (PC per kg)	8.33	0.50	12.57	–	2.40	m	14.97
240 mm girth	–	0.50	12.57	–	3.84	m	16.41
300 mm girth	–	0.50	12.57	–	4.80	m	17.37
Stepped flashings; wedging into grooves; Pvf2 finish							
180 mm girth	–	1.00	25.13	–	2.88	m	28.01
270 mm girth	–	1.00	25.13	–	4.32	m	29.45
Sundries							
provision of square batten roll at 500 mm centres (per m)	–	0.20	5.03	–	1.57	m	6.60
Standing seam aluminium roofing							
Kalzip; 65 mm seam, 400 cover width, Ref BS AW 3004 standard natural aluminium, stucco embossed finish, 0.9 mm thick; ST Clips fixed with stainless steel fasteners; 37 Plus 180 mm Glassfibre Insulation compressed to 165 mm (0.25 U-value); vapour control layer, clear reinforced polyethelyne 530MNs/g all laps sealed; Liner Sheets, profiled steel, 1000 mm cover width, bright white polyester paint finish Ref TR35/200S, 0.7 mm thick, fixed with stainless steel fasteners.							
roof coverings (twin skin construction); pitch not less than 1.5°; fixed to cold rolled purlins (not included	–	–	–	–	–	m²	79.01

17 SHEET ROOF COVERINGS

Item	PC £	Labour hours	Labour £	Plant £	Material £	Unit	Total rate £
17.06 ALUMINIUM SHEET COVERINGS/ FLASHINGS – cont							
Standing seam aluminium roofing – cont							
Eaves details							
40 × 20 mm extruded aluminium drip angle fixed to Kalzip sheet using aluminium blind sealed rivets; black solid rubber eaves filler blocks; ST clips fixed with stainless steel fasteners	–	–	–	–	–	m	**25.60**
0.90 mm thick stucco embossed natural aluminium external eaves closure; 375 mm girth twice bent	–	–	–	–	–	m	**10.74**
0.70 mm thick bright white polyester liner sheet closure internal flashing; 200 mm girth once bent with stainless steel fasteners, black solid rubber profiled liner small flute filler sealed top and bottom with sealant tape	–	–	–	–	–	m	**12.37**
Verge details							
extruded aluminium gable end channel fixed to Kalzip seam using aluminium blind rivets; extruded aluminium gable end clips fixed to St Clips with stainless steel fasteners; extruded aluminium gable tolerence clip hooked over gable end channel	–	–	–	–	–	m	**20.88**
0.90 mm thick stucco embossed natural aluminium external verge closure 600 mm girth four times bent, fixed to extruded aluminium gable tolerance clip and vertical cladding with stainless steel fasteners, black profiled filler blocks to vertical clddding	–	–	–	–	–	m	**20.98**
0.70 mm thick bright white polyester liner sheet closure internal flashing 200 mm girth once bent fixed with stainless steel fasteners, black solid rubber profiled filler blocks sealed top and bottom with sealant tape	–	–	–	–	–	m	**12.59**
Duo-Ridge details							
2 nr extruded aluminium zed sections fixed to Kalzip seams using aluminium blind sealed rivets; 2 nr natural aluminium stucco embossed U Type ridge closures fixed to Kalzip seams using aluminium blind sealed rivets; 2 nr black solid rubber ridge filler blocks, 2 nr ST clips fixed with stainless steel fasteners; fix seam of Kaizip sheet to ST clips using aluminium blind sealed rivets (for fixed point); turn up Kalzip 400 sheets both sides	–	–	–	–	–	m	**40.44**
0.90 mm thick stucco embossed natural aluminium external ridge closure; 600 mm girth three times bent, fixed to U Type Ridge closure with stainless steel fasteners	–	–	–	–	–	m	**16.06**
0.70 mm thick bright white polyester liner sheet closure flashing 600 mm girth once bent fixed with stainless steel fasteners, black solid rubber profiled filler blocks sealed top and bottom with sealant tape	–	–	–	–	–	m	**14.02**

17 SHEET ROOF COVERINGS

Item	PC £	Labour hours	Labour £	Plant £	Material £	Unit	Total rate £
Accessories							
smooth curving Kalzip sheets	–	–	–	–	–	m²	9.30
crimp curving liner (below 52.5 m convex radius)	–	–	–	–	–	sheet	15.68
polyster coating Kalzip sheets	–	–	–	–	–	m²	4.84
PvDF coating Kalzip sheets	–	–	–	–	–	m²	5.59
vapour control layer, foil encapsulated polythene 4300MNs/g	–	–	–	–	–	m²	0.77
200 mm thick thermal insulation quilt	–	–	–	–	–	m²	0.25
30 mm thick semi-rigid acoustic insulation slab	–	–	–	–	–	m²	5.80
1.0 mm flashings etc.; fixing/wedging into grooves							
flashing; 500 mm girth	–	–	–	–	–	m	17.14
flashing; 750 mm girth	–	–	–	–	–	m	22.62
flashing; 1000 mm girth	–	–	–	–	–	m	31.82
1.2 mm flashings etc.; fixing/wedging into grooves							
flashing; 500 mm girth	–	–	–	–	–	m	18.84
flashing; 750 mm girth	–	–	–	–	–	m	24.89
flashing; 1000 mm girth	–	–	–	–	–	m	31.46
1.4 mm flashings etc.; fixing/wedging into grooves							
flashing; 500 mm girth	–	–	–	–	–	m	21.67
flashing; 750 mm girth	–	–	–	–	–	m	28.61
flashing; 1000 mm girth	–	–	–	–	–	m	43.00
Aluminium Alumasc Skyline coping system; polyester powder coated							
Coping; fixing straps plugged and screwed to brickwork							
362 mm wide; for parapet wall 241–300 mm wide	–	0.50	12.57	–	42.04	m	54.61
90° angle	–	0.25	6.28	–	84.11	nr	90.39
90° tee junction	–	0.35	8.80	–	87.26	nr	96.06
stop end	–	0.15	3.77	–	58.76	nr	62.53
stop end upstand	–	0.20	5.03	–	60.07	nr	65.10
17.07 COPPER STRIP SHEET COVERINGS/ FLASHINGS							
Copper roofing; BS EN 504; on and including Geotec underlay							
The following rates are based upon nett deck or wall areas							
Roof and dormer coverings							
0.6 mm thick roof coverings; mill finish							
flat (in wood roll construction) (Prime Cost £ per kg)	8.11	2.20	55.29	–	70.79	m²	126.08
eaves detail ED1	–	0.40	10.05	–	8.63	m	18.68
abutment upstands at perimeters	–	0.66	16.58	–	4.32	m	20.90
pitched over 3° (in standing seam construction)	–	1.70	42.72	–	57.84	m²	100.56
vertical (in angled or flat seam construction)	–	1.80	45.24	–	57.84	m²	103.08
0.6 mm thick dormer coverings; mill finish							
flat (in wood roll construction)	8.11	3.20	80.42	–	70.79	m²	151.21
eaves detail ED1	–	0.40	10.05	–	8.63	m	18.68
pitched over 3° (in standing seam construction)	–	2.50	62.83	–	57.84	m²	120.67
vertical (in angled or flat seam construction)	–	2.70	67.86	–	57.84	m²	125.70

17 SHEET ROOF COVERINGS

Item	PC £	Labour hours	Labour £	Plant £	Material £	Unit	Total rate £
17.07 COPPER STRIP SHEET COVERINGS/ FLASHINGS – cont							
Roof and dormer coverings – cont							
0.6 mm thick roof coverings; oxide finish							
flat (in wood roll construction) (Prime Cost £ per kg)	9.95	2.20	55.29	–	85.86	m²	**141.15**
eaves detail ED1	–	0.40	10.05	–	10.59	m	**20.64**
abutment upstands at perimeters	–	0.66	16.58	–	5.29	m	**21.87**
pitched over 3° (in standing seam construction)	–	1.70	42.72	–	70.45	m²	**113.17**
vertical (in angled or flat seam construction)	–	1.60	40.21	–	70.45	m²	**110.66**
0.6 mm thick dormer coverings; oxide finish							
flat (in wood roll construction)	9.95	3.00	75.40	–	85.86	m²	**161.26**
eaves detail ED1	–	0.40	10.05	–	10.59	m	**20.64**
pitched over 3° (in standing seam construction)	–	2.50	62.83	–	70.45	m²	**133.28**
vertical (in angled or flat seam construction)	–	2.70	67.86	–	70.45	m²	**138.31**
0.6 mm thick roof coverings; KME pre-patinated finish							
flat (in wood roll construction)	83.20	2.20	55.29	–	126.96	m²	**182.25**
eaves detail ED1	–	0.40	10.05	–	16.11	m	**26.16**
abutment upstands at perimeters	–	0.66	16.58	–	8.06	m	**24.64**
pitched over 3° (in standing seam construction)	–	1.70	42.72	–	102.80	m²	**145.52**
vertical (in angled or flat seam construction)	–	1.80	45.24	–	102.80	m²	**148.04**
0.6 mm thick dormer coverings; KME pre-patinated finish							
flat (in wood roll construction)	83.20	3.00	75.40	–	126.96	m²	**202.36**
eaves detail ED1	–	0.40	10.05	–	16.11	m	**26.16**
pitched over 3° (in standing seam construction)	–	2.50	62.83	–	102.80	m²	**165.63**
vertical (in angled or flat seam construction)	–	2.70	67.86	–	102.80	m²	**170.66**
0.7 mm thick roof coverings; mill finish							
flat (in wood roll construction) (Prime Cost £ per kg)	8.11	2.00	50.26	–	80.21	m²	**130.47**
eaves detail ED1	–	0.40	10.05	–	9.89	m	**19.94**
abutment upstands at perimeters	–	0.66	16.58	–	4.95	m	**21.53**
pitched over 3° (in standing seam construction)	–	1.50	37.70	–	65.38	m²	**103.08**
vertical (in angled or flat seam construction)	–	1.60	40.21	–	65.38	m²	**105.59**
0.7 mm thick dormer coverings; mill finish							
flat (in wood roll construction)	8.11	3.00	75.40	–	80.21	m²	**155.61**
eaves detail ED1	–	0.40	10.05	–	9.89	m	**19.94**
pitched over 3° (in standing seam construction)	–	2.50	62.83	–	65.34	m²	**128.17**
vertical (in angled or flat seam construction)	–	2.70	67.86	–	65.30	m²	**133.16**
0.7 mm thick roof coverings; oxide finish							
flat (in wood roll construction) (Prime Cost £ per kg)	9.95	2.00	50.26	–	108.96	m²	**159.22**
eaves detail ED1	–	0.40	10.05	–	12.13	m	**22.18**
abutment upstands at perimeters	–	0.66	16.58	–	6.06	m	**22.64**
pitched over 3° (in standing seam construction)	–	1.50	37.70	–	88.74	m²	**126.44**
vertical (in angled or flat seam construction)	–	1.60	40.21	–	79.12	m²	**119.33**
0.7 mm thick dormer coverings; oxide finish							
flat (in wood roll construction)	9.95	3.00	75.40	–	97.41	m²	**172.81**
eaves detail ED1	–	0.40	10.05	–	12.13	m	**22.18**
pitched over 3° (in standing seam construction)	–	2.50	62.83	–	79.12	m²	**141.95**
vertical (in angled or flat seam construction)	–	2.70	67.86	–	79.12	m²	**146.98**

17 SHEET ROOF COVERINGS

Item	PC £	Labour hours	Labour £	Plant £	Material £	Unit	Total rate £
0.7 mm thick roof coverings; KME pre-patinated finish							
flat (in wood roll construction)	95.75	2.20	55.29	–	145.75	m²	**201.04**
eaves detail ED1	–	0.40	10.05	–	18.53	m	**28.58**
abutment upstands at perimeters	–	0.66	16.58	–	9.27	m	**25.85**
pitched over 3° (in standing seam construction)	–	1.70	42.72	–	117.95	m²	**160.67**
vertical (in angled or flat seam construction)	–	1.80	45.24	–	117.95	m²	**163.19**
0.7 mm thick dormer coverings; KME pre-patinated finish							
flat (in wood roll construction)	95.75	3.00	75.40	–	145.75	m²	**221.15**
eaves detail ED1	–	0.40	10.05	–	18.53	m	**28.58**
pitched over 3° (in standing seam construction)	–	2.50	62.83	–	117.95	m²	**180.78**
vertical (in angled or flat seam construction)	–	2.70	67.86	–	117.95	m²	**185.81**
0.6 mm thick copper flashings, etc.							
Flashings; wedging into grooves; mill finish							
150 mm girth (Prime Cost £ per kg)	8.52	0.50	12.57	–	5.17	m	**17.74**
240 mm girth	–	0.50	12.57	–	10.35	m	**22.92**
300 mm girth	–	0.50	12.57	–	12.93	m	**25.50**
Stepped flashings; wedging into grooves; mill finish							
180 mm girth	–	1.00	25.13	–	7.76	m	**32.89**
270 mm girth	–	1.00	25.13	–	11.64	m	**36.77**
Flashings; wedging into grooves; oxide finish							
150 mm girth (Prime Cost £ per kg)	10.44	0.50	12.57	–	7.94	m	**20.51**
240 mm girth	–	0.50	12.57	–	12.70	m	**25.27**
300 mm girth	–	0.50	12.57	–	15.88	m	**28.45**
Stepped flashings; wedging into grooves; oxide finish							
180 mm girth	–	1.00	25.13	–	9.53	m	**34.66**
270 mm girth	–	1.00	25.13	–	14.29	m	**39.42**
Flashings; wedging into grooves; KME pre-patinated finish							
150 mm girth (Prime Cost £ per kg)	87.36	0.50	12.57	–	12.08	m	**24.65**
240 mm girth	–	0.50	12.57	–	19.33	m	**31.90**
300 mm girth	–	0.50	12.57	–	24.17	m	**36.74**
Stepped flashings; wedging into grooves; KME pre-patinated finish							
180 mm girth	–	1.00	25.13	–	14.50	m	**39.63**
270 mm girth	–	1.00	25.13	–	21.75	m	**46.88**
0.7 mm thick copper flashings, etc.							
Flashings; wedging into grooves; mill finish							
150 mm girth (Prime Cost £ per kg)	8.11	0.50	12.57	–	7.41	m	**19.98**
240 mm girth	–	0.50	12.57	–	11.86	m	**24.43**
300 mm girth	–	0.50	12.57	–	14.82	m	**27.39**
Stepped flashings; wedging into grooves; mill finish							
180 mm girth	–	1.00	25.13	–	8.89	m	**34.02**
270 mm girth	–	1.00	25.13	–	13.34	m	**38.47**
Flashings; wedging into grooves; oxide finish							
150 mm girth (Prime Cost £ per kg)	9.95	0.50	12.57	–	9.09	m	**21.66**
240 mm girth	–	0.50	12.57	–	14.55	m	**27.12**
300 mm girth	–	0.50	12.57	–	18.19	m	**30.76**

17 SHEET ROOF COVERINGS

Item	PC £	Labour hours	Labour £	Plant £	Material £	Unit	Total rate £
17.07 COPPER STRIP SHEET COVERINGS/ FLASHINGS – cont							
0.7 mm thick copper flashings, etc. – cont							
Stepped flashings; wedging into grooves; oxide finish							
180 mm girth	–	1.00	25.13	–	10.91	m	**36.04**
270 mm girth	–	1.00	25.13	–	16.37	m	**41.50**
Flashings; wedging into grooves; KME pre-patinated finish							
150 mm girth (Prime Cost £ per kg)	95.75	0.50	12.57	–	13.90	m	**26.47**
240 mm girth	–	0.50	12.57	–	22.24	m	**34.81**
300 mm girth	–	0.50	12.57	–	27.80	m	**40.37**
Stepped flashings; wedging into grooves; KME pre-patinated finish							
180 mm girth	–	1.00	25.13	–	16.68	m	**41.81**
270 mm girth	–	1.00	25.13	–	25.02	m	**50.15**
Sundries							
provision of square batten roll at 500 mm centres (per m)	–	0.20	5.03	–	1.57	m	**6.60**
17.08 ZINC STRIP SHEET COVERINGS/ FLASHINGS							
Zinc roofing; CP 143–5; on and including Delta Trella underlay							
The following rates are based upon nett deck or wall areas							
Natural Bright Rheinzink							
Roof, dormer and wall coverings							
0.7 mm thick roof coverings							
flat (in wood roll construction) (Prime Cost £ per kg)	3.95	2.00	50.26	–	37.77	m²	**88.03**
eaves detail ED1	–	0.40	10.05	–	6.37	m	**16.42**
abutment upstands at perimeters	–	0.66	16.58	–	3.19	m	**19.77**
pitched over 3° (in standing seam construction)	–	1.50	37.70	–	31.37	m²	**69.07**
0.7 mm thick dormer coverings							
flat (in wood roll construction) (Prime Cost £ per kg)	3.95	3.00	75.40	–	37.77	m²	**113.17**
eaves detail ED1	–	0.40	10.05	–	6.37	m	**16.42**
pitched over 3° (in standing seam construction)	–	2.50	62.83	–	31.37	m²	**94.20**
0.8 mm thick wall coverings							
vertical (in angled or flat seam construction)	–	1.60	40.21	–	35.14	m²	**75.35**
0.8 mm thick dormer coverings							
vertical (in angled or flat seam construction)	–	2.70	67.86	–	35.25	m²	**103.11**

17 SHEET ROOF COVERINGS

Item	PC £	Labour hours	Labour £	Plant £	Material £	Unit	Total rate £
0.8 mm thick zinc flashings, etc.; Natural Bright Rheinzink							
Flashings; wedging into grooves							
150 mm girth	–	0.50	12.57	–	3.66	m	**16.23**
240 mm girth	–	0.50	12.57	–	5.86	m	**18.43**
300 mm girth	–	0.50	12.57	–	7.32	m	**19.89**
Stepped flashings; wedging into grooves							
180 mm girth	–	1.00	25.13	–	4.39	m	**29.52**
270 mm girth	–	1.00	25.13	–	6.59	m	**31.72**
Integral box gutter							
900 mm girth; 2 × bent; 2 × welted	–	2.00	50.26	–	30.26	m	**80.52**
Valley gutter							
600 mm girth; 2 × bent; 2 × welted	–	1.50	37.70	–	18.15	m	**55.85**
Hips and ridges							
450 mm girth; 2 × bent; 2 × welted	–	2.00	50.26	–	10.97	m	**61.23**
Natural Bright Rheinzink PRO							
Roof, dormer and wall coverings							
0.7 mm thick roof coverings							
flat (in wood roll construction) (Prime Cost £ per kg)	5.38	2.00	50.26	–	51.29	m²	**101.55**
eaves detail ED1	–	0.40	10.05	–	8.66	m	**18.71**
abutment upstands at perimeters	–	0.66	16.58	–	4.33	m	**20.91**
pitched over 3° (in standing seam construction)	–	1.50	37.70	–	42.61	m²	**80.31**
0.7 mm thick dormer coverings							
flat (in wood roll construction) (Prime Cost £ per kg)	5.38	3.00	75.40	–	51.29	m²	**126.69**
eaves detail ED1	–	0.40	10.05	–	8.66	m	**18.71**
pitched over 3° (in standing seam construction)	–	2.50	62.83	–	42.60	m²	**105.43**
0.8 mm thick wall coverings							
vertical (in angled or flat seam construction)	–	1.60	40.21	–	47.73	m²	**87.94**
0.8 mm thick dormer coverings							
vertical (in angled or flat seam construction)	–	2.70	67.86	–	47.73	m²	**115.59**
0.8 mm thick zinc flashings, etc.; Natural Bright Rheinzink PRO							
Flashings; wedging into grooves							
150 mm girth	–	0.50	12.57	–	4.97	m	**17.54**
240 mm girth	–	0.50	12.57	–	7.96	m	**20.53**
300 mm girth	–	0.50	12.57	–	9.94	m	**22.51**
Stepped flashings; wedging into grooves							
180 mm girth	–	1.00	25.13	–	5.96	m	**31.09**
270 mm girth	–	1.00	25.13	–	8.95	m	**34.08**
Integral box gutter							
900 mm girth; 2 × bent; 2 × welted	–	2.00	50.26	–	41.10	m	**91.36**
Valley gutter							
600 mm girth; 2 × bent; 2 × welted	–	1.50	37.70	–	24.65	m	**62.35**
Hips and ridges							
450 mm girth; 2 × bent; 2 × welted	–	2.00	50.26	–	14.91	m	**65.17**

17 SHEET ROOF COVERINGS

Item	PC £	Labour hours	Labour £	Plant £	Material £	Unit	Total rate £
17.08 ZINC STRIP SHEET COVERINGS/ FLASHINGS – cont							
Pre-weathered Rheinzink							
Roof, dormer and wall coverings							
0.7 mm thick roof coverings; pre-weathered Rheinzink							
flat (in wood roll construction) (Prime Cost £ per kg)	4.56	2.00	50.26	–	43.63	m²	**93.89**
eaves detail ED1	–	0.40	10.05	–	7.36	m	**17.41**
abutment upstands at perimeters	–	0.66	16.58	–	3.68	m	**20.26**
pitched over 3° (in standing seam construction)	–	1.50	37.70	–	36.24	m²	**73.94**
0.7 mm thick dormer coverings; pre-weathered Rheinzink							
flat (in wood roll construction) (Prime Cost £ per kg)	4.56	3.00	75.40	–	43.63	m²	**119.03**
eaves detail ED1	–	0.40	10.05	–	7.36	m	**17.41**
pitched over 3° (in standing seam construction)	–	2.50	62.83	–	36.24	m²	**99.07**
0.8 mm thick wall coverings; pre-weathered Rheinzink							
vertical (in angled or flat seam construction)	–	1.60	40.21	–	40.59	m²	**80.80**
0.8 mm thick dormer coverings; pre-weathered Rheinzink							
vertical (in angled or flat seam construction)	–	2.70	67.86	–	40.59	m²	**108.45**
0.8 mm thick zinc flashings, etc.; pre-weathered Rheinzink							
Flashings; wedging into grooves							
150 mm girth	–	0.50	12.57	–	4.23	m	**16.80**
240 mm girth	–	0.50	12.57	–	6.77	m	**19.34**
300 mm girth	–	0.50	12.57	–	8.46	m	**21.03**
Stepped flashings; wedging into grooves							
180 mm girth	–	1.00	25.13	–	5.07	m	**30.20**
270 mm girth	–	1.00	25.13	–	7.61	m	**32.74**
Integral box gutter							
900 mm girth; 2 × bent; 2 × welted	–	2.00	50.26	–	34.96	m	**85.22**
Valley gutter							
600 mm girth; 2 × bent; 2 × welted	–	1.50	37.70	–	20.97	m	**58.67**
Hips and ridges							
450 mm girth; 2 × bent; 2 × welted	–	2.00	50.26	–	12.68	m	**62.94**
Pre-weathered Rheinzink PRO							
Roof, dormer and wall coverings							
0.7 mm thick roof coverings							
flat (in wood roll construction) (Prime Cost £ per kg)	5.98	2.00	50.26	–	57.15	m²	**107.41**
eaves detail ED1	–	0.40	10.05	–	9.65	m	**19.70**
abutment upstands at perimeters	–	0.66	16.58	–	4.82	m	**21.40**
pitched over 3° (in standing seam construction)	–	1.50	37.70	–	47.48	m²	**85.18**

17 SHEET ROOF COVERINGS

Item	PC £	Labour hours	Labour £	Plant £	Material £	Unit	Total rate £
0.7 mm thick dormer coverings							
flat (in wood roll construction) (Prime Cost £ per kg)	5.98	3.00	75.40	–	57.15	m²	**132.55**
eaves detail ED1	–	0.40	10.05	–	9.65	m	**19.70**
pitched over 3° (in standing seam construction)	–	2.50	62.83	–	47.48	m²	**110.31**
0.8 mm thick wall coverings							
vertical (in angled or flat seam construction)	–	1.60	40.21	–	53.18	m²	**93.39**
0.8 mm thick dormer coverings							
vertical (in angled or flat seam construction)	–	2.70	67.86	–	53.18	m²	**121.04**
0.8 mm thick zinc flashings, etc.; pre-weathered Rheinzink PRO							
Flashings; wedging into grooves							
150 mm girth	–	0.50	12.57	–	5.54	m	**18.11**
240 mm girth	–	0.50	12.57	–	8.87	m	**21.44**
300 mm girth	–	0.50	12.57	–	11.08	m	**23.65**
Stepped flashings; wedging into grooves							
180 mm girth	–	1.00	25.13	–	6.64	m	**31.77**
270 mm girth	–	1.00	25.13	–	9.97	m	**35.10**
Integral box gutter							
900 mm girth; 2 × bent; 2 × welted	–	–	–	–	–	m	**45.80**
Valley gutter							
600 mm girth; 2 × bent; 2 × welted	–	1.50	37.70	–	27.47	m	**65.17**
Hips and ridges							
450 mm girth; 2 × bent; 2 × welted	–	2.00	50.26	–	16.61	m	**66.87**
VM Natural Bright							
Roof, dormer and wall coverings							
0.7 mm thick roof coverings							
flat (in wood roll construction) (Prime Cost £ per kg)	3.95	2.00	50.26	–	37.77	m²	**88.03**
eaves detail ED1	–	0.40	10.05	–	6.37	m	**16.42**
abutment upstands at perimeters	–	0.66	16.58	–	3.19	m	**19.77**
pitched over 3° (in standing seam construction)	–	1.50	37.70	–	31.37	m²	**69.07**
0.7 mm thick dormer coverings							
flat (in wood roll construction) (Prime Cost £ per kg)	3.95	3.00	75.40	–	37.77	m²	**113.17**
eaves detail ED1	–	0.40	10.05	–	6.37	m	**16.42**
pitched over 3° (in standing seam construction)	–	2.50	62.83	–	31.37	m²	**94.20**
0.8 mm thick wall coverings							
vertical (in angled or flat seam construction)	–	1.60	40.21	–	35.14	m²	**75.35**
0.8 mm thick dormer coverings							
vertical (in angled or flat seam construction)	–	2.70	67.86	–	35.14	m²	**103.00**
0.8 mm thick zinc flashings, etc.; VM Natural Bright							
Flashings; wedging into grooves							
150 mm girth	–	0.50	12.57	–	3.66	m	**16.23**
240 mm girth	–	0.50	12.57	–	5.86	m	**18.43**
300 mm girth	–	0.50	12.57	–	7.32	m	**19.89**
Stepped flashings; wedging into grooves							
180 mm girth	–	1.00	25.13	–	4.39	m	**29.52**
270 mm girth	–	1.00	25.13	–	6.59	m	**31.72**

17 SHEET ROOF COVERINGS

Item	PC £	Labour hours	Labour £	Plant £	Material £	Unit	Total rate £
17.08 ZINC STRIP SHEET COVERINGS/ FLASHINGS – cont							
0.8 mm thick zinc flashings, etc. – cont							
Integral box gutter							
900 mm girth; 2 × bent; 2 × welted	–	2.00	50.26	–	30.26	m	**80.52**
Valley gutter							
600 mm girth; 2 × bent; 2 × welted	–	1.50	37.70	–	18.15	m	**55.85**
Hips and ridges							
450 mm girth; 2 × bent; 2 × welted	–	2.00	50.26	–	9.96	m	**60.22**
VM Natural Bright PLUS							
Roof, dormer and wall coverings							
0.7 mm thick roof coverings							
flat (in wood roll construction) (Prime Cost £ per kg)	5.38	2.00	50.26	–	51.29	m²	**101.55**
eaves detail ED1	–	0.40	10.05	–	8.66	m	**18.71**
abutment upstands at perimeters	–	0.66	16.58	–	4.33	m	**20.91**
pitched over 3° (in standing seam construction)	–	1.50	37.70	–	42.61	m²	**80.31**
0.7 mm thick dormer coverings							
flat (in wood roll construction) (Prime Cost £ per kg)	5.38	3.00	75.40	–	51.29	m²	**126.69**
eaves detail ED1	–	0.40	10.05	–	8.66	m	**18.71**
pitched over 3° (in standing seam construction)	–	2.50	62.83	–	42.61	m²	**105.44**
0.8 mm thick wall coverings							
vertical (in angled or flat seam construction)	–	1.60	40.21	–	47.73	m²	**87.94**
0.8 mm thick dormer coverings							
vertical (in angled or flat seam construction)	–	2.70	67.86	–	47.73	m²	**115.59**
0.8 mm thick zinc flashings, etc.; VM Natural Bright PLUS							
Flashings; wedging into grooves							
150 mm girth	–	0.50	12.57	–	4.97	m	**17.54**
240 mm girth	–	0.50	12.57	–	7.96	m	**20.53**
300 mm girth	–	0.50	12.57	–	9.94	m	**22.51**
Stepped flashings; wedging into grooves							
180 mm girth	–	1.00	25.13	–	5.96	m	**31.09**
270 mm girth	–	1.00	25.13	–	8.95	m	**34.08**
Integral box gutter							
900 mm girth; 2 × bent; 2 × welted	–	2.00	50.26	–	41.10	m	**91.36**
Valley gutter							
600 mm girth; 2 × bent; 2 × welted	–	1.50	37.70	–	24.65	m	**62.35**
Hips and ridges							
450 mm girth; 2 × bent; 2 × welted	–	2.00	50.26	–	14.91	m	**65.17**
VM Quartz (pre-weathered)							
Roof, dormer and wall coverings							
0.7 mm thick roof coverings							
flat (in wood roll construction) (Prime Cost £ per kg)	4.90	2.00	50.26	–	46.90	m²	**97.16**
eaves detail ED1	–	0.40	10.05	–	7.91	m	**17.96**
abutment upstands at perimeters	–	0.66	16.58	–	3.96	m	**20.54**
pitched over 3° (in standing seam construction)	–	1.50	37.70	–	38.96	m²	**76.66**

17 SHEET ROOF COVERINGS

Item	PC £	Labour hours	Labour £	Plant £	Material £	Unit	Total rate £
0.7 mm thick dormer coverings							
flat (in wood roll construction) (Prime Cost £ per kg)	4.90	3.00	75.40	–	46.90	m²	122.30
eaves detail ED1	–	0.40	10.05	–	7.91	m	17.96
pitched over 3° (in standing seam construction)	–	2.50	62.83	–	38.96	m²	101.79
0.8 mm thick wall coverings							
vertical (in angled or flat seam construction)	–	1.60	40.21	–	43.64	m²	83.85
0.8 mm thick dormer coverings							
vertical (in angled or flat seam construction)	–	2.70	67.86	–	43.64	m²	111.50
0.8 mm thick zinc flashings, etc.; VM Quartz (pre-weathered)							
Flashings; wedging into grooves							
150 mm girth	–	0.50	12.57	–	4.54	m	17.11
240 mm girth	–	0.50	12.57	–	7.28	m	19.85
300 mm girth	–	0.50	12.57	–	9.09	m	21.66
Stepped flashings; wedging into grooves							
180 mm girth	–	1.00	25.13	–	5.45	m	30.58
270 mm girth	–	1.00	25.13	–	8.18	m	33.31
Integral box gutter							
900 mm girth; 2 × bent; 2 × welted	–	2.00	50.26	–	37.58	m	87.84
Valley gutter							
600 mm girth; 2 × bent; 2 × welted	–	1.50	37.70	–	22.54	m	60.24
Hips and ridges							
450 mm girth; 2 × bent; 2 × welted	–	2.00	50.26	–	13.63	m	63.89
VM Quartz (pre-weathered) PLUS							
Roof, dormer and wall coverings							
0.7 mm thick roof coverings							
flat (in wood roll construction) (Prime Cost £ per kg)	6.32	2.00	50.26	–	60.65	m²	110.91
eaves detail ED1	–	0.40	10.05	–	10.24	m	20.29
abutment upstands at perimeters	–	0.66	16.58	–	5.12	m	21.70
pitched over 3° (in standing seam construction)	–	1.50	37.70	–	50.39	m²	88.09
0.7 mm thick dormer coverings							
flat (in wood roll construction) (Prime Cost £ per kg)	6.32	3.00	75.40	–	60.65	m²	136.05
eaves detail ED1	–	0.40	10.05	–	10.24	m	20.29
pitched over 3° (in standing seam construction)	–	2.50	62.83	–	50.39	m²	113.22
0.8 mm thick wall coverings							
vertical (in angled or flat seam construction)	–	1.60	40.21	–	56.43	m²	96.64
0.8 mm thick dormer coverings							
vertical (in angled or flat seam construction)	–	2.70	67.86	–	56.43	m²	124.29
0.8 mm thick zinc flashings, etc.; VM Quartz (pre-weathered) PLUS							
Flashings; wedging into grooves							
150 mm girth	–	0.50	12.57	–	5.88	m	18.45
240 mm girth	–	0.50	12.57	–	9.41	m	21.98
300 mm girth	–	0.50	12.57	–	11.75	m	24.32

17 SHEET ROOF COVERINGS

Item	PC £	Labour hours	Labour £	Plant £	Material £	Unit	Total rate £
17.08 ZINC STRIP SHEET COVERINGS/ FLASHINGS – cont							
0.8 mm thick zinc flashings, etc. – cont							
Stepped flashings; wedging into grooves							
180 mm girth	–	1.00	25.13	–	7.05	m	**32.18**
270 mm girth	–	1.00	25.13	–	10.58	m	**35.71**
Integral box gutter							
900 mm girth; 2 × bent; 2 × welted	–	2.00	50.26	–	48.60	m	**98.86**
Valley gutter							
600 mm girth; 2 × bent; 2 × welted	–	1.50	37.70	–	29.15	m	**66.85**
Hips and ridges							
450 mm girth; 2 × bent; 2 × welted	–	2.00	50.26	–	17.62	m	**67.88**
Sundries and accessories							
Klober breather membrane/underlay	–	0.20	5.03	–	4.59	m^2	**9.62**
Delta Trela Chestwig underlay	–	0.20	5.03	–	7.85	m^2	**12.88**
Delta Trela Football Studs underlay	–	0.20	5.03	–	1.45	m^2	**6.48**
Trapezoidal batten roll at 500 mm centres (per m)	–	0.20	5.03	–	1.45	m	**6.48**
Zinflash; 0.6 mm thick lead look flashing (no patination oil required)							
Flashings; wedging into grooves							
150 mm girth	–	0.50	12.57	–	6.49	m	**19.06**
250 mm girth	–	0.50	12.57	–	10.80	m	**23.37**
300 mm girth	–	0.50	12.57	–	12.97	m	**25.54**
380 mm girth	–	0.50	12.57	–	16.43	m	**29.00**
450 mm girth	–	0.50	12.57	–	19.45	m	**32.02**
Stepped flashings; wedging into grooves							
150 mm girth	–	1.00	25.13	–	6.49	m	**31.62**
250 mm girth	–	1.00	25.13	–	10.80	m	**35.93**
300 mm girth	–	1.00	25.13	–	12.97	m	**38.10**
380 mm girth	–	1.00	25.13	–	16.43	m	**41.56**
450 mm girth	–	1.00	25.13	–	19.45	m	**44.58**
17.09 STAINLESS STEEL SHEET COVERINGS/ FLASHINGS							
Terne-coated stainless steel roofing; Associated Lead Mills Ltd; or other equal and approved: on and including Metmatt underlay							
The following rates are based upon nett deck or wall areas							
Roof, dormer and wall coverings in Uginox grade 316; marine							
0.4 mm thick roof coverings							
flat (in wood roll construction) (Prime Cost £ per kg)	9.18	2.00	50.26	–	43.89	m^2	**94.15**
eaves detail ED1	–	0.40	10.05	–	5.18	m	**15.23**
abutment upstands at perimeters	–	0.66	16.58	–	2.59	m	**19.17**
pitched over 3° (in standing seam construction)	–	1.50	37.70	–	36.12	m^2	**73.82**

17 SHEET ROOF COVERINGS

Item	PC £	Labour hours	Labour £	Plant £	Material £	Unit	Total rate £
0.5 mm thick dormer coverings							
flat (in wood roll construction) (Prime Cost £ per kg)	8.57	3.00	75.40	–	50.91	m²	126.31
eaves detail ED1	–	0.40	10.05	–	4.83	m	14.88
pitched over 3° (in standing seam construction)	–	2.50	62.83	–	41.74	m²	104.57
0.5 mm thick wall coverings							
vertical (in angled or flat seam construction)	8.57	1.60	40.21	–	41.74	m²	81.95
vertical (with Coulisseau joint construction)	–	2.50	62.83	–	43.09	m²	105.92
0.5 mm thick Uginox grade 316 flashings, etc.							
Flashings; wedging into grooves							
150 mm girth (Prime Cost £ per kg)	8.57	0.50	12.57	–	4.59	m	17.16
240 mm girth	–	0.50	12.57	–	7.35	m	19.92
300 mm girth	–	0.50	12.57	–	9.18	m	21.75
Stepped flashings; wedging into grooves							
180 mm girth	–	1.00	25.13	–	5.51	m	30.64
270 mm girth	–	1.00	25.13	–	8.26	m	33.39
Fan apron							
250 mm girth	–	0.50	12.57	–	7.65	m	20.22
Integral box gutter							
900 mm girth; 2 × bent; 2 × welted	–	2.00	50.26	–	31.52	m	81.78
Valley gutter							
600 mm girth; 2 × bent; 2 × welted	–	1.50	37.70	–	22.40	m	60.10
Hips and ridges							
450 mm girth; 2 × bent; 2 × welted	–	2.00	50.26	–	13.77	m	64.03
Roof, dormer and wall coverings in Ugitop grade 304							
0.4 mm thick roof coverings							
flat (in wood roll construction) (Prime Cost £ per kg)	6.94	2.00	50.26	–	37.67	m²	87.93
eaves detail ED1	–	0.40	10.05	–	3.92	m	13.97
abutment upstands at perimeters	–	0.66	16.58	–	1.96	m	18.54
pitched over 3° (in standing seam construction)	–	1.50	37.70	–	28.53	m²	66.23
0.5 mm thick dormer coverings							
flat (in wood roll construction) (Prime Cost £ per kg)	6.60	3.00	75.40	–	40.36	m²	115.76
eaves detail ED1	–	0.40	10.05	–	3.72	m	13.77
pitched over 3° (in standing seam construction)	–	2.50	62.83	–	33.29	m²	96.12
0.5 mm thick wall coverings							
vertical (in angled or flat seam construction)	–	1.60	40.21	–	33.29	m²	73.50
vertical (with Coulisseau joint construction)	–	2.50	62.83	–	34.33	m²	97.16
0.5 mm thick Ugitop grade 304 flashings, etc.							
Flashings; wedging into grooves							
150 mm girth (Prime Cost £ per kg)	6.60	0.50	12.57	–	3.38	m	15.95
240 mm girth	–	0.50	12.57	–	5.95	m	18.52
300 mm girth	–	0.50	12.57	–	7.43	m	20.00
Stepped flashings; wedging into grooves							
180 mm girth	–	1.00	25.13	–	4.46	m	29.59
270 mm girth	–	1.00	25.13	–	6.69	m	31.82

17 SHEET ROOF COVERINGS

Item	PC £	Labour hours	Labour £	Plant £	Material £	Unit	Total rate £
17.09 STAINLESS STEEL SHEET COVERINGS/ FLASHINGS – cont							
0.5 mm thick Ugitop grade 304 flashings, etc. – cont							
Fan apron							
250 mm girth	–	0.50	12.57	–	6.19	m	**18.76**
Integral box gutter							
900 mm girth; 2 × bent; 2 × welted	–	2.00	50.26	–	25.52	m	**75.78**
Valley gutter							
600 mm girth; 2 × bent; 2 × welted	–	1.50	37.70	–	18.13	m	**55.83**
Hips and ridges							
450 mm girth; 2 × bent; 2 × welted	–	2.00	50.26	–	11.15	m	**61.41**
Roof, dormer and wall coverings in Ugitop grade 316							
0.4 mm thick roof coverings							
flat (in wood roll construction) (Prime Cost £ per kg)	8.57	2.00	50.26	–	45.28	m²	**95.54**
eaves detail ED1	–	0.40	10.05	–	4.83	m	**14.88**
abutment upstands at perimeters	–	0.66	16.58	–	2.41	m	**18.99**
pitched over 3° (in standing seam construction)	–	1.50	37.70	–	34.01	m²	**71.71**
0.5 mm thick dormer coverings							
flat (in wood roll construction) (Prime Cost £ per kg)	8.16	3.00	75.40	–	48.78	m²	**124.18**
eaves detail ED1	–	0.40	10.05	–	4.60	m	**14.65**
pitched over 3° (in standing seam construction)	–	2.50	62.83	–	40.03	m²	**102.86**
0.5 mm thick wall coverings							
vertical (in angled or flat seam construction)	–	1.60	40.21	–	40.03	m²	**80.24**
vertical (with Coulisseau joint construction)	–	2.50	62.83	–	34.33	m²	**97.16**
0.5 mm thick Ugitop grade 316 flashings, etc.							
Flashings; wedging into grooves							
150 mm girth	–	0.50	12.57	–	4.59	m	**17.16**
240 mm girth	–	0.50	12.57	–	7.35	m	**19.92**
300 mm girth	–	0.50	12.57	–	9.18	m	**21.75**
Stepped flashings; wedging into grooves							
180 mm girth	–	1.00	25.13	–	5.51	m	**30.64**
270 mm girth	–	1.00	25.13	–	8.26	m	**33.39**
Fan apron							
250 mm girth	–	0.50	12.57	–	7.65	m	**20.22**
Integral box gutter							
900 mm girth; 2 × bent; 2 × welted	–	2.00	50.26	–	31.52	m	**81.78**
Valley gutter							
600 mm girth; 2 × bent; 2 × welted	–	1.50	37.70	–	22.40	m	**60.10**
Hips and ridges							
450 mm girth; 2 × bent; 2 × welted	–	2.00	50.26	–	13.77	m	**64.03**
Sundries							
provision of square batten roll at 500 mm centres (per m)	–	0.20	5.03	–	1.57	m	**6.60**

18 TILE AND SLATE ROOF AND WALL COVERINGS

Item	PC £	Labour hours	Labour £	Plant £	Material £	Unit	Total rate £
MATERIAL RATES							
Clay tiles; plain, interlocking and pantiles							
Dreadnought							
Handformed Classic Staffordshire Blue	–	–	–	–	742.63	1000	742.63
Handformed Classic Purple Brown	–	–	–	–	742.63	1000	742.63
Handformed Classic Bronze	–	–	–	–	1044.01	1000	1044.01
Handformed Classic Deep Red	–	–	–	–	1011.77	1000	1011.77
Red smooth/sandfaced	–	–	–	–	645.81	1000	645.81
Country brown smooth/sandfaced	–	–	–	–	699.68	1000	699.68
Brown Antique smooth/sandfaced	–	–	–	–	635.10	1000	635.10
Blue/Dark Heather	–	–	–	–	742.63	1000	742.63
Sandtoft pantiles							
Bridgewater Double Roman	–	–	–	–	5975.44	1000	5975.44
Gaelic	–	–	–	–	3048.90	1000	3048.90
Arcadia	–	–	–	–	1672.31	1000	1672.31
William Blyth pantiles							
Celtic (French)	–	–	–	–	1302.95	1000	1302.95
Concrete tiles; plain and interlocking							
Marley Eternit roof tiles							
Anglia	–	–	–	–	734.18	1000	734.18
Ashmore	–	–	–	–	884.67	1000	884.67
Duo Modern	–	–	–	–	1107.35	1000	1107.35
Pewter Mendip	–	–	–	–	1263.81	1000	1263.81
Malvern	–	–	–	–	1179.66	1000	1179.66
Plain	–	–	–	–	421.27	1000	421.27
Redland roof tiles							
Redland 49	–	–	–	–	699.68	1000	699.68
Mini Stoneworld	–	–	–	–	1091.08	1000	1091.08
Grovebury	–	–	–	–	3726.54	1000	3726.54

18.01 PLAIN TILING

NOTE: The following items of tile roofing unless otherwise described, include for conventional fixing assuming normal exposure with appropriate nails and/or rivets or clips to pressure impregnated softwood battens fixed with galvanized nails; prices also include for all bedding and pointing at verges, beneath ridge tiles, etc.

Clay interlocking plain tiles; Sandtoft 20/20 natural red faced or other equal; 75 mm lap; on 25 mm × 38 mm battens and type 1F reinforced underlay

Item	PC £	Labour hours	Labour £	Plant £	Material £	Unit	Total rate £
Tiles 370 mm × 223 mm (Prime Cost £ per 1000)	821.70	–	–	–	821.70	1000	821.70
roof coverings	12.33	0.50	11.32	–	18.92	m²	30.24
extra over coverings for							
fixing every tile	–	0.02	0.45	–	1.26	m²	1.71
double course at eaves	–	0.28	6.34	–	15.09	m	21.43
verges; extra single undercloak course of plain tiles	–	0.28	6.34	–	7.67	m	14.01
open valleys; cutting both sides	–	0.17	3.85	–	3.55	m	7.40
dry ridge tiles	–	0.56	12.68	–	19.25	m	31.93
dry hips; cutting both sides	–	0.69	15.63	–	15.07	m	30.70
holes for pipes and the like	–	0.19	4.31	–	–	nr	4.31

18 TILE AND SLATE ROOF AND WALL COVERINGS

Item	PC £	Labour hours	Labour £	Plant £	Material £	Unit	Total rate £
18.01 PLAIN TILING – cont							
Clay pantiles; Sandtoft Old English; red sand faced or other equal; 75 mm lap; on 25 mm × 38 mm battens and type 1F reinforced underlay							
Pantiles 342 mm × 241 mm (Prime Cost £ per 1000)	1405.80	–	–	–	1405.80	1000	**1405.80**
roof coverings	22.49	0.50	11.32	–	30.55	m²	**41.87**
extra over coverings for							
fixing every tile	–	0.02	0.45	–	3.14	m²	**3.59**
double course at eaves	–	0.31	7.02	–	7.13	m	**14.15**
verges; extra single undercloak course of plain tiles	–	0.28	6.34	–	18.03	m	**24.37**
open valleys; cutting both sides	–	0.17	3.85	–	6.08	m	**9.93**
ridge tiles; tile slips	–	0.56	12.68	–	55.47	m	**68.15**
hips; cutting both sides	–	0.69	15.63	–	61.54	m	**77.17**
holes for pipes and the like	–	0.19	4.31	–	–	nr	**4.31**
Clay pantiles; William Blyth's Lincoln natural or other equal; 75 mm lap; on 19 mm × 38 mm battens and type 1F reinforced underlay							
Pantiles 343 mm × 280 mm (Prime Cost £ per 1000)	1518.00	–	–	–	1518.00	1000	**1518.00**
roof coverings	23.53	0.50	11.32	–	31.67	m²	**42.99**
extra over coverings for							
fixing every tile	–	0.02	0.45	–	3.14	m²	**3.59**
other colours	–	–	–	–	2.02	m²	**2.02**
double course at eaves	–	0.31	7.02	–	7.61	m	**14.63**
verges; extra single undercloak course of plain tiles	–	0.28	6.34	–	15.70	m	**22.04**
open valleys; cutting both sides	–	0.17	3.85	–	6.57	m	**10.42**
ridge tiles; tile slips	–	0.56	12.68	–	34.43	m	**47.11**
hips; cutting both sides	–	0.69	15.63	–	41.00	m	**56.63**
holes for pipes and the like	–	0.19	4.31	–	–	nr	**4.31**
Clay plain tiles; Hinton, Perry and Davenhill Dreadnought smooth red machine-made or other equal; on 19 mm × 38 mm battens and type 1F reinforced underlay							
Tiles 265 mm × 165 mm (Prime Cost £ per 1000)	627.00	–	–	–	627.00	1000	**627.00**
roof coverings; to 64 mm lap	37.62	0.50	11.32	–	53.76	m²	**65.08**
wall coverings; to 38 mm lap	33.23	0.75	16.98	–	46.72	m²	**63.70**
extra over coverings for							
ornamental tiles	–	–	–	–	51.26	m²	**51.26**
double course at eaves	–	0.23	5.21	–	10.42	m	**15.63**
verges	–	0.28	6.34	–	15.87	m	**22.21**
swept valleys; cutting both sides	–	0.60	13.59	–	10.85	m	**24.44**
bonnet hips; cutting both sides	–	0.74	16.76	–	137.23	m	**153.99**
external vertical angle tiles; supplementary nail fixings	–	0.37	8.38	–	101.66	m	**110.04**
half round ridge tiles	–	0.56	12.68	–	20.11	m	**32.79**
holes for pipes and the like	–	0.19	4.31	–	–	nr	**4.31**

18 TILE AND SLATE ROOF AND WALL COVERINGS

Item	PC £	Labour hours	Labour £	Plant £	Material £	Unit	Total rate £
Concrete plain tiles; BS EN 490 group A; on 25 mm × 38 mm battens and type 1F reinforced underlay							
Tiles 267 mm × 165 mm (Prime Cost £ per 1000)	360.00	–	–	–	360.00	1000	**360.00**
roof coverings; to 64 mm lap	21.60	0.25	5.67	–	36.43	m²	**42.10**
wall coverings; to 38 mm lap	19.08	0.30	6.80	–	31.42	m²	**38.22**
extra over coverings for							
ornamental tiles	–	–	–	–	36.79	m²	**36.79**
double course at eaves	–	0.23	5.21	–	4.62	m	**9.83**
verges	–	0.31	7.02	–	1.56	m	**8.58**
swept valleys; cutting both sides	–	0.60	13.59	–	44.04	m	**57.63**
bonnet hips; cutting both sides	–	0.74	16.76	–	44.21	m	**60.97**
external vertical angle tiles; supplementary nail fixings	–	0.37	8.38	–	34.09	m	**42.47**
half round ridge tiles	–	0.46	10.42	–	10.05	m	**20.47**
third round hip tiles; cutting both sides	–	0.46	10.42	–	12.38	m	**22.80**
holes for pipes and the like	–	0.19	4.31	–	–	nr	**4.31**

18.02 INTERLOCKING TILING

NOTE: The following items of tile roofing unless otherwise described, include for conventional fixing assuming normal exposure with appropriate nails and/or rivets or clips to pressure impregnated softwood battens fixed with galvanized nails; prices also include for all bedding and pointing at verges, beneath ridge tiles, etc.

Item	PC £	Labour hours	Labour £	Plant £	Material £	Unit	Total rate £
Concrete interlocking tiles; Marley Eternit Anglia granule finish tiles or other equal; 75 mm lap; on 25 mm × 38 mm battens and type 1F reinforced underlay							
Tiles 387 mm × 230 mm (Prime Cost £ per 1000)	621.70	–	–	–	621.70	1000	**621.70**
roof coverings	9.76	0.42	9.52	–	16.02	m²	**25.54**
extra over coverings for							
fixing every tile	–	0.02	0.45	–	0.53	m²	**0.98**
eaves; eaves filler	–	0.04	0.91	–	13.67	m	**14.58**
verges; 150 mm wide asbestos free strip undercloak	–	0.21	4.76	–	2.43	m	**7.19**
valley trough tiles; cutting both sides	–	0.51	11.55	–	31.51	m	**43.06**
segmental ridge tiles; tile slips	–	0.51	11.55	–	16.63	m	**28.18**
segmental hip tiles; tile slips; cutting both sides	–	0.65	14.72	–	18.64	m	**33.36**
dry ridge tiles; segmental including batten sections; unions and filler pieces	–	0.28	6.34	–	23.33	m	**29.67**
segmental mono-ridge tiles	–	0.51	11.55	–	25.83	m	**37.38**
gas ridge terminal	–	0.46	10.42	–	84.57	nr	**94.99**
holes for pipes and the like	–	0.19	4.31	–	–	nr	**4.31**

18 TILE AND SLATE ROOF AND WALL COVERINGS

Item	PC £	Labour hours	Labour £	Plant £	Material £	Unit	Total rate £
18.02 INTERLOCKING TILING – cont							
Concrete interlocking tiles; Marley Eternit							
Ludlow Major granule finish tiles or other equal;							
75 mm lap; on 25 mm × 38 mm battens and type							
1F reinforced underlay							
Tiles 420 mm × 330 mm (Prime Cost £ per 1000)	1065.80	–	–	–	1065.80	1000	**1065.80**
roof coverings	10.97	0.32	7.25	–	15.76	m²	**23.01**
extra over coverings for							
fixing every tile	–	0.02	0.45	–	0.53	m²	**0.98**
eaves; eaves filler	–	0.04	0.91	–	0.46	m	**1.37**
verges; 150 mm wide asbestos free strip							
undercloak	–	0.21	4.76	–	2.43	m	**7.19**
dry verge system; extruded white PVC	–	0.14	3.17	–	14.52	m	**17.69**
segmental ridge cap to dry verge	–	0.02	0.45	–	4.80	m	**5.25**
valley trough tiles; cutting both sides	–	0.51	11.55	–	32.48	m	**44.03**
segmental ridge tiles	–	0.46	10.42	–	10.87	m	**21.29**
segmental hip tiles; cutting both sides	–	0.60	13.59	–	14.33	m	**27.92**
dry ridge tiles; segmental including batten							
sections; unions and filler pieces	–	0.28	6.34	–	23.37	m	**29.71**
segmental mono-ridge tiles	–	0.46	10.42	–	22.47	m	**32.89**
gas ridge terminal	–	0.46	10.42	–	84.57	nr	**94.99**
holes for pipes and the like	–	0.19	4.31	–	–	nr	**4.31**
Concrete interlocking tiles; Marley Eternit							
Mendip granule finish double pantiles or other							
equal; 75 mm lap; on 22 mm × 38 mm battens and							
type 1F reinforced underlay							
Tiles 420 mm × 330 mm (Prime Cost £ per 1000)	999.20	–	–	–	999.20	1000	**999.20**
roof coverings (PC £ per 1000)	10.18	0.32	7.25	–	14.95	m²	**22.20**
extra over coverings for							
fixing every tile	–	0.02	0.45	–	0.53	m²	**0.98**
eaves; eaves filler	–	0.02	0.45	–	13.35	m	**13.80**
verges; 150 mm wide asbestos free strip							
undercloak	–	0.21	4.76	–	2.43	m	**7.19**
dry verge system; extruded white PVC	–	0.14	3.17	–	14.52	m	**17.69**
segmental ridge cap to dry verge	–	0.02	0.45	–	4.80	m	**5.25**
valley trough tiles; cutting both sides	–	0.51	11.55	–	32.33	m	**43.88**
segmental ridge tiles	–	0.51	11.55	–	16.63	m	**28.18**
segmental hip tiles; cutting both sides	–	0.65	14.72	–	19.87	m	**34.59**
dry ridge tiles; segmental including batten							
sections; unions and filler pieces	–	0.28	6.34	–	23.37	m	**29.71**
segmental mono-ridge tiles	–	0.46	10.42	–	25.36	m	**35.78**
gas ridge terminal	–	0.46	10.42	–	84.57	nr	**94.99**
holes for pipes and the like	–	0.19	4.31	–	–	nr	**4.31**

18 TILE AND SLATE ROOF AND WALL COVERINGS

Item	PC £	Labour hours	Labour £	Plant £	Material £	Unit	Total rate £
Concrete interlocking tiles; Marley Eternit Modern smooth finish tiles or other equal; 75 mm lap; on 25 mm × 38 mm battens and type 1F reinforced underlay							
Tiles 420 mm × 220 mm (Prime Cost £ per 1000)	888.20	–	–	–	888.20	1000	888.20
roof coverings	9.23	0.32	7.25	–	14.37	m²	21.62
extra over coverings for							
fixing every tile	–	0.02	0.45	–	0.53	m²	0.98
verges; 150 wide asbestos free strip undercloak	–	0.21	4.76	–	2.43	m	7.19
dry verge system; extruded white PVC	–	0.19	4.31	–	14.52	m	18.83
Modern ridge cap to dry verge	–	0.02	0.45	–	4.80	m	5.25
valley trough tiles; cutting both sides	–	0.51	11.55	–	32.08	m	43.63
Modern ridge tiles	–	0.46	10.42	–	13.53	m	23.95
Modern hip tiles; cutting both sides	–	0.60	13.59	–	16.42	m	30.01
dry ridge tiles; Modern; including batten sections; unions and filler pieces	–	0.28	6.34	–	26.04	m	32.38
Modern mono-ridge tiles	–	0.46	10.42	–	22.47	m	32.89
gas ridge terminal	–	0.46	10.42	–	84.57	nr	94.99
holes for pipes and the like	–	0.19	4.31	–	–	nr	4.31
Concrete interlocking tiles; Marley Eternit Ecologic Ludlow Major granule finish tiles or other equal; 75 mm lap; on 25 mm × 38 mm battens and type 1F reinforced underlay							
Tiles 420 mm × 330 mm (Prime Cost £ per 1000)	1054.70	–	–	–	1054.70	1000	1054.70
roof coverings	10.34	0.32	7.25	–	15.65	m²	22.90
extra over coverings for							
fixing every tile	–	0.02	0.45	–	0.53	m²	0.98
eaves; eaves filler	–	0.04	0.91	–	0.46	m	1.37
verges; 150 mm wide asbestos free strip undercloak	–	0.21	4.76	–	2.43	m	7.19
dry verge system; extruded white PVC	–	0.14	3.17	–	14.52	m	17.69
segmental ridge cap to dry verge	–	0.02	0.45	–	4.80	m	5.25
valley trough tiles; cutting both sides	–	0.51	11.55	–	32.44	m	43.99
segmental ridge tiles	–	0.46	10.42	–	10.87	m	21.29
segmental hip tiles; cutting both sides	–	0.60	13.59	–	14.29	m	27.88
dry ridge tiles; segmental including batten sections; unions and filler pieces	–	0.28	6.34	–	23.37	m	29.71
segmental mono-ridge tiles	–	0.46	10.42	–	22.47	m	32.89
gas ridge terminal	–	0.46	10.42	–	84.57	nr	94.99
holes for pipes and the like	–	0.19	4.31	–	–	nr	4.31

18 TILE AND SLATE ROOF AND WALL COVERINGS

Item	PC £	Labour hours	Labour £	Plant £	Material £	Unit	Total rate £
18.02 INTERLOCKING TILING – cont							
Concrete interlocking tiles; Marley Eternit Wessex smooth finish tiles or other equal; 75 mm lap; on 25 mm × 38 mm battens and type 1F reinforced underlay							
Tiles 413 mm × 330 mm (Prime Cost £ per 1000)	1554.30	–	–	–	1554.30	1000	**1554.30**
roof coverings	15.39	0.32	7.25	–	21.50	m²	**28.75**
extra over coverings for							
fixing every tile	–	0.02	0.45	–	0.53	m²	**0.98**
verges; 150 mm wide asbestos free strip undercloak	–	0.21	4.76	–	2.43	m	**7.19**
dry verge system; extruded white PVC	–	0.19	4.31	–	14.52	m	**18.83**
Modern ridge cap to dry verge	–	0.02	0.45	–	4.80	m	**5.25**
valley trough tiles; cutting both sides	–	0.51	11.55	–	33.53	m	**45.08**
Modern ridge tiles	–	0.46	10.42	–	13.53	m	**23.95**
Modern hip tiles; cutting both sides	–	0.60	13.59	–	18.57	m	**32.16**
dry ridge tiles; Modern; including batten sections; unions and filler pieces	–	0.28	6.34	–	26.00	m	**32.34**
Modern mono-ridge tiles	–	0.46	10.42	–	22.47	m	**32.89**
gas ridge terminal	–	0.46	10.42	–	84.57	nr	**94.99**
holes for pipes and the like	–	0.19	4.31	–	–	nr	**4.31**
Concrete interlocking tiles; Fenland Pantile smooth finish pantiles or other equal; 75 mm lap; on 25 mm × 38 mm battens and type 1F reinforced underlay							
Tiles 381 mm × 229 mm (Prime Cost £ per 1000)	684.00	–	–	–	684.00	1000	**684.00**
roof coverings	12.78	0.42	9.52	–	18.69	m²	**28.21**
extra over coverings for							
fixing every tile	–	0.04	0.91	–	0.22	m²	**1.13**
eaves; eaves filler	–	0.04	0.91	–	1.51	m	**2.42**
verges; extra single undercloak course of plain tiles	–	0.28	6.34	–	9.02	m	**15.36**
valley trough tiles; cutting both sides	–	0.56	12.68	–	41.76	m	**54.44**
universal ridge tiles	–	0.46	10.42	–	18.32	m	**28.74**
universal hip tiles; cutting both sides	–	0.60	13.59	–	22.02	m	**35.61**
universal gas flue ridge tile	–	0.46	10.42	–	92.98	nr	**103.40**
universal ridge vent tile with 110 mm dia. adaptor	–	0.50	11.32	–	119.69	nr	**131.01**
holes for pipes and the like	–	0.19	4.31	–	–	nr	**4.31**

18 TILE AND SLATE ROOF AND WALL COVERINGS

Item	PC £	Labour hours	Labour £	Plant £	Material £	Unit	Total rate £
Concrete interlocking tiles; Redland Renown granule finish tiles or other equal and approved; 418 mm × 330 mm; to 75 mm lap; on 25 mm × 38 mm battens and type 1F reinforced underlay							
Tiles 418 mm × 330 mm (Prime Cost £ per 1000)	856.90	–	–	–	856.90	1000	856.90
roof coverings	8.31	0.32	7.25	–	13.92	m²	21.17
extra over coverings for							
fixing every tile	–	0.02	0.45	–	0.26	m²	0.71
verges; extra single undercloak course of plain tiles	–	0.23	5.21	–	5.00	m	10.21
cloaked verge system	–	0.14	3.17	–	11.38	m	14.55
valley trough tiles; cutting both sides	–	0.51	11.55	–	40.83	m	52.38
universal ridge tiles	–	0.46	10.42	–	18.32	m	28.74
universal hip tiles; cutting both sides	–	0.60	13.59	–	21.10	m	34.69
dry ridge system; universal ridge tiles	–	0.23	5.21	–	47.29	m	52.50
universal half round mono-pitch ridge tiles	–	0.51	11.55	–	38.98	m	50.53
universal gas flue ridge tile	–	0.46	10.42	–	92.98	nr	103.40
universal ridge vent tile with 110 mm dia. adaptor	–	0.46	10.42	–	119.69	nr	130.11
holes for pipes and the like	–	0.19	4.31	–	–	nr	4.31
Concrete interlocking tiles; Redland Regent granule finish bold roll tiles or other equal; 75 mm lap; on 25 mm × 38 mm battens and type 1F reinforced underlay							
Tiles 418 mm × 332 mm (Prime Cost £ per 1000)	909.10	–	–	–	909.10	1000	909.10
roof coverings	8.82	0.32	7.25	–	14.46	m²	21.71
extra over coverings for							
fixing every tile	–	0.03	0.68	–	0.82	m²	1.50
eaves; eaves filler	–	0.04	0.91	–	1.23	m	2.14
verges; extra single undercloak course of plain tiles	–	0.23	5.21	–	4.07	m	9.28
cloaked verge system	–	0.14	3.17	–	11.23	m	14.40
valley trough tiles; cutting both sides	–	0.51	11.55	–	41.00	m	52.55
universal ridge tiles	–	0.46	10.42	–	18.32	m	28.74
universal hip tiles; cutting both sides	–	0.60	13.59	–	21.28	m	34.87
dry ridge system; universal ridge tiles	–	0.23	5.21	–	59.32	m	64.53
universal half round mono-pitch ridge tiles	–	0.51	11.55	–	38.98	m	50.53
universal gas flue ridge tile	–	0.46	10.42	–	92.98	nr	103.40
universal ridge vent tile with 110 mm dia. adaptor	–	0.46	10.42	–	119.69	nr	130.11
holes for pipes and the like	–	0.19	4.31	–	–	nr	4.31

18 TILE AND SLATE ROOF AND WALL COVERINGS

Item	PC £	Labour hours	Labour £	Plant £	Material £	Unit	Total rate £
18.02 INTERLOCKING TILING – cont							
Concrete interlocking slates; Redland Mini Stonewold concrete slates or other equal; 75 mm lap; on 25 mm × 38 mm battens and type 1F reinforced underlay							
Slates 418 mm × 334 mm (Prime Cost £ per 1000)	1059.30	–	–	–	1059.30	1000	**1059.30**
roof coverings	11.33	0.32	7.25	–	17.35	m²	**24.60**
extra over coverings for							
fixing every tile	–	0.02	0.45	–	3.25	m²	**3.70**
eaves; eaves filler	–	0.02	0.45	–	4.49	m	**4.94**
verges; extra single undercloak course of plain tiles	–	0.28	6.34	–	5.28	m	**11.62**
ambi-dry verge system	–	0.19	4.31	–	19.15	m	**23.46**
ambi-dry verge eave/ridge end piece	–	0.02	0.45	–	6.86	m	**7.31**
valley trough tiles; cutting both sides	–	0.51	11.55	–	41.32	m	**52.87**
universal angle ridge tiles	–	0.46	10.42	–	16.54	m	**26.96**
universal hip tiles; cutting both sides	–	0.60	13.59	–	34.78	m	**48.37**
dry ridge system; universal angle ridge tiles	–	0.23	5.21	–	25.55	m	**30.76**
universal mono-pitch angle ridge tiles	–	0.51	11.55	–	32.32	m	**43.87**
universal gas flue angle ridge tile	–	0.46	10.42	–	113.87	nr	**124.29**
universal angle ridge vent tile with 110 mm dia. adaptor	–	0.46	10.42	–	145.32	nr	**155.74**
holes for pipes and the like	–	0.19	4.31	–	–	nr	**4.31**
Concrete interlocking slates; Redland Richmond 10 Slates laid broken bonded, normally from right to left; smooth finish tiles or other equal; 75 mm lap; on 25 mm × 38 mm battens and type 1F reinforced underlay							
Slates 418 mm × 330 mm (Prime Cost £ per 1000)	1679.90	–	–	–	1679.90	1000	**1679.90**
roof coverings	17.99	0.32	7.25	–	24.62	m²	**31.87**
extra over coverings for							
fixing every tile	–	0.02	0.45	–	1.02	m²	**1.47**
eaves; eaves filler	–	0.02	0.45	–	4.20	m	**4.65**
verges; extra single undercloak course of plain tiles	–	0.23	5.21	–	5.45	m	**10.66**
ambi-dry verge system	–	0.19	4.31	–	19.15	m	**23.46**
ambi-dry verge eave/ridge end piece	–	0.02	0.45	–	6.86	m	**7.31**
universal valley trough tiles; cutting both sides	–	0.56	12.68	–	53.47	m	**66.15**
universal hip tiles; cutting both sides	–	0.60	13.59	–	21.99	m	**35.58**
universal angle ridge tiles	–	0.46	10.42	–	16.54	m	**26.96**
dry ridge system; universal angle ridge tiles	–	0.23	5.21	–	25.55	m	**30.76**
universal mono-pitch angle ridge tiles	–	0.51	11.55	–	32.32	m	**43.87**
gas ridge terminal	–	0.46	10.42	–	112.32	nr	**122.74**
ridge vent with 110 mm dia. flexible adaptor	–	0.46	10.42	–	143.78	nr	**154.20**
holes for pipes and the like	–	0.19	4.31	–	–	nr	**4.31**

18 TILE AND SLATE ROOF AND WALL COVERINGS

Item	PC £	Labour hours	Labour £	Plant £	Material £	Unit	Total rate £
Concrete interlocking slates; Redland Stonewold II smooth finish tiles or other equal; 75 mm lap; on 25 mm × 38 mm battens and type 1F reinforced underlay							
Slates 430 mm × 380 mm (Prime Cost £ per 1000)	5620.50	–	–	–	5620.50	1000	5620.50
roof coverings	54.52	0.32	7.25	–	65.08	m²	72.33
extra over coverings for							
fixing every tile	–	0.02	0.45	–	3.25	m²	3.70
eaves; eaves filler	–	0.02	0.45	–	4.49	m	4.94
verges; extra single undercloak course of plain tiles	–	0.28	6.34	–	5.63	m	11.97
ambi-dry verge system	–	0.19	4.31	–	19.15	m	23.46
ambi-dry verge eave/ridge end piece	–	0.02	0.45	–	6.86	m	7.31
valley trough tiles; cutting both sides	–	0.51	11.55	–	56.29	m	67.84
universal angle ridge tiles	–	0.46	10.42	–	16.54	m	26.96
universal hip tiles; cutting both sides	–	0.60	13.59	–	34.78	m	48.37
dry ridge system; universal angle ridge tiles	–	0.23	5.21	–	25.55	m	30.76
universal mono-pitch angle ridge tiles	–	0.51	11.55	–	32.32	m	43.87
universal gas flue angle ridge tile	–	0.46	10.42	–	113.87	nr	124.29
universal angle ridge vent tile with 110 mm dia. adaptor	–	0.46	10.42	–	145.32	nr	155.74
holes for pipes and the like	–	0.19	4.31	–	–	nr	4.31
Concrete interlocking slates; Redland Cambrian Slates made from 60% recycled Welsh slate or other equal; 50 mm lap; on 25 mm × 38 mm battens and type 1F reinforced underlay							
Slates 300 mm × 336 mm (Prime Cost £ per 1000)	2279.10	–	–	–	2279.10	1000	2279.10
roof coverings	36.24	0.32	7.25	–	48.06	m²	55.31
extra over coverings for							
fixing every tile	–	0.02	0.45	–	1.43	m²	1.88
eaves; eaves filler	–	0.02	0.45	–	4.99	m	5.44
verges; extra single undercloak course of plain tiles	–	0.23	5.21	–	14.48	m	19.69
ambi-dry verge system	–	0.19	4.31	–	19.15	m	23.46
ambi-dry verge eave/ridge end piece	–	0.02	0.45	–	6.86	m	7.31
universal valley trough tiles; cutting both sides	–	0.56	12.68	–	53.47	m	66.15
universal hip tiles; cutting both sides	–	0.60	13.59	–	23.59	m	37.18
universal angle ridge tiles	–	0.46	10.42	–	16.54	m	26.96
dry ridge system; universal angle ridge tiles	–	0.23	5.21	–	25.55	m	30.76
universal mono-pitch angle ridge tiles	–	0.51	11.55	–	32.32	m	43.87
gas ridge terminal	–	0.46	10.42	–	112.32	nr	122.74
ridge vent with 110 mm dia. flexible adaptor	–	0.46	10.42	–	143.78	nr	154.20
holes for pipes and the like	–	0.19	4.31	–	–	nr	4.31

18 TILE AND SLATE ROOF AND WALL COVERINGS

Item	PC £	Labour hours	Labour £	Plant £	Material £	Unit	Total rate £
18.02 INTERLOCKING TILING – cont							
Sundries and accessories							
Hip irons							
galvanized mild steel; fixing with screws	–	0.09	2.04	–	3.41	nr	**5.45**
Rytons Clip strip or other equal continuous soffit ventilator							
51 mm wide; plastic	–	0.28	6.34	–	1.90	m	**8.24**
Rytons over fascia ventilator or other equal and approved; continuous eaves ventilator							
40 mm wide; plastic	–	0.09	2.04	–	0.97	m	**3.01**
Rytons roof ventilator or other equal; to suit rafters at 600 mm centres							
250 mm deep × 43 mm high; plastic	–	0.09	2.04	–	2.47	m	**4.51**
Rytons push and lock ventilators or other equal; circular							
83 mm dia.; plastic	–	0.04	0.93	–	0.92	nr	**1.85**
Fixing only							
lead soakers (supply cost not included)	–	0.07	1.76	–	–	nr	**1.76**
Pressure impregnated softwood counter battens; 25 mm × 50 mm							
450 mm centres	–	0.06	1.36	–	3.22	m²	**4.58**
600 mm centres	–	0.04	0.91	–	2.44	m²	**3.35**
UNDERLAYS							
Underlay; BS EN 13707 type 1B; bitumen felt weighing 14 kg/10 m²; 75 mm laps							
to sloping or vertical surfaces	0.70	0.02	0.45	–	1.15	m²	**1.60**
Underlay; BS EN 13707 type 1F; reinforced bitumen felt weighing 22.50 kg/10 m²; 75 mm laps							
to sloping or vertical surfaces	0.88	0.02	0.45	–	1.34	m²	**1.79**
Underlay; Visqueen HP vapour barrier or other equal; multi-layer reinforce LDPE membrane with an aluminium core							
to sloping or vertical surfaces	0.44	0.02	0.45	–	0.96	m²	**1.41**
Underlay; Visqueen HP vapour barrier or other equal; multi-layer reinforce LDPE membrane with an aluminium core							
to sloping or vertical surfaces	0.44	0.02	0.45	–	0.96	m²	**1.41**
Underlay; reinforced vapour permeable membrane; 75 mm laps; Klober Permo Ecovent or similar							
to sloping or vertical surfaces	1.59	0.02	0.45	–	2.08	m²	**2.53**
Underlay; Klober Permo air breathable reinforced vapour permeable membrane; 75 mm laps							
to sloping or vertical surfaces	2.39	0.03	0.57	–	2.89	m²	**3.46**

18 TILE AND SLATE ROOF AND WALL COVERINGS

Item	PC £	Labour hours	Labour £	Plant £	Material £	Unit	Total rate £
18.03 FIBRE CEMENT SLATING							
Fibre cement artificial slates; Eternit Garsdale/ E2000T or other equal to 75 mm lap; on 19 mm × 50 mm battens and type 1F reinforced underlay							
Coverings; 600 mm × 600 mm slates							
roof coverings	–	0.60	13.59	–	23.88	m²	**37.47**
wall coverings	–	0.74	16.76	–	25.74	m²	**42.50**
Coverings; 600 mm × 300 mm slates							
roof coverings	–	0.46	10.42	–	24.03	m²	**34.45**
wall coverings	–	0.60	13.59	–	24.03	m²	**37.62**
Extra over slate coverings for							
double course at eaves	–	0.23	5.21	–	5.78	m	**10.99**
verges; extra single undercloak course	–	0.31	7.02	–	1.13	m	**8.15**
open valleys; cutting both sides	–	0.19	4.31	–	4.51	m	**8.82**
stop end	–	0.09	2.04	–	12.46	nr	**14.50**
roll top ridge tiles	–	0.56	12.68	–	40.15	m	**52.83**
stop end	–	0.09	2.04	–	22.21	nr	**24.25**
mono-pitch ridge tiles	–	0.46	10.42	–	49.66	m	**60.08**
stop end	–	0.09	2.04	–	58.85	nr	**60.89**
duo-pitch ridge tiles	–	0.46	10.42	–	47.56	m	**57.98**
stop end	–	0.09	2.04	–	38.33	nr	**40.37**
half round hip tiles; cutting both sides	–	0.19	4.31	–	82.16	m	**86.47**
holes for pipes and the like	–	0.19	4.31	–	–	nr	**4.31**
18.04 NATURAL SLATING							
NOTE: The following items of slate roofing unless otherwise described, include for conventional fixing assuming normal exposure with appropriate nails and/or rivets or clips to pressure impregnated softwood battens fixed with galvanized nails; prices also include for all bedding and pointing at verges; beneath verge tiles etc.							
Natural slates; First quality Spanish blue grey; uniform size; to 75 mm lap; on 25 mm × 50 mm battens and type 1F reinforced underlay							
Coverings; 400 mm × 250 mm slates (Prime Cost £ per 1000)	1500.00	–	–	–	1545.00	1000	**1545.00**
roof coverings	36.75	1.20	27.18	–	50.40	m²	**77.58**
wall coverings	36.75	1.06	24.01	–	50.40	m²	**74.41**
Coverings; 500 mm × 250 mm slates (Prime Cost £ per 1000)	1390.00	–	–	–	1431.70	1000	**1431.70**
roof coverings	25.99	1.20	27.18	–	36.45	m²	**63.63**
wall coverings	25.99	0.88	19.93	–	36.45	m²	**56.38**
Coverings; 500 mm × 375 mm slates (Prime Cost £ per 1000)	3880.00	–	–	–	3996.40	1000	**3996.40**
roof coverings	48.89	1.20	27.18	–	59.74	m²	**86.92**
wall coverings	48.89	0.69	15.63	–	59.74	m²	**75.37**

18 TILE AND SLATE ROOF AND WALL COVERINGS

Item	PC £	Labour hours	Labour £	Plant £	Material £	Unit	Total rate £
18.04 NATURAL SLATING – cont							
Natural slates – cont							
Extra over coverings for							
double course at eaves	–	0.28	6.34	–	15.43	m	**21.77**
verges; extra single undercloak course	–	0.39	8.84	–	8.50	m	**17.34**
open valleys; cutting both sides	–	0.20	4.53	–	33.57	m	**38.10**
blue/black glass reinforced concrete 152 mm half round ridge tiles	–	0.46	10.42	–	16.87	m	**27.29**
blue/black glass reinforced concrete 125 mm × 125 mm plain angle ridge tiles	–	0.46	10.42	–	16.87	m	**27.29**
mitred hips; cutting both sides	–	0.20	4.53	–	33.57	m	**38.10**
blue/black glass reinforced concrete 152 mm half round hip tiles; cutting both sides	–	0.65	14.72	–	50.44	m	**65.16**
blue/black glass reinforced concrete 125 mm × 125 mm plain angle hip tiles; cutting both sides	–	0.65	14.72	–	50.43	m	**65.15**
holes for pipes and the like	–	0.19	4.31	–	–	nr	**4.31**
Natural slates; Welsh blue grey; 7 mm nominal thickness; uniform size; to 75 mm lap; on 25 mm × 50 mm battens and type 1F reinforced underlay							
Coverings; 400 mm × 250 mm slates (Prime Cost £ per 1000)	3200.00	–	–	–	3296.00	1000	**3296.00**
roof coverings	78.40	0.70	15.85	–	95.45	m²	**111.30**
wall coverings	78.40	1.00	22.65	–	95.45	m²	**118.10**
Coverings; 500 mm × 250 mm slate (Prime Cost £ per 1000)	5330.00	–	–	–	5489.90	1000	**5489.90**
roof coverings	99.67	0.60	13.59	–	116.13	m²	**129.72**
wall coverings	99.67	0.80	18.12	–	116.13	m²	**134.25**
Coverings; 500 mm × 300 mm slates (Prime Cost £ per 1000)	8780.00	–	–	–	9043.40	1000	**9043.40**
roof coverings	136.97	0.60	13.59	–	156.47	m²	**170.06**
wall coverings	136.97	0.75	16.98	–	156.47	m²	**173.45**
Coverings; 600 mm × 300 mm slates (Prime Cost £ per 1000)	10500.00	–	–	–	10815.00	1000	**10815.00**
roof coverings	132.30	0.50	11.32	–	149.95	m²	**161.27**
wall coverings	132.30	0.65	14.72	–	149.95	m²	**164.67**
Extra over coverings for							
double course at eaves	–	0.25	5.67	–	39.06	m	**44.73**
verges; extra single undercloak course	–	0.35	7.93	–	22.81	m	**30.74**
open valleys; cutting both sides	–	0.20	4.53	–	90.85	m	**95.38**
blue/black glazed ware 152 mm half round ridge tiles	–	0.46	10.42	–	20.71	m	**31.13**
blue/black glazed ware 125 mm × 125 mm plain angle ridge tiles	–	0.46	10.42	–	23.93	m	**34.35**
mitred hips; cutting both sides	–	0.20	4.53	–	90.85	m	**95.38**
blue/black glazed ware 152 mm half round hip tiles; cutting both sides	–	0.65	14.72	–	111.56	m	**126.28**
blue/black glazed ware 125 mm × 125 mm plain angle hip tiles; cutting both sides	–	0.65	14.72	–	114.77	m	**129.49**
holes for pipes and the like	–	0.19	4.31	–	–	nr	**4.31**

18 TILE AND SLATE ROOF AND WALL COVERINGS

Item	PC £	Labour hours	Labour £	Plant £	Material £	Unit	Total rate £
Natural slates; Westmoreland green; random lengths; 457 mm–229 mm proportionate widths to 75 mm lap; in diminishing courses; on 25 mm × 50 mm battens and type 1F underlay							
Coverings (Prime Cost £ per tonne; approximately 17.5 m²)	3150.00	–	–	–	3244.50	tonne	**3244.50**
roof coverings	179.55	1.00	22.65	–	203.49	m²	**226.14**
wall coverings	179.55	1.30	29.45	–	203.49	m²	**232.94**
Extra over coverings for							
double course at eaves	–	0.60	13.59	–	35.79	m	**49.38**
verges; extra single undercloak course slates							
152 mm wide	–	0.67	15.17	–	31.03	m	**46.20**
holes for pipes and the like	–	0.25	5.67	–	–	nr	**5.67**
18.05 NATURAL OR ARTIFICIAL STONE SLATING							
Reconstructed stone slates; Hardrow Slates or similar; standard colours; 75 mm lap; on 25 mm × 50 mm battens and type 1F reinforced underlay							
Coverings; 457 mm × 305 mm slates (Prime Cost £ per 1000)	1854.00	–	–	–	1909.62	1000	**1909.62**
roof coverings	31.89	0.74	16.76	–	44.04	m²	**60.80**
wall coverings	31.89	0.93	21.06	–	44.04	m²	**65.10**
Coverings; 457 mm × 457 mm slates (Prime Cost £ per 1000)	2421.00	–	–	–	2493.63	1000	**2493.63**
roof coverings	27.84	0.60	13.59	–	39.35	m²	**52.94**
wall coverings	27.84	0.79	17.89	–	39.35	m²	**57.24**
Extra over 457 mm × 305 mm coverings for							
double course at eaves	–	0.28	6.34	–	8.23	m	**14.57**
verges; pointed	–	0.39	8.84	–	–0.85	m	**7.99**
open valleys; cutting both sides	–	0.20	4.53	–	20.05	m	**24.58**
ridge tiles	–	0.46	10.42	–	53.83	m	**64.25**
hip tiles; cutting both sides	–	0.65	14.72	–	40.65	m	**55.37**
holes for pipes and the like	–	0.19	4.31	–	–	nr	**4.31**
Reconstructed stone slates; Bradstone Cotswold style or similar; random lengths 550 mm–300 mm; proportional widths; to 80 mm lap; in diminishing courses; on 25 mm × 50 mm battens and type 1F reinforced underlay							
Roof coverings							
All-in rate inclusive of eaves and verges (Prime Cost £ per m²)	40.17	0.97	21.97	–	53.00	m²	**74.97**
Extra over coverings for							
open valleys/mitred hips; cutting both sides	–	0.42	9.52	–	19.55	m²	**29.07**
ridge tiles	–	0.61	13.81	–	26.33	m	**40.14**
hip tiles; cutting both sides	–	0.97	21.97	–	44.22	m	**66.19**
holes for pipes and the like	–	0.28	6.34	–	–	nr	**6.34**

18 TILE AND SLATE ROOF AND WALL COVERINGS

Item	PC £	Labour hours	Labour £	Plant £	Material £	Unit	Total rate £
18.05 NATURAL OR ARTIFICIAL STONE SLATING – cont							
Reconstructed stone slates; Bradstone Moordale style or similar; random lengths 550 mm–450 mm; proportional widths; to 80 mm lap; in diminishing course; on 25 mm × 50 mm battens and type 1F reinforced underlay							
Roof coverings							
all-in rate inclusive of eaves and verges (Prime Cost £ per m²)	37.80	0.97	21.97	–	50.43	m²	**72.40**
Extra over coverings for							
open valleys/mitred hips; cutting both sides	–	0.42	9.52	–	18.40	m²	**27.92**
ridge tiles	–	0.61	13.81	–	26.33	m	**40.14**
holes for pipes and the like	–	0.28	6.34	–	–	nr	**6.34**
18.06 TIMBER SHINGLES							
Red Cedar sawn shingles preservative treated; uniform length 400 mm; to 125 mm gauge; on 25 mm × 38 mm battens and type 1F reinforced underlay							
Shingles; uniform length 400 mm (Prime Cost £ per bundle)	93.24	–	–	–	96.04	bundle	**96.04**
roof coverings; 125 mm gauge (2.28 m² per bundle)	40.93	0.97	21.97	–	54.92	m²	**76.89**
wall coverings; 190 mm gauge (3.47 m²/bundle)	26.85	0.74	16.76	–	36.52	m²	**53.28**
Extra over for							
double course at eaves	–	0.19	4.31	–	5.55	m	**9.86**
open valleys; cutting both sides	–	0.19	4.31	–	10.84	m	**15.15**
preformed ridge capping	–	0.28	6.34	–	12.90	m	**19.24**
preformed hip capping; cutting both sides	–	0.46	10.42	–	23.72	m	**34.14**
double starter course to cappings	–	0.09	2.04	–	12.72	m	**14.76**
holes for pipes and the like	–	0.14	3.17	–	–	nr	**3.17**

19 WATERPROOFING

Item	PC £	Labour hours	Labour £	Plant £	Material £	Unit	Total rate £
19.01 MASTIC ASPHALT ROOFING							
Mastic asphalt to BS 6925 Type R988							
20 mm thick two coat coverings; felt isolating membrane; to concrete (or timber) base; flat or to falls or slopes not exceeding 10° from horizontal							
over 300 mm wide	–	–	–	–	–	m²	25.56
225 mm–300 mm wide	–	–	–	–	–	m²	39.47
150 mm–225 mm wide	–	–	–	–	–	m²	46.11
not exceeding 150 mm wide	–	–	–	–	–	m²	59.35
Add to the above for covering with							
10 mm thick limestone chippings in hot bitumen	–	–	–	–	–	m²	4.17
coverings with solar reflective paint	–	–	–	–	–	m²	4.69
300 mm × 300 mm × 8 mm g.r.p. tiles in hot bitumen	–	–	–	–	–	m²	70.68
Cutting to line; jointing to old asphalt	–	–	–	–	–	m	8.09
13 mm thick two coat skirtings to brickwork base							
not exceeding 150 mm girth	–	–	–	–	–	m	17.42
150 mm–225 mm girth	–	–	–	–	–	m	20.00
225 mm–300 mm girth	–	–	–	–	–	m	24.47
13 mm thick three coat skirtings; expanded metal lathing reinforcement nailed to timber base							
not exceeding 150 mm girth	–	–	–	–	–	m	29.28
150 mm–225 mm girth	–	–	–	–	–	m	34.88
225 mm–300 mm girth	–	–	–	–	–	m	40.82
13 mm thick two coat fascias to concrete base							
not exceeding 150 mm girth	–	–	–	–	–	m	17.42
150 mm–225 mm girth	–	–	–	–	–	m	20.00
20 mm thick two coat linings to channels to concrete base							
not exceeding 150 mm girth	–	–	–	–	–	m	38.30
150 mm–225 mm girth	–	–	–	–	–	m	43.55
225 mm–300 mm girth	–	–	–	–	–	m	44.81
20 mm thick two coat lining to cesspools							
250 mm × 150 mm × 150 mm deep	–	–	–	–	–	nr	37.53
Collars around pipes, standards and like members	–	–	–	–	–	nr	26.85
Accessories							
Eaves trim; extruded aluminium alloy; working asphalt into trim							
Alutrim; type A roof edging or other equal and approved	–	–	–	–	–	m	16.00
extra; angle	–	–	–	–	–	nr	8.96
Roof screed ventilator – aluminium alloy							
Extr-aqua-vent or other equal and approved; set on screed over and including dished sinking; working collar around ventilator	–	–	–	–	–	nr	31.01

19 WATERPROOFING

Item	PC £	Labour hours	Labour £	Plant £	Material £	Unit	Total rate £
19.01 MASTIC ASPHALT ROOFING – cont							
Bituminous lightweight insulating roof screeds							
Bit-Ag or similar roof screed or other equal and approved; to falls or cross-falls; bitumen felt vapour barrier; over 300 mm wide							
75 mm (average) thick	–	–	–	–	–	m^2	**56.67**
100 mm (average) thick	–	–	–	–	–	m^2	**71.82**
19.02 APPLIED LIQUID TANKING							
Tanking and damp-proofing							
Synthaprufe or other equal and approved; blinding with sand; horizontal on slabs							
two coats	–	0.19	3.21	–	3.79	m^2	**7.00**
three coats	–	0.26	4.40	–	5.61	m^2	**10.01**
One coat Vandex Super 0.75 kg/m^2 slurry or other equal and approved; one consolidating coat of Vandex BB75 1 kg/m^2 slurry or other equal and approved; horizontal on beds							
over 225 mm wide	–	0.32	5.41	–	5.09	m^2	**10.50**
Intergritank; Methacrylate resin based structural waterproofing membrane; in two separate colour coded coats; minimumm 2 mm overalll dry film finish; on a primed substrate. Typical project size between 250 m^2 and 10,000 m^2, over 250 mm wide							
100 m^2–499 m^2	–	–	–	–	–	m^2	**60.71**
500 m^2–1999 m^2	–	–	–	–	–	m^2	**50.11**
over 2000 m^2	–	–	–	–	–	m^2	**40.47**
19.03 MASTIC ASPHALT TANKING AND DAMP-PROOF MEMBRANES							
Mastic asphalt to BS 6925 Type T 1097							
13 mm thick one coat coverings to concrete base; flat; subsequently covered							
over 300 mm wide	–	–	–	–	–	m^2	**19.20**
225 mm–300 mm wide	–	–	–	–	–	m^2	**55.15**
150 mm–225 mm wide	–	–	–	–	–	m^2	**60.49**
not exceeding 150 mm wide	–	–	–	–	–	m^2	**75.56**
20 mm thick two coat coverings to concrete base; flat; subsequently covered							
over 300 mm wide	–	–	–	–	–	m^2	**24.16**
225 mm–300 mm wide	–	–	–	–	–	m^2	**49.78**
150 mm–225 mm wide	–	–	–	–	–	m^2	**69.66**
not exceeding 150 mm wide	–	–	–	–	–	m^2	**81.40**
30 mm thick three coat coverings to concrete base; flat; subsequently covered							
over 300 mm wide	–	–	–	–	–	m^2	**38.78**
225 mm–300 mm wide	–	–	–	–	–	m^2	**79.93**
150 mm–225 mm wide	–	–	–	–	–	m^2	**86.73**
not exceeding 150 mm wide	–	–	–	–	–	m^2	**105.67**

19 WATERPROOFING

Item	PC £	Labour hours	Labour £	Plant £	Material £	Unit	Total rate £
13 mm thick two coat coverings to brickwork base; vertical; subsequently covered							
over 300 mm wide	–	–	–	–	–	m²	**53.36**
225 mm–300 mm wide	–	–	–	–	–	m²	**76.70**
150 mm–225 mm wide	–	–	–	–	–	m²	**82.83**
not exceeding 150 mm wide	–	–	–	–	–	m²	**108.25**
20 mm thick three coat coverings to brickwork base; vertical; subsequently covered							
over 300 mm wide	–	–	–	–	–	m²	**86.32**
225 mm–300 mm wide	–	–	–	–	–	m²	**103.34**
150 mm–225 mm wide	–	–	–	–	–	m²	**113.40**
not exceeding 150 mm wide	–	–	–	–	–	m²	**147.02**
Turning into groove 20 mm deep	–	–	–	–	–	m	**1.04**
Internal angle fillets; subsequently covered	–	–	–	–	–	m	**6.01**

19.04 FLEXIBLE SHEET TANKING AND DAMP-PROOF MEMBRANES

Sheet tanking

Item	PC £	Labour hours	Labour £	Plant £	Material £	Unit	Total rate £
Preprufe pre-applied self-adhesive waterproofing membranes for use below concrete slabs or behind concrete walls. HPE film with a pressure sensitive adhesive and weather resistant protective coating Preprufe 300R heavy duty grade for use below slabs and on rafts							
over 300 mm wide; horizontal	–	0.10	1.69	–	21.03	m²	**22.72**
not exceeding 300 mm wide; horizontal	–	0.11	1.85	–	8.82	m	**10.67**
Preprufe 300R Plus heavy duty grade for use below slabs and on rafts designed to accept placing of heavy reinforcement							
over 300 mm wide; horizontal	–	0.11	1.85	–	22.31	m²	**24.16**
not exceeding 300 mm wide; horizontal	–	0.12	2.04	–	8.92	m	**10.96**
Preprufe 160R thinner grade for use blindside, zero property line applications against soil retention							
over 300 mm wide; horizontal	–	0.10	1.69	–	18.48	m²	**20.17**
not exceeding 300 mm wide; horizontal	–	0.11	1.85	–	7.75	m	**9.60**
Preprufe 160R Plus thinner grade for use blindside, zero property line applications against soil retention. Vertical use ony							
over 300 mm wide; horizontal	–	0.11	1.85	–	19.61	m²	**21.46**
not exceeding 300 mm wide; horizontal	–	0.12	2.03	–	7.85	m	**9.88**
Preprufe 800PA reinforced cross laminated HDPE film for use below ground car parks, basements, underground reservoirs and tanks							
over 300 mm wide; horizontal	–	0.15	2.53	–	11.89	m²	**14.42**
not exceeding 300 mm wide; horizontal	–	0.17	2.79	–	4.99	m	**7.78**
Visqueen self-adhesive damp-proof membrane							
over 300 mm wide; horizontal	–	–	–	–	–	m²	**7.97**
not exceeding 300 mm wide; horizontal	–	–	–	–	–	m	**3.07**

19 WATERPROOFING

Item	PC £	Labour hours	Labour £	Plant £	Material £	Unit	Total rate £
19.04 FLEXIBLE SHEET TANKING AND DAMP-PROOF MEMBRANES – cont							
Sheet tanking – cont							
Tanking primer for self-adhesive DPM							
over 300 mm wide; horizontal	–	–	–	–	–	m²	5.52
not exceeding 300 mm wide; horizontal	–	–	–	–	–	m	2.52
Bituthene sheeting or other equal and approved; lapped joints; horizontal on slabs							
8000 grade	–	0.10	1.69	–	28.15	m²	29.84
Bituthene sheeting or other equal and approved; lapped joints; dressed up vertical face of concrete							
8000 grade	–	0.17	2.87	–	28.15	m²	31.02
RIW Structureseal tanking and damp-proof membrane; or other equal and approved							
over 300 mm wide; horizontal	–	–	–	–	–	m²	8.59
Structureseal Fillet							
40 mm × 40 mm	–	–	–	–	–	m	6.51
Ruberoid Plasfrufe 2000SA self-adhesive damp-proof membrane							
over 300 mm wide; horizontal	–	–	–	–	–	m²	15.17
not exceeding 300 mm wide; horizontal	–	–	–	–	–	m	5.82
extra for 50 mm thick sand blinding	–	–	–	–	–	m²	2.67
Servipak protection board or other equal; butt jointed; taped joints; to horizontal surfaces;							
3 mm thick	–	0.14	2.37	–	13.79	m²	16.16
6 mm thick	–	0.14	2.37	–	15.76	m²	18.13
12 mm thick	–	0.19	3.21	–	26.99	m²	30.20
Servipak protection board or other equal; butt jointed; taped joints; to vertical surfaces							
3 mm thick	–	0.19	3.21	–	13.79	m²	17.00
6 mm thick	–	0.19	3.21	–	15.76	m²	18.97
12 mm thick	–	0.23	3.88	–	26.99	m²	30.87
19.05 SPECIALIST WATERPROOF RENDERING							
Sika waterproof rendering or other equal; steel trowelled							
20 mm work to walls; three coat; to concrete base							
over 300 mm wide	–	–	–	–	–	m²	39.10
not exceeding 300 mm wide	–	–	–	–	–	m²	59.23
25 mm work to walls; three coat; to concrete base							
over 300 mm wide	–	–	–	–	–	m²	46.20
not exceeding 300 mm wide	–	–	–	–	–	m²	71.07
40 mm work to walls; four coat; to concrete base							
over 300 mm wide	–	–	–	–	–	m²	68.12
not exceeding 300 mm wide	–	–	–	–	–	m²	106.62

19 WATERPROOFING

Item	PC £	Labour hours	Labour £	Plant £	Material £	Unit	Total rate £
Sto External render only system; comprising glassfibre mesh reinforcement embedded in 10 mm Sto Levell Cote with Sto Armat Classic Basecoat Render and Stolit K 1.5 Decorative Topcoat Render (white)							
15 mm thick work to walls; two coats; to brickwork or blockwork base							
over 300 mm wide	–	–	–	–	–	m²	60.04
extra over for							
bellcast bead	–	–	–	–	–	m	5.72
external angle with PVC mesh angle bead	–	–	–	–	–	m	5.27
internal angle with Sto Armor angle	–	–	–	–	–	m	5.27
render stop bead	–	–	–	–	–	m	5.27
K-Rend render or similar through-colour render system							
18 mm thick work to walls; two coats; to brickwork or blockwork base; first coat 8 mm standard base coat; second coat 10 mm K-rend silicone WP/FT							
over 300 mm wide	–	–	–	–	–	m²	74.33

20 PROPRIETARY LININGS AND PARTITIONS

Item	PC £	Labour hours	Labour £	Plant £	Material £	Unit	Total rate £
20.01 PARTITIONS AND WALLS							
Gyproc metal stud proprietary partitions or other equal; metal Gypframe C studs at 600 mm centres; floor and ceiling channels							
73 mm partition; 48 mm studs and channels; one layer of 12.5 mm Gyproc Wallboard each side; joints filled with joint filler and joint tape to receive direct decoration							
average height 2.00 m	–	0.63	15.58	–	18.44	m²	**34.02**
average height 3.00 m	–	0.58	14.54	–	16.87	m²	**31.41**
Rate per m (SMM7 measurement rule)							
height 2.10 m–2.40 m	–	1.50	37.39	–	44.24	m	**81.63**
height 2.40 m–2.70 m	–	1.69	42.13	–	47.62	m	**89.75**
height 2.70 m–3.00 m	–	1.75	43.62	–	50.60	m	**94.22**
78 mm partition; 48 mm studs and channels; one layer of 15 mm Gyproc Wallboard each side; joints filled with joint filler and joint tape to receive direct decoration							
average height 2.00 m	–	0.65	16.20	–	19.88	m²	**36.08**
average height 3.00 m	–	0.61	15.20	–	18.31	m²	**33.51**
Rate per m (SMM7 measurement rule)							
height 2.10 m–2.40 m	–	1.57	39.26	–	47.71	m	**86.97**
height 2.40 m–2.70 m	–	1.77	44.11	–	51.53	m	**95.64**
height 2.70 m–3.00 m	–	1.99	49.60	–	61.62	m	**111.22**
95 mm partition; 70 mm studs and channels; one layer of 12.5 mm Gyproc Wallboard each side; joints filled with joint filler and joint tape to receive direct decoration							
average height 2.00 m	–	0.63	15.58	–	19.12	m²	**34.70**
average height 3.00 m	–	0.58	14.54	–	17.66	m²	**32.20**
Rate per m (SMM7 measurement rule)							
height 2.10 m–2.40 m	–	1.50	37.39	–	45.88	m	**83.27**
height 2.40 m–2.70 m	–	1.69	42.13	–	49.49	m	**91.62**
height 2.70 m–3.00 m	–	1.75	43.62	–	52.99	m	**96.61**
height 3.00 m–3.30 m	–	2.00	49.85	–	56.50	m	**106.35**
100 mm partition; 70 mm studs and channels; one layer of 15 mm Gyproc Wallboard each side; joints filled with joint filler and joint tape to receive direct decoration							
average height 2.00 m	–	0.65	16.20	–	20.53	m²	**36.73**
average height 3.00 m	–	0.61	15.20	–	21.26	m²	**36.46**
Rate per m (SMM7 measurement rule)							
height 2.10 m–2.40 m	–	1.57	39.26	–	49.35	m	**88.61**
height 2.40 m–2.70 m	–	1.77	44.11	–	53.40	m	**97.51**
height 2.70 m–3.00 m	–	1.99	49.60	–	63.78	m	**113.38**
height 3.00 m–3.30 m	–	2.10	52.34	–	67.36	m	**119.70**

20 PROPRIETARY LININGS AND PARTITIONS

Item	PC £	Labour hours	Labour £	Plant £	Material £	Unit	Total rate £
120 mm partition; 70 mm studs and channels; two layers of 12.5 mm Gyproc Wallboard each side; joints filled with joint filler and joint tape to receive direct decoration							
average height 2.00 m	–	0.73	18.20	–	26.02	m²	**44.22**
average height 3.00 m	–	0.68	17.04	–	21.78	m²	**38.82**
Rate per m (SMM7 measurement rule)							
height 2.10 m–2.40 m	–	1.75	43.62	–	62.44	m	**106.06**
height 2.40 m–2.70 m	–	1.99	49.60	–	68.12	m	**117.72**
height 2.70 m–3.00 m	–	2.05	51.10	–	73.70	m	**124.80**
height 3.00 m–3.30 m	–	2.50	62.31	–	79.27	m	**141.58**
130 mm partition; 70 mm studs and channels; two layers of 15 mm Gyproc Wallboard each side; joints filled with joint filler and joint tape to receive direct decoration							
average height 2.00 m	–	0.80	19.94	–	28.91	m²	**48.85**
average height 3.00 m	–	0.75	18.69	–	27.56	m²	**46.25**
Rate per m (SMM7 measurement rule)							
height 2.10 m–2.40 m	–	1.82	45.36	–	69.39	m	**114.75**
height 2.40 m–2.70 m	–	2.07	51.59	–	53.69	m	**105.28**
height 2.70 m–3.00 m	–	2.29	57.08	–	88.83	m	**145.91**
height 3.00 m–3.30 m	–	2.40	59.82	–	94.91	m	**154.73**
Gypwall metal stud proprietary partitions or other equal; metal GypWall Acoustic C studs at 600 mm centres; deep flange floor and ceiling channels							
95 mm partition; 70 mm studs and channels; one layer of 12.5 mm Gyproc SoundBloc each side; joints filled with joint filler and joint tape to receive direct decoration							
average height 2.00 m	–	0.63	15.58	–	29.06	m²	**44.64**
average height 3.00 m	–	0.58	14.54	–	27.66	m²	**42.20**
Rate per m (SMM7 measurement rule)							
height 2.10 m–2.40 m	–	1.50	37.39	–	69.72	m	**107.11**
height 2.40 m–2.70 m	–	1.69	42.13	–	75.67	m	**117.80**
height 2.70 m–3.00 m	–	1.75	43.62	–	82.97	m	**126.59**
100 mm partition; 70 mm studs and channels; one layer of 15 mm Gyproc SoundBloc each side; joints filled with joint filler and joint tape to receive direct decoration							
average height 2.00 m	–	0.63	15.58	–	31.06	m²	**46.64**
average height 3.00 m	–	0.58	14.54	–	29.67	m²	**44.21**
Rate per m (SMM7 measurement rule)							
height 2.10 m–2.40 m	–	1.57	39.26	–	74.56	m	**113.82**
height 2.40 m–2.70 m	–	1.77	44.11	–	81.12	m	**125.23**
height 2.70 m–3.00 m	–	1.99	49.60	–	87.18	m	**136.78**

20 PROPRIETARY LININGS AND PARTITIONS

Item	PC £	Labour hours	Labour £	Plant £	Material £	Unit	Total rate £
20.01 PARTITIONS AND WALLS – cont							
Gypwall metal stud proprietary partitions or other equal – cont							
120 mm partition; 70 mm studs and channels; two layers of 12.5 mm Gyproc SoundBloc each side; joints filled with joint filler and joint tape to receive direct decoration							
average height 2.00 m	–	0.80	19.94	–	39.47	m²	**59.41**
average height 3.00 m	–	0.75	18.69	–	38.08	m²	**56.77**
Rate per m (SMM7 measurement rule)							
height 2.10 m–2.40 m	–	2.00	49.85	–	95.14	m	**144.99**
height 2.40 m–2.70 m	–	2.20	54.84	–	104.10	m	**158.94**
height 2.70 m–3.00 m	–	2.30	57.33	–	114.56	m	**171.89**
130 mm partition; 70 mm studs and channels; two layers of 15 mm Gyproc SoundBloc each side; joints filled with joint filler and joint tape to receive direct decoration							
average height 2.00 m	–	0.80	19.94	–	43.51	m²	**63.45**
average height 3.00 m	–	0.75	18.69	–	42.11	m²	**60.80**
Rate per m (SMM7 measurement rule)							
height 2.10 m–2.40 m	–	2.00	49.85	–	104.82	m	**154.67**
height 2.40 m–2.70 m	–	2.20	54.84	–	115.00	m	**169.84**
height 2.70 m–3.00 m	–	2.30	57.33	–	126.66	m	**183.99**
Gyproc metal stud proprietary partitions or other equal; metal Gypframe C studs at 600 mm centres; floor and ceiling channels							
95 mm partition; 70 mm studs and channels; one layer of 12.5 mm Gyproc Fireline each side; joints filled with joint filler and joint tape to receive direct decoration							
average height 2.00 m	–	0.80	19.94	–	36.90	m²	**56.84**
average height 3.00 m	–	0.75	18.69	–	35.51	m²	**54.20**
Rate per m (SMM7 measurement rule)							
height 2.10 m–2.40 m	–	1.50	37.39	–	66.65	m	**104.04**
height 2.40 m–2.70 m	–	1.70	42.37	–	72.21	m	**114.58**
height 2.70 m–3.00 m	–	1.75	43.62	–	77.28	m	**120.90**
100 mm partition; 70 mm studs and channels; one layer of 15 mm Gyproc Fireline each side; joints filled with joint filler and joint tape to receive direct decoration							
average height 2.00 m	–	0.80	19.94	–	40.62	m²	**60.56**
average height 3.00 m	–	0.75	18.69	–	39.22	m²	**57.91**
Rate per m (SMM7 measurement rule)							
height 2.10 m–2.40 m	–	1.50	37.39	–	71.10	m	**108.49**
height 2.40 m–2.70 m	–	1.70	42.37	–	77.23	m	**119.60**
height 2.70 m–3.00 m	–	1.75	43.62	–	82.85	m	**126.47**

20 PROPRIETARY LININGS AND PARTITIONS

Item	PC £	Labour hours	Labour £	Plant £	Material £	Unit	Total rate £
120 mm partition; 70 mm studs and channels; two layers of 12.5 mm Gyproc Fireline each side; joints filled with joint filler and joint tape to receive direct decoration							
average height 2.00 m	–	1.10	27.42	–	37.02	m²	**64.44**
average height 3.00 m	–	1.15	28.66	–	34.25	m²	**62.91**
Rate per m (SMM7 measurement rule)							
height 2.10 m–2.40 m	–	2.00	49.85	–	88.84	m	**138.69**
height 2.40 m–2.70 m	–	2.20	54.84	–	97.18	m	**152.02**
height 2.70 m–3.00 m	–	2.30	57.33	–	102.73	m	**160.06**
130 mm partition; 70 mm studs and channels; two layers of 15 mm Gyproc Fireline each side; joints filled with joint filler and joint tape to receive direct decoration							
average height 2.00 m	–	1.10	27.42	–	40.73	m²	**68.15**
average height 3.00 m	–	1.15	28.66	–	37.96	m²	**66.62**
Rate per m (SMM7 measurement rule)							
height 2.10 m–2.40 m	–	2.00	49.85	–	97.75	m	**147.60**
height 2.40 m–2.70 m	–	2.20	54.84	–	107.20	m	**162.04**
height 2.70 m–3.00 m	–	2.30	57.33	–	115.79	m	**173.12**
Gypwall Rapid/db Plus metal stud housing partitioning system; or other equal; floor and ceiling channels plugged and screwed to concrete or nailed to timber							
75 mm partition; 43 mm studs at 600 mm centres; one layer of 15 mm SoundBloc Rapid each side; joints filled with joint filler and joint tape to receive direct decoration							
average height 2.00 m	–	0.63	14.46	–	19.23	m²	**33.69**
average height 3.00 m	–	0.83	19.28	–	24.70	m²	**43.98**
Rate per m (SMM7 measurement rule)							
height 2.10 m–2.40 m	–	1.50	34.70	–	46.16	m	**80.86**
height 2.40 m–2.70 m	–	1.75	40.49	–	51.29	m	**91.78**
height 2.70 m–3.00 m	–	2.00	46.27	–	56.55	m	**102.82**
Angles, corners and similar							
T-junctions	–	0.17	4.05	–	5.45	m	**9.50**
splayed corners	–	0.13	2.89	–	3.66	m	**6.55**
corners	–	0.10	2.32	–	1.86	m	**4.18**
fair ends	–	0.11	2.54	–	0.96	m	**3.50**
Cavity insulation to stud partitions pinned to board							
25 mm Isover APR							
average height 2.00 m	–	0.06	1.44	–	1.26	m	**2.70**
average height 3.00	–	0.07	1.62	–	1.26	m	**2.88**
Rate per m (SMM7 measurement rule)							
height 2.10 m–2.40 m	–	0.15	3.47	–	3.01	m	**6.48**
height 2.40 m–2.70 m	–	0.16	3.70	–	3.39	m	**7.09**
height 3.00 m–3.30 m	–	0.17	3.93	–	4.14	m	**8.07**

20 PROPRIETARY LININGS AND PARTITIONS

Item	PC £	Labour hours	Labour £	Plant £	Material £	Unit	Total rate £
20.01 PARTITIONS AND WALLS – cont							
Cavity insulation to stud partitions pinned to board – cont							
50 mm Isover APR							
average height 2.00 m	–	0.06	1.44	–	2.06	m	3.50
average height 3.00 m	–	0.07	1.62	–	2.06	m	3.68
Rate per m (SMM7 measurement rule)							
height 2.10 m–2.40 m	–	0.15	3.47	–	4.94	m	8.41
height 2.40 m–2.70 m	–	0.16	3.70	–	5.56	m	9.26
height 3.00 m–3.30 m	–	0.17	3.93	–	6.80	m	10.73
Terrazzo faced partitions; polished on two faces							
Precast reinforced terrazzo faced WC partitions							
38 mm thick; over 300 mm wide	–	–	–	–	–	m²	337.45
50 mm thick; over 300 mm wide	–	–	–	–	–	m²	346.30
Wall post; once rebated							
64 mm × 102 mm	–	–	–	–	–	m	154.89
64 mm × 152 mm	–	–	–	–	–	m	168.17
Centre post; twice rebated							
64 mm × 102 mm	–	–	–	–	–	m	162.64
64 mm × 152 mm	–	–	–	–	–	m	179.23
Lintel; once rebated							
64 mm × 102 mm	–	–	–	–	–	m	168.17
Pair of brass topped plates or sockets cast into posts for fixings (not included)	–	–	–	–	–	nr	38.72
Brass indicator bolt lugs cast into posts for fixings (not included)	–	–	–	–	–	nr	16.59
DEMOUNTABLE PARTITIONS							
Fire-resistant concertina partition							
Insulated panel and two-hour fire wall system for warehouses etc., comprising white polyester coated galvanized steel frame and 0.55 mm galvanized steel panels either side of rockwool infill							
100 mm thick wall: 31 Rw dB acoustic rating	–	–	–	–	–	m²	50.59
150 mm thick wall: 31 Rw dB acoustic rating	–	–	–	–	–	m²	54.38
intumescent mastic sealant; bedding frames at perimeter of metal fire walls	–	–	–	–	–	m	4.44
Room divider moveable wall							
Laminated both sides top hung movable acoustic panel wall with concealed uPVC vertical edge profiles, 1106 m × 3000 mm panels and two point panel support system							
105 mm thick wall: 47 Rw dB acoustic rating	–	–	–	–	–	m²	467.92
105 mm thick wall: 50 Rw dB acoustic rating	–	–	–	–	–	m²	505.85
105 mm thick wall: 53 Rw dB acoustic rating	–	–	–	–	–	m²	543.80

20 PROPRIETARY LININGS AND PARTITIONS

Item	PC £	Labour hours	Labour £	Plant £	Material £	Unit	Total rate £
20.02 LININGS TO WALLS							
SUPPLY ONLY SHEET LINING – ALTERNATIVE MATERIAL PRICES							
Fibreboard; 19 mm Decorative faced							
Ash	–	–	–	–	11.95	m²	**11.95**
Beech	–	–	–	–	11.42	m²	**11.42**
Oak	–	–	–	–	11.77	m²	**11.77**
Edgings; self-adhesive							
22 mm Ash	–	–	–	–	0.34	m	**0.34**
22 mm Beech	–	–	–	–	0.34	m	**0.34**
22 mm Oak	–	–	–	–	0.34	m	**0.34**
Chipboard standard grade							
12 mm	–	–	–	–	2.25	m²	**2.25**
18 mm	–	–	–	–	3.18	m²	**3.18**
22 mm	–	–	–	–	3.89	m²	**3.89**
25 mm	–	–	–	–	4.46	m²	**4.46**
Chipboard; melamine faced							
15 mm	–	–	–	–	3.32	m²	**3.32**
18 mm	–	–	–	–	3.61	m²	**3.61**
Medium density fibreboard; external quality							
6 mm	–	–	–	–	4.69	m²	**4.69**
9 mm	–	–	–	–	6.23	m²	**6.23**
19 mm	–	–	–	–	10.10	m²	**10.10**
25 mm	–	–	–	–	14.09	m²	**14.09**
Wallboard plank							
9.5 mm	–	–	–	–	3.09	m²	**3.09**
12.5 mm	–	–	–	–	3.09	m²	**3.09**
15 mm	–	–	–	–	3.75	m²	**3.75**
Moisture-resistant board							
12.5 mm	–	–	–	–	5.26	m²	**5.26**
15 mm	–	–	–	–	6.33	m²	**6.33**
Fireline board							
12.5 mm	–	–	–	–	4.15	m²	**4.15**
15 mm	–	–	–	–	5.00	m²	**5.00**
Wall linings; Gyproc GypLyner IWL walling system or other equal; comprising 48 mm wide metal I stud frame; 50 mmm wide metal C stud floor and head channels; plugged and screwed to concrete							
62.5 mm lining; outer skin of 12.50 mm thick tapered edge wallboard one side; joints filled with joint filler and joint tape to receive direct decoration							
average height 2.00 m	–	0.53	12.26	–	9.39	m²	**21.65**
average height 3.00 m	–	0.53	12.26	–	9.23	m²	**21.49**
average height 4.00 m	–	0.53	12.26	–	9.20	m²	**21.46**
Rate per m (SMM7 measurement rule)							
height 2.10 m–2.40 m	–	1.30	30.08	–	21.48	m	**51.56**
height 2.40 m–2.70 m	–	1.45	33.55	–	25.37	m	**58.92**
height 2.70 m–3.00 m	–	1.60	37.02	–	27.70	m	**64.72**

Prices for Measured Works

20 PROPRIETARY LININGS AND PARTITIONS

Item	PC £	Labour hours	Labour £	Plant £	Material £	Unit	Total rate £
20.02 LININGS TO WALLS – cont							
Wall linings – cont							
Rate per m (SMM7 measurement rule) – cont							
height 3.00 m–3.30 m	–	1.75	40.49	–	30.09	m	**70.58**
height 3.30 m–3.60 m	–	1.90	43.96	–	32.54	m	**76.50**
height 3.60 m–3.90 m	–	2.05	47.43	–	34.99	m	**82.42**
height 3.90 m–4.20 m	–	2.20	50.90	–	37.44	m	**88.34**
65 mm lining; outer skin of 15.00 mm thick tapered edge wallboard one side; joints filled with joint filler and joint tape to receive direct decoration							
average height 2.00 m	–	0.53	12.26	–	10.42	m²	**22.68**
average height 3.00 m	–	0.53	12.26	–	10.09	m²	**22.35**
average height 4.00 m	–	0.53	12.26	–	9.69	m²	**21.95**
Rate per m (SMM7 measurement rule)							
height 2.10 m–2.40 m	–	1.30	30.08	–	23.13	m	**53.21**
height 2.40 m–2.70 m	–	1.45	33.55	–	27.23	m	**60.78**
height 2.70 m–3.00 m	–	1.60	37.02	–	29.77	m	**66.79**
height 3.00 m–3.30 m	–	1.75	40.49	–	32.37	m	**72.86**
height 3.30 m–3.60 m	–	1.90	43.96	–	35.03	m	**78.99**
height 3.60 m–3.90 m	–	2.05	47.43	–	37.69	m	**85.12**
height 3.90 m–4.20 m	–	2.20	50.90	–	40.35	m	**91.25**
62.5 mm lining; outer skin of 12.50 mm thick tapered edge Fireline board one side; joints filled with joint filler and joint tape to receive direct decoration							
average height 2.00 m	–	0.53	12.26	–	10.71	m²	**22.97**
average height 3.00 m	–	0.53	12.26	–	10.22	m²	**22.48**
average height 4.00 m	–	0.53	12.26	–	9.99	m²	**22.25**
Rate per m (SMM7 measurement rule)							
height 2.10 m–2.40 m	–	1.30	30.25	–	24.16	m	**54.41**
height 2.40 m–2.70 m	–	1.45	33.72	–	28.39	m	**62.11**
height 2.70 m–3.00 m	–	1.60	37.19	–	31.05	m	**68.24**
height 3.00 m–3.30 m	–	1.75	40.66	–	33.78	m	**74.44**
height 3.30 m–3.60 m	–	1.90	44.14	–	36.56	m	**80.70**
height 3.60 m–3.90 m	–	2.05	47.61	–	39.36	m	**86.97**
height 3.90 m–4.20 m	–	2.20	51.08	–	42.15	m	**93.23**
65 mm partition; outer skin of 15.00 mm thick tapered edge Fireline board one side; joints filled with joint filler and joint tape to receive direct decoration							
average height 2.00 m	–	0.53	12.26	–	11.60	m²	**23.86**
average height 3.00 m	–	0.53	12.26	–	11.10	m²	**23.36**
average height 4.00 m	–	0.53	12.26	–	10.88	m²	**23.14**
Rate per m (SMM7 measurement rule)							
height 2.10 m–2.40 m	–	1.30	30.25	–	26.29	m	**56.54**
height 2.40 m–2.70 m	–	1.45	33.72	–	30.78	m	**64.50**
height 2.70 m–3.00 m	–	1.60	37.19	–	33.71	m	**70.90**
height 3.00 m–3.30 m	–	1.75	40.66	–	36.71	m	**77.37**
height 3.30 m–3.60 m	–	1.90	44.14	–	39.76	m	**83.90**
height 3.60 m–3.90 m	–	2.05	47.61	–	42.81	m	**90.42**
height 3.90 m–4.20 m	–	2.20	51.08	–	45.87	m	**96.95**

20 PROPRIETARY LININGS AND PARTITIONS

Item	PC £	Labour hours	Labour £	Plant £	Material £	Unit	Total rate £
62.5 mm lining; outer skin of 12.50 mm thick tapered edge wallboard one side; filling cavity with 50 mm Isover insulation; wallboard joints filled with joint filler and joint tape to receive direct decoration							
average height 2.00 m	–	0.65	15.04	–	11.43	m²	**26.47**
average height 3.00 m	–	0.65	15.04	–	10.92	m²	**25.96**
average height 4.00 m	–	0.65	15.04	–	10.78	m²	**25.82**
Rate per m (SMM7 measurement rule)							
height 2.10 m–2.40 m	–	1.42	33.03	–	25.98	m	**59.01**
height 2.40 m–2.70 m	–	1.60	37.19	–	30.42	m	**67.61**
height 2.70 m–3.00 m	–	1.65	38.17	–	33.31	m	**71.48**
height 3.00 m–3.30 m	–	1.92	44.49	–	36.27	m	**80.76**
height 3.30 m–3.60 m	–	2.08	48.30	–	39.27	m	**87.57**
height 3.60 m–3.90 m	–	2.25	52.12	–	42.29	m	**94.41**
height 3.90 m–4.20 m	–	2.41	55.94	–	45.31	m	**101.25**
65 mm lining; outer skin of 15.00 mm thick tapered edge wallboard one side; filling cavity with 50 mm Isover slabs; wallboard joints filled with joint filler and joint tape to receive direct decoration							
average height 2.00 m	–	0.65	15.04	–	12.12	m²	**27.16**
average height 3.00 m	–	0.65	15.04	–	11.61	m²	**26.65**
average height 4.00 m	–	0.65	15.04	–	11.47	m²	**26.51**
Rate per m (SMM7 measurement rule)							
height 2.10 m–2.40 m	–	1.42	33.03	–	27.63	m	**60.66**
height 2.40 m–2.70 m	–	1.60	37.19	–	32.29	m	**69.48**
height 2.70 m–3.00 m	–	1.65	38.17	–	35.38	m	**73.55**
height 3.00 m–3.30 m	–	1.92	44.49	–	38.54	m	**83.03**
height 3.30 m–3.60 m	–	2.08	48.30	–	41.77	m	**90.07**
height 3.60 m–3.90 m	–	2.25	52.12	–	44.98	m	**97.10**
height 3.90 m–4.20 m	–	2.41	55.94	–	48.20	m	**104.14**
Gypsum plasterboard; fixing on dabs or with nails; joints filled with joint filler and joint tape to receive direct decoration; to softwood base (measured elsewhere)							
Plain grade tapered edge wallboard							
9.50 mm board to walls							
average height 2.00 m	–	0.16	3.70	–	4.69	m²	**8.39**
average height 3.00 m	–	0.16	3.58	–	4.70	m²	**8.28**
average height 4.00 m	–	0.15	3.47	–	4.70	m²	**8.17**
Rate per m (SMM7 measurement rule)							
wall height 2.40 m–2.70 m	–	0.43	10.12	–	12.67	m	**22.79**
wall height 2.70 m–3.00 m	–	0.46	10.83	–	14.07	m	**24.90**
wall height 3.00 m–3.30 m	–	0.50	11.74	–	15.48	m	**27.22**
wall height 3.30 m–3.60 m	–	0.59	13.83	–	16.97	m	**30.80**
9.50 mm board to reveals and soffits of openings and recesses							
not exceeding 300 mm wide	–	0.14	3.33	–	2.39	m	**5.72**
300 mm–600 mm wide	–	0.22	5.18	–	3.54	m	**8.72**

20 PROPRIETARY LININGS AND PARTITIONS

Item	PC £	Labour hours	Labour £	Plant £	Material £	Unit	Total rate £
20.02 LININGS TO WALLS – cont							
Plain grade tapered edge wallboard – cont							
9.50 mm board to faces of columns – 4 nr faces							
not exceeding 300 mm total girth	–	0.20	4.62	–	3.68	m	**8.30**
300 mm–600 mm total girth	–	0.20	4.62	–	4.88	m	**9.50**
600 mm–900 mm total girth	–	0.20	4.62	–	6.10	m	**10.72**
900 mm–1200 mm total girth	–	0.20	4.62	–	7.30	m	**11.92**
12.50 mm board to walls							
average height 2.00 m	–	0.16	3.70	–	6.55	m²	**10.25**
average height 3.00 m	–	0.16	3.58	–	6.55	m²	**10.13**
average height 4.00 m	–	0.15	3.47	–	6.55	m²	**10.02**
Rate per m (SMM7 measurement rule)							
wall height 2.40 m–2.70 m	–	0.45	10.41	–	17.70	m	**28.11**
wall height 2.70 m–3.00 m	–	0.50	11.57	–	19.65	m	**31.22**
wall height 3.00 m–3.30 m	–	0.55	12.72	–	21.74	m	**34.46**
wall height 3.30 m–3.60 m	–	0.60	13.88	–	23.67	m	**37.55**
12.50 mm board to reveals and soffits of openings and recesses							
not exceeding 300 mm wide	–	0.19	4.57	–	2.96	m	**7.53**
300 mm–600 mm wide	–	0.37	8.92	–	4.67	m	**13.59**
12.50 mm board to faces of columns – 4 nr faces							
not exceeding 300 mm total girth	–	0.30	6.94	–	4.18	m	**11.12**
300 mm–600 mm total girth	–	0.30	6.94	–	5.91	m	**12.85**
600 mm–900 mm total girth	–	0.30	6.94	–	7.62	m	**14.56**
900 mm–1200 mm total girth	–	0.30	6.94	–	9.34	m	**16.28**
Tapered edge wallboard TEN							
12.50 mm board to walls							
average height 2.00 m	–	0.16	3.70	–	4.85	m²	**8.55**
average height 3.00 m	–	0.16	3.58	–	4.85	m²	**8.43**
average height 4.00 m	–	0.15	3.47	–	4.85	m²	**8.32**
Rate per m (SMM7 measurement rule)							
wall height 2.40 m–2.70 m	–	0.47	11.05	–	13.12	m	**24.17**
wall height 2.70 m–3.00 m	–	0.56	13.23	–	14.57	m	**27.80**
wall height 3.00 m–3.30 m	–	0.65	15.40	–	16.04	m	**31.44**
wall height 3.30 m–3.60 m	–	0.78	18.49	–	17.57	m	**36.06**
12.50 mm board to reveals and soffits of openings and recesses							
not exceeding 300 mm wide	–	0.19	4.57	–	2.44	m	**7.01**
300 mm–600 mm wide	–	0.37	8.92	–	3.65	m	**12.57**
12.50 mm board to faces of columns – 4 nr faces							
not exceeding 300 mm total girth	–	0.25	6.23	–	3.80	m	**10.03**
300 mm–600 mm total girth	–	0.20	4.62	–	5.01	m	**9.63**
600 mm–900 mm total girth	–	0.20	4.62	–	6.22	m	**10.84**
900 mm–1200 mm total girth	–	0.20	4.62	–	7.45	m	**12.07**

20 PROPRIETARY LININGS AND PARTITIONS

Item	PC £	Labour hours	Labour £	Plant £	Material £	Unit	Total rate £
Tapered edge plank							
19 mm plank to walls							
average height 2.00 m	–	0.16	3.70	–	8.03	m²	**11.73**
average height 3.00 m	–	0.16	3.58	–	8.03	m²	**11.61**
average height 4.00 m	–	0.15	3.47	–	8.03	m²	**11.50**
Rate per m (SMM7 measurement rule)							
wall height 2.40 m–2.70 m	–	1.02	24.67	–	21.70	m	**46.37**
wall height 2.70 m–3.00 m	–	1.20	29.02	–	24.11	m	**53.13**
wall height 3.00 m–3.30 m	–	1.30	31.51	–	26.52	m	**58.03**
wall height 3.30 m–3.60 m	–	1.53	37.01	–	29.02	m	**66.03**
19 mm plank to reveals and soffits of openings and recesses							
not exceeding 300 mm wide	–	0.20	4.81	–	3.40	m	**8.21**
300 mm–600 mm wide	–	0.42	10.07	–	5.55	m	**15.62**
19 mm plank to faces of columns – 4 nr faces							
not exceeding 300 mm total girth	–	0.80	19.94	–	4.74	m	**24.68**
300 mm–600 mm total girth	–	0.80	18.51	–	6.92	m	**25.43**
600 mm–900 mm total girth	–	0.80	18.51	–	9.11	m	**27.62**
900 mm–1200 mm total girth	–	0.80	18.51	–	11.28	m	**29.79**
ThermaLine Plus Board (0.19 W/mK)							
27 mm board to walls							
average height 2.00 m	–	0.16	3.70	–	14.69	m²	**18.39**
average height 3.00 m	–	0.16	3.58	–	14.69	m²	**18.27**
average height 4.00 m	–	0.15	3.47	–	14.69	m²	**18.16**
Rate per m (SMM7 measurement rule)							
wall height 2.40 m–2.70 m	–	1.06	24.52	–	39.68	m	**64.20**
wall height 2.70 m–3.00 m	–	1.23	28.46	–	44.08	m	**72.54**
wall height 3.00 m–3.30 m	–	1.34	31.00	–	48.49	m	**79.49**
wall height 3.30 m–3.60 m	–	1.62	37.48	–	52.98	m	**90.46**
27 mm board to reveals and soffits of openings and recesses							
not exceeding 300 mm wide	–	0.21	4.86	–	5.40	m	**10.26**
300 mm–600 mm wide	–	0.43	9.95	–	9.55	m	**19.50**
27 mm board to faces of columns – 4 nr faces							
not exceeding 300 mm total girth	–	0.52	12.03	–	10.79	m	**22.82**
300 mm–600 mm total girth	–	0.65	15.04	–	10.79	m	**25.83**
600 mm–900 mm total girth	–	0.75	17.36	–	19.11	m	**36.47**
900 mm–1200 mm total girth	–	1.00	23.13	–	19.12	m	**42.25**
48 mm board to walls							
average height 2.00 m	–	0.16	3.70	–	20.00	m²	**23.70**
average height 3.00 m	–	0.16	3.58	–	20.00	m²	**23.58**
average height 4.00 m	–	0.15	3.47	–	20.00	m²	**23.47**
Rate per m (SMM7 measurement rule)							
wall height 2.40 m–2.70 m	–	1.06	24.52	–	54.03	m	**78.55**
wall height 2.70 m–3.00 m	–	1.30	30.08	–	60.04	m	**90.12**
wall height 3.00 m–3.30 m	–	1.43	33.08	–	66.04	m	**99.12**
wall height 3.30 m–3.60 m	–	1.71	39.56	–	72.13	m	**111.69**

20 PROPRIETARY LININGS AND PARTITIONS

Item	PC £	Labour hours	Labour £	Plant £	Material £	Unit	Total rate £
20.02 LININGS TO WALLS – cont							
ThermaLine Plus Board (0.19 W/mK) – cont							
48 mm board to reveals and soffits of openings and recesses							
not exceeding 300 mm wide	–	0.23	5.33	–	6.99	m	**12.32**
300 mm–600 mm wide	–	0.46	10.64	–	12.74	m	**23.38**
48 mm board to faces of columns – 4 nr faces							
not exceeding 300 mm total girth	–	0.52	12.03	–	13.98	m	**26.01**
300 mm–600 mm total girth	–	0.65	15.04	–	13.98	m	**29.02**
600 mm–900 mm total girth	–	0.75	17.36	–	25.48	m	**42.84**
900 mm–1200 mm total girth	–	1.00	23.13	–	25.49	m	**48.62**
ThermaLine Super boards							
50 mm board to walls							
average height 2.00 m	–	0.40	9.25	–	26.78	m²	**36.03**
average height 3.00 m	–	0.43	9.95	–	26.78	m²	**36.73**
average height 4.00 m	–	0.47	10.99	–	26.78	m²	**37.77**
Rate per m (SMM7 measurement rule)							
wall height 2.40 m–2.70 m	–	1.06	24.52	–	72.31	m	**96.83**
wall height 2.70 m–3.00 m	–	1.30	30.08	–	80.34	m	**110.42**
wall height 3.00 m–3.30 m	–	1.43	33.08	–	88.37	m	**121.45**
wall height 3.30 m–3.60 m	–	1.71	39.56	–	96.49	m	**136.05**
50 mm board to reveals and soffits of openings and recesses							
not exceeding 300 mm wide	–	0.52	12.03	–	9.02	m	**21.05**
300 mm–600 mm wide	–	0.46	10.64	–	16.80	m	**27.44**
50 mm board to faces of columns – 4 nr faces							
not exceeding 300 mm total girth	–	0.52	12.03	–	10.27	m	**22.30**
300–600 mm total girth	–	0.65	15.04	–	18.06	m	**33.10**
600 mm–900 mm total girth	–	0.75	17.36	–	25.84	m	**43.20**
900 mm–1200 mm total girth	–	1.00	23.13	–	33.63	m	**56.76**
60 mm board to walls							
average height 2.00 m	–	0.40	9.25	–	31.47	m²	**40.72**
average height 3.00 m	–	0.43	9.95	–	31.47	m²	**41.42**
average height 4.00 m	–	0.47	10.99	–	31.47	m²	**42.46**
Rate per m (SMM7 measurement rule)							
wall height 2.40 m–2.70 m	–	1.06	24.52	–	84.94	m	**109.46**
wall height 2.70 m–3.00 m	–	1.30	30.08	–	94.39	m	**124.47**
wall height 3.00 m–3.30 m	–	1.43	33.08	–	103.82	m	**136.90**
wall height 3.30 m–3.60 m	–	1.71	39.56	–	113.35	m	**152.91**
70 mm board to walls							
average height 2.00 m	–	0.40	9.25	–	37.19	m²	**46.44**
average height 3.00 m	–	0.43	9.95	–	37.19	m²	**47.14**
average height 4.00 m	–	0.47	10.99	–	37.19	m²	**48.18**
Rate per m (SMM7 measurement rule)							
wall height 2.40 m–2.70 m	–	1.06	24.52	–	100.44	m	**124.96**
wall height 2.70 m–3.00 m	–	1.30	30.08	–	111.59	m	**141.67**
wall height 3.00 m–3.30 m	–	1.43	33.08	–	122.76	m	**155.84**
wall height 3.30 m–3.60 m	–	1.71	39.56	–	133.99	m	**173.55**

20 PROPRIETARY LININGS AND PARTITIONS

Item	PC £	Labour hours	Labour £	Plant £	Material £	Unit	Total rate £
70 mm board to reveals and soffits of openings and recesses							
not exceeding 300 mm wide	–	0.23	5.33	–	12.14	m	**17.47**
300 mm–600 mm wide	–	0.46	10.64	–	23.05	m	**33.69**
70 mm board to faces of columns – 4 nr faces							
not exceeding 300 mm total girth	–	0.52	12.03	–	13.40	m	**25.43**
300–600 mm total girth	–	0.65	15.04	–	24.30	m	**39.34**
600 mm–900 mm total girth	–	0.75	17.36	–	35.22	m	**52.58**
900 mm–1200 mm total girth	–	1.00	23.13	–	46.12	m	**69.25**
80 mm board to walls							
average height 2.00 m	–	0.40	9.25	–	43.00	m²	**52.25**
average height 3.00 m	–	0.43	9.95	–	43.00	m²	**52.95**
average height 4.00 m	–	0.47	10.99	–	43.00	m²	**53.99**
Rate per m (SMM7 measurement rule)							
wall height 2.40 m–2.70 m	–	1.06	24.52	–	116.10	m	**140.62**
wall height 2.70 m–3.00 m	–	1.30	30.08	–	129.01	m	**159.09**
wall height 3.00 m–3.30 m	–	1.43	33.08	–	141.91	m	**174.99**
wall height 3.30 m–3.60 m	–	1.71	39.56	–	154.89	m	**194.45**
80 mm board to reveals and soffits of openings and recesses							
not exceeding 300 mm wide	–	0.23	5.33	–	13.88	m	**19.21**
300 mm–600 mm wide	–	0.46	10.64	–	26.53	m	**37.17**
80 mm board to faces of columns – 4 nr faces							
not exceeding 300 mm total girth	–	0.52	12.03	–	15.14	m	**27.17**
300–600 mm total girth	–	0.65	15.04	–	27.78	m	**42.82**
600 mm–900 mm total girth	–	0.75	17.36	–	40.44	m	**57.80**
900 mm–1200 mm total girth	–	1.00	23.13	–	53.10	m	**76.23**
90 mm board to walls							
average height 2.00 m	–	0.40	9.25	–	45.46	m²	**54.71**
average height 3.00 m	–	0.43	9.95	–	45.46	m²	**55.41**
average height 4.00 m	–	0.47	10.99	–	45.46	m²	**56.45**
Rate per m (SMM7 measurement rule)							
wall height 2.40 m–2.70 m	–	1.06	24.52	–	122.77	m	**147.29**
wall height 2.70 m–3.00 m	–	1.30	30.08	–	136.40	m	**166.48**
wall height 3.00 m–3.30 m	–	1.43	33.08	–	150.05	m	**183.13**
wall height 3.30 m–3.60 m	–	1.71	39.56	–	163.77	m	**203.33**
90 mm board to reveals and soffits of openings and recesses							
not exceeding 300 mm wide	–	0.23	5.33	–	14.63	m	**19.96**
300 mm–600 mm wide	–	0.46	10.64	–	28.02	m	**38.66**
90 mm board to faces of columns – 4 nr faces							
not exceeding 300 mm total girth	–	0.52	12.03	–	15.88	m	**27.91**
300–600 mm total girth	–	0.65	15.04	–	29.26	m	**44.30**
600 mm–900 mm total girth	–	0.75	17.36	–	42.66	m	**60.02**
900 mm–1200 mm total girth	–	1.00	23.13	–	56.05	m	**79.18**

20 PROPRIETARY LININGS AND PARTITIONS

Item	PC £	Labour hours	Labour £	Plant £	Material £	Unit	Total rate £
20.02 LININGS TO WALLS – cont							
Kingspan Kooltherm K18 insulated plasterboard; fixing with nails; joints filled with joint filler and joint tape to receive direct decoration; to softwood base (measured elsewhere)							
12.5 mm plasterboard bonded to CFC/HCFC free rigid phenolic insulation; overall thickness (0.21 W/mK)							
37.5 mm thick panel	–	0.23	5.20	–	25.12	m²	**30.32**
42.5 mm thick panel	–	0.23	5.20	–	16.21	m²	**21.41**
52.5 mm thick panel	–	0.25	5.79	–	20.31	m²	**26.10**
62.5 mm thick panel	–	0.25	5.79	–	18.03	m²	**23.82**
72.5 mm thick panel	–	0.25	5.79	–	21.63	m²	**27.42**
Plasterboard jointing system; filling joint with jointing compounds							
To walls and ceilings							
to suit 9.50 mm or 12.50 mm thick boards	–	0.09	2.08	–	2.74	m	**4.82**
Angle trim; plasterboard edge support system							
To walls and ceilings							
to suit 9.50 mm or 12.50 mm thick boards	–	0.09	2.08	–	2.59	m	**4.67**
Gyproc SoundBloc tapered edge plasterboard with higher density core; fixing on dabs or with nails; joints filled with joint filler and joint tape to receive direct decoration; to softwood base							
12.50 mm board to walls							
average height 2.00 m	–	0.36	8.33	–	5.82	m²	**14.15**
average height 3.00 m	–	0.38	8.80	–	5.82	m²	**14.62**
average height 4.00 m	–	0.40	9.25	–	5.91	m²	**15.16**
Rate per m (SMM7 measurement rule)							
wall height 2.40 m–2.70 m	–	0.97	22.44	–	15.70	m	**38.14**
wall height 2.70 m–3.00 m	–	1.11	25.68	–	17.44	m	**43.12**
wall height 3.00 m–3.30 m	–	1.32	30.54	–	19.19	m	**49.73**
wall height 3.30 m–3.60 m	–	1.43	33.08	–	21.01	m	**54.09**
12.50 mm board to ceilings							
over 300 mm wide	–	0.41	9.49	–	5.81	m²	**15.30**
15.00 mm board to walls							
average height 2.00 m	–	0.36	8.33	–	6.78	m²	**15.11**
average height 3.00 m	–	0.38	8.80	–	6.78	m²	**15.58**
average height 4.00 m	–	0.40	9.25	–	6.87	m²	**16.12**
Rate per m (SMM7 measurement rule)							
wall height 2.40 m–2.70 m	–	1.00	23.13	–	18.31	m	**41.44**
wall height 2.70 m–3.00 m	–	1.14	26.38	–	20.35	m	**46.73**
wall height 3.00 m–3.30 m	–	1.27	29.39	–	22.39	m	**51.78**
wall height 3.30 m–3.60 m	–	1.46	33.77	–	24.50	m	**58.27**
15.00 mm board to reveals and soffits of openings and recesses							
not exceeding 300 mm wide	–	0.20	4.62	–	3.02	m	**7.64**
300 mm–600 mm wide	–	0.38	8.80	–	4.80	m	**13.60**
15.00 mm board to ceilings							
over 300 mm wide	–	0.43	9.95	–	6.79	m²	**16.74**

20 PROPRIETARY LININGS AND PARTITIONS

Item	PC £	Labour hours	Labour £	Plant £	Material £	Unit	Total rate £
Two layers of gypsum plasterboard; plain grade square and tapered edge wallboard; fixing on dabs or with nails; joints filled with joint filler and joint tape; top layer to receive direct decoration; to softwood base							
19 mm two layer board to walls							
average height 2.00 m	–	0.48	11.10	–	7.95	m²	19.05
average height 3.00 m	–	0.51	11.80	–	7.95	m²	19.75
average height 4.00 m	–	0.54	12.49	–	8.05	m²	20.54
Rate per m (SMM7 measurement rule)							
wall height 2.40 m–2.70 m	–	1.30	30.08	–	21.48	m	51.56
wall height 2.70 m–3.00 m	–	1.48	34.24	–	23.85	m	58.09
wall height 3.00 m–3.30 m	–	1.67	38.64	–	26.24	m	64.88
wall height 3.30 m–3.60 m	–	1.94	44.89	–	28.72	m	73.61
19 mm two layer board to reveals and soffits of openings and recesses							
not exceeding 300 mm wide	–	0.28	6.48	–	3.38	m	9.86
300 mm–600 mm wide	–	0.56	12.96	–	5.51	m	18.47
19 mm two layer board to faces of columns – 4 nr							
not exceeding 300 mm total girth	–	0.50	11.57	–	5.74	m	17.31
300–600 mm total girth	–	0.69	15.97	–	6.76	m	22.73
600 mm–900 mm total girth	–	0.75	17.36	–	8.89	m	26.25
900 mm–1200 mm total girth	–	1.00	23.13	–	10.59	m	33.72
25 mm two layer board to walls							
average height 2.00 m	–	0.48	11.10	–	9.81	m²	20.91
average height 3.00 m	–	0.51	11.80	–	9.81	m²	21.61
average height 4.00 m	–	0.54	12.49	–	9.91	m²	22.40
Rate per m (SMM7 measurement rule)							
wall height 2.40 m–2.70 m	–	1.39	32.16	–	26.50	m	58.66
wall height 2.70 m–3.00 m	–	1.57	36.33	–	29.44	m	65.77
wall height 3.00 m–3.30 m	–	1.76	40.72	–	32.39	m	73.11
wall height 3.30 m–3.60 m	–	2.04	47.19	–	35.41	m	82.60
25 mm two layer board to reveals and soffits of openings and recesses							
not exceeding 300 mm wide	–	0.28	6.48	–	3.93	m	10.41
300 mm–600 mm wide	–	0.56	12.96	–	6.62	m	19.58
25 mm two layer board to faces of columns – 4 nr							
not exceeding 300 mm total girth	–	0.35	8.10	–	5.17	m	13.27
300–600 mm total girth	–	0.69	15.97	–	7.88	m	23.85
600 mm–900 mm total girth	–	1.00	23.13	–	10.57	m	33.70
900 mm–1200 mm total girth	–	1.25	28.96	–	13.21	m	42.17
Gyproc Dri-Wall dry lining system or other equal or approved; plain grade tapered edge wallboard; fixed to walls with adhesive; joints filled with joint filler and joint tape; to receive direct decoration							
9.50 mm board to walls							
average height 2.00 m	–	0.41	9.51	–	6.24	m²	15.75
average height 3.00 m	–	0.43	10.02	–	6.24	m²	16.26
average height 4.00 m	–	0.51	11.71	–	6.24	m²	17.95

20 PROPRIETARY LININGS AND PARTITIONS

Item	PC £	Labour hours	Labour £	Plant £	Material £	Unit	Total rate £
20.02 LININGS TO WALLS – cont							
Gyproc Dri-Wall dry lining system or other equal or approved – cont							
Rate per m (SMM7 measurement rule)							
wall height 2.40 m–2.70 m	–	1.11	25.68	–	16.84	m	**42.52**
wall height 2.70 m–3.00 m	–	1.28	29.61	–	18.69	m	**48.30**
wall height 3.00 m–3.30 m	–	1.43	33.08	–	20.54	m	**53.62**
wall height 3.30 m–3.60 m	–	1.67	38.64	–	22.47	m	**61.11**
9.50 mm board to reveals and soffits of openings and recesses							
not exceeding 300 mm wide	–	0.23	5.50	–	2.83	m	**8.33**
300 mm–600 mm wide	–	0.46	11.00	–	4.43	m	**15.43**
9.50 mm board to faces of columns – 4 nr faces							
300 mm total girth	–	0.45	10.41	–	2.88	m	**13.29**
300–600 mm total girth	–	0.58	13.42	–	5.65	m	**19.07**
600 mm–900 mm total girth	–	0.75	17.36	–	6.87	m	**24.23**
900 mm–1200 mm total girth	–	0.95	21.98	–	8.02	m	**30.00**
Angle; with joint tape bedded and covered with Jointex or other equal							
internal	–	0.05	1.25	–	0.62	m	**1.87**
external	–	0.11	2.74	–	0.62	m	**3.36**
Gyproc Dri-Wall M/F dry lining system or other equal; mild steel channel fixed to walls with adhesive; tapered edge wallboard screwed to channel; joints filled with joint filler and joint tape							
9.50 mm board to walls							
average height 2.00 m	–	0.55	12.72	–	9.51	m²	**22.23**
average height 3.00 m	–	0.57	13.31	–	9.50	m²	**22.81**
average height 4.00 m	–	0.62	14.27	–	9.52	m²	**23.79**
Rate per m (SMM7 measurement rule)							
wall height 2.40 m–2.70 m	–	1.48	34.24	–	25.66	m	**59.90**
wall height 2.70 m–3.00 m	–	1.69	40.36	–	28.48	m	**68.84**
wall height 3.00 m–3.30 m	–	1.90	43.96	–	31.34	m	**75.30**
wall height 3.30 m–3.60 m	–	2.22	51.37	–	34.28	m	**85.65**
9.50 mm board to reveals and soffits of openings and recesses							
not exceeding 300 mm wide	–	0.23	5.50	–	2.49	m	**7.99**
300 mm–600 mm wide	–	0.46	11.00	–	3.76	m	**14.76**
12.50 mm board to walls							
average height 2.00 m	–	0.55	12.72	–	11.36	m²	**24.08**
average height 3.00 m	–	0.57	13.31	–	11.35	m²	**24.66**
average height 4.00 m	–	0.62	14.27	–	11.37	m²	**25.64**
Rate per m (SMM7 measurement rule)							
wall height 2.40 m–2.70 m	–	1.48	34.24	–	30.66	m	**64.90**
wall height 2.70 m–3.00 m	–	1.69	40.36	–	34.05	m	**74.41**
wall height 3.00 m–3.30 m	–	1.90	43.96	–	37.46	m	**81.42**
wall height 3.30 m–3.60 m	–	2.22	51.37	–	40.95	m	**92.32**
12.50 mm board to reveals and soffits of openings and recesses							
not exceeding 300 mm wide	–	0.23	5.50	–	3.05	m	**8.55**
300 mm–600 mm wide	–	0.46	11.00	–	4.87	m	**15.87**

20 PROPRIETARY LININGS AND PARTITIONS

Item	PC £	Labour hours	Labour £	Plant £	Material £	Unit	Total rate £
Megadeco wallboard; fixing on dabs or with screws; joints filled with joint filler and joint tape to receive direct decoration; to softwood							
12.50 mm board to walls							
average height 2.00 m	–	0.36	8.33	–	5.58	m²	**13.91**
average height 3.00 m	–	0.38	8.77	–	5.58	m²	**14.35**
average height 4.00 m	–	0.40	9.25	–	5.58	m²	**14.83**
Rate per m (SMM7 measurement rule)							
wall height 2.40 m–2.70 m	–	0.97	22.44	–	18.50	m	**40.94**
wall height 2.70 m–3.00 m	–	1.11	26.93	–	20.56	m	**47.49**
wall height 3.00 m–3.30 m	–	1.25	28.92	–	22.62	m	**51.54**
wall height 3.30 m–3.60 m	–	1.43	33.08	–	24.67	m	**57.75**
12.50 mm board to ceilings							
over 300 mm wide	–	0.41	9.85	–	6.85	m²	**16.70**
15 mm board to walls							
average height 2.00 m	–	0.38	8.74	–	6.85	m²	**15.59**
average height 3.00 m	–	0.40	9.20	–	6.85	m²	**16.05**
average height 4.00 m	–	0.42	9.71	–	6.85	m²	**16.56**
Rate per m (SMM7 measurement rule)							
wall height 2.40 m–2.70 m	–	1.02	23.57	–	18.50	m	**42.07**
wall height 2.70 m–3.00 m	–	1.17	28.28	–	20.56	m	**48.84**
wall height 3.00 m–3.30 m	–	1.31	30.36	–	22.62	m	**52.98**
wall height 3.30 m–3.60 m	–	1.50	34.74	–	24.67	m	**59.41**
15 mm board to ceilings							
over 300 mm wide	–	0.43	10.33	–	6.85	m²	**17.18**
Gypsum cladding; Glasroc Firecase S board or other equal; fixed with adhesive; joints pointed in adhesive							
25 mm thick column linings, faces = 4; 2 hour fire protection rating							
not exceeding 300 mm girth	–	0.30	6.94	–	22.43	m	**29.37**
300 mm–600 mm girth	–	0.30	6.94	–	30.46	m	**37.40**
600 mm–900 mm girth	–	0.45	10.41	–	38.49	m	**48.90**
900 mm–1200 mm girth	–	0.55	12.72	–	46.51	m	**59.23**
1200 mm–1500 mm girth	–	0.60	13.88	–	54.54	m	**68.42**
30 mm thick beam linings, faces = 3; 2 hour fire protection rating							
not exceeding 300 mm girth	–	0.60	13.88	–	20.41	m	**34.29**
300 mm–600 mm girth	–	0.60	13.88	–	29.49	m	**43.37**
600 mm–900 mm girth	–	0.75	17.36	–	38.55	m	**55.91**
900 mm–1200 mm girth	–	0.90	20.83	–	47.63	m	**68.46**
1200 mm–1500 mm girth	–	1.20	27.76	–	56.70	m	**84.46**

Prices for Measured Works

20 PROPRIETARY LININGS AND PARTITIONS

Item	PC £	Labour hours	Labour £	Plant £	Material £	Unit	Total rate £
20.02 LININGS TO WALLS – cont							
Vermiculite gypsum cladding; Vermiculux board or other equal; fixed with adhesive; joints pointed in adhesive							
25 mm thick column linings, faces = 4; 2 hour fire protection rating							
not exceeding 300 mm girth	–	0.20	4.62	–	13.38	m	**18.00**
300 mm–600 mm girth	–	0.30	6.94	–	26.76	m	**33.70**
600 mm–900 mm girth	–	0.45	10.41	–	39.73	m	**50.14**
900 mm–1200 mm girth	–	0.60	13.88	–	52.97	m	**66.85**
1200 mm–1500 mm girth	–	0.70	16.19	–	66.08	m	**82.27**
30 mm thick beam linings, faces = 3; 2 hour fire protection rating							
not exceeding 300 mm girth	–	0.50	11.57	–	16.84	m	**28.41**
300 mm–600 mm girth	–	0.60	13.88	–	33.68	m	**47.56**
600 mm–900 mm girth	–	0.90	20.83	–	53.71	m	**74.54**
900 mm–1200 mm girth	–	1.00	23.13	–	66.71	m	**89.84**
1200 mm–1500 mm girth	–	1.20	27.76	–	83.29	m	**111.05**
55 mm thick column linings, faces = 4 ; 4 hour fire protection rating							
not exceeding 300 mm girth	–	0.60	13.88	–	35.70	m	**49.58**
300 mm–600 mm girth	–	0.75	17.36	–	71.40	m	**88.76**
600 mm–900 mm girth	–	0.80	18.51	–	106.70	m	**125.21**
900 mm–1200 mm girth	–	1.00	23.13	–	142.26	m	**165.39**
1200 mm–1500 mm girth	–	1.02	23.60	–	177.70	m	**201.30**
60 mm thick beam linings, faces = 3; 4 hour fire protection rating							
not exceeding 300 mm girth	–	0.60	13.88	–	38.57	m	**52.45**
300 mm–600 mm girth	–	0.75	17.36	–	77.14	m	**94.50**
600 mm–900 mm girth	–	0.80	18.51	–	115.31	m	**133.82**
900 mm–1200 mm girth	–	1.00	23.13	–	153.75	m	**176.88**
1200 mm–1500 mm girth	–	1.02	23.60	–	192.05	m	**215.65**
Add to the above rates for working at height							
for work 3.50 m–5.00 m high	–	–	–	–	–	%	**2.50**
for work 5.00 m–6.50 m high	–	–	–	–	–	%	**5.00**
for work 6.50 m–8.00 m high	–	–	–	–	–	%	**12.50**
for work over 8.00 m high	–	–	–	–	–	%	**17.50**
Cutting and fitting around steel joints, angles, trunking, ducting, ventilators, pipes, tubes, etc.							
not exceeding 0.30 m girth	–	0.28	6.48	–	–	nr	**6.48**
0.30 m–1 m girth	–	0.37	8.56	–	–	nr	**8.56**
1 m–2 m girth	–	0.51	11.80	–	–	nr	**11.80**
over 2 m girth	–	0.42	9.71	–	–	m	**9.71**

20 PROPRIETARY LININGS AND PARTITIONS

Item	PC £	Labour hours	Labour £	Plant £	Material £	Unit	Total rate £
Supalux or other equal; sanded finish							
Lining to walls 6 mm thick							
over 600 mm wide	25.65	0.31	7.17	–	26.74	m²	33.91
not exceeding 600 mm wide	–	0.38	8.80	–	16.07	m	24.87
Lining to walls 9 mm thick							
over 600 mm wide	33.65	0.33	7.63	–	34.98	m²	42.61
not exceeding 600 mm wide	–	0.38	8.80	–	21.00	m	29.80
Lining to walls 12 mm thick							
over 600 mm wide	43.73	0.37	8.56	–	45.36	m²	53.92
not exceeding 600 mm wide	–	0.44	10.18	–	27.23	m	37.41
Masterboard or other equal; sanded finish							
Lining to walls 6 mm thick							
over 600 mm wide	12.83	0.31	7.17	–	13.53	m²	20.70
not exceeding 600 mm wide	–	0.38	8.80	–	8.14	m	16.94
Lining to walls 9 mm thick							
over 600 mm wide	32.46	0.33	7.63	–	33.75	m²	41.38
not exceeding 600 mm wide	–	0.38	8.80	–	20.27	m	29.07
Lining to walls 12 mm thick							
over 600 mm wide	33.89	0.40	9.25	–	35.23	m²	44.48
not exceeding 600 mm wide	–	0.38	8.80	–	24.65	m	33.45
Monolux 500 or other equal; 6 mm × 50 mm							
Supalux cover fillets or other equal one side							
Lining to walls 19 mm thick							
over 600 mm wide	289.48	0.65	15.04	–	301.97	m²	317.01
not exceeding 600 mm wide	–	0.92	21.28	–	183.50	m	204.78
Lining to walls 25 mm thick							
over 600 mm wide	369.21	0.69	15.97	–	384.09	m²	400.06
not exceeding 600 mm wide	221.52	0.98	22.67	–	232.78	m	255.45
Cement particle consisting of a mixture of wood fibre, cement and additive, Class 0, Cembrit or similar rigid high performance board as a sheathing							
Lining to walls 8 mm thick							
over 600 mm wide	7.87	0.65	15.04	–	11.91	m²	26.95
not exceeding 600 mm wide	–	0.92	21.28	–	9.47	m	30.75
Lining to walls 10 mm thick							
over 600 mm wide	9.20	0.65	15.04	–	13.28	m²	28.32
not exceeding 600 mm wide	–	0.92	21.28	–	10.29	m	31.57
Lining to walls 12 mm thick							
over 600 mm wide	9.15	0.65	15.04	–	13.23	m²	28.27
not exceeding 600 mm wide	–	0.92	21.28	–	10.26	m	31.54
Lining to walls 16 mm thick							
over 600 mm wide	5.02	0.65	15.04	–	8.97	m²	24.01
not exceeding 600 mm wide	–	0.92	21.28	–	7.70	m	28.98
Lining to walls 18 mm thick							
over 600 mm wide	6.46	0.65	15.04	–	10.46	m²	25.50
not exceeding 600 mm wide	–	0.92	21.28	–	8.60	m	29.88

20 PROPRIETARY LININGS AND PARTITIONS

Item	PC £	Labour hours	Labour £	Plant £	Material £	Unit	Total rate £
20.02 LININGS TO WALLS – cont							
Fibre cement board consisting of fibre cement sheet; smooth finish for paint; A1 Non-combustible; high performance board							
Lining to walls 6 mm thick							
over 600 mm wide	7.25	0.65	15.04	–	11.27	m²	**26.31**
not exceeding 600 mm wide	–	0.92	21.28	–	9.08	m	**30.36**
Lining to walls 9 mm thick							
over 600 mm wide	11.31	0.65	15.04	–	15.45	m²	**30.49**
not exceeding 600 mm wide	–	0.92	21.28	–	11.60	m	**32.88**
Lining to walls 12 mm thick							
over 600 mm wide	15.00	0.65	15.04	–	19.25	m²	**34.29**
not exceeding 600 mm wide	–	0.92	21.28	–	13.87	m	**35.15**
Glass reinforced gypsum Glasroc Multi-board or other equal and approved; fixing with nails; joints filled with joint filler and joint tape; finishing with Jointex or other equal and approved to receive decoration; to softwood base							
10 mm board to walls							
average wall height 2.00 m	–	0.37	8.49	–	32.91	m²	**41.40**
average wall height 3.00 m	–	0.36	8.42	–	32.91	m²	**41.33**
average wall height 4.00 m	–	0.36	8.33	–	32.91	m²	**41.24**
Rate per m (SMM7 measurement rule)							
wall height 2.40 m–2.70 m	–	0.93	21.52	–	88.84	m	**110.36**
wall height 2.70 m–3.00 m	–	1.06	24.52	–	98.70	m	**123.22**
wall height 3.00 m–3.30 m	–	1.07	24.68	–	108.58	m	**133.26**
wall height 3.30 m–3.60 m	–	1.39	32.16	–	118.53	m	**150.69**
12.50 mm board to walls							
average wall height 2.00 m	–	0.39	8.91	–	43.02	m²	**51.93**
average wall height 3.00 m	–	0.38	8.84	–	43.02	m²	**51.86**
average wall height 4.00 m	–	0.38	8.74	–	43.02	m²	**51.76**
Rate per m (SMM7 measurement rule)							
wall height 2.40 m–2.70 m	–	0.97	22.44	–	116.17	m	**138.61**
wall height 2.70 m–3.00 m	–	1.11	25.68	–	129.08	m	**154.76**
wall height 3.00 m–3.30 m	–	1.25	28.92	–	141.99	m	**170.91**
wall height 3.30 m–3.60 m	–	1.14	26.44	–	154.97	m	**181.41**
Blockboard (Birch faced)							
Lining to walls 18 mm thick							
over 600 wide	7.56	0.46	10.64	–	8.22	m²	**18.86**
not exceeding 600 wide	–	0.60	13.88	–	5.00	m	**18.88**
holes for pipes and the like	–	0.04	0.93	–	–	nr	**0.93**
Chipboard (plain)							
Lining to walls 12 mm thick							
over 600 mm wide	2.35	0.35	8.10	–	2.84	m²	**10.94**
not exceeding 600 mm wide	–	0.40	9.25	–	1.61	m	**10.86**
holes for pipes and the like	–	0.02	0.46	–	–	nr	**0.46**

20 PROPRIETARY LININGS AND PARTITIONS

Item	PC £	Labour hours	Labour £	Plant £	Material £	Unit	Total rate £
Lining to walls 15 mm thick							
over 600 mm wide	3.28	0.37	8.56	–	3.80	m²	12.36
not exceeding 600 mm wide	–	0.44	10.18	–	2.35	m	12.53
holes for pipes and the like	–	0.03	0.69	–	–	nr	0.69
Lining to walls 18 mm thick							
over 600 mm wide	3.32	0.39	9.02	–	3.94	m²	12.96
not exceeding 600 mm wide	–	0.50	11.57	–	2.37	m	13.94
holes for pipes and the like	–	0.04	0.93	–	–	nr	0.93
Fire-retardant chipboard/MDF; Class 1 spread of flame							
Lining to walls 12 mm thick							
over 600 mm wide	–	0.35	8.10	–	17.93	m²	26.03
not exceeding 600 mm wide	–	0.40	9.25	–	10.83	m	20.08
holes for pipes and the like	–	0.02	0.46	–	–	nr	0.46
Lining to walls 15 mm thick							
over 600 mm wide	–	0.35	8.10	–	16.37	m²	24.47
not exceeding 600 mm wide	–	0.40	9.25	–	9.89	m	19.14
holes for pipes and the like	–	0.02	0.46	–	–	nr	0.46
Lining to walls 18 mm thick							
over 600 mm wide	–	0.39	9.02	–	18.57	m²	27.59
not exceeding 600 mm wide	–	0.50	11.57	–	11.21	m	22.78
holes for pipes and the like	–	0.04	0.93	–	–	nr	0.93
Lining to walls 25 mm thick							
over 600 mm wide	–	0.41	9.49	–	27.54	m²	37.03
not exceeding 600 mm wide	–	0.56	12.96	–	16.59	m	29.55
holes for pipes and the like	–	0.05	1.15	–	–	nr	1.15
Chipboard melamine faced; white matt finish; laminated masking strips							
Lining to walls 15 mm thick							
over 600 mm wide	4.23	0.97	22.44	–	5.11	m²	27.55
not exceeding 600 mm wide	–	1.26	29.15	–	3.15	m	32.30
holes for pipes and the like	–	0.06	1.39	–	–	nr	1.39
Insulation board to BS EN 622							
Lining to walls 12 mm thick							
over 600 mm wide	2.25	0.22	5.09	–	2.74	m²	7.83
not exceeding 600 mm wide	–	0.26	6.02	–	1.71	m	7.73
holes for pipes and the like	–	0.01	0.23	–	–	nr	0.23
Plywood (Eastern European); internal quality							
Lining to walls 4 mm thick							
over 600 mm wide	2.72	0.34	7.87	–	3.22	m²	11.09
not exceeding 600 mm wide	–	0.44	10.18	–	2.00	m	12.18
Lining to walls 6 mm thick							
over 600 mm wide	3.93	0.37	8.56	–	4.47	m²	13.03
not exceeding 600 mm wide	–	0.48	11.10	–	2.75	m	13.85
Lining to walls 12 mm thick							
over 600 mm wide	7.30	0.43	9.95	–	7.94	m²	17.89
not exceeding 600 mm wide	–	0.56	12.96	–	4.83	m	17.79
Lining to walls 18 mm thick							
over 600 mm wide	10.65	0.46	10.64	–	11.40	m²	22.04
not exceeding 600 mm wide	–	0.60	13.88	–	6.90	m	20.78

20 PROPRIETARY LININGS AND PARTITIONS

Item	PC £	Labour hours	Labour £	Plant £	Material £	Unit	Total rate £
20.02 LININGS TO WALLS – cont							
Plywood (Eastern European); external quality							
Lining to walls 4 mm thick							
over 600 mm wide	6.73	0.34	7.87	–	7.35	m²	**15.22**
not exceeding 600 mm wide	–	0.44	10.18	–	4.48	m	**14.66**
Lining to walls 6.5 mm thick							
over 600 mm wide	7.53	0.37	8.56	–	8.18	m²	**16.74**
not exceeding 600 mm wide	–	0.48	11.10	–	4.97	m	**16.07**
Lining to walls 9 mm thick							
over 600 mm wide	9.69	0.40	9.25	–	10.40	m²	**19.65**
not exceeding 600 mm wide	–	0.52	12.03	–	6.30	m	**18.33**
Lining to walls 12 mm thick							
over 600 mm wide	12.10	0.43	9.95	–	12.89	m²	**22.84**
not exceeding 600 mm wide	–	0.56	12.96	–	7.80	m	**20.76**
holes for pipes and the like	–	0.03	0.69	–	–	nr	**0.69**
Extra over linings fixed with screws	–	0.10	2.32	–	0.08	m²	**2.40**
Internal quality American Cherry veneered plywood; 6 mm thick							
Lining to walls							
over 600 mm wide	8.82	0.41	9.49	–	9.40	m²	**18.89**
not exceeding 600 mm wide	–	0.54	12.49	–	5.77	m	**18.26**
Glazed hardboard to BS EN 622; on and including 38 mm × 38 mm sawn softwood framing							
3.20 mm thick panel							
to side of bath	–	1.67	38.64	–	10.99	nr	**49.63**
to end of bath	–	0.65	15.04	–	3.59	nr	**18.63**
RIGID SHEET ACOUSTIC PANEL LININGS							
Perforated steel acoustic wall panels; Eckel type HD EFP or other equal and approved; polyurethene enamel finish; fibrous glass acoustic insulation							
Walls							
average height 3.00 m; fixed to timber or masonry	–	–	–	–	–	m²	**212.87**
Acoustic Panel linings; Troldekt Ultrafine 1200 mm × 600 mm × 25 mm on and including 50 mm × 70 mm timber framework at 600 mm centres both ways and supported from the roof with 50 mm × 50 mm hangers 400 mm–600 mm long							
Ceilings							
allow for ceiling tiles to be clipped to prevent wind uplift when external doors are opened	–	–	–	–	–	m²	**117.47**
perimeter edge detail with 50 mm × 50 mm support batten	–	–	–	–	–	m	**7.45**

20 PROPRIETARY LININGS AND PARTITIONS

Item	PC £	Labour hours	Labour £	Plant £	Material £	Unit	Total rate £
20.03 LININGS TO CEILINGS							
Plaster; one coat Thistle board finish or similar; steel trowelled 3 mm work to ceilings; one coat on and including gypsum plasterboard; fixing with nails; 3 mm joints filled with plaster and jute scrim cloth; to softwood base; plain grade baseboard or lath with rounded edges							
9.50 mm thick boards to ceilings							
over 600 mm wide	–	0.89	14.96	–	5.47	m²	**20.43**
over 600 mm wide; 3.50 m–5.00 m high	–	1.03	17.44	–	5.47	m²	**22.91**
not exceeding 600 mm wide	–	0.43	7.64	–	1.65	m	**9.29**
9.50 mm thick boards to ceilings; in staircase areas or plant rooms							
over 600 mm wide	–	0.98	16.55	–	5.47	m²	**22.02**
not exceeding 600 mm wide	–	0.47	8.35	–	1.65	m	**10.00**
9.50 mm thick boards to isolated beams							
over 600 mm wide	–	1.05	17.80	–	5.47	m²	**23.27**
not exceeding 600 mm wide	–	0.50	8.89	–	1.65	m	**10.54**
12.50 mm thick boards to ceilings							
over 600 mm wide	–	0.95	16.02	–	5.47	m²	**21.49**
over 600 mm wide; 3.50 m–5.00 m high	–	1.06	17.97	–	5.47	m²	**23.44**
not exceeding 600 mm wide	–	0.45	8.00	–	1.65	m	**9.65**
12.50 mm thick boards to ceilings; in staircase areas or plant rooms							
over 600 mm wide	–	1.06	17.97	–	5.47	m²	**23.44**
not exceeding 600 mm wide	–	0.51	9.06	–	1.65	m	**10.71**
12.50 mm thick boards to isolated beams							
over 600 mm wide	–	1.15	19.57	–	5.47	m²	**25.04**
not exceeding 600 mm wide	–	0.56	9.96	–	1.65	m	**11.61**

Prices for Measured Works

21 CLADDING AND COVERING

Item	PC £	Labour hours	Labour £	Plant £	Material £	Unit	Total rate £
21.01 CURTAIN WALLING AND GLAZED ROOFING							
Stick curtain walling system; proprietary solution from system supplier; e.g. Schuco or equivalent							
Polyester powder coated solid colour matt finish or natural anodized finish mullions spaced 1.5 m apart and spanning typical storey height of 3.8 m. Floor to ceiling glass sealed units with 8.8 mm low-e coated laminated inner pane, air filled cavity and 8 mm clear monolithic heat strengthened outer pane, retained by external pressure plates and caps. Rates to include glass fronted solid spandrel panels, all brackets, membranes, fire stopping between floors, trade contractor preliminaries, including external access equipment							
flat system	–	–	–	–	–	m²	**522.75**
extra over for							
neutral selective high performance coating on surface #2 in lieu of low-e coating on surface #3 (G-values circa 0.25–0.30); for assisting in solar control	–	–	–	–	–	m²	**42.99**
inner laminated glass to be heat strengthened laminated to mitigate thermal fracture risk	–	–	–	–	–	m²	**42.99**
outer glass to be heat strengthened laminated in lieu of monolithic heat strengthened	–	–	–	–	–	m²	**42.99**
ceramic fritting glass on surface #2 for visual and/or performance requirements (minimal G-value improvement)	–	–	–	–	–	m²	**58.09**
flush glass finish without external face caps, achieved by concealed toggle fixings locating within perimeter channels within sealed units including silicone sealing between glass panes	–	–	–	–	–	m²	**58.09**
typical coping detail, including pressed aluminium profiles, membranes, seals, etc.	–	–	–	–	–	m	**313.65**
typical sill detail, including pressed aluminium profiles, membranes, seals, etc.	–	–	–	–	–	m	**255.57**
intermediate transoms (per transom)	–	–	–	–	–	m	**55.76**

21 CLADDING AND COVERING

Item	PC £	Labour hours	Labour £	Plant £	Material £	Unit	Total rate £
Unitized curtain walling system; proprietary solution from system supplier e.g. Schuco or equivalent alternatively pre-designed system solution from specialist facade contractor Polyester powder coated solid colour matt finish or natural anodized curtain walling Element widths of 1.5 m spanning typical storey height of 3.8 m. Floor to ceiling glass sealed units with 8.8 mm low-e coated laminated inner pane, air filled cavity and 8 mm monolithic heat strengthed outer pane, retained by external beading system. Rates include 1.1 m solid spandrel panels, all brackets, membranes, fire stopping between floors, trade contractor preliminaries, including external access equipment							
flat system	–	–	–	–	–	m²	**1080.36**
extra over for							
neutral selective high performance coating on surface #2 in lieu of low-e coating on surface #3 (G-values circa 0.25–0.30); for assisting in solar control	–	–	–	–	–	m²	**42.99**
inner laminated glass to be heat strengthened laminated to mitigate thermal fracture risk	–	–	–	–	–	m²	**42.99**
flush glass finish without external face caps, achieved by carrier frames with glass sealed units factory silicone bonded; often referred to as SSG (Structural Silicone Glazing)	–	–	–	–	–	m²	**69.70**
typical coping detail, including pressed aluminium profiles, membranes, seals, etc.	–	–	–	–	–	m	**313.65**
typical sill detail, including pressed aluminium profiles, membranes, seals, etc.	–	–	–	–	–	m	**255.57**
Other curtain walling systems/costs Unitized curtain walling system; bespoke project specific solution via specialist facade contractor mostly based in mainland Europe.							
generally as described for system supplier solution but profiles catered to specific performance and visual requirements, thus additional design development. Note: These rates are subject to currency fluctuations between £ and €.	–	–	–	–	–	m²	**1080.36**
flush glass finish finish (SEG) assuming direct bonding in he factory but with base chassis design to accommodate carrier frame for glass replacement, avoiding site bending	–	–	–	–	–	m²	**34.86**
project specific unitized curtain walling generally requires project specific performance testing. This rate is for a single wall type.	–	–	–	–	–	nr	**104549.80**

21 CLADDING AND COVERING

Item	PC £	Labour hours	Labour £	Plant £	Material £	Unit	Total rate £
21.01 CURTAIN WALLING AND GLAZED ROOFING – cont							
Other curtain walling systems/costs – cont							
Unitized curtain walling system – cont							
visual mock-ups are often required for project specific unitized curtain walling solutions and in cases for proprietary unitized and stick curtain walling projects. This rate is for a single wall type.	–	–	–	–	–	nr	34849.93
all curtain walling should be site hose tested. The rate depends upon the quantum of joints to be tested, generally 5%. Assume 5 days @ £1000 extra over for	–	–	–	–	–	nr	6969.99
neutral selective high performance coating on surface #2 in lieu of low-e coating on surface #3 (G-values circa 0.25–0.30); for assisting in solar control	–	–	–	–	–	m²	42.99
inner laminated glass to be heat strengthened laminated to mitigate thermal fracture risk	–	–	–	–	–	m²	42.99
outer glass to be heat strengthened laminated in lieu of monolithic heat strengthened	–	–	–	–	–	m²	42.99
alternative solutions for achieving Part L and comfort criteria, include increasing the amount of solidity	–	–	–	–	–	m²	104.53
Patent glazing; aluminium alloy bars 2.55 m long at 622 mm centres; fixed to supports							
Roof cladding							
single glazed with 6.4 mm laminated glass	–	–	–	–	–	m²	175.93
single glazed with 7 mm thick Georgian wired cast glass	–	–	–	–	–	m²	181.99
thermally broken and double glazed with low-e clear toughened and laminated double glazed units; aluminium finished RAL matt colour	–	–	–	–	–	m²	479.25
Opening roof vents							
600 mm × 900 mm top hung opening roof vent; manually operated	–	–	–	–	–	nr	539.93
600 mm × 900 mm top hung opening roof vent; electrically operated	–	–	–	–	–	nr	691.59
Skylight							
self-supporting hipped or gable ended lantern/ skylight thermally broken and double glazed with low-e clear toughened and laminated double glazed units; aluminium finished RAL matt colour	–	–	–	–	–	m²	958.51
Associated code 4 lead flashings							
top flashing; 210 mm girth	–	–	–	–	–	m	75.21
bottom flashing; 240 mm girth	–	–	–	–	–	m	86.14
end flashing; 300 mm girth	–	–	–	–	–	m	94.64

21 CLADDING AND COVERING

Item	PC £	Labour hours	Labour £	Plant £	Material £	Unit	Total rate £
Wall cladding							
single glazed with 6.4 mm laminated glass	–	–	–	–	–	m²	**174.71**
single glazed with 7 mm thick Georgian wired cast glass	–	–	–	–	–	m²	**194.13**
thermally broken and double glazed with low-e clear toughened and laminated double glazed units; aluminium finished RAL matt colour	–	–	–	–	–	m²	**499.88**
Extra for aluminium alloy perimeter members							
38 mm × 38 mm × 3 mm angle jamb	–	–	–	–	–	m	**26.70**
pressed sill member	–	–	–	–	–	m	**53.39**
pressed channel head and PVC case	–	–	–	–	–	m	**53.39**
21.02 RIGID SHEET CLADDING							
Resoplan sheet or other equal and approved; Eternit UK Ltd; flexible neoprene gasket joints; fixing with stainless steel screws and coloured caps							
6 mm thick cladding to walls							
over 300 mm wide	–	1.94	44.89	–	71.23	m²	**116.12**
not exceeding 300 mm wide	–	0.65	15.04	–	26.07	m	**41.11**
Eternit 2000 Glasal sheet or other equal and approved; Eternit UK Ltd; flexible neoprene gasket joints; fixing with stainless steel screws and coloured caps							
7.50 mm thick cladding to walls							
over 300 mm wide	–	1.94	44.89	–	62.76	m²	**107.65**
not exceeding 300 mm wide	–	0.65	15.04	–	23.53	m	**38.57**
external angle trim	–	0.09	2.08	–	12.13	m	**14.21**
7.50 mm thick cladding to eaves, verge soffit boards, fascia boards or the like							
100 mm wide	–	0.46	10.64	–	12.02	m	**22.66**
200 mm wide	–	0.56	12.96	–	17.77	m	**30.73**
300 mm wide	–	0.65	15.04	–	23.53	m	**38.57**
21.03 WEATHERBOARDING							
Prodema ProdEX high density resin-bonded cellulose fibre weatherboarding panels; including secondary supports and fixing							
Walls							
8 mm panels face fixed on to timber battens	–	–	–	–	–	m²	**200.83**
8 mm panels face fixed on to aluminium rails	–	–	–	–	–	m²	**227.96**
8 mm panels adhesive fixed on to timber battens or aluminium rails	–	–	–	–	–	m²	**244.26**
10 mm panels secret fixed on to helping hand aluminium system	–	–	–	–	–	m²	**298.54**

21 CLADDING AND COVERING

Item	PC £	Labour hours	Labour £	Plant £	Material £	Unit	Total rate £
21.04 PROFILED SHEET CLADDING							
Galvanized steel strip troughed sheets; Corus Products or other equal							
Roof cladding or decking; sloping not exceeding 50°; fixing to steel purlins with plastic headed self-tapping screws							
0.7 mm thick; 46 profile	–	–	–	–	–	m²	14.21
0.7 mm thick; 60 profile	–	–	–	–	–	m²	15.56
0.7 mm thick; 100 profile	–	–	–	–	–	m²	16.91
Galvanized steel strip troughed sheets; PMF Strip Mill Products or other equal							
Roof cladding; sloping not exceeding 50°; fixing to steel purlins with plastic headed self-tapping screws							
0.7 mm thick type HPS200 13.5/3 corrugated	–	–	–	–	–	m²	16.02
0.7 mm thick type HPS200 R32/1000	–	–	–	–	–	m²	14.55
0.7 mm thick type Arcline 40; plasticol finished	–	–	–	–	–	m²	21.38
extra over last for aluminium roof cladding or decking	–	–	–	–	–	m²	8.79
Accessories for roof cladding							
HPS200 Drip flashing; 250 mm girth	–	–	–	–	–	m	5.12
HPS200 Ridge flashing; 375 mm girth	–	–	–	–	–	m	6.50
HPS200 Gable flashing; 500 mm girth	–	–	–	–	–	m	8.26
HPS200 Internal angle; 625 mm girth	–	–	–	–	–	m	9.52
GRP transluscent rooflights; factory assembled							
Rooflight; vertical fixing to steel purlins (measured elsewhere)							
double skin; class 3 over 1	–	–	–	–	–	m²	49.63
triple skin; class 3 over 1	–	–	–	–	–	m²	55.14
Lightweight galvanized steel roof tiles; Decra Roof Systems; or other equal; coated finish							
Zinc coated tile panels, dry fixed							
Classic profile; clay or concrete tile appearance	–	0.23	5.21	–	20.16	m²	25.37
Stratos profile; slate or concrete tiles appearance	–	0.23	5.21	–	25.04	m²	30.25
Accessories for roof cladding							
pitched D ridge	–	0.09	2.04	–	11.94	m	13.98
barge cover (handed)	–	0.09	2.04	–	11.76	m	13.80
in line air vent	–	0.09	2.04	–	68.78	nr	70.82
in line soil vent	–	0.09	2.04	–	68.78	nr	70.82
gas flue terminal	–	0.19	4.31	–	139.05	nr	143.36
Kingspan KS100RW composite roof panels for roof pitches greater than 4° (after deflection)							
External coating XL Forte (steel), internal coating bright white polyester (steel)							
80 mm thick panel; U-value 0.25 W/m² K	–	–	–	–	–	m²	43.01
115 mm thick panel; U-value 0.18 W/m² K	–	–	–	–	–	m²	45.17
150 mm thick panel; U-value 0.14 W/m² K	–	–	–	–	–	m²	60.65
KS1000PC polycarbonate rooflight	–	–	–	–	–	m²	13.24

21 CLADDING AND COVERING

Item	PC £	Labour hours	Labour £	Plant £	Material £	Unit	Total rate £
Associated flashings							
eaves – 200 mm girth	–	–	–	–	–	m	**3.31**
ridge – 620 mm girth	–	–	–	–	–	m	**19.85**
hip – 620 mm girth	–	–	–	–	–	m	**19.85**
Kingspan KS1000 KZ composite standing seam roof panels for roof pitches greater than 1.5° (after deflection)							
External Coating XL Forte (Steel) Internal Coating Bright White Polyester (steel)							
90 mm thick panel; U-value 0.25 W/m^2 K	–	–	–	–	–	m^2	**69.47**
110 mm thick panel; U-value 0.20 W/m^2 K	–	–	–	–	–	m^2	**74.99**
125 mm thick panel; U-value 0.18 W/m^2 K	–	–	–	–	–	m^2	**78.30**
Associated flashings							
eaves – 200 mm girth	–	–	–	–	–	m	**3.31**
ridge – 620 mm girth	–	–	–	–	–	m	**19.85**
hip – 620 mm girth	–	–	–	–	–	m	**19.85**
Fibre cement corrugated sheets; Eternit 2000 or other equal							
Roof cladding; sloping not exceeding 50°; fixing to steel purlins with hook bolts							
Profile 3; natural grey	–	0.23	5.21	–	17.56	m^2	**22.77**
Profile 3; coloured	–	0.23	5.21	–	23.19	m^2	**28.40**
Profile 6; natural grey	–	0.28	6.34	–	16.11	m^2	**22.45**
Profile 6; coloured	–	0.28	6.34	–	17.21	m^2	**23.55**
Profile 6; natural grey; insulated 100 mm glass fibre infill; white finish steel lining panel	–	0.46	10.42	–	31.35	m^2	**41.77**
Profile 6; coloured; insulated 100 mm glass fibre infill; white finish steel lining panel	–	0.46	10.42	–	32.01	m^2	**42.43**
Accessories; to Profile 3 cladding; natural grey							
eaves filler	–	0.09	2.04	–	11.71	m	**13.75**
external corner piece	–	0.11	2.49	–	8.65	m	**11.14**
apron flashing	–	0.11	2.49	–	11.71	m	**14.20**
plain wing or close fitting two piece adjustable capping to ridge	–	0.16	3.63	–	10.98	m	**14.61**
ventilating two piece adjustable capping to ridge	–	0.16	3.63	–	16.88	m	**20.51**
Accessories; to Profile 6 cladding; natural grey							
eaves filler	–	0.09	2.04	–	7.06	m	**9.10**
external corner piece	–	0.11	2.49	–	8.03	m	**10.52**
apron flashing	–	0.11	2.49	–	7.84	m	**10.33**
underglazing flashing	–	0.11	2.49	–	10.33	m	**12.82**
plain cranked crown to ridge	–	0.16	3.63	–	21.00	m	**24.63**
plain wing or close fitting two piece adjustable capping to ridge	–	0.16	3.63	–	14.11	m	**17.74**
ventilating two piece adjustable capping to ridge	–	0.16	3.63	–	18.05	m	**21.68**

21 CLADDING AND COVERING

Item	PC £	Labour hours	Labour £	Plant £	Material £	Unit	Total rate £
21.05 WALL CLADDING – METAL							
Extended, hard skinned, foamed PVC-UE profiled sections; Swish Celuka or other equal and approved; Class 1 fire-rated to BS 476; Part 7; in white finish							
Wall cladding; vertical; fixing to timber							
100 mm shiplap profiles; Code 001	–	0.35	8.10	–	71.17	m²	**79.27**
150 mm shiplap profiles; Code 002	–	0.32	7.41	–	63.07	m²	**70.48**
125 mm feather-edged profiles; Code C208	–	0.34	7.87	–	68.72	m²	**76.59**
vertical angles	–	0.19	4.40	–	7.31	m	**11.71**
raking cutting	–	0.14	3.23	–	–	m	**3.23**
Kingspan KS1000RW Composite Wall Panels							
External coating XL Forte (steel) internal coating bright white polyester (steel)							
60 mm thick; U-value 0.35 W/m²K	–	–	–	–	–	m²	**51.80**
80 mm thick; U-value 0.26 W/m²K	–	–	–	–	–	m²	**54.52**
100 mm thick; U-value 0.20 W/m²K	–	–	–	–	–	m²	**57.92**
Kingspan KS1000MR Composite Wall Panels							
External coating Spectum (steel) internal coating bright white polyester (steel)							
60 mm thick; U-value 0.35 W/m²K	–	–	–	–	–	m²	**92.68**
80 mm thick; U-value 0.26 W/m²K	–	–	–	–	–	m²	**99.50**
100 mm thick; U-value 0.20 W/m²K	–	–	–	–	–	m²	**106.32**
Kingspan KS900MR Composite Wall Panels							
External coating Spectum (steel) internal coating bright white polyester (steel)							
60 mm thick; U-value 0.35 W/m²K	–	–	–	–	–	m²	**106.32**
80 mm thick; U-value 0.26 W/m²K	–	–	–	–	–	m²	**114.49**
100 mm thick; U-value 0.20 W/m²K	–	–	–	–	–	m²	**122.66**
Kingspan KS600MR Composite Wall Panels							
External coating Sprectum (steel) internal coating bright white polyester (steel)							
60 mm thick; U-value 0.35 W/m²K	–	–	–	–	–	m²	**143.11**
80 mm thick; U-value 0.26 W/m²K	–	–	–	–	–	m²	**155.38**
100 mm thick; U-value 0.20 W/m²K	–	–	–	–	–	m²	**167.64**
Kingspan KS1000 Optimo FLAT Composite Wall Panels							
External coating Spectum (steel) internal coating bright white polyester (steel)							
60 mm thick; U-value 0.35 W/m²K	–	–	–	–	–	m²	**155.38**
80 mm thick; U-value 0.26 W/m²K	–	–	–	–	–	m²	**163.56**
100 mm thick; U-value 0.20 W/m²K	–	–	–	–	–	m²	**171.74**

21 CLADDING AND COVERING

Item	PC £	Labour hours	Labour £	Plant £	Material £	Unit	Total rate £
Kingspan KS900 Optimo FLAT Composite Wall Panels							
External coating Spectrum (steel) internal coating bright white polyester (steel)							
60 mm thick; U-value 0.35 W/m²K	–	–	–	–	–	m²	177.18
80 mm thick; U-value 0.26 W/m²K	–	–	–	–	–	m²	185.36
100 mm thick; U-value 0.20 W/m²K	–	–	–	–	–	m²	193.52
Kingspan KS600 Optimo FLAT Composite Wall Panels							
External coating Spectrum (steel) internal coating bright white polyester (steel)							
60 mm thick; U-value 0.35 W/m²K	–	–	–	–	–	m²	231.71
80 mm thick; U-value 0.26 W/m²K	–	–	–	–	–	m²	245.33
100 mm thick; U-value 0.20 W/m²K	–	–	–	–	–	m²	258.97
Associated flashings							
0.7 mm pre-coated steel							
cill; 300 mm girth	–	–	–	–	–	m	24.53
composite top hat	–	–	–	–	–	m	28.61
Miscellaneous items							
panel bearers at 1500 mm centres	–	–	–	–	–	nr	10.90
preformed corners to horizontally laid panels	–	–	–	–	–	nr	122.66
21.06 WALL CLADDING – RAINSCREEN							
Timber							
Western Red Cedar tongued and grooved wall cladding on and including treated softwrood battens on breather mambrane, 10 mm Eternit Blueclad board and 50 mm insulation board; the whole fixed to Metsec frame system; including sealing all joints etc.							
26 mm thick cladding to walls; boards laid horizontally	–	–	–	–	–	m²	113.81
Aluminium							
Reynobond rainscreen cladding; aluminium composite material cassettes with thermoplastic cores, back ventilated, including insulation, vapour control membrane and aluminium support system							
4 mm thick cladding; fixed to walls	–	–	–	–	–	m²	202.35
Clay							
Terracotta clay rainscreen cladding; including insulation, vapour control membrane and aluminium support system							
400 × 200 × 30 mm tile cladding; fixed to walls	–	–	–	–	–	m²	366.75

22 GENERAL JOINERY

Item	PC £	Labour hours	Labour £	Plant £	Material £	Unit	Total rate £
22.01 SOFTWOOD							
Wrought softwood							
Skirtings, picture rails, dado rails and the like; chamfered or ogee							
19 mm × 44 mm; chamfered	–	0.09	2.08	–	1.63	m	**3.71**
19 mm × 44 mm; ogee	–	0.09	2.08	–	1.25	m	**3.33**
19 mm × 69 mm; chamfered	–	0.09	2.08	–	2.43	m	**4.51**
19 mm × 69 mm; ogee	–	0.09	2.08	–	1.84	m	**3.92**
19 mm × 94 mm; chamfered	–	0.09	2.08	–	3.23	m	**5.31**
19 mm × 94 mm; ogee	–	0.09	2.08	–	2.43	m	**4.51**
19 mm × 144 mm; ogee	–	0.11	2.54	–	3.61	m	**6.15**
19 mm × 169 mm; ogee	–	0.11	2.54	–	4.19	m	**6.73**
25 mm × 50 mm; ogee	–	0.09	2.08	–	2.07	m	**4.15**
25 mm × 69 mm; chamfered	–	0.09	2.08	–	3.13	m	**5.21**
25 mm × 94 mm; chamfered	–	0.09	2.08	–	4.18	m	**6.26**
25 mm × 144 mm; chamfered	–	0.11	2.54	–	6.29	m	**8.83**
25 mm × 144 mm; ogee	–	0.11	2.54	–	4.68	m	**7.22**
25 mm × 169 mm; ogee	–	0.11	2.54	–	5.45	m	**7.99**
25 mm × 219 mm; ogee	–	0.13	3.01	–	6.99	m	**10.00**
returned ends	–	0.14	3.23	–	–	nr	**3.23**
mitres	–	0.09	2.08	–	–	nr	**2.08**
Architraves, cover fillets and the like; bull nosed; chamfered or ogee							
13 mm × 25 mm; bull nosed	–	0.11	2.54	–	0.68	m	**3.22**
13 mm × 50 mm; ogee	–	0.11	2.54	–	0.89	m	**3.43**
16 mm × 32 mm; half bull nosed	–	0.11	2.54	–	0.95	m	**3.49**
16 mm × 38 mm; ogee	–	0.11	2.54	–	0.84	m	**3.38**
16 mm × 50 mm; ogee	–	0.11	2.54	–	1.04	m	**3.58**
19 mm × 50 mm; chamfered	–	0.11	2.54	–	2.11	m	**4.65**
19 mm × 63 mm; chamfered	–	0.11	2.54	–	2.62	m	**5.16**
19 mm × 69 mm; chamfered	–	0.11	2.54	–	2.43	m	**4.97**
25 mm × 44 mm; chamfered	–	0.11	2.54	–	2.42	m	**4.96**
25 mm × 50 mm; ogee	–	0.11	2.54	–	1.51	m	**4.05**
25 mm × 63 mm; chamfered	–	0.11	2.54	–	3.37	m	**5.91**
25 mm × 69 mm; chamfered	–	0.11	2.54	–	3.68	m	**6.22**
32 mm × 88 mm; ogee	–	0.11	2.54	–	3.14	m	**5.68**
38 mm × 38 mm; ogee	–	0.11	2.54	–	1.71	m	**4.25**
50 mm × 50 mm; ogee	–	0.11	2.54	–	2.81	m	**5.35**
returned ends	–	0.14	3.23	–	–	nr	**3.23**
mitres	–	0.09	2.08	–	–	nr	**2.08**
Stops; screwed on							
16 mm × 38 mm	–	0.09	2.08	–	0.69	m	**2.77**
16 mm × 50 mm	–	0.09	2.08	–	0.91	m	**2.99**
19 mm × 38 mm	–	0.09	2.08	–	0.81	m	**2.89**
25 mm × 38 mm	–	0.09	2.08	–	1.06	m	**3.14**
25 mm × 50 mm	–	0.09	2.08	–	1.37	m	**3.45**

22 GENERAL JOINERY

Item	PC £	Labour hours	Labour £	Plant £	Material £	Unit	Total rate £
Glazing beads and the like							
13 mm × 16 mm	–	0.04	0.93	–	1.98	m	**2.91**
13 mm × 19 mm	–	0.04	0.93	–	2.35	m	**3.28**
13 mm × 25 mm	–	0.04	0.93	–	3.10	m	**4.03**
13 mm × 25 mm; screwed	–	0.09	2.08	–	3.10	m	**5.18**
13 mm × 25 mm; fixing with brass cups and screws	–	0.12	2.78	–	3.17	m	**5.95**
16 mm × 25 mm; screwed	–	0.08	1.85	–	3.81	m	**5.66**
16 mm quadrant	–	0.04	0.93	–	2.14	m	**3.07**
19 mm quadrant or scotia	–	0.04	0.93	–	5.96	m	**6.89**
19 mm × 36 mm; screwed	–	0.04	0.93	–	6.52	m	**7.45**
25 mm × 38 mm; screwed	–	0.04	0.93	–	9.05	m	**9.98**
25 mm quadrant or scotia	–	0.04	0.93	–	5.96	m	**6.89**
38 mm scotia	–	0.04	0.93	–	13.77	m	**14.70**
50 mm scotia	–	0.04	0.93	–	23.84	m	**24.77**
Isolated shelves, worktops, seats and the like							
19 mm × 150 mm	–	0.15	3.47	–	3.86	m	**7.33**
19 mm × 200 mm	–	0.20	4.62	–	5.33	m	**9.95**
25 mm × 150 mm	–	0.15	3.47	–	4.41	m	**7.88**
25 mm × 200 mm	–	0.20	4.62	–	6.27	m	**10.89**
32 mm × 150 mm	–	0.15	3.47	–	5.15	m	**8.62**
32 mm × 200 mm	–	0.20	4.62	–	7.02	m	**11.64**
Isolated shelves, worktops, seats and the like; cross-tongued joints							
19 mm × 300 mm	–	0.26	6.02	–	15.87	m	**21.89**
19 mm × 450 mm	–	0.31	7.17	–	23.94	m	**31.11**
19 mm × 600 mm	–	0.37	8.56	–	30.97	m	**39.53**
25 mm × 300 mm	–	0.26	6.02	–	17.04	m	**23.06**
25 mm × 450 mm	–	0.31	7.17	–	25.81	m	**32.98**
25 mm × 600 mm	–	0.37	8.56	–	33.58	m	**42.14**
32 mm × 300 mm	–	0.26	6.02	–	18.05	m	**24.07**
32 mm × 450 mm	–	0.31	7.17	–	27.44	m	**34.61**
32 mm × 600 mm	–	0.37	8.56	–	35.79	m	**44.35**
Isolated shelves, worktops, seats and the like; slatted with 50 wide slats at 75 mm centres							
19 mm thick	–	0.60	13.88	–	39.07	m	**52.95**
25 mm thick	–	0.60	13.88	–	39.93	m	**53.81**
32 mm thick	–	0.60	13.88	–	40.69	m	**54.57**
Window boards, nosings, bed moulds and the like; rebated and rounded							
19 mm × 75 mm	–	0.17	3.93	–	5.48	m	**9.41**
19 mm × 150 mm	–	0.19	4.40	–	6.76	m	**11.16**
19 mm × 225 mm; in one width	–	0.24	5.55	–	8.33	m	**13.88**
19 mm × 300 mm; cross-tongued joints	–	0.28	6.48	–	19.07	m	**25.55**
25 mm × 75 mm	–	0.17	3.93	–	5.79	m	**9.72**
25 mm × 150 mm	–	0.19	4.40	–	7.40	m	**11.80**
25 mm × 225 mm; in one width	–	0.24	5.55	–	9.32	m	**14.87**
25 mm × 300 mm; cross-tongued joints	–	0.28	6.48	–	20.61	m	**27.09**
32 mm × 75 mm	–	0.17	3.93	–	6.06	m	**9.99**
32 mm × 150 mm	–	0.19	4.40	–	7.96	m	**12.36**
32 mm × 225 mm; in one width	–	0.24	5.55	–	10.17	m	**15.72**

22 GENERAL JOINERY

Item	PC £	Labour hours	Labour £	Plant £	Material £	Unit	Total rate £
22.01 SOFTWOOD – cont							
Wrought softwood – cont							
Window boards, nosings, bed moulds and the like – cont							
32 mm × 300 mm; cross-tongued joints	–	0.28	6.48	–	21.91	m	**28.39**
38 mm × 75 mm	–	0.17	3.93	–	6.62	m	**10.55**
38 mm × 150 mm	–	0.19	4.40	–	9.16	m	**13.56**
38 mm × 225 mm; in one width	–	0.24	5.55	–	11.87	m	**17.42**
38 mm × 300 mm; cross-tongued joints	–	0.28	6.48	–	24.52	m	**31.00**
returned and fitted ends	–	0.14	3.23	–	–	nr	**3.23**
Handrails; mopstick							
50 mm dia.	–	0.23	5.33	–	2.83	m	**8.16**
Handrails; rounded							
44 mm × 50 mm	–	0.23	5.33	–	3.18	m	**8.51**
50 mm × 75 mm	–	0.25	5.79	–	5.41	m	**11.20**
63 mm × 87 mm	–	0.28	6.48	–	7.16	m	**13.64**
75 mm × 100 mm	–	0.32	7.41	–	9.97	m	**17.38**
Handrails; pigs ear							
44 mm × 50 mm	–	0.23	5.33	–	3.35	m	**8.68**
50 mm × 75 mm	–	0.25	5.79	–	5.71	m	**11.50**
63 mm × 87 mm	–	0.28	6.48	–	7.54	m	**14.02**
75 mm × 100 mm	–	0.32	7.41	–	10.49	m	**17.90**
Sundries on softwood							
Extra over fixing with nails for							
gluing and pinning	–	0.02	0.40	–	0.04	m	**0.44**
masonry nails	–	0.02	0.41	–	0.13	m	**0.54**
steel screws	–	0.02	0.39	–	0.10	m	**0.49**
self-tapping screws	–	0.02	0.40	–	0.11	m	**0.51**
steel screws; gluing	–	0.03	0.68	–	0.10	m	**0.78**
steel screws; sinking; filling heads	–	0.04	0.85	–	0.10	m	**0.95**
steel screws; sinking; pellating over	–	0.08	1.86	–	0.10	m	**1.96**
brass cups and screws	–	0.10	2.31	–	0.27	m	**2.58**
Extra over for							
countersinking	–	0.01	0.34	–	–	m	**0.34**
pellating	–	0.07	1.62	–	–	m	**1.62**
Head or nut in softwood							
let in flush	–	0.04	0.85	–	–	nr	**0.85**
Head or nut; in hardwood							
let in flush	–	0.06	1.28	–	–	nr	**1.28**
let in over; pellated	–	0.13	2.99	–	–	nr	**2.99**

22 GENERAL JOINERY

Item	PC £	Labour hours	Labour £	Plant £	Material £	Unit	Total rate £
22.02 HARDWOOD							
Selected Sapele							
Skirtings, picture rails, dado rails and the like;							
chamfered or ogee							
19 mm × 44 mm; chamfered	4.51	0.13	3.01	–	4.86	m	**7.87**
19 mm × 44 mm; ogee	2.58	0.13	3.01	–	2.87	m	**5.88**
19 mm × 69 mm; chamfered	7.08	0.13	3.01	–	7.51	m	**10.52**
19 mm × 69 mm; ogee	4.04	0.13	3.01	–	4.38	m	**7.39**
19 mm × 94 mm; chamfered	9.64	0.13	3.01	–	10.15	m	**13.16**
19 mm × 94 mm; ogee	5.50	0.13	3.01	–	5.88	m	**8.89**
19 mm × 144 mm; ogee	8.43	0.15	3.47	–	8.89	m	**12.36**
19 mm × 169 mm; ogee	9.89	0.15	3.47	–	10.39	m	**13.86**
25 mm × 44 mm; ogee	5.94	0.13	3.01	–	6.33	m	**9.34**
25 mm × 69 mm; chamfered	9.32	0.13	3.01	–	9.81	m	**12.82**
25 mm × 94 mm; chamfered	12.69	0.13	3.01	–	13.29	m	**16.30**
25 mm × 144 mm; chamfered	19.44	0.15	3.47	–	20.23	m	**23.70**
25 mm × 144 mm; ogee	11.09	0.15	3.47	–	11.63	m	**15.10**
25 mm × 169 mm; ogee	13.01	0.15	3.47	–	13.62	m	**17.09**
25 mm × 219 mm; ogee	16.87	0.17	3.93	–	17.58	m	**21.51**
returned ends	–	0.20	4.62	–	–	nr	**4.62**
mitres	–	0.14	3.23	–	–	nr	**3.23**
Architraves, cover fillets and the like; bull nosed;							
chamfered or ogee							
13 mm × 25 mm; bull nosed	1.82	0.15	3.47	–	2.09	m	**5.56**
13 mm × 50 mm; ogee	2.60	0.15	3.47	–	2.89	m	**6.36**
16 mm × 32 mm; bull nosed	2.86	0.15	3.47	–	3.16	m	**6.63**
16 mm × 38 mm; ogee	2.44	0.15	3.47	–	2.72	m	**6.19**
16 mm × 50 mm; ogee	3.20	0.15	3.47	–	3.51	m	**6.98**
19 mm × 50 mm; chamfered	4.75	0.15	3.47	–	5.10	m	**8.57**
19 mm × 63 mm; chamfered	5.99	0.15	3.47	–	6.39	m	**9.86**
19 mm × 69 mm; chamfered	6.55	0.15	3.47	–	6.96	m	**10.43**
25 mm × 44 mm; chamfered	5.50	0.15	3.47	–	5.88	m	**9.35**
25 mm × 50 mm; ogee	5.00	0.15	3.47	–	5.36	m	**8.83**
25 mm × 63 mm; chamfered	7.88	0.15	3.47	–	8.33	m	**11.80**
25 mm × 69 mm; chamfered	8.63	0.15	3.47	–	9.09	m	**12.56**
32 mm × 88 mm; ogee	11.26	0.15	3.47	–	11.81	m	**15.28**
38 mm × 38 mm; ogee	5.78	0.15	3.47	–	6.17	m	**9.64**
50 mm × 50 mm; ogee	10.00	0.15	3.47	–	10.51	m	**13.98**
returned ends	–	0.20	4.62	–	–	nr	**4.62**
mitres	–	0.14	3.23	–	–	nr	**3.23**
Stops; screwed on							
16 mm × 38 mm	2.98	0.14	3.23	–	3.07	m	**6.30**
16 mm × 50 mm	3.92	0.14	3.23	–	4.04	m	**7.27**
19 mm × 38 mm	3.54	0.14	3.23	–	3.65	m	**6.88**
25 mm × 38 mm	4.65	0.14	3.23	–	4.79	m	**8.02**
25 mm × 50 mm	6.12	0.14	3.23	–	6.30	m	**9.53**

22 GENERAL JOINERY

Item	PC £	Labour hours	Labour £	Plant £	Material £	Unit	Total rate £
22.02 HARDWOOD – cont							
Selected Sapele – cont							
Glazing beads and the like							
13 mm × 16 mm	0.80	0.06	1.39	–	0.82	m	**2.21**
13 mm × 19 mm	0.95	0.06	1.39	–	0.98	m	**2.37**
13 mm × 25 mm	1.25	0.06	1.39	–	1.29	m	**2.68**
13 mm × 25 mm; screwed	1.25	0.10	2.32	–	1.29	m	**3.61**
13 mm × 25 mm; fixing with brass cups and screws	1.25	0.20	4.62	–	1.29	m	**5.91**
16 mm × 25 mm; screwed	1.54	0.06	1.39	–	1.59	m	**2.98**
16 mm quadrant	0.87	0.06	1.39	–	0.90	m	**2.29**
19 mm quadrant or scotia	1.39	0.06	1.39	–	1.43	m	**2.82**
19 mm × 36 mm; screwed	2.63	0.10	2.32	–	2.71	m	**5.03**
25 mm × 38 mm; screwed	3.65	0.11	2.54	–	3.76	m	**6.30**
25 mm quadrant or scotia	2.40	0.06	1.39	–	2.47	m	**3.86**
38 mm scotia	5.54	0.06	1.39	–	5.71	m	**7.10**
50 mm scotia	9.60	0.06	1.39	–	9.89	m	**11.28**
Isolated shelves; worktops, seats and the like							
19 mm × 150 mm	8.51	0.20	4.62	–	8.77	m	**13.39**
19 mm × 200 mm	10.06	0.28	6.48	–	10.36	m	**16.84**
25 mm × 150 mm	9.97	0.20	4.62	–	10.27	m	**14.89**
25 mm × 200 mm	11.98	0.28	6.48	–	12.34	m	**18.82**
32 mm × 150 mm	11.26	0.20	4.62	–	11.60	m	**16.22**
32 mm × 200 mm	13.61	0.28	6.48	–	14.02	m	**20.50**
Isolated shelves, worktops, seats and the like; cross-tongued joints							
19 mm × 300 mm	23.03	0.35	8.10	–	23.72	m	**31.82**
19 mm × 450 mm	36.04	0.42	9.71	–	37.12	m	**46.83**
19 mm × 600 mm	47.87	0.51	11.80	–	49.31	m	**61.11**
25 mm × 300 mm	25.86	0.35	8.10	–	26.64	m	**34.74**
25 mm × 450 mm	40.62	0.42	9.71	–	41.84	m	**51.55**
25 mm × 600 mm	53.99	0.51	11.80	–	55.61	m	**67.41**
32 mm × 300 mm	28.27	0.35	8.10	–	29.12	m	**37.22**
32 mm × 450 mm	44.52	0.42	9.71	–	45.86	m	**55.57**
32 mm × 600 mm	59.19	0.51	11.80	–	60.97	m	**72.77**
Isolated shelves, worktops, seats and the like; slatted with 50 wide slats at 75 mm centres							
19 mm thick	66.34	0.80	18.51	–	69.13	m²	**87.64**
25 mm thick	71.02	0.80	18.51	–	73.94	m²	**92.45**
32 mm thick	75.03	0.80	18.51	–	78.07	m²	**96.58**
Window boards, nosings, bed moulds and the like; rebated and rounded							
19 mm × 75 mm	7.19	0.22	5.09	–	7.88	m	**12.97**
19 mm × 150 mm	10.43	0.25	5.79	–	11.23	m	**17.02**
19 mm × 225 mm; in one width	12.80	0.33	7.63	–	13.67	m	**21.30**
19 mm × 300 mm; cross-tongued joints	26.08	0.37	8.56	–	27.35	m	**35.91**
25 mm × 75 mm	7.92	0.22	5.09	–	8.64	m	**13.73**
25 mm × 150 mm	11.61	0.25	5.79	–	12.44	m	**18.23**
25 mm × 225 mm; in one width	15.25	0.33	7.63	–	16.19	m	**23.82**

22 GENERAL JOINERY

Item	PC £	Labour hours	Labour £	Plant £	Material £	Unit	Total rate £
25 mm × 300 mm; cross-tongued joints	30.30	0.37	8.56	–	31.69	m	**40.25**
32 mm × 75 mm	8.59	0.22	5.09	–	9.32	m	**14.41**
32 mm × 150 mm	12.92	0.25	5.79	–	13.78	m	**19.57**
32 mm × 225 mm; in one width	17.23	0.33	7.63	–	18.23	m	**25.86**
32 mm × 300 mm; cross-tongued joints	33.33	0.37	8.56	–	34.80	m	**43.36**
returned and fitted ends	–	0.21	4.86	–	–	nr	**4.86**
Handrails; rounded							
44 mm × 50 mm	17.43	0.31	7.17	–	17.95	m	**25.12**
50 mm × 75 mm	29.71	0.33	7.63	–	30.60	m	**38.23**
63 mm × 87 mm	39.29	0.37	8.56	–	40.47	m	**49.03**
75 mm × 100 mm	54.67	0.42	9.71	–	56.31	m	**66.02**
Handrails; pigs ear							
44 mm × 50 mm	7.96	0.31	7.17	–	8.20	m	**15.37**
50 mm × 75 mm	13.58	0.33	7.63	–	13.99	m	**21.62**
63 mm × 87 mm	17.96	0.37	8.56	–	18.50	m	**27.06**
75 mm × 100 mm	24.98	0.42	9.71	–	25.73	m	**35.44**
Sundries on hardwood							
Extra over fixing with nails for							
gluing and pinning	–	0.02	0.40	–	0.04	m	**0.44**
masonry nails	–	0.02	0.41	–	0.13	m	**0.54**
steel screws	–	0.02	0.39	–	0.10	m	**0.49**
self-tapping screws	–	0.02	0.40	–	0.11	m	**0.51**
steel screws; gluing	–	0.03	0.68	–	0.10	m	**0.78**
steel screws; sinking; filling heads	–	0.04	0.85	–	0.10	m	**0.95**
steel screws; sinking; pellating over	–	0.08	1.86	–	0.10	m	**1.96**
brass cups and screws	–	0.10	2.31	–	0.27	m	**2.58**
Extra over for							
countersinking	–	0.01	0.34	–	–	m	**0.34**
pellating	–	0.07	1.62	–	–	m	**1.62**
Head or nut in softwood							
let in flush	–	0.04	0.85	–	–	nr	**0.85**
Head or nut in hardwood							
let in flush	–	0.06	1.28	–	–	nr	**1.28**
let in over; pellated	–	0.13	2.99	–	–	nr	**2.99**
22.03 MEDIUM DENSITY FIBREBOARD							
Medium density fibreboard							
Skirtings, picture rails, dado rails and the like;							
chamfered or ogee; white primed							
18 mm × 50 mm; chamfered	–	0.09	2.08	–	2.24	m	**4.32**
18 mm × 50 mm; ogee	–	0.09	2.08	–	1.76	m	**3.84**
18 mm × 75 mm; chamfered	–	0.09	2.08	–	3.25	m	**5.33**
18 mm × 75 mm; ogee	–	0.09	2.08	–	2.53	m	**4.61**
18 mm × 100 mm; chamfered	–	0.09	2.08	–	4.26	m	**6.34**
18 mm × 100 mm; ogee	–	0.09	2.08	–	3.31	m	**5.39**
18 mm × 150 mm; ogee	–	0.11	2.54	–	4.85	m	**7.39**
18 mm × 175 mm; ogee	–	0.11	2.54	–	5.63	m	**8.17**
22 mm × 100 mm; chamfered	–	0.09	2.08	–	6.33	m	**8.41**
25 mm × 50 mm; ogee	–	0.09	2.08	–	3.04	m	**5.12**
25 mm × 75 mm; chamfered	–	0.09	2.08	–	4.44	m	**6.52**

22 GENERAL JOINERY

Item	PC £	Labour hours	Labour £	Plant £	Material £	Unit	Total rate £
22.03 MEDIUM DENSITY FIBREBOARD – cont							
Medium density fibreboard – cont							
Skirtings, picture rails, dado rails and the like – cont							
25 mm × 100 mm; chamfered	–	0.09	2.08	–	5.84	m	7.92
25 mm × 150 mm; chamfered	–	0.11	2.54	–	8.65	m	11.19
25 mm × 150 mm; ogee	–	0.11	2.54	–	6.65	m	9.19
25 mm × 175 mm; ogee	–	0.11	2.54	–	7.74	m	10.28
25 mm × 225 mm; ogee	–	0.13	3.01	–	9.89	m	12.90
returned ends	–	0.14	3.23	–	–	nr	3.23
mitres	–	0.09	2.08	–	–	nr	2.08
Architraves, cover fillets and the like; bull nosed; chamfered or ogee							
12 mm × 25 mm; bullnosed	–	0.11	2.54	–	0.89	m	3.43
12 mm × 50 mm; ogee	–	0.11	2.54	–	1.75	m	4.29
15 mm × 32 mm; bull nosed	–	0.11	2.54	–	1.30	m	3.84
15 mm × 38 mm; ogee	–	0.11	2.54	–	1.68	m	4.22
15 mm × 50 mm; ogee	–	0.11	2.54	–	2.13	m	4.67
18 mm × 50 mm; chamfered	–	0.11	2.54	–	2.24	m	4.78
18 mm × 63 mm; chamfered	–	0.11	2.54	–	2.76	m	5.30
18 mm × 75 mm; chamfered	–	0.11	2.54	–	3.25	m	5.79
25 mm × 44 mm; chamfered	–	0.11	2.54	–	2.69	m	5.23
25 mm × 50 mm; ogee	–	0.11	2.54	–	3.04	m	5.58
25 mm × 63 mm; chamfered	–	0.11	2.54	–	3.76	m	6.30
25 mm × 75 mm; chamfered	–	0.11	2.54	–	4.44	m	6.98
30 mm × 88 mm; ogee	–	0.11	2.54	–	6.98	m	9.52
38 mm × 38 mm; ogee	–	0.11	2.54	–	3.91	m	6.45
50 mm × 50 mm; ogee	–	0.11	2.54	–	6.62	m	9.16
returned ends	–	0.14	3.23	–	–	nr	3.23
mitres	–	0.09	2.08	–	–	nr	2.08
Stops; screwed on							
15 mm × 38 mm	–	0.09	2.08	–	2.79	m	4.87
15 mm × 50 mm	–	0.09	2.08	–	3.66	m	5.74
18 mm × 38 mm	–	0.09	2.08	–	3.30	m	5.38
25 mm × 38 mm	–	0.09	2.08	–	4.33	m	6.41
25 mm × 50 mm	–	0.09	2.08	–	5.67	m	7.75
Glazing beads and the like							
12 mm × 16 mm	–	0.04	0.93	–	0.19	m	1.12
12 mm × 19 mm	–	0.04	0.93	–	0.22	m	1.15
12 mm × 25 mm	–	0.04	0.93	–	0.29	m	1.22
12 mm × 25 mm; screwed	–	0.15	3.47	–	0.29	m	3.76
12 mm × 25 mm; fixing with brass cups and screws	–	0.16	3.58	–	0.29	m	3.87
15 mm × 25 mm; screwed	–	0.04	0.93	–	0.44	m	1.37
15 mm quadrant	–	0.04	0.93	–	0.22	m	1.15
18 mm quadrant or scotia	–	0.04	0.93	–	0.31	m	1.24
18 mm × 36 mm; screwed	–	0.04	0.93	–	0.70	m	1.63
25 mm × 38 mm; screwed	–	0.04	0.93	–	1.00	m	1.93
25 mm quadrant or scotia	–	0.04	0.93	–	0.60	m	1.53
38 mm scotia	–	0.04	0.93	–	1.38	m	2.31
50 mm scotia	–	0.04	0.93	–	2.38	m	3.31

22 GENERAL JOINERY

Item	PC £	Labour hours	Labour £	Plant £	Material £	Unit	Total rate £
Medium density fibreboard; Sapele veneered one side; 18 mm thick							
Window boards and the like; rebated; hardwood lipped on one edge							
18 mm × 200 mm	–	0.25	5.79	–	18.67	m	**24.46**
18 mm × 250 mm	–	0.28	6.48	–	19.64	m	**26.12**
18 mm × 300 mm	–	0.31	7.17	–	20.13	m	**27.30**
18 mm × 350 mm	–	0.33	7.63	–	21.60	m	**29.23**
returned and fitted ends	–	0.20	4.62	–	3.63	nr	**8.25**
Isolated shelves, worktops, seats and the like							
18 mm × 150 mm	–	0.15	3.47	–	3.71	m	**7.18**
18 mm × 200 mm	–	0.20	4.62	–	3.92	m	**8.54**
25 mm × 150 mm	–	0.15	3.47	–	4.28	m	**7.75**
25 mm × 200 mm	–	0.20	4.62	–	4.57	m	**9.19**
30 mm × 150 mm	–	0.15	3.47	–	5.99	m	**9.46**
30 mm × 200 mm	–	0.20	4.62	–	6.64	m	**11.26**
Isolated shelves, worktops, seats and the like; cross-tongued joints							
18 mm × 300 mm	–	0.26	6.02	–	12.14	m	**18.16**
18 mm × 450 mm	–	0.31	7.17	–	13.85	m	**21.02**
18 mm × 600 mm	–	0.37	8.56	–	23.16	m	**31.72**
25 mm × 300 mm	–	0.26	6.02	–	12.79	m	**18.81**
25 mm × 450 mm	–	0.31	7.17	–	15.65	m	**22.82**
25 mm × 600 mm	–	0.37	8.56	–	22.58	m	**31.14**
30 mm × 300 mm	–	0.26	6.02	–	14.75	m	**20.77**
30 mm × 450 mm	–	0.31	7.17	–	17.57	m	**24.74**
30 mm × 600 mm	–	0.37	8.56	–	25.56	m	**34.12**
Isolated shelves, worktops, seats and the like; slatted with 50 wide slats at 75 mm centres							
18 mm thick	–	0.60	13.88	–	38.36	m	**52.24**
25 mm thick	–	0.60	13.88	–	40.67	m	**54.55**
30 mm thick	–	0.60	13.88	–	42.79	m	**56.67**
Window boards, nosings, bed moulds and the like; rebated and rounded							
18 mm × 75 mm	–	0.17	3.93	–	4.12	m	**8.05**
18 mm × 150 mm	–	0.19	4.40	–	4.62	m	**9.02**
18 mm × 225 mm	–	0.24	5.55	–	5.02	m	**10.57**
18 mm × 300 mm	–	0.28	6.48	–	5.52	m	**12.00**
25 mm × 75 mm	–	0.17	3.93	–	4.28	m	**8.21**
25 mm × 150 mm	–	0.19	4.40	–	5.02	m	**9.42**
25 mm × 225 mm	–	0.24	5.55	–	5.60	m	**11.15**
25 mm × 300 mm	–	0.28	6.48	–	6.28	m	**12.76**
30 mm × 75 mm	–	0.17	3.93	–	5.80	m	**9.73**
30 mm × 150 mm	–	0.19	4.40	–	7.12	m	**11.52**
30 mm × 225 mm	–	0.24	5.55	–	8.15	m	**13.70**
30 mm × 300 mm	–	0.28	6.48	–	9.36	m	**15.84**
38 mm × 75 mm	–	0.17	3.93	–	6.55	m	**10.48**
38 mm × 150 mm	–	0.19	4.40	–	8.15	m	**12.55**
38 mm × 225 mm	–	0.24	5.55	–	9.36	m	**14.91**
38 mm × 300 mm	–	0.28	6.48	–	10.83	m	**17.31**
returned and fitted ends	–	–	–	–	1.28	nr	**1.28**

Prices for Measured Works

22 GENERAL JOINERY

Item	PC £	Labour hours	Labour £	Plant £	Material £	Unit	Total rate £
22.03 MEDIUM DENSITY FIBREBOARD – cont							
Medium density fibreboard; American White Ash veneered one side; 18 mm thick							
Window boards and the like; rebated; hardwood lipped on one edge							
18 mm × 200 mm	–	0.25	5.79	–	19.41	m	**25.20**
18 mm × 250 mm	–	0.28	6.48	–	20.63	m	**27.11**
18 mm × 300 mm	–	0.31	7.17	–	21.22	m	**28.39**
18 mm × 350 mm	–	0.33	7.63	–	23.02	m	**30.65**
returned and fitted ends	–	0.20	4.62	–	3.63	nr	**8.25**
Pin-boards; medium board							
Sundeala A pin-board or other equal and approved; fixed with adhesive to backing (not included); over 300 mm wide							
6 mm thick	–	0.56	12.96	–	7.07	m²	**20.03**
9 mm thick Colourboard	–	0.56	12.96	–	12.10	m²	**25.06**
22.04 FRAMED PANEL CUBICLE PARTITIONS							
Toilet cubicle partitions and IPS systems; Amwells or other equal and approved; standard colours and ironmongery; assembling and screwing to floor and wall							
Axis MFC standard cubic system, for use in small offices, retail and community halls etc.; standard cubicle set; 800 mm × 1500 mm × 1980 mm high per cubicle, with polished aluminium framing; 19 mm melamine-faced chipboard divisions and doors							
one cubicle set; 2 nr panels; 1 nr door	–	7.00	175.92	–	382.91	nr	**558.83**
range of 3 cubicle sets; 4 nr panels; 3 nr doors	–	21.50	540.34	–	920.57	nr	**1460.91**
range of 6 cubicle sets; 7 nr panels; 6 nr doors	–	42.50	1068.11	–	1727.06	nr	**2795.17**
reduction of 1 nr panel for end unit adjoining side wall	–	–	–	–	–87.50	nr	**–87.50**
Splash SGL cubic system for heavy use environments like schools, swimming pools and prisons; standard cubicle set; 800 mm × 1500 mm × 1980 mm high per cubicle, with polished aluminium framing; 19 mm melamine-faced chipboard divisions and doors							
one cubicle set; 2 nr panels; 1 nr door	–	7.00	175.92	–	787.40	nr	**963.32**
range of 3 cubicle sets; 4 nr panels; 3 nr doors	–	21.50	540.34	–	1816.83	nr	**2357.17**
range of 6 cubicle sets; 7 nr panels; 6 nr door	–	42.50	1068.11	–	3360.95	nr	**4429.06**
reduction of 1 nr panel for end unit adjoining side wall	–	–	–	–	–242.02	nr	**–242.02**

22 GENERAL JOINERY

Item	PC £	Labour hours	Labour £	Plant £	Material £	Unit	Total rate £
Minima stylish stainless steel framed cubicle system in MFC, HPL, SGL, Real Wood veneer or Glass; 800 mm × 1500 mm × 2100 mm high per cubicle, with satin polished stainless steel framing; 18 mm high pressure laminated (HPL) chipboard divisions and doors							
one cubicle set; 2 nr panels; 1 nr door	–	7.00	175.92	–	920.34	nr	1096.26
range of 3 cubicle sets; 4 nr panels; 3 nr doors	–	21.50	540.34	–	2051.34	nr	2591.68
range of 6 cubicle set; 7 nr panels; 6 nr doors	–	42.50	1068.11	–	3747.83	nr	4815.94
reduction of 1 nr panel for end unit adjoining side wall	–	–	–	–	−204.52	nr	−204.52
Urban flush fronted, floor to ceiling option cubicle system in either 13 mm or 20 mm panels, material options HPL, SGL or Real Wood Veneer: cubicle set; 800 mm × 1500 mm × 2400 mm high per cubicle, with sating finished stainless steel ironmongery; 20 mm divisions, doors and pilasters							
one cubicle set; 2 nr panels; 1 nr door	–	8.00	201.06	–	1400.96	nr	1602.02
range of 3 cubicle set; 4 nr panels; 3 nr doors	–	23.50	590.60	–	3446.16	nr	4036.76
range of 6 cubicle sets; 7 nr panels; 6 nr doors	–	46.80	1176.18	–	6513.97	nr	7690.15
reduction of 1 nr panel for end unit adjoining side wall	–	–	–	–	−271.56	nr	−271.56
Sylan high end specification flush fronted system floor to ceiling 44 mm doors, no visible fixings. Finishes in Real wood Veneer, HPL, high Gloss paint and lacquer; cubicle set; 800 mm × 1500 mm × 2400 mm high per cubicle, with satin finished stainless steel ironmongery; 30 mm high pressure laminated (HPL) chipboard divisions and 44 mm solid cored real wood veneered doors and pilasters							
one cubicle set; 2 nr panels; 1 nr door	–	10.00	251.32	–	2604.23	nr	2855.55
range of 3 cubicle set; 4 nr panels; 3 nr doors	–	30.00	753.96	–	6134.47	nr	6888.43
range of 6 cubicle sets; 7 nr panels; 6 nr doors	–	60.00	1507.92	–	11429.28	nr	12937.20
reduction of 1 nr panel for end unit adjoining side wall	–	–	–	–	−452.22	nr	−452.22
IPS panel systems; 12.50 mm melamine faced chipboard; fixed to existing timber subframe							
IPS back panel system to accomodate urinals; to conceal pipework; approx 2.70 m high × 2.10 m wide	–	–	–	–	–	nr	1030.93
IPS back panel system to accomodate wash hand basins; to conceal pipework; approx 2.70 m high × 5.70 m wide	–	–	–	–	–	nr	2032.49
IPS back panel system to accomodate wash hand basins; to conceal pipework; approx 2.70 m high × 2.10 m wide	–	–	–	–	–	nr	1148.43

22 GENERAL JOINERY

Item	PC £	Labour hours	Labour £	Plant £	Material £	Unit	Total rate £
22.05 ASSOCIATED METALWORK							
Metalwork; mild steel							
Angle section bearers; for building in							
90 mm × 90 mm × 6 mm	–	0.31	7.73	–	8.65	m	**16.38**
120 mm × 120 mm × 8 mm	–	0.32	7.97	–	12.90	m	**20.87**
200 mm × 150 mm × 12 mm	–	0.37	9.22	–	32.36	m	**41.58**
Metalwork; mild steel; galvanized							
Waterbars; groove in timber							
6 mm × 30 mm	–	0.46	10.64	–	6.06	m	**16.70**
6 mm × 40 mm	–	0.46	10.64	–	7.61	m	**18.25**
6 mm × 50 mm	–	0.46	10.64	–	5.61	m	**16.25**
Angle section bearers; for building in							
90 mm × 90 mm × 6 mm	–	0.31	7.73	–	17.33	m	**25.06**
120 mm × 120 mm × 8 mm	–	0.32	7.97	–	30.60	m	**38.57**
200 mm × 150 mm × 12 mm	–	0.37	9.22	–	68.18	m	**77.40**
Dowels; mortice in timber							
8 mm dia. × 100 mm long	–	0.04	0.93	–	0.82	nr	**1.75**
10 mm dia. × 50 mm long	–	0.04	0.93	–	1.29	nr	**2.22**
Cramps							
25 mm × 3 mm × 230 mm girth; one end bent, holed and screwed to softwood; other end fishtailed for building in	–	0.06	1.39	–	1.94	nr	**3.33**
Metalwork; stainless steel							
Angle section bearers; for building in							
90 mm × 90 mm × 6 mm	–	0.31	7.73	–	37.98	m	**45.71**
120 mm × 120 mm × 8 mm	–	0.32	7.97	–	67.04	m	**75.01**
200 mm × 150 mm × 12 mm	–	0.37	9.22	–	147.04	m	**156.26**

23 WINDOWS, SCREENS AND LIGHTS

Item	PC £	Labour hours	Labour £	Plant £	Material £	Unit	Total rate £
23.01 WINDOWS, SCREENS AND LIGHTS							
SUPPLY ONLY PRICES							
NOTE: The following supply only prices are for purpose-made components, to which fixings, sealants etc. labour and overheads and profit need to be added, before they may be used to arrive at a guide price for a complete window. The reader is then referred to the subsequent SUPPLY AND FIX pages for fixing costs based on the overall window size.							
Purpose-made window casements; treated wrought softwood							
Casements; rebated; moulded							
44 mm thick	–	–	–	–	57.08	m²	57.08
57 mm thick	–	–	–	–	59.93	m²	59.93
Casements; rebated; moulded; in medium panes							
44 mm thick	–	–	–	–	91.24	m²	91.24
57 mm thick	–	–	–	–	95.07	m²	95.07
Casements; rebated; moulded; with semi-circular head							
44 mm thick	–	–	–	–	119.29	m²	119.29
57 mm thick	–	–	–	–	123.03	m²	123.03
Casements; rebated; moulded; to bullseye window							
44 mm thick; 600 mm dia.	–	–	–	–	189.10	nr	189.10
44 mm thick; 900 mm dia.	–	–	–	–	225.28	nr	225.28
57 mm thick; 600 mm dia.	–	–	–	–	198.02	nr	198.02
57 mm thick; 900 mm dia.	–	–	–	–	27.24	nr	27.24
Fitting and hanging casements (in factory)							
square or rectangular	–	–	–	–	13.35	nr	13.35
semi-circular	–	–	–	–	21.68	nr	21.68
bullseye	–	–	–	–	27.24	nr	27.24
Purpose-made window casements; selected Sapele							
Casements; rebated; moulded							
44 mm thick	–	–	–	–	64.86	m²	64.86
57 mm thick	–	–	–	–	71.07	m²	71.07
Casements; rebated; moulded; in medium panes							
44 mm thick	–	–	–	–	105.06	m²	105.06
57 mm thick	–	–	–	–	113.35	m²	113.35
Casements; rebated; moulded with semi-circular head							
44 mm thick	–	–	–	–	132.33	m²	132.33
57 mm thick	–	–	–	–	140.43	m²	140.43
Casements; rebated; moulded; to bullseye window							
44 mm thick; 600 mm dia.	–	–	–	–	233.88	nr	233.88
44 mm thick; 900 mm dia.	–	–	–	–	281.46	nr	281.46
57 mm thick; 600 mm dia.	–	–	–	–	253.17	nr	253.17
57 mm thick; 900 mm dia.	–	–	–	–	306.04	nr	306.04
Fitting and hanging casements (in factory)							
square or rectangular	–	–	–	–	14.45	nr	14.45
semi-circular	–	–	–	–	23.90	nr	23.90
bullseye	–	–	–	–	30.56	nr	30.56

23 WINDOWS, SCREENS AND LIGHTS

Item	PC £	Labour hours	Labour £	Plant £	Material £	Unit	Total rate £
23.01 WINDOWS, SCREENS AND LIGHTS – cont							
Purpose-made window frames; treated wrought softwood							
Frames; rounded; rebated check grooved							
44 mm × 69 mm	–	–	–	–	15.52	m	**15.52**
44 mm × 94 mm	–	–	–	–	16.27	m	**16.27**
44 mm × 119 mm	–	–	–	–	17.01	m	**17.01**
57 mm × 94 mm	–	–	–	–	17.05	m	**17.05**
69 mm × 144 mm	–	–	–	–	22.45	m	**22.45**
90 mm × 140 mm	–	–	–	–	31.89	m	**31.89**
Mullions and transoms; twice rounded, rebated and check grooved							
57 mm × 69 mm	–	–	–	–	18.07	m	**18.07**
57 mm × 94 mm	–	–	–	–	18.99	m	**18.99**
69 mm × 94 mm	–	–	–	–	21.47	m	**21.47**
69 mm × 144 mm	–	–	–	–	31.55	m	**31.55**
Sill; sunk weathered, rebated and grooved							
69 mm × 94 mm	–	–	–	–	37.68	m	**37.68**
69 mm × 144 mm	–	–	–	–	39.91	m	**39.91**
Purpose-made window frames; selected Sapele							
Frames; rounded; rebated check grooved							
44 mm × 69 mm	–	–	–	–	19.81	m	**19.81**
44 mm × 94 mm	–	–	–	–	21.31	m	**21.31**
44 mm × 119 mm	–	–	–	–	22.80	m	**22.80**
57 mm × 94 mm	–	–	–	–	24.94	m	**24.94**
69 mm × 144 mm	–	–	–	–	35.73	m	**35.73**
90 mm × 140 mm	–	–	–	–	50.75	m	**50.75**
Mullions and transoms; twice rounded, rebated and check grooved							
57 mm × 69 mm	–	–	–	–	22.52	m	**22.52**
57 mm × 94 mm	–	–	–	–	26.19	m	**26.19**
69 mm × 94 mm	–	–	–	–	31.39	m	**31.39**
69 mm × 144 mm	–	–	–	–	47.74	m	**47.74**
Sill; sunk weathered, rebated and grooved							
69 mm × 94 mm	–	–	–	–	44.49	m	**44.49**
69 mm × 144 mm	–	–	–	–	49.32	m	**49.32**

23 WINDOWS, SCREENS AND LIGHTS

Item	PC £	Labour hours	Labour £	Plant £	Material £	Unit	Total rate £
SUPPLY AND FIX PRICES							
Standard windows; treated wrought softwood;							
Jeld-Wen or other equal							
Casement windows; factory glazed with low-e 24 mm							
double glazing (U-value = 1.6 W/m^{2}K); with 140 mm							
wide softwood sills; opening casements and ventilators							
hung on rustproof hinges; fitted with aluminized							
lacquered finish casement stays and fasteners							
488 mm × 750 mm; ref LEWN07V	291.86	0.65	15.04	–	300.77	nr	**315.81**
488 mm × 900 mm; ref LEWN09V	296.27	0.74	17.12	–	305.31	nr	**322.43**
630 mm × 750 mm; ref LEW107C	266.54	0.74	17.12	–	274.69	nr	**291.81**
630 mm × 750 mm; ref LEW107V	306.61	0.74	17.12	–	315.95	nr	**333.07**
630 mm × 900 mm; ref LEW109V	311.77	0.83	19.20	–	321.28	nr	**340.48**
630 mm × 900 mm; ref LEW109CH	279.13	0.74	17.12	–	287.66	nr	**304.78**
630 mm × 1050 mm; ref LEW110C	292.22	0.93	21.52	–	301.18	nr	**322.70**
630 mm × 1050 mm; ref LEW110V	321.63	0.74	17.12	–	331.48	nr	**348.60**
915 mm × 900 mm; ref LEW2NO9W	384.67	1.02	23.60	–	396.36	nr	**419.96**
915 mm × 1050 mm; ref LEW2N1OW	397.40	1.06	24.52	–	409.53	nr	**434.05**
915 mm × 1200 mm; ref LEW2N12W	410.14	1.11	25.68	–	422.65	nr	**448.33**
915 mm × 1350 mm; ref LEW2N13W	422.87	1.25	28.92	–	435.76	nr	**464.68**
915 mm × 1500 mm; ref LEW2N15W	457.27	1.30	30.08	–	471.24	nr	**501.32**
1200 mm × 750 mm; ref LEW2O7C	394.28	1.06	24.52	–	406.30	nr	**430.82**
1200 mm × 750 mm; ref LEW2O7CV	496.99	1.06	24.52	–	512.10	nr	**536.62**
1200 mm × 900 mm; ref LEW2O9C	415.18	1.11	25.68	–	427.83	nr	**453.51**
1200 mm × 900 mm; ref LEW2O9W	428.87	1.11	25.68	–	441.94	nr	**467.62**
1200 mm × 900 mm; ref LEW2O9CV	513.21	1.11	25.68	–	528.81	nr	**554.49**
1200 mm × 1050 mm; ref LEW210C	436.57	1.25	28.92	–	449.92	nr	**478.84**
1200 mm × 1050 mm; ref LEW210W	444.47	1.25	28.92	–	458.06	nr	**486.98**
1200 mm × 1050 mm; ref LEW210T	534.61	1.25	28.92	–	550.91	nr	**579.83**
1200 mm × 1050 mm; ref LEW210CV	744.82	1.25	28.92	–	767.41	nr	**796.33**
1200 mm × 1200 mm; ref LEW212C	458.88	1.34	31.00	–	472.96	nr	**503.96**
1200 mm × 1200 mm; ref LEW212W	460.09	1.34	31.00	–	474.20	nr	**505.20**
1200 mm × 1200 mm; ref LEW212TX	589.57	1.34	31.00	–	607.57	nr	**638.57**
1200 mm × 1200 mm; ref LEW212CV	556.92	1.34	31.00	–	573.94	nr	**604.94**
1200 mm × 1350 mm; ref LEW213W	475.70	1.43	33.08	–	490.28	nr	**523.36**
1200 mm × 1350 mm; ref LEW213CV	586.68	1.43	33.08	–	604.59	nr	**637.67**
1200 mm × 1500 mm; ref LEW215W	520.48	1.57	36.33	–	536.44	nr	**572.77**
1770 mm × 750 mm; ref LEW307CC	582.13	1.30	30.08	–	599.90	nr	**629.98**
1770 mm × 900 mm; ref LEW309CC	614.05	1.57	36.33	–	632.78	nr	**669.11**
1770 mm × 1050 mm; ref LEW310C	540.36	1.62	37.48	–	556.88	nr	**594.36**
1770 mm × 1050 mm; ref LEW310T	638.75	1.57	36.33	–	658.22	nr	**694.55**
1770 mm × 1050 mm; ref LEW310CC	646.98	1.30	30.08	–	666.70	nr	**696.78**
1770 mm × 1050 mm; ref LEW310CW	651.18	1.30	30.08	–	671.02	nr	**701.10**
1770 mm × 1200 mm; ref LEW312C	568.42	1.67	38.64	–	585.83	nr	**624.47**
1770 mm × 1200 mm; ref LEW312T	665.39	1.67	38.64	–	685.71	nr	**724.35**
1770 mm × 1200 mm; ref LEW312CC	681.71	1.67	38.64	–	702.52	nr	**741.16**
1770 mm × 1200 mm; ref LEW312CW	672.55	1.67	38.64	–	693.09	nr	**731.73**
1770 mm × 1200 mm; ref LEW312CVC	779.55	1.67	38.64	–	803.29	nr	**841.93**
1770 mm × 1350 mm; ref LEW313CC	731.37	1.76	40.72	–	753.66	nr	**794.38**
1770 mm × 1350 mm; ref LEW313CW	603.95	1.76	40.72	–	622.42	nr	**663.14**
1770 mm × 1350 mm; ref LEW313CVC	829.21	1.76	40.72	–	854.45	nr	**895.17**

23 WINDOWS, SCREENS AND LIGHTS

Item	PC £	Labour hours	Labour £	Plant £	Material £	Unit	Total rate £
23.01 WINDOWS, SCREENS AND LIGHTS – cont							
Standard windows – cont							
Casement windows – cont							
1770 mm × 1500 mm; ref LEW315T	768.91	1.85	42.80	–	792.34	nr	**835.14**
2340 mm × 1050 mm; ref LEW410CWC	863.08	1.80	41.64	–	889.33	nr	**930.97**
2340 mm × 1200 mm; ref LEW412CWC	903.58	1.90	43.96	–	931.09	nr	**975.05**
2340 mm × 1350 mm; ref LEW413CWC	958.98	2.04	47.19	–	988.20	nr	**1035.39**
Top-hung casement windows; factory glazed with low-e 24 mm double glazing (U-value = 1.6 W/m² K); with 140 mm wide softwood sills; opening casements and ventilators hung on rustproof hinges; fitted with aluminized lacquered finish casement stays							
630 mm × 750 mm; ref LEW107 A	274.64	0.74	17.12	–	283.03	nr	**300.15**
630 mm × 900 mm; ref LEW109 A	287.23	0.83	19.20	–	296.00	nr	**315.20**
630 mm × 1050 mm; ref LEW110 A	294.05	0.93	21.52	–	303.07	nr	**324.59**
915 mm × 750 mm; ref LEW2N07 A	342.51	0.97	22.44	–	352.94	nr	**375.38**
915 mm × 900 mm; ref LEW2N09 A	357.97	1.02	23.60	–	368.86	nr	**392.46**
915 mm × 1050 mm; ref LEW2N10 A	367.67	1.06	24.52	–	378.91	nr	**403.43**
915 mm × 1350 mm; ref LEW2N13 AS	383.13	1.25	28.92	–	394.88	nr	**423.80**
1200 mm × 750 mm; ref LEW207 A	381.67	1.06	24.52	–	393.33	nr	**417.85**
1200 mm × 900 mm; ref LEW209 A	400.01	1.11	25.68	–	412.21	nr	**437.89**
1200 mm × 1050 mm; ref LEW210 A	412.58	1.25	28.92	–	425.21	nr	**454.13**
1200 mm × 1200 mm; ref LEW212 A	430.93	1.34	31.00	–	444.11	nr	**475.11**
1200 mm × 1350 mm; ref LEW213 AS	488.15	1.43	33.08	–	503.09	nr	**536.17**
1200 mm × 1500 mm; ref LEW215 AS	513.74	1.57	36.33	–	529.45	nr	**565.78**
1770 mm × 1050 mm; ref LEW310 AE	638.75	1.57	36.33	–	658.22	nr	**694.55**
1770 mm × 1200 mm; ref LEW312 AE	665.39	1.67	38.64	–	685.66	nr	**724.30**
Standard Oak windows; Jeld-Wen or other equal; factory applied preservative stain base coat							
Side-hung casement windows; factory glazed with low-e 24 mm double glazing (U-value 1.4 W/m² K); 45 mm × 140 mm hardwood sills; weather stripping; opening sashes on canopy hinges; fitted with fasteners; brown finish ironmongery							
630 mm × 750 mm; ref LEW107C	570.00	0.88	20.36	–	587.25	nr	**607.61**
630 mm × 900 mm; ref LEW109C	612.00	1.11	25.68	–	630.51	nr	**656.19**
630 mm × 1050 mm; ref LEW110C	648.00	1.20	27.76	–	667.59	nr	**695.35**
630 mm × 1195 mm; ref LEW112C	686.40	1.20	27.76	–	707.15	nr	**734.91**
1195 mm × 1195 mm; ref LEW212C	1008.00	1.39	32.16	–	1038.45	nr	**1070.61**
1195 mm × 1345 mm; ref LEW213C	1008.00	1.39	32.16	–	1038.45	nr	**1070.61**

23 WINDOWS, SCREENS AND LIGHTS

Item	PC £	Labour hours	Labour £	Plant £	Material £	Unit	Total rate £
Casement windows; factory glazed with low-e 24 mm double glazing (U-value = 1.4 W/m² K); 45 mm × 140 mm hardwood sills; weather stripping; opening sashes on canopy hinges; fitted with fasteners; brown finish ironmongery							
630 mm × 900 mm; ref LEW109 AH	618.00	0.88	20.36	–	636.69	nr	**657.05**
630 mm × 1050 mm; ref LEW110 AH	648.00	1.20	27.76	–	667.65	nr	**695.41**
915 mm × 900 mm; ref LEW2N09 AH	777.00	1.39	32.16	–	800.46	nr	**832.62**
915 mm × 1050 mm; ref LEW2N10 AH	810.00	1.48	34.24	–	834.51	nr	**868.75**
1200 mm × 895 mm; ref LEW213 ASH	888.00	1.67	38.64	–	914.85	nr	**953.49**
1200 mm × 1050 mm; ref LEW210 AH	924.00	1.57	36.33	–	951.93	nr	**988.26**
1770 mm × 1050 mm; ref LEW310 AEH	1200.00	1.80	41.64	–	1236.26	nr	**1277.90**
Purpose-made double-hung sash windows; treated wrought softwood							
Cased frames of 100 mm × 25 mm grooved inner linings; 114 mm × 25 mm grooved outer linings; 125 mm × 38 mm twice rebated head linings; 125 mm × 32 mm twice rebated grooved pulley stiles; 150 mm × 13 mm linings; 50 mm × 19 mm parting slips; 25 mm × 19 mm inside beads; 150 mm × 75 mm Oak twice sunk weathered throated sill; 50 mm thick rebated and moulded sashes; moulded horns							
over 1.25 m² each; both sashes in medium panes; including spiral spring balances	417.16	2.08	48.12	–	507.63	m²	**555.75**
up to 1.25 m² each; both sashes in medium panes; including spiral spring balances with cased mullions	474.17	2.31	53.45	–	566.35	m²	**619.80**
Purpose-made double-hung sash windows; selected Sapele							
Cased frames of 100 mm × 25 mm grooved inner linings; 114 mm × 25 mm grooved outer linings; 125 mm × 38 mm twice rebated head linings; 125 mm × 32 mm twice rebated grooved pulley stiles; 150 mm × 13 mm linings; 50 mm × 19 mm parting slips; 25 mm × 19 mm inside beads; 150 mm × 75 mm Oak twice sunk weathered throated sill; 50 mm thick rebated and moulded sashes; moulded horns							
over 1.25 m² each; both sashes in medium panes; including spiral sash balances	464.26	2.78	64.31	–	556.14	m²	**620.45**
over 1.25 m² each; both sashes in medium panes; including spiral sash balances with cased mullions	498.16	3.08	71.26	–	591.06	m²	**662.32**

23 WINDOWS, SCREENS AND LIGHTS

Item	PC £	Labour hours	Labour £	Plant £	Material £	Unit	Total rate £
23.01 WINDOWS, SCREENS AND LIGHTS – cont							
Clement EB24 range of factory finished steel fixed light; casement and fanlight windows and doors; with a U-value of 2.0 W/m²K (part L compliant); to EN ISO 9001 2000; polyester powder coated; factory glazed with low-e double glazing; fixed in position; including lugs plugged and screwed to brickwork or blockwork							
Basic fixed light including easy-glaze snap-on beads							
508 mm × 292 mm	210.76	2.00	50.26	–	217.19	nr	**267.45**
508 mm × 457 mm	229.92	2.00	50.26	–	236.92	nr	**287.18**
508 mm × 628 mm	249.08	2.00	50.26	–	256.71	nr	**306.97**
508 mm × 923 mm	287.41	2.00	50.26	–	296.23	nr	**346.49**
508 mm × 1218 mm	325.72	2.50	62.83	–	335.75	nr	**398.58**
Basic 'Tilt and Turn' window; including easy-glaze snap-on beads							
508 mm × 292 mm	498.16	2.00	50.26	–	513.21	nr	**563.47**
508 mm × 457 mm	517.33	2.00	50.26	–	532.95	nr	**583.21**
508 mm × 628 mm	536.48	2.00	50.26	–	552.73	nr	**602.99**
508 mm × 923 mm; including fixed light	632.28	2.00	50.26	–	651.45	nr	**701.71**
508 mm × 1218 mm; including fixed light	670.60	2.50	62.83	–	690.98	nr	**753.81**
Basic casement; including easy-glaze snap-on beads							
508 mm × 628 mm	593.97	2.00	50.26	–	611.93	nr	**662.19**
508 mm × 923 mm	632.28	2.00	50.26	–	651.45	nr	**701.71**
508 mm × 1218 mm	670.60	2.50	62.83	–	690.98	nr	**753.81**
Double door							
1143 mm × 2057 mm	3755.38	3.50	87.96	–	3868.39	nr	**3956.35**
Extra for							
pressed steel sills; to suit above windows	57.49	0.50	11.57	–	59.27	m	**70.84**
G + bar	114.95	–	–	–	118.40	m	**118.40**
simulated leaded light	114.95	–	–	–	118.40	m	**118.40**
Thermally broken composite double glazed aluminium/timber windows; Velfac 200 or other approved; with a maximum glazing U-value of 1.5 W/m²K; argon filled cavity; low-e glazing with laminated glass unless otherwise specified; including multipoint espagnolette locking mechanisms and other ironmongery							
Standard fixed casement windows							
900 mm × 900 mm single fixed pane; low-e glass 6/14/4	231.00	2.70	62.47	–	243.97	nr	**306.44**
900 mm × 2000 mm single fixed pane; low-e glass 6/14/4	488.00	5.94	137.42	–	511.72	nr	**649.14**
1200 mm × 1200 mm single fixed pane; low-e glass 6/14/4	315.00	4.75	109.89	–	332.12	nr	**442.01**
1200 mm × 2200 mm three fixed panes; low-e glass 6/14/4	630.00	7.90	182.77	–	659.74	nr	**842.51**
2200 mm × 2200 mm single fixed pane; low-e glass 6/12/6	780.00	12.00	277.63	–	818.91	nr	**1096.54**

23 WINDOWS, SCREENS AND LIGHTS

Item	PC £	Labour hours	Labour £	Plant £	Material £	Unit	Total rate £
Outward opening standard sash casement windows							
900 mm × 900 mm top-hung sash; low-e glass 6/14/4	546.00	2.67	61.77	–	568.42	nr	630.19
900 mm × 2200 mm with small top-hung sash; fixed lower pane; low-e glass 6/14/4	546.00	5.90	136.50	–	567.82	nr	704.32
900 mm × 3000 mm with small top-hung sash; fixed lower pane; low-e glass 6/14/4	690.00	8.00	185.08	–	722.86	nr	907.94
1600 mm × 1600 mm with two side-hung sashes; low-e glass 4/16/4	640.00	2.60	60.15	–	668.91	nr	729.06
1600 mm × 1600 mm with two side-hung projecting sashes; low-e glass 6/14/4	710.00	2.60	60.15	–	736.23	nr	796.38
1800 mm × 900 mm with two side-hung projecting sashes; low-e glass 6/14 E	475.00	5.50	127.25	–	492.69	nr	619.94
1800 mm × 3000 mm with two side-hung projecting sashes; two top-hung sashes; low-e glass 6/14/4	1460.00	15.00	347.04	–	1519.37	nr	1866.41
2000 mm × 1600 mm with one side-hung sash next to a top-hung projecting sash over a fixed sash; low-e glass 6/16/4	780.00	9.50	219.79	–	811.41	nr	1031.20
1200 mm × 2200 mm with fixed lower sash and top-hung projecting upper sash; lower low-e glass/toughened 4/14/4; upper low-e glass 6/14/4	620.00	7.90	182.77	–	650.28	nr	833.05
1200 mm × 2200 mm with fixed lower fully reversible upper sash; low-e glass/toughened 6/14/4; upper low-e glass 6/14/4	675.00	7.90	182.77	–	706.93	nr	889.70
1800 mm × 900 mm with two side-hung projecting sashes; low-e glass 6/14/4; 60 minute fire integrity	710.00	5.50	127.25	–	737.68	nr	864.93
Outward opening standard doors							
1800 mm × 2200 mm French casement patio door; low-e toughened glass 4/16/4; deadlock; handles; cylinder	3150.00	13.00	300.76	–	3259.96	nr	3560.72
Allternative cavity fill							
extra for Krypton cavity fill (over Argon)	–	–	–	–	30.90	m²	30.90
uPVC windows; Profile 22 or other equal and approved; reinforced where appropriate with aluminium alloy; including standard ironmongery; sills and factory glazed with low-e 24 mm double glazing; fixed in position; including lugs plugged and screwed to brickwork or blockwork							
Casement/fixed light; including e.p.d.m. glazing gaskets and weather seals							
630 mm × 900 mm; ref P109C	71.05	2.50	62.83	–	73.39	nr	136.22
630 mm × 1200 mm; ref P112V	82.74	2.50	62.83	–	85.48	nr	148.31
1200 mm × 1200 mm; ref P212C	127.16	2.50	62.83	–	131.28	nr	194.11
1770 mm × 1200 mm; ref P312CC	151.69	3.00	75.40	–	156.55	nr	231.95

23 WINDOWS, SCREENS AND LIGHTS

Item	PC £	Labour hours	Labour £	Plant £	Material £	Unit	Total rate £
23.01 WINDOWS, SCREENS AND LIGHTS – cont							
uPVC windows – cont							
Casement/fixed light; including vents; e.p.d.m.							
glazing gaskets and weather seals							
630 mm × 900 mm; ref P109V	87.11	2.50	62.83	–	89.93	nr	**152.76**
630 mm × 1200 mm; ref P112C	91.52	2.50	62.83	–	94.52	nr	**157.35**
1200 mm × 1200 mm; ref P212W	136.19	2.50	62.83	–	140.58	nr	**203.41**
1200 mm × 1200 mm; ref P212CV	172.15	2.50	62.83	–	177.62	nr	**240.45**
1770 mm × 1200 mm; ref P312WW	162.87	3.00	75.40	–	168.07	nr	**243.47**
Secured by Design accreditation							
Casement/fixed light; including e.p.d.m. glazing							
gaskets and weather seals							
630 mm × 900 mm; ref P109C	74.09	2.50	62.83	–	76.52	nr	**139.35**
630 mm × 1200 mm; ref P112V	85.78	2.50	62.83	–	88.61	nr	**151.44**
1200 mm × 1200 mm; ref P212C	130.19	2.50	62.83	–	134.40	nr	**197.23**
1770 mm × 1200 mm; ref P312CC	154.72	3.00	75.40	–	159.67	nr	**235.07**
Casement/fixed light; including vents; e.p.d.m.							
glazing gaskets and weather seals							
630 mm × 900 mm; ref P109V	90.14	2.50	62.83	–	93.05	nr	**155.88**
630 mm × 1200 mm; ref P112C	94.55	2.50	62.83	–	97.64	nr	**160.47**
1200 mm × 1200 mm; ref P212W	139.21	2.50	62.83	–	143.70	nr	**206.53**
1200 mm × 1200 mm; ref P212CV	178.22	2.50	62.83	–	183.88	nr	**246.71**
1770 mm × 1200 mm; ref P312WW	165.92	3.00	75.40	–	171.21	nr	**246.61**
WER A rating							
Casement/fixed light; including e.p.d.m. glazing							
gaskets and weather seals							
630 mm × 900 mm; ref P109C	72.03	2.50	62.83	–	74.40	nr	**137.23**
630 mm × 1200 mm; ref P112V	84.12	2.50	62.83	–	86.90	nr	**149.73**
1200 mm × 1200 mm; ref P212C	130.32	2.50	62.83	–	134.54	nr	**197.37**
1770 mm × 1200 mm; ref P312CC	156.83	3.00	75.40	–	161.84	nr	**237.24**
Casement/fixed light; including vents; e.p.d.m.							
glazing gaskets and weather seals							
630 mm × 900 mm; ref P109V	89.08	2.50	62.83	–	91.96	nr	**154.79**
630 mm × 1200 mm; ref P112C	93.66	2.50	62.83	–	96.73	nr	**159.56**
1200 mm × 1200 mm; ref P212W	140.13	2.50	62.83	–	144.64	nr	**207.47**
1200 mm × 1200 mm; ref P212CV	175.68	2.50	62.83	–	181.26	nr	**244.09**
1770 mm × 1200 mm; ref P312WW	168.50	3.00	75.40	–	173.86	nr	**249.26**
WER C rating							
Casement/fixed light; including e.p.d.m. glazing							
gaskets and weather seals							
630 mm × 900 mm; ref P109C	71.05	2.50	62.83	–	73.39	nr	**136.22**
630 mm × 1200 mm; ref P112V	82.74	2.50	62.83	–	85.48	nr	**148.31**
1200 mm × 1200 mm; ref P212C	127.16	2.50	62.83	–	131.28	nr	**194.11**
1770 mm × 1200 mm; ref P312CC	151.69	3.00	75.40	–	156.55	nr	**231.95**

23 WINDOWS, SCREENS AND LIGHTS

Item	PC £	Labour hours	Labour £	Plant £	Material £	Unit	Total rate £
Casement/fixed light; including vents; e.p.d.m. glazing gaskets and weather seals							
630 mm × 900 mm; ref P109V	87.11	2.50	62.83	–	89.93	nr	**152.76**
630 mm × 1200 mm; ref P112C	91.52	2.50	62.83	–	94.52	nr	**157.35**
1200 mm × 1200 mm; ref P212W	136.19	2.50	62.83	–	140.58	nr	**203.41**
1200 mm × 1200 mm; ref P212CV	172.15	2.50	62.83	–	177.62	nr	**240.45**
1770 mm × 1200 mm; ref P312WW	162.87	3.00	75.40	–	168.07	nr	**243.47**
Colour finish uPVC windows							
Casement/fixed light; including e.p.d.m. glazing gaskets and weather seals							
630 mm × 900 mm; ref P109C	100.12	2.50	62.83	–	103.33	nr	**166.16**
630 mm × 1200 mm; ref P112V	113.50	2.50	62.83	–	117.16	nr	**179.99**
1200 mm × 1200 mm; ref P212C	172.80	2.50	62.83	–	178.29	nr	**241.12**
1770 mm × 1200 mm; ref P312CC	207.21	3.00	75.40	–	213.74	nr	**289.14**
Casement/fixed light; including vents; e.p.d.m. glazing gaskets and weather seals							
630 mm × 900 mm; ref P109V	118.16	2.50	62.83	–	121.91	nr	**184.74**
630 mm × 1200 mm; ref P112C	123.74	2.50	62.83	–	127.71	nr	**190.54**
1200 mm × 1200 mm; ref P212W	183.49	2.50	62.83	–	189.30	nr	**252.13**
1200 mm × 1200 mm; ref P212CV	228.37	2.50	62.83	–	235.53	nr	**298.36**
1770 mm × 1200 mm; ref P312WW	217.70	2.50	62.83	–	224.54	nr	**287.37**
uPVC windows; Profile 22 or other equal and approved; reinforced where appropriate with aluminium alloy; in refurbishment work, including standard ironmongery; sills and factory glazed with low-e 24 mm double glazing; removing existing windows and fixing new in position; including lugs plugged and screwed to brickwork or blockwork							
Casement/fixed light; including e.p.d.m. glazing gaskets and weather seals							
630 mm × 900 mm; ref P109C	71.05	5.00	125.66	–	73.39	nr	**199.05**
630 mm × 1200 mm; ref P112V	91.52	5.00	125.66	–	94.52	nr	**220.18**
1200 mm × 1200 mm; ref P212C	127.16	6.00	150.79	–	131.28	nr	**282.07**
1770 mm × 1200 mm; ref P312CC	151.69	6.50	163.36	–	156.55	nr	**319.91**
Casement/fixed light; including vents; e.p.d.m. glazing gaskets and weather seals							
630 mm × 900 mm; ref P109V	87.11	5.00	125.66	–	89.93	nr	**215.59**
630 mm × 1200 mm; ref P112C	82.74	5.50	138.23	–	85.48	nr	**223.71**
1200 mm × 1200 mm; ref P212W	136.19	6.00	150.79	–	140.58	nr	**291.37**
1200 mm × 1200 mm; ref P212CV	172.15	6.00	150.79	–	177.62	nr	**328.41**
1770 mm × 1200 mm; ref P312WW	162.87	6.50	163.36	–	168.07	nr	**331.43**

23 WINDOWS, SCREENS AND LIGHTS

Item	PC £	Labour hours	Labour £	Plant £	Material £	Unit	Total rate £
23.01 WINDOWS, SCREENS AND LIGHTS – cont							
Aluminium windows; Schuco AWS 50 (or similar) proprietary system or equal and approved							
Polyester powder coated solid colour matt finish or natural anodized window system of glass sealed units with 6.4 mm low-e coated laminated inner pane, air filled cavity and 6 mm clear annealed outer pane. Rates to include all brackets, membranes, cills, silicone seals, trade contractor preliminaries, including external access equipment							
ribbon construction windows 1.5 m high	–	–	–	–	–	m²	**609.07**
punched hole windows fixing into prepared apertures by others	–	–	–	–	–	m²	**653.85**
Extra for							
1.25 m wide × 1.5 m high opening vents, assuming tilt and turn operation	–	–	–	–	–	m²	**194.39**
neutral selective high performance coating in lieu of low-e, for assisting in solar control	–	–	–	–	–	m²	**50.34**
outer glass pane to be toughened and heat soak tested or heat strengthened in lieu of annealed	–	–	–	–	–	m²	**35.95**
inner laminated glass to be toughened and heat soak tested laminated, or heat strengthened laminated	–	–	–	–	–	m²	**70.69**
23.02 ROOFLIGHTS							
Rooflights, skylights, roof windows and frames; pre-glazed; treated Nordic Red Pine and aluminium trimmed Velux windows or other equal; type U flashings and soakers (for tiles and pantiles up to 45 mm deep), and sealed double glazing unit (trimming opening not included)							
Roof windows; U-value = 1.3 W/m²K; size							
550 mm × 780 mm	306.18	1.85	42.80	–	385.59	nr	**428.39**
550 mm × 980 mm	323.82	2.08	48.12	–	405.06	nr	**453.18**
660 mm × 1180 mm	375.90	2.31	53.45	–	467.82	nr	**521.27**
780 mm × 980 mm	350.70	2.31	53.45	–	441.86	nr	**495.31**
780 mm × 1180 mm	396.90	2.78	64.31	–	493.43	nr	**557.74**
780 mm × 1400 mm	438.90	2.31	53.45	–	541.85	nr	**595.30**
940 mm × 1600 mm	541.80	2.78	64.31	–	662.15	nr	**726.46**
1140 mm × 1180 mm	527.10	2.78	64.31	–	643.11	nr	**707.42**
1340 mm × 1400 mm	614.25	2.78	64.31	–	746.50	nr	**810.81**
extra for electric powered windows plugged in to adjacent power supply	–	–	–	–	324.45	nr	**324.45**

23 WINDOWS, SCREENS AND LIGHTS

Item	PC £	Labour hours	Labour £	Plant £	Material £	Unit	Total rate £
Rooflights, skylights, roof windows and frames; uPVC; plugged and screwed to concrete; or screwed to timber							
Rooflight; Cox Suntube range or other equal and approved; double skin polycarbonate dome							
230 mm dia.; for flat roof using felt or membrane	294.29	2.50	57.83	–	303.75	nr	**361.58**
230 mm dia.; for up to 30° pitch roof with standard tiles	325.39	3.00	69.41	–	335.66	nr	**405.07**
230 mm dia.; for up to 30° pitch roof with bold roll tiles	302.06	3.00	69.41	–	311.69	nr	**381.10**
300 mm dia.; for flat roof using felt or membrane	477.52	2.50	57.83	–	492.47	nr	**550.30**
300 mm dia.; for up to 30° pitch roof with standard tiles	477.52	3.00	69.41	–	492.36	nr	**561.77**
300 mm dia.; for up to 30° pitch roof with bold roll tiles	511.93	3.00	69.41	–	527.85	nr	**597.26**
Rooflight; Cox Galaxy range or other equal and approved; double skin polycarbonate dome only; fitting to existing kerb							
600 mm × 600 mm	136.59	1.50	34.70	–	141.04	nr	**175.74**
900 mm × 900 mm	252.08	1.75	40.49	–	260.10	nr	**300.59**
1200 mm × 1800 mm	739.59	2.00	46.27	–	762.35	nr	**808.62**
Rooflight; Cox Galaxy range or other equal and approved; triple skin polycarbonate dome only; fitting to existing kerb							
600 mm × 600 mm rooflight	211.00	1.50	34.70	–	217.67	nr	**252.37**
900 mm × 900 mm rooflight	378.68	1.75	40.49	–	390.49	nr	**430.98**
1200 mm × 1800 mm rooflight	1144.93	2.00	46.27	–	1179.84	nr	**1226.11**
Rooflight; Cox Trade range or other equal and approved; double skin polycarbonate fixed light dome on 150 mm PVC upstand							
600 mm × 600 mm	234.32	2.00	46.27	–	241.75	nr	**288.02**
900 mm × 900 mm	378.68	2.25	52.06	–	390.55	nr	**442.61**
1200 mm × 1800 mm	749.59	2.50	57.83	–	772.71	nr	**830.54**
Rooflight; Cox Trade range or other equal and approved; double skin polycarbonate manual opening light dome on 150 mm PVC upstand							
600 mm × 600 mm	366.47	2.50	57.83	–	378.32	nr	**436.15**
900 mm × 900 mm	530.81	2.75	63.62	–	547.70	nr	**611.32**
1200 mm × 1800 mm	998.34	3.10	71.72	–	1029.38	nr	**1101.10**
Rooflight; Cox Trade range or other equal and approved; double skin polycarbonate electric opening light dome on 150 mm PVC upstand (not including adjacent power supply)							
600 mm × 600 mm	622.99	2.75	63.62	–	642.53	nr	**706.15**
900 mm × 900 mm	829.54	3.00	69.41	–	855.39	nr	**924.80**
1200 mm × 1800 mm	1391.46	3.50	80.98	–	1434.30	nr	**1515.28**
Rooflight; Cox Trade range or other equal and approved; triple skin polycarbonate fixed light dome on 150 mm PVC upstand							
600 mm × 600 mm	275.40	2.00	46.27	–	284.06	nr	**330.33**
900 mm × 900 mm	453.08	2.00	46.27	–	467.07	nr	**513.34**
1200 m × 1800 mm	918.39	2.50	57.83	–	946.57	nr	**1004.40**

23 WINDOWS, SCREENS AND LIGHTS

Item	PC £	Labour hours	Labour £	Plant £	Material £	Unit	Total rate £
23.02 ROOFLIGHTS – cont							
Rooflights, skylights, roof windows and frames – cont							
Rooflight; Cox Trade range or other equal and approved; triple skin polycarbonate manual opening light dome on 150 mm PVC upstand							
600 mm × 600 mm	407.55	2.50	57.83	–	420.63	nr	**478.46**
900 mm × 900 mm	605.21	2.50	57.83	–	624.22	nr	**682.05**
1200 m × 1800 mm	1167.14	3.10	71.72	–	1203.24	nr	**1274.96**
Rooflight; Cox Trade range or other equal and approved; triple skin polycarbonate electric opening light dome on 150 mm PVC upstand							
600 mm × 600 mm light	664.07	2.75	63.62	–	684.85	nr	**748.47**
900 mm × 900 mm light	903.94	2.75	63.62	–	931.91	nr	**995.53**
1200 mm × 1800 mm light	1560.26	3.50	80.98	–	1608.15	nr	**1689.13**
Rooflight; Cox 2000 range or other equal and approved double skin polycarbonate fixed light dome on 235 mm solid core PVC upstand							
600 mm × 600 mm	749.59	2.00	46.27	–	772.48	nr	**818.75**
900 mm × 900 mm	1016.62	2.50	57.83	–	1047.52	nr	**1105.35**
1200 mm × 1800 mm	2711.58	3.00	69.41	–	2793.39	nr	**2862.80**
Rooflight; Cox 2000 range or other equal and approved double skin polycarbonate manual opening light dome on 235 mm solid core PVC upstand							
600 mm × 600 mm	1029.45	2.50	57.83	–	1061.19	nr	**1119.02**
900 mm × 900 mm	1296.48	3.00	69.41	–	1336.23	nr	**1405.64**
1200 mm × 1800 mm	3249.07	3.60	83.29	–	3347.46	nr	**3430.75**
Rooflight; Cox 2000 range or other equal and approved double skin polycarbonate electric opening light dome on 235 mm solid core PVC upstand (not including adjacent power supply)							
600 mm × 600 mm	1354.82	2.75	63.62	–	1396.32	nr	**1459.94**
900 mm × 900 mm	1621.84	3.25	75.19	–	1671.36	nr	**1746.55**
1200 mm × 1800 mm	3657.74	4.00	92.55	–	3768.38	nr	**3860.93**
Rooflight; Cox 2000 range or other equal and approved triple skin polycarbonate fixed light dome on 235 mm solid core PVC upstand							
600 mm × 600 mm	1009.44	2.00	46.27	–	1040.12	nr	**1086.39**
900 mm × 900 mm	1474.80	2.50	57.83	–	1519.45	nr	**1577.28**
1200 mm × 1800 mm	3622.34	3.00	69.41	–	3731.46	nr	**3800.87**
Rooflight; Cox 2000 range or other equal and approved triple skin polycarbonate manual opening light dome on 235 mm solid core PVC upstand							
600 mm × 600 mm	1289.30	2.50	57.83	–	1328.83	nr	**1386.66**
900 mm × 900 mm	1754.66	3.00	69.41	–	1808.15	nr	**1877.56**
1200 mm × 1800 mm	4159.83	3.60	83.29	–	4285.53	nr	**4368.82**

23 WINDOWS, SCREENS AND LIGHTS

Item	PC £	Labour hours	Labour £	Plant £	Material £	Unit	Total rate £
Rooflight; Cox 2000 range or other equal and approved triple skin polycarbonate electric opening light dome on 235 mm solid core PVC upstand (not incluing adjacent power supply)							
600 mm × 600 mm	1614.66	2.75	63.62	–	1663.95	nr	**1727.57**
900 mm × 900 mm	2080.03	3.25	75.19	–	2143.29	nr	**2218.48**
1200 mm × 1800 mm	4568.49	4.00	92.55	–	4706.46	nr	**4799.01**

23.03 LOUVRES

Louvres, Brise Soleils and frames; polyester powder coated aluminium; fixing in position including brackets
Brise soleil, to mitigate the effects of solar gain. This rate assumes a single natural anodized extruded aluminium fin, with brackets and orientated either horizontally or vertically. The quantity of fins per storey height should be calculated to achieve desired shading

Item	PC £	Labour hours	Labour £	Plant £	Material £	Unit	Total rate £
300 mm deep	–	–	–	–	–	m	**174.26**

24 DOORS, SHUTTERS AND HATCHES

Item	PC £	Labour hours	Labour £	Plant £	Material £	Unit	Total rate £
24.01 EXTERNAL DOORS AND FRAMES							
EXTERNAL DOOR FRAMES							
External softwood door frame composite standard joinery sets							
External door frame composite set; 56 mm × 78 mm wide (finished); for external doors							
762 mm × 1981 mm × 44 mm	71.62	0.75	17.36	–	73.97	nr	**91.33**
813 mm × 1981 mm × 44 mm	73.52	0.75	17.36	–	75.93	nr	**93.29**
838 mm × 1981 mm × 44 mm	73.04	0.75	17.36	–	75.44	nr	**92.80**
External door frame composite set; 56 mm × 78 mm wide (finished); with 45 mm × 140 mm (finished) hardwood sill; for external doors							
686 mm × 1981 mm × 44 mm	88.75	1.00	23.13	–	91.62	nr	**114.75**
762 mm × 1981 mm × 44 mm	91.88	1.00	23.13	–	94.84	nr	**117.97**
838 mm × 1981 mm × 44 mm	108.33	1.00	23.13	–	111.79	nr	**134.92**
826 mm × 2040 mm × 44 mm	95.01	1.00	23.13	–	98.07	nr	**121.20**
EXTERNAL DOORS							
NOTE: door ironmogery is not included unless stated							
Doors; standard matchboarded; wrought softwood							
Matchboarded, framed, ledged and braced doors; 44 mm thick overall; 19 mm thick tongued, grooved and V-jointed boarding; one side vertical boarding							
762 mm × 1981 mm	62.33	1.67	38.64	–	64.20	nr	**102.84**
838 mm × 1981 mm	68.12	1.67	38.64	–	70.16	nr	**108.80**
Flush door; external quality; skeleton or cellular core; plywood faced both sides; lipped all round							
762 mm × 1981 mm × 54 mm	79.70	1.62	37.48	–	82.09	nr	**119.57**
838 mm × 1981 mm × 54 mm	81.42	1.62	37.48	–	83.86	nr	**121.34**
Fire doors							
Flush door; half-hour fire-resisting; external quality; skeleton or cellular core; plywood faced both sides; lipped on all four edges							
762 mm × 1981 mm × 44 mm	167.94	1.71	39.56	–	172.98	nr	**212.54**
838 mm × 1981 mm × 44 mm	169.65	1.71	39.56	–	174.74	nr	**214.30**
Flush door; half-hour fire-resisting; external quality with 6 mm Georgian wired polished plate glass opening; skeleton or cellular core; plywood faced both sides; lipped on all four edges; including glazing beads							
762 mm × 1981 mm × 44 mm	167.94	1.71	39.56	–	172.98	nr	**212.54**
838 mm × 1981 mm × 44 mm	169.65	1.71	39.56	–	174.74	nr	**214.30**
726 mm × 2040 mm × 44 mm	167.94	1.71	39.56	–	172.98	nr	**212.54**
826 mm × 2040 mm × 44 mm	169.73	1.71	39.56	–	174.82	nr	**214.38**
926 mm × 2040 mm × 44 mm	179.09	1.71	39.56	–	184.46	nr	**224.02**

24 DOORS, SHUTTERS AND HATCHES

Item	PC £	Labour hours	Labour £	Plant £	Material £	Unit	Total rate £
External softwood door frame composite standard joinery sets							
External door frame composite set; 56 mm × 78 mm wide (finished); for external doors							
762 mm × 1981 mm × 44 mm	71.62	0.75	17.36	–	73.97	nr	91.33
813 mm × 1981 mm × 44 mm	73.52	0.75	17.36	–	75.93	nr	93.29
838 mm × 1981 mm × 44 mm	73.04	0.75	17.36	–	75.44	nr	92.80
Doorsets; Anti-Vandal Security door and frame units; Bastion Security Ltd or other equal; to BS 5051; factory primed; fixing with frame anchors to masonry; cutting mortices; external 46 mm thick insulated door with birch grade plywood; sheet steel bonded into door core; 2 mm thick polyester coated laminate finish; hardwood lippings all edges; 95 mm × 65 mm hardwood frame; polyester coated standard ironmongery; weather stripping all round; low projecting aluminium threshold; plugging; screwing							
for 980 mm × 2100 mm structural opening; single doorsets; panic bolt	–	–	–	–	–	nr	2441.25
for 1830 mm × 2100 mm structural opening; double doorsets; panic bolt	–	–	–	–	–	nr	3181.50
Doorsets; galvanized steel door and frame units; treated softwood frame, primed hardwood sill; fixing in position; plugged and screwed to brickwork or blockwork							
Door and frame							
838 mm × 1981 mm	–	2.78	64.31	–	536.57	nr	600.88
Doorsets; steel security door and frame; Hormann or other equal; including ironmongery, weather seals and all necessary fixing accessories							
Horman ref E55–1 doorset							
to suit structural opening 1100 × 2105 mm; fire-rating 30 minutes; acoustic rating 38dB; including stainless steel ironmongery	–	–	–	–	–	nr	2213.15
to suit structural opening 2000 × 2105 mm; fire-rating 30 minutes; acoustic rating 38dB; including stainless steel ironmongery	–	–	–	–	–	nr	3288.09
Doorsets; steel bullet-resistant door and frame units; Wormald Doors or other equal; Medite laquered panels; ironmongery							
Door and frame							
1000 mm × 2060 mm overall; fixed to masonry	–	–	–	–	–	nr	4615.99

24 DOORS, SHUTTERS AND HATCHES

Item	PC £	Labour hours	Labour £	Plant £	Material £	Unit	Total rate £
24.01 EXTERNAL DOORS AND FRAMES – cont							
uPVC doors; Profile 22 or other equal; reinforced where appropriate with aluminium alloy; including standard ironmongery; thresholds and factory glazed with low-e 24 mm double glazing; fixed in position; including lugs plugged and screwed to brickwork or blockwork							
Fixed light; including e.p.d.m. glazing gaskets and weather seals							
2100 mm × 900 mm uPVC door with midrail; half glazed	338.47	4.00	100.53	–	348.85	nr	**449.38**
2100 mm × 900 mm uPVC door with midrail; glazed	334.08	3.00	75.40	–	344.36	nr	**419.76**
Secured by Design accrediation							
Fixed light; including e.p.d.m. glazing gaskets and weather seals							
2100 mm × 900 mm uPVC door with midrail; half glazed	385.56	2.00	50.26	–	397.35	nr	**447.61**
2100 mm × 900 mm uPVC door with midrail; glazed	389.91	3.00	75.40	–	401.86	nr	**477.26**
WER A rating							
Fixed light; including e.p.d.m. glazing gaskets and weather seals							
2100 mm × 900 mm uPVC door with midrail; half glazed	347.66	4.00	100.53	–	358.32	nr	**458.85**
2100 mm × 900 mm uPVC door with midrail; glazed	339.30	3.00	75.40	–	349.74	nr	**425.14**
WER C rating							
Fixed light; including e.p.d.m. glazing gaskets and weather seals							
2100 mm × 900 mm uPVC door with midrail; half glazed	338.47	4.00	100.53	–	348.85	nr	**449.38**
2100 mm × 900 mm uPVC door with midrail; glazed	334.08	3.00	75.40	–	344.36	nr	**419.76**
Colour finish							
Fixed light; including e.p.d.m. glazing gaskets and weather seals							
2100 mm × 900 mm uPVC door with midrail; half glazed	518.22	4.00	100.53	–	533.99	nr	**634.52**
2100 mm × 900 mm uPVC door with midrail; glazed	413.95	3.00	75.40	–	426.63	nr	**502.03**
Sliding/folding; aluminium double glazed sliding patio doors; white acrylic finish; with and including 18 thick annealed double glazing; fixed in position; including lugs plugged and screwed to brickwork or blockwork							
Patio doors							
1800 mm × 2100 mm	1595.00	16.00	402.11	–	1643.46	nr	**2045.57**
2400 mm × 2100 mm	1625.00	16.00	402.11	–	1674.36	nr	**2076.47**
2700 mm × 2100 mm	1925.00	16.00	402.11	–	1983.36	nr	**2385.47**

24 DOORS, SHUTTERS AND HATCHES

Item	PC £	Labour hours	Labour £	Plant £	Material £	Unit	Total rate £
24.02 REVOLVING DOORS							
Automatic revolving doors; structural glass							
ASSA ABLOY Automatic Glass Revolving Door System providing hands free operation and Hi-Tech functionality. Modern modular design comprising four door leaves, curved glazed outer walls, glass ceiling/canopy, complete with in-floor mounted operator and imbedded Configurable Digital Control. System including normal installation to comply with BS7036–0–2014 / BS-EN16005, complete with the installer's Declaration of Conformity in accordance with Machinery Regulation							
ASSA ABLOY RD300 structural glass Automatic Revolving Door; 3 or 4 wing 1800 mm dia., 2200 mm internal height	–	–	–	–	–	nr	43178.63
ASSA ABLOY RD300 structural glass Automatic Revolving Door; 3 or 4 wing 2100 mm dia., 2200 mm internal height	–	–	–	–	–	nr	44112.22
ASSA ABLOY RD300 structural glass Automatic Revolving Door; 3 or 4 wing 2400 mm dia., 2200 mm internal height	–	–	–	–	–	nr	45045.81
ASSA ABLOY RD300 structural glass Automatic Revolving Door; 3 or 4 wing 2700 mm dia., 2200 mm internal height	–	–	–	–	–	nr	45746.01
ASSA ABLOY RD300 structural glass Automatic Revolving Door; 3 or 4 wing 3000 mm dia., 2200 mm internal height	–	–	–	–	–	nr	46679.60
Automatic revolving doors; framed							
ASSA ABLOY Framed Automatic Revolving Door System providing hands free operation and ESCAPE functionality. Modern modular design comprising four collapsible door leaves with electromechanical breakout, framed curved glazed outer walls, radially segmented ceiling and semi-circular dustcovers, complete with operator and imbedded Configurable Digital Control. System including normal installation to comply with BS7036–0–2014 / BS-EN16005, complete with the installer's Declaration of Conformity in accordance with Machinery Regulations							
RD3 / RD4 Framed Automatic Revolving Door; 3 or 4 wing 1800 mm dia., 2200 mm internal height	–	–	–	–	–	nr	28941.35
RD3 / RD4 Framed Automatic Revolving Door; 3 or 4 wing 2100 mm dia., 2200 mm internal height	–	–	–	–	–	nr	29524.85
RD3 / RD4 Framed Automatic Revolving Door; 3 or 4 wing 2400 mm dia., 2200 mm internal height	–	–	–	–	–	nr	30108.34

24 DOORS, SHUTTERS AND HATCHES

Item	PC £	Labour hours	Labour £	Plant £	Material £	Unit	Total rate £
24.02 REVOLVING DOORS – cont							
Automatic revolving doors – cont							
ASSA ABLOY Framed Automatic Revolving Door System providing hands free operation and ESCAPE functionality – cont							
RD3/RD4 Framed Automatic Revolving Door; 3 or 4 wing 2700 mm dia., 2200 mm internal height	–	–	–	–	–	nr	30691.84
RD3/RD4 Framed Automatic Revolving Door; 3 or 4 wing 3000 mm dia., 2200 mm internal height	–	–	–	–	–	nr	31275.33
RD3/RD4 Framed Automatic Revolving Door; 3 or 4 wing 3300 mm dia., 2200 mm internal height	–	–	–	–	–	nr	34192.81
RD3/RD4 Framed Automatic Revolving Door; 3 or 4 wing 3600 mm dia., 2200 mm internal height	–	–	–	–	–	nr	34776.30
Access control revolving doors; framed							
ASSA ABLOY Framed Automatic Revolving Door System providing controlled single or bi-directional controlled access. Modern modular design comprising four door leaves, framed curved glazed outer walls, radially segmented ceiling and semi-circular dustcovers, complete with operator, anti-tailgate and imbedded Configurable Digital Control. System including normal installation to comply with BS7036–0–2014 / BS-EN16005, complete with the installer's Declaration of Conformity in accordance with Machinery Regulations							
RD4-A1 Framed Access Control Revolving Door; 4 wing 1800 mm dia., 2200 mm internal height	–	–	–	–	–	nr	30691.84
RD4-A2 Framed Access & Egress Control Revolving Door; 4 wing 1800 mm dia., 2200 mm internal height	–	–	–	–	–	nr	31275.33
RD4-A1 Framed Access Control Revolving Door; 4 wing 2100 mm dia., 2200 mm internal height	–	–	–	–	–	nr	31275.33
RD4-A2 Framed Access & Egress Control Revolving Door; 4 wing 2100 mm dia., 2200 mm internal height	–	–	–	–	–	nr	31858.83
RD4-A1 Framed Access Control Revolving Door; 4 wing 2400 mm dia., 2200 mm internal height	–	–	–	–	–	nr	31858.83
RD4-A2 Framed Access & Egress Control Revolving Door; 4 wing 2400 mm dia., 2200 mm internal height	–	–	–	–	–	nr	32442.32

24 DOORS, SHUTTERS AND HATCHES

Item	PC £	Labour hours	Labour £	Plant £	Material £	Unit	Total rate £
Manually operated revolving doors; framed							
ASSA ABLOY Intelligent Manually Operated Revolving Door System providing Comfort, Convenience and Safety automatically at the user's pace. Modern modular design comprising four door leaves, framed curved glazed outer walls, radially segmented ceiling and semi-circular dustcovers, complete with electro-mechanical operator and Configurable Digital Control. System including normal installation to comply with BS7036–0–2014 / BS-EN16005, complete with the installer's Declaration of Conformity in accordance with Machinery Regulations							
RD100 Framed Intelligent Manually Operated Revolving Door; 3 or 4 wing 1800 mm dia., 2600 mm internal height	–	–	–	–	–	nr	25440.38
RD100 Framed Intelligent Manually Operated Automatic Revolving Door; 3 or 4 wing 2100 mm dia., 2600 mm internal height	–	–	–	–	–	nr	26023.88
RD100 Framed Intelligent Manually Operated Automatic Revolving Door; 3 or 4 wing 2400 mm dia., 2600 mm internal height	–	–	–	–	–	nr	26607.37
RD100 Framed Intelligent Manually Operated Automatic Revolving Door; 3 or 4 wing 2700 mm dia., 2600 mm internal height	–	–	–	–	–	nr	27190.87
RD100 Framed Intelligent Manually Operated Automatic Revolving Door; 3 or 4 wing 3000 mm dia., 2600 mm internal height	–	–	–	–	–	nr	27774.36
Automatic revolving doors; framed							
ASSA ABLOY Framed Automatic Revolving Door System providing hands free operation and safe adaptive Manual Speed control. Modern modular design comprising four door leaves, curved glazed outer walls, radially segmented ceiling and semi-circular dustcovers, complete with operator and imbedded Configurable Digital Control. System including normal installation to comply with BS7036–0–2014 / BS-EN16005, complete with the installer's Declaration of Conformity in accordance with Machinery Regulations							
RD150 Framed Automatic Revolving Door; 3 or 4 wing 1800 mm dia., 2200 mm internal height	–	–	–	–	–	nr	27540.96
RD150 Framed Automatic Revolving Door; 3 or 4 wing 2100 mm dia., 2200 mm internal height	–	–	–	–	–	nr	28124.46
RD150 Framed Automatic Revolving Door; 3 or 4 wing 2400 mm dia., 2200 mm internal height	–	–	–	–	–	nr	28707.95
RD150 Framed Automatic Revolving Door; 3 or 4 wing 2700 mm dia., 2200 mm internal height	–	–	–	–	–	nr	29291.45
RD150 Framed Automatic Revolving Door; 3 or 4 wing 3000 mm dia., 2200 mm internal height	–	–	–	–	–	nr	29874.94

24 DOORS, SHUTTERS AND HATCHES

Item	PC £	Labour hours	Labour £	Plant £	Material £	Unit	Total rate £
24.02 REVOLVING DOORS – cont							
Automatic high capacity revolving doors; 2-wing ASSA ABLOY UniTurn providing safe hands free operation with capacity for multiple occupants and trolleys per compartment. Modern modular design with two centre passage manually pivoting break out door leaves arranged between two rotating showcases. Curved glazed outer walls and canopy enclose a whole rotating radially segmented ceiling, fixed semi-circular dustcovers and load bearing peripheral drive with imbedded Configurable Digital Control. System including normal installation to comply with BS7036–0–2014 / BS-EN16005, complete with the installer's Declaration of Conformity in accordance with Machinery Regulations							
UniTurn High Capacity Automatic Revolving Door; 2 wing 3600 mm dia., 2200 mm internal height	–	–	–	–	–	nr	60683.48
UniTurn High Capacity Automatic Revolving Door; 2 wing 4200 mm dia., 2200 mm internal height	–	–	–	–	–	nr	63367.56
UniTurn High Capacity Automatic Revolving Door; 2 wing 4800 mm dia., 2200 mm internal height	–	–	–	–	–	nr	66051.63
UniTurn High Capacity Automatic Revolving Door; 2 wing 5400 mm dia., 2200 mm internal height	–	–	–	–	–	nr	68735.71

24 DOORS, SHUTTERS AND HATCHES

Item	PC £	Labour hours	Labour £	Plant £	Material £	Unit	Total rate £
Automatic high capacity revolving doors; 3-wing, panic exit							
ASSA ABLOY RD3L System providing safe hands free operation, capacity for multiple occupants with trolleys. Modern modular design comprising three fixed door leaves arranged around the centre shaft and electro-mechanical emergency break out leaves folding around secondary pivot points. Curved glazed outer walls and canopy enclose a whole rotating radially segmented ceiling, fixed semi-circular dustcovers and load compensating peripheral drive with imbedded Configurable Digital Control. System including normal installation to comply with BS7036–0–2014 / BS-EN16005, complete with the installer's Declaration of Conformity in accordance with Machinery Regulations							
RD3L High Capacity Automatic Revolving Door; 3 wing 4200 mm dia., 2600 mm internal height. For up to five occupants per compartment	–	–	–	–	–	nr	**64067.75**
RD3L High Capacity Automatic Revolving Door; 3 wing 4800 mm dia., 2600 mm internal height. For up to seven occupants per compartment	–	–	–	–	–	nr	**66285.03**
RD3L High Capacity Automatic Revolving Door; 3 wing 5400 mm dia., 2600 mm internal height. For up to nine occupants per compartment	–	–	–	–	–	nr	**68385.61**
RD3L High Capacity Automatic Revolving Door; 3 wing 6200 mm dia., 2600 mm internal height. For up to twelve occupants per compartment	–	–	–	–	–	nr	**70486.20**

24 DOORS, SHUTTERS AND HATCHES

Item	PC £	Labour hours	Labour £	Plant £	Material £	Unit	Total rate £
24.03 SLIDING DOOR ENTRANCES							
Sliding door entrances (including doors and screens)							
Automatic Sliding Door EN1627: RC2 Anti-Burglar Security Entrance System. ASSA ABLOY SL510 P RC2 Anti-Burglar Sliding Door System complete with 100 mm operator providing automated operation of the sliding door leaves. RAL powder coated or anodized with insulated P4 A security glass. Providing full retraction of the doors to maximize the clear open width.. Operator should optimize hold open time in relation to changing traffic flow. The door should change opening width in relation to changing traffic flow. The timeframe for change should be flexible and possible to program. The electromechanical lock must block all active door leaves separately. Third Party Verified system EPDs to be a demand. System including normal installation to comply with BS7036–0–2014 / BS-EN16005, complete with the installer's Declaration of Conformity in accordance with Machinery Regulations							
ASSA ABLOY SL510 P EN1627: RC2 1P; Single Door Leaf and 100 mm Operator with Pocket Safety Screen 1200 mm wide × 2300 mm high	–	–	–	–	–	nr	6895.85
ASSA ABLOY SL510 P EN1627: RC2 2PL/R; Single Door Leaf with One Sidescreen and 100 mm Operator with Pocket Safety Screen 2000 mm wide × 2300 mm high	–	–	–	–	–	nr	7956.75
ASSA ABLOY SL510 P EN1627: RC2 2P; Double Bi-Parting Door Leaves and 100 mm Operator with Pocket Safety Screens 2000 mm wide × 2300 mm high	–	–	–	–	–	nr	9017.65
ASSA ABLOY SL510 P EN1627: RC2 4P; Double Bi-Parting Door Leaves with Two Sidescreens and 100 mm Operator with Pocket Safety Screens 4000 mm wide × 2300 mm high	–	–	–	–	–	nr	12200.35

24 DOORS, SHUTTERS AND HATCHES

Item	PC £	Labour hours	Labour £	Plant £	Material £	Unit	Total rate £
Automatic Sliding Door EN1627: RC3 Anti-Burglar Security Entrance System. ASSA ABLOY SL510 P RC3 Anti-Burglar Sliding Door System complete with 100 mm operator providing automated operation of the sliding door leaves. RAL powder coated or anodized with insulated P5 A security glass. Providing full retraction of the doors to maximize the clear open width. Operator should optimize hold open time in relation to changing traffic flow. The door should change opening width in relation to changing traffic flow. The timeframe for change should be flexible and possible to program. The electromechanical lock must block all active door leaves separately. Third Party Verified system EPDs to be a demand. System including normal installation to comply with BS7036–0–2014 / BS-EN16005, complete with the installer's Declaration of Conformity in accordance with Machinery Regulations							
ASSA ABLOY SL510 P RC3 1P; Single Door Leaf and 100 mm Operator with Pocket Safety Screen 1200 mm wide × 2300 mm high	–	–	–	–	–	nr	–
ASSA ABLOY SL510 P RC3 2PL/R; Single Door Leaf with One Sidescreen and 100 mm Operator with Pocket Safety Screen 2000 mm wide × 2300 mm high	–	–	–	–	–	nr	8487.20
ASSA ABLOY SL510 P RC3 2P; Double Bi-Parting Door Leaves and 100 mm Operator with Pocket Safety Screens 2000 mm high × 2300 mm high	–	–	–	–	–	nr	9548.10
ASSA ABLOY SL510 P RC3 4P; Double Bi-Parting Door Leaves with Two Sidescreens and 100 mm Operator with Pocket Safety Screens 4000 mm wide × 2300 mm high	–	–	–	–	–	nr	12730.80

24 DOORS, SHUTTERS AND HATCHES

Item	PC £	Labour hours	Labour £	Plant £	Material £	Unit	Total rate £
24.03 SLIDING DOOR ENTRANCES – cont							
Sliding door entrances (including doors and screens) – cont							
Automatic Sliding Door Rugged Framed Entrance System. ASSA ABLOY SL510 F Sliding Door System complete with 100 mm operator providing automated operation of the sliding door leaves. RAL powder coated or anodized with insulated glass. Providing full retraction of the doors to maximize the clear open width. Operator should optimize hold open time in relation to changing traffic flow. The door should change opening width in relation to changing traffic flow. The timeframe for change should be flexible and possible to program. The electromechanical lock must block all active door leaves separately. Third Party Verified system EPDs to be a demand. System including normal installation to comply with BS7036–0–2014 / BS-EN16005, complete with the installer's Declaration of Conformity in accordance with Machinery Regulations							
ASSA ABLOY SL510 F 1P; Single Door Leaf and 100 mm Operator with Pocket Safety Screen 1200 mm wide × 2300 mm high	–	–	–	–	–	nr	**3713.15**
ASSA ABLOY SL510 F 2PL/R; Single Door Leaf with One Sidescreen and 100 mm Operator with Pocket Safety Screen 2000 mm wide × 2300 mm high	–	–	–	–	–	nr	**4774.05**
ASSA ABLOY SL510 F 2P; Double Bi-Parting Door Leaves and 100 mm Operator with Pocket Safety Screens 2000 mm wide × 2300 mm high	–	–	–	–	–	nr	**5092.32**
ASSA ABLOY SL510 F 4P; Double Bi-Parting Door Leaves with Two Sidescreens and 100 mm Operator with Pocket Safety Screens 4000 mm wide × 2300 mm high	–	–	–	–	–	nr	**6365.40**

24 DOORS, SHUTTERS AND HATCHES

Item	PC £	Labour hours	Labour £	Plant £	Material £	Unit	Total rate £
Automatic Sliding Door Narrow Framed Entrance System. ASSA ABLOY SL510 SE Sliding Door System complete with 100 mm operator providing automated operation of the sliding door leaves. RAL powder coated or anodized with insulated glass. Providing full retraction of the doors to maximize the clear open width. Operator should optimize hold open time in relation to changing traffic flow. The door should change opening width in relation to changing traffic flow. The timeframe for change should be flexible and possible to program. The electromechanical lock must block all active door leaves separately. Third Party Verified system EPDs to be a demand. System including normal installation to comply with BS7036–0–2014 / BS-EN16005, complete with the installer's Declaration of Conformity in accordance with Machinery Regulations							
ASSA ABLOY SL510 SE 1P; Single Door Leaf and 100 mm Operator with Pocket Safety Screen 1200 mm wide × 2300 mm high	–	–	–	–	–	nr	3713.15
ASSA ABLOY SL510 SE 2PL/R; Single Door Leaf with One Sidescreen and 100 mm Operator with Pocket Safety Screen 2000 mm wide × 2300 mm high	–	–	–	–	–	nr	4774.05
ASSA ABLOY SL510 SE 2P; Double Bi-Parting Door Leaves and 100 mm Operator with Pocket Safety Screens 2000 mm wide × 2300 mm high	–	–	–	–	–	nr	5092.32
ASSA ABLOY SL510 SE 4P; Double Bi-Parting Door Leaves with Two Sidescreens and 100 mm Operator with Pocket Safety Screens 4000 mm wide × 2300 mm high	–	–	–	–	–	nr	6365.40

24 DOORS, SHUTTERS AND HATCHES

Item	PC £	Labour hours	Labour £	Plant £	Material £	Unit	Total rate £
24.04 SWING DOORS							
Swing door operators (mounted onto doors supplied by others)							
Automatic Low Kinetic Energy Swing Door Operator. ASSA ABLOY SW100 Surface Mounted Door Operator for internal doors. The SW100 is an electromechanical operator and has been developed to comply with European standards and requirements for low-energy power operated doors. Mounted onto a suitable transom by others. Door and frame by others. Fire-rated upto 60 min. EPD verified by 3rd Party							
ASSA ABLOY SW100–1 Single Door Leaf Low Energy Door Operator; Low risk users environment	–	–	–	–	–	nr	1856.58
ASSA ABLOY SW100–1 Single Door Leaf Low Energy Door Operator; High risk users environment	–	–	–	–	–	nr	1962.66
ASSA ABLOY SW100–2 Double Door Leaf Low Energy Door Operator; Low risk users environment	–	–	–	–	–	nr	2546.16
ASSA ABLOY SW100–2 Double Door Leaf Low Energy Door Operator; High risk users environment	–	–	–	–	–	nr	2970.52
ASSA ABLOY SW300 70 mm Surface Mounted Door Operator for external and internal doors. The SW300 is an electromechanical operator and has been developed to comply with European standards and requirements for low and high energy power operated doors. Door can be used manually with power assist or by push pad(s). Includes intelligent control for wind and stack pressure management. Mounted onto a suitable transom by others. Door and frame by others. Fire-rated upto 60 min. EPD verified by 3rd Party							
ASSA ABLOY SW300–1 Single Door Leaf Door Operator; Low risk users environment	–	–	–	–	–	nr	2333.98
ASSA ABLOY SW300–1 Single Door Leaf Door Operator; High risk users environment	–	–	–	–	–	nr	2546.16
ASSA ABLOY SW300- 2 Double Door Leaf Door Operator; Low risk users environment	–	–	–	–	–	nr	3182.70
ASSA ABLOY SW300- 2 Double Door Leaf Door Operator; High risk users environment	–	–	–	–	–	nr	3607.06

24 DOORS, SHUTTERS AND HATCHES

Item	PC £	Labour hours	Labour £	Plant £	Material £	Unit	Total rate £
ASSA ABLOY SW300 70 mm Surface Mounted Door Operator for external and internal doors. The SW300 is an electromechanical operator and has been developed to comply with European standards and requirements for low and high energy power operated doors. Allows the door to open manually with power assist in both directions with one direction automatically by push pad(s). Includes intelligent control for wind and stack pressure management. Mounted onto a suitable transom by others. Door and frame by others. Fire-rated upto 60 min. EPD verified by 3rd Party							
ASSA ABLOY SW300–1DA Single Door Leaf Door Operator; Low risk users environment	–	–	–	–	–	nr	2440.07
ASSA ABLOY SW300–1DA Single Door Leaf Door Operator; High risk users environment	–	–	–	–	–	nr	2652.25
ASSA ABLOY SW300- 2DA Double Door Leaf Door Operator; Low risk users environment	–	–	–	–	–	nr	3394.88
ASSA ABLOY SW300- 2DA Double Door Leaf Door Operator; High risk users environment	–	–	–	–	–	nr	3819.24
Swing door systems (including doors and screens)							
ASSA ABLOY SW200 I, anti-ligature, digitally controlled and intelligent complete swing door system. Includes built-in anti-finger trap door leaf mounted into a tight sealed frame. Complete with an overhead concealed swing door operator and ball-raced heavy duty bottom pivot. Door automatically opens inwards or outwards and can also be opened manually with a maximum of 22N opening force. Tested to 1 m cycles. Verified with 3rd party EPD documentation. System including normal installation to comply with BS7036–0/ EN16005							
ASSA ABLOY SW200 I–1 Single Door Leaf Concealed In-Head Swing Door Entrance; High risk users environment	–	–	–	–	–	nr	4561.87
ASSA ABLOY SW200 I–1 Double Door Leaf Concealed In-Head Swing Door Entrance; High risk users environment	–	–	–	–	–	nr	6153.22

24 DOORS, SHUTTERS AND HATCHES

Item	PC £	Labour hours	Labour £	Plant £	Material £	Unit	Total rate £
24.04 SWING DOORS – cont							
Swing door systems (including doors and screens) – cont ASSA ABLOY SW200 B Balance Door System, consisting of a liner frame, one door leaf and transom. The Balance Door System is very effective in windy conditions, combining the benefits of a swing and sliding door system. The door system should be fully functional in wind loads according to table 1013665. A perfect solution for narrow passages or in corridors. 3rd party EPD verification							
ASSA ABLOY SW200 B–1 Single Door Leaf Balance Door Entrance; High risk users environment	–	–	–	–	–	nr	5304.50
ASSA ABLOY SW200 B–2 Double Door Leaf Balance Door Entrance; High risk users environment	–	–	–	–	–	nr	8487.20
24.05 SHUTTER DOORS AND PANEL DOORS							
ASSA ABLOY Entrance Systems Ltd; Overhead Sectional Insulated Doors ASSA ABLOY; Thermal transmittance (EN12428) = 0.43 W/m^2 K (Installed Door 1.1 W/m^2 K); Water penetration (EN12425) – Class 3; Air permeability (EN12426) – Class 3; Wind-Load (EN12424) – Class 3 reinforced steel panel; spring break device; versatile track system; Single phase electrically operated; manual override, 'deadman' control, pre-coated panel in a standard colour; 1 row DAOP vision panels							
ASSA ABLOY 1042P, Standard Lift; 2500 mm × 3000 mm; 42 mm thick insulated sandwich panels	–	–	–	–	–	nr	2328.05
ASSA ABLOY OH1042P, Standard Lift; 3000 mm × 3000 mm; 42 mm thick insulated sandwich panels	–	–	–	–	–	nr	3156.25
ASSA ABLOY OH1042P, Standard Lift; 4000 mm × 5000 mm; 42 mm thick insulated sandwich panels	–	–	–	–	–	nr	4040.00
ASSA ABLOY OH1042P, Standard Lift; 6000 mm × 6000 mm; 42 mm thick insulated sandwich panels	–	–	–	–	–	nr	6726.60
ASSA ABLOY OH1042P, Standard Lift; 8000 mm × 6000 mm; 42 mm thick insulated sandwich panels	–	–	–	–	–	nr	9696.00

24 DOORS, SHUTTERS AND HATCHES

Item	PC £	Labour hours	Labour £	Plant £	Material £	Unit	Total rate £
Insulated overhead sectional speed doors; ASSA ABLOY; Thermal transmittance (EN12428) = 0.43 W/m² K (Installed Door 1.1 W/m² K); Water penetration (EN12425) – Class 3; Air permeability (EN12426) – Class 3; Wind-Load (EN12424) – Class 3 reinforced steel panel; spring break device; versatile track system; Single phase electrically operated; manual override,'deadman' control, pre-coated panel in a standard colour; 1 row DAOP vision panels							
ASSA ABLOY OH1042S, Standard Lift; 2500 mm × 3000 mm; 42 mm thick insulated sandwich panels	–	–	–	–	–	nr	3838.00
ASSA ABLOY OH1042S, Standard Lift; 3000 mm × 3000 mm; 42 mm thick insulated sandwich panels	–	–	–	–	–	nr	4040.00
ASSA ABLOY OH1042S, Standard Lift; 4000 mm × 5000 mm; 42 mm thick insulated sandwich panels	–	–	–	–	–	nr	5050.00
ASSA ABLOY Entrance Systems Ltd; High Speed Doors							
ASSA ABLOY RR3000 ISO – a low noise high speed insulated secure door, fabricated using 50 mm thick thermally insulated laths. The anti-cold bridge laths are covered in a hard wearing steel skin. A doubled sealing system links the laths ensuring excellent wind and water tightness and a smooth operation of the door. Patented V-Drive Technology supports the door curtain movement, thereby ensuring longevity. The laths are wound onto winding discs without contact and travel within a specially designed channel guide. Using this technology, the door can be opened and closed very quickly, with a soft start and stop feature with minimal wear. Should accidental damage occur then each lath can be replaced individually, rapidly and easily thanks to their individual suspension from the rubber drive belt. Optionally double glazed window laths can be integrated within the door curtain. Door can complete up to 5 opening/closing cycles per minute. Opening speed up to 2.1 m/s closing at 0.7 m/s							
ASSA ABLOY RR3000 ISO, external 3000 mm × 3000 mm	–	–	–	–	–	nr	14948.00
ASSA ABLOY RR3000 ISO, external 5000 mm × 5000 mm	–	–	–	–	–	nr	20301.00
ASSA ABLOY RR3000 ISO, external 7000 mm × 6000 mm	–	–	–	–	–	nr	23230.00

24 DOORS, SHUTTERS AND HATCHES

Item	PC £	Labour hours	Labour £	Plant £	Material £	Unit	Total rate £
24.05 SHUTTER DOORS AND PANEL DOORS – cont							
ASSA ABLOY Entrance Systems Ltd – cont							
High speed internal fabric door; ASSA ABLOY; breakaway facility with auto repair system; electrically operated frequency inverter motor with opening speeds to 2.4 m/s; lifetime expectations 1,000,000 cycles; galvanized steel guide bodies; soft bottom edge for safety; high tensile fabric curtain; stop and return safety-edge, safety photo cells, vision strip, push button operation to both sides							
ASSA ABLOY HS9020G internal; one row windows 2000 mm × 2500 mm	–	–	–	–	–	nr	**4741.95**
ASSA ABLOY HS9010P internal; one row windows 3000 mm × 3000 mm	–	–	–	–	–	nr	**7527.53**
ASSA ABLOY HS9010P internal; one row windows 4000 mm × 4000 mm	–	–	–	–	–	nr	**8029.50**
High Speed external fabric doors; ASSA ABLOY; breakaway facility with auto repair system; air permeability (EN12426) Class 2; Wind load resistance Class 4; lifetime expectations 1,000,000 cycles; electrically operated frequency inverter motor, opening speeds up to 2.4 m/s; galvanized steel guide bodies; soft bottom edge for safety; high tensile fabric curtain; stop and return safety-edge, safety photo cells, push button operation to both sides, vision strip							
ASSA ABLOY HS8010P external; one row windows 3000 mm × 3000 mm	–	–	–	–	–	nr	**8019.40**
ASSA ABLOY HS8010P external; one row windows 4000 mm × 4500 mm	–	–	–	–	–	nr	**9292.00**
ASSA ABLOY HS8010P external; one row windows 5500 mm × 5500 mm	–	–	–	–	–	nr	**9847.50**

24 DOORS, SHUTTERS AND HATCHES

Item	PC £	Labour hours	Labour £	Plant £	Material £	Unit	Total rate £
ASSA ABLOY Entrance Systems Ltd; High Speed Clean Room Doors							
Fraunhofer Institute certified high vertical opening door. Certified according to ISO 5 and GMP class C. Compliant to the European norms and regulations in cleanroom applications according to DIN EN ISO 14644–1 and GMP. V2 A stainless steel. The low particle emissions and sealing properties are supported by the unique clean room design. Hidden containment of all mechanical and electronic parts solution for controlled environmental areas. The standard smooth clear PVC curtain has vertical coloured fabric stripes for optimal transparency, person recognition and cleaning. Optional anti-static curtain available, some restrictions in performance, size or certification may apply. Standard alternative would be a silicon free curtain. Opening/closing speed: Up to 2.0 / 1.0 m/s. Leakage: max. $20 \, m^3/h$ at a pressure difference of 25 Pa, at a door size of $3 \, m^2$							
ASSA ABLOY RR300 Clean internal; clear PVC curtain 2000 mm × 2500 mm	–	–	–	–	–	nr	8752.66
ASSA ABLOY RR300 Clean internal; clear PVC curtain 3000 mm × 3000 mm	–	–	–	–	–	nr	9268.77
DOCK LEVELLERS							
Swingdock leveller from ASSA ABLOY Entrance Systems; electrohydraulic operation with two lifting cylinders; integrated frame to suit various pit applications; Control for door and leveller from one communal control panel, including interlocking of leveller/door; hold-to-run operation; mains power switch; load bearing capacity 60 kN in line with EN1398; colour RAL5010; toe guards, warning signs							
ASSA ABLOY DL6010S 500 mm swing lip; 2500 mm length × 2000 mm wide × 600 mm height	–	–	–	–	–	nr	3312.80
ASSA ABLOY DL6010S 500 mm swing lip; 4000 mm length × 2000 mm wide × 1100 mm height	–	–	–	–	–	nr	3984.45

24 DOORS, SHUTTERS AND HATCHES

Item	PC £	Labour hours	Labour £	Plant £	Material £	Unit	Total rate £
24.05 SHUTTER DOORS AND PANEL DOORS – cont							
DOCK LEVELLERS – cont							
Teledock leveller from ASSA ABLOY Entrance Systems; electrohydraulic operation with two lifting cylinders; one lip ram cylinder; movable telescopic lip; integrated frame to suit various pit applications; Control for door and leveller from one communal control panel, including interlocking of leveller/door; hold-to-run operation; mains power switch; load bearing capacity 60kN in line with EN1398; colour RAL5010; toe guards, warning signs							
ASSA ABLOY DL6020T 500 mm telescopic lip; 2500 mm length × 2000 mm wide	–	–	–	–	–	nr	**3519.85**
ASSA ABLOY DL6020T 1000 mm telescopic lip; 4000 mm length × 2000 mm wide	–	–	–	–	–	nr	**4398.55**
DOCK SHELTERS							
Curtain mechanical shelter; ASSA ABLOY Entrance Systems; extruded aluminium frame; spring loaded-frame connected with parallel bracing arms; 2 side curtains, one top curtain; double-layered high quality polyester, coated on both sides, longitudinal strength: approx. 750 N, transverse strength: approx. 900 N; 4500 mm above outside yard level, 600 mm projection, 1000 mm top curtain, top curtain to have one slit at the contact point on each side, integrated rain-channels							
ASSA ABLOY DS6060 A; 3200 mm height × 3250 mm width	–	–	–	–	–	nr	**1242.30**
ASSA ABLOY DS6060 A; 600 mm projection, 1000 mm top curtain; 3400 mm height × 3250 width	–	–	–	–	–	nr	**1343.30**
ASSA ABLOY DS6060 A; 600 mm projection, 1000 mm top curtain; 3400 mm height × 3450 width	–	–	–	–	–	nr	**1446.32**
Inflatable mechanical shelter; ASSA ABLOY Entrance Systems; hot dipped galvanized surface treatment, polyester painted, corrosion protection class 3; top bag with polyester fabric panels, 1000 mm extension; side bags with polyester fabric panels, 650 mm extension, strong blower for straight movement of top and side bags; side section with steel guards; colour from standard range							
ASSA ABLOY DS6070B; 3755 mm height × 3600 width × 770 mm depth	–	–	–	–	–	nr	**4141.00**
ASSA ABLOY DS6070B; 4055 mm height × 3600 width × 770 mm depth	–	–	–	–	–	nr	**4242.00**
ASSA ABLOY DS6070B; 4555 mm height × 3600 width × 770 mm depth	–	–	–	–	–	nr	**4451.07**

24 DOORS, SHUTTERS AND HATCHES

Item	PC £	Labour hours	Labour £	Plant £	Material £	Unit	Total rate £
LOAD HOUSES							
Stand-alone Load House; ASSA ABLOY Entrance Systems; a complete loading unit attached externally to building; modular steel skin/insulated cladding; two wall elements, one roof element; hot-dipped galvanized steel frame; installed at 90°/ 45° to building; NL 2000/3000 mm; NW3300/ 3600 mm; DH 950/1300 mm; 4° roof slope; drainpipe, gutter; standard colour range							
ASSA ABLOY LH6081L Load house,6010SA Swingdock c/w Autodock platform; 2000 mm length × 3300 mm width × 3495 mm height	–	–	–	–	–	nr	6726.60
ASSA ABLOY LH6081L Load house, 6010SA Swingdock c/w Autodock platform; 3000 mm length × 3600 mm width × 3575 mm height	–	–	–	–	–	nr	8489.05
Rolling shutters and collapsible gates; steel counter shutters; push-up, self-coiling; polyester power coated; fixing by bolting							
Shutter doors							
3000 mm × 1000 mm	–	–	–	–	–	nr	1565.30
4000 mm × 1000 mm; in two panels	–	–	–	–	–	nr	2706.66
Rolling shutters with collapsible gates; galvanized steel; one hour fire-resisting; self-coiling; activated by fusible link; fixing with bolts							
Shutter doors							
1000 mm × 2750 mm	–	–	–	–	–	nr	1869.66
1500 mm × 2750 mm	–	–	–	–	–	nr	1972.92
2400 mm × 2750 mm	–	–	–	–	–	nr	2358.81
Translucent GRP stacking door (78% translucency); U-value = 2.6 W/m² K; electrically operated; Envirodoor Ltd; fully enclosed aluminium track system with SAA finish; manual over-ride, lock interlock, stop and return safety cage, deadmans down button, anti-flip device, photoelectric cell and beam deflectors; fixing by bolting; standard panel finishes							
Stacking doors							
HT40 stacking door 2500 mm × 3000 mm; 40 mm thick × 500 mm high twin walled GRP translucent panels	–	–	–	–	–	nr	6083.13
HT40 stacking door 4500 mm × 6000 mm; 40 mm thick × 500 mm high twin walled translucent panels	–	–	–	–	–	nr	11304.94
HT60-N stacking door 6000 mm × 6000 mm: 60 mm thick × 500 mm high twinn walled translucent panels	–	–	–	–	–	nr	17226.56

24 DOORS, SHUTTERS AND HATCHES

Item	PC £	Labour hours	Labour £	Plant £	Material £	Unit	Total rate £
24.05 SHUTTER DOORS AND PANEL DOORS – cont							
Translucent GRP stacking door (78% translucency) – cont							
Stacking doors – cont							
HT60-H stacking door 7000 mm × 8000 mm: 60 mm thick × 1000 mm high twin walled translucent panels	–	–	–	–	–	nr	**24224.86**
HT80–400 stacking door 10000 mm × 10000 mm; 80 mm thick × 1000 mm high twin walled translucent panels	–	–	–	–	–	nr	**59862.30**
Doors; galvanized steel up and over type garage doors; Catnic Horizon 90 or other equal and approved; spring counter balanced; fixed to timber frame (not included)							
Garage door							
2135 mm × 1980 mm	454.40	3.70	87.27	–	477.30	nr	**564.57**
2135 mm × 2135 mm	497.60	3.70	87.27	–	522.66	nr	**609.93**
2400 mm × 2135 mm	628.80	3.70	87.27	–	660.45	nr	**747.72**
3965 mm × 2135 mm	1396.80	5.55	130.89	–	1467.06	nr	**1597.95**
Grilles; Galaxy nylon rolling counter grille or other equal and approved; Bolton Brady Ltd; colour, off-white; self-coiling; fixing by bolting							
Grilles							
3000 mm × 1000 mm	–	–	–	–	–	nr	**1260.00**
4000 mm × 1000 mm	–	–	–	–	–	nr	**2020.00**
Sliding/folding partitions; Alco Beldan Ltd or equal and approved							
Sliding/folding partitions							
ref. NW100 Moveable Wall; 5000 mm (wide) × 2495 mm (high) comprising 4 nr. 954 mm (wide) standard panels and 1 nr. 954 mm (wide) telescopic panel; sealing; fixing	–	–	–	–	–	nr	**15750.00**
24.06 INTERNAL DOORS AND FRAMES							
DOOR FRAMES AND LININGS							
Internal softwood door lining set; with loose stops							
32 mm × 115 mm wide (finished) set; with loose stops; for internal doors							
686 / 762 / 838 wide doors	14.58	0.70	16.19	–	15.22	nr	**31.41**
32 mm × 138 mm wide (finished) set; with loose stops; for internal doors							
686 / 762 / 838 wide doors	16.66	0.70	16.19	–	17.36	nr	**33.55**

24 DOORS, SHUTTERS AND HATCHES

Item	PC £	Labour hours	Labour £	Plant £	Material £	Unit	Total rate £
Internal softwood fire door door lining set; with loose stops							
38 mm × 115 mm wide (finished) set; with loose stops; for internal doors							
686 / 762 / 838 wide doors	21.96	0.70	16.19	–	22.81	nr	**39.00**
38 mm × 138 mm wide (finished) set; with loose stops; for internal doors							
686 / 762 / 838 wide doors	26.07	0.70	16.19	–	27.05	nr	**43.24**
Door frames and door linings, sets; purpose-made; wrought softwood							
Jambs and heads; as linings							
32 mm × 63 mm	–	0.16	3.70	–	6.02	m	**9.72**
32 mm × 100 mm	–	0.16	3.70	–	6.77	m	**10.47**
32 mm × 140 mm	–	0.16	3.70	–	7.21	m	**10.91**
Jambs and heads; as frames; rebated, rounded and grooved							
44 mm × 75 mm	–	0.16	3.70	–	9.68	m	**13.38**
44 mm × 100 mm	–	0.16	3.70	–	10.40	m	**14.10**
44 mm × 115 mm	–	0.16	3.70	–	10.45	m	**14.15**
44 mm × 140 mm	–	0.19	4.40	–	10.96	m	**15.36**
57 mm × 100 mm	–	0.19	4.40	–	11.12	m	**15.52**
57 mm × 125 mm	–	0.19	4.40	–	11.75	m	**16.15**
69 mm × 88 mm	–	0.19	4.40	–	11.32	m	**15.72**
69 mm × 100 mm	–	0.19	4.40	–	12.13	m	**16.53**
69 mm × 125 mm	–	0.20	4.62	–	12.88	m	**17.50**
69 mm × 150 mm	–	0.20	4.62	–	13.64	m	**18.26**
94 mm × 100 mm	–	0.23	5.33	–	18.39	m	**23.72**
94 mm × 150 mm	–	0.23	5.33	–	21.44	m	**26.77**
Mullions and transoms; in linings							
32 mm × 63 mm	–	0.11	2.54	–	8.01	m	**10.55**
32 mm × 100 mm	–	0.11	2.54	–	8.79	m	**11.33**
32 mm × 140 mm	–	0.11	2.54	–	9.16	m	**11.70**
Mullions and transoms; in frames; twice rebated, rounded and grooved							
44 mm × 75 mm	–	0.11	2.54	–	12.14	m	**14.68**
44 mm × 100 mm	–	0.11	2.54	–	12.66	m	**15.20**
44 mm × 115 mm	–	0.11	2.54	–	12.66	m	**15.20**
44 mm × 140 mm	–	0.13	3.01	–	13.15	m	**16.16**
57 mm × 100 mm	–	0.13	3.01	–	13.32	m	**16.33**
57 mm × 125 mm	–	0.13	3.01	–	13.95	m	**16.96**
69 mm × 88 mm	–	0.13	3.01	–	13.15	m	**16.16**
69 mm × 100 mm	–	0.13	3.01	–	13.92	m	**16.93**
Add 5% to the above material prices for selected softwood for staining							

24 DOORS, SHUTTERS AND HATCHES

Item	PC £	Labour hours	Labour £	Plant £	Material £	Unit	**Total rate £**
24.06 INTERNAL DOORS AND FRAMES – cont							
Door frames and door linings, sets; purpose-made; medium density fireboard							
Jambs and heads; as linings							
18 mm × 126 mm	–	0.16	3.70	–	7.23	m	**10.93**
22 mm × 126 mm	–	0.16	3.70	–	7.50	m	**11.20**
25 mm × 126 mm	–	0.16	3.70	–	7.64	m	**11.34**
Door frames and door linings, sets; purpose-made; selected Sapele							
Jambs and heads; as linings							
32 mm × 63 mm	9.02	0.21	4.86	–	9.34	m	**14.20**
32 mm × 100 mm	11.14	0.21	4.86	–	11.53	m	**16.39**
32 mm × 140 mm	12.19	0.21	4.86	–	12.66	m	**17.52**
Jambs and heads; as frames; rebated, rounded and grooved							
44 mm × 75 mm	14.78	0.21	4.86	–	15.27	m	**20.13**
44 mm × 100 mm	16.83	0.21	4.86	–	17.39	m	**22.25**
44 mm × 115 mm	17.40	0.21	4.86	–	18.01	m	**22.87**
44 mm × 140 mm	18.23	0.25	5.79	–	18.88	m	**24.67**
57 mm × 100 mm	18.64	0.25	5.79	–	19.30	m	**25.09**
57 mm × 125 mm	20.40	0.25	5.79	–	21.11	m	**26.90**
69 mm × 88 mm	18.62	0.25	5.79	–	19.23	m	**25.02**
69 mm × 100 mm	20.72	0.25	5.79	–	21.44	m	**27.23**
69 mm × 125 mm	22.81	0.28	6.48	–	23.60	m	**30.08**
69 mm × 150 mm	24.90	0.28	6.48	–	25.75	m	**32.23**
94 mm × 100 mm	29.09	0.28	6.48	–	30.07	m	**36.55**
94 mm × 150 mm	35.67	0.28	6.48	–	36.84	m	**43.32**
Mullions and transoms; in linings							
32 mm × 63 mm	11.32	0.15	3.47	–	11.66	m	**15.13**
32 mm × 100 mm	13.43	0.15	3.47	–	13.83	m	**17.30**
32 mm × 140 mm	14.49	0.15	3.47	–	14.92	m	**18.39**
Mullions and transoms; in frames; twice rebated, rounded and grooved							
44 mm × 75 mm	18.19	0.15	3.47	–	18.74	m	**22.21**
44 mm × 100 mm	19.57	0.15	3.47	–	20.16	m	**23.63**
44 mm × 115 mm	20.14	0.15	3.47	–	20.74	m	**24.21**
44 mm × 140 mm	20.98	0.17	3.93	–	21.61	m	**25.54**
57 mm × 100 mm	21.40	0.17	3.93	–	22.04	m	**25.97**
57 mm × 125 mm	23.16	0.17	3.93	–	23.85	m	**27.78**
69 mm × 88 mm	20.98	0.17	3.93	–	21.61	m	**25.54**
69 mm × 100 mm	23.61	0.17	3.93	–	24.32	m	**28.25**
Sills; once sunk weathered; once rebated, three times grooved							
63 mm × 175 mm	52.25	0.31	7.17	–	53.82	m	**60.99**
75 mm × 125 mm	50.33	0.31	7.17	–	51.84	m	**59.01**
75 mm × 150 mm	52.71	0.31	7.17	–	54.29	m	**61.46**

24 DOORS, SHUTTERS AND HATCHES

Item	PC £	Labour hours	Labour £	Plant £	Material £	Unit	Total rate £
Door frames and door linings, sets; European Oak							
Sills; once sunk weathered; once rebated, three times grooved							
63 mm × 175 mm	85.89	0.31	7.17	–	88.47	m	**95.64**
75 mm × 125 mm	84.81	0.31	7.17	–	87.35	m	**94.52**
75 mm × 150 mm	92.72	0.31	7.17	–	95.50	m	**102.67**
Fire-resisting door frame; internal and external; fitted with 15 mm × 4 mm intumescent strips; 12 mm deep rebates; screwed to masonry/ concrete							
Softwood frames; no sill – open in or out. Door size							
762 mm × 1981 mm × 44 mm; FD30; intemescent strip only	88.10	0.75	17.36	–	90.95	nr	**108.31**
838 mm × 1981 mm × 44 mm; FD30; intemescent strip only	90.51	0.75	17.36	–	93.43	nr	**110.79**
826 mm × 2040 mm × 44 mm; FD30; intemescent strip only	103.86	0.75	17.36	–	107.18	nr	**124.54**
762 mm × 1981 mm × 44 mm; FD30; intemescent strip/smoke seal	93.74	0.75	17.36	–	96.76	nr	**114.12**
838 mm × 1981 mm × 44 mm; FD30; intemescent strip/smoke seal	95.80	0.75	17.36	–	98.88	nr	**116.24**
826 mm × 2040 mm × 44 mm; FD30; intemescent strip/smoke seal	95.80	0.75	17.36	–	98.88	nr	**116.24**
Hardwood frames; no sill – open in or out. Door size							
762 mm × 1981 mm × 44 mm; FD60; intemescent strip/smoke seal	177.70	0.75	17.36	–	183.23	nr	**200.59**
838 mm × 1981 mm × 44 mm; FD60; intemescent strip/smoke seal	181.70	0.75	17.36	–	187.35	nr	**204.71**
826 mm × 2040 mm × 44 mm; FD60; intemescent strip/smoke seal	184.45	0.75	17.36	–	190.18	nr	**207.54**
INTERNAL DOORS							
NOTE: Door frames, linings and ironmogery is not included.							
Doors; standard flush; softwood composition							
Moulded panel doors; white based coated facings suitable for paint finish only; two, four or six panel options							
533 mm × 1981 mm × 35 mm	35.06	0.75	17.36	–	36.11	nr	**53.47**
610 mm × 1981 mm × 35 mm	35.06	0.75	17.36	–	36.11	nr	**53.47**
762 mm × 1981 mm × 35 mm	35.06	0.75	17.36	–	36.11	nr	**53.47**
838 mm × 1981 mm × 35 mm	37.87	0.75	17.36	–	39.01	nr	**56.37**
914 mm × 1981 mm × 35 mm	42.07	0.75	17.36	–	43.33	nr	**60.69**
526 mm × 2040 mm × 40 mm	35.06	0.75	17.36	–	36.11	nr	**53.47**
626 mm × 2040 mm × 40 mm	35.06	0.75	17.36	–	36.11	nr	**53.47**
726 mm × 2040 mm × 40 mm	35.06	0.75	17.36	–	36.11	nr	**53.47**
826 mm × 2040 mm × 40 mm	37.87	0.75	17.36	–	39.01	nr	**56.37**
926 mm × 2040 mm × 40 mm	42.07	0.75	17.36	–	43.33	nr	**60.69**

24 DOORS, SHUTTERS AND HATCHES

Item	PC £	Labour hours	Labour £	Plant £	Material £	Unit	Total rate £
24.06 INTERNAL DOORS AND FRAMES – cont							
Doors – cont							
Flush door; internal quality; skeleton or cellular core; faced both sides; lipped on two long edges; Jeld-Wen paint grade veneer or other equal							
457 mm × 1981 mm × 35 mm	52.96	1.16	26.84	–	54.55	nr	**81.39**
533 mm × 1981 mm × 35 mm	53.96	1.16	26.84	–	55.58	nr	**82.42**
610 mm × 1981 mm × 35 mm	53.96	1.16	26.84	–	55.58	nr	**82.42**
686 mm × 1981 mm × 35 mm	53.96	1.16	26.84	–	55.58	nr	**82.42**
762 mm × 1981 mm × 35 mm	53.96	1.16	26.84	–	55.58	nr	**82.42**
838 mm × 1981 mm × 35 mm	56.10	1.16	26.84	–	57.78	nr	**84.62**
864 mm × 1981 mm × 35 mm	57.24	1.16	26.84	–	58.96	nr	**85.80**
426 mm × 2040 mm × 40 mm	57.83	1.16	26.84	–	59.56	nr	**86.40**
526 mm × 2040 mm × 40 mm	57.83	1.16	26.84	–	59.56	nr	**86.40**
626 mm × 2040 mm × 40 mm	57.83	1.16	26.84	–	59.56	nr	**86.40**
726 mm × 2040 mm × 40 mm	59.03	1.16	26.84	–	60.80	nr	**87.64**
826 mm × 2040 mm × 40 mm	67.24	1.16	26.84	–	69.26	nr	**96.10**
926 mm × 2040 mm × 40 mm	69.11	1.16	26.84	–	71.18	nr	**98.02**
Flush door; internal quality; skeleton or cellular core; chipboard veneered; faced both sides; lipped on two long edges; Jeld-Wen Sapele veneered or other equal and approved							
457 mm × 1981 mm × 35 mm	76.80	1.25	28.92	–	79.10	nr	**108.02**
533 mm × 1981 mm × 35 mm	76.80	1.25	28.92	–	79.10	nr	**108.02**
610 mm × 1981 mm × 35 mm	76.80	1.25	28.92	–	79.10	nr	**108.02**
686 mm × 1981 mm × 35 mm	82.01	1.25	28.92	–	84.47	nr	**113.39**
762 mm × 1981 mm × 35 mm	82.01	1.25	28.92	–	84.47	nr	**113.39**
838 mm × 1981 mm × 35 mm	87.21	1.25	28.92	–	89.83	nr	**118.75**
526 mm × 2040 mm × 40 mm	82.01	1.25	28.92	–	84.47	nr	**113.39**
626 mm × 2040 mm × 40 mm	87.21	1.25	28.92	–	89.83	nr	**118.75**
726 mm × 2040 mm × 40 mm	87.21	1.25	28.92	–	89.83	nr	**118.75**
826 mm × 2040 mm × 40 mm	92.42	1.25	28.92	–	95.19	nr	**124.11**
926 mm × 2040 mm × 40 mm	97.63	1.25	28.92	–	100.56	nr	**129.48**
Doors; purpose-made panelled; wrought softwood							
Panelled doors; one open panel for glass; including glazing beads							
686 mm × 1981 mm × 44 mm	101.47	1.62	37.48	–	104.51	nr	**141.99**
762 mm × 1981 mm × 44 mm	102.30	1.62	37.48	–	105.37	nr	**142.85**
838 mm × 1981 mm × 44 mm	103.15	1.62	37.48	–	106.24	nr	**143.72**
Panelled doors; two open panel for glass; including glazing beads							
686 mm × 1981 mm × 44 mm	141.86	1.62	37.48	–	146.12	nr	**183.60**
762 mm × 1981 mm × 44 mm	143.09	1.62	37.48	–	147.38	nr	**184.86**
838 mm × 1981 mm × 44 mm	144.32	1.62	37.48	–	148.65	nr	**186.13**
Panelled doors; four 19 mm thick plywood panels; mouldings worked on solid both sides							
686 mm × 1981 mm × 44 mm	215.43	1.62	37.48	–	221.89	nr	**259.37**
762 mm × 1981 mm × 44 mm	218.22	1.62	37.48	–	224.77	nr	**262.25**
838 mm × 1981 mm × 44 mm	221.01	1.62	37.48	–	227.64	nr	**265.12**

24 DOORS, SHUTTERS AND HATCHES

Item	PC £	Labour hours	Labour £	Plant £	Material £	Unit	Total rate £
Panelled doors; six 25 mm thick panels raised and fielded; mouldings worked on solid both sides							
686 mm × 1981 mm × 44 mm	396.40	1.94	44.89	–	408.29	nr	**453.18**
762 mm × 1981 mm × 44 mm	400.36	1.94	44.89	–	412.37	nr	**457.26**
838 mm × 1981 mm × 44 mm	404.31	1.94	44.89	–	416.44	nr	**461.33**
rebated edges beaded	–	–	–	–	2.45	m	**2.45**
rounded edges or heels	–	–	–	–	0.57	m	**0.57**
weatherboard fixed to bottom rail	–	0.23	5.33	–	8.71	m	**14.04**
stopped groove for weatherboard	–	–	–	–	2.79	m	**2.79**
Doors; purpose-made panelled; selected Sapele							
Panelled doors; one open panel for glass; including glazing beads							
686 mm × 1981 mm × 44 mm	137.38	2.31	53.45	–	141.50	nr	**194.95**
762 mm × 1981 mm × 44 mm	139.23	2.31	53.45	–	143.41	nr	**196.86**
838 mm × 1981 mm × 44 mm	141.11	2.31	53.45	–	145.34	nr	**198.79**
686 mm × 1981 mm × 57 mm	146.83	2.54	58.76	–	151.23	nr	**209.99**
762 mm × 1981 mm × 57 mm	149.05	2.54	58.76	–	153.52	nr	**212.28**
838 mm × 1981 mm × 57 mm	151.24	2.54	58.76	–	155.78	nr	**214.54**
Panelled doors; 250 mm wide cross tongued intermediate rail; two open panels for glass; mouldings worked on the solid one side; 19 mm × 13 mm beads one side; fixing with brass cups and screws							
686 mm × 1981 mm × 44 mm	210.01	2.31	53.45	–	216.31	nr	**269.76**
762 mm × 1981 mm × 44 mm	213.55	2.31	53.45	–	219.96	nr	**273.41**
838 mm × 1981 mm × 44 mm	223.94	2.31	53.45	–	230.66	nr	**284.11**
686 mm × 1981 mm × 57 mm	223.94	2.54	58.76	–	230.66	nr	**289.42**
762 mm × 1981 mm × 57 mm	228.13	2.54	58.76	–	234.97	nr	**293.73**
838 mm × 1981 mm × 57 mm	232.43	2.54	58.76	–	239.40	nr	**298.16**
Panelled doors; four panels; (19 mm thick for 44 mm doors, 25 mm thick for 57 mm doors); mouldings worked on solid both sides							
686 mm × 1981 mm × 44 mm	295.10	2.31	53.45	–	303.95	nr	**357.40**
762 mm × 1981 mm × 44 mm	318.11	2.31	53.45	–	327.65	nr	**381.10**
838 mm × 1981 mm × 44 mm	333.84	2.31	53.45	–	343.86	nr	**397.31**
686 mm × 1981 mm × 57 mm	301.07	2.54	58.76	–	310.10	nr	**368.86**
762 mm × 1981 mm × 57 mm	325.97	2.54	58.76	–	335.75	nr	**394.51**
838 mm × 1981 mm × 57 mm	307.04	2.54	58.76	–	316.25	nr	**375.01**
Panelled doors; 150 mm wide stiles in one width; 430 mm wide cross tongued bottom rail; six panels raised and fielded one side; (19 mm thick for 44 mm doors, 25 mm thick for 57 mm doors); mouldings worked on solid both sides							
686 mm × 1981 mm × 44 mm	503.76	2.31	53.45	–	518.87	nr	**572.32**
762 mm × 1981 mm × 44 mm	555.93	2.31	53.45	–	572.61	nr	**626.06**
838 mm × 1981 mm × 44 mm	566.48	2.31	53.45	–	583.47	nr	**636.92**
686 mm × 1981 mm × 57 mm	536.44	2.54	58.76	–	552.53	nr	**611.29**
762 mm × 1981 mm × 57 mm	591.51	2.54	58.76	–	609.26	nr	**668.02**
838 mm × 1981 mm × 57 mm	604.96	2.54	58.76	–	623.11	nr	**681.87**
rebated edges beaded	–	–	–	–	3.12	m	**3.12**
rounded edges or heels	–	–	–	–	0.83	m	**0.83**
weatherboard fixed to bottom rail	–	0.31	7.17	–	11.79	m	**18.96**
stopped groove for weatherboard	–	–	–	–	2.89	m	**2.89**

24 DOORS, SHUTTERS AND HATCHES

Item	PC £	Labour hours	Labour £	Plant £	Material £	Unit	Total rate £
24.06 INTERNAL DOORS AND FRAMES – cont							
Internal white foiled moisture-resistant MDF door lining composite standard joinery set							
22 mm × 77 mm wide (finished) set; with loose stops; for internal doors							
610 mm × 1981 mm × 35 mm	10.96	0.70	16.19	–	11.49	nr	**27.68**
686 mm × 1981 mm × 35 mm	10.91	0.70	16.19	–	11.44	nr	**27.63**
762 mm × 1981 mm × 35 mm	10.91	0.70	16.19	–	11.44	nr	**27.63**
838 mm × 1981 mm × 35 mm	12.02	0.70	16.19	–	12.59	nr	**28.78**
864 mm × 1981 mm × 35 mm	10.91	0.70	16.19	–	11.44	nr	**27.63**
22 mm × 150 mm wide (finished) set; with loose stops; for internal doors							
610 mm × 1981 mm × 35 mm	12.07	0.70	16.19	–	12.63	nr	**28.82**
686 mm × 1981 mm × 35 mm	12.02	0.70	16.19	–	12.59	nr	**28.78**
762 mm × 1981 mm × 35 mm	12.02	0.70	16.19	–	12.59	nr	**28.78**
838 mm × 1981 mm × 35 mm	12.02	0.70	16.19	–	12.59	nr	**28.78**
864 mm × 1981 mm × 35 mm	12.02	0.70	16.19	–	12.59	nr	**28.78**
FIRE DOORS							
Flush door; half-hour fire-resisting (FD30); chipboard veneered; faced both sides; lipped on two long edges; Jeld-Wen paint grade veneer or other equal and approved							
610 mm × 1981 mm × 44 mm	75.40	1.62	37.48	–	77.66	nr	**115.14**
686 mm × 1981 mm × 44 mm	75.40	1.62	37.48	–	77.66	nr	**115.14**
762 mm × 1981 mm × 44 mm	75.40	1.62	37.48	–	77.66	nr	**115.14**
838 mm × 1981 mm × 44 mm	77.53	1.62	37.48	–	79.86	nr	**117.34**
526 mm × 2040 mm × 44 mm	79.27	1.62	37.48	–	81.65	nr	**119.13**
626 mm × 2040 mm × 44 mm	79.27	1.62	37.48	–	81.65	nr	**119.13**
726 mm × 2040 mm × 44 mm	80.47	1.62	37.48	–	82.88	nr	**120.36**
826 mm × 2040 mm × 44 mm	88.68	1.62	37.48	–	91.34	nr	**128.82**
Moulded panel doors; half-hour fire-resisting (FD30); white based coated facings suitable for paint finish only; two, four or six panel options; Premdor Fireshield or other equal and approved							
526 mm × 2040 mm × 44 mm	86.95	1.10	25.45	–	89.56	nr	**115.01**
626 mm × 2040 mm × 44 mm	86.95	1.10	25.45	–	89.56	nr	**115.01**
726 mm × 2040 mm × 44 mm	297.30	1.10	25.45	–	306.22	nr	**331.67**
826 mm × 2040 mm × 44 mm	91.86	1.10	25.45	–	94.62	nr	**120.07**
926 mm × 2040 mm × 44 mm	98.17	1.10	25.45	–	101.12	nr	**126.57**
Flush door; half-hour fire-resisting (FD30); faced both sides; lipped on two long edges; Jeld-Wen Sapele veneered or other equal and approved							
610 mm × 1981 mm × 44 mm	91.83	1.71	39.56	–	94.58	nr	**134.14**
686 mm × 1981 mm × 44 mm	97.03	1.71	39.56	–	99.94	nr	**139.50**
762 mm × 1981 mm × 44 mm	97.03	1.71	39.56	–	99.94	nr	**139.50**
838 mm × 1981 mm × 44 mm	102.24	1.71	39.56	–	105.31	nr	**144.87**
726 mm × 2040 mm × 44 mm	102.24	1.71	39.56	–	105.31	nr	**144.87**
826 mm × 2040 mm × 44 mm	107.45	1.71	39.56	–	110.67	nr	**150.23**
926 mm × 2040 mm × 44 mm	112.65	1.71	39.56	–	116.03	nr	**155.59**

24 DOORS, SHUTTERS AND HATCHES

Item	PC £	Labour hours	Labour £	Plant £	Material £	Unit	Total rate £
Flush door; half-hour fire-resisting (FD30); chipboard for painting; hardwood lipping two long edges; Premdor or other equal							
526 mm × 2040 mm × 44 mm	54.70	1.62	37.48	–	56.34	nr	**93.82**
626 mm × 2040 mm × 44 mm	54.70	1.62	37.48	–	56.34	nr	**93.82**
726 mm × 2040 mm × 44 mm	54.70	1.62	37.48	–	56.34	nr	**93.82**
826 mm × 2040 mm × 44 mm	56.10	1.62	37.48	–	57.78	nr	**95.26**
926 mm × 2040 mm × 44 mm	63.10	1.62	37.48	–	64.99	nr	**102.47**
826 mm × 2040 mm × 44 mm; single side vision panel 150 mm × 700 mm; factory fitted clear fire-rated glass	163.38	1.71	39.56	–	168.28	nr	**207.84**
826 mm × 2040 mm × 44 mm; two side vision panels 150 mm × 700 mm; factory fitted clear fire-rated glass	225.08	1.71	39.56	–	231.83	nr	**271.39**
926 mm × 2040 mm × 44 mm; single side vision panel 150 mm × 700 mm; factory fitted clear fire-rated glass	170.39	1.71	39.56	–	175.50	nr	**215.06**
926 mm × 2040 mm × 44 mm; two side vision panels 150 mm × 700 mm; factory fitted clear fire-rated glass	232.10	1.71	39.56	–	239.06	nr	**278.62**
Flush door; half-hour fire-resisting (FD30); White Oak veneer; hardwood lipping all edges; Premdor Fireshield or other equal and approved;							
526 mm × 2040 mm × 44 mm	92.56	1.62	37.48	–	95.34	nr	**132.82**
626 mm × 2040 mm × 44 mm	92.56	1.62	37.48	–	95.34	nr	**132.82**
726 mm × 2040 mm × 44 mm	92.56	1.62	37.48	–	95.34	nr	**132.82**
826 mm × 2040 mm × 44 mm	96.06	1.62	37.48	–	98.94	nr	**136.42**
926 mm × 2040 mm × 44 mm	103.78	1.62	37.48	–	106.89	nr	**144.37**
826 mm × 2040 mm × 44 mm; single side vision panel 508 mm × 1649 mm; factory fitted clear fire-rated glass	213.16	1.71	39.56	–	219.55	nr	**259.11**
826 mm × 2040 mm × 44 mm; two side vision panels 150 mm × 775 mm and 150 mm × 700 mm; factory fitted clear fire-rated glass	232.10	1.71	39.56	–	239.06	nr	**278.62**
926 mm × 2040 mm × 44 mm; single side vision panel 508 mm × 1649 mm; factory fitted clear fire-rated glass	220.88	1.71	39.56	–	227.51	nr	**267.07**
926 mm × 2040 mm × 44 mm; two side vision panels 150 mm × 775 mm and 150 mm × 700 mm; factory fitted clear fire-rated glass	239.81	1.71	39.56	–	247.00	nr	**286.56**
Flush door; one-hour fire-resisting (FD60); chipboard for painting; hardwood lipping two long edges; Premdor Firemaster or other equal and approved							
626 mm × 2040 mm × 54 mm	203.35	1.71	39.56	–	209.45	nr	**249.01**
726 mm × 2040 mm × 54 mm	203.35	1.71	39.56	–	209.45	nr	**249.01**
826 mm × 2040 mm × 54 mm	225.08	1.71	39.56	–	231.83	nr	**271.39**
926 mm × 2040 mm × 54 mm	232.10	1.71	39.56	–	239.06	nr	**278.62**

24 DOORS, SHUTTERS AND HATCHES

Item	PC £	Labour hours	Labour £	Plant £	Material £	Unit	Total rate £
24.06 INTERNAL DOORS AND FRAMES – cont							
FIRE DOORS – cont							
Moulded panel doors; half-hour fire-resisting (FD60); white based coated facings suitable for paint finish only; two, four or six panel options; Premdor Firemaster or other equal and approved							
626 mm × 2040 mm × 54 mm	273.47	1.71	39.56	–	281.67	nr	**321.23**
726 mm × 2040 mm × 54 mm	273.47	1.71	39.56	–	281.67	nr	**321.23**
826 mm × 2040 mm × 54 mm	279.78	1.71	39.56	–	288.17	nr	**327.73**
926 mm × 2040 mm × 54 mm	288.89	1.71	39.56	–	297.56	nr	**337.12**
Flush door; one-hour fire-resisting (FD60); White Oak veneer; hardwood lipping all edges; Premdor Firemaster or other equal and approved							
626 mm × 2040 mm × 54 mm	229.29	1.94	44.89	–	236.17	nr	**281.06**
726 mm × 2040 mm × 54 mm	229.29	1.94	44.89	–	236.17	nr	**281.06**
826 mm × 2040 mm × 54 mm	242.61	1.94	44.89	–	249.89	nr	**294.78**
926 mm × 2040 mm × 54 mm	250.32	1.94	44.89	–	257.83	nr	**302.72**
Flush door; one-hour fire-resisting (FD60); Steamed Beech veneer; hardwood lipping all edges; Premdor Firemaster or other equal and approved							
626 mm × 2040 mm × 54 mm	230.69	1.94	44.89	–	237.61	nr	**282.50**
726 mmx 2040 mm × 54 mm	230.69	1.94	44.89	–	237.61	nr	**282.50**
826 mm × 2040 mm × 54 mm	246.12	1.94	44.89	–	253.50	nr	**298.39**
926 mm × 2040 mm × 54 mm	253.83	1.94	44.89	–	261.44	nr	**306.33**
HYGENIC DOORSETS							
Altro Whiterock flush doors complete with frame and stops, leaf and integral hinges. NOTE EXCLUDING IRONMONGERY							
946 mm × 2040 mm single leaf doorset; non fire-rated	1118.58	3.00	69.41	–	1152.34	nr	**1221.75**
946 mm × 2040 mm single leaf doorset; FD30	1189.65	3.00	69.41	–	1225.55	nr	**1294.96**
946 mm × 2040 mm single leaf doorset; FD60	1459.51	3.00	69.41	–	1503.50	nr	**1572.91**
1350 mm × 2040 mm leaf and & half doorset; non fire-rated	1611.95	4.00	92.55	–	1660.57	nr	**1753.12**
1350 mm × 2040 mm leaf & half doorset; FD30	1709.80	4.00	92.55	–	1761.35	nr	**1853.90**
1350 mm × 2040 mm leaf and half doorset; FD60	2163.00	4.00	92.55	–	2228.15	nr	**2320.70**
1800 mm × 2040 mm double leaf doorset; non fire-rated	1781.90	6.00	138.81	–	1835.67	nr	**1974.48**
1800 mm × 2040 mm double leaf doorset; FD30	1946.70	6.00	138.81	–	2005.41	nr	**2144.22**
1800 mm × 2040 mm double leaf doorset; FD60	2379.30	6.00	138.81	–	2450.99	nr	**2589.80**
Intumescent strips							
Factory installed intumescent smoke seals to fire doors; 3 edges, jambs and head; site fixed intumescent strips	–	–	–	–	29.05	door	**29.05**
fire and smoke intumescent strips							
15 mm × 4 mm – FD30 doors	–	0.15	3.47	–	0.80	m	**4.27**
fire and smoke intumescent strips							
20 mm × 4 mm – FD60 doors	–	0.15	3.47	–	1.13	m	**4.60**

24 DOORS, SHUTTERS AND HATCHES

Item	PC £	Labour hours	Labour £	Plant £	Material £	Unit	Total rate £
Bedding and pointing frames							
Pointing wood frames or sills with mastic							
one side	–	0.09	1.60	–	1.22	m	**2.82**
both sides	–	0.19	3.38	–	2.44	m	**5.82**
Pointing wood frames or sills with polysulphide sealant							
one side	–	0.09	1.60	–	3.09	m	**4.69**
both sides	–	0.19	3.38	–	6.19	m	**9.57**
Bedding wood frames in cement mortar (1:3) and point							
one side	–	0.07	1.57	–	0.17	m	**1.74**
both sides	–	0.09	2.02	–	0.22	m	**2.24**
one side in mortar; other side in mastic	–	0.19	3.84	–	1.38	m	**5.22**
24.07 ASSOCIATED IRONMONGERY							
NOTE: Ironmongery is largely a matter of selection and specification and prices vary considerably; indicative prices for reasonable quantities of good quality ironmongery are given below.							
Allgood Ltd or other equal; to softwood							
Bolts							
75 × 35 mm Modric anodized aluminium straight barrel bolt	9.63	0.30	6.94	–	9.92	nr	**16.86**
150 × 35 mm Modric anodized aluminium straight barrel bolt	11.00	0.30	6.94	–	11.33	nr	**18.27**
75 × 35 mm Modric anodized aluminium necked barrel bolt	10.79	0.30	6.94	–	11.11	nr	**18.05**
150 × 35 mm Modric anodized aluminium necked barrel bolt	13.79	0.30	6.94	–	14.20	nr	**21.14**
11 mm Easiclean socket for wood or stone	4.71	0.10	2.32	–	4.85	nr	**7.17**
Security hinge bolt chubb WS12	9.75	0.50	11.57	–	10.04	nr	**21.61**
203 × 19 × 11 mm lever action flush bolt set, with coil spring and intumescent pack for FD30 and FD60 fire doors	33.11	0.60	13.88	–	34.10	nr	**47.98**
609 × 19 mm lever action flush bolt set, with coil sprin and intumescent pack for FD30 and FD60 fire doors	96.51	0.60	13.88	–	99.41	nr	**113.29**
Stainless steel indicating bolt complete with outside indicator and emergency release	69.16	0.60	13.88	–	71.23	nr	**85.11**
Catches							
Magnetic catch	0.48	0.20	4.62	–	0.49	nr	**5.11**
Door closers and furniture							
13 mm stainless steel rebate component for 7104/08/78/79/86	28.70	0.60	13.88	–	29.56	nr	**43.44**
70 × 70 mm Modric anodized aluminium electrically powered hold open wall magnet (excluding power supply and connection)	134.97	0.40	9.25	–	139.02	nr	**148.27**
Modric anodized aluminium bathroom configuration with quadaxial assembly, turn, release and optional indicator	52.59	0.80	18.51	–	54.17	nr	**72.68**

24 DOORS, SHUTTERS AND HATCHES

Item	PC £	Labour hours	Labour £	Plant £	Material £	Unit	Total rate £
24.07 ASSOCIATED IRONMONGERY – cont							
Allgood Ltd or other equal – cont							
Door closers and furniture – cont							
Concealed jamb door closer check action	157.44	1.00	23.13	–	162.16	nr	**185.29**
75 × 57 × 170 mm Modric anodized aluminium door coordinator for pairs of rebated leaves, CE Marked to BS EN1158 3–5–3/5–1–1–0	34.83	0.80	18.51	–	35.87	nr	**54.38**
290 × 48 × 50 mm Modric anodized aluminium rectangular overhead door closer with adjustable power and adjustable backcheck intumescent protected arm heavy duty U.L. & certifire listed & CE Marked to BS EN1154 4–8–2/4–1–1–3 and kite-marked	101.15	1.00	23.13	–	104.18	nr	**127.31**
Stainless steel overhead door closer. Projecting armset, Power EN 2–5, CE marked, c/w Backcheck, Latch action and Speed control. Max door width 1100 mm, Max door weight 100 kg	100.86	1.00	23.13	–	103.89	nr	**127.02**
288 × 45 × 32 mm fully concealed overhead door closer complete with track and arm for single action doors, adjustable power, latch action and backcheck. Certifire approved	168.98	0.80	18.51	–	174.05	nr	**192.56**
75 × 45 mm heavy duty floor pivot set with thrust roller bearing 200 kg load capacity. Complete with forged steel intumescent protected double action strap with 10 mm height adjustment and matching cover plate	257.50	2.30	53.21	–	265.22	nr	**318.43**
Cavalier floor spring unit with cover plate and loose box for concrete, adjustable power 25 mm offset strap & top centre, intumescent pack. Certfire listed	273.10	2.30	53.21	–	281.29	nr	**334.50**
Double action pivot set for door maximum width 1100 mm and maximum weight 80 kg	86.21	2.30	53.21	–	88.80	nr	**142.01**
Surface vertical rod push bar panic bolt, reversible, to suit doors 2500x1100 mm maximum, silver finish, CE marked to EN1125 class 3–7–5–1–1–3–2–2–A	144.07	1.50	34.70	–	148.39	nr	**183.09**
Rim push bar panic latch, reversible, to suit doors 1100 mm wide maximum, silver finish, CE marked to EN1125 class 3–7–5–1–1–3–2–2–A	102.72	1.30	30.08	–	105.80	nr	**135.88**
76 × 51 × 13 mm adjustable heavy roller catch, stainless steel forend and strike complete with satin nickel plate roller bolt	8.12	0.60	13.88	–	8.36	nr	**22.24**
External access device for use with XX10280/2 panic hardware to suit door thickness 45–55 mm, complete with SS3006N lever, SS755 rose, SS796 profile escutcheon and spindle. For use with MA7420 A51 or MA7420 A55 profile cylinders	34.12	1.30	30.08	–	35.14	nr	**65.22**
142 × 22 mm Ø Concealed jamb door closer light duty	18.27	0.80	18.51	–	18.82	nr	**37.33**

24 DOORS, SHUTTERS AND HATCHES

Item	PC £	Labour hours	Labour £	Plant £	Material £	Unit	Total rate £
80 × 40 × 45 mm Emergency release door stop with holdback facility	86.21	1.00	23.13	–	88.80	nr	**111.93**
Modric anodized aluminium quadaxial lever MA3503 assembly tested to BS EN1906 4/7/-/1/1/4/0/U	31.96	0.80	18.51	–	32.92	pair	**51.43**
Modric anodized aluminium quadaxial lever MA3502 assembly tested to BS EN1906 4/7/-/1/1/4/0/U	34.07	0.80	18.51	–	35.09	pair	**53.60**
Modric anodized aluminium quadaxial lever assembly tested to BS EN1906 4/7/-/1/1/4/0/U with Biocote® antibacterial protection	61.44	0.80	18.51	–	63.28	pair	**81.79**
Modric stainless steel quadaxial lever assembly tested to BS EN1906 4/7/-/1/1/4/0/U	58.49	0.80	18.51	–	60.24	pair	**78.75**
152 × 38 × 13 mm Modric anodized aluminium security door chain leather covered	45.58	0.40	9.25	–	46.95	nr	**56.20**
50 Ø × 3 mm Modric anodized aluminium circular covered rose for profile cylinder	4.48	0.10	2.32	–	4.61	nr	**6.93**
50 Ø × 3 mm Modric anodized aluminium circular covered rose with indicator and emergency release	9.24	0.15	3.47	–	9.52	nr	**12.99**
50 Ø × 3 mm Modric anodized aluminium circular covered rose with heavy turn, 5–8 mm spindle	16.21	0.15	3.47	–	16.70	nr	**20.17**
Budget lock escutcheon – satin stainless steel 316	8.83	0.10	2.32	–	9.09	nr	**11.41**
50 Ø × 3 mm Stainless steel circular covered rose for profile cylinder	7.18	0.10	2.32	–	7.40	nr	**9.72**
50 Ø × 3 mm Stainless steel circular covered rose with indicator and emergency release	9.67	0.15	3.47	–	9.96	nr	**13.43**
50 Ø × 3 mm Stainless steel circular covered rose with heavy turn, 5–8 mm spindle	19.94	0.15	3.47	–	20.54	nr	**24.01**
330 × 76 × 1.6 mm Modric anodized aluminium push plate	3.53	0.15	3.47	–	3.64	nr	**7.11**
330 × 76 × 1.6 mm Stainless steel push plate	8.18	0.15	3.47	–	8.43	nr	**11.90**
800 × 150 × 1.5 mm Modric anodized aluminium kicking plate, drilled and countersunk with screws	8.70	0.25	5.79	–	8.96	nr	**14.75**
900 × 150 × 1.5 mm Modric anodized aluminium kicking plate, drilled and countersunk with screws	9.79	0.25	5.79	–	10.08	nr	**15.87**
1000 × 150 × 1.5 mm Modric anodized aluminium kicking plate, drilled & countersunk with screws	10.86	0.25	5.79	–	11.19	nr	**16.98**
800 × 150 × 1.5 mm Stainless steel kicking plate, drilled and countersunk with screws	15.39	0.25	5.79	–	15.85	nr	**21.64**
900 × 150 × 1.5 mm Stainless steel kicking plate, drilled and countersunk with screws	17.31	0.25	5.79	–	17.83	nr	**23.62**
1000 × 150 × 1.5 mm Stainless steel kicking plate, drilled & countersunk with screws	19.24	0.25	5.79	–	19.82	nr	**25.61**
610 × 70 × 19 mm Ø Modric anodized aluminium grab handle bolt through fixing for doors 10 to 55 mm thick	31.45	0.40	9.25	–	32.39	nr	**41.64**
400 × 19 mm Ø Stainless steel D line straight pull handle with M8 threaded holes, fixing centres 300 mm	48.66	0.33	7.63	–	55.47	nr	**63.10**

Prices for Measured Works

24 DOORS, SHUTTERS AND HATCHES

Item	PC £	Labour hours	Labour £	Plant £	Material £	Unit	Total rate £
24.07 ASSOCIATED IRONMONGERY – cont							
Allgood Ltd or other equal – cont							
Hinges							
100 × 75 × 3 mm Stainless steel triple knuckle concealed twin bearings, button tipped butt hinges, jig drilled for metal doors/frames, complete with M6 × 12MT 'undercut' machine screws, stainless steel 316 CE marked to EN1935	19.30	0.25	5.79	–	19.88	pair	**25.67**
100 × 100 × 3 mm Stainless steel triple knuckle concealed twin Newton bearings, button tipped hinges, jig drilled, stainless steel grade 316 CE marked to EN1935	34.30	0.25	5.79	–	35.33	pair	**41.12**
Latches							
Modric anodized aluminium round cylinder for rim night latch, 2 keyed satin nickel plated	27.15	0.40	9.25	–	27.96	nr	**37.21**
93 × 75 mm Cylinder rim non-deadlocking night latch case only 60 mm backset	23.26	0.40	9.25	–	23.96	nr	**33.21**
71 series mortice latch, case only, low friction latchbolt, griptight follower, heavy spring for levers. Radius forend and sq strike. CE marked to BS EN12209 3/X/8/1/0G/-/B/02/0	17.58	0.80	18.51	–	18.11	nr	**36.62**
Modric anodized aluminium latch configuration with quadaxial assembly	31.75	0.80	18.51	–	32.70	nr	**51.21**
Modric anodized aluminium Nightlatch configuration with quadaxial assembly and single cylinder	58.91	0.80	18.51	–	60.68	nr	**79.19**
Locks							
44 mm case Bright zinc plated steel mortice budget lock with slotted strike plate 33 mm backset	29.85	0.80	18.51	–	30.75	nr	**49.26**
76 × 58 mm b/s Stainless steel cubicle mortice deadlock with 8 mm follower	14.14	0.80	18.51	–	14.56	nr	**33.07**
'A' length European profile double cylinder lock, 2 keyed satin nickel plated	27.84	0.80	18.51	–	28.68	nr	**47.19**
'A' length European profile cylinder and large turn, 2 keyed satin nickel plated	31.63	0.80	18.51	–	32.58	nr	**51.09**
'A' length European profile cylinder and large turn, 2 keyed under master key, satin nickel plated	29.11	0.80	18.51	–	29.98	nr	**48.49**
'A' length European profile single cylinder, 2 keyed satin nickel plated	21.53	0.80	18.51	–	22.18	nr	**40.69**
'A' length European profile single cylinder, 2 keyed under master key, satin nickel plated	21.53	0.80	18.51	–	22.18	nr	**40.69**

24 DOORS, SHUTTERS AND HATCHES

Item	PC £	Labour hours	Labour £	Plant £	Material £	Unit	Total rate £
93 × 60 mm b/s 71 series profile cylinder mortice deadlock, case only. Single throw 22 mm deadbolt. Radius forend and square strike. CE marked to BS EN12209 3/X/8/1/0/G/4/B/A/0/0	17.58	0.80	18.51	–	18.11	nr	36.62
92 × 60 mm b/s 71 series bathroom lock, case only, low friction latchbolt, griptight follower, heavy spring for levers, twin 8 mm followers at 78 mm centres. Radius forend and square strike. CE marked to BS EN12209 3/X/8/0/0/G-/B/0/2/0	20.84	0.80	18.51	–	21.47	nr	39.98
93 × 60 mm b/s 71 series profile cylinder mortice lock, case only, low friction latchbolt, griptight follower. Heavy spring for levers, 22 mm throw deadbolt, cylinder withdraws bolt bolts. Radius forend and square strike. CE marked to BS EN12209 3/X/8/1/0G/4/B/A2/0	20.84	0.80	18.51	–	21.47	nr	39.98
92 × 60 mm b/s71 series profile cylinder emergency lock, case only. Low friction latchbolt, griptight follower, heavy spring for lever, single throw 22 mm deadbolt, lever can withdraw both bolts. Radius forend and strike	68.56	0.80	18.51	–	70.62	nr	89.13
Modric anodized aluminium lock configuration with quadaxial assembly and cylinder with turn	90.75	0.80	18.51	–	93.47	nr	111.98
Sundries							
76 mm Ø Modric anodized aluminium circular sex symbol male	4.92	0.08	1.85	–	5.07	nr	6.92
76 mm Ø Modric anodized aluminium circular symbol fire door keep locked	4.92	0.08	1.85	–	5.07	nr	6.92
76 mm Ø Modric anodized aluminium circular symbol fire door keep shut	4.92	0.10	2.32	–	5.07	nr	7.39
38 × 47 mm Ø Modric anodized aluminium heavy circular floor door stop with cover	9.96	0.10	2.32	–	10.26	nr	12.58
38 × 47 mm Ø Stainless steel heavy circular floor door stop with cover	17.79	0.10	2.32	–	18.32	nr	20.64
63 × 19 mm Ø Modric anodized aluminium Circular heavy duty skirting buffer with thief-resistant insert	6.51	0.10	2.32	–	6.71	nr	9.03
102 × 25 mm Ø Stainless steel circular heavy duty skirting buffer with thief-resistant insert	11.28	0.10	2.32	–	11.62	nr	13.94
152 mm Cabin hook satin chrome on brass	23.60	0.15	3.47	–	24.31	nr	27.78
14 mm Ø × 145 × 94 mm Toilet roll holder, length 145 mm, colour white, satin stainless steel 316	65.68	0.15	3.47	–	69.43	nr	72.90
Towel rail with bushes, fixing centres 450 mm, satin stainless steel 316	89.27	0.25	5.79	–	97.29	nr	103.08
Toilet brush holder with toilet brush, with bushes, satin stainless steel 316	138.62	0.20	4.62	–	146.34	nr	150.96
Bathline 850 mm lift up support rail	209.97	0.75	17.36	–	216.27	set	233.63
Bathline 600 × 95 × 35 mm support rail with concealed fixing roses	107.62	0.50	11.57	–	110.85	set	122.42
Bathline 400 × 250 × 35 mm backrest rail with concealed fixing roses	107.62	0.50	11.57	–	110.85	set	122.42

24 DOORS, SHUTTERS AND HATCHES

Item	PC £	Labour hours	Labour £	Plant £	Material £	Unit	Total rate £
24.07 ASSOCIATED IRONMONGERY – cont							
Allgood Ltd or other equal; to hardwood							
Bolts							
75 × 35 mm Modric anodized aluminium straight barrel bolt	9.63	0.40	9.25	–	9.92	nr	**19.17**
150 × 35 mm Modric anodized aluminium straight barrel bolt	11.00	0.40	9.25	–	11.33	nr	**20.58**
75 × 35 mm Modric anodized aluminium necked barrel bolt	10.79	0.40	9.25	–	11.11	nr	**20.36**
150 × 35 mm Modric anodized aluminium necked barrel bolt	13.79	0.40	9.25	–	14.20	nr	**23.45**
11 mm Easiclean socket for wood or stone	4.71	0.15	3.47	–	4.85	nr	**8.32**
Security hinge bolt chubb WS12	9.75	0.65	15.04	–	10.04	nr	**25.08**
203 × 19 × 11 mm lever action flush bolt set with coil spring and intumescent pack for FD30 and FD60 fire doors	33.11	0.80	18.51	–	34.10	nr	**52.61**
609 × 19 mm lever action flush bolt set with coil spring and intumescent pack for FD30 and FD60 fire doors	96.51	0.80	18.51	–	99.41	nr	**117.92**
Stainless steel indicating bolt complete with outside indicator and emergency release	69.16	0.80	18.51	–	71.23	nr	**89.74**
Catches							
Magnetic catch	0.48	0.25	5.79	–	0.49	nr	**6.28**
Door closers and furniture							
13 mm stainless steel rebate component for 7104/08/78/79/86	28.70	0.80	18.51	–	29.56	nr	**48.07**
70 × 70 mm Modric anodized aluminium electrically powered hold open wall magnet. CE marked to BS EN1155:1997 & A1:2002 3–5–6/3–1–1–3	134.97	0.55	12.72	–	139.02	nr	**151.74**
Modric anodized aluminium bathroom configuration with quadaxial assembly, turn, release and optional indicator	52.59	1.05	24.29	–	54.17	nr	**78.46**
495 × 17.5 × 16 mm concealed overhead door restraining stay with aluminium channel. Stainless steel plated arm and bracket with adjustable friction slide. Block and spring buffer to cussion door opening. Not for use with door closing devices	27.07	1.35	31.23	–	27.88	nr	**59.11**
Concealed jamb door closer check action	157.44	1.35	31.23	–	162.16	nr	**193.39**
75 × 57 × 170 mm Modric anodized aluminium door coordinator for pairs of rebated leaves, CE Marked to BS EN1158 3–5–3/5–1–1–0	34.83	1.05	24.29	–	35.87	nr	**60.16**
290 × 48 × 50 mm Modric anodized aluminium rectangular overhead door closer with adjustable power and adjustable backcheck intumescent protected arm heavy duty U.L. & certifire listed & CE Marked to BS EN1154 4–8–2/4–1–1–3 and kite-marked	101.15	1.35	31.23	–	104.18	nr	**135.41**

24 DOORS, SHUTTERS AND HATCHES

Item	PC £	Labour hours	Labour £	Plant £	Material £	Unit	Total rate £
Stainless steel overhead door closer. Projecting armset, Power EN 2–5, CE marked, c/w Backcheck, Latch action and Speed control. Max door width 1100 mm, Max door weight 100 kg	100.86	1.35	31.23	–	103.89	nr	**135.12**
288 × 45 × 32 mm fully concealed overhead door closer complete with track and arm for single action doors, adjustable power, latch action and backcheck. Certifire approved	168.98	1.05	24.29	–	174.05	nr	**198.34**
75 × 45 mm heavy duty floor pivot set with thrust roller bearing 200 kg load capacity. Complete with forged steel intumescent protected double action strap with 10 mm height adjustment and matching cover plate	257.50	3.05	70.57	–	265.22	nr	**335.79**
Cavalier floor spring unit with cover plate and loose box for concrete, adjustable power 25 mm offset strap & top centre with intumescent pack. Certfire listed	273.10	2.30	53.21	–	281.29	nr	**334.50**
Double action pivot set for door maximum width 1100 mm and maximum weight 80 kg	86.21	3.05	70.57	–	88.80	nr	**159.37**
Surface vertical rod push bar panic bolt, reversible, to suit doors 2500x1100 mm maximum, silver finish, CE marked to EN1125 class 3–7–5–1–1–3–2–2–A	144.07	2.00	46.27	–	148.39	nr	**194.66**
Rim push bar panic latch, reversible, to suit doors 1100 mm wide maximum, silver finish, CE marked to EN1125 class 3–7–5–1–1–3–2–2–A	102.72	1.75	40.49	–	105.80	nr	**146.29**
76 × 51 × 13 mm adjustable heavy roller catch, stainless steel forend and strike complete with satin nickel roller bolt satin chrome	8.12	0.80	18.51	–	8.36	nr	**26.87**
External access device for use with XX10280/2 panic hardware to suit door thickness 45–55 mm, complete with SS3006N lever, SS755 rose, SS796 profile escutcheon and spindle.For use with MA7420 A51 or MA7420 A55 profile cylinders	34.12	1.75	40.49	–	35.14	nr	**75.63**
142 × 22 mm Ø Concealed jamb door closer light duty	18.27	1.05	24.29	–	18.82	nr	**43.11**
80 × 40 × 45 mm Emergency release door stop with holdback facility	86.21	1.35	31.23	–	88.80	nr	**120.03**
Modric anodized aluminium quadaxial lever MA3503 assembly tested to BS EN1906 4/7/-/1/1/4/0/U	31.96	1.05	24.29	–	32.92	pair	**57.21**
Modric anodized aluminium quadaxial lever MA3502 assembly Tested to BS EN1906 4/7/-/1/1/4/0/U	34.07	1.05	24.29	–	35.09	pair	**59.38**
Modric anodized aluminium quadaxial lever assembly Tested to BS EN1906 4/7/-/1/1/4/0/U with Biocote® antibacterial protection	61.44	1.05	24.29	–	63.28	pair	**87.57**
Modric stainless steel quadaxial lever assembly tested to BS EN1906 4/7/-/1/1/4/0/U	58.49	1.05	24.29	–	60.24	pair	**84.53**
152 × 38 × 13 mm Modric anodized aluminium security door chain leather covered	45.58	0.55	12.72	–	46.95	nr	**59.67**

24 DOORS, SHUTTERS AND HATCHES

Item	PC £	Labour hours	Labour £	Plant £	Material £	Unit	Total rate £
24.07 ASSOCIATED IRONMONGERY – cont							
Allgood Ltd or other equal – cont							
Door closers and furniture – cont							
50 Ø × 3 mm Modric anodized aluminium circular covered rose for profile cylinder	4.48	0.15	3.47	–	4.61	nr	**8.08**
50 Ø × 3 mm Modric anodized aluminium circular covered rose with indicator and emergency release	9.24	0.20	4.62	–	9.52	nr	**14.14**
50 Ø × 3 mm Modric anodized aluminium circular covered rose with heavy turn, 5–8 mm spindle	16.21	0.20	4.62	–	16.70	nr	**21.32**
Budget lock escutcheon – satin stainless steel 316	8.83	0.15	3.47	–	9.09	nr	**12.56**
50 Ø × 3 mm Stainless steel circular covered rose for profile cylinder	7.18	0.15	3.47	–	7.40	nr	**10.87**
50 Ø × 3 mm Stainless steel circular covered rose with indicator and emergency release	9.67	0.20	4.62	–	9.96	nr	**14.58**
50 Ø × 3 mm Stainless steel circular covered rose with heavy turn, 5–8 mm spindle	19.94	0.20	4.62	–	20.54	nr	**25.16**
330 × 76 × 1.6 mm Modric anodized aluminium push plate	3.53	0.20	4.62	–	3.64	nr	**8.26**
330 × 76 × 1.6 mm Stainless steel push plate	8.18	0.20	4.62	–	8.43	nr	**13.05**
800 × 150 × 1.5 mm Modric anodized aluminium kicking plate, drilled and countersunk with screws	8.70	0.35	8.10	–	8.96	nr	**17.06**
900 × 150 × 1.5 mm Modric anodized aluminium kicking plate, drilled and countersunk with screws	9.79	0.35	8.10	–	10.08	nr	**18.18**
1000 × 150 × 1.5 mm Modric anodized aluminium kicking plate, drilled & countersunk with screws	10.86	0.35	8.10	–	11.19	nr	**19.29**
800 × 150 × 1.5 mm Stainless steel kicking plate, drilled and countersunk with screws	15.39	0.35	8.10	–	15.85	nr	**23.95**
900 × 150 × 1.5 mm Stainless steel kicking plate, drilled and countersunk with screws	17.31	0.35	8.10	–	17.83	nr	**25.93**
1000 × 150 × 1.5 mm Stainless steel kicking plate, drilled & countersunk with screws	19.24	0.35	8.10	–	19.82	nr	**27.92**
610 × 70 × 19 mm Ø Modric anodized aluminium grab handle bolt through fixing for doors 10 to 55 mm thick	31.45	0.55	12.72	–	32.39	nr	**45.11**
400 × 19 mm Ø Stainless steel D line straight pull handle with M8 threaded holes, fixing centres 300 mm	48.66	0.45	10.41	–	55.47	nr	**65.88**

24 DOORS, SHUTTERS AND HATCHES

Item	PC £	Labour hours	Labour £	Plant £	Material £	Unit	Total rate £
Hinges							
100 × 75 × 3 mm Stainless steel triple knuckle concealed twin Newton bearings, button tipped butt hinges, jig drilled for metal doors/frames, complete with M6 × 12MT 'undercut' machine screws, stainless steel 316 CE marked to EN1935	19.30	0.35	8.10	–	19.88	pair	**27.98**
100 × 100 × 3 mm Stainless steel triple knuckle concealed twin Newton bearings, button tipped hinges, jig drilled, stainless steel grade 316 CE marked to EN1935	34.30	0.35	8.10	–	35.33	pair	**43.43**
Latches							
Modric anodized aluminium round cylinder for rim night latch, 2 keyed satin nickel plated	27.15	0.55	12.72	–	27.96	nr	**40.68**
93 × 75 mm Cylinder rim non-deadlocking night latch case only 60 mm backset	23.26	0.55	12.72	–	23.96	nr	**36.68**
71 series mortice latch, case only, low friction latchbolt, griptight follower, heavy spring for levers. Radius forend and sq strike. CE marked to BS EN12209 3/X/8/1/0G/-/B/02/0	17.58	1.05	24.29	–	18.11	nr	**42.40**
Modric anodized aluminium latch configuration with quadaxial assembly	31.75	1.05	24.29	–	32.70	nr	**56.99**
Modric anodized aluminium Nightlatch configuration with quadaxial assembly and single cylinder	58.91	1.05	24.29	–	60.68	nr	**84.97**
Locks							
44 mm case Bright zinc plated steel mortice budget lock with slotted strike plate 33 mm backset	29.85	1.05	24.29	–	30.75	nr	**55.04**
76 × 58 mm b/s Stainless steel cubicle mortice deadlock with 8 mm follower	14.14	1.05	24.29	–	14.56	nr	**38.85**
'A' length European profile double cylinder lock, 2 keyed satin nickel plated	27.84	1.05	24.29	–	28.68	nr	**52.97**
'A' length European profile cylinder and large turn, 2 keyed satin nickel plated	31.63	1.05	24.29	–	32.58	nr	**56.87**
'A' length European profile cylinder and large turn, 2 keyed under master key, satin nickel plated	29.11	1.05	24.29	–	29.98	nr	**54.27**
'A' length European profile single cylinder, 2 keyed satin nickel plated	21.53	1.05	24.29	–	22.18	nr	**46.47**
'A' length European profile single cylinder, 2 keyed under master key, satin nickel plated	21.53	1.05	24.29	–	22.18	nr	**46.47**

24 DOORS, SHUTTERS AND HATCHES

Item	PC £	Labour hours	Labour £	Plant £	Material £	Unit	Total rate £
24.07 ASSOCIATED IRONMONGERY – cont							
Allgood Ltd or other equal – cont							
Locks – cont							
93 × 60 mm b/s 71 series profile cylinder mortice deadlock, case only. Single throw 22 mm deadbolt. Radius forend and square strike. CE marked to BS EN12209 3/X/8/1/0/G/4/B/A/0/0	17.58	1.05	24.29	–	18.11	nr	**42.40**
92 × 60 mm b/s 71 series bathroom lock, case only, low friction latchbolt, griptight follower, heavy spring for levers, twin 8 mm followers at 78 mm centres. Radius forend and square strike. CE marked to BS EN12209 3/X/8/0/0/G-/B/0/2/0	20.84	1.05	24.29	–	21.47	nr	**45.76**
93 × 60 mm b/s 71 series profile cylinder mortice lock, case only, low friction latchbolt, griptight follower. Heavy spring for levers, 22 mm throw deadbolt, cylinder withdraws bolt bolts. Radius forend and square strike. CE marked to BS EN12209 3/X/8/1/0G/4/B/A2/0	20.84	1.05	24.29	–	21.47	nr	**45.76**
92 × 60 mm b/s 71 series profile cylinder emergency lock, case only. Low friction latchbolt, griptight follower, heavy spring for lever, single throw 22 mm deadbolt, lever can withdraw both bolts. Radius forend and strike	68.56	1.05	24.29	–	70.62	nr	**94.91**
Modric anodized aluminium lock configuration with quadaxial assembly and cylinder with turn	90.75	1.05	24.29	–	93.47	nr	**117.76**
Sundries							
76 mm Ø Modric anodized aluminium circular sex symbol male	4.92	0.10	2.32	–	5.07	nr	**7.39**
76 mm Ø Modric anodized aluminium circular symbol fire door keep locked	4.92	0.10	2.32	–	5.07	nr	**7.39**
76 mm Ø Modric anodized aluminium circular symbol fire door keep shut	4.92	0.15	3.47	–	5.07	nr	**8.54**
38 × 47 mm Ø Modric anodized aluminium heavy circular floor door stop with cover	9.96	0.15	3.47	–	10.26	nr	**13.73**
38 × 47 mm Ø Stainless steel heavy circular floor door stop with cover	17.79	0.15	3.47	–	18.32	nr	**21.79**
63 × 19 mm Ø Modric anodized aluminium circular heavy duty skirting buffer with thief-resistant insert	6.51	0.15	3.47	–	6.71	nr	**10.18**
102 × 25 mm Ø Stainless steel circular heavy duty skirting buffer with thief-resistant insert	11.28	0.15	3.47	–	11.62	nr	**15.09**
152 mm Cabin hook satin chrome on brass	23.60	0.20	4.62	–	24.31	nr	**28.93**
14 mm Ø × 145 × 94 mm Toilet roll holder, length 145 mm, colour white, satin stainless steel 316	65.68	0.20	4.62	–	69.43	nr	**74.05**
Towel rail with bushes, fixing centres 450 mm, satin stainless steel 316	89.27	0.35	8.10	–	97.29	nr	**105.39**
Toilet brush holder with toilet brush, with bushes, satin stainless steel 316	138.62	0.25	5.79	–	146.34	nr	**152.13**

24 DOORS, SHUTTERS AND HATCHES

Item	PC £	Labour hours	Labour £	Plant £	Material £	Unit	Total rate £
Bathline 850 mm lift up support rail	209.97	0.75	17.36	–	216.27	set	**233.63**
Bathline 600 × 95 × 35 mm support rail with concealed fixing roses	107.62	0.50	11.57	–	110.85	set	**122.42**
Bathline 400 × 250 × 35 mm backrest rail with concealed fixing roses	107.62	0.50	11.57	–	110.85	set	**122.42**
ASSA ABLOY Ltd typical door ironmongery sets These are standard doorsets compilations which ASSA ABLOY most commonly supply							
Classroom doorset (fire-rated)	–	2.00	46.27	–	424.27	nr	**470.54**
Ironmongery schedule and cost							
UNION PowerLOAD 603 bushed bearing butt hinges (1.5 pair) with a fixed in and radius corners, 100 mm × 88 mm, satin stainless steel, grade 13 BS EN 1935	9.60	–	–	–	–	nr	–
ASSA ABLOY DC700 A overhead cam motion backcheck closure with slide arm size 2–6. Full cover standard silver	163.84	–	–	–	–	nr	–
UNION keyULTRA euro profile key and turn cylinder, turn/key, 35 mm/35 mm , satin chrome, BS EN 1303 Security Grade 2 when fitted with security escutcheon (sold separately)	39.09	–	–	–	–	nr	–
Union 2S21 euro-profile sashlock, 72 mm centresm, 55 mm backset, satin stainless steel, Grade 3 category of use	11.89	–	–	–	–	nr	–
Union 1000 01 style, 19 mm dia. lever on 8 mm round rose, sprung, satin stainless steel grade 304, bolt through connections	8.20	–	–	–	–	pair	–
7012F ASSA finger guard 2.05 m push side, silver	97.22	–	–	–	–	nr	–
UNION 1000EE Euro Escutcheon, 54 mm dia. × 8 mm projection, satin stainless steel	0.65	–	–	–	–	nr	–
Kick Plate 900 mm × 150 mm, satin stainless steel softened & finished edges with radius corners	18.14	–	–	–	–	nr	–
'Fire Door Keep Shut' 75 mm dia. sign, satin stainless steel	5.05	–	–	–	–	nr	–
UNION DS100 Floor Mounted Door Stop, half moon design, 45 mm dia. × 24 mm high, satin stainless steel	2.46	–	–	–	–	nr	–
Non-locking office doorset (fire-rated)	–	2.00	46.27	–	269.46	nr	**315.73**
Ironmongery schedule and cost							
UNION PowerLOAD 603 bushed bearing butt hinges (1.5 pair) with a fixed pin and radius corners, 100 mm × 88 mm, satin stainless steel, grade 13 BS EN 1935	28.79	–	–	–	–	nr	–
ASSA ABLOY DC700 A overhead cam motion backcheck closure with slide arm size 2–6. Full cover standard silver	163.84	–	–	–	–	nr	–

24 DOORS, SHUTTERS AND HATCHES

Item	PC £	Labour hours	Labour £	Plant £	Material £	Unit	Total rate £
24.07 ASSOCIATED IRONMONGERY – cont							
Non-locking office doorset (fire-rated) – cont							
Ironmongery schedule and cost – cont							
UNION keyULTRA euro profile key and turn cylinder, turn/key, 35 mm/35 mm, satin chrome, BS EN 1303 Security Grade 2 when fitted with security escutcheon (sold separately)	39.09	–	–	–	–	nr	–
UNION 2C23 DIN mortice latch, 55 mm backset, satin stainless steel, radius forend, Grade 3 Category of Use, BS EN 12209, CE marked	11.94	–	–	–	–	nr	–
Union 1000 01 style, 19 mm dia. lever on 8 mm round rose, sprung, satin stainless steel grade 304, bolt through connections	8.20	–	–	–	–	pair	–
Kick Plate 900 mm × 150 mm, satin stainless steel softened & finished edges with radius corners	18.14	–	–	–	–	nr	–
'Fire Door Keep Shut' 75 mm dia. sign, satin stainless steel	5.05	–	–	–	–	nr	–
UNION DS100 Floor Mounted Door Stop, half moon design, 45 mm dia. × 24 mm high, satin stainless steel	2.46	–	–	–	–	nr	–
Office/Store locking doorset (fire-rated)	–	2.00	46.27	–	311.01	nr	**357.28**
Ironmongery schedule and cost							
UNION PowerLOAD 603 bushed bearing butt hinges (1.5 pair) with a fixed pin and radius corners, 100 mm × 88 mm, satin stainless steel, grade 13 BS EN 1935	28.79	–	–	–	–	nr	–
ASSA ABLOY DC700 A overhead cam motion backcheck closure with slide arm size 2–6. Full cover standard silver	163.84	–	–	–	–	nr	–
UNION keyULTRA euro profile key and turn cylinder, turn/key, 35 mm/35 mm, satin chrome, BS EN 1303 Security Grade 2 when fitted with security escutcheon (sold separately)	39.09	–	–	–	–	nr	–
Union 2S21 euro-profile sashlock, 72 mm centresm, 55 mm backset, satin stainless steel, Grade 3 category of use	11.89	–	–	–	–	nr	–
Union 1000 01 style, 19 mm dia. lever on 8 mm round rose, sprung, satin stainless steel grade 304, bolt through connections	8.20	–	–	–	–	pair	–
Kick Plate 200 mm × 1.5 mm, satin stainless steel softened & finished edges with radius corners	18.14	–	–	–	–	nr	–
'Fire Door Keep Shut' 75 mm dia. sign, satin stainless steel	5.05	–	–	–	–	nr	–
UNION DS100 Floor Mounted Door Stop, half moon design, 45 mm dia. × 24 mm high, satin stainless steel	2.46	–	–	–	–	nr	–

24 DOORS, SHUTTERS AND HATCHES

Item	PC £	Labour hours	Labour £	Plant £	Material £	Unit	Total rate £
Common room locking doorset (fire-rated)	–	2.50	57.83	–	433.56	nr	**491.39**
Ironmongery schedule and cost							
UNION PowerLOAD 603 bushed bearing butt hinges (1.5 pair) with a fixed pin and radius corners, 100 mm × 88 mm, satin stainless steel, Grade 13 BS EN 1935	28.79	–	–	–	–	nr	–
ASSA ABLOY DC700 A overhead cam motion backcheck closure with slide arm size 2–6. Full cover standard silver	163.84	–	–	–	–	nr	–
UNION keyULTRA euro profile key and turn cylinder, turn/key (Large), 35 mm/35 mm, 6pin, satin chrome, BS EN 1303 Security Grade 2 when fitted with security escutcheon (sold separately)	51.82	–	–	–	–	nr	–
UNION 2C22 DIN euro-profile deadlock, 55 mm backset, satin stainless steel, radius forend, Grade 3 Category of Use, BS EN 12209, CE marked	17.37	–	–	–	–	nr	–
UNION 01 style pull handle 19 mm dia. × 425 mm centres, satin stainless steel, Grade 304, bolt through fixing	5.22	–	–	–	–	nr	–
Kick Plate 200 mm × 1.5 mm, satin stainless steel softened & finished edges with radius corners	18.14	–	–	–	–	nr	–
Push Plate 450 mm × 100 mm × 1.5 mm satin stainless steel, softened & finished edges with radius corners	2.05	–	–	–	–	nr	–
7012F ASSA finger guard 2.05 m push side, silver	97.22	–	–	–	–	nr	–
UNION 1000EE Euro Escutcheon, 54 mm dia. × 8 mm projection, satin stainless steel	0.65	–	–	–	–	nr	–
UNION DS100 Floor Mounted Door Stop, Half Moon Design, 45 mm Dia. × 24 mm high, Satin Stainless Steel	2.46	–	–	–	–	nr	–
'Fire Door Keep Shut' 75 mm dia. sign, satin stainless steel	5.05	–	–	–	–	nr	–
PULL or PUSH signage, 75 mm dia., satin stainless steel	2.24	–	–	–	–	nr	–
Maintenance/Plant room doorset (fire-rated)	–	1.75	40.49	–	297.41	nr	**337.90**
Ironmongery schedule and cost							
1 ½ pairs Union Powerload 603 brushed bearing butt hinge with a fixed pin; 100 mm × 88 mm; satin stainless steel grade 13 BS EN 1935	28.79	–	–	–	–	nr	–
ASSA ABLOY DC700 A overhead cam motion backcheck closure with slide arm size 2–6. Full cover standard silver	163.84	–	–	–	–	nr	–
UNION 2C22 DIN euro-profile deadlock, 55 mm backset, satin stainless steel, radius forend, Grade 3 Category of Use, BS EN 12209, CE marked	17.37	–	–	–	–	nr	–

24 DOORS, SHUTTERS AND HATCHES

Item	PC £	Labour hours	Labour £	Plant £	Material £	Unit	Total rate £
24.07 ASSOCIATED IRONMONGERY – cont							
Maintenance/Plant room doorset (fire-rated) – cont							
Ironmongery schedule and cost – cont							
UNION keyULTRA euro profile single cylinder, 35 mm/35 mm, satin chrome, BS EN 1303 Security Grade 2 when fitted with security escutcheon	51.82	–	–	–	–	nr	–
Kick Plate 900 mm × 150 mm, satin stainless steel softened & finished edges with Radius Corner	18.14	–	–	–	–	nr	–
'Fire Door Keep Locked' 75 mm Dia. Sign, Satin Stainless Steel	5.05	–	–	–	–	nr	–
Euro Rim Cylinder Pull, stainless steel	0.61	–	–	–	–	nr	–
UNION 1000EE Euro Escutcheon, 54 mm dia. × 8 mm projection, satin stainless steel	0.65	–	–	–	–	nr	–
UNION DS100 Floor Mounted Door Stop, half moon design, 45 mm Dia. × 24 mm high, satin stainless steel	2.46	–	–	–	–	nr	–
Standard bathroom doorset (Unisex)	–	1.78	41.29	–	236.79	nr	**278.08**
Ironmongery schedule and cost							
1 ½ pairs Union Powerload 603 brushed bearing butt hinge with a fixed pin; 100 mm × 88 mm; satin stainless steel Grade 13 BS EN 1935	28.79	–	–	–	–	nr	–
ASSA ABLOY DC700 A overhead cam motion backcheck closure with slide arm size 2–6. Full cover standard silver	163.84	–	–	–	–	nr	–
UNION 2C27 DIN bathroom lock, 78 mm centres, 55 mm backset, satin stainless steel, radius forend, Grade 3 Category of Use, BS EN 12209, CE marked	12.35	–	–	–	–	nr	–
Union 1000 01 style, 19 mm dia. lever on 8 mm round rose, sprung, satin stainless steel grade 304, bolt through connections	8.20	–	–	–	–	pair	–
UNION 1000ER emergency release, 54 mm dia. × 8 mm projection, satin stainless steel	3.94	–	–	–	–	nr	–
UNION 1000T standard turn, 54 mm dia. × 8 mm projection, satin stainless steel	5.86	–	–	–	–	nr	–
UNION DS100 Floor Mounted Door Stop, half moon design, 45 mm dia. × 24 mm high, satin stainless steel	2.46	–	–	–	–	nr	–
UNION US200 Unisex Symbol, 75 mm dia., satin stainless steel, Grade 304	2.24	–	–	–	–	nr	–
UNION HC100 hat & coat hook with a buffer, satin stainless steel	2.21	–	–	–	–	nr	–

24 DOORS, SHUTTERS AND HATCHES

Item	PC £	Labour hours	Labour £	Plant £	Material £	Unit	Total rate £
Accessible toilet doorset	–	2.55	59.00	–	102.58	nr	**161.58**
Ironmongery schedule and cost							
1 ½ pairs Union Powerload 603 brushed bearing butt hinge with a fixed pin; 100 mm × 88 mm; satin stainless steel Grade 13 BS EN 1935	28.79	–	–	–	–	nr	–
UNION 2C27 DIN bathroom lock, 78 mm centres, 55 mm backset, satin stainless steel, radius forend, Grade 3 Category of Use, BS EN 12209, CE marked	12.35	–	–	–	–	nr	–
Union 1000 01 style, 19 mm dia. lever on 8 mm round rose, sprung, satin stainless steel grade 304, bolt through connections	8.20	–	–	–	–	pair	–
Kick plates 900 mm × 150 mm, stainless steel	18.14	–	–	–	–	nr	–
UNION WS200 Disabled Symbol, 75 mm dia., satin stainless steel, Grade 304	2.24	–	–	–	–	nr	–
UNION 1000ER emergency release, 54 mm dia. × 8 mm projection, satin stainless steel	3.94	–	–	–	–	nr	–
UNION 1000LT Large Turn, 54 mm dia. × 8 mm projection, satin stainless steel	3.14	–	–	–	–	nr	–
UNION DS100 Floor Mounted Door Stop, half moon design, 45 mm dia. × 24 mm high, satin stainless steel	2.46	–	–	–	–	nr	–
UNION HC100 hat & coat hook with a buffer, satin stainless steel	2.21	–	–	–	–	nr	–
Single Lobby/Corridor doorset (fire-rated)	–	2.25	52.06	–	360.94	nr	**413.00**
Ironmongery schedule and cost							
UNION PowerLOAD 603 bushed bearing butt hinges (1.5 pair) with a fixed pin and radius corners, 100 mm × 88 mm, satin stainless steel, Grade 13 BS EN 1935	28.79	–	–	–	–	nr	–
UNION 01 style pull handle 19 mm dia. × 425 mm centres, satin stainless steel, Grade 304, bolt through fixing	5.22	–	–	–	–	nr	–
ASSA ABLOY DC700 A overhead cam motion backcheck closure with slide arm size 2–6. Full cover standard silver	163.84	–	–	–	–	nr	–
7012F ASSA finger guard 2.05 m push side, silver	97.22	–	–	–	–	nr	–
Kick Plate 900 mm × 150 mm, satin stainless steel softened & finished edges with radius corners	18.14	–	–	–	–	nr	–
'Fire Door Keep Shut' 75 mm dia. sign, satin stainless steel	5.05	–	–	–	–	nr	–
PULL or PUSH signage, 75 mm dia., satin stainless steel	2.24	–	–	–	–	nr	–
UNION DS100 Floor Mounted Door Stop, half moon design, 45 mm dia. × 24 mm high, satin stainless steel	2.46	–	–	–	–	nr	–

24 DOORS, SHUTTERS AND HATCHES

Item	PC £	Labour hours	Labour £	Plant £	Material £	Unit	Total rate £
24.07 ASSOCIATED IRONMONGERY – cont							
Double lobby/corridor doorset (fire-rated)	–	4.00	92.55	–	745.50	nr	**838.05**
Ironmongery schedule and cost							
UNION PowerLOAD 603 bushed bearing butt hinges (3 pair) with a fixed pin and radius corners, 100 mm × 88 mm, satin stainless steel, Grade 13 BS EN 1935	57.58	–	–	–	–	nr	–
UNION 8899 Electro Mag closer hold open/swing free, fixed strength size 4, painted silver, app 1, 61 and 66, BS EN 1155, CE Marked, Certifire approved. Supplied with a curve cover c/w matching arms.	175.31	–	–	–	–	nr	–
UNION 01 style pull handle 19 mm dia. × 425 mm centres, satin stainless steel, Grade 304, bolt through fixing	5.22	–	–	–	–	nr	–
Push Plate 450 mm × 100 mm × 1.5 mm satin stainless steel, softened & finished edges with radius corners	2.05	–	–	–	–	nr	–
7012F ASSA finger guard 2.05 m push side, silver	97.22	–	–	–	–	nr	–
Kick Plate 900 mm × 150 mm, satin stainless steel softened & finished edges with radius corners	18.14	–	–	–	–	nr	–
'Fire Door Keep Shut' 75 mm dia. sign, satin stainless steel	5.05	–	–	–	–	nr	–
PULL or PUSH signage, 75 mm dia., satin stainless steel	2.24	–	–	–	–	nr	–
UNION DS100 Floor Mounted Door Stop, half moon design, 45 mm dia. × 24 mm high, satin stainless steel	2.46	–	–	–	–	nr	–
External rebated double fire escape doorset	–	4.00	92.55	–	1633.85	nr	**1726.40**
Ironmongery schedule and cost							
UNION PowerLOAD 603 bushed bearing butt hinges (1.5 pair) with a fixed pin and radius corners, 100 mm × 88 mm, satin stainless steel, Grade 13 BS EN 1935	28.79	–	–	–	–	nr	–
UNION satin stainless steel, Ball Bearing Dog Bolt Hinge	145.77	–	–	–	–	nr	–
ASSA ABLOY DC700 A overhead cam motion backcheck closure with slide arm size 2–6. Full cover standard silver	163.84	–	–	–	–	nr	–

24 DOORS, SHUTTERS AND HATCHES

Item	PC £	Labour hours	Labour £	Plant £	Material £	Unit	Total rate £
UNION 883T panic exit double rebated doorset, 3 point locking with pullman latches, touch bar design, 900 mm maximum door width, silver, EN1125, CE marked	205.88	–	–	–	–	nr	–
UNION 885 outside access device (OAD), lever handle to suit 880 series only, silver, uses a UNION 2X50 5 pin euro profile cylinder to be ordered separately	50.19	–	–	–	–	nr	–
UNION keyULTRA euro profile single cylinder, 35 mm/35 mm, satin chrome, BS EN 1303 Security Grade 2 when fitted with security escutcheon	51.82	–	–	–	–	nr	–
Generic door door limiting stays	44.31	–	–	–	–	nr	–
Kick Plate 900 mm × 150 mm, satin stainless steel softened & finished edges with radius corners	18.14	–	–	–	–	nr	–
'Fire Door Keep Shut' 75 mm dia. sign, satin stainless steel	5.05	–	–	–	–	nr	–
SP(102)FA1201 – Aluminium	29.77	–	–	–	–	nr	–
Double aluminium entrance doorset	–	5.00	115.68	–	750.50	nr	**866.18**
Ironmongery schedule and cost							
Adams Rite SENTINAL Radius Weatherstrip Faceplate, satin anodized aluminium	5.97	–	–	–	–	nr	–
Adams Rite 4700 Series Faceplate Radius w/s, satin anodized aluminium	4.99	–	–	–	–	nr	–
Adams Rite SENTINEL 6 Deadlock, 30 mm Backset BS EN 12209	16.24	–	–	–	–	nr	–
ADAMS RITE SENTINEL Armoured Strike Plate	5.69	–	–	–	–	nr	–
ADAMS RITE SENTINEL Radius Armoured Trim Strike, satin anodized aluminium	10.98	–	–	–	–	nr	–
ADAMS RITE 4596 Euro Paddle Handle, Pull to Left 45 mm Door Satin Anodized Aluminium	83.53	–	–	–	–	nr	–
ADAMS RITE 4750 Deadlatch EuroProfile 28 mm backset body only – left hand	43.73	–	–	–	–	nr	–
ADAMS RITE 2 Point Latch Push to Left, satin anodized aluminium	162.28	–	–	–	–	nr	–
ADAMS RITE Transom ARC–51N EN3 Spring Closer HO side load 70 mm pivot	80.43	–	–	–	–	nr	–
ADAMS RITE 7108 12vDC Fail Safe Escape Electric Release, Satin Anodized Aluminium. Specifer to confirm depth of section as could clash with mechanism of 4781.	190.40	–	–	–	–	nr	–
ADAMS RITE Sentinal Euro Profile Security Escutcheon	43.98	–	–	–	–	nr	–

24 DOORS, SHUTTERS AND HATCHES

Item	PC £	Labour hours	Labour £	Plant £	Material £	Unit	Total rate £
24.07 ASSOCIATED IRONMONGERY – cont							
External toilet lobby doorset (fire-rated)	–	2.50	57.83	–	435.19	nr	**493.02**
Ironmongery schedule and cost							
1 ½ pairs Union Powerload 603 brushed bearing butt hinge with a fixed pin; 100 mm × 88 mm; satin stainless steel Grade 13 BS EN 1935	28.79	–	–	–	–	nr	–
ASSA ABLOY DC700 A overhead cam motion backcheck closure with slide arm size 2–6. Full cover standard silver	163.84	–	–	–	–	nr	–
UNION 2C22 DIN euro-profile deadlock, 55 mm backset, satin stainless steel, radius forend, Grade 3 Category of Use, BS EN 12209, CE marked	17.37	–	–	–	–	nr	–
UNION 01 style pull handle 19 mm dia. × 425 mm centres, satin stainless steel, Grade 304, bolt through fixing	5.22	–	–	–	–	nr	–
UNION keyULTRA euro profile single cylinder, 35 mm/35 mm, satin chrome, BS EN 1303 Security Grade 2 when fitted with security escutcheon	51.82	–	–	–	–	nr	–
Push Plate 450 mm × 100 mm × 1.5 mm satin stainless steel, softened & finished edges with radius corners	2.05	–	–	–	–	nr	–
Kick Plate 900 mm × 150 mm, satin stainless steel softened & finished edges with radius corner	18.14	–	–	–	–	nr	–
7012F ASSA finger guard 2.05 m push side, silver	97.22	–	–	–	–	nr	–
'Fire Door Keep Locked' 75 mm dia. sign, satin stainless steel	5.05	–	–	–	–	nr	–
UNION 1000EE Euro Escutcheon, 54 mm dia. × 8 mm projection, satin stainless steel	0.65	–	–	–	–	nr	–
UNION DS100 Floor Mounted Door Stop, half moon design, 45 mm dia. × 24 mm high, satin stainless steel	2.46	–	–	–	–	nr	–
PULL or PUSH signage, 75 mm dia., satin stainless steel	2.24	–	–	–	–	nr	–
UNION US200 Unisex Symbol, 75 mm dia., satin stainless steel, Grade 304	2.24	–	–	–	–	nr	–
Internal toilet lobby doorset (non fire-rated)	–	2.00	46.27	–	250.41	nr	**296.68**
Ironmongery schedule and cost							
1 ½ pairs Union Powerload 603 brushed bearing butt hinge with a fixed pin; 100 mm × 88 mm; satin stainless steel Grade 13 BS EN 1935	28.79	–	–	–	–	nr	–
ASSA ABLOY DC700 A overhead cam motion backcheck closure with slide arm size 2–6. Full cover standard silver	163.84	–	–	–	–	nr	–
UNION 01 style pull handle 19 mm dia. × 425 mm centres, satin stainless steel, Grade 304, bolt through fixing	5.22	–	–	–	–	nr	–
Push Plate 450 mm × 100 mm × 1.5 mm satin stainless steel, softened & finished edges with radius corners	2.05	–	–	–	–	nr	–

24 DOORS, SHUTTERS AND HATCHES

Item	PC £	Labour hours	Labour £	Plant £	Material £	Unit	Total rate £
Kick Plate 900 mm × 150 mm, satin stainless steel softened & finished edges with radius corners	18.14	–	–	–	–	nr	–
UNION DS100 Floor Mounted Door Stop, half moon design, 45 mm dia. × 24 mm high, satin stainless steel	2.46	–	–	–	–	nr	–
PULL or PUSH signage, 75 mm dia., satin stainless steel	2.24	–	–	–	–	nr	–
Double classroom store doorset (non fire-rated)	–	1.50	34.70	–	364.75	nr	**399.45**
Ironmongery schedule and cost							
UNION PowerLOAD 603 bushed bearing butt hinges (3 pair) with a fixed pin and radius corners, 100 mm × 88 mm, satin stainless steel, Grade 13 BS EN 1935	28.79	–	–	–	–	nr	–
Union 2S21 euro-profile sashlock, 72 mm centresm, 55 mm backset, satin stainless steel, Grade 3 category of use	11.89	–	–	–	–	nr	–
UNION keyULTRA euro profile key and turn cylinder, turn/key (Large), 35 mm/35 mm, 6pin, satin chrome, BS EN 1303 Security Grade 2 when fitted with security escutcheon (sold separately)	51.82	–	–	–	–	nr	–
Union 1000 01 style, 19 mm dia. lever on 8 mm round rose, sprung, satin stainless steel grade 304, bolt through connections	8.20	–	–	–	–	pair	–
UNION 1000EE Euro Escutcheon, 54 mm dia. × 8 mm projection, Satin Stainless Steel	0.65	–	–	–	–	nr	–
Kick Plate 900 mm × 150 mm, satin stainless steel softened & finished edges with radius corners	18.14	–	–	–	–	nr	–
UNION 8060 aluminium lever action flush bolt 200 mm satin stainless steel satin chrome	7.79	–	–	–	–	pair	–
Sliding door gear; Hillaldam Coburn Ltd or other equal and approved; Commercial/Light industrial; for top hung timber/metal doors, weight not exceeding 365 kg							
Sliding door gear							
bottom guide; fixed to concrete in groove	28.68	0.46	10.64	–	29.54	m	**40.18**
top track	38.99	0.23	5.33	–	40.16	m	**45.49**
detachable locking bar	48.78	0.31	7.17	–	50.24	nr	**57.41**
hangers; timber doors	78.70	0.46	10.64	–	81.06	nr	**91.70**
hangers; metal doors	50.29	0.46	10.64	–	51.80	nr	**62.44**
head brackets; open, soffit fixing; screwing to timber	10.03	0.32	7.41	–	10.35	nr	**17.76**
head brackets; open, side fixing; bolting to masonry	10.53	0.46	10.64	–	12.01	nr	**22.65**
door guide to timber door	8.82	0.23	5.33	–	9.08	nr	**14.41**
door stop; rubber buffers; to masonry	38.89	0.69	15.97	–	40.06	nr	**56.03**
drop bolt; screwing to timber	34.00	0.46	10.64	–	35.02	nr	**45.66**
bow handle; to timber	13.48	0.23	5.33	–	13.88	nr	**19.21**
Sundries							
20 mm rubber door stop; plugged and screwed to concrete	2.20	0.09	2.08	–	2.27	nr	**4.35**

25 STAIRS, WALKWAYS AND BALUSTRADES

Item	PC £	Labour hours	Labour £	Plant £	Material £	Unit	Total rate £
25.01 STAIRS							
Standard staircases; wrought softwood (parana pine)							
Stairs; 25 mm thick treads with rounded nosings; 9 mm thick plywood risers; 32 mm thick strings; bullnose bottom tread; 50 mm × 75 mm hardwood handrail; 32 mm square plain balusters; 100 mm square plain newel posts							
straight flight; 838 mm wide; 2676 mm going; 2600 mm rise; with two newel posts	–	6.48	149.92	–	808.39	nr	**958.31**
dogleg staircase; 838 mm wide; 2676 mm going; 2600 mm rise; with two newel posts; quarter space landing third riser from top	–	6.48	149.92	–	873.28	nr	**1023.20**
dogleg staircase; 838 mm wide; 2676 mm going; 2600 mm rise; with two newel posts; half space landing third riser from top	–	7.40	171.21	–	1041.99	nr	**1213.20**
Hardwood staircases; purpose-made; assembled at works							
Fixing only complete staircase including landings, balustrades, etc.							
plugging and screwing to brickwork or blockwork	–	13.88	321.12	–	3.17	nr	**324.29**
Spiral staircases, balustrades and handrails; mild steel; galvanized and polyester powder coated							
Staircase							
2080 mm dia. × 3695 mm high; 18 nr treads; 16 mm dia. intermediate balusters; 1040 mm × 1350 mm landing unit with matching balustrade both sides; fixing with 16 mm dia. resin anchors to masonry at landing and with 12 mm dia. expanding bolts to concrete at base	–	–	–	–	–	nr	**20000.00**
The following are supply only prices for purpose-made staircase components in selected Sapele supplied as part of an assembled staircase and may be used to arrive at a guide price for a complete hardwood staircase							
Board landings; cross-tongued joints; 100 mm × 50 mm sawn softwood bearers							
25 mm thick	–	–	–	–	104.43	m²	**104.43**
32 mm thick	–	–	–	–	117.38	m²	**117.38**
Treads; cross-tongued joints and risers; rounded nosings; tongued, grooved, glued and blocked together; one 175 mm × 50 mm sawn softwood carriage							
25 mm treads; 19 mm risers	–	–	–	–	209.88	m²	**209.88**
ends; quadrant	–	–	–	–	63.93	nr	**63.93**
ends; housed to hardwood	–	–	–	–	1.18	nr	**1.18**
32 mm treads; 25 mm risers	–	–	–	–	217.42	m²	**217.42**
ends; quadrant	–	–	–	–	82.17	nr	**82.17**
ends; housed to hardwood	–	–	–	–	1.18	nr	**1.18**

25 STAIRS, WALKWAYS AND BALUSTRADES

Item	PC £	Labour hours	Labour £	Plant £	Material £	Unit	Total rate £
Winders; cross-tongued joints and risers in one width; rounded nosings; tongued, grooved glued and blocked together; one 175 mm × 50 mm sawn softwood carriage							
25 mm treads; 19 mm risers	–	–	–	–	291.74	m²	**291.74**
32 mm treads; 25 mm risers	–	–	–	–	298.61	m²	**298.61**
wide ends; housed to hardwood	–	–	–	–	2.36	nr	**2.36**
narrow ends; housed to hardwood	–	–	–	–	1.77	nr	**1.77**
Closed strings; in one width; 230 mm wide; rounded twice							
32 mm thick	–	–	–	–	38.43	m	**38.43**
38 mm thick	–	–	–	–	41.85	m	**41.85**
50 mm thick	–	–	–	–	46.63	m	**46.63**
Closed strings; cross-tongued joints; 280 mm wide; once rounded							
32 mm thick	–	–	–	–	49.88	m	**49.88**
extra for short ramp	–	–	–	–	25.63	nr	**25.63**
38 mm thick	–	–	–	–	54.41	m	**54.41**
extra for short ramp	–	–	–	–	29.13	nr	**29.13**
50 mm thick	–	–	–	–	60.72	m	**60.72**
extra for short ramp	–	–	–	–	36.06	nr	**36.06**
The following labours are irrespective of timber width							
ends; fitted	–	–	–	–	1.51	nr	**1.51**
ends; framed	–	–	–	–	8.94	nr	**8.94**
extra for tongued heading joint	–	–	–	–	4.41	nr	**4.41**
Closed strings; ramped; crossed tongued joints 280 mm wide; once rounded							
32 mm thick	–	–	–	–	49.88	m	**49.88**
44 mm thick	–	–	–	–	54.41	m	**54.41**
57 mm thick	–	–	–	–	60.72	m	**60.72**
Apron linings; in one width 230 mm wide							
19 mm thick	–	–	–	–	13.26	m	**13.26**
25 mm thick	–	–	–	–	15.64	m	**15.64**
The following are supply only prices for purpose-made staircase components in selected American Oak; supplied as part of an assembled staircase							
Board landings; cross-tongued joints; 100 mm × 50 mm sawn softwood bearers							
25 mm thick	–	–	–	–	154.06	m²	**154.06**
32 mm thick	–	–	–	–	186.03	m²	**186.03**
Treads; cross-tongued joints and risers; rounded nosings; tongued, grooved, glued and blocked together; one 175 mm × 50 mm sawn softwood carriage							
25 mm treads; 19 mm risers	–	–	–	–	255.75	m²	**255.75**
ends; quadrant	–	–	–	–	127.76	nr	**127.76**
ends; housed to hardwood	–	–	–	–	1.56	nr	**1.56**
32 mm treads; 25 mm risers	–	–	–	–	293.05	m²	**293.05**
ends; quadrant	–	–	–	–	157.26	nr	**157.26**
ends; housed to hardwood	–	–	–	–	1.56	nr	**1.56**

25 STAIRS, WALKWAYS AND BALUSTRADES

Item	PC £	Labour hours	Labour £	Plant £	Material £	Unit	Total rate £
25.01 STAIRS – cont							
The following are supply only prices for purpose-made staircase components in selected American Oak – cont							
Winders; cross-tongued joints and risers in one width; rounded nosings; tongued, grooved glued and blocked together; one 175 mm × 50 mm sawn softwood carriage							
25 mm treads; 19 mm risers	–	–	–	–	324.19	m²	**324.19**
32 mm treads; 25 mm risers	–	–	–	–	353.72	m²	**353.72**
wide ends; housed to hardwood	–	–	–	–	3.15	nr	**3.15**
narrow ends; housed to hardwood	–	–	–	–	2.37	nr	**2.37**
Closed strings; in one width; 230 mm wide; rounded twice							
32 mm thick	–	–	–	–	60.42	m	**60.42**
44 mm thick	–	–	–	–	69.74	m	**69.74**
57 mm thick	–	–	–	–	95.74	m	**95.74**
Closed strings; cross-tongued joints; 280 mm wide; once rounded							
32 mm thick	–	–	–	–	76.77	m	**76.77**
extra for short ramp	–	–	–	–	43.90	nr	**43.90**
38 mm thick	–	–	–	–	88.96	m	**88.96**
extra for short ramp	–	–	–	–	50.00	nr	**50.00**
50 mm thick	–	–	–	–	121.99	m	**121.99**
extra for short ramp	–	–	–	–	66.51	nr	**66.51**
Closed strings; ramped; crossed tongued joints 280 mm wide; once rounded							
32 mm thick	–	–	–	–	88.29	m	**88.29**
44 mm thick	–	–	–	–	102.32	m	**102.32**
57 mm thick	–	–	–	–	140.28	m	**140.28**
Apron linings; in one width 230 mm wide							
19 mm thick	–	–	–	–	20.57	m	**20.57**
25 mm thick	–	–	–	–	25.21	m	**25.21**
Handrails; rounded							
40 mm × 50 mm	–	–	–	–	16.67	m	**16.67**
50 mm × 75 mm	–	–	–	–	21.36	m	**21.36**
57 mm × 87 mm	–	–	–	–	32.09	m	**32.09**
69 mm × 100 mm	–	–	–	–	43.20	m	**43.20**
Handrails; moulded							
40 mm × 50 mm	–	–	–	–	18.30	m	**18.30**
50 mm × 75 mm	–	–	–	–	22.98	m	**22.98**
57 mm × 87 mm	–	–	–	–	33.70	m	**33.70**
69 mm × 100 mm	–	–	–	–	44.79	m	**44.79**
Add to above for							
grooved once	–	–	–	–	0.82	m	**0.82**
ends; framed	–	–	–	–	8.26	nr	**8.26**
ends; framed on rake	–	–	–	–	10.46	nr	**10.46**

25 STAIRS, WALKWAYS AND BALUSTRADES

Item	PC £	Labour hours	Labour £	Plant £	Material £	Unit	Total rate £
Heading joints to handrail; mitred or raked							
overall size not exceeding 50 mm × 75 mm	–	–	–	–	44.09	nr	**44.09**
overall size not exceeding 69 mm × 100 mm	–	–	–	–	52.33	nr	**52.33**
Knee piece to handrail; mitred or raked							
overall size not exceeding 69 mm × 100 mm	–	–	–	–	93.68	nr	**93.68**
Balusters; stiffeners							
25 mm × 25 mm	–	–	–	–	3.81	m	**3.81**
32 mm × 32 mm	–	–	–	–	4.80	m	**4.80**
44 mm × 44 mm	–	–	–	–	7.58	m	**7.58**
ends; housed	–	–	–	–	1.92	nr	**1.92**
Sub-rails							
32 mm × 63 mm	–	–	–	–	10.27	m	**10.27**
ends; framed joint to newel	–	–	–	–	8.26	nr	**8.26**
Knee rails							
32 mm × 140 mm	–	–	–	–	17.83	m	**17.83**
ends; framed joint to newel	–	–	–	–	8.26	nr	**8.26**
Newel posts							
44 mm × 94 mm; half newel	–	–	–	–	13.37	m	**13.37**
69 mm × 69 mm	–	–	–	–	22.79	m	**22.79**
94 mm × 94 mm	–	–	–	–	56.88	m	**56.88**
Newel caps; splayed on four sides							
62.50 mm × 125 mm × 50 mm	–	–	–	–	11.47	nr	**11.47**
100 mm × 100 mm × 50 mm	–	–	–	–	12.12	nr	**12.12**
125 mm × 125 mm × 50 mm	–	–	–	–	13.33	nr	**13.33**
25.02 WALKWAYS							
Flooring; metalwork							
Chequer plate flooring; galvanized mild steel; over 300 mm wide; bolted to steel supports							
6 mm thick	–	–	–	–	–	m²	**316.15**
8 mm thick	–	–	–	–	–	m²	**328.31**
Open mesh flooring; galvanized; over 300 mm wide; bolted to steel supports							
8 mm thick	–	–	–	–	–	m²	**316.15**
Surface treatment							
At works							
galvanizing	–	–	–	–	–	tonne	**419.51**
shotblasting	–	–	–	–	–	m²	**5.47**
touch up primer and one coat of two pack epoxy zinc phosphate or chromate primer	–	–	–	–	–	m²	**10.95**
25.03 BALUSTRADES							
Standard balustrades; wrought softwood							
Landing balustrade; 50 mm × 75 mm hardwood handrail; 32 mm square plain balusters; one end of handrail jointed to newel post; other end built into wall; balusters housed in at bottom (newel post and mortices both not included)							
3.00 m long	–	3.70	85.60	–	140.59	nr	**226.19**

25 STAIRS, WALKWAYS AND BALUSTRADES

Item	PC £	Labour hours	Labour £	Plant £	Material £	Unit	Total rate £
25.03 BALUSTRADES – cont							
Steel							
Balustrades; galvanized mild steel CHS posts and top rail, with one infill rail							
1100 mm high	–	–	–	–	–	m	**255.35**
Balustrades; painted mild steel flat bar posts and CHS top rail, with 3 nr. stainless steel infills							
1100 mm high	–	–	–	–	–	m	**364.80**
bends	–	–	–	–	–	nr	**32.83**
ends	–	–	–	–	–	nr	**21.88**
Stainless steel							
Balustrades; stainless steel flat bar posts and circular handrail, with 3 nr. stainless steel infills							
1100 mm high	–	–	–	–	–	m	**437.75**
Balustrades; stainless steel 50 mm Ø posts and circular handrail, with 10 mm thick toughened glass infill panels							
1100 mm high	–	–	–	–	–	m	**729.58**
Glass							
Balustrades; laminated glass; with stainless steel cap channel to top and including all necessary support fixings							
1100 mm high	–	–	–	–	–	m	**729.58**
Softwood handrails							
Handrails; rounded							
40 mm × 50 mm	–	–	–	–	14.41	m	**14.41**
50 mm × 75 mm	–	–	–	–	17.37	m	**17.37**
57 mm × 87 mm	–	–	–	–	20.38	m	**20.38**
69 mm × 100 mm	–	–	–	–	25.33	m	**25.33**
Handrails; moulded							
40 mm × 50 mm	–	–	–	–	16.02	m	**16.02**
50 mm × 75 mm	–	–	–	–	18.98	m	**18.98**
57 mm × 87 mm	–	–	–	–	22.00	m	**22.00**
69 mm × 100 mm	–	–	–	–	26.94	m	**26.94**
Add to above for							
grooved once	–	–	–	–	0.71	m	**0.71**
ends; framed	–	–	–	–	6.73	nr	**6.73**
ends; framed on rake	–	–	–	–	8.26	nr	**8.26**
Heading joints to handrail; mitred or raked							
overall size not exceeding 50 mm × 75 mm	–	–	–	–	33.06	nr	**33.06**
overall size not exceeding 69 mm × 100 mm	–	–	–	–	41.33	nr	**41.33**
Knee piece to handrail; mitred or raked							
overall size not exceeding 69 mm × 100 mm	–	–	–	–	88.17	nr	**88.17**
Balusters; stiffeners							
25 mm × 25 mm	–	–	–	–	3.70	m	**3.70**
32 mm × 32 mm	–	–	–	–	4.18	m	**4.18**
44 mm × 44 mm	–	–	–	–	5.50	m	**5.50**
ends; housed	–	–	–	–	1.66	nr	**1.66**

25 STAIRS, WALKWAYS AND BALUSTRADES

Item	PC £	Labour hours	Labour £	Plant £	Material £	Unit	Total rate £
Sub-rails							
32 mm × 63 mm	–	–	–	–	8.50	m	**8.50**
ends; framed joint to newel	–	–	–	–	7.17	nr	**7.17**
Knee rails							
32 mm × 140 mm	–	–	–	–	14.08	m	**14.08**
ends; framed joint to newel	–	–	–	–	7.17	nr	**7.17**
Newel posts							
44 mm × 94 mm; half newel	–	–	–	–	9.78	m	**9.78**
69 mm × 69 mm	–	–	–	–	10.61	m	**10.61**
94 mm × 94 mm	–	–	–	–	21.69	m	**21.69**
Newel caps; splayed on four sides							
62.50 mm × 125 mm × 50 mm	–	–	–	–	10.36	nr	**10.36**
100 mm × 100 mm × 50 mm	–	–	–	–	10.56	nr	**10.56**
125 mm × 125 mm × 50 mm	–	–	–	–	11.09	nr	**11.09**
Metal handrails							
Wall rails; painted mild steel CHS wall rail; with wall rose bracket							
42 mm dia.	–	–	–	–	–	m	**103.36**
Wall rails; stainless steel circular wall rail; with wall rose bracket							
42 mm dia.	–	–	–	–	–	m	**133.75**
25.04 LADDERS							
Loft ladders; fixing with screws to timber lining (not included)							
Loft ladders							
Youngman Easiway 3 section aluminium ladder; 2.3 m to 3.0 m ceiling height	–	0.93	20.89	–	150.09	nr	**170.98**
Youngman Eco folding 3 section timber ladder; 2.8 m ceiling height	–	2.00	44.92	–	210.09	nr	**255.01**
Youngman Spacemeaker aluminium sliding ladder; 2.6 m ceiling height	–	1.25	28.08	–	90.09	nr	**118.17**
Youngman Deluxe 2 section aluminium ladder; 3.25 m ceiling height; spring assisted	–	2.50	56.15	–	300.09	nr	**356.24**
Access ladders; mild steel							
Ladders							
400 mm wide; 3850 mm long (overall); 12 mm dia. rungs; 65 mm × 15 mm strings; 50 mm × 5 mm safety hoops; fixing with expanded bolts; to masonry; mortices; welded fabrication	–	–	–	–	–	nr	**2362.50**

27 GLAZING

Item	PC £	Labour hours	Labour £	Plant £	Material £	Unit	Total rate £
MATERIAL RATES							
SUPPLY ONLY RATES							
Ordinary transluscent/patterned glass							
3 mm	–	–	–	–	35.56	m²	**35.56**
4 mm	–	–	–	–	37.74	m²	**37.74**
5 mm	–	–	–	–	45.88	m²	**45.88**
6 mm	–	–	–	–	50.37	m²	**50.37**
Obscured ground sheet glass – patterned							
4 mm white	–	–	–	–	53.34	m²	**53.34**
6 mm white	–	–	–	–	58.71	m²	**58.71**
Rough cast							
6 mm	–	–	–	–	50.00	m²	**50.00**
Ordinary Georgian wired							
7 mm cast	–	–	–	–	50.99	m²	**50.99**
6 mm polish	–	–	–	–	78.42	m²	**78.42**
Cetuff toughened; float							
4 mm	–	–	–	–	41.76	m²	**41.76**
5 mm	–	–	–	–	55.36	m²	**55.36**
6 mm	–	–	–	–	60.96	m²	**60.96**
10 mm	–	–	–	–	100.03	m²	**100.03**
Clear laminated; safety							
4.40 mm	–	–	–	–	49.78	m²	**49.78**
6.40 mm	–	–	–	–	59.43	m²	**59.43**
27.01 GENERAL GLAZING							
SUPPLY AND FIX PRICES							
NOTE: The following measured rates are provided by a glazing contractor and assume in excess of 500 m², within 20 miles of the suppliers branch.							
Standard plain glass; BS EN 14449; clear float; panes area 0.15 m²–4.00 m²							
3 mm thick; glazed with							
screwed beads	–	–	–	–	–	m²	**58.66**
4 mm thick; glazed with							
screwed beads	–	–	–	–	–	m²	**62.22**
5 mm thick; glazed with							
screwed beads	–	–	–	–	–	m²	**75.78**
6 mm thick; glazed with							
screwed beads	–	–	–	–	–	m²	**83.00**
Standard plain glass; BS EN 14449; obscure patterned; panes area 0.15 m²–4.00 m²							
4 mm thick; glazed with							
screwed beads	–	–	–	–	–	m²	**87.98**
6 mm thick; glazed with							
screwed beads	–	–	–	–	–	m²	**96.82**

27 GLAZING

Item	PC £	Labour hours	Labour £	Plant £	Material £	Unit	Total rate £
Standard plain glass; BS EN 14449; rough cast; panes area 0.15 m²–4.00 m²							
6 mm thick; glazed with							
screwed beads	–	–	–	–	–	m²	79.61
Standard plain glass; BS EN 14449; Georgian wired cast; panes area 0.15 m²–4.00 m²							
7 mm thick; glazed with							
screwed beads	–	–	–	–	–	m²	81.13
extra for lining up wired glass	–	–	–	–	–	m²	4.20
Standard plain glass; BS EN 14449; Georgian wired polished; panes area 0.15 m²–4.00 m²							
6 mm thick; glazed with							
screwed beads	–	–	–	–	–	m²	125.90
extra for lining up wired glass	–	–	–	–	–	m²	4.20
Factory-made double hermetically sealed units; to wood or metal with screwed or clipped beads							
Two panes; BS EN 14449; clear float glass; 4 mm thick; 6 mm air space							
0.35 m²–2.00 m²	–	–	–	–	–	m²	156.54
Two panes; BS 952; clear float glass; 6 mm thick; 6 mm air space							
0.35 m²–2.0 m²	–	–	–	–	–	m²	182.31
2.00 m²–4.00 m²	–	–	–	–	–	m²	274.12
Factory-made double hermetically sealed units; with inner pane of Pilkington's K low emissivity coated glass; to wood or metal with screwed or clipped beads							
Two panes; BS EN 14449; clear float glass; 4 mm thick; 6 mm air space							
0.35 m²–2.00 m²	–	–	–	–	–	m²	190.22
Two panes; BS EN 14449; clear float glass; 6 mm thick; 6 mm air space							
0.35 m²–2.0 m²	–	–	–	–	–	m²	221.71
2.00 m²–4.00 m²	–	–	–	–	–	m²	333.44
Factory-made triple hermetically sealed units; with inner pane of Pilkington's K low emissivity coated glass; to wood or metal with screwed or clipped beads							
Three panes; BS EN 14449; clear float glass; 4 mm thick; 6 mm air spaces							
0.35 m²–2.00 m²	–	–	–	–	–	m²	306.59
Three panes; BS EN 14449; clear float glass; 6 mm thick; 6 mm air spaces							
0.35 m²–2.0 m²	–	–	–	–	–	m²	356.90
2.00 m²–4.00 m²	–	–	–	–	–	m²	536.88

27 GLAZING

Item	PC £	Labour hours	Labour £	Plant £	Material £	Unit	Total rate £
27.02 FIRE RESISTING GLAZING							
Special glass; BS EN 14449; Pyran half-hour fire-resisting glass or other equal							
6.50 mm thick rectangular panes; glazed with screwed hardwood beads and Sealmaster Fireglaze intumescent compound or other equal and approved to rebated frame							
300 mm × 400 mm pane	–	0.74	18.60	–	63.30	nr	**81.90**
400 mm × 800 mm pane	–	0.92	23.12	–	165.50	nr	**188.62**
500 mm × 1400 mm pane	–	1.48	37.19	–	354.22	nr	**391.41**
600 mm × 1800 mm pane	–	1.86	46.74	–	559.26	nr	**606.00**
Special glass; BS EN 14449; Pyrostop one-hour fire-resisting glass or other equal							
15 mm thick regular panes; glazed with screwed hardwood beads and Sealmaster Fireglaze intumescent liner and compound or other equal and approved both sides							
300 mm × 400 mm pane	–	2.22	55.80	–	116.28	nr	**172.08**
400 mm × 800 mm pane	–	2.78	69.86	–	238.84	nr	**308.70**
500 mm × 1400 mm pane	–	3.70	92.99	–	503.20	nr	**596.19**
600 mm × 1800 mm pane	–	4.62	116.11	–	759.34	nr	**875.45**
27.03 TOUGHENED AND LAMINATED GLAZING							
Special glass; BS EN 14449; toughened clear float; panes area 0.15 m² – 4.00 m²							
4 mm thick; glazed with							
screwed beads	–	–	–	–	–	m²	**62.39**
5 mm thick; glazed with							
screwed beads	–	–	–	–	–	m²	**82.86**
6 mm thick; glazed with							
screwed beads	–	–	–	–	–	m²	**91.15**
10 mm thick; glazed with							
screwed beads	–	–	–	–	–	m²	**151.26**
Special glass; BS EN 14449; clear laminated safety glass; panes area 0.15 m² – 4.00 m²							
4.40 mm thick; glazed with							
screwed beads	–	–	–	–	–	m²	**83.68**
6.40 mm thick; glazed with							
screwed beads	–	–	–	–	–	m²	**99.88**
Special glass; BS EN 14449; clear laminated security glass							
7.50 mm thick regular panes; glazed with screwed hardwood beads and Intergens intumescent strip							
300 mm × 400 mm pane	–	0.74	18.96	–	37.46	nr	**56.42**
400 mm × 800 mm pane	–	0.92	23.57	–	97.32	nr	**120.89**
500 mm × 1400 mm pane	–	1.48	37.92	–	210.59	nr	**248.51**
600 mm × 1800 mm pane	–	1.92	49.19	–	321.18	nr	**370.37**

27 GLAZING

Item	PC £	Labour hours	Labour £	Plant £	Material £	Unit	Total rate £
27.04 MIRRORS AND LOUVRES							
Mirror panels; BS EN 14449; silvered; insulation backing							
4 mm thick float; fixing with adhesive							
1000 mm × 1000 mm	–	–	–	–	–	nr	**62.26**
1000 mm × 2000 mm	–	–	–	–	–	nr	**124.63**
1000 mm × 4000 mm	–	–	–	–	–	nr	**454.27**
Glass louvres; BS EN 14449; with long edges ground or smooth							
6 mm thick float							
150 mm wide	–	–	–	–	–	m	**31.03**
7 mm thick Georgian wired cast							
150 mm wide	–	–	–	–	–	m	**48.84**
6 mm thick Georgian wire polished							
150 mm wide	–	–	–	–	–	m	**68.62**
27.05 ADDITIONAL LABOURS							
Drill holes in glass							
Drill holes 6 mm to 15 mm dia. to glass thickness:							
not exceeding 6 mm thick	–	–	–	–	4.03	nr	**4.03**
not exceeding 10 mm thick	–	–	–	–	5.18	nr	**5.18**
not exceeding 12 mm thick	–	–	–	–	6.43	nr	**6.43**
not exceeding 19 mm thick	–	–	–	–	8.10	nr	**8.10**
not exceeding 25 mm thick	–	–	–	–	10.05	nr	**10.05**
Drill holes 16 mm to 38 mm dia. to glass thickness:							
not exceeding 6 mm thick	–	–	–	–	5.79	nr	**5.79**
not exceeding 10 mm thick	–	–	–	–	7.65	nr	**7.65**
not exceeding 12 mm thick	–	–	–	–	9.14	nr	**9.14**
not exceeding 19 mm thick	–	–	–	–	11.52	nr	**11.52**
not exceeding 25 mm thick	–	–	–	–	14.24	nr	**14.24**
Drill holes over 38 mm dia. to glass thickness:							
not exceeding 6 mm thick	–	–	–	–	11.52	nr	**11.52**
not exceeding 10 mm thick	–	–	–	–	13.88	nr	**13.88**
not exceeding 12 mm thick	–	–	–	–	16.44	nr	**16.44**
not exceeding 19 mm thick	–	–	–	–	20.23	nr	**20.23**
not exceeding 25 mm thick	–	–	–	–	25.25	nr	**25.25**
Other works to glass							
Curved cutting to glass							
to 4 mm thick panes	–	–	–	–	5.54	m	**5.54**
to 6 mm thick panes	–	–	–	–	5.54	m	**5.54**
to 6 mm thick wired panes	–	–	–	–	8.45	m	**8.45**
intumescant paste to glazed panels for die doors; per side treated	–	–	–	–	11.70	m	**11.70**
imitation washleather/black velvet bedding to edge of glass	–	–	–	–	2.02	m	**2.02**

28 FLOOR, WALL, CEILING AND ROOF FINISHES

Item	PC £	Labour hours	Labour £	Plant £	Material £	Unit	Total rate £
28.01 FLOORS; SCREEDS (CEMENT: SAND; CONCRETE; GRANOLITHIC)							
Cement and sand (1:3) screeds; steel trowelled							
Work to floors; one coat level; to concrete base; screeded; over 600 mm wide							
25 mm thick	–	0.41	10.22	0.44	0.52	m²	**11.18**
50 mm thick	–	0.52	12.96	0.44	1.49	m²	**14.89**
75 mm thick	–	0.66	16.45	0.44	2.25	m²	**19.14**
100 mm thick	–	0.70	17.45	0.44	3.00	m²	**20.89**
Add to the above for work to falls and crossfalls and to slopes							
not exceeding 15° from horizontal	–	0.02	0.49	–	–	m²	**0.49**
over 15° from horizontal	–	0.09	2.25	–	–	m²	**2.25**
water repellent additive incorporated in the mix	–	0.02	0.49	–	0.57	m²	**1.06**
oil repellent additive incorporated in the mix	–	0.07	1.74	–	4.94	m²	**6.68**
Fine concrete (1: 4–5) levelling screeds; steel trowelled							
Work to floors; one coat; level; to concrete base; over 600 mm wide							
50 mm thick	–	0.52	12.96	–	5.55	m²	**18.51**
75 mm thick	–	0.66	16.45	–	8.33	m²	**24.78**
extra over last for isolation joint to perimeter	–	0.20	4.99	–	3.11	m	**8.10**
Early drying floor screed; RMC Mortars Readyscreed; or other equal and approved; steel trowelled							
Work to floors; one coat; level; to concrete base; over 600 mm wide							
100 mm thick	–	0.70	17.45	–	13.01	m²	**30.46**
100 mm thick with galvanized chicken wire anticrack reinforcement	–	0.71	17.70	–	14.08	m²	**31.78**
Granolithic paving; cement and granite chippings 5 to dust (1:1:2); steel trowelled							
Work to floors; one coat; level; laid on concrete while green; bonded; over 600 mm wide							
25 mm thick	–	0.75	18.69	0.44	4.32	m²	**23.45**
38 mm thick	–	0.80	19.94	0.44	6.89	m²	**27.27**
Work to floors; two coat; laid on hacked concrete with slurry; over 600 mm wide							
50 mm thick	–	1.20	29.91	0.44	8.61	m²	**38.96**
75 mm thick	–	1.50	37.39	0.44	13.78	m²	**51.61**
Work to landings; one coat; level; laid on concrete while green; bonded; over 600 mm wide							
25 mm thick	–	0.80	19.94	0.44	4.32	m²	**24.70**
38 mm thick	–	1.00	24.93	0.44	6.89	m²	**32.26**

28 FLOOR, WALL, CEILING AND ROOF FINISHES

Item	PC £	Labour hours	Labour £	Plant £	Material £	Unit	Total rate £
Work to landings; two coat; laid on hacked concrete with slurry; over 600 mm wide							
50 mm thick	–	2.00	49.85	0.44	8.61	m²	**58.90**
75 mm thick	–	2.50	62.31	0.44	13.78	m²	**76.53**
Add to the above over 600 mm wide for							
liquid hardening additive incorporated in the mix	–	0.04	1.00	–	0.79	m²	**1.79**
oil-repellent additive incorporated in the mix	–	0.07	1.74	–	4.94	m²	**6.68**
25 mm work to treads; one coat; to concrete base							
225 mm wide	–	0.83	20.69	–	10.71	m	**31.40**
275 mm wide	–	0.83	20.69	–	12.00	m	**32.69**
returned end	–	0.17	4.23	–	–	nr	**4.23**
13 mm skirtings; rounded top edge and coved bottom junction; to brickwork or blockwork base							
75 mm wide on face	–	0.51	12.71	–	0.52	m	**13.23**
150 mm wide on face	–	0.69	17.20	–	9.42	m	**26.62**
ends; fair	–	0.04	1.00	–	–	nr	**1.00**
angles	–	0.06	1.49	–	–	nr	**1.49**
13 mm outer margin to stairs; to follow profile of and with rounded nosing to treads and risers; fair edge and arris at bottom, to concrete base							
75 mm wide	–	0.83	20.69	–	5.14	m	**25.83**
angles	–	0.06	1.49	–	–	nr	**1.49**
13 mm wall string to stairs; fair edge and arris on top; coved bottom junction with treads and risers; to brickwork or blockwork base							
275 mm (extreme) wide	–	0.74	18.45	–	8.99	m	**27.44**
ends	–	0.04	1.00	–	–	nr	**1.00**
angles	–	0.06	1.49	–	–	nr	**1.49**
ramps	–	0.07	1.74	–	–	nr	**1.74**
ramped and wreathed corners	–	0.09	2.25	–	–	nr	**2.25**
13 mm outer string to stairs; rounded nosing on top at junction with treads and risers; fair edge and arris at bottom; to concrete base							
300 mm (extreme) wide	–	0.74	18.45	–	11.13	m	**29.58**
ends	–	0.04	1.00	–	–	nr	**1.00**
angles	–	0.06	1.49	–	–	nr	**1.49**
ramps	–	0.07	1.74	–	–	nr	**1.74**
ramps and wreathed corners	–	0.09	2.25	–	–	nr	**2.25**
19 mm thick skirtings; rounded top edge and coved bottom junction; to brickwork or blockwork base							
75 mm wide on face	–	0.51	12.71	–	9.42	m	**22.13**
150 mm wide on face	–	0.69	17.20	–	14.56	m	**31.76**
ends; fair	–	0.04	1.00	–	–	nr	**1.00**
angles	–	0.06	1.49	–	–	nr	**1.49**
19 mm riser; one rounded nosing; to concrete base							
150 mm high; plain	–	0.83	20.69	–	8.14	m	**28.83**
150 mm high; undercut	–	0.83	20.69	–	8.14	m	**28.83**
180 mm high; plain	–	0.83	20.69	–	11.13	m	**31.82**
180 mm high; undercut	–	0.83	20.69	–	11.13	m	**31.82**

28 FLOOR, WALL, CEILING AND ROOF FINISHES

Item	PC £	Labour hours	Labour £	Plant £	Material £	Unit	Total rate £
28.02 FLOORS; SCREEDS (RESIN)							
Resins can be difficult to price as there may be site-specific variables, such as substrate condition and moisture content. There are numerous variables of thicknesses and surface textures, but we have provided cost guides for the most commonly used products. The installed cost will reduce by amounts increasing to approx. 10% as the overall area increases above 200 m^2 toward 4,000 m^2 and rise significantly as the area reduces toward 20 m^2							
Latex self-levelling floor screeds; steel trowelled							
Work to floors; level; to concrete base; over 600 mm wide							
3 mm thick; one coat	–	–	–	–	–	m^2	**8.74**
5 mm thick; two coats	–	–	–	–	–	m^2	**10.93**
Resin flooring; Altro epoxy and polyurethane resin flooring system, level, to concrete; sand cement screed or securely bonded 25 mm plywood							
Work to floors; level; to concrete base; over 600 mm wide and over 300 m^2 total area							
AltroCoat water-based epoxy coating; 140 micron thick for two coats	–	0.45	6.80	–	8.00	m^2	**14.80**
AltroTect solvent free high build epoxy coating; 350 micron thick for two coats	–	0.45	6.80	–	9.74	m^2	**16.54**
AltroSeal UVR chemically resistant polyurethane coating; 140 micron thick for two coats	–	0.45	6.80	–	9.24	m^2	**16.04**
AltroFlow EP, epoxy self-smoothing resin screed; 2 mm thick including primer	–	0.90	13.59	–	22.11	m^2	**35.70**
AltroFlow PU, polyurethane self-smoothing resin screed; 2 mm thickness including primer	–	0.90	13.59	–	14.62	m^2	**28.21**
AltroFlow PU, polyurethane self-smoothing resin screed; 4 mm thickness including primer	–	0.90	13.59	–	24.54	m^2	**38.13**
Altro Flexiflow, flexible polyurethame self-smoothing resin screed; 2 mm thickness including primer	–	0.90	13.59	–	25.46	m^2	**39.05**
Altro TB screed, trowel applied multi-coloured epoxy screed; 3 mm thickness including primer	–	1.20	18.12	–	26.49	m^2	**44.61**
AltroScreed Quartz EP, trowel applied multi-coloured epoxy screed; 4 mm thickness including primer	–	1.20	18.12	–	28.58	m^2	**46.70**
Altro MultiScreed, trowel applied flecked epoxy screed; 4 mm thickness including primer	–	1.20	18.12	–	26.28	m^2	**44.40**
AltroGrip EP, multi-layer, textured, slip-resistant, plain coloured; 4 mm thickness including primer	–	1.20	18.12	–	27.11	m^2	**45.23**

28 FLOOR, WALL, CEILING AND ROOF FINISHES

Item	PC £	Labour hours	Labour £	Plant £	Material £	Unit	Total rate £
AltroGrip PU, multi-layer, textured, slip-resistant, plain coloured; 4 mm thickness including primer	–	1.20	18.12	–	21.82	m²	**39.94**
AltroCrete PU Excel, heavy duty polyurethane screed; 6 mm thickness including primer	–	1.20	18.12	–	29.49	m²	**47.61**
AltroCrete PU Excel, heavy duty polyurethane screed; 8 mm thickness including primer	–	1.20	18.12	–	33.90	m²	**52.02**
Perimeter site-formed resin cove, 100 mm high, 40 mm radius	–	–	–	–	29.41	m	**29.41**
Isocrete K screeds or other equal; steel trowelled							
Work to floors; level; to concrete base; over 600 mm wide							
35 mm thick; plus polymer bonder coat	–	–	–	–	–	m²	**17.83**
40 mm thick	–	–	–	–	–	m²	**16.47**
45 mm thick	–	–	–	–	–	m²	**17.40**
50 mm thick	–	–	–	–	–	m²	**18.33**
Work to floors; to falls or cross-falls; to concrete base; over 600 mm wide							
55 mm (average) thick	–	–	–	–	–	m²	**19.25**
60 mm (average) thick	–	–	–	–	–	m²	**20.21**
65 mm (average) thick	–	–	–	–	–	m²	**21.14**
75 mm (average) thick	–	–	–	–	–	m²	**23.01**
90 mm (average) thick	–	–	–	–	–	m²	**25.84**
Isocrete K screeds; quick drying; or other equal and approved; steel trowelled							
Work to floors; level or to floors n.e. 15° frojm the horizontal; to concrete base; over 600 mm wide							
55 mm thick	–	–	–	–	–	m²	**24.97**
75 mm thick	–	–	–	–	–	m²	**31.18**
Isocrete pumpable Self-Level Plus screeds; or other equal and approved; protected with Corex type polythene; knifed off prior to laying floor finish; flat smooth finish							
Work to floors; level or to floors n.e. 15° frojm the horizontal; to concrete base; over 600 mm wide							
20 mm thick	–	–	–	–	–	m²	**30.40**
50 mm thick	–	–	–	–	–	m²	**40.55**
28.03 FLOORS; MASTIC ASPHALT							
Mastic asphalt flooring; black							
20 mm thick; one coat coverings; felt isolating membrane; to concrete base; flat							
over 300 mm wide	–	–	–	–	–	m²	**21.56**
225 mm–300 mm wide	–	–	–	–	–	m²	**40.05**
150 mm–225 mm wide	–	–	–	–	–	m²	**44.00**
not exceeding 150 mm wide	–	–	–	–	–	m²	**53.86**

28 FLOOR, WALL, CEILING AND ROOF FINISHES

Item	PC £	Labour hours	Labour £	Plant £	Material £	Unit	Total rate £
28.03 FLOORS; MASTIC ASPHALT – cont							
Mastic asphalt flooring – cont							
25 mm thick; one coat coverings; felt isolating membrane; to concrete base; flat							
over 300 mm wide	–	–	–	–	–	m²	25.02
225 mm–300 mm wide	–	–	–	–	–	m²	42.72
150 mm–225 mm wide	–	–	–	–	–	m²	46.56
not exceeding 150 mm wide	–	–	–	–	–	m²	56.46
20 mm three coat skirtings to brickwork base							
not exceeding 150 mm girth	–	–	–	–	–	m	22.03
150 mm–225 mm girth	–	–	–	–	–	m	26.93
225 mm–300 mm girth	–	–	–	–	–	m	31.82
Mastic asphalt flooring; acid-resisting; black							
20 mm thick; one coat coverings; felt isolating membrane; to concrete base flat							
over 300 mm wide	–	–	–	–	–	m²	25.25
225 mm–300 mm wide	–	–	–	–	–	m²	46.19
150 mm–225 mm wide	–	–	–	–	–	m²	47.70
not exceeding 150 mm wide	–	–	–	–	–	m²	57.55
25 mm thick; one coat coverings; felt isolating membrane; to concrete base; flat							
over 300 mm wide	–	–	–	–	–	m²	29.84
225 mm–300 mm wide	–	–	–	–	–	m²	47.48
150 mm–225 mm wide	–	–	–	–	–	m²	51.39
not exceeding 150 mm wide	–	–	–	–	–	m²	61.27
20 mm thick; three coat skirtings to brickwork base							
not exceeding 150 mm girth	–	–	–	–	–	m	22.27
150 mm–225 mm girth	–	–	–	–	–	m	25.93
225 mm–300 mm girth	–	–	–	–	–	m	29.43
Mastic asphalt flooring; red							
20 mm thick; one coat coverings; felt isolating membrane; to concrete base; flat							
over 300 mm wide	–	–	–	–	–	m²	35.34
225 mm–300 mm wide	–	–	–	–	–	m²	58.37
150 mm–225 mm wide	–	–	–	–	–	m²	63.06
not exceeding 150 mm wide	–	–	–	–	–	m²	75.42
20 mm thick; three coat skirtings to brickwork base							
not exceeding 150 mm girth	–	–	–	–	–	m	27.73
150 mm–225 mm girth	–	–	–	–	–	m	35.34

28 FLOOR, WALL, CEILING AND ROOF FINISHES

Item	PC £	Labour hours	Labour £	Plant £	Material £	Unit	Total rate £
28.04 FLOORS; EDGE FIXED CARPETING; CARPET TILES							
Fitted carpeting; Wilton wool/nylon or other equal and approved; 80/20 velvet pile; heavy domestic plain							
Work to floors							
over 600 mm wide	42.00	0.37	6.25	–	47.59	m²	**53.84**
Work to treads and risers							
not exceeding 600 mm wide	25.20	0.74	12.50	–	28.55	m	**41.05**
Fitted carpet; Forbo Flooring; Flocked flooring							
Work to floors							
Flotex classic textile; 2.0 m wide roll	25.23	0.45	10.41	–	28.28	m²	**38.69**
Flotex HD textile; 2.0 m wide roll	27.00	0.45	10.41	–	30.24	m²	**40.65**
Fitted carpeting; Gradus woven polypropylene tufted loop							
Work to floors over 600 mm wide							
Bodega tufted loop pile	17.29	0.37	6.25	–	19.59	m²	**25.84**
Genus tufted cut pile	37.81	0.37	6.25	–	42.84	m²	**49.09**
Work to treads and risers							
not exceeding 600 mm wide	22.68	0.74	12.50	–	25.70	m	**38.20**
Carpet tiles; Forbo Flooring; 500 mm × 500 mm carpet tiles; adhesive applied to subfloor							
Work to floors:							
Teviot; low level loop pile; 2.50 mm thick	15.59	0.45	10.41	–	17.61	m²	**28.02**
Barcode; low level loop pile; 2.50 mm thick	20.77	0.45	10.41	–	23.34	m²	**33.75**
Create Space 2; random lay, batchless	22.18	0.45	10.41	–	24.91	m²	**35.32**
Arran; textured loop; 4.00 mm thick	29.83	0.45	10.41	–	33.37	m²	**43.78**
Alignment; textured cut and loop; 4.00 mm thick	28.23	0.45	10.41	–	31.61	m²	**42.02**
Underlay to carpeting							
Work to floors							
over 600 mm wide	3.35	0.07	1.18	–	3.63	m²	**4.81**
raking cutting	–	0.07	1.00	–	–	m	**1.00**
Carpet tiles to concrete or timber subfloor							
Work to floors over 600 mm wide							
Heuga 530 heavy duty loop pile	20.80	0.28	6.98	–	22.50	m²	**29.48**
Gradus Bodega floor tiles 500 mm × 500 mm, loop pile	21.14	0.37	6.25	–	23.95	m²	**30.20**
Gradus Adventure floor tiles 500 mm × 500 mm; loop pile	35.44	0.33	5.63	–	40.15	m²	**45.78**
Gradus Cityscene floor tiles 500 mm × 500 mm; loop pile	37.70	0.33	5.63	–	42.71	m²	**48.34**
Work to treads and risers not exceeding 600 mm wide							
not exceeding 600 mm wide	12.68	0.74	12.50	–	14.37	m	**26.87**

28 FLOOR, WALL, CEILING AND ROOF FINISHES

Item	PC £	Labour hours	Labour £	Plant £	Material £	Unit	Total rate £
28.04 FLOORS; EDGE FIXED CARPETING; CARPET TILES – cont							
Sundries							
Carpet gripper fixed to floor; standard edging							
22 mm wide	–	0.04	0.58	–	0.44	m	**1.02**
Stair nosings; aluminium; Gradus or equivalent							
Medium duty hard aluminium alloy stair tread							
nosings; plugged and screwed in concrete							
56 mm × 32 mm; ref AS11	12.36	0.23	4.09	–	13.12	m	**17.21**
84 mm × 32 mm; ref AS12	17.13	0.28	4.97	–	18.16	m	**23.13**
Heavy duty aluminium alloy stair tread nosings;							
plugged and screwed to concrete							
48 mm × 38 mm; ref HE1	14.43	0.28	4.97	–	15.31	m	**20.28**
82 mm × 38 mm; ref HE2	20.03	0.32	5.69	–	21.23	m	**26.92**
Door entrance mats							
Entrance mat systems; aluminium wiper bar; laying							
in position; closed construction;12 mm thick							
Gradus Topguard; 900 mm × 550 mm; single							
wipers	170.64	0.46	6.57	–	175.76	nr	**182.33**
Gradus Topguard; 1200 mm × 750 mm; single							
wipers	307.15	0.46	6.57	–	316.36	nr	**322.93**
Gradus Topguard; 2400 mm × 1200 mm; single							
wipers	982.87	0.93	13.30	–	1012.36	nr	**1025.66**
Gradus Topguard; 900 mm × 550 mm; double							
wipers	197.17	0.46	6.57	–	203.09	nr	**209.66**
Gradus Topguard; 1200 mm × 750 mm; double							
wipers	354.90	0.46	6.57	–	365.55	nr	**372.12**
Gradus Topguard; 2400 mm × 1200 mm; double							
wipers	1135.68	0.93	13.30	–	1169.75	nr	**1183.05**
Gradus Topguard; single wipers	358.34	0.50	7.15	–	369.09	m²	**376.24**
Gradus Topguard; double wipers	414.05	0.50	7.15	–	426.47	m²	**433.62**
Entrance mat systems; aluminium wiper bar; laying							
in position; closed construction;18 mm thick							
Gradus Topguard; 900 mm × 550 mm; single							
wipers	177.02	0.46	6.57	–	182.33	nr	**188.90**
Gradus Topguard; 1200 mm × 750 mm; single							
wipers	318.63	0.46	6.57	–	328.19	nr	**334.76**
Gradus Topguard; 2400 mm × 1200 mm; single							
wipers	1019.62	0.93	13.30	–	1050.21	nr	**1063.51**
Gradus Topguard; 900 mm × 550 mm; double							
wipers	203.95	0.46	6.57	–	210.07	nr	**216.64**
Gradus Topguard; 1200 mm × 750 mm; double							
wipers	367.11	0.46	6.57	–	378.12	nr	**384.69**
Gradus Topguard; 2400 mm × 1200 mm; double							
wipers	1174.75	0.93	13.30	–	1209.99	nr	**1223.29**
Gradus Topguard; single wipers	371.74	0.50	7.15	–	382.89	m²	**390.04**
Gradus Topguard; double wipers	428.29	0.50	7.15	–	441.14	m²	**448.29**

28 FLOOR, WALL, CEILING AND ROOF FINISHES

Item	PC £	Labour hours	Labour £	Plant £	Material £	Unit	Total rate £
Entrance mat systems; aluminium wiper bar; laying in position; open construction;12 mm thick							
Gradus Topguard; 900 mm × 550 mm; single wipers	183.52	0.46	6.57	–	189.03	nr	**195.60**
Gradus Topguard; 1200 mm × 750 mm; single wipers	330.34	0.46	6.57	–	340.25	nr	**346.82**
Gradus Topguard; 2400 mm × 1200 mm; single wipers	1057.10	0.93	13.30	–	1088.81	nr	**1102.11**
Gradus Topguard; 900 mm × 550 mm; double wipers	206.46	0.46	6.57	–	212.65	nr	**219.22**
Gradus Topguard; 1200 mm × 750 mm; double wipers	371.62	0.46	6.57	–	382.77	nr	**389.34**
Gradus Topguard; 2400 mm × 1200 mm; double wipers	1189.19	0.93	13.30	–	1224.87	nr	**1238.17**
Gradus Topguard; single wipers	385.40	0.50	7.15	–	396.96	m²	**404.11**
Gradus Topguard; double wipers	433.56	0.50	7.15	–	446.57	m²	**453.72**
Entrance mat systems; aluminium wiper bar; laying in position; open 5 mm construction;18 mm thick							
Gradus Topguard; 900 mm × 550 mm; single wipers	193.95	0.46	6.57	–	199.77	nr	**206.34**
Gradus Topguard; 1200 mm × 750 mm; single wipers	349.12	0.46	6.57	–	359.59	nr	**366.16**
Gradus Topguard; 2400 mm × 1200 mm; single wipers	1117.17	0.93	13.30	–	1150.69	nr	**1163.99**
Gradus Topguard; 900 mm × 550 mm; double wipers	218.57	0.46	6.57	–	225.13	nr	**231.70**
Gradus Topguard; 1200 mm × 750 mm; double wipers	393.43	0.46	6.57	–	405.23	nr	**411.80**
Gradus Topguard; 2400 mm × 1200 mm; double wipers	1258.98	0.93	13.30	–	1296.75	nr	**1310.05**
Gradus Topguard; single wipers	407.30	0.50	7.15	–	419.52	m²	**426.67**
Gradus Topguard; double wipers	459.00	0.50	7.15	–	472.77	m²	**479.92**
Entrance mat systems; brass wiper bar; laying in position; closed construction; 18 mm thick							
Gradus Topguard; 900 mm × 550 mm; single wipers	300.34	0.46	6.57	–	309.35	nr	**315.92**
Gradus Topguard; 1200 mm × 750 mm; single wipers	540.61	0.46	6.57	–	556.83	nr	**563.40**
Gradus Topguard; 2400 mm × 1200 mm; single wipers	1729.95	0.93	13.30	–	1781.85	nr	**1795.15**
Gradus Topguard; 900 mm × 550 mm; double wipers	307.87	0.46	6.57	–	317.11	nr	**323.68**
Gradus Topguard; 1200 mm × 750 mm; double wipers	554.17	0.46	6.57	–	570.80	nr	**577.37**
Gradus Topguard; 2400 mm × 1200 mm; double wipers	1773.36	0.93	13.30	–	1826.56	nr	**1839.86**
Gradus Topguard; single wipers	630.71	0.50	7.15	–	649.63	m²	**656.78**
Gradus Topguard; double wipers	646.54	0.50	7.15	–	665.94	m²	**673.09**

28 FLOOR, WALL, CEILING AND ROOF FINISHES

Item	PC £	Labour hours	Labour £	Plant £	Material £	Unit	Total rate £
28.04 FLOORS; EDGE FIXED CARPETING; CARPET TILES – cont							
Door entrance mats – cont							
Entrance mat systems; brass wiper bar; laying in position; 5 mm open construction; 18 mm thick							
Gradus Topguard; 900 mm × 550 mm; single wipers	324.37	0.46	6.57	–	334.10	nr	**340.67**
Gradus Topguard; 1200 mm × 750 mm; single wipers	583.87	0.46	6.57	–	601.39	nr	**607.96**
Gradus Topguard; 2400 mm × 1200 mm; single wipers	1868.39	0.93	13.30	–	1924.44	nr	**1937.74**
Gradus Topguard; 900 mm × 550 mm; double wipers	350.96	0.46	6.57	–	361.49	nr	**368.06**
Gradus Topguard; 1200 mm × 750 mm; double wipers	631.73	0.46	6.57	–	650.68	nr	**657.25**
Gradus Topguard; 2400 mm × 1200 mm; double wipers	2021.53	0.93	13.30	–	2082.18	nr	**2095.48**
Gradus Topguard; single wipers	681.18	0.50	7.15	–	701.62	m²	**708.77**
Gradus Topguard; double wipers	737.02	0.50	7.15	–	759.13	m²	**766.28**
Other systems							
Coral Classic textile secondary and circulation matting system; 2.0 m wide roll	–	0.46	6.57	–	44.73	m²	**51.30**
Coral Brush Activ secondary and circulation matting system; 2.0 m wide roll	–	0.46	6.57	–	50.63	m²	**57.20**
Coral Duo secondary matting system; 2.0 m wide roll	–	0.46	6.57	–	50.63	m²	**57.20**
Matwells							
Polished stainless steel matwell; angle rim with lugs; bedding in screed							
Stainless steel frame bed in screed 1100 mm × 1400 mm	157.97	0.93	13.30	–	162.71	nr	**176.01**
Stainless steel frame bed in screed 500 mm × 1300 mm	113.74	1.00	14.30	–	117.15	nr	**131.45**
Stainless steel frame bed in screed 1000 mm × 1600 mm	164.29	2.00	28.59	–	169.22	nr	**197.81**
Stainless steel frame bed in screed 1100 mm × 1400 mm	157.97	1.20	17.16	–	162.71	nr	**179.87**
Stainless steel frame bed in screed 1800 mm × 3000 mm	303.31	1.50	21.44	–	312.41	nr	**333.85**
Polished aluminium matwell; angle rim with lugs brazed on; bedding in screed							
900 mm × 550 mm; constructed with 25 mm × 25 mm × 3 mm angle	25.88	0.93	13.30	–	26.66	nr	**39.96**
1200 mm × 750 mm; constructed with 34 mm × 26 mm × 6 mm angle	41.77	1.00	14.30	–	43.02	nr	**57.32**
2400 mm × 1200 mm; constructed with 50 mm × 50 mm × 6 mm angle	75.03	1.50	21.44	–	77.28	nr	**98.72**

28 FLOOR, WALL, CEILING AND ROOF FINISHES

Item	PC £	Labour hours	Labour £	Plant £	Material £	Unit	Total rate £
25 mm × 31 mm × 6 mm thick polished brass matwell; comprising angle rim with lugs brazed on; bedding in screed							
900 mm × 550 mm	84.40	0.93	13.30	–	86.93	nr	**100.23**
1200 mm × 750 mm	113.51	1.00	14.30	–	116.92	nr	**131.22**
2400 mm x1200 mm	209.55	1.50	21.44	–	215.84	nr	**237.28**
28.05 FLOORS; GRANITE; MARBLE; SLATE; TERRAZZO							
Clay floor quarries; level bedding 10 mm thick and jointing in cement and sand (1:3); butt joints; straight both ways; flush pointing with grout; to cement and sand base							
Work to floors; over 600 mm wide							
150 mm × 150 mm × 12 mm thick; heatherbrown	29.28	0.74	18.45	–	48.15	m²	**66.60**
150 mm × 150 mm × 12 mm thick; red	29.28	0.74	18.45	–	48.15	m²	**66.60**
200 mm × 200 mm × 12 mm thick; heatherbrown	29.40	0.60	14.96	–	48.28	m²	**63.24**
Works to floors; in staircase areas or plant rooms							
150 mm × 150 mm × 12 mm thick; heatherbrown	29.28	0.83	20.69	–	48.15	m²	**68.84**
150 mm × 150 mm × 12 mm thick; red	30.75	0.83	20.69	–	48.15	m²	**68.84**
200 mm × 200 mm × 12 mm thick; heatherbrown	29.40	0.69	17.20	–	48.28	m²	**65.48**
Work to floors; not exceeding 600 mm wide							
150 mm × 150 mm × 12 mm thick; heatherbrown	–	0.37	9.22	–	11.37	m	**20.59**
150 mm × 150 mm × 12 mm thick; red	–	0.37	9.22	–	11.37	m	**20.59**
200 mm × 200 mm × 12 mm thick; heatherbrown	–	0.31	7.73	–	11.19	m	**18.92**
fair square cutting against flush edges of existing finishes	–	0.11	1.85	–	2.60	m	**4.45**
raking cutting	–	0.19	3.25	–	2.91	m	**6.16**
cutting around pipes; not exceeding 0.30 m girth	–	0.14	2.49	–	–	nr	**2.49**
extra for cutting and fitting into recessed manhole cover 600 mm × 600 mm	–	0.93	16.53	–	–	nr	**16.53**
Work to sills; round edge tiles							
150 mm × 150 mm × 12 mm thick; interior; heatherbrown or red	–	0.31	7.73	–	11.86	m	**19.59**
150 mm × 150 mm × 12 mm thick; exterior; heatherbrown or red	–	0.32	7.97	–	15.25	m	**23.22**
fitted end	–	0.14	2.49	–	–	nr	**2.49**
Coved skirtings; 150 mm high; rounded top edge							
150 mm × 150 mm × 12 mm thick; ref. CBTR; heatherbrown or red	–	0.23	5.74	–	29.87	m	**35.61**
ends	–	0.04	0.71	–	–	nr	**0.71**
angles	–	0.14	2.49	–	6.36	nr	**8.85**

28 FLOOR, WALL, CEILING AND ROOF FINISHES

Item	PC £	Labour hours	Labour £	Plant £	Material £	Unit	Total rate £
28.05 FLOORS; GRANITE; MARBLE; SLATE; TERRAZZO – cont							
50 mm × 50 mm × 5.50 mm thick slip-resistant mosaic floor tiles, in 300 mm × 300 mm sheets, fixing with adhesive; butt joints; straight both ways; flush pointing with white grout; to cement and sand base							
Work to floors							
over 600 mm wide	93.71	1.76	43.87	–	98.84	m²	**142.71**
not exceeding 600 mm wide	–	1.38	34.40	–	60.00	m	**94.40**
Dakota mahogany granite cladding; polished finish; jointed and pointed in coloured mortar (1:2:8)							
20 mm work to floors; level; to cement and sand base							
over 600 mm wide	–	–	–	–	–	m²	**420.40**
20 mm × 300 mm treads; plain nosings	–	–	–	–	–	m	**240.24**
raking, cutting	–	–	–	–	–	m	**42.05**
polished edges	–	–	–	–	–	m	**540.52**
birdsmouth	–	–	–	–	–	m	**54.05**
Riven Welsh slate floor tiles; level; bedding 10 mm thick and jointing in cement and sand (1:3); butt joints; straight both ways; flush pointing with coloured mortar; to cement and sand base							
Work to floors; over 600 mm wide							
regular size and patter; 12 mm–15 mm thick	31.50	0.56	13.96	–	57.11	m²	**71.07**
Work to floors; not exceeding 600 mm wide							
regular size and patter; 12 mm–15 mm thick	–	0.56	13.96	–	34.60	m	**48.56**
Riven Chinese or Spanish slate floor tiles; level; bedding 10 mm thick and jointing in cement and sand (1:3); butt joints; straight both ways; flush pointing with coloured mortar; to cement and sand base							
Work to floors; over 600 mm wide							
regular size and patter; 12 mm–15 mm thick	18.00	0.56	13.96	–	41.82	m²	**55.78**
Work to floors; not exceeding 600 mm wide							
regular size and patter; 12 mm–15 mm thick	–	0.56	13.96	–	34.60	m	**48.56**
Roman Travertine marble cladding; polished finish; jointed and pointed in coloured mortar (1:2:8)							
20 mm thick work to floors; level; to cement and sand base							
over 600 mm wide	–	–	–	–	–	m²	**276.26**
20 mm × 300 mm treads; plain nosings	–	–	–	–	–	m	**165.76**
raking cutting	–	–	–	–	–	m	**32.43**
polished edges	–	–	–	–	–	m	**30.03**
birdsmouth	–	–	–	–	–	m	**60.05**

28 FLOOR, WALL, CEILING AND ROOF FINISHES

Item	PC £	Labour hours	Labour £	Plant £	Material £	Unit	Total rate £
Terrazzo tiles; BS EN 13748; aggregate size random ground grouted and polished to 80's grit finish; standard colour range; 3 mm joints symmetrical layout; bedding in 42 mm cement semi-dry mix (1:4); grouting with neat matching cement							
300 mm × 300 mm × 28 mm (nominal) Terrazzo tile units; hydraulically pressed, mechanically vibrated, steam cured; to floors on concrete base (not included); sealed with penetrating case hardener or other equal and approved; 2 coats applied immediately after final polishing							
plain; laid level	–	–	–	–	–	m²	**52.46**
plain; to slopes exceeding 15° from horizontal	–	–	–	–	–	m²	**63.95**
to small areas/toilets	–	–	–	–	–	m²	**120.08**
Accessories							
plastic division strips; 6 mm × 38 mm; set into floor tiling above crack inducing joints, to the nearest full tile module	–	–	–	–	–	m	**3.61**
Specially made terrazzo precast units; BS EN 13748–1; aggregate size random; standard colour range; 3 mm joints; grouting with neat matching cement							
Standard tread and riser square combined terrazzo units (with riser cast down) or other equal and approved; 280 mm wide; 150 mm high; 40 mm thick; machine-made; vibrated and fully machine polished; incorporating 1 nr. Ferodo anti-slip insert ref. OT40D or other equal and approved cast-in during manufacture; one end polished only							
fixed with cement: sand (1:4) mortar on prepared backgrounds (not included); grouted in neat tinted cement; wiped clean on completion of fixing	–	–	–	–	–	m	**274.73**
Standard tread square terrazzo units or other equal and approved; 40 mm thick; 280 mm wide; factory polished; incorporating 1 nr. Ferodo anti-slip insert ref. OT40D or other equal and approved							
fixed with cement: sand (1:4) mortar on prepared backgrounds (not included); grouted in neat tinted cement; wiped clean on completion of fixing	–	–	–	–	–	m	**165.30**
extra over for 55 × 55 mm contrasting colour to step nosing	–	–	–	–	–	m	**57.77**
Standard riser square terrazzo units or other equal and approved; 40 mm thick; 150 mm high; factory polished							
fixed with cemnt: sand (1:4) mortar on prepared backgrounds (not included); grouted in neat tinted cement; wiped clean on completion of fixing	–	–	–	–	–	m	**102.58**

28 FLOOR, WALL, CEILING AND ROOF FINISHES

Item	PC £	Labour hours	Labour £	Plant £	Material £	Unit	Total rate £
28.05 FLOORS; GRANITE; MARBLE; SLATE; TERRAZZO – cont							
Specially made terrazzo precast units – cont							
Standard coved terrazzo skirting units or other equal and approved; 904 mm long; 150 mm high; nominal finish; 23 mm thick; with square top edge							
fixed with cement: sand (1:4) mortar on prepared backgrounds (by others); grouted in neat tinted cement; wiped clean on completion of fixing	–	–	–	–	–	m	**96.35**
extra over for special internal/external angle pieces to match	–	–	–	–	–	m	**27.37**
extra over for special polished ends	–	–	–	–	–	nr	**9.02**
28.06 FLOORS; CORK; RUBBER; VINYL							
Linoleum sheet; Forbo Flooring; laid level; fixing with adhesive; welded joints; to cement and sand base							
Work to floors; over 600 mm wide							
Marmoleum Real marbled sheet, 2.0 m wide × 2.50 mm thick	16.12	0.37	9.22	–	17.79	m²	**27.01**
Marmoleum Modular marbled tile 333 mm × 333 mm × 3.20 mm thick	21.49	0.45	11.22	–	23.59	m²	**34.81**
Decibel 17db marbled acoustic sheet, 2.0 m wide; 3.50 mm thick	21.49	0.45	11.22	–	23.59	m²	**34.81**
High performance vinyl sheet; Forbo Flooring Eternal; laid level with welded seams; fixing with adhesive; to cement and sand base							
Work to floors; over 600 mm wide							
2.0 m wide roll × 2.00 mm thick	16.27	0.35	8.72	–	17.94	m²	**26.66**
Slip-resistant vinyl sheet; Forbo Flooring Surestep; laid level with welded seams; fixing with adhesive; to cement and sand base							
Work to floors; over 600 mm wide							
2.0 m wide roll × 2.00 mm thick	16.27	0.37	9.22	–	17.94	m²	**27.16**
Anti-static vinyl tiles; Forbo Flooring Colorex SD; laid level; fixing with adhesive; to cement and sand base							
Work to floors; over 600 mm wide							
615 mm × 615 mm tiles × 2.00 mm thick	22.73	0.50	12.46	–	24.94	m²	**37.40**
Acoustic vinyl sheet; Forbo Flooring Sarlon Traffic 19db; level with welded seams; fixing with adhesive; to cement and sand base							
Work to floors; over 600 mm wide							
2.0 m wide roll × 3.40 mm thick	17.36	0.37	9.22	–	19.13	m²	**28.35**

28 FLOOR, WALL, CEILING AND ROOF FINISHES

Item	PC £	Labour hours	Labour £	Plant £	Material £	Unit	Total rate £
Vinyl sheet; Altro range; with welded seams; level; fixing with adhesive where appropriate; to cement and sand base							
Work to floors; over 600 mm wide and over 200 m² total area							
2.00 mm thick; Walkway 20 and Walkway Plus 20	12.91	0.56	13.96	–	15.78	m²	**29.74**
2.00 mm thick; Marine 20	21.22	0.55	13.71	–	25.28	m²	**38.99**
2.00 mm thick; Aquarius	21.22	0.55	13.71	–	25.28	m²	**38.99**
2.00 mm thick; Suprema II	22.79	0.55	13.71	–	26.72	m²	**40.43**
2.00 mm thick; Wood Safety	17.10	0.55	13.71	–	20.82	m²	**34.53**
2.20 mm thick; Xpresslay and Xpresslay Plus (NB No adhesive required)	15.62	0.44	10.97	–	16.89	m²	**27.86**
2.50 mm thick; Classic 25	23.33	0.60	14.96	–	27.04	m²	**42.00**
2.50 mm thick; Designer 25	23.33	0.60	14.96	–	27.04	m²	**42.00**
2.50 mm thick; Unity 25	25.34	0.60	14.96	–	29.21	m²	**44.17**
3.50 mm thick; Stronghold 30	27.70	0.65	16.20	–	32.32	m²	**48.52**
Gradus loosleay LVT vinyl floor covering; 4.5 mm thick; medium traffic areas							
Work to floors; over 600 mm wide and over 200 m² total area							
Creation 55 vinyl tiles; 500 mm × 500 mm	0.57	0.65	15.73	–	2.88	m²	**18.61**
Creation 55 vinyl planks; 229 mm × 1220 mm	0.57	0.65	15.73	–	2.88	m²	**18.61**
Homogeneous vinyl sheet; Marleyflor Plus or other equal; level; with welded seams; fixing with adhesive; level; to cement and sand base							
Work to floors; over 600 mm wide							
2.00 mm thick	7.59	0.42	10.46	–	8.67	m²	**19.13**
100 mm high skirtings	–	0.11	2.74	–	2.12	m	**4.86**
Safety sheet; Marleyflor Granite Multisafe or other equal; level; with welded seams; fixing with adhesive; level; to cement and sand base							
Work to floors; over 600 mm wide							
2.00 mm thick	15.07	0.42	10.46	–	16.77	m²	**27.23**
Vinyl sheet; Marleyflor Omnisports or other equal; level; with welded seams; fixing with adhesive; level; to cement and sand base							
Work to floors; over 600 mm wide							
7.65 mm thick; Pro	30.01	0.90	22.43	–	32.92	m²	**55.35**
8.75 mm thick; Competition	35.17	1.00	24.93	–	38.51	m²	**63.44**
Vinyl semi-flexible tiles; Marleyflor Homogeneous tiles range; level; fixing with adhesive; butt joints; straight both ways; to cement and sand base							
Work to floors; over 600 mm wide							
300 mm × 300 mm × 2.00 mm thick; Vylon Plus	6.42	0.23	5.74	–	7.41	m²	**13.15**
500 mm × 500 mm × 2.00 mm thick; Marleyflor Plus	7.88	0.20	4.99	–	8.99	m²	**13.98**

Prices for Measured Works

28 FLOOR, WALL, CEILING AND ROOF FINISHES

Item	PC £	Labour hours	Labour £	Plant £	Material £	Unit	Total rate £
28.06 FLOORS; CORK; RUBBER; VINYL – cont							
Vinyl tiles; Polyflex Plus; level; fixing with adhesive; butt joints; straight both ways; to cement and sand base							
Work to floors; over 600 mm wide							
300 mm × 300 mm × 2.00 mm thick	6.63	0.23	5.74	–	7.64	m²	**13.38**
Vinyl tiles; Polyflex Camaro; level; fixing with adhesive; butt joints; straight both ways; to cement and sand base							
Work to floors; over 600 mm wide							
300 mm × 300 mm × 2.00 mm thick	14.94	0.30	7.48	–	16.62	m²	**24.10**
Vinyl tiles; Polyflor XL; level; fixing with adhesive; butt joints; straight both ways; to cement and sand base							
Work to floors; over 600 mm wide							
300 mm × 300 mm × 2.00 mm thick	8.28	0.32	7.97	–	9.42	m²	**17.39**
Vinyl sheet; Polyflor XL; level; fixing with adhesive; butt joints; straight both ways; to cement and sand base							
Work to floors; over 600 mm wide							
2.00 mm thick	5.69	0.28	6.86	–	6.62	m²	**13.48**
Vinyl sheet; Polysafe Standard; level; fixing with adhesive; welded seams; to cement and sand base							
Work to floors; over 600 mm wide							
2.00 mm thick	10.44	0.28	6.86	–	11.76	m²	**18.62**
2.50 mm thick	15.39	0.28	6.86	–	17.12	m²	**23.98**
Vinyl sheet; Polysafe Hydro; level; fixing with adhesive; welded seams; to cement and sand base							
Work to floors; over 600 mm wide							
2.00 mm thick	13.49	0.28	6.86	–	15.06	m²	**21.92**
Luxury mineral vinyl tiles; Marley I D Naturelle; level; fixing with adhesive; butt joints; straight both ways; to cement and sand base							
Work to floors; over 600 mm wide							
330 mm × 330 mm × 2.00 mm thick	10.53	0.23	5.74	–	11.86	m²	**17.60**
Acoustic vinyl tiles; Marley Tapiflex 243; level; fixing with adhesive; butt joints; straight both ways; to cement and sand base							
Work to floors; over 600 mm wide							
500 mm × 500 mm × 2.00 mm thick	14.21	0.20	4.99	–	15.84	m²	**20.83**

28 FLOOR, WALL, CEILING AND ROOF FINISHES

Item	PC £	Labour hours	Labour £	Plant £	Material £	Unit	Total rate £
Linoleum tiles; Marley Veneto XF; level; fixing with adhesive; butt joints; straight both ways; to cement and sand base							
Work to floors; over 600 mm wide							
500 mm × 500 mm × 2.50 mm thick	16.85	0.20	4.99	–	18.68	m²	**23.67**
Rubber studded tiles; Altro Mondopave; level; fixing with adhesive; butt joints; straight to cement and sand base							
Work to floors; over 600 mm wide							
500 mm × 500 mm × 2.50 mm thick; type MRB; black	26.76	0.60	14.96	–	31.47	m²	**46.43**
500 mm × 500 mm × 4.00 mm thick; type MRB; black	31.40	0.60	14.96	–	36.47	m²	**51.43**
Work to landings; over 600 mm wide							
500 mm × 500 mm × 4.00 mm thick; type MRB; black	31.40	0.74	18.45	–	39.84	m²	**58.29**
Work to stairs							
tread; 275 mm wide	–	0.46	11.46	–	11.87	m	**23.33**
riser; 180 mm wide	–	0.56	13.96	–	8.30	m	**22.26**
Cork tiles; natural finish, unsealed; level; fixing with adhesive; butt joints; straight both ways; to cement and sand base							
Work to floors; over 600 mm wide							
300 mm × 300 mm × 8.00 mm thick; contract quality	22.42	0.40	9.97	–	29.15	m²	**39.12**
300 mm × 300 mm × 6.00 mm thick; high density	18.13	0.40	9.97	–	24.29	m²	**34.26**
300 mm × 300 mm × 4.00 mm thick; domestic	15.79	0.37	9.22	–	21.63	m²	**30.85**
300 mm × 300 mm × 3.20 mm thick; domestic	12.85	0.37	9.22	–	18.30	m²	**27.52**
Cork tiles; PVC surfaced for heaviest wear areas; level; fixing with adhesive; butt joints; straight both ways; to cement and sand base							
600 mm × 300 mm × 3.20 mm thick	36.15	0.37	9.22	–	40.97	m²	**50.19**
Sundry floor sheeting underlays							
For floor finishings; over 600 mm wide							
building paper to BS 1521; class A; 75 mm lap (laying only)	–	0.05	0.71	–	–	m²	**0.71**
3.20 mm thick hardboard	–	0.19	4.74	–	1.74	m²	**6.48**
6.00 mm thick plywood	–	0.28	6.98	–	8.29	m²	**15.27**
Skirtings; plastic; Gradus or equivalent							
Set-in skirtings							
100 mm high; ref. SI100	–	0.11	2.74	–	3.13	m	**5.87**
150 mm high; ref. SI150	–	0.22	5.48	–	5.12	m	**10.60**
Set-on skirtings							
100 mm high; ref. SO100	–	0.22	5.48	–	2.52	m	**8.00**
150 mm high; ref. SO150	–	0.40	9.97	–	4.78	m	**14.75**

28 FLOOR, WALL, CEILING AND ROOF FINISHES

Item	PC £	Labour hours	Labour £	Plant £	Material £	Unit	Total rate £
28.06 FLOORS; CORK; RUBBER; VINYL – cont							
Stair nosings; aluminium; Gradus or equivalent							
Medium duty hard aluminium alloy stair tread							
nosings; plugged and screwed in concrete							
56 mm × 32 mm; ref AS11	12.67	0.23	4.09	–	13.12	m	**17.21**
84 mm × 32 mm; ref AS12	17.56	0.28	4.97	–	18.16	m	**23.13**
Heavy duty aluminium alloy stair tread nosings;							
plugged and screwed to concrete							
48 mm × 38 mm; ref HE1	14.79	0.28	4.97	–	15.31	m	**20.28**
82 mm × 38 mm; ref HE2	20.53	0.32	5.69	–	21.23	m	**26.92**
28.07 FLOORS; WOOD BLOCK							
Wood blocks; Havwoods or other equal and							
approved; 25 mm thick; level; laid to							
herringbone pattern with 2 block borderl; fixing							
with adhesive; to cement: sand base; sanded							
and sealed							
Work to floors; over 300 mm wide							
Merbau	–	–	–	–	–	m²	**134.87**
Iroko	–	–	–	–	–	m²	**134.87**
American Oak	–	–	–	–	–	m²	**141.35**
European Oak	–	–	–	–	–	m²	**147.83**
Add to wood block flooring over 300 mm wide for							
buff; one coat seal	–	–	–	–	–	m²	**5.18**
buff; two coats seal	–	–	–	–	–	m²	**8.43**
sand; three coats for seal or oil	–	–	–	–	–	m²	**23.34**
28.08 FLOORS; RAISED ACCESS FLOORS							
Raised flooring system; laid on or fixed to							
concrete floor							
Full access system; 150 mm high overall; pedestal							
supports							
PSA light grade; steel finish	–	–	–	–	–	m²	**37.99**
PSA medium grade; steel finish	–	–	–	–	–	m²	**37.99**
PSA heavy grade; steel finish	–	–	–	–	–	m²	**54.63**
Extra for							
factory applied needlepunch carpet	–	–	–	–	–	m²	**17.81**
factory applied anti-static vinyl	–	–	–	–	–	m²	**29.68**
factory applied black PVC edge strips	–	–	–	–	–	m	**5.40**
ramps; 3.00 m × 1.40 m (no finish)	–	–	–	–	–	nr	**836.40**
steps (no finish)	–	–	–	–	–	m	**52.27**
forming cut-out for electrical boxes	–	–	–	–	–	nr	**4.75**
supply and lay protection to raised floor;							
2440 × 1220 polypropylene sheets with taped							
joints	–	–	–	–	–	m²	**2.07**

28 FLOOR, WALL, CEILING AND ROOF FINISHES

Item	PC £	Labour hours	Labour £	Plant £	Material £	Unit	Total rate £
28.09 WALLS; PLASTERED; RENDERED; ROUGHCAST							
Plaster; one coat Thistle board finish or other equal; steel trowelled							
3 mm thick work to walls or ceilings; one coat; to plasterboard base							
over 600 mm wide	–	0.25	6.23	–	0.06	m²	**6.29**
over 600 mm wide; in staircase areas or plant rooms	–	0.35	8.72	–	0.06	m²	**8.78**
not exceeding 600 mm wide	–	0.15	3.74	–	0.04	m	**3.78**
Plaster; one coat Thistle board finish, 3 mm work to walls or ceilings; one coat on and including gypsum plasterboard; fixing with nails; 3 mm joints filled with plaster scrim; to softwood base; plain grade baseboard or lath with rounded edges							
9.50 mm thick boards to walls							
over 600 mm wide	–	0.97	16.38	–	5.47	m²	**21.85**
not exceeding 600 mm wide	–	0.37	6.58	–	1.65	m	**8.23**
9.50 mm thick boards to walls; in staircase areas or plant rooms							
over 600 mm wide	–	1.06	17.97	–	5.47	m²	**23.44**
not exceeding 600 mm wide	–	0.46	8.18	–	1.65	m	**9.83**
9.50 mm thick boards to isolated columns							
over 600 mm wide	–	1.06	17.97	–	5.47	m²	**23.44**
not exceeding 600 mm wide	–	0.56	9.96	–	1.65	m	**11.61**
12.50 mm thick boards to walls; in staircase areas or plant rooms							
over 600 mm wide	–	1.12	19.04	–	5.47	m²	**24.51**
not exceeding 600 mm wide	–	0.50	8.89	–	1.65	m	**10.54**
12.50 mm thick boards to isolated columns							
over 600 mm wide	–	1.12	19.04	–	5.47	m²	**24.51**
not exceeding 600 mm wide	–	0.59	10.49	–	1.65	m	**12.14**
Cement and sand (1:3) beds and backings							
10 mm thick work to walls; one coat; to brickwork or blockwork base							
over 600 mm wide	–	0.60	14.96	0.44	1.02	m²	**16.42**
not exceeding 600 mm wide	–	0.35	8.72	0.22	0.59	m	**9.53**
13 mm thick; work to walls; two coats; to brickwork or blockwork base							
over 600 mm wide	–	0.80	19.94	0.44	1.05	m²	**21.43**
not exceeding 600 mm wide	–	0.40	9.97	0.22	0.59	m	**10.78**
15 mm thick work to walls; two coats; to brickwork or blockwork base							
over 600 mm wide	–	0.90	22.43	0.44	1.12	m²	**23.99**
not exceeding 600 mm wide	–	0.40	9.97	0.22	0.67	m	**10.86**

28 FLOOR, WALL, CEILING AND ROOF FINISHES

Item	PC £	Labour hours	Labour £	Plant £	Material £	Unit	Total rate £
28.09 WALLS; PLASTERED; RENDERED; ROUGHCAST – cont							
Cement and sand (1:3); steel trowelled							
13 mm thick work to walls; two coats; to brickwork or blockwork base							
over 600 mm wide	–	0.80	19.94	0.44	1.05	m²	**21.43**
not exceeding 600 mm wide	–	0.40	9.97	0.22	0.59	m	**10.78**
16 mm thick work to walls; two coats; to brickwork or blockwork base							
over 600 mm wide	–	0.90	22.43	0.44	1.12	m²	**23.99**
not exceeding 600 mm wide	–	0.40	9.97	0.22	0.72	m	**10.91**
19 mm thick work to walls; two coats; to brickwork or blockwork base							
over 600 mm wide	–	0.95	23.68	0.44	1.25	m²	**25.37**
not exceeding 600 mm wide	–	0.40	9.97	0.22	0.94	m	**11.13**
Add to above							
over 600 mm wide in water repellent cement	–	0.10	2.49	–	1.44	m²	**3.93**
finishing coat in colour cement	–	0.35	8.72	0.22	0.96	m²	**9.90**
Cement–lime–sand (1:2:9); steel trowelled							
19 mm thick work to walls; two coats; to brickwork or blockwork base							
over 600 mm wide	–	0.95	23.68	0.44	0.96	m²	**25.08**
not exceeding 600 mm wide	–	0.40	9.97	0.22	4.49	m	**14.68**
Cement–lime–sand (1:1:6); steel trowelled							
13 mm thick work to walls; two coats; to brickwork or blockwork base							
over 600 mm wide	–	0.95	23.68	0.44	1.17	m²	**25.29**
not exceeding 600 mm wide	–	0.40	9.97	0.22	4.49	m	**14.68**
Add to the above over 600 mm wide for							
waterr epellent cement	–	0.10	2.49	–	1.44	m²	**3.93**
Plaster; first 11 mm coat of Thistle Hardwall plaster; second 2 mm finishing coat of Thistle Multi-Finish plaster; steel trowelled							
13 mm thick work to walls; two coats; to brickwork or blockwork base							
over 600 mm wide	–	0.50	12.46	–	1.70	m²	**14.16**
over 600 mm wide; in staircase areas, plant rooms and similar restricted areas	–	0.60	14.96	–	1.70	m²	**16.66**
not exceeding 600 mm wide	–	0.25	6.23	–	1.02	m	**7.25**
13 mm thick work to isolated brickwork or blockwork columns; two coats							
over 600 mm wide	–	1.00	24.93	–	1.70	m²	**26.63**
not exceeding 600 mm wide	–	0.50	12.46	–	1.02	m	**13.48**

28 FLOOR, WALL, CEILING AND ROOF FINISHES

Item	PC £	Labour hours	Labour £	Plant £	Material £	Unit	Total rate £
Plaster; first 8 mm or 11 mm coat of Thistle Bonding plaster; second 2 mm finishing coat of Thistle Multi-Finish plaster; steel trowelled finish							
13 mm thick work to walls; two coats; to concrete base							
over 600 mm wide	–	0.50	12.46	–	1.71	m²	**14.17**
over 600 mm wide; in staircase areas, plant rooms and similar restricted areas	–	0.60	14.96	–	1.71	m²	**16.67**
not exceeding 600 mm wide	–	0.25	6.23	–	1.01	m	**7.24**
13 mm thick work to isolated piers or columns; two coats; to concrete base							
over 600 mm wide	–	1.00	24.93	–	1.71	m²	**26.64**
not exceeding 600 mm wide	–	0.50	12.46	–	1.03	m	**13.49**
13 mm thick work to vertical face of metal lathing arch former							
not exceeding 0.50 m² per side	–	0.75	18.69	–	0.03	nr	**18.72**
0.50 m²–1 m² per side	–	1.00	24.93	–	0.05	nr	**24.98**
Plaster; first coat of Limelite renovating plaster; finishing coat of Limelite finishing plaster; or other equal							
13 mm thick work to walls; two coats; to brickwork or blockwork base							
over 600 mm wide	–	0.90	22.43	–	2.07	m²	**24.50**
over 600 mm wide; in staircase areas, plant rooms and similar restricted areas	–	1.00	24.93	–	2.07	m²	**27.00**
not exceeding 600 mm wide	–	0.45	11.22	–	1.19	m	**12.41**
Dubbing out existing walls with plaster; average 6 mm thick							
over 600 mm wide	–	0.25	6.23	–	0.06	m²	**6.29**
not exceeding 600 mm wide	–	0.15	3.74	–	0.04	m	**3.78**
Dubbing out existing walls with undercoat plaster; average 12 mm thick							
over 600 mm wide	–	0.50	12.46	–	0.12	m²	**12.58**
not exceeding 600 mm wide	–	0.25	6.23	–	0.07	m	**6.30**
Plaster; first coat of Thistle X-ray plaster or other equal; finishing coat of Thistle X-ray finishing plaster or other equal and approved; steel trowelled							
17 mm thick work to walls; two coats; to brickwork or blockwork base							
over 600 mm wide	–	–	–	–	–	m²	**78.39**
over 600 mm wide; in staircase areas or plant rooms	–	–	–	–	–	m²	**84.21**
not exceeding 600 mm wide	–	–	–	–	–	m	**31.33**
17 mm thick work to isolated columns; two coats							
over 600 mm wide	–	–	–	–	–	m²	**127.15**
not exceeding 600 mm wide	–	–	–	–	–	m	**50.83**

28 FLOOR, WALL, CEILING AND ROOF FINISHES

Item	PC £	Labour hours	Labour £	Plant £	Material £	Unit	Total rate £
28.09 WALLS; PLASTERED; RENDERED; ROUGHCAST – cont							
Plaster; one coat Thistle projection spray applied plaster or other equal; steel trowelled							
13 mm thick work to walls; one coat; to brickwork or blockwork base							
over 600 mm wide	–	0.15	3.74	1.14	10.38	m²	**15.26**
over 600 mm wide; in staircase areas or plant rooms	–	0.20	4.99	1.29	10.38	m²	**16.66**
not exceeding 600 mm wide	–	0.10	2.49	0.71	5.97	m	**9.17**
Cemrend self-coloured render or other equal; one coat; to brickwork or blockwork base							
20 mm thick work to walls; to brickwork or blockwork base							
over 600 mm wide	–	–	–	–	–	m²	**43.69**
not exceeding 600 mm wide	–	–	–	–	–	m	**28.13**
Tyrolean decorative rendering or similar; 13 mm thick first coat of cement–lime–sand (1:1:6); finishing three coats of Cullamix or other equal and approved; applied with approved hand operated machine external							
To walls; four coats; to brickwork or blockwork base							
over 600 mm wide	–	–	–	–	–	m²	**40.16**
not exceeding 600 mm wide	–	–	–	–	–	m	**20.06**
Drydash (pebbledash) finish of chippings on and including cement–lime–sand (1:2:9) backing							
18 mm thick work to walls; two coats; to brickwork or blockwork base. Over 600 mm wide areas							
Glenarm White chippings	2.51	1.45	36.14	0.44	5.53	m²	**42.11**
Derbyshire Spar/Red Granite chippings	3.76	1.45	36.14	0.44	6.82	m²	**43.40**
Black and White, Red and White Spar chippings	4.23	1.45	36.14	0.44	7.29	m²	**43.87**
Calcined Flint, Snowdrop chippings	5.00	1.45	36.14	0.44	8.10	m²	**44.68**
Barleycorn, Durite chippings	5.41	1.45	36.14	0.44	8.51	m²	**45.09**
Beige Marble, Buff Quartz chippings	6.44	1.45	36.14	0.44	9.58	m²	**46.16**
Accessories							
Expamet render beads or other equal and approved; white PVC nosings; to brickwork or blockwork base							
external stop bead; 20 mm thick	–	0.07	1.25	–	3.80	m	**5.05**
Expamet render beads or other equal and approved; stainless steel; to brickwork or blockwork base							
stop bead; 20 mm thick	–	0.07	1.25	–	3.11	m	**4.36**
stop bead; 20 mm thick	–	0.07	1.25	–	3.11	m	**4.36**

28 FLOOR, WALL, CEILING AND ROOF FINISHES

Item	PC £	Labour hours	Labour £	Plant £	Material £	Unit	Total rate £
Expamet plaster beads or other equal and approved; to brickwork or blockwork base							
angle bead; 13 mm thick	–	0.08	1.42	–	0.59	m	2.01
architrave bead; 13 mm thick	–	0.10	1.78	–	3.72	m	5.50
stop bead; 10 mm thick	–	0.07	1.25	–	0.78	m	2.03
stop bead; 13 mm thick	–	0.07	1.25	–	0.78	m	2.03
movement bead; 13 mm thick	–	0.09	1.60	–	12.63	m	14.23
Expamet plaster beads or other equal and approved; stainless steel; to brickwork or blockwork base							
angle bead; 20 mm thick	–	0.08	1.42	–	0.81	m	2.23
stop bead; 10 mm thick	–	0.07	1.25	–	3.11	m	4.36
stop bead; 13 mm thick	–	0.07	1.25	–	3.11	m	4.36
Expamet thin coat plaster beads or other equal and approved; galvanized steel; to timber base							
angle bead; 3 mm thick	–	0.07	1.25	–	0.59	m	1.84
stop bead; 3 mm thick	–	0.06	1.07	–	0.56	m	1.63
stop bead; 6 mm thick	–	0.06	1.07	–	0.58	m	1.65

28.10 WALLS; INSULATED RENDER

Sto Therm Classic M-system insulation render
70 mm EPS insulation fixed with adhesive to SFS structure (measured separately) with horizontal PVC intermediate track and vertical T-spines; with glassfibre mesh reinforcement embedded in Sto Armat Classic Basecoat Render and Stolit K 1.5 Decorative Topcoat Render (white)

Item	PC £	Labour hours	Labour £	Plant £	Material £	Unit	Total rate £
over 300 mm wide	–	–	–	–	–	m²	71.27

70 mm EPS insulation mechanically fixed to SFS structure (measured separately) with horizontal PVC intermediate track and vertical T-spines; with glassfibre mesh reinforcement embedded in Sto Armat Classic Basecoat Render and Stolit K 1.5 Decorative Topcoat Render (white)

Item	PC £	Labour hours	Labour £	Plant £	Material £	Unit	Total rate £
over 300 mm wide	–	–	–	–	–	m²	78.42
rendered heads and reveals not exceeding 100 mm wide; including angle beads	–	–	–	–	–	m	19.70
Extra for							
aluminium starter track at base of insulated render system	–	–	–	–	–	m	11.99
external angle with PVC mesh angle bead	–	–	–	–	–	m	5.27
internal angle with Sto Armor angle	–	–	–	–	–	m	5.27
render stop bead	–	–	–	–	–	m	5.27
Sto seal tape to all vertical abutments	–	–	–	–	–	m	4.90
Sto Armor mat HD mesh reinforcement to areas prone to physical damage (e.g. 1800 mm high adjoining floor level)							
over 300 mm wide	–	–	–	–	–	m²	15.76

28 FLOOR, WALL, CEILING AND ROOF FINISHES

Item	PC £	Labour hours	Labour £	Plant £	Material £	Unit	Total rate £
28.11 WALLS; CORK; CERAMIC TILING; MARBLE							
Glazed ceramic wall tiles; fixing with adhesive; butt joints; straight both ways; flush pointing with white grout; to plaster base. Ranges and prices vary enormously for tiles are toonumerous to schedule here. Prices are for good quality tiles							
Work to walls; over 600 mm wide							
152 mm × 152 mm × 5.50 mm thick; white	18.90	0.56	13.96	–	22.33	m²	**36.29**
152 mm × 152 mm × 5.50 mm thick; light colours	23.10	0.56	13.96	–	26.87	m²	**40.83**
152 mm × 152 mm × 5.50 mm thick; dark colours	31.50	0.56	13.96	–	35.96	m²	**49.92**
200 mm × 100 mm × 6.50 mm thick; white and light colours	21.00	0.56	13.96	–	24.61	m²	**38.57**
250 mm × 200 mm × 7.00 mm thick; white and light colours	26.25	0.56	13.96	–	30.28	m²	**44.24**
Work to walls; in staircase areas or plant rooms							
152 mm × 152 mm × 5.50 mm thick; white	18.90	0.62	15.45	–	22.33	m²	**37.78**
Work to walls; not exceeding 600 mm wide							
152 mm × 152 mm × 5.50 mm thick; white	11.34	0.56	13.96	–	13.35	m	**27.31**
152 mm × 152 mm × 5.50 mm thick; light colours	13.86	0.56	13.96	–	16.68	m	**30.64**
152 mm × 152 mm × 5.50 mm thick; dark colours	18.90	0.56	13.96	–	21.16	m	**35.12**
200 mm × 100 mm × 6.50 mm thick; white and light colours	12.60	0.56	13.96	–	14.71	m	**28.67**
250 mm × 200 mm × 7.00 mm thick; white and light colours	15.75	0.46	11.46	–	18.12	m	**29.58**
cutting around pipes; not exceeding 0.30 m girth	–	0.09	1.60	–	–	nr	**1.60**
Work to sills; 150 mm wide; rounded edge tiles							
152 mm × 152 mm × 5.50 mm thick; white	2.83	0.23	5.74	–	3.34	m	**9.08**
fitted end	–	0.09	1.60	–	–	nr	**1.60**
198 mm × 64.50 mm × 6 mm thick wall tiles; fixing with adhesive; butt joints; straight both ways; flush pointing with white grout; to plaster base							
Work to walls							
over 600 mm wide	30.00	1.67	41.62	–	35.89	m²	**77.51**
not exceeding 600 mm wide	18.00	1.30	32.40	–	21.48	m	**53.88**
20 mm × 20 mm × 5.50 mm thick glazed mosaic wall tiles; fixing with adhesive; butt joints; straight both ways; flush pointing with white grout; to plaster base							
Work to walls							
over 600 mm wide	35.00	1.76	43.87	–	41.66	m²	**85.53**
not exceeding 600 mm wide	21.00	1.38	34.40	–	25.87	m	**60.27**

28 FLOOR, WALL, CEILING AND ROOF FINISHES

Item	PC £	Labour hours	Labour £	Plant £	Material £	Unit	Total rate £
Dakota mahogany granite cladding; polished finish; jointed and pointed in coloured mortar (1:2:8)							
20 mm thick work to walls; to cement and sand base							
over 600 mm wide	–	–	–	–	–	m²	**432.42**
not exceeding 600 mm wide	–	–	–	–	–	m	**192.18**
40 mm thick work to walls; to cement and sand base							
over 600 mm wide	–	–	–	–	–	m²	**720.70**
not exceeding 600 mm wide	–	–	–	–	–	m	**324.32**
Roman Travertine marble cladding; polished finish; jointed and pointed in coloured mortar (1:2:8)							
20 mm thick work to walls; to cement and sand base							
over 600 mm wide	–	–	–	–	–	m²	**327.91**
not exceeding 600 mm wide	–	–	–	–	–	m	**150.15**
40 mm thick work to walls; to cement and sand base							
over 600 mm wide	–	–	–	–	–	m²	**451.64**
not exceeding 600 mm wide	–	–	–	–	–	m	**204.20**
Cork tiles; fixing with adhesive; butt joints; straight both ways; to plastered walls							
3 mm thick							
6.00 mm thick; pinboard 2 tone	6.66	0.45	11.22	–	10.35	m²	**21.57**
3.00 mm thick; standard natural finish	11.79	0.40	9.97	–	16.17	m²	**26.14**
4 mm thick							
4.00 mm thick; character natural finish	15.84	0.40	9.97	–	20.75	m²	**30.72**
28.12 WALLS; WATERPROOF PANELLING							
Waterproof panelling system with moisture-resistant MDF core and high pressure laminate backing.and decorative laminate face							
Supply and fit Showerwall 10.5 mm thick panels complete with aluminium trims, sealants and adhesive							
White gloss finish fixed to substrate with adhesive	–	0.25	5.79	–	63.54	m²	**69.33**
Perganmon Marble gloss finish fixed to substrate with adhesive	–	0.25	5.79	–	70.11	m²	**75.90**
28.13 CEILINGS; PLASTERED; RENDERED							
Plaster; first 8 mm or 11 mm coat of Thistle Bonding plaster; second 2 mm finishing coat of Thistle Multi-Finish plaster; steel trowelled finish							
10 mm thick work to ceilings; two coats; to concrete base							
over 600 mm wide	–	0.50	12.46	–	0.14	m²	**12.60**
over 600 wide; 3.50 m–5.00 m high	–	0.75	18.69	–	0.14	m²	**18.83**
over 600 mm wide; in staircase areas, plant rooms and similar restricted areas	–	0.75	18.69	–	0.14	m²	**18.83**
not exceeding 600 mm wide	–	0.33	8.30	–	0.07	m	**8.37**

28 FLOOR, WALL, CEILING AND ROOF FINISHES

Item	PC £	Labour hours	Labour £	Plant £	Material £	Unit	Total rate £
28.13 CEILINGS; PLASTERED; RENDERED – cont							
Plaster – cont							
10 mm thick work to isolated beams; two coats; to concrete base							
over 600 mm wide	–	1.10	27.42	–	0.14	m²	**27.56**
over 600 mm wide; 3.50 m–5.00 m high	–	1.20	29.91	–	0.14	m²	**30.05**
not exceeding 600 mm wide	–	0.50	12.46	–	0.07	m	**12.53**
Plaster; first 11 mm coat of Thistle Bonding plaster; second coat 2 mm finishing coat of Thistle Multi-Finish plaster; steel trowelled							
13 mm thick work to ceilings; to metal lathing base							
over 600 mm wide	–	0.50	12.46	–	0.14	m²	**12.60**
over 600 mm wide; in staircase areas, plant rooms and similar restricted areas	–	0.75	18.69	–	0.14	m²	**18.83**
not exceeding 600 mm wide	–	0.33	8.30	–	0.07	m	**8.37**
Plaster; one coat Thistle board finish or similar; steel trowelled 3 mm work to ceilings; one coat on and including gypsum plasterboard; fixing with nails; 3 mm joints filled with plaster and jute scrim cloth; to softwood base; plain grade baseboard or lath with rounded edges							
9.50 mm thick boards to ceilings							
over 600 mm wide	–	0.89	14.96	–	5.47	m²	**20.43**
over 600 mm wide; 3.50 m–5.00 m high	–	1.03	17.44	–	5.47	m²	**22.91**
not exceeding 600 mm wide	–	0.43	7.64	–	1.65	m	**9.29**
9.50 mm thick boards to ceilings; in staircase areas or plant rooms							
over 600 mm wide	–	0.98	16.55	–	5.47	m²	**22.02**
not exceeding 600 mm wide	–	0.47	8.35	–	1.65	m	**10.00**
9.50 mm thick boards to isolated beams							
over 600 mm wide	–	1.05	17.80	–	5.47	m²	**23.27**
not exceeding 600 mm wide	–	0.50	8.89	–	1.65	m	**10.54**
12.50 mm thick boards to ceilings							
over 600 mm wide	–	0.95	16.02	–	5.47	m²	**21.49**
over 600 mm wide; 3.50 m–5.00 m high	–	1.06	17.97	–	5.47	m²	**23.44**
not exceeding 600 mm wide	–	0.45	8.00	–	1.65	m	**9.65**
12.50 mm thick boards to ceilings; in staircase areas or plant rooms							
over 600 mm wide	–	1.06	17.97	–	5.47	m²	**23.44**
not exceeding 600 mm wide	–	0.51	9.06	–	1.65	m	**10.71**
12.50 mm thick boards to isolated beams							
over 600 mm wide	–	1.15	19.57	–	5.47	m²	**25.04**
not exceeding 600 mm wide	–	0.56	9.96	–	1.65	m	**11.61**

28 FLOOR, WALL, CEILING AND ROOF FINISHES

Item	PC £	Labour hours	Labour £	Plant £	Material £	Unit	Total rate £
Plaster; one coat Thistle board finish or other equal; steel trowelled							
3 mm thick work to walls or ceilings; one coat; to plasterboard base							
over 600 mm wide	–	0.25	6.23	–	0.06	m²	**6.29**
over 600 mm wide; in staircase areas, plant rooms and similar restricted areas	–	0.35	8.72	–	0.06	m²	**8.78**
not exceeding 600 mm wide	–	0.15	3.74	–	0.04	m	**3.78**
Cement–lime–sand (1:1:6); steel trowelled							
19 mm thick work to ceilings; three coats; to metal lathing base							
over 600 mm wide	–	0.95	23.68	0.44	0.96	m²	**25.08**
not exceeding 600 mm wide	–	0.40	9.97	0.22	4.49	m	**14.68**
Gyproc cove or other equal and approved; fixing with adhesive; filling and pointing joints with plaster							
Cove							
125 mm girth	–	0.19	3.38	–	1.31	m	**4.69**
angles	–	0.03	0.54	–	0.70	nr	**1.24**
28.14 METAL MESH LATHING FOR PLASTERED COATINGS							
Arch frames							
Preformed galvanized expanded steel semi-circular arch-frames; Expamet or other equal; to suit walls up to 230 mm thick							
for 760 mm opening; ref ESC 30	73.28	0.46	7.37	–	75.48	nr	**82.85**
for 920 mm opening; ref ESC 36	93.14	0.46	7.37	–	95.93	nr	**103.30**
for 1220 mm opening; ref ESC 48	125.93	0.46	7.37	–	129.71	nr	**137.08**
Lathing; Expamet BB expanded metal lathing or other equal; 50 mm laps							
Mesh linings to ceilings; fixing with staples; to softwood base; over 300 mm wide							
ref BB263; 0.500 mm thick; galvanized steel	3.05	0.56	8.98	–	3.14	m²	**12.12**
ref BB264; 0.675 mm thick; stainless steel	14.46	0.56	8.98	–	14.89	m²	**23.87**
Mesh linings to ceilings; fixing with wire; to steelwork; over 300 mm wide							
ref BB263; 0.500 mm thick; galvanized steel	–	0.59	9.48	–	3.14	m²	**12.62**
ref BB264; 0.675 mm thick; stainless steel	–	0.59	9.48	–	14.89	m²	**24.37**
Mesh linings to ceilings; fixing with wire; to steelwork; not exceeding 300 mm wide							
ref BB263; 0.500 mm thick; galvanized steel	–	0.37	5.91	–	3.14	m²	**9.05**
ref BB264; 0.675 mm thick; stainless steel	–	0.37	5.91	–	14.89	m²	**20.80**
raking cutting	–	0.19	3.38	–	–	m	**3.38**
cutting and fitting around pipes; not exceeding 0.30 m girth	–	0.28	4.97	–	–	nr	**4.97**

28 FLOOR, WALL, CEILING AND ROOF FINISHES

Item	PC £	Labour hours	Labour £	Plant £	Material £	Unit	Total rate £
28.14 METAL MESH LATHING FOR PLASTERED COATINGS – cont							
Lathing; Expamet Riblath or Spreedpro or other equal stiffened expanded metal lathing; 50 mm laps; 2.50 m × 0.60 m sheets							
Rigid rib mesh lining to walls; fixing with nails; to softwood base; over 300 mm wide							
Riblath ref 269; galvanized steel	8.19	0.46	7.37	–	8.58	m²	**15.95**
Riblath ref 267; stanless steel	15.81	0.46	7.37	–	16.43	m²	**23.80**
Rigid rib mesh lining to walls; fixing with nails; to softwood base; not exceeding 300 mm wide							
Riblath ref 269; galvanized steel	–	0.28	4.49	–	2.63	m	**7.12**
Riblath ref 267; stanless steel	–	0.28	4.49	–	4.99	m	**9.48**
Rigid rib mesh lining to ceilings; fixing with wire; to steelwork; over 300 mm wide							
Riblath ref 269; galvanized steel	–	0.59	9.48	–	9.05	m²	**18.53**
Riblath ref 267; stanless steel	–	0.59	9.48	–	16.90	m²	**26.38**
28.15 SPRAYED MINERAL FIBRE COATINGS							
Prepare and apply by spray Mandolite CP2 fire protection or other equal and approved on structural steel/metalwork							
16 mm thick (one hour) fire protection							
to walls and columns	–	–	–	–	–	m²	**12.37**
to ceilings and beams	–	–	–	–	–	m²	**13.65**
to isolated metalwork	–	–	–	–	–	m²	**27.20**
22 mm thick (one and a half hour) fire protection							
to walls and columns	–	–	–	–	–	m²	**14.39**
to ceilings and beams	–	–	–	–	–	m²	**15.95**
to isolated metalwork	–	–	–	–	–	m²	**31.92**
28 mm thick (two hour) fire protection							
to walls and columns	–	–	–	–	–	m²	**16.87**
to ceilings and beams	–	–	–	–	–	m²	**18.43**
to isolated metalwork	–	–	–	–	–	m²	**36.83**
52 mm thick (four hour) fire protection							
to walls and columns	–	–	–	–	–	m²	**25.50**
to ceilings and beams	–	–	–	–	–	m²	**28.40**
to isolated metalwork	–	–	–	–	–	m²	**56.52**
Prepare and apply by spray; cementitious Pyrok WF26 render or other equal and approved; on expanded metal lathing (not included)							
15 mm thick							
to ceilings and beams	–	–	–	–	–	m²	**39.15**

28 FLOOR, WALL, CEILING AND ROOF FINISHES

Item	PC £	Labour hours	Labour £	Plant £	Material £	Unit	Total rate £
28.16 TIMBER FLOOR AND WALL LININGS							
Wrought softwood							
Boarding to internal walls; tongued and grooved and V-jointed							
12 mm × 100 mm boards	–	0.74	17.12	–	22.89	m²	40.01
16 mm × 100 mm boards	–	0.74	17.12	–	24.72	m²	41.84
19 mm × 100 mm boards	–	0.74	17.12	–	28.16	m²	45.28
19 mm × 125 mm boards	–	0.69	15.97	–	25.85	m²	41.82
19 mm × 125 mm boards; chevron pattern	–	1.11	25.68	–	25.85	m²	51.53
25 mm × 125 mm boards	–	0.69	15.97	–	22.19	m²	38.16
12 mm × 100 mm boards; knotty pine	–	0.74	17.12	–	13.83	m²	30.95
Wood strip; 22 mm thick; Junckers All in Beech Sylva Sport Premium pretreated or other equal and approved; tongued and grooved joints; on bearers etc.; level fixing to cement and sand base							
Strip flooring; over 300 mm wide							
on 45 × 45 mm blue bat bearers	–	–	–	–	–	m²	76.68
on 10 mm Pro Foam	–	–	–	–	–	m²	87.59
on Uno bat 50 mm bearers	–	–	–	–	–	m²	80.22
on New Era levelling system	–	–	–	–	–	m²	82.82
on Uno bat 62 mm bearers	–	–	–	–	–	m²	81.52
on Duo bat 110 mm bearers	–	–	–	–	–	m²	136.81
Wood strip; 22 mm thick; Junckers pretreated or other equal and approved flooring systems; tongued and grooved joints; on bearers etc.; level fixing to cement and sand base							
Strip flooring; over 300 mm wide							
Sylva Squash Beech untreated on blue bat bearers	–	–	–	–	–	m²	80.40
Classic Beech clip system	–	–	–	–	–	m²	106.52
Harmoni Oak clip system	–	–	–	–	–	m²	111.54
Classic Beech on blue bat bearers	–	–	–	–	–	m²	107.65
Harmoni Oak on blue bat bearers	–	–	–	–	–	m²	110.23
Unfinished wood strip; 22 mm thick; Havwoods or other equal and approved; tongued and grooved joints; secret fixed; laid on semi-sprung bearers; fixing to cement and sand base; sanded and sealed							
Strip flooring; over 300 mm wide							
Prime Iroko	–	–	–	–	–	m²	103.74
Prime Maple	–	–	–	–	–	m²	103.74
American Oak	–	–	–	–	–	m²	113.72

29 DECORATION

Item	PC £	Labour hours	Labour £	Plant £	Material £	Unit	Total rate £
MATERIAL RATES							
BASIC PAINT PRICES – MATERIAL ONLY							
Paints							
matt emulsion	–	–	–	–	2.68	5 litre	**2.68**
gloss	–	–	–	–	23.50	5 litre	**23.50**
Weathershield gloss	–	–	–	–	59.51	5 litre	**59.51**
Weathershield undercoat	–	–	–	–	61.01	5 litre	**61.01**
Sandtex masonry paint – brilliant white	–	–	–	–	18.39	5 litre	**18.39**
Sandtex masonry paint – coloured	–	–	–	–	47.24	5 litre	**47.24**
Primer/undercoats							
oil-based undercoat	–	–	–	–	29.33	5 litre	**29.33**
acrylic	–	–	–	–	29.33	5 litre	**29.33**
red oxide	–	–	–	–	36.77	5 litre	**36.77**
water-based	–	–	–	–	27.71	5 litre	**27.71**
zinc phosphate	–	–	–	–	29.33	5 litre	**29.33**
MDF primer	–	–	–	–	66.46	5 litre	**66.46**
knotting solution	–	–	–	–	94.86	5 litre	**94.86**
Special paints							
solar reflective aluminium	–	–	–	–	42.31	5 litre	**42.31**
Coo-Var anti-graffiti coating; 2 pack; clear	–	–	–	–	149.41	5 litre	**149.41**
bituminous emulsion blacking paint	–	–	–	–	44.35	5 litre	**44.35**
Stains and preservatives							
Cuprinaol – clear	–	–	–	–	44.83	5 litre	**44.83**
Cuprinaol – boiled linseed oil	–	–	–	–	114.48	5 litre	**114.48**
Sadolin – Extra	–	–	–	–	62.37	5 litre	**62.37**
Sadolin – New Base	–	–	–	–	28.90	5 litre	**28.90**
Sikkens – Cetol HLS	–	–	–	–	76.83	5 litre	**76.83**
Sikkens – Cetol TSI	–	–	–	–	106.68	5 litre	**106.68**
Sikkens – Cetol Filter 7	–	–	–	–	115.26	5 litre	**115.26**
Protim Solignum – Architectural	–	–	–	–	92.65	5 litre	**92.65**
Protim Solignum – Green	–	–	–	–	92.65	5 litre	**92.65**
Protim Solignum – Cedar	–	–	–	–	92.65	5 litre	**92.65**
Varnishes							
polyurethane	–	–	–	–	58.59	5 litre	**58.59**
29.01 PAINTING AND CLEAR FINISHES – EXTERNAL							
NOTE: The following prices include for preparing surfaces. Painting woodwork also includes for knotting prior to applying the priming coat and for all stopping of nail holes etc.							
Two coats of cement paint, Sandtex Matt or other equal and approved							
Brick or block walls							
over 300 mm girth	–	0.26	4.62	–	2.36	m²	**6.98**
Cement render or concrete walls							
over 300 mm girth	–	0.23	4.09	–	1.56	m²	**5.65**
Roughcast walls							
over 300 mm girth	–	0.40	7.11	–	1.56	m²	**8.67**

29 DECORATION

Item	PC £	Labour hours	Labour £	Plant £	Material £	Unit	Total rate £
One coat sealer and two coats of external grade emulsion paint, Dulux Weathershield or other equal and approved							
Brick or block walls							
over 300 mm girth	–	0.43	7.64	–	10.26	m²	**17.90**
Cement render or concrete walls							
over 300 mm girth	–	0.35	6.22	–	6.84	m²	**13.06**
Concrete soffits							
over 300 mm girth	–	0.40	7.11	–	6.84	m²	**13.95**
One coat sealer (applied by brush) and two coats of external grade emulsion paint, Dulux Weathershield or other equal and approved (spray applied)							
Roughcast							
over 300 mm girth	–	0.29	5.16	–	13.99	m²	**19.15**
One coat sealer and two coats of anti-graffiti paint (spray applied)							
Brick or block walls							
over 300 mm girth	–	0.01	0.10	0.54	5.37	m²	**6.01**
Cement render or concrete walls							
over 300 mm girth	–	0.01	0.10	0.54	5.76	m²	**6.40**
2.5 mm of Vandalene or similar anti-climb paint (spray applied)							
General surfaces							
over 300 mm girth	–	0.01	0.10	0.54	5.25	m²	**5.89**
Two coats solar reflective aluminium paint; on bituminous roofing							
General surfaces							
over 300 mm girth	–	0.44	7.82	–	14.32	m²	**22.14**
Touch up primer; two undercoats and one finishing coat of gloss oil paint; on wood surfaces							
General surfaces							
over 300 mm girth	–	0.35	6.22	–	4.12	m²	**10.34**
isolated surfaces not exceeding 300 mm girth	–	0.15	2.67	–	1.11	m	**3.78**
isolated areas not exceeding 1.00 m²; irrespective of girth	–	0.27	4.80	–	2.21	nr	**7.01**
Glazed windows and screens							
panes; area not exceeding 0.10 m²	–	0.59	10.49	–	3.65	m²	**14.14**
panes; area 0.10 m²–0.50 m²	–	0.59	10.49	–	3.12	m²	**13.61**
panes; area 0.50 m²–1.00 m²	–	0.47	8.35	–	2.69	m²	**11.04**
panes; area over 1.00 m²	–	0.35	6.22	–	2.21	m²	**8.43**
Glazed windows and screens; multi-coloured work							
panes; area not exceeding 0.10 m²	–	0.68	12.09	–	3.65	m²	**15.74**
panes; area 0.10 m²–0.50 m²	–	0.55	9.77	–	3.17	m²	**12.94**
panes; area 0.50 m²–1.00 m²	–	0.47	8.35	–	2.69	m²	**11.04**
panes; area over 1.00 m²	–	0.41	7.29	–	2.21	m²	**9.50**

29 DECORATION

Item	PC £	Labour hours	Labour £	Plant £	Material £	Unit	Total rate £
29.01 PAINTING AND CLEAR FINISHES – EXTERNAL – cont							
Knot; one coat primer; two undercoats and one finishing coat of gloss oil paint; on wood surfaces							
General surfaces							
over 300 mm girth	–	0.46	8.18	–	3.71	m²	**11.89**
isolated surfaces not exceeding 300 mm girth	–	0.19	3.38	–	5.37	m	**8.75**
isolated areas not exceeding 1.00 m²; irrespective of girth	–	0.35	6.22	–	2.30	nr	**8.52**
Glazed windows and screens							
panes; area not exceeding 0.10 m²	–	0.78	13.86	–	4.13	m²	**17.99**
panes; area 0.10 m²–0.50 m²	–	0.62	11.02	–	3.69	m²	**14.71**
panes; area 0.50 m²–1.00 m²	–	0.55	9.77	–	2.84	m²	**12.61**
panes; area over 1.00 m²	–	0.46	8.18	–	2.00	m²	**10.18**
Glazed windows and screens; multi-coloured work							
panes; area not exceeding 0.10 m²	–	0.89	15.82	–	4.13	m²	**19.95**
panes; area 0.10 m²–0.50 m²	–	0.72	12.80	–	3.71	m²	**16.51**
panes; area 0.50 m²–1.00 m²	–	0.64	11.38	–	2.84	m²	**14.22**
panes; area over 1.00 m²	–	0.54	9.60	–	2.00	m²	**11.60**
Touch up primer; two undercoats and one finishing coat of gloss oil paint; on iron or steel surfaces							
General surfaces							
over 300 mm girth	–	0.35	6.22	–	2.21	m²	**8.43**
isolated surfaces not exceeding 300 mm girth	–	0.14	2.49	–	0.63	m	**3.12**
isolated areas not exceeding 1.00 m²; irrespective of girth	–	0.26	4.62	–	1.28	nr	**5.90**
Glazed windows and screens							
panes; area not exceeding 0.10 m²	–	0.59	10.49	–	2.35	m²	**12.84**
panes; area 0.10 m²–0.50 m²	–	0.47	8.35	–	2.04	m²	**10.39**
panes; area 0.50 m²–1.00 m²	–	0.41	7.29	–	1.74	m²	**9.03**
panes; area over 1.00 m²	–	0.35	6.22	–	1.43	m²	**7.65**
Structural steelwork							
over 300 mm girth	–	0.40	7.11	–	2.37	m²	**9.48**
Members of roof trusses							
over 300 mm girth	–	0.54	9.60	–	2.68	m²	**12.28**
Ornamental railings and the like; each side measured overall							
over 300 mm girth	–	0.60	10.67	–	2.77	m²	**13.44**
Eaves gutters							
over 300 mm girth	–	0.64	11.38	–	2.66	m²	**14.04**
not exceeding 300 mm girth	–	0.25	4.45	–	1.14	m	**5.59**
Pipes or conduits							
over 300 mm girth	–	0.54	9.60	–	2.66	m²	**12.26**
not exceeding 300 mm girth	–	0.21	3.73	–	0.92	m	**4.65**

29 DECORATION

Item	PC £	Labour hours	Labour £	Plant £	Material £	Unit	Total rate £
One coat primer; two undercoats and one finishing coat of gloss oil paint; on iron or steel surfaces							
General surfaces							
over 300 mm girth	–	0.43	7.64	–	2.60	m²	**10.24**
isolated surfaces not exceeding 300 mm girth	–	0.18	3.20	–	0.69	m	**3.89**
isolated areas not exceeding 1.00 m²; irrespective of girth	–	0.32	5.69	–	1.37	nr	**7.06**
Glazed windows and screens							
panes; area not exceeding 0.10 m²	–	0.71	12.62	–	2.44	m²	**15.06**
panes; area 0.10 m²–0.50 m²	–	0.56	9.96	–	2.14	m²	**12.10**
panes; area 0.50 m²–1.00 m²	–	0.50	8.89	–	1.83	m²	**10.72**
panes; area over 1.00 m²	–	0.43	7.64	–	1.37	m²	**9.01**
Structural steelwork							
over 300 mm girth	–	0.48	8.53	–	2.75	m²	**11.28**
Members of roof trusses							
over 300 mm girth	–	0.64	11.38	–	3.05	m²	**14.43**
Ornamental railings and the like; each side measured overall							
over 300 mm girth	–	0.72	12.80	–	3.05	m²	**15.85**
Eaves gutters							
over 300 mm girth	–	0.76	13.51	–	2.88	m²	**16.39**
not exceeding 300 mm girth	–	0.31	5.51	–	1.01	m	**6.52**
Pipes or conduits							
over 300 mm girth	–	0.64	11.38	–	2.88	m²	**14.26**
not exceeding 300 mm girth	–	0.25	4.45	–	5.01	m	**9.46**
One coat of Hammerite paint or other equal; on iron or steel surfaces							
General surfaces							
over 300 mm girth	–	0.15	2.67	–	2.78	m²	**5.45**
isolated surfaces not exceeding 300 mm girth	–	0.08	1.42	–	0.88	m	**2.30**
isolated areas not exceeding 1.00 m²; irrespective of girth	–	0.11	1.96	–	1.60	nr	**3.56**
Glazed windows and screens							
panes; area not exceeding 0.10 m²	–	0.25	4.45	–	2.06	m²	**6.51**
panes; area 0.10 m²–0.50 m²	–	0.19	3.38	–	2.34	m²	**5.72**
panes; area 0.50 m²–1.00 m²	–	0.18	3.20	–	2.11	m²	**5.31**
panes; area over 1.00 m²	–	0.15	2.67	–	2.11	m²	**4.78**
Structural steelwork							
over 300 mm girth	–	0.17	3.02	–	2.55	m²	**5.57**
Members of roof trusses							
over 300 mm girth	–	0.23	4.09	–	2.78	m²	**6.87**
Ornamental railings and the like; each side measured overall							
over 300 mm girth	–	0.26	4.62	–	2.78	m²	**7.40**
Eaves gutters							
over 300 mm girth	–	0.27	4.80	–	3.01	m²	**7.81**
not exceeding 300 mm girth	–	0.08	1.42	–	1.48	m	**2.90**
Pipes or conduits							
over 300 mm girth	–	0.26	4.62	–	2.55	m²	**7.17**
not exceeding 300 mm girth	–	0.08	1.42	–	1.21	m	**2.63**

29 DECORATION

Item	PC £	Labour hours	Labour £	Plant £	Material £	Unit	Total rate £
29.01 PAINTING AND CLEAR FINISHES – EXTERNAL – cont							
Two coats of creosote; on wood surfaces							
General surfaces							
over 300 mm girth	–	0.16	2.84	–	5.58	m²	**8.42**
isolated surfaces not exceeding 300 mm girth	–	0.01	0.10	–	2.99	m	**3.09**
Two coats of Solignum wood preservative or other equal and approved; on wood surfaces							
General surfaces							
over 300 mm girth	–	0.14	2.49	–	4.82	m²	**7.31**
isolated surfaces not exceeding 300 mm girth	–	0.05	0.89	–	1.42	m	**2.31**
Three coats of polyurethane; on wood surfaces							
General surfaces							
over 300 mm girth	–	0.29	5.16	–	3.90	m²	**9.06**
isolated surfaces not exceeding 300 mm girth	–	0.11	1.96	–	1.96	m	**3.92**
isolated areas not exceeding 1.00 m²;							
irrespective of girth	–	0.21	3.73	–	2.26	nr	**5.99**
Two coats of New Base primer or other equal and approved; and two coats of Extra or other equal and approved; Sadolin Ltd; pigmented; on wood surfaces							
General surfaces							
over 300 mm girth	–	0.43	7.64	–	4.53	m²	**12.17**
isolated surfaces not exceeding 300 mm girth	–	0.13	2.31	–	0.80	m	**3.11**
Glazed windows and screens							
panes; area not exceeding 0.10 m²	–	0.71	12.62	–	3.25	m²	**15.87**
panes; area 0.10 m²–0.50 m²	–	0.57	10.14	–	3.08	m²	**13.22**
panes; area 0.50 m²–1.00 m²	–	0.50	8.89	–	2.89	m²	**11.78**
panes; area over 1.00 m²	–	0.43	7.64	–	2.35	m²	**9.99**
Two coats Sikkens Cetol Filter 7 exterior stain or other equal and approved; on wood surfaces							
General surfaces							
over 300 mm girth	–	0.20	3.55	–	5.83	m²	**9.38**
isolated surfaces not exceeding 300 mm girth	–	0.09	1.60	–	2.05	m	**3.65**
isolated areas not exceeding 1.00 m²;							
irrespective of girth	–	0.14	2.49	–	2.99	nr	**5.48**

29 DECORATION

Item	PC £	Labour hours	Labour £	Plant £	Material £	Unit	Total rate £
29.02 PAINTING AND CLEAR FINISHES – INTERNAL							
NOTE: The following prices include for preparing surfaces. Painting woodwork also includes for knotting prior to applying the priming coat and for all stopping of nail holes etc.							
Touch up primer; two undercoats and one finishing coat of gloss oil paint; on wood surfaces							
General surfaces							
over 300 mm girth	–	0.35	6.22	–	4.12	m²	**10.34**
isolated surfaces not exceeding 300 mm girth	–	0.15	2.67	–	1.11	m	**3.78**
isolated areas not exceeding 1.00 m²;							
irrespective of girth	–	0.27	4.80	–	2.21	nr	**7.01**
One coat primer; on wood surfaces before fixing							
General surfaces							
over 300 mm girth	–	0.08	1.42	–	2.02	m²	**3.44**
isolated surfaces not exceeding 300 mm girth	–	0.02	0.36	–	0.72	m	**1.08**
isolated areas not exceeding 1.00 m²;							
irrespective of girth	–	0.06	1.07	–	0.48	nr	**1.55**
One coat polyurethane sealer; on wood surfaces before fixing							
General surfaces							
over 300 mm girth	–	0.10	1.78	–	1.36	m²	**3.14**
isolated surfaces not exceeding 300 mm girth	–	0.03	0.54	–	0.49	m	**1.03**
isolated areas not exceeding 1.00 m²;							
irrespective of girth	–	0.08	1.42	–	0.63	nr	**2.05**
One coat of Sikkens Cetol HLS stain or other equal; on wood surfaces before fixing							
General surfaces							
over 300 mm girth	–	0.11	1.96	–	1.46	m²	**3.42**
isolated surfaces not exceeding 300 mm girth	–	0.03	0.54	–	1.05	m	**1.59**
isolated areas not exceeding 1.00 m²;							
irrespective of girth	–	0.08	1.42	–	0.69	nr	**2.11**
One coat of Sikkens Cetol TS interior stain or other equal; on wood surfaces before fixing							
General surfaces							
over 300 mm girth	–	0.11	1.96	–	1.96	m²	**3.92**
isolated surfaces not exceeding 300 mm girth	–	0.03	0.54	–	0.91	m	**1.45**
isolated areas not exceeding 1.00 m²;							
irrespective of girth	–	0.08	1.42	–	0.93	nr	**2.35**

29 DECORATION

Item	PC £	Labour hours	Labour £	Plant £	Material £	Unit	Total rate £
29.02 PAINTING AND CLEAR FINISHES – INTERNAL – cont							
One coat Cuprinol clear wood preservative or other equal; on wood surfaces before fixing							
General surfaces							
over 300 mm girth	–	0.08	1.42	–	1.14	m²	**2.56**
isolated surfaces not exceeding 300 mm girth	–	0.02	0.36	–	0.42	m	**0.78**
isolated areas not exceeding 1.00 m²; irrespective of girth	–	0.05	0.89	–	0.53	nr	**1.42**
One coat HCC Protective Coatings Ltd Permacor urethane alkyd gloss finishing coat or other equal; on previously primed steelwork							
Members of roof trusses							
over 300 mm girth	–	0.15	2.67	–	0.67	m²	**3.34**
Two coats emulsion paint							
Brick or block walls							
over 300 mm girth	–	0.21	3.73	–	0.37	m²	**4.10**
Cement render or concrete							
over 300 mm girth	–	0.20	3.55	–	0.36	m²	**3.91**
isolated surfaces not exceeding 300 mm girth	–	0.10	1.78	–	0.19	m	**1.97**
Plaster walls or plaster/plasterboard ceilings							
over 300 mm girth	–	0.18	3.20	–	0.37	m²	**3.57**
over 300 mm girth; in multi colours	–	0.24	4.26	–	0.38	m²	**4.64**
over 300 mm girth; in staircase areas	–	0.21	3.73	–	0.37	m²	**4.10**
cutting in edges on flush surfaces	–	0.08	1.42	–	–	m	**1.42**
Plaster/plasterboard ceilings							
over 300 mm girth; 3.50 m–5.00 m high	–	0.21	3.73	–	0.37	m²	**4.10**
One mist and two coats emulsion paint							
Brick or block walls							
over 300 mm girth	–	0.19	3.38	–	0.47	m²	**3.85**
Cement render or concrete							
over 300 mm girth	–	0.19	3.38	–	0.46	m²	**3.84**
Plaster walls or plaster/plasterboard ceilings							
over 300 mm girth	–	0.18	3.20	–	0.46	m²	**3.66**
over 300 mm girth; in multi colours	–	0.25	4.45	–	0.46	m²	**4.91**
over 300 mm girth; in staircase areas	–	0.21	3.73	–	0.46	m²	**4.19**
cutting in edges on flush surfaces	–	0.09	1.60	–	–	m	**1.60**
Plaster/plasterboard ceilings							
over 300 mm girth; 3.50 m–5.00 m high	–	0.21	3.73	–	0.46	m²	**4.19**
One mist Supermatt; one full Supermatt and one full coat of quick drying Acrylic Eggshell							
Brick or block walls							
over 300 mm girth	–	0.19	3.38	–	2.06	m²	**5.44**

29 DECORATION

Item	PC £	Labour hours	Labour £	Plant £	Material £	Unit	Total rate £
Cement render or concrete							
over 300 mm girth	–	0.19	3.38	–	1.92	m²	**5.30**
Plaster walls or plaster/plasterboard ceilings							
over 300 mm girth	–	0.18	3.20	–	1.92	m²	**5.12**
over 300 mm girth; in multi colours	–	0.25	4.45	–	1.92	m²	**6.37**
over 300 mm girth; in staircase areas	–	0.21	3.73	–	1.92	m²	**5.65**
cutting in edges on flush surfaces	–	0.09	1.60	–	–	m	**1.60**
Plaster/plasterboard ceilings							
over 300 mm girth; 3.50 m–5.00 m high	–	0.21	3.73	–	1.92	m²	**5.65**
Clean and prepare surfaces, one coat Keim dilution, one coat primer and two coats of Keim Ecosil paint							
Brick or block walls							
over 300 mm girth	–	0.30	5.34	–	5.82	m²	**11.16**
Cement render or concrete							
over 300 mm girth	–	0.30	5.34	–	5.10	m²	**10.44**
Plaster walls or plaster/plasterboard ceilings							
over 300 mm girth	–	0.30	5.34	–	5.10	m²	**10.44**
over 300 mm girth; in staircase areas	–	0.30	5.34	–	5.10	m²	**10.44**
cutting in edges on flush surfaces	–	0.09	1.60	–	–	m	**1.60**
Plaster/plasterboard ceilings							
over 300 mm girth; 3.50 m–5.00 m high	–	0.25	4.45	–	5.38	m²	**9.83**
One coat Tretol No 10 Sealer or other equal; two coats Tretol sprayed Supercover Spraytone emulsion paint or other equal and approved							
Plaster walls or plaster/plasterboard ceilings							
over 300 mm girth	–	0.25	6.23	–	0.32	m²	**6.55**
Textured plastic; Artex or other equal finish							
Plasterboard ceilings							
over 300 mm girth	–	0.19	3.38	–	3.84	m²	**7.22**
Touch up primer; one undercoat and one finishing coat of gloss oil paint; on wood surfaces							
General surfaces							
over 300 mm girth	–	0.20	3.55	–	3.46	m²	**7.01**
isolated surfaces not exceeding 300 mm girth	–	0.08	1.42	–	1.22	m	**2.64**
isolated areas not exceeding 1.00 m²; irrespective of girth	–	0.18	3.20	–	1.88	nr	**5.08**
Glazed windows and screens							
panes; area not exceeding 0.10 m²	–	0.38	6.76	–	3.13	m²	**9.89**
panes; area 0.10 m²–0.50 m²	–	0.31	5.51	–	2.31	m²	**7.82**
panes; area 0.50 m²–1.00 m²	–	0.26	4.62	–	1.92	m²	**6.54**
panes; area over 1.00 m²	–	0.23	4.09	–	1.65	m²	**5.74**

29 DECORATION

Item	PC £	Labour hours	Labour £	Plant £	Material £	Unit	Total rate £
29.02 PAINTING AND CLEAR FINISHES – INTERNAL – cont							
Knot; one coat primer; stop; one undercoat and one finishing coat of gloss oil paint; on wood surfaces							
General surfaces							
over 300 mm girth	–	0.33	5.87	–	4.09	m²	**9.96**
isolated surfaces not exceeding 300 mm girth	–	0.13	2.31	–	1.45	m	**3.76**
isolated areas not exceeding 0.50 m²; irrespective of girth	–	0.25	4.45	–	2.79	nr	**7.24**
Glazed windows and screens							
panes; area not exceeding 0.10 m²	–	0.56	9.96	–	4.09	m²	**14.05**
panes; area 0.10 m²–0.50 m²	–	0.45	8.00	–	3.40	m²	**11.40**
panes; area 0.50 m²–1.00 m²	–	0.40	7.11	–	3.40	m²	**10.51**
panes; area over 1.00 m²	–	0.33	5.87	–	2.59	m²	**8.46**
One coat primer; one undercoat and one finishing coat of gloss oil paint							
Plaster surfaces							
over 300 mm girth	–	0.30	5.34	–	5.23	m²	**10.57**
One coat primer; two undercoats and one finishing coat of gloss oil paint							
Plaster surfaces							
over 300 mm girth	–	0.40	7.11	–	7.23	m²	**14.34**
One coat primer; two undercoats and one finishing coat of eggshell paint							
Plaster surfaces							
over 300 mm girth	–	0.40	7.11	–	7.29	m²	**14.40**
Touch up primer; one undercoat and one finishing coat of gloss paint; on iron or steel surfaces							
General surfaces							
over 300 mm girth	–	0.23	4.09	–	2.98	m²	**7.07**
isolated surfaces not exceeding 300 mm girth	–	0.09	1.60	–	1.04	m	**2.64**
isolated areas not exceeding 0.50 m²; irrespective of girth	–	0.18	3.20	–	1.65	nr	**4.85**
Glazed windows and screens							
panes; area not exceeding 0.10 m²	–	0.38	6.76	–	3.10	m²	**9.86**
panes; area 0.10 m²–0.50 m²	–	0.31	5.51	–	2.32	m²	**7.83**
panes; area 0.50 m²–1.00 m²	–	0.26	4.62	–	1.85	m²	**6.47**
panes; area over 1.00 m²	–	0.23	4.09	–	1.59	m²	**5.68**
Structural steelwork							
over 300 mm girth	–	0.25	4.45	–	3.13	m²	**7.58**
Members of roof trusses							
over 300 mm girth	–	0.34	6.05	–	3.61	m²	**9.66**

29 DECORATION

Item	PC £	Labour hours	Labour £	Plant £	Material £	Unit	Total rate £
Ornamental railings and the like; each side measured overall							
over 300 mm girth	–	0.40	7.11	–	3.91	m²	**11.02**
Iron or steel radiators							
over 300 mm girth	–	0.23	4.09	–	3.31	m²	**7.40**
Pipes or conduits							
over 300 mm girth	–	0.34	6.05	–	3.46	m²	**9.51**
not exceeding 300 mm girth	–	0.13	2.31	–	1.19	m	**3.50**
One coat primer; one undercoat and one finishing coat of gloss oil paint; on iron or steel surfaces							
General surfaces							
over 300 mm girth	–	0.30	5.34	–	3.01	m²	**8.35**
isolated surfaces not exceeding 300 mm girth	–	0.12	2.13	–	1.88	m	**4.01**
isolated areas not exceeding 0.50 m²; irrespective of girth	–	0.23	4.09	–	3.02	nr	**7.11**
Glazed windows and screens							
panes; area not exceeding 0.10 m²	–	0.50	8.89	–	4.92	m²	**13.81**
panes; area 0.10 m²–0.50 m²	–	0.40	7.11	–	3.99	m²	**11.10**
panes; area 0.50 m²–1.00 m²	–	0.34	6.05	–	3.43	m²	**9.48**
panes; area over 1.00 m²	–	0.30	5.34	–	3.02	m²	**8.36**
Structural steelwork							
over 300 mm girth	–	0.33	5.87	–	4.86	m²	**10.73**
Members of roof trusses							
over 300 mm girth	–	0.45	8.00	–	5.08	m²	**13.08**
Ornamental railings and the like; each side measured overall							
over 300 mm girth	–	0.51	9.06	–	6.03	m²	**15.09**
Iron or steel radiators							
over 300 mm girth	–	0.30	5.34	–	5.08	m²	**10.42**
Pipes or conduits							
over 300 mm girth	–	0.45	8.00	–	5.08	m²	**13.08**
not exceeding 300 mm girth	–	0.18	3.20	–	1.68	m	**4.88**
Two coats of bituminous paint; on iron or steel surfaces							
General surfaces							
over 300 mm girth	–	0.23	4.09	–	2.13	m²	**6.22**
Inside of galvanized steel cistern							
over 300 mm girth	–	0.34	6.05	–	2.59	m²	**8.64**
Two coats bituminous paint; first coat blinded with clean sand prior to second coat; on concrete surfaces							
General surfaces							
over 300 mm girth	–	0.79	14.05	–	4.91	m²	**18.96**

29 DECORATION

Item	PC £	Labour hours	Labour £	Plant £	Material £	Unit	Total rate £
29.02 PAINTING AND CLEAR FINISHES – INTERNAL – cont							
Mordant solution; one coat HCC Protective Coatings Ltd Permacor Alkyd MIO or other equal; one coat Permatex Epoxy Gloss finishing coat or other equal and approved on galvanized steelwork							
Structural steelwork							
over 300 mm girth	–	0.44	7.82	–	3.59	m^2	**11.41**
One coat HCC Protective Coatings Ltd Epoxy Zinc Primer or other equal and approved; two coats Permacor Alkyd MIO or other equal; one coat Permacor Epoxy Gloss finishing coat or other equal and approved on steelwork							
Structural steelwork							
over 300 mm girth	–	0.63	11.20	–	7.30	m^2	**18.50**
Steel protection; HCC Protective Coatings Ltd Unitherm or other equal; two coats to steelwork							
Structural steelwork							
over 300 mm girth	–	0.99	17.60	–	2.61	m^2	**20.21**
Two coats of Dulux Floorshield floor paint; on screeded concrete surfaces							
General surfaces							
over 300 mm girth	–	0.25	4.45	–	9.03	m^2	**13.48**
Nitoflor Hardtop STD floor hardener and dust proofer or other equal; Fosroc Expandite Ltd; two coats; on concrete surfaces							
General surfaces							
over 300 mm girth	–	0.24	3.43	–	0.13	m^2	**3.56**
Two coats of boiled linseed oil; on hardwood surfaces							
General surfaces							
over 300 mm girth	–	0.18	3.20	–	8.82	m^2	**12.02**
isolated surfaces not exceeding 300 mm girth	–	0.07	1.25	–	2.79	m	**4.04**
isolated areas not exceeding 0.50 m^2;							
irrespective of girth	–	0.13	2.31	–	5.14	nr	**7.45**
Two coats polyurethane varnish; on wood surfaces							
General surfaces							
over 300 mm girth	–	0.18	3.20	–	2.41	m^2	**5.61**
isolated surfaces not exceeding 300 mm girth	–	0.07	1.25	–	0.91	m	**2.16**
isolated areas not exceeding 0.50 m^2;							
irrespective of girth	–	0.13	2.31	–	0.34	nr	**2.65**

29 DECORATION

Item	PC £	Labour hours	Labour £	Plant £	Material £	Unit	Total rate £
Three coats polyurethane varnish; on wood surfaces							
General surfaces							
over 300 mm girth	–	0.26	4.62	–	3.62	m²	**8.24**
isolated surfaces not exceeding 300 mm girth	–	0.10	1.78	–	1.16	m	**2.94**
isolated areas not exceeding 0.50 m²;							
irrespective of girth	–	0.19	3.38	–	2.03	nr	**5.41**
One undercoat; and one finishing coat; of Albi clear flame retardant surface coating or other equal; on wood surfaces							
General surfaces							
over 300 mm girth	–	0.34	6.05	–	6.73	m²	**12.78**
isolated surfaces not exceeding 300 mm girth	–	0.14	2.49	–	2.36	m	**4.85**
isolated areas not exceeding 0.50 m²;							
irrespective of girth	–	0.19	3.38	–	5.17	nr	**8.55**
Two undercoats; and one finishing coat; of Albi clear flame retardant surface coating or other equal; on wood surfaces							
General surfaces							
over 300 mm girth	–	0.40	7.11	–	8.54	m²	**15.65**
isolated surfaces not exceeding 300 mm girth	–	0.20	3.55	–	3.44	m	**6.99**
isolated areas not exceeding 0.50 m²;							
irrespective of girth	–	0.33	5.87	–	4.60	nr	**10.47**
Seal and wax polish; dull gloss finish on wood surfaces							
General surfaces							
over 300 mm girth	–	0.50	12.46	–	0.36	m²	**12.82**
isolated surfaces not exceeding 300 mm girth	–	0.25	6.23	–	0.07	m	**6.30**
isolated areas not exceeding 0.50 m²;							
irrespective of girth	–	0.35	8.72	–	0.16	nr	**8.88**
One coat of Sadolin Extra or other equal; clear or pigmented; one further coat of Holdex clear interior silk matt lacquer or similar							
General surfaces							
over 300 mm girth	–	0.25	4.45	–	5.99	m²	**10.44**
isolated surfaces not exceeding 300 mm girth	–	0.10	1.78	–	2.81	m	**4.59**
isolated areas not exceeding 0.50 m²;							
irrespective of girth	–	0.20	3.55	–	2.90	nr	**6.45**
Glazed windows and screens							
panes; area not exceeding 0.10 m²	–	0.42	7.47	–	3.42	m²	**10.89**
panes; area 0.10 m²–0.50 m²	–	0.33	5.87	–	3.18	m²	**9.05**
panes; area 0.50 m²–1.00 m²	–	0.29	5.16	–	2.96	m²	**8.12**
panes; area over 1.00 m²	–	0.25	4.45	–	2.81	m²	**7.26**

Prices for Measured Works

29 DECORATION

Item	PC £	Labour hours	Labour £	Plant £	Material £	Unit	Total rate £
29.02 PAINTING AND CLEAR FINISHES – **INTERNAL – cont**							
Two coats of Sadolin Extra or other equal; clear **or pigmented; two further coats of PV67 clear** **interior silk matt lacquer or similar**							
General surfaces							
over 300 mm girth	–	0.40	7.11	–	10.93	m²	**18.04**
isolated surfaces not exceeding 300 mm girth	–	0.16	2.84	–	5.46	m	**8.30**
isolated areas not exceeding 0.50 m²;							
irrespective of girth	–	0.30	5.34	–	6.22	nr	**11.56**
Glazed windows and screens							
panes; area not exceeding 0.10 m²	–	0.66	11.73	–	6.67	m²	**18.40**
panes; area 0.10 m²–0.50 m²	–	0.52	9.25	–	6.22	m²	**15.47**
panes; area 0.50 m²–1.00 m²	–	0.45	8.00	–	5.77	m²	**13.77**
panes; area over 1.00 m²	–	0.40	7.11	–	5.46	m²	**12.57**
Two coats of Sikkens Cetol TSI interior stain or **other equal; on wood surfaces**							
General surfaces							
over 300 mm girth	–	0.19	3.38	–	3.44	m²	**6.82**
isolated surfaces not exceeding 300 mm girth	–	0.08	1.42	–	1.25	m	**2.67**
isolated areas not exceeding 0.50 m²;							
irrespective of girth	–	0.13	2.31	–	1.91	nr	**4.22**
Body in and wax polish; dull gloss finish; on **hardwood surfaces**							
General surfaces							
over 300 mm girth	–	0.60	14.96	–	0.16	m²	**15.12**
isolated surfaces not exceeding 300 mm girth	–	0.30	7.48	–	0.07	m	**7.55**
isolated areas not exceeding 0.50 m²;							
irrespective of girth	–	0.40	9.97	–	0.07	nr	**10.04**
Stain; body in and wax polish; dull gloss finish; **on hardwood surfaces**							
General surfaces							
over 300 mm girth	–	0.75	18.69	–	0.16	m²	**18.85**
isolated surfaces not exceeding 300 mm girth	–	0.30	7.48	–	0.16	m	**7.64**
isolated areas not exceeding 0.50 m²;							
irrespective of girth	–	0.50	12.46	–	0.16	nr	**12.62**
Seal; two coats of synthetic resin lacquer; **decorative flatted finish; wire down, wax and** **burnish; on wood surfaces**							
General surfaces							
over 300 mm girth	–	1.00	24.93	–	1.02	m²	**25.95**
isolated surfaces not exceeding 300 mm girth	–	0.40	9.97	–	0.73	m	**10.70**
isolated areas not exceeding 0.50 m²;							
irrespective of girth	–	0.65	16.20	–	0.73	nr	**16.93**

29 DECORATION

Item	PC £	Labour hours	Labour £	Plant £	Material £	Unit	Total rate £
Stain; body in and fully French polish; full gloss finish; on hardwood surfaces							
General surfaces							
over 300 mm girth	–	1.10	27.42	–	2.10	m²	**29.52**
isolated surfaces not exceeding 300 mm girth	–	0.50	12.46	–	1.11	m	**13.57**
isolated areas not exceeding 0.50 m²;							
irrespective of girth	–	0.75	18.69	–	2.07	nr	**20.76**
Stain; fill grain and fully French polish; full gloss finish; on hardwood surfaces							
General surfaces							
over 300 mm girth	–	1.60	39.88	–	3.33	m²	**43.21**
isolated surfaces not exceeding 300 mm girth	–	0.60	14.96	–	1.11	m	**16.07**
isolated areas not exceeding 0.50 m²;							
irrespective of girth	–	1.00	24.93	–	2.07	nr	**27.00**
Stain black; body in and fully French polish; ebonized finish; on hardwood surfaces							
General surfaces							
over 300 mm girth	–	1.75	43.62	–	3.81	m²	**47.43**
isolated surfaces not exceeding 300 mm girth	–	0.80	19.94	–	1.11	m	**21.05**
isolated areas not exceeding 0.50 m²;							
irrespective of girth	–	1.30	32.40	–	2.07	nr	**34.47**
29.03 DECORATIVE PAPERS OR FABRICS							
Lining paper; and hanging							
Plaster walls or columns							
over 1.00 m² (PC £ per roll)	3.00	0.19	3.38	–	0.42	m²	**3.80**
Plaster ceilings or beams							
over 1.00 m² girth (PC £ per roll)	3.00	0.23	4.09	–	0.42	m²	**4.51**
Decorative paper-backed vinyl wallpaper; and hanging. Because of the enormous range of papers available we have simply allowed a prime cost amount to purchase the paper							
Plaster walls or columns							
over 1.00 m² girth (PC £ per roll)	30.00	0.23	4.09	–	5.24	m²	**9.33**
PVC wall lining; Altro Whiterock; or other equal; fixed directly to plastered brick or blockwork with Altrofix adhesive							
Work to walls over 300 mm wide; over 100 m² total area							
Standard white	19.53	1.25	22.22	–	27.33	m²	**49.55**
Satins	29.03	1.25	22.22	–	37.59	m²	**59.81**
Chameleon	48.82	1.25	22.22	–	59.00	m²	**81.22**

Prices for Measured Works

30 SUSPENDED CEILINGS

Item	PC £	Labour hours	Labour £	Plant £	Material £	Unit	Total rate £
30.01 SUSPENDED CEILINGS							
Gyproc M/F suspended ceiling system or other equal approved; hangers plugged and screwed to concrete soffit, 900 mm × 1800 mm × 12.50 mm tapered edge wallboard infill; joints filled with joint filler and taped to receive direct decoration							
Lining to ceilings; hangers 150 mm–500 mm long							
over 600 mm wide	–	–	–	–	–	m²	35.93
not exceeding 600 mm wide in isolated strips	–	–	–	–	–	m	36.42
Edge treatments							
20 × 20 mm SAS perimeter shadow gap; screwed to plasterboard	–	–	–	–	–	m	6.54
20 × 20 mm SAS shadow gap around 450 mm dia. column; including 15 × 44 mm batten plugged and screwed to concrete	–	–	–	–	–	nr	77.68
Vertical bulkhead; including additional hangers							
over 600 mm wide	–	–	–	–	–	m²	45.36
not exceeding 600 mm wide in isolated strips	–	–	–	–	–	m	44.74
Rockfon, or other equal suspended concealed ceiling system; 400 mm to 600 mm long hangers plugged and screwed to concrete soffit							
Lining to ceilings; 600 mm × 600 mm × 20 mm Sonar suspended ceiling tiles							
over 600 mm wide	–	–	–	–	–	m²	45.99
not exceeding 600 mm wide in isolated strips	–	–	–	–	–	m	27.12
edge trim; shadow-line trim	–	–	–	–	–	m	5.37
Vertical bulkhead, as upstand to rooflight well; including additional hangers; perimeter trim							
300 mm × 600 mm wide	–	–	–	–	–	m	50.34
Ecophon, or other equal; Z demountable suspended concealed ceiling system; 400 mm long hangers plugged and screwed to concrete soffit							
Lining to ceilings; 600 mm × 600 mm × 20 mm Gedina ET15 suspended ceiling tiles							
over 600 mm wide	–	–	–	–	–	m²	39.91
not exceeding 600 mm wide in isolated strips	–	–	–	–	–	m	25.17
edge trim; shadow-line trim	–	–	–	–	–	m	4.97
Vertical bulkhead, as upstand to rooflight well; including additional hangers; perimeter trim							
300 mm × 600 mm wide	–	–	–	–	–	m	46.33
Lining to ceilings; 600 mm × 600 mm × 20 mm Hygiene Performance washable suspended ceiling tiles							
over 600 mm wide	–	–	–	–	–	m²	53.29
not exceeding 600 mm wide in isolated strips	–	–	–	–	–	m	43.69
edge trim; shadow-line trim	–	–	–	–	–	m	7.02

30 SUSPENDED CEILINGS

Item	PC £	Labour hours	Labour £	Plant £	Material £	Unit	Total rate £
Vertical bulkhead, as upstand to rooflight well; including additional hangers; perimeter trim							
not exceeding 600 mm wide in isolated strips	–	–	–	–	–	m	48.10
Lining to ceilings; 1200 mm × 1200 mm × 20 mm Focus DG suspended ceiling tiles							
over 600 mm wide	–	–	–	–	–	m²	48.34
not exceeding 600 mm wide in isolated strips	–	–	–	–	–	m	28.65
edge trim; shadow-line trim	–	–	–	–	–	m	4.97
Vertical bulkhead, as upstand to rooflight well; including additional hangers; perimeter trim							
300 mm × 600 mm wide	–	–	–	–	–	m	48.81
Z demountable suspended ceiling system or other equal; hangers plugged and screwed to concrete soffit, 600 mm × 600 mm × 19 mm Echostop glass reinforced fibrous plaster lightweight plain bevelled edge tiles							
Lining to ceilings; hangers 150 mm–500 mm long							
over 600 mm wide	–	–	–	–	–	m²	98.78
not exceeding 600 mm wide in isolated strips	–	–	–	–	–	m	70.44
Concealed galvanized steel suspension system; hangers plugged and screwed to concrete soffit, Burgess white stove enamelled perforated mild steel tiles 600 mm × 600 mm							
Lining to ceilings; hangers average 150 mm–500 mm long							
over 600 mm wide	–	–	–	–	–	m²	47.49
not exceeding 600 mm wide in isolated strips	–	–	–	–	–	m	41.64
Semi-concealed galvanized steel Trulok suspension system or other equal; hangers plugged and screwed to concrete; Armstrong Microlook Ultima Plain 600 mm × 600 mm × 18 mm mineral ceiling tiles							
Linings to ceilings; hangers 500 mm–1000 mm long							
over 600 mm wide	–	–	–	–	–	m²	38.53
over 600 mm wide; 3.50 m–5.00 m high	–	–	–	–	–	m²	44.04
over 600 mm wide; in staircase areas or plant rooms	–	–	–	–	–	m²	44.04
not exceeding 600 mm wide in isolated strips	–	–	–	–	–	m	38.53
cutting and fitting around modular downlighter including yoke	–	–	–	–	–	nr	17.97
24 mm × 19 mm white finished angle edge trim	–	–	–	–	–	m	4.44
Vertical bulkhead; including additional hangers							
over 600 mm wide	–	–	–	–	–	m²	55.10
not exceeding 600 mm wide in isolated strips	–	–	–	–	–	m	49.68

30 SUSPENDED CEILINGS

Item	PC £	Labour hours	Labour £	Plant £	Material £	Unit	Total rate £
30.01 SUSPENDED CEILINGS – cont							
Metal; SAS system 330; EMAC suspension system; 100 mm Omega C profiles at 1500 mm centres filled in with 1400 mm × 250 mm perforated metal tiles with 18 mm thick × 80 kg/ m³ density foil wrapped tissue-faced acoustic pad adhered above; ceiling to achieve 40dB with 0.7 absorption coefficient							
Linings to ceilings; hangers 500 mm–700 mm long							
over 600 mm wide	–	–	–	–	–	m²	54.37
not exceeding 600 mm wide in isolated strips	–	–	–	–	–	m	27.74
cutting and reinforcing to receive a recessed light maximum 1300 mm × 500 mm.	–	–	–	–	–	nr	15.53
edge trim; to perimeter	–	–	–	–	–	m	13.87
edge trim around 450 mm dia. column;	–	–	–	–	–	nr	53.26
Galvanized steel suspension system; hangers plugged and screwed to concrete soffit, Luxalon stove enamelled aluminium linear panel ceiling, type 80B or other equal and approved, complete with mineral insulation							
Linings to ceilings; hangers 500 mm–700 mm long							
over 600 mm wide	–	–	–	–	–	m²	89.65
not exceeding 600 mm wide in isolated strips	–	–	–	–	–	m	43.46
Armstrong suspended ceiling; assume large rooms over 250 m²							
Mineral fibre							
mineral fibre; basic range; Cortega, Tatra, Academy; exposed grid	–	–	–	–	–	m²	23.64
mineral fibre; medium quality; Corline; exposed grid	–	–	–	–	–	m²	34.30
mineral fibre; medium quality; Corline; concealed grid	–	–	–	–	–	m²	51.10
mineral fibre; medium quality; Corline; silhouette grid	–	–	–	–	–	m²	63.87
mineral fibre; Specific; Bioguard; exposed grid	–	–	–	–	–	m²	28.39
mineral fibre; Specific; Bioguard; clean room grid	–	–	–	–	–	m²	44.94
mineral fibre; Specific; Hygiene, Cleanroom; exposed grid	–	–	–	–	–	m²	61.50
Open cell							
metal open cell; Cellio Global White; exposed grid	–	–	–	–	–	m²	57.96
metal open cell; Cellio Black; exposed grid	–	–	–	–	–	m²	69.79
Wood veneer							
wood veneer board; Microlook 8; plain; exposed grid	–	–	–	–	–	m²	75.70
wood veneer; plain; exposed grid	–	–	–	–	–	m²	104.08
wood veneers; perforated; exposed grid	–	–	–	–	–	m²	179.78
wood veneers; plain; concealed grid	–	–	–	–	–	m²	171.49
wood veneers; perforated; concealed grid	–	–	–	–	–	m²	236.56

30 SUSPENDED CEILINGS

Item	PC £	Labour hours	Labour £	Plant £	Material £	Unit	Total rate £
Acoustic panel linings; Troldekt Ultrafine 1200 mm × 600 mm × 25 mm on and including 50 mm × 70 mm timber framework at 600 mm centres both ways and supported from the roof with 50 mm × 50 mm hangers 400 mm–600 mm long							
Ceilings							
allow for ceiling tiles to be clipped to prevent wind uplift when external doors are opened	–	–	–	–	–	m²	**117.47**
perimeter edge detail with 50 mm × 50 mm support batten	–	–	–	–	–	m	**7.45**
30.02 SOLID SUSPENDED CEILINGS							
Gypsum plasterboard; BS EN 520; plain grade tapered edge wallboard; fixing on dabs or with nails; joints left open to receive Artex finish or other equal and approved; to softwood base							
9.50 mm board to ceilings							
over 600 mm wide	–	0.23	5.33	–	3.25	m²	**8.58**
9.50 mm board to beams							
girth not exceeding 600 mm	–	0.28	6.48	–	1.96	m²	**8.44**
girth 600 mm–1200 mm	–	0.37	8.56	–	3.90	m²	**12.46**
12.50 mm board to ceilings							
over 300 mm wide	–	0.31	7.17	–	3.27	m²	**10.44**
12.50 mm board to beams							
girth not exceeding 600 mm	–	0.28	6.48	–	1.97	m²	**8.45**
girth 600 mm–1200 mm	–	0.37	8.56	–	3.91	m²	**12.47**
Gypsum plasterboard; fixing on dabs or with nails; joints filled with joint tape and filler to receive direct decoration; to softwood base (measured elsewhere)							
Plain grade tapered edge wallboard							
9.50 mm board to ceilings							
over 600 mm wide	–	0.29	6.89	–	4.69	m²	**11.58**
9.50 mm board to faces of beams – 3 nr							
not exceeding 600 mm total girth	–	0.46	11.00	–	4.78	m	**15.78**
600 mm–1200 mm total girth	–	0.67	15.94	–	7.10	m	**23.04**
1200 mm–1800 mm total girth	–	0.75	17.89	–	9.41	m	**27.30**
12.50 mm board to ceilings							
over 600 mm wide	–	0.31	7.35	–	6.55	m²	**13.90**
12.50 mm board to faces of beams – 3 nr							
not exceeding 600 mm total girth	–	0.46	11.00	–	5.90	m	**16.90**
600 mm–1200 mm total girth	–	0.67	15.94	–	9.33	m	**25.27**
1200 mm–1800 mm total girth	–	0.75	17.89	–	12.76	m	**30.65**
external angle; with joint tape bedded and covered with Jointex or other equal and approved	–	0.11	2.74	–	0.62	m	**3.36**

30 SUSPENDED CEILINGS

Item	PC £	Labour hours	Labour £	Plant £	Material £	Unit	Total rate £
30.02 SOLID SUSPENDED CEILINGS – cont							
Tapered edge wallboard TEN							
12.50 mm board to ceilings							
over 600 mm wide	–	0.41	9.85	–	4.86	m²	**14.71**
12.50 mm board to faces of beams – 3 nr							
not exceeding 600 mm total girth	–	0.56	13.49	–	4.88	m	**18.37**
600 mm–1200 mm total girth	–	1.02	24.67	–	7.29	m	**31.96**
1200 mm–1800 mm total girth	–	1.30	31.60	–	9.71	m	**41.31**
external angle; with joint tape bedded and							
covered with Jointex or other equal and approved	–	0.11	2.74	–	0.62	m	**3.36**
Tapered edge plank							
19 mm plank to ceilings							
over 600 mm wide	–	0.43	10.31	–	8.03	m²	**18.34**
19 mm plank to faces of beams – 3 nr							
not exceeding 600 mm total girth	–	0.60	14.42	–	6.79	m	**21.21**
600 mm–1200 mm total girth	–	1.06	25.60	–	11.11	m	**36.71**
1200 mm–1800 mm total girth	–	1.34	32.53	–	15.43	m	**47.96**
ThermaLine Plus boards							
27 mm board to ceilings							
over 600 mm wide	–	0.46	10.64	–	14.70	m²	**25.34**
27 mm board to faces of beams – 3 nr							
not exceeding 600 mm total girth	–	0.56	12.96	–	10.78	m	**23.74**
600 mm–1200 mm total girth	–	1.06	24.52	–	19.10	m	**43.62**
1200 mm–1800 mm total girth	–	1.43	33.08	–	27.42	m	**60.50**
48 mm board to ceilings							
over 600 mm wide	–	0.49	11.34	–	20.01	m²	**31.35**
48 mm board to faces of beams – 3 nr							
not exceeding 600 mm total girth	–	0.58	13.42	–	14.00	m	**27.42**
600 mm–1200 mm total girth	–	1.17	27.07	–	25.50	m	**52.57**
1200 mm–1800 mm total girth	–	1.57	36.33	–	37.02	m	**73.35**
ThermaLine Super boards							
50 mm board to ceilings							
over 600 mm wide	–	0.49	11.34	–	26.78	m²	**38.12**
50 mm board to faces of beams – 3 nr							
not exceeding 600 mm total girth	–	0.58	13.42	–	18.06	m	**31.48**
600 mm–1200 mm total girth	–	1.17	27.07	–	33.63	m	**60.70**
1200 mm–1800 mm total girth	–	1.57	36.33	–	49.20	m	**85.53**
60 mm board to ceilings							
over 600 mm wide	–	0.49	11.34	–	37.19	m²	**48.53**
60 mm board to faces of beams – 3 nr							
not exceeding 600 mm total girth	–	0.58	13.42	–	24.31	m	**37.73**
600 mm–1200 mm total girth	–	1.17	27.07	–	46.12	m	**73.19**
1200 mm–1800 mm total girth	–	1.57	36.33	–	67.95	m	**104.28**

31 INSULATION, FIRE STOPPING AND FIRE PROTECTION

Item	PC £	Labour hours	Labour £	Plant £	Material £	Unit	Total rate £
MATERIAL RATES							
Altrernative insulation supply only prices, delivered in full loads							
Crown FrameTherm Roll 40 (Thermal conductivity 0.040 W/mK)							
90 mm thick	–	–	–	–	2.34	m²	**2.34**
140 mm thick	–	–	–	–	3.58	m²	**3.58**
Crown FrameTherm Roll 32 (Thermal conductivity 0.032 W/mK)							
90 mm thick	–	–	–	–	6.88	m²	**6.88**
140 mm thick	–	–	–	–	12.83	m²	**12.83**
Crown Factoryclad 40 (Thermal conductivity 0.040 W/mK)							
180 mm thick	–	–	–	–	3.24	m²	**3.24**
100 mm thick	–	–	–	–	1.53	m²	**1.53**
Thermafleece CosyWool Roll (0.039 W/mK); 75% sheep's wool, 15% recycled polyester and 10% polyester binder with a high recycled content							
50 mm thick	–	–	–	–	4.03	m²	**4.03**
75 mm thick	–	–	–	–	5.99	m²	**5.99**
100 mm thick	–	–	–	–	7.85	m²	**7.85**
140 mm thick	–	–	–	–	11.09	m²	**11.09**
Thermafleece TF35 high density wool insulating batts (0.035 W/mK); 60% British wool, 30% recycled polyester and 10% polyester binder with a high recycled content							
50 mm thick	–	–	–	–	6.29	m²	**6.29**
75 mm thick	–	–	–	–	9.80	m²	**9.80**
31.01 INSULATION							
SUPPLY AND FIX PRICES							
Sisalkraft building papers/vapour barriers or other equal and approved							
Building paper; 150 mm laps; fixed to softwood							
Moistop grade 728 (class A1F)	–	0.08	1.35	–	1.50	m²	**2.85**
Vapour barrier/reflective insulation 150 mm laps; fixed to softwood							
Insulex grade 714; single sided	–	0.08	1.35	–	1.78	m²	**3.13**
Insulation board underlays							
Vapour barrier							
reinforced; metal lined	–	–	–	–	–	m²	**15.96**
Rockwool; Duorock flat insulation board							
140 mm thick (0.25 U-value)	–	–	–	–	–	m²	**43.74**

31 INSULATION, FIRE STOPPING AND FIRE PROTECTION

Item	PC £	Labour hours	Labour £	Plant £	Material £	Unit	Total rate £
31.01 INSULATION – cont							
Insulation board underlays – cont							
Kingspan Thermaroof TR21 zero OPD urethene insulation board							
50 mm thick	–	–	–	–	–	m²	**26.88**
90 mm thick	–	–	–	–	–	m²	**46.32**
100 mm thick	–	–	–	–	–	m²	**51.45**
Wood fibre boards; impregnated; density 220–350 kg/m³							
12.70 mm thick	–	–	–	–	–	m²	**7.25**
Mat or quilt insulation							
Glass fibre roll; Crown Loft Roll 44 (Thermal conductivity 0.044 W/mK) or other equal; laid loose							
100 mm thick	1.45	0.09	1.52	–	1.49	m²	**3.01**
150 mm thick	2.19	0.10	1.69	–	2.26	m²	**3.95**
200 mm thick	3.03	0.11	1.85	–	3.12	m²	**4.97**
Glass fibre quilt; Isover Modular roll (Thermal conductivity 0.043 W/mK) or other equal and approved; laid loose							
100 mm thick	1.42	0.09	1.52	–	1.46	m²	**2.98**
150 mm thick	2.12	0.10	1.69	–	2.18	m²	**3.87**
200 mm thick	3.14	0.15	2.53	–	3.23	m²	**5.76**
Mineral fibre quilt; Isover Acoustic Partition Roll (APR 1200) or other equal; pinned vertically to softwood							
25 mm thick	1.19	0.08	1.35	–	1.49	m²	**2.84**
50 mm thick	1.95	0.09	1.52	–	2.28	m²	**3.80**
75 mm thick	3.27	0.10	1.69	–	3.64	m²	**5.33**
100 mm thick	4.23	0.15	2.53	–	4.62	m²	**7.15**
Crown Dritherm Cavity Slab 37 (Thermal conductivity 0.037 W/mK) glass fibre batt or other equal; as full or partial cavity fill; including cutting and fitting around wall ties and retaining discs							
50 mm thick	2.17	0.12	2.03	–	2.50	m²	**4.53**
75 mm thick	3.69	0.13	2.19	–	4.07	m²	**6.26**
100 mm thick	4.42	0.14	2.37	–	4.83	m²	**7.20**
Crown Dritherm Cavity Slab 34 (Thermal conductivity 0.034 W/mK) glass fibre batt or other equal; as full or partial cavity fill; including cutting and fitting around wall ties and retaining discs							
65 mm thick	2.37	0.12	2.03	–	2.71	m²	**4.74**
75 mm thick	2.75	0.13	2.19	–	3.10	m²	**5.29**
85 mm thick	4.17	0.13	2.19	–	4.56	m²	**6.75**
100 mm thick	4.17	0.14	2.37	–	4.56	m²	**6.93**

31 INSULATION, FIRE STOPPING AND FIRE PROTECTION

Item	PC £	Labour hours	Labour £	Plant £	Material £	Unit	Total rate £
Crown Dritherm Cavity Slab 32 (Thermal conductivity 0.032 W/mK) glass fibre batt or other equal; as full or partial cavity fill; including cutting and fitting around wall ties and retaining discs							
65 mm thick	3.08	0.12	2.03	–	3.44	m²	**5.47**
75 mm thick	3.59	0.13	2.19	–	3.97	m²	**6.16**
85 mm thick	4.07	0.13	2.19	–	4.46	m²	**6.65**
100 mm thick	4.69	0.14	2.37	–	5.10	m²	**7.47**
Crown Frametherm Roll 40 (Thermal conductivity 0.040 W/mK) glass fibre semi-rigid or rigid batt or other equal; pinned vertically in timber frame construction							
90 mm thick	4.04	0.14	2.37	–	4.16	m²	**6.53**
140 mm thick	5.85	0.16	2.70	–	6.03	m²	**8.73**
Crown Rafter Roll 32 (Thermal conductivity 0.032 W/mK) glass fibre flanged building roll; pinned vertically or to slope between timber framing							
50 mm thick	3.91	0.13	2.19	–	4.03	m²	**6.22**
75 mm thick	5.59	0.14	2.37	–	5.76	m²	**8.13**
100 mm thick	7.19	0.15	2.53	–	7.41	m²	**9.94**
Board or slab insulation							
Kay-Cel (Thermal conductivity 0.033 W/mK) expanded polystyrene board standard grade SD/N or other equal; fixed with adhesive							
25 mm thick	–	0.14	3.23	–	3.55	m²	**6.78**
50 mm thick	–	0.16	3.70	–	6.16	m²	**9.86**
75 mm thick	–	0.18	4.16	–	8.71	m²	**12.87**
100 mm thick	–	0.19	4.40	–	11.30	m²	**15.70**
KIngspan Kooltherm K8 (Thermal conductivity 0.022 W/mK) Premium performance rigid thermoset modified resin insulant manufactured with a blowing agent that has zero Ozone Depletion Potential (ODP) and low Global Warming Potential (GWP) or other equal; as partial cavity fill; including cutting and fitting around wall ties and retaining discs							
40 mm thick	–	0.15	3.47	–	14.86	m²	**18.33**
50 mm thick	–	0.15	3.47	–	14.86	m²	**18.33**
60 mm thick	–	0.15	3.47	–	17.84	m²	**21.31**
75 mm thick	–	0.15	3.47	–	17.63	m²	**21.10**
Kingspan Thermawall TW50 (Thermal conductivity 0.022 W/mK) High performance rigid thermoset polyisocyanurate (PIR) insulant or other equal and approved; as partial cavity fill; including cutting and fitting around wall ties and retaining discs							
25 mm thick	–	0.17	3.93	–	10.06	m²	**13.99**
50 mm thick	–	0.17	3.93	–	13.16	m²	**17.09**
75 mm thick	–	0.18	4.16	–	19.18	m²	**23.34**
100 mm thick	–	0.19	4.40	–	24.86	m²	**29.26**

31 INSULATION, FIRE STOPPING AND FIRE PROTECTION

Item	PC £	Labour hours	Labour £	Plant £	Material £	Unit	Total rate £
31.01 INSULATION – cont							
Board or slab insulation – cont							
Kingspan Thermafloor TF70 (Thermal conductivity 0.022 W/mK) high performance rigid urethane floor insulation for solid concrete and suspended ground floors							
50 mm thick	–	0.17	4.05	–	9.72	m²	**13.77**
75 mm thick	–	0.17	4.05	–	14.41	m²	**18.46**
100 mm thick	–	0.17	4.05	–	17.99	m²	**22.04**
125 mm thick	–	0.17	4.05	–	24.96	m²	**29.01**
150 mm thick	–	0.20	4.62	–	29.22	m²	**33.84**
Ravatherm XPS X 500 SL (Thermal conductivity 0.033 W/mK) extruded polystyrene foam or other equal and approved							
50 mm thick	–	0.46	10.64	–	8.54	m²	**19.18**
75 mm thick	–	0.46	10.64	–	12.56	m²	**23.20**
100 mm thick	–	0.46	10.64	–	17.07	m²	**27.71**
160 mm thick	–	0.48	11.32	–	34.80	m²	**46.12**
Ravatherm XPS blocks or other equal; cold bridging insulation fixed with adhesive to brick, block or concrete base							
Insulation to walls							
50 mm thick	–	0.33	7.63	–	12.49	m²	**20.12**
75 mm thick	–	0.35	8.10	–	16.33	m²	**24.43**
100 mm thick	–	0.37	8.56	–	19.68	m²	**28.24**
165 mm thick	–	0.40	9.25	–	28.75	m²	**38.00**
Insulation to isolated columns							
50 mm thick	–	0.41	9.49	–	12.49	m²	**21.98**
75 mm thick	–	0.43	9.95	–	16.33	m²	**26.28**
Insulation to ceilings							
50 mm thick	–	0.36	8.33	–	12.49	m²	**20.82**
75 mm thick	–	0.39	9.02	–	16.33	m²	**25.35**
100 mm thick	–	0.42	9.71	–	19.68	m²	**29.39**
165 mm thick	–	0.45	10.41	–	28.75	m²	**39.16**
Insulation to isolated beams							
50 mm thick	–	0.43	9.95	–	12.49	m²	**22.44**
75 mm thick	–	0.46	10.64	–	16.33	m²	**26.97**
Pittsburgh Company Foamglas cellular glass insulation with PC 56 component adhesive as required							
Flat roof insulation; Foamglas T3+; Thermal conductivity 0.036 W/mK; compressive strength > 500 kPa size 600 mm × 450 mm slabs							
80 mm thick slab with adhered joints	43.48	1.20	26.89	–	53.27	m²	**80.16**
100 mm thick slab with adhered joints	53.24	1.20	26.89	–	63.32	m²	**90.21**
150 mm thick slab with adhered joints	79.87	1.30	29.14	–	90.75	m²	**119.89**
180 mm thick slab with adhered joints	95.84	1.40	31.37	–	107.20	m²	**138.57**

31 INSULATION, FIRE STOPPING AND FIRE PROTECTION

Item	PC £	Labour hours	Labour £	Plant £	Material £	Unit	Total rate £
Flat roof insulation; Foamglas T4+; Thermal conductivity 0.041 W/mK; compressive strength > 600 kPa size 600 mm × 450 mm slabs							
80 mm thick slab with adhered joints	42.59	1.20	26.89	–	52.35	m²	79.24
100 mm thick slab with adhered joints	53.23	1.20	26.89	–	63.31	m²	90.20
150 mm thick slab with adhered joints	79.87	1.30	29.14	–	90.75	m²	119.89
180 mm thick slab with adhered joints	95.84	1.40	31.37	–	107.20	m²	138.57
Wall insulation slabs; Foamglas F; Thermal conductivity 0.038 W/mK; 600 mm × 450 mm slabs							
60 mm thick slab with adhered joints	34.75	1.25	28.02	–	44.28	m²	72.30
100 mm thick slab with adhered joints	72.38	1.25	28.02	–	83.04	m²	111.06
150 mm thick slab with adhered joints	108.58	1.25	28.02	–	120.31	m²	148.33
Ecotherm Eco-Versal General Purpose Insulation Board; PIR rigid thermal insulation board; Thermal conductivity 0.022 W/mK. For pitched roofs, ceilings, floors and walls							
30 mm thick slab with adhered joints	5.78	1.20	26.89	–	14.44	m²	41.33
50 mm thick slab with adhered joints	7.94	1.20	26.89	–	16.67	m²	43.56
100 mm thick slab with adhered joints	14.20	1.30	29.14	–	23.11	m²	52.25
150 mm thick slab with adhered joints	22.10	1.40	31.37	–	31.25	m²	62.62
Sheepswool mix							
Thermafleece TF35 high density wool insulating batts (0.035 W/mK); 60% British wool, 30% recycled polyester and 10% polyester binder with a high recycled content							
50 mm thick	–	0.35	8.10	–	6.29	m²	14.39
75 mm thick	–	0.38	8.67	–	9.80	m²	18.47
Thermafleece EcoRoll (0.039 W/mK); 75% sheep's wool, 15% recycled polyester and 10% polyester binder with a high recycled content							
50 mm thick	–	0.15	2.53	–	4.03	m²	6.56
75 mm thick	–	0.14	2.37	–	5.99	m²	8.36
100 mm thick	–	0.15	2.53	–	7.85	m²	10.38
140 mm thick	–	0.17	2.96	–	11.09	m²	14.05

31 INSULATION, FIRE STOPPING AND FIRE PROTECTION

Item	PC £	Labour hours	Labour £	Plant £	Material £	Unit	Total rate £
31.01 INSULATION – cont							
Tapered insulation board underlays							
Tapered insulation £/m² prices can vary dramatically depending upon the factors which determine the scheme layout; these primarily being gutter/outlet locations and the length of fall involved. The following guide assumes a U-value of 0.18 W/m² K as a benchmark.							
As the required insulation value will vary from project to project the required U-value should be determined by calculating the buildings energy consumption at the design stage. Due to tapered insulation scheme prices varying by project, the following prices are indicative. Please contact a specialist for a project specific quotation. U-value must be calculated in accordance with BSENISO 6946:2007 Annex C							
Tapered PIR (Polyisocyanurate) boards; bedded in hot bitumen							
effective thickness achieving 0.18 W/m² K	32.46	–	–	–	–	m²	**68.66**
minimum thickness achieving 0.18 W/m² K	38.61	–	–	–	–	m²	**76.22**
Tapered PIR (Polyisocyanurate) board; mechanically fastened							
effective thickness achieving 0.18 W/m² K	32.46	–	–	–	–	m²	**71.67**
Tapered Aspire; Hybrid EPS/PIR boards; bedded in hot bitumen							
effective thickness achieving 0.18 W/m² K	30.21	–	–	–	–	m²	**68.66**
minimum thickness achieving 0.18 W/m² K	36.37	–	–	–	–	m²	**76.22**
Tapered PIR (Polyisocyanurate) board; mechanically fastened							
effective thickness achieving 0.18 W/m² K	30.21	–	–	–	–	m²	**71.67**
minimum thickness achieving 0.18 W/m² K	36.37	–	–	–	–	m²	**79.23**
Tapered Rockwool boards; bedded in hot bitumen							
effective thickness achieving 0.18 W/m² K	57.97	–	–	–	–	m²	**101.00**
minimum thickness achieving 0.18 W/m² K	71.63	–	–	–	–	m²	**110.09**
Tapered Rockwool boards; mechanically fastened							
effective thickness achieving 0.18 W/m² K	57.97	–	–	–	–	m²	**104.02**
minimum thickness achieving 0.18 W/m² K	71.63	–	–	–	–	m²	**112.85**
Tapered EPS (Expanded polystyrene) boards; bedded in hot bitumen							
effective thickness achieving 0.18 W/m² K	26.53	–	–	–	–	m²	**61.17**
minimum thickness achieving 0.18 W/m² K	32.46	–	–	–	–	m²	**68.66**
Tapered EPS (Expanded polystyrene) boards; mechanicaly fastened							
effective thickness achieving 0.18 W/m² K	26.53	–	–	–	–	m²	**64.09**
minimum thickness achieving 0.18 W/m² K	32.46	–	–	–	–	m²	**71.66**

31 INSULATION, FIRE STOPPING AND FIRE PROTECTION

Item	PC £	Labour hours	Labour £	Plant £	Material £	Unit	Total rate £
Insulation board overlays							
Dow Roofmate SL extruded polystyrene foam boards or other equal and approved; Thermal conductivity – 0.028 W/mK							
50 mm thick	–	–	–	–	–	m²	**17.79**
140 mm thick	–	–	–	–	–	m²	**31.06**
160 mm thick	–	–	–	–	–	m²	**33.65**
Dow Roofmate LG extruded polystyrene foam boards or other equal and approved; Thermal conductivity – 0.028 W/mK							
80 mm thick	–	–	–	–	–	m²	**63.37**
100 mm thick	–	–	–	–	–	m²	**67.91**
120 mm thick	–	–	–	–	–	m²	**72.52**
31.02 FIRE STOPPING							
Fire stopping							
Cape Firecheck channel; intumescent coatings on cut mitres; fixing with brass cups and screws							
19 mm × 44 mm or 19 mm × 50 mm	–	0.56	12.96	–	24.11	m	**37.07**
Sealmaster intumescent fire and smoke seals or other equal; pinned into groove in timber							
type N30; for single leaf half hour door	–	0.28	6.48	–	12.11	m	**18.59**
type N60; for single leaf one hour door	–	0.31	7.17	–	18.72	m	**25.89**
type IMN or IMP; for meeting or pivot stiles of pair of one hour doors; per stile	–	0.31	7.17	–	18.72	m	**25.89**
intumescent plugs in timber; including boring	–	0.09	2.08	–	0.48	nr	**2.56**
Rockwool fire stops or other equal; between top of brick/block wall and concrete soffit							
75 mm deep × 100 mm wide	–	0.08	1.85	–	11.41	m	**13.26**
75 mm deep × 150 mm wide	–	0.10	2.32	–	17.12	m	**19.44**
75 mm deep × 200 mm wide	–	0.12	2.78	–	22.82	m	**25.60**
90 mm deep × 100 mm wide	–	0.10	2.32	–	13.73	m	**16.05**
90 mm deep × 150 mm wide	–	0.12	2.78	–	20.60	m	**23.38**
90 mm deep × 200 mm wide	–	0.14	3.23	–	27.46	m	**30.69**
Fire protection compound							
Quelfire QF4, fire protection compound or other equal; filling around pipes, ducts and the like; including all necessary formwork							
300 mm × 300 mm × 250 mm; pipes – 2	–	0.93	17.82	–	17.36	nr	**35.18**
500 mm × 500 mm × 250 mm; pipes – 2	–	1.16	21.64	–	52.06	nr	**73.70**
Fire barriers							
Rockwool fire barrier or other equal; between top of suspended ceiling and concrete soffit							
75 mm layer × 900 mm wide; one hour	–	0.56	12.96	–	78.44	m²	**91.40**
90 mm layer × 900 mm wide; two hour	–	1.10	25.45	–	99.25	m²	**124.70**

31 INSULATION, FIRE STOPPING AND FIRE PROTECTION

Item	PC £	Labour hours	Labour £	Plant £	Material £	Unit	Total rate £
31.02 FIRE STOPPING – cont							
Fire barriers – cont							
Corofil C144 fire barrier to edge of slab; fixed with non-flammable contact adhesive							
to suit void 30 mm wide × 100 mm deep; one hour	–	–	–	–	–	m	**15.49**
Lamatherm fire barrier or other equal; to void below raised access floors							
75 mm thick × 300 mm high; half hour	–	0.17	3.93	–	11.76	m	**15.69**
75 mm thick × 600 mm high; half hour	–	0.17	3.93	–	25.78	m	**29.71**
90 mm thick × 300 mm high; half hour	–	0.17	3.93	–	16.52	m	**20.45**
90 mm thick × 600 mm high; half hour	–	0.17	3.93	–	34.40	m	**38.33**

32 FURNITURE, FITTINGS AND EQUIPMENT

Item	PC £	Labour hours	Labour £	Plant £	Material £	Unit	Total rate £
32.01 GENERAL FIXTURES, FURNISHES AND EQUIPMENT							
SUPPLY ONLY PRICES							
NOTE: The fixing of general fixtures will vary considerably dependent upon the size of the fixture and the method of fixing employed. Prices for fixing like sized kitchen fittings may be suitable for certain fixtures, although adjustment to those rates will almost invariably be necessary.							
The following supply only prices are for purpose-made fittings components in various materials supplied as part of an assembled fitting and therefore may be used to arrive at a guide price for a complete fitting.							
Fitting components; medium density fibreboard							
Backs, fronts, sides or divisions; over 300 mm wide							
12 mm thick	–	–	–	–	25.51	m²	**25.51**
18 mm thick	–	–	–	–	27.03	m²	**27.03**
25 mm thick	–	–	–	–	30.10	m²	**30.10**
Shelves or worktops; over 300 mm wide							
18 mm thick	–	–	–	–	27.03	m²	**27.03**
25 mm thick	–	–	–	–	30.10	m²	**30.10**
Flush doors; lipped on four edges							
450 mm × 750 mm × 18 mm	–	–	–	–	39.15	nr	**39.15**
450 mm × 750 mm × 25 mm	–	–	–	–	39.90	nr	**39.90**
600 mm × 900 mm × 18 mm	–	–	–	–	46.22	nr	**46.22**
600 mm × 900 mm × 25 mm	–	–	–	–	47.42	nr	**47.42**
Fitting components; moisture-resistant medium density fibreboard							
Backs, fronts, sides or divisions; over 300 mm wide							
12 mm thick	–	–	–	–	28.55	m²	**28.55**
18 mm thick	–	–	–	–	31.61	m²	**31.61**
25 mm thick	–	–	–	–	34.66	m²	**34.66**
Shelves or worktops; over 300 mm wide							
18 mm thick	–	–	–	–	31.61	m²	**31.61**
25 mm thick	–	–	–	–	34.66	m²	**34.66**
Flush doors; lipped on four edges							
450 mm × 750 mm × 18 mm	–	–	–	–	39.90	nr	**39.90**
450 mm × 750 mm × 25 mm	–	–	–	–	41.02	nr	**41.02**
600 mm × 900 mm × 18 mm	–	–	–	–	47.42	nr	**47.42**
600 mm × 900 mm × 25 mm	–	–	–	–	49.25	nr	**49.25**
Fitting components; medium density fibreboard; melamine faced both sides							
Backs, fronts, sides or divisions; over 300 mm wide							
12 mm thick	–	–	–	–	33.65	m²	**33.65**
18 mm thick	–	–	–	–	37.48	m²	**37.48**

32 FURNITURE, FITTINGS AND EQUIPMENT

Item	PC £	Labour hours	Labour £	Plant £	Material £	Unit	Total rate £
32.01 GENERAL FIXTURES, FURNISHES AND EQUIPMENT – cont							
Fitting components – cont							
Shelves or worktops; over 300 mm wide							
18 mm thick	–	–	–	–	37.48	m²	**37.48**
Flush doors; lipped on four edges							
450 mm × 750 mm × 18 mm	–	–	–	–	28.01	nr	**28.01**
600 mm × 900 mm × 25 mm	–	–	–	–	35.52	nr	**35.52**
Fitting components; medium density fibreboard; formica faced both sides							
Backs, fronts, sides or divisions; over 300 mm wide							
12 mm thick	–	–	–	–	102.48	m²	**102.48**
18 mm thick	–	–	–	–	106.57	m²	**106.57**
Shelves or worktops; over 300 mm wide							
18 mm thick	–	–	–	–	106.57	m²	**106.57**
Flush doors; lipped on four edges							
450 mm × 750 mm × 18 mm	–	–	–	–	56.42	nr	**56.42**
600 mm × 900 mm × 25 mm	–	–	–	–	57.65	nr	**57.65**
Fitting components; wrought softwood							
Backs, fronts, sides or divisions; cross-tongued joints; over 300 mm wide							
25 mm thick	–	–	–	–	48.98	m²	**48.98**
Shelves or worktops; cross-tongued joints; over 300 mm wide							
25 mm thick	–	–	–	–	48.98	m²	**48.98**
Bearers							
19 mm × 38 mm	–	–	–	–	2.56	m	**2.56**
25 mm × 50 mm	–	–	–	–	2.84	m	**2.84**
44 mm × 44 mm	–	–	–	–	3.04	m	**3.04**
44 mm × 75 mm	–	–	–	–	3.49	m	**3.49**
Bearers; framed; to backs, fronts or sides							
19 mm × 38 mm	–	–	–	–	5.87	m	**5.87**
25 mm × 50 mm	–	–	–	–	6.36	m	**6.36**
50 mm × 50 mm	–	–	–	–	8.20	m	**8.20**
50 mm × 75 mm	–	–	–	–	9.40	m	**9.40**
Add 5% to the above material prices for selected softwood staining							
Fitting components; selected Sapele							
Backs, fronts, sides or divisions; cross-tongued joints; over 300 mm wide							
25 mm thick	–	–	–	–	77.10	m²	**77.10**
Shelves or worktops; cross-tongued joints; over 300 mm wide							
25 mm thick	–	–	–	–	77.10	m²	**77.10**

32 FURNITURE, FITTINGS AND EQUIPMENT

Item	PC £	Labour hours	Labour £	Plant £	Material £	Unit	Total rate £
Bearers							
19 mm × 38 mm	–	–	–	–	3.86	m	3.86
25 mm × 50 mm	–	–	–	–	4.82	m	4.82
50 mm × 50 mm	–	–	–	–	5.41	m	5.41
50 mm × 75 mm	–	–	–	–	7.03	m	7.03
Bearers; framed; to backs, fronts or sides							
19 mm × 38 mm	–	–	–	–	7.47	m	7.47
25 mm × 50 mm	–	–	–	–	8.28	m	8.28
50 mm × 50 mm	–	–	–	–	11.09	m	11.09
50 mm × 75 mm	–	–	–	–	13.91	m	13.91
Fitting components; Iroko							
Backs, fronts, sides or divisions; cross-tongued joints; over 300 mm wide							
25 mm thick	–	–	–	–	84.98	m²	84.98
Shelves or worktops; cross-tongued joints; over 300 mm wide							
25 mm thick	–	–	–	–	84.98	m²	84.98
Draining boards; cross-tongued joints; over 300 mm wide							
25 mm thick	–	–	–	–	106.45	m²	106.45
stopped flutes	–	–	–	–	5.51	m	5.51
grooves; cross-grain	–	–	–	–	0.81	m	0.81
Bearers							
19 mm × 38 mm	–	–	–	–	4.22	m	4.22
25 mm × 50 mm	–	–	–	–	5.40	m	5.40
50 mm × 50 mm	–	–	–	–	6.13	m	6.13
50 mm × 75 mm	–	–	–	–	8.09	m	8.09
Bearers; framed; to backs, fronts or sides							
19 mm × 38 mm	–	–	–	–	7.73	m	7.73
25 mm × 50 mm	–	–	–	–	8.59	m	8.59
50 mm × 50 mm	–	–	–	–	11.62	m	11.62
50 mm × 75 mm	–	–	–	–	15.08	m	15.08
Lockers and cupboards; Welconstruct Distribution or other equal and approved							
Standard clothes lockers; steel body and door within reinforced 19G frame, powder coated finish, cam locks							
1 compartment; placing in position							
300 mm × 300 mm × 1800 mm	–	0.23	3.29	–	61.03	nr	64.32
380 mm × 380 mm × 1800 mm	–	0.23	3.29	–	87.34	nr	90.63
450 mm × 450 mm × 1800 mm	–	0.28	4.01	–	89.17	nr	93.18
Compartment lockers; steel body and door within reinforced 19G frame, powder coated finish, cam locks							
2 compartments; placing in position							
300 mm × 300 mm × 1800 mm	–	0.23	3.29	–	70.58	nr	73.87
380 mm × 380 mm × 1800 mm	–	0.23	3.29	–	85.09	nr	88.38
450 mm × 450 mm × 1800 mm	–	0.28	4.01	–	92.28	nr	96.29

32 FURNITURE, FITTINGS AND EQUIPMENT

Item	PC £	Labour hours	Labour £	Plant £	Material £	Unit	Total rate £
32.01 GENERAL FIXTURES, FURNISHES AND EQUIPMENT – cont							
Lockers and cupboards – cont							
4 compartments; placing in position							
300 mm × 300 mm × 1800 mm	–	0.23	3.29	–	82.98	nr	**86.27**
380 mm × 380 mm × 1800 mm	–	0.23	3.29	–	99.61	nr	**102.90**
450 mm × 450 mm × 1800 mm	–	0.28	4.01	–	99.61	nr	**103.62**
Timber clothes lockers; veneered MDF finish, routed door, cam locks							
1 compartment; placing in position							
380 mm × 380 mm × 1830 mm	–	0.28	4.01	–	217.35	nr	**221.36**
4 compartments; placing in position							
380 mm × 380 mm × 1830 mm	–	0.28	4.01	–	323.97	nr	**327.98**
Vandal-resistant lockers							
1030 high mm × 370 mm × 560 mm; one compartment	–	0.23	3.29	–	230.43	nr	**233.72**
1930 mm × 370 mm × 560 mm; two compartments	–	0.23	3.29	–	357.42	nr	**360.71**
850 mm × 740 mm × 560 mm; 2 high × 2 wide	–	0.23	3.29	–	469.22	nr	**472.51**
1930 mm × 740 mm × 560 mm; 5 high × 2 wide	–	0.23	3.29	–	1062.31	nr	**1065.60**
Shelving support systems; The Welconstruct Company or other equal standard duty; maximum bayload of 2000 kg							
Shelving support systems; steel body; stove enamelled finish; assembling							
Open initial bay; 5 shelves; placing in position							
1000 mm × 300 mm × 1850 mm	–	0.69	11.66	–	169.13	nr	**180.79**
1000 mm × 600 mm × 1850 mm	–	0.69	11.66	–	214.43	nr	**226.09**
Open extension bay; 5 shelves; placing in position							
1000 mm × 300 mm × 1850 mm	–	0.83	14.03	–	109.83	nr	**123.86**
1000 mm × 600 mm × 1850 mm	–	0.83	14.03	–	150.81	nr	**164.84**
Closed initial bay; 5 shelves; placing in position							
1000 mm × 300 mm × 1850 mm	–	0.69	11.66	–	237.10	nr	**248.76**
1000 mm × 600 mm × 1850 mm	–	0.69	11.66	–	307.04	nr	**318.70**
Closed extension bay; 5 shelves; placing in position							
1000 mm × 300 mm × 1850 mm	–	0.83	14.03	–	176.17	nr	**190.20**
1000 mm × 600 mm × 1850 mm	–	0.83	14.03	–	229.53	nr	**243.56**
Extra for pair of doors; fixing in position							
1000 mm × 1850 mm	–	0.75	12.67	–	333.23	nr	**345.90**
Cloakroom racks; The Welconstruct Company or other equal and approved							
Cloakroom racks; 40 mm × 40 mm square tube framing, polyester powder coated finish; beech slatted seats and rails to one side only; placing in position							
1675 mm × 325 mm × 1500 mm; 5 nr coat hooks	–	0.30	5.07	–	379.53	nr	**384.60**
1825 mm × 325 mm × 1500 mm; 15 nr coat hangers	–	0.30	5.07	–	434.90	nr	**439.97**

32 FURNITURE, FITTINGS AND EQUIPMENT

Item	PC £	Labour hours	Labour £	Plant £	Material £	Unit	Total rate £
Extra for							
shoe baskets	–	–	–	–	85.47	nr	**85.47**
mesh bottom shelf	–	–	–	–	59.69	nr	**59.69**
Cloakroom racks; 40 mm × 40 mm square tube framing, polyester powder coated finish; beech slatted seats and rails to both sides; placing in position							
1675 mm × 600 mm × 1500 mm; 10 nr coat hooks	–	0.40	6.76	–	520.47	nr	**527.23**
1825 mm × 600 mm × 1500 mm; 30 nr coat hangers	–	0.40	6.76	–	542.61	nr	**549.37**
Extra for							
shoe baskets	–	–	–	–	106.80	nr	**106.80**
mesh bottom shelf	–	–	–	–	72.58	nr	**72.58**
6 mm thick rectangular glass mirrors; silver backed; fixed with chromium plated domed headed screws; to background requiring plugging							
Mirror with polished edges							
365 mm × 254 mm	12.17	0.74	13.15	–	12.86	nr	**26.01**
400 mm × 300 mm	15.87	0.74	13.15	–	16.68	nr	**29.83**
560 mm × 380 mm	27.50	0.83	14.76	–	28.65	nr	**43.41**
640 mm × 460 mm	35.94	0.93	16.53	–	37.35	nr	**53.88**
Mirror with bevelled edges							
365 mm × 254 mm	21.68	0.74	13.15	–	22.66	nr	**35.81**
400 mm × 300 mm	25.39	0.74	13.15	–	26.48	nr	**39.63**
560 mm × 380 mm	42.31	0.83	14.76	–	43.91	nr	**58.67**
640 mm × 460 mm	52.88	0.93	16.53	–	54.80	nr	**71.33**
Internal blinds							
Roller blinds; Luxaflex EOS type 10 roller; Compact Fabric; plain type material; 1219 mm drop; fixing with screws							
1016 mm wide	50.33	0.93	13.30	–	51.84	nr	**65.14**
2031 mm wide	74.30	1.45	20.73	–	76.53	nr	**97.26**
2843 mm wide	92.28	1.97	28.16	–	95.05	nr	**123.21**
Roller blinds; Luxaflex EOS type 10 roller; Compact Fabric; fire-resisting material; 1219 mm drop; fixing with screws							
1016 mm wide	65.92	0.93	13.30	–	67.90	nr	**81.20**
2031 mm wide	98.27	1.45	20.73	–	101.22	nr	**121.95**
2843 mm wide	124.64	1.97	28.16	–	128.38	nr	**156.54**
Roller blinds; Luxaflex EOS type 10 roller; Light-resistant; blackout material; 1219 mm drop; fixing with screws							
1016 mm wide	85.09	0.93	13.30	–	87.64	nr	**100.94**
2031 mm wide	142.61	1.45	20.73	–	146.89	nr	**167.62**
2843 mm wide	192.95	1.97	28.16	–	198.74	nr	**226.90**

32 FURNITURE, FITTINGS AND EQUIPMENT

Item	PC £	Labour hours	Labour £	Plant £	Material £	Unit	Total rate £
32.01 GENERAL FIXTURES, FURNISHES AND EQUIPMENT – cont							
Internal blinds – cont							
Roller blinds; Luxaflex Lite-master Crank Op; 100% blackout; 1219 mm drop; fixing with screws							
1016 mm wide	237.29	1.96	28.02	–	244.41	nr	**272.43**
2031 mm wide	317.58	2.75	39.32	–	327.11	nr	**366.43**
2843 mm wide	409.87	3.53	50.47	–	422.17	nr	**472.64**
Vertical louvre blinds; 89 mm wide louvres; Luxaflex EOS type; Florida Fabric; 1219 mm drop; fixing with screws							
1016 mm wide	68.31	0.82	11.72	–	70.36	nr	**82.08**
2031 mm wide	104.27	1.30	18.58	–	107.40	nr	**125.98**
3046 mm wide	142.61	1.77	25.31	–	146.89	nr	**172.20**
Vertical louvre blinds; 127 mm wide louvres; Luxaflex EOS type; Florida Fabric; 1219 mm drop; fixing with screws							
1016 mm wide	57.52	0.88	12.58	–	59.25	nr	**71.83**
2031 mm wide	87.50	1.35	19.30	–	90.13	nr	**109.43**
3046 mm wide	117.45	1.81	25.87	–	120.97	nr	**146.84**
Venetian blinds; 80 mm wide louvres; Levolux 480 type; solid slat; stock colour; manual hand crank; 2700 mm drop; standard brackets fixed to suitable grounds							
875 mm wide	–	–	–	–	–	nr	**230.00**
1150 mm wide	–	–	–	–	–	nr	**260.00**
1500 mm wide	–	–	–	–	–	nr	**300.00**
2500 mm wide	–	–	–	–	–	nr	**400.00**
3000 mm wide	–	–	–	–	–	nr	**435.00**
3500 mm wide	–	–	–	–	–	nr	**500.00**
4000 mm wide	–	–	–	–	–	nr	**610.00**
Door entrance mats							
Entrance mats; single aluminium wiper bar; laying in position; 12 mm thick							
Nuway Tuftiguard Plain 12 mm depth; 500 mm × 1300 mm	–	0.50	7.15	–	255.00	nr	**262.15**
Nuway Tuftiguard Plain 12 mm depth; 1000 mm × 1600 mm	–	0.65	9.29	–	582.85	nr	**592.14**
Nuway Tuftiguard Plain 12 mm depth; 1800 mm × 3000 mm	–	1.25	17.87	–	1967.11	nr	**1984.98**
Nuway Tuftiguard Classic 12 mm depth; 500 mm × 1300 mm	–	0.50	7.15	–	280.52	nr	**287.67**
Nuway Tuftiguard Classic 12 mm depth; 1000 mm × 1600 mm	–	0.65	9.29	–	641.20	nr	**650.49**
Nuway Tuftiguard Classic 12 mm depth; 1800 mm × 3000 mm	–	1.25	17.87	–	2164.02	nr	**2181.89**

32 FURNITURE, FITTINGS AND EQUIPMENT

Item	PC £	Labour hours	Labour £	Plant £	Material £	Unit	Total rate £
Entrance mats; double aluminium wiper bar; laying in position; 18 mm thick							
Gradus Topguard; 900 mm × 550 mm	–	0.46	6.57	–	182.33	nr	**188.90**
Gradus Topguard; 1200 mm × 750 mm	–	0.46	6.57	–	328.19	nr	**334.76**
Gradus Topguard; 2400 mm × 1200 mm	–	0.93	13.30	–	1050.21	nr	**1063.51**
Nuway Tuftiguard Plain 17 mm depth; 500 mm × 1300 mm	–	0.55	7.86	–	309.08	nr	**316.94**
Nuway Tuftiguard Plain 17 mm depth; 1000 mm × 1600 mm	–	0.72	10.36	–	706.47	nr	**716.83**
Nuway Tuftiguard Plain 17 mm depth; 1800 mm × 3000 mm	–	1.38	19.66	–	2384.33	nr	**2403.99**
Nuway Tuftiguard Classic 17 mm depth; 500 mm × 1300 mm	–	0.55	7.86	–	340.39	nr	**348.25**
Nuway Tuftiguard Classic 17 mm depth; 1000 mm × 1600 mm	–	0.72	10.36	–	778.03	nr	**788.39**
Nuway Tuftiguard Classic 17 mm depth; 1800 mm × 3000 mm	–	1.38	19.66	–	2625.87	nr	**2645.53**
Coral Classic textile secondary and circulation matting system for moisture removal; 2.0 m wide roll	–	0.46	6.57	–	45.81	m²	**52.38**
Coral Brush Activ secondary and circulation matting system for soil and moisture removal; 2.0 m wide roll	–	0.46	6.57	–	51.73	m²	**58.30**
Coral Duo secondary matting system ribbed; 2.0 m wide roll	–	0.46	6.57	–	51.73	m²	**58.30**
Matwells							
Polished stainless steel matwell; angle rim with lugs; bedding in screed							
stainless steel frame bed in screed 500 mm × 1300 mm	113.74	0.93	13.30	–	117.15	nr	**130.45**
stainless steel frame bed in screed 1000 mm × 1600 mm	101.10	0.93	13.30	–	104.13	nr	**117.43**
stainless steel frame bed in screed 1100 mm × 1400 mm	157.97	1.00	14.30	–	162.71	nr	**177.01**
stainless steel frame bed in screed 1800 mm × 3000 mm	303.31	2.00	28.59	–	312.41	nr	**341.00**
Polished aluminium matwell; angle rim with lugs brazed on; bedding in screed							
900 mm × 550 mm; constructed with 25 × 25 × 3 mm angle	25.88	0.93	13.30	–	26.66	nr	**39.96**
1200 mm × 750 mm; constructed with 34 × 26 × 6 mm angle	41.77	1.00	14.30	–	43.02	nr	**57.32**
2400 mm × 1200 mm; constructed with 50 × 50 × 6 mm angle	75.03	1.50	21.44	–	77.28	nr	**98.72**

32 FURNITURE, FITTINGS AND EQUIPMENT

Item	PC £	Labour hours	Labour £	Plant £	Material £	Unit	Total rate £
32.01 GENERAL FIXTURES, FURNISHES AND EQUIPMENT – cont							
Matwells – cont							
Polished brass matwell; comprising angle rim with lugs brazed on; bedding in screed							
900 mm × 550 mm; constructed with							
25 × 25 × 5 mm angle	84.40	0.93	13.30	–	86.93	nr	**100.23**
1200 mm × 750 mm; constructed with							
38 × 38 × 6 mm angle	113.51	1.00	14.30	–	116.92	nr	**131.22**
2400 mm x1200 mm; constructed with							
38 × 38 × 6 mm angle	209.55	1.50	21.44	–	215.84	nr	**237.28**
32.02 KITCHEN FITTINGS							
NOTE: Kitchen fittings vary considerably. PC supply prices for reasonable quantities for a moderately priced standard range of kitchen fittings have been shown.							
Kitchen units: Supplying and fixing to backgrounds requiring plugging; including any pre-assembly							
Wall units							
300 mm × 300 mm × 720 mm	150.00	1.11	19.16	–	150.16	nr	**169.32**
500 mm × 300 mm × 720 mm	180.00	1.16	20.02	–	180.16	nr	**200.18**
600 mm × 300 mm × 720 mm	190.00	1.30	22.44	–	190.16	nr	**212.60**
800 mm × 300 mm × 720 mm	290.00	1.48	25.54	–	290.16	nr	**315.70**
Floor units with drawers							
500 mm × 600 mm × 870 mm	265.00	1.16	20.02	–	265.16	nr	**285.18**
600 mm × 600 mm × 870 mm	290.00	1.30	22.44	–	290.16	nr	**312.60**
1000 mm × 600 mm × 870 mm	440.00	1.57	27.10	–	440.16	nr	**467.26**
Sink units (excluding sink top)							
1000 mm × 600 mm × 870 mm	400.00	1.48	25.54	–	400.16	nr	**425.70**
Kitchen worktop: Laminated worktops; single rolled edge; prices include for fixing							
38 mm thick; 600 mm wide	105.00	0.37	6.39	–	105.06	m	**111.45**
extra for forming hole for inset sink	–	0.69	11.91	–	–	nr	**11.91**
extra for jointing strip at corner intersection of worktops	–	0.14	2.42	–	15.00	nr	**17.42**
extra for butt and scribe joint at corner intersection of worktops	–	4.16	71.80	–	–	nr	**71.80**

32 FURNITURE, FITTINGS AND EQUIPMENT

Item	PC £	Labour hours	Labour £	Plant £	Material £	Unit	Total rate £
32.03 SANITARY APPLIANCES AND FITTINGS							
Sinks; Armitage Shanks or equal							
Sinks; white glazed fireclay; BS 6465; pointing all round with silicone sealant							
Belfast sink; 46 cm × 38 cm × 21 cm; pair of Nuastyle 21 basin taps with dual indices, chrome handle; wall mounts 38 mm slotted waste, chain and plug, screw stay; pair of 40.5 cm aluminium alloy build-in brackets with 35.5 cm studs	407.87	2.78	69.86	–	532.16	nr	**602.02**
Belfast sink; 61 cm × 38 cm × 21 cm; pair of Nuastyle 21 basin taps with dual indices, chrome handle; wall mounts; 38 mm slotted waste, chain and plug, screw stay; pair of 40.5 cm aluminium alloy build-in brackets with 35.5 cm studs	409.51	2.78	69.86	–	534.21	nr	**604.07**
Belfast drainer sink; 61 cm × 38 cm × 21 cm; pair of Nuastyle 21 basin taps with dual indices, chrome handle; wall mounts; 38 mm slotted waste, chain and plug, screw stay; pair of 40.5 cm aluminium alloy build-in brackets with 35.5 cm studs	520.98	2.78	69.86	–	649.36	nr	**719.22**
Lavatory basins; Armitage Shanks or equal							
Basins; white vitreous china; BS 6465 Part 3; pointing all round with silicone sealant							
Portman 21 40 cm basin; with overflow, chain hole and two tapholes; pair of Nuastyle 21 basin taps with dual indices; slotted basin waste with plastic plug, chain waste and plug; 32 × 75 mm seal plastic standard bottle trap; pair of Portman concealed brackets with waste support; Isovalve 15 mm plastic servicing valve with outlet for copper	95.55	2.13	53.53	–	168.28	nr	**221.81**
Portman 21 50 cm basin; with overflow, chain hole and two tapholes; pair of Nuastyle 21 basin taps with dual indices; slotted basin waste with plastic plug, chain waste and plug; 32 × 75 mm seal plastic standard bottle trap; pair of Portman concealed brackets with waste support; Isovalve 15 mm plastic servicing valve with outlet for copper	216.18	2.13	53.53	–	292.76	nr	**346.29**
Portman 21 60 cm basin; with overflow, chain hole and two tapholes; pair of Nuastyle 21 basin taps with dual indices; slotted basin waste with plastic plug, chain waste and plug; 32 × 75 mm seal plastic standard bottle trap; pair of Portman concealed brackets with waste support; Isovalve 15 mm plastic servicing valve with outlet for copper	285.25	2.13	53.53	–	364.13	nr	**417.66**

32 FURNITURE, FITTINGS AND EQUIPMENT

Item	PC £	Labour hours	Labour £	Plant £	Material £	Unit	Total rate £
32.03 SANITARY APPLIANCES AND FITTINGS – cont							
Drinking fountains; Armitage Shanks or equal							
White vitreous china fountains; pointing all round with silicone selant							
Aqualon wall mounted drinking fountain; Aqualon self-closing valve with fittings and plastic waste; 32 × 75 mm seal plastic standard bottle trap	535.75	2.31	58.05	–	567.89	nr	**625.94**
Polished stainless steel fountains; pointing all round with silicone selant							
Purita wall mounted drinking fountain with self-closing valve and fittings; 32 mm unslotted basin strainer waste	424.20	2.31	58.05	–	437.73	nr	**495.78**
Purita pedestal mounted drinking fountain 90 cm high with self-closing valve and fittings; 32 mm unslotted basin strainer waste	1092.51	2.78	69.86	–	1127.87	nr	**1197.73**
Baths; Armitage Shanks or equal							
Pointing all round with silicone selant							
Sandringham acrylic rectangular bath with chrome plated grips and two tapholes; standard pair of standard bath taps with chrome handles; bath chain waste with plastic plug and overflow; cast brass P trap with plain outlet and overflow connection; 170 cm long × 70 cm wide; white or coloured	265.35	3.50	87.96	–	273.31	nr	**361.27**
Nisa lowline heavy gauge steel rectangular bath with chrome plated grips and two tapholes; standard pair of standard bath taps with chrome handles; bath chain waste with plastic plug and overflow; cast brass P trap with plain outlet and overflow connection; 170 cm long × 70 cm wide; white or coloured	673.58	3.50	87.96	–	693.79	nr	**781.75**
Water closets; Armitage Shanks or equal							
White vitreous china pans and cisterns; pointing all round base with silicone sealant							
Wentworth close coupled washdown closet pan with horizontal outlet; Orion 3 plastic toilet seat and cover; Panketa pan connector 14° finned; Universal close coupled bottom inlet cistern with syphon	305.49	3.05	76.65	–	327.88	nr	**404.53**
extra for; Panketa pan connector 90° finned	–	–	–	–	3.10	nr	**3.10**
Cameo close coupled washdown closet pan with horizontal outlet; Accolade/Cameo plastic toilet seat and cover; Panketa pan connector 14° finned; Cameo 6 litre close coupled cistern with dual flush valve	517.83	3.05	76.65	–	546.60	nr	**623.25**
extra for; Panketa pan connector 90° finned	–	–	–	–	3.10	nr	**3.10**

32 FURNITURE, FITTINGS AND EQUIPMENT

Item	PC £	Labour hours	Labour £	Plant £	Material £	Unit	Total rate £
Wall urinals; Armitage Shanks or equal							
White vitreous china bowls and cisterns; pointing all round with silicone sealant							
Single Sanura 40 cm urinal bowl; Sanura top inlet spreader; pair of wall hangers for urinal bowl; 38 mm plastic domed waste; 38 × 75 mm seal plastic standard bottle trap; Conceala 4½ litres capacity auto cistern and cover; Sanura concealed flushpipe for single urinal bowl; screwing	373.47	3.70	92.99	–	425.32	nr	**518.31**
Single Sanura 40 cm urinal bowl; Sanura top inlet spreader; pair of wall hangers for urinal bowl; 38 mm plastic domed waste; 38 × 75 mm seal plastic standard bottle trap; Mura 4½ litres capacity auto cistern and cover; Sanura/Mura exposed flushpipe for single urinal bowl	409.54	3.70	92.99	–	462.46	nr	**555.45**
Single Sanura 50 cm urinal bowl; Sanura top inlet spreader; pair of wall hangers for urinal bowl; 38 mm plastic domed waste; 38 × 75 mm seal plastic standard bottle trap; Conceala 4½ litres capacity auto cistern and cover; Sanura concealed flushpipe for single urinal bowl	504.86	3.70	92.99	–	560.81	nr	**653.80**
Single Sanura 50 cm urinal bowl; Sanura top inlet spreader; pair of wall hangers for urinal bowl; 38 mm plastic domed waste; 38 × 75 mm seal plastic standard bottle trap; Mura 4½ litres capacity auto cistern and cover; Sanura/Mura exposed flushpipe for single urinal bowl	540.93	3.70	92.99	–	597.97	nr	**690.96**
Range of 2 nr Sanura 40 cm urinal bowls; Sanura top inlet spreader; pairs of wall hangers for urinal bowls; 38 mm plastic domed wastes; 38 × 75 mm seal plastic standard bottle traps; Conceala 9 litres capacity auto cistern and cover; Sanura concealed flushpipe for range of 2 nr urinal bowls	620.14	6.95	174.67	–	720.01	nr	**894.68**
Range of 2 nr Sanura 50 cm urinal bowls; Sanura top inlet spreader; pairs of wall hangers for urinal bowls; 38 mm plastic domed wastes; 38 × 75 mm seal plastic standard bottle traps; Conceala 9 litres capacity auto cistern and cover; Sanura concealed flushpipe for range of 2 nr urinal bowls	882.92	6.95	174.67	–	991.02	nr	**1165.69**
Range of 3 nr Sanura 40 cm urinal bowls; Sanura top inlet spreader; pairs of wall hangers for urinal bowls; 38 mm plastic domed wastes; 38 × 75 mm seal plastic standard bottle traps; Conceala 9 litres capacity auto cistern and cover; Sanura concealed flushpipe for range of 3 nr urinal bowls	857.49	10.15	255.09	–	1005.13	nr	**1260.22**

32 FURNITURE, FITTINGS AND EQUIPMENT

Item	PC £	Labour hours	Labour £	Plant £	Material £	Unit	Total rate £
32.03 SANITARY APPLIANCES AND FITTINGS – cont							
Wall urinals – cont							
White vitreous china bowls and cisterns – cont							
Range of 3 nr Sanura 50 cm urinal bowls; Sanura top inlet spreader; pairs of wall hangers for urinal bowls; 38 mm plastic domed wastes; 38 × 75 mm seal plastic standard bottle traps; Conceala 9 litres capacity auto cistern and cover; Sanura concealed flushpipe for range of 3 nr urinal bowls	1251.67	10.15	255.09	–	1411.12	nr	**1666.21**
Range of 4 nr Sanura 40 cm urinal bowls; Sanura top inlet spreader; pairs of wall hangers for urinal bowls; 38 mm plastic domed wastes; 38 × 75 mm seal plastic standard bottle traps; Conceala 9 litres capacity auto cistern and cover; Sanura concealed flushpipe for range of 4 nr urinal bowls; screwing	1104.49	13.40	336.77	–	1300.17	nr	**1636.94**
Range of 4 nr Sanura 50 cm urinal bowls; Sanura top inlet spreader; pairs of wall hangers for urinal bowls; 38 mm plastic domed wastes; 38 × 75 mm seal plastic standard bottle traps; Conceala 9 litres capacity auto cistern and cover; Sanura concealed flushpipe for range of 4 nr urinal bowls	1630.05	13.40	336.77	–	1842.19	nr	**2178.96**
Range of 5 nr Sanura 40 cm urinal bowls; Sanura top inlet spreader; pairs of wall hangers for urinal bowls; 38 mm plastic domed wastes; 38 × 75 mm seal plastic standard bottle traps; Conceala 9 litres capacity auto cistern and cover; Sanura concealed flushpipe for range of 5 nr urinal bowls	1341.96	16.65	418.45	–	1585.40	nr	**2003.85**
Range of 5 nr Sanura 50 cm urinal bowls; Sanura top inlet spreader; pairs of wall hangers for urinal bowls; 38 mm plastic domed wastes; 38 × 75 mm seal plastic standard bottle traps; Conceala 9 litres capacity auto cistern and cover; Sanura concealed flushpipe for range of 5 nr urinal bowls	1998.92	16.65	418.45	–	2262.93	nr	**2681.38**
White vitreous china division panels; pointing all round with silicone sealant							
urinal division with screw and hanger	112.68	0.70	17.59	–	117.96	nr	**135.55**
Bidets; Armitage Shanks or equal							
Tiffany back to wall bidet with one taphole; vitreous china; chromium plated pop-up waste and mixer tap with hand wheels; 58 cm × 39 cm; white or coloured	598.03	3.50	87.96	–	615.97	nr	**703.93**
Shower tray and fittings							
Simplicity shower tray; acrylic; with outlet and grated waste; chain and plug; bedding and pointing in waterproof cement mortar							
760 mm × 760 mm ; white or coloured	94.91	3.00	75.40	–	97.76	nr	**173.16**

32 FURNITURE, FITTINGS AND EQUIPMENT

Item	PC £	Labour hours	Labour £	Plant £	Material £	Unit	Total rate £
Shower fitting; riser pipe with mixing valve and shower rose; chromium plated; plugging and screwing mixing valve and pipe bracket							
15 mm dia. riser pipe; 127 mm dia. shower rose	510.64	5.00	125.66	–	525.96	nr	**651.62**
Corner fitting shower enclosure							
Bliss flat top hinged door with front panel and clear glass side panel	1005.35	3.00	42.89	–	1035.51	nr	**1078.40**
Miscellaneous fittings; Magrini Ltd or equal							
Vertical nappy changing unit							
ref KBCS; screwing	–	0.60	10.67	–	232.49	nr	**243.16**
Horizontal nappy changing unit							
ref KBHS; screwing	–	0.60	10.67	–	232.49	nr	**243.16**
Stay Safe baby seat							
ref MX33; screwing to wall	–	0.55	9.77	–	87.19	nr	**96.96**
Miscellaneous fittings; Pressalit Ltd or equal; stainless steel							
Grab rails							
300 mm long ref RT100000; screw fix to wall	–	0.50	8.89	–	51.94	nr	**60.83**
450 mm long ref RT101000; screw fix to wall	–	0.50	8.89	–	60.06	nr	**68.95**
600 mm long ref RT102000; screw fix to wall	–	0.50	8.89	–	68.90	nr	**77.79**
800 mm long ref RT103000; screw fix to wall	–	0.50	8.89	–	77.47	nr	**86.36**
1000 mm long ref RT104000; screw fix to wall	–	0.50	8.89	–	89.32	nr	**98.21**
Angled grab rails							
900 mm long, angled 135° ref RT110000; screw fix to wall	–	0.50	8.89	–	124.07	nr	**132.96**
1300 mm long, angled 90° ref RT119000; screw fix to wall	–	0.75	13.34	–	175.85	nr	**189.19**
Hinged grab rails							
600 mm long ref R3016000; screw fix to wall	–	0.35	6.22	–	182.36	nr	**188.58**
600 mm long with spring counter balance ref RF016000; screw fix to wall	–	0.35	6.22	–	254.49	nr	**260.71**
850 mm long ref R3010000; screw fix to wall	–	0.35	6.22	–	221.43	nr	**227.65**
850 mm long with spring counter balance ref RF010000; screw fix to wall	–	0.35	6.22	–	273.15	nr	**279.37**
Shower seat; wall mounted; with padded seat and back	–	1.50	26.67	–	270.29	nr	**296.96**
Document M pack; flush lever; WC seat; steel grab rails	–	4.00	89.85	–	256.00	n	**345.85**

32.04 NOTICES AND SIGNS

Plain script; in gloss oil paint; on painted or varnished surfaces

Item	PC £	Labour hours	Labour £	Plant £	Material £	Unit	Total rate £
Capital letters; lower case letters or numerals							
per coat; per 25 mm high	–	0.09	1.60	–	–	nr	**1.60**

33 DRAINAGE ABOVE GROUND

Item	PC £	Labour hours	Labour £	Plant £	Material £	Unit	Total rate £
33.01 RAINWATER INSTALLATIONS							
Aluminium							
Pipes and fittings; Heritage cast; with or without ears cast on; polyester powder coated finish							
63 mm dia. pipes; plugged and screwed	21.48	0.34	6.79	–	24.84	m	**31.63**
extra for							
fittings with one end	14.64	0.20	4.00	–	14.04	nr	**18.04**
fittings with two ends	15.61	0.39	7.79	–	14.27	nr	**22.06**
fittings with three ends	20.41	0.56	11.19	–	21.89	nr	**33.08**
shoe	14.64	0.20	4.00	–	15.89	nr	**19.89**
bend	15.61	0.39	7.79	–	16.13	nr	**23.92**
single branch	20.41	0.56	11.19	–	21.89	nr	**33.08**
offset 228 projection	36.08	0.39	9.02	–	37.99	nr	**47.01**
offset 304 projection	40.21	0.39	7.79	–	42.23	nr	**50.02**
access pipe	44.54	0.39	7.79	–	45.90	nr	**53.69**
connection to clay pipes; cement and sand (1:2) joint	–	0.14	2.79	–	0.15	nr	**2.94**
75 mm dia. pipes; plugged and screwed	25.02	0.37	7.39	–	28.69	m	**36.08**
extra for							
shoe	20.13	0.23	4.59	–	21.56	nr	**26.15**
bend	19.77	0.42	8.38	–	20.41	nr	**28.79**
single branch	24.55	0.60	11.98	–	26.16	nr	**38.14**
offset 228 projection	39.88	0.42	8.38	–	41.90	nr	**50.28**
offset 304 projection	44.11	0.42	8.38	–	46.26	nr	**54.64**
access pipe	48.67	0.42	8.38	–	50.15	nr	**58.53**
connection to clay pipes; cement and sand (1:2) joint	–	0.16	3.19	–	0.15	nr	**3.34**
100 mm dia. pipes; plugged and screwed	42.59	0.42	8.38	–	47.69	m	**56.07**
extra for							
shoe	24.26	0.26	5.19	–	25.82	nr	**31.01**
bend	27.51	0.46	9.19	–	28.41	nr	**37.60**
single branch	32.90	0.69	13.78	–	34.79	nr	**48.57**
offset 228 projection	46.14	0.46	9.19	–	47.56	nr	**56.75**
offset 304 projection	51.22	0.46	9.19	–	52.79	nr	**61.98**
access pipe	57.67	0.46	9.19	–	59.44	nr	**68.63**
connection to clay pipes; cement and sand (1:2) joint	–	0.19	3.79	–	0.15	nr	**3.94**
75 mm × 75 mm square pipes; plugged and screwed	47.91	0.34	6.79	–	53.44	m	**60.23**
extra for							
fittings with one end	46.39	0.20	4.00	–	44.45	nr	**48.45**
fittings with two ends	116.88	0.39	7.79	–	116.29	nr	**124.08**
fittings with three ends	116.88	0.56	11.19	–	121.25	nr	**132.44**
shoe	46.39	0.20	4.00	–	48.60	nr	**52.60**
bend	50.08	0.39	7.79	–	51.63	nr	**59.42**
single branch	116.88	0.56	11.19	–	121.25	nr	**132.44**
offset 75 mm projection	64.13	0.39	9.02	–	66.88	nr	**75.90**
access pipe	73.35	0.39	7.79	–	75.58	nr	**83.37**
connection to clay pipes; cement and sand (1:2) joint	–	0.14	2.79	–	0.15	nr	**2.94**

33 DRAINAGE ABOVE GROUND

Item	PC £	Labour hours	Labour £	Plant £	Material £	Unit	Total rate £
100 mm × 75 mm square pipes; plugged and screwed	56.26	0.37	7.39	–	62.47	m	**69.86**
extra for							
shoe	50.16	0.23	4.59	–	52.49	nr	**57.08**
bend	56.78	0.46	9.19	–	58.56	nr	**67.75**
single branch	126.24	0.60	11.98	–	130.90	nr	**142.88**
offset 75 mm projection	73.05	0.42	8.38	–	76.07	nr	**84.45**
access pipe	78.44	0.42	8.38	–	80.81	nr	**89.19**
connection to clay pipes; cement and sand (1:2) joint	–	0.16	3.19	–	0.15	nr	**3.34**
100 mm × 100 mm square pipes; plugged and screwed	68.01	0.42	8.38	–	75.18	m	**83.56**
extra for							
shoe	56.10	0.26	5.19	–	58.62	nr	**63.81**
bend	130.90	0.46	9.19	–	134.89	nr	**144.08**
single branch	65.96	0.69	13.78	–	68.83	nr	**82.61**
offset 75 mm projection	84.02	0.46	9.19	–	86.57	nr	**95.76**
access pipe	85.95	0.46	9.19	–	88.56	nr	**97.75**
connection to clay pipes; cement and sand (1:2) joint	–	0.19	3.79	–	0.15	nr	**3.94**
Roof outlets; circular aluminium; with domed grating; joint to pipe; vertical spigot outlet							
50 mm dia.	118.13	0.56	12.68	–	121.67	nr	**134.35**
75 mm dia.	150.37	0.60	13.59	–	154.88	nr	**168.47**
100 mm dia.	190.40	0.65	14.72	–	196.11	nr	**210.83**
150 mm dia.	237.32	0.69	15.63	–	244.44	nr	**260.07**
Balcony outlets; circular aluminium; with flat grating; joint to pipe							
50 mm dia.	118.13	0.56	12.68	–	121.67	nr	**134.35**
75 mm dia.	150.37	0.60	13.59	–	154.88	nr	**168.47**
100 mm dia.	190.40	0.65	14.72	–	196.11	nr	**210.83**
150 mm dia.	237.32	0.69	15.63	–	244.44	nr	**260.07**
PVC balloon grating							
110 mm dia.	7.65	0.06	1.36	–	7.88	nr	**9.24**
63 mm dia.	4.80	0.06	1.36	–	4.94	nr	**6.30**
Aluminium gutters and fittings; polyester powder coated finish							
100 mm half round gutters; on brackets; screwed to timber	29.73	0.32	7.41	–	38.04	m	**45.45**
extra for							
stop end	7.86	0.15	3.47	–	15.08	nr	**18.55**
running outlet	17.41	0.31	7.17	–	19.80	nr	**26.97**
stop end outlet	15.51	0.15	3.47	–	22.96	nr	**26.43**
angle	16.10	0.31	7.17	–	18.47	nr	**25.64**
113 mm half round gutters; on brackets; screwed to timber	31.12	0.32	7.41	–	39.55	m	**46.96**
extra for							
stop end	8.27	0.15	3.47	–	15.56	nr	**19.03**
running outlet	18.99	0.31	7.17	–	21.42	nr	**28.59**
stop end outlet	17.80	0.15	3.47	–	25.32	nr	**28.79**
angle	18.15	0.31	7.17	–	15.11	nr	**22.28**

33 DRAINAGE ABOVE GROUND

Item	PC £	Labour hours	Labour £	Plant £	Material £	Unit	Total rate £
33.01 RAINWATER INSTALLATIONS – cont							
Aluminium gutters and fittings – cont							
125 mm half round gutters; on brackets; screwed to timber	34.97	0.37	8.56	–	47.56	m	**56.12**
extra for							
stop end	10.07	0.17	3.93	–	21.09	nr	**25.02**
running outlet	20.58	0.32	7.41	–	23.06	nr	**30.47**
stop end outlet	18.87	0.17	3.93	–	30.17	nr	**34.10**
angle	20.16	0.32	7.41	–	22.63	nr	**30.04**
100 mm ogee gutters; on brackets; screwed to timber	37.08	0.34	7.87	–	49.03	m	**56.90**
extra for							
stop end	8.31	0.16	3.70	–	9.98	nr	**13.68**
running outlet	20.44	0.32	7.41	–	21.68	nr	**29.09**
stop end outlet	20.44	0.16	3.70	–	31.03	nr	**34.73**
angle	17.25	0.32	7.41	–	11.93	nr	**19.34**
112 mm ogee gutters; on brackets; screwed to timber	41.24	0.39	9.02	–	53.95	m	**62.97**
extra for							
stop end	8.86	0.16	3.70	–	10.56	nr	**14.26**
running outlet	20.69	0.32	7.41	–	21.94	nr	**29.35**
stop end outlet	20.69	0.16	3.70	–	31.87	nr	**35.57**
angle	20.53	0.32	7.41	–	21.79	nr	**29.20**
125 mm ogee gutters; on brackets; screwed to timber	45.54	0.39	9.02	–	59.35	m	**68.37**
extra for							
stop end	9.71	0.18	4.16	–	11.43	nr	**15.59**
running outlet	22.60	0.34	7.87	–	23.92	nr	**31.79**
stop end outlet	22.60	0.18	4.16	–	34.71	nr	**38.87**
angle	23.93	0.34	7.87	–	17.33	nr	**25.20**
Cast iron; primed finish only							
Pipes and fittings; ears cast on; joints							
65 mm pipes; primed; nailed to masonry	45.51	0.48	9.59	–	47.45	m	**57.04**
extra for							
shoe	38.99	0.30	5.99	–	38.05	nr	**44.04**
bend	23.88	0.53	10.59	–	22.48	nr	**33.07**
single branch	50.11	0.67	13.38	–	49.03	nr	**62.41**
offset 225 mm projection	42.52	0.53	10.59	–	38.87	nr	**49.46**
offset 305 mm projection	49.84	0.53	10.59	–	45.47	nr	**56.06**
connection to clay pipes; cement and sand (1:2) joint	–	0.14	2.79	–	0.16	nr	**2.95**
75 mm pipes; primed; nailed to masonry	45.51	0.51	10.19	–	47.76	m	**57.95**
extra for							
shoe	38.99	0.32	6.39	–	38.55	nr	**44.94**
bend	28.97	1.11	24.07	–	28.23	nr	**52.30**
single branch	55.25	0.69	13.78	–	55.33	nr	**69.11**
offset 225 mm projection	42.52	0.56	11.19	–	39.38	nr	**50.57**
offset 305 mm projection	52.28	0.56	11.19	–	48.49	nr	**59.68**
connection to clay pipes; cement and sand (1:2) joint	–	0.16	3.19	–	0.16	nr	**3.35**

33 DRAINAGE ABOVE GROUND

Item	PC £	Labour hours	Labour £	Plant £	Material £	Unit	Total rate £
100 mm pipes; primed; nailed to masonry	61.13	0.56	11.19	–	64.19	m	**75.38**
extra for							
shoe	51.77	0.37	7.39	–	51.37	nr	**58.76**
bend	40.92	0.60	11.98	–	40.19	nr	**52.17**
single branch	64.38	0.74	14.78	–	64.68	nr	**79.46**
offset 225 mm projection	83.47	0.60	11.98	–	80.24	nr	**92.22**
offset 305 mm projection	85.10	0.60	11.98	–	80.66	nr	**92.64**
connection to clay pipes; cement and sand (1:2) joint	–	0.19	3.79	–	0.15	nr	**3.94**
100 mm × 75 mm rectangular pipes; primed; nailing to masonry	131.26	0.56	11.19	–	136.43	m	**147.62**
extra for							
shoe	155.99	0.37	7.39	–	152.93	nr	**160.32**
offset 225 mm projection	200.52	0.37	7.39	–	190.68	nr	**198.07**
offset 305 mm projection	231.06	0.37	7.39	–	219.44	nr	**226.83**
connection to clay pipes; cement and sand (1:2) joint	–	0.19	3.79	–	0.15	nr	**3.94**
Rainwater head; 225 mm × 125 mm × 125 mm, rectangular; for pipes							
65 mm dia.	119.83	0.53	10.59	–	125.06	nr	**135.65**
75 mm dia.	119.83	0.56	11.19	–	125.57	nr	**136.76**
Rainwater head; 280 mm × 130 mm × 130 mm, rectangular; for pipes							
100 mm dia.	165.46	0.60	11.98	–	173.50	nr	**185.48**
Rainwater head; flat back, octagonal; for pipes							
65 mm dia.	86.12	0.53	10.59	–	90.34	nr	**100.93**
75 mm dia.	86.12	0.56	11.19	–	90.85	nr	**102.04**
Gutters and fittings; Primed finish only							
100 mm half round gutters; primed; on brackets; screwed to timber	23.26	0.37	8.56	–	31.68	m	**40.24**
extra for							
stop end	5.60	0.16	3.70	–	9.12	nr	**12.82**
running outlet	16.20	0.32	7.41	–	15.53	nr	**22.94**
angle	16.66	0.32	7.41	–	18.96	nr	**26.37**
115 mm half round gutters; primed; on brackets; screwed to timber	24.24	0.37	8.56	–	32.82	m	**41.38**
extra for							
stop end	7.22	0.16	3.70	–	10.79	nr	**14.49**
running outlet	17.66	0.32	7.41	–	17.03	nr	**24.44**
angle	17.10	0.32	7.41	–	19.32	nr	**26.73**
125 mm half round gutters; primed; on brackets; screwed to timber	28.38	0.42	9.71	–	37.14	m	**46.85**
extra for							
stop end	7.22	0.19	4.40	–	10.94	nr	**15.34**
running outlet	20.14	0.37	8.56	–	19.31	nr	**27.87**
angle	20.14	0.37	8.56	–	21.81	nr	**30.37**
150 mm half round gutters; primed; on brackets; screwed to timber	44.64	0.46	10.64	–	54.40	m	**65.04**

33 DRAINAGE ABOVE GROUND

Item	PC £	Labour hours	Labour £	Plant £	Material £	Unit	Total rate £
33.01 RAINWATER INSTALLATIONS – cont							
Cast iron – cont							
Gutters and fittings – cont							
extra for							
stop end	10.06	0.20	4.62	–	17.81	nr	**22.43**
running outlet	34.93	0.42	9.71	–	33.28	nr	**42.99**
angle	36.86	0.42	9.71	–	37.58	nr	**47.29**
100 mm ogee gutters; primed; on brackets;							
screwed to timber	25.94	0.39	9.02	–	34.62	m	**43.64**
extra for							
stop end	5.72	0.17	3.93	–	12.34	nr	**16.27**
running outlet	17.67	0.34	7.87	–	16.97	nr	**24.84**
angle	17.35	0.34	7.87	–	19.78	nr	**27.65**
115 mm ogee gutters; primed; on brackets;							
screwed to timber	28.54	0.39	9.02	–	37.34	m	**46.36**
extra for							
stop end	7.41	0.17	3.93	–	14.15	nr	**18.08**
running outlet	18.80	0.34	7.87	–	17.92	nr	**25.79**
angle	18.80	0.34	7.87	–	20.82	nr	**28.69**
125 mm ogee gutters; primed; on brackets;							
screwed to timber	29.94	0.43	9.95	–	39.57	m	**49.52**
extra for							
stop end	7.41	0.19	4.40	–	14.95	nr	**19.35**
running outlet	20.51	0.39	9.02	–	19.71	nr	**28.73**
angle	20.51	0.39	9.02	–	23.20	nr	**32.22**
Cast iron; pre-painted finish; black standard							
Pipes and fittings; ears cast on; joints							
65 mm pipes; primed; nailed to masonry	53.73	0.48	9.59	–	55.91	m	**65.50**
extra for							
shoe	45.05	0.30	5.99	–	43.61	nr	**49.60**
bend	29.57	0.53	10.59	–	27.67	nr	**38.26**
single branch	55.90	0.67	13.38	–	53.97	nr	**67.35**
offset 225 mm projection	48.77	0.53	10.59	–	44.13	nr	**54.72**
offset 305 mm projection	56.23	0.53	10.59	–	50.70	nr	**61.29**
connection to clay pipes; cement and sand							
(1:2) joint	–	0.14	2.79	–	0.16	nr	**2.95**
75 mm pipes; primed; nailed to masonry	53.73	0.51	10.19	–	56.22	m	**66.41**
extra for							
shoe	45.05	0.32	6.39	–	44.11	nr	**50.50**
bend	34.83	1.11	24.07	–	33.59	nr	**57.66**
single branch	61.06	0.69	13.78	–	60.31	nr	**74.09**
offset 225 mm projection	48.77	0.56	11.19	–	44.63	nr	**55.82**
offset 305 mm projection	58.72	0.56	11.19	–	53.77	nr	**64.96**
connection to clay pipes; cement and sand							
(1:2) joint	–	0.16	3.19	–	0.16	nr	**3.35**
100 mm pipes; primed; nailed to masonry	75.50	0.56	11.19	–	79.00	m	**90.19**

33 DRAINAGE ABOVE GROUND

Item	PC £	Labour hours	Labour £	Plant £	Material £	Unit	Total rate £
extra for							
shoe	57.26	0.37	7.39	–	55.84	nr	**63.23**
bend	47.11	0.60	11.98	–	45.38	nr	**57.36**
single branch	71.50	0.74	14.78	–	70.25	nr	**85.03**
offset 225 mm projection	90.79	0.60	11.98	–	85.71	nr	**97.69**
offset 305 mm projection	90.79	0.60	11.98	–	84.15	nr	**96.13**
connection to clay pipes; cement and sand (1:2) joint	–	0.19	3.79	–	0.15	nr	**3.94**
100 mm × 75 mm rectangular pipes; primed; nailing to masonry	131.26	0.56	11.19	–	136.43	m	**147.62**
extra for							
shoe	155.99	0.37	7.39	–	152.93	nr	**160.32**
offset 225 mm projection	200.52	0.37	7.39	–	190.68	nr	**198.07**
offset 305 mm projection	231.06	0.37	7.39	–	219.44	nr	**226.83**
connection to clay pipes; cement and sand (1:2) joint	–	0.19	3.79	–	0.15	nr	**3.94**
Rainwater head; 225 mm × 125 mm × 125 mm, rectangular; for pipes							
65 mm dia.	131.87	0.53	10.59	–	137.46	nr	**148.05**
75 mm dia.	131.87	0.56	11.19	–	137.97	nr	**149.16**
Rainwater head; 280 mm × 130 mm × 130 mm, rectangular; for pipes							
100 mm dia.	177.91	0.60	11.98	–	186.34	nr	**198.32**
Rainwater head; flat back, octagonal; for pipes							
65 mm dia.	93.75	0.53	10.59	–	98.20	nr	**108.79**
75 mm dia.	93.75	0.56	11.19	–	98.70	nr	**109.89**
Gutters and fittings; Pre-painted finish; black standard							
100 mm half round gutters; pre-painted; on brackets; screwed to timber	29.04	0.37	8.56	–	39.13	m	**47.69**
extra for							
stop end	8.53	0.16	3.70	–	12.82	nr	**16.52**
running outlet	19.38	0.32	7.41	–	18.32	nr	**25.73**
angle	19.78	0.32	7.41	–	22.55	nr	**29.96**
115 mm half round gutters; pre-painted; on brackets; screwed to timber	31.77	0.37	8.56	–	42.64	m	**51.20**
extra for							
stop end	9.91	0.16	3.70	–	14.51	nr	**18.21**
running outlet	20.73	0.32	7.41	–	19.57	nr	**26.98**
angle	20.25	0.32	7.41	–	23.14	nr	**30.55**
125 mm half round gutters; pre-painted; on brackets; screwed to timber	10.41	0.42	9.71	–	21.23	m	**30.94**
extra for							
stop end	9.91	0.19	4.40	–	14.88	nr	**19.28**
running outlet	23.09	0.37	8.56	–	22.69	nr	**31.25**
angle	–	0.37	8.56	–	25.69	nr	**34.25**
150 mm half round gutters; pre-painted; on brackets; screwed to timber	54.24	0.46	10.64	–	65.65	m	**76.29**

33 DRAINAGE ABOVE GROUND

Item	PC £	Labour hours	Labour £	Plant £	Material £	Unit	Total rate £
33.01 RAINWATER INSTALLATIONS – cont							
Cast iron – cont							
Gutters and fittings – cont							
extra for							
stop end	14.15	0.20	4.62	–	23.26	nr	**27.88**
running outlet	49.20	0.42	9.71	–	47.18	nr	**56.89**
angle	58.38	0.42	9.71	–	59.31	nr	**69.02**
100 mm ogee gutters; pre-painted; on brackets; screwed to timber	35.29	0.39	9.02	–	45.67	m	**54.69**
extra for							
stop end	8.60	0.17	3.93	–	16.60	nr	**20.53**
running outlet	20.58	0.34	7.87	–	19.20	nr	**27.07**
angle	20.38	0.34	7.87	–	22.57	nr	**30.44**
115 mm ogee gutters; pre-painted; on brackets; screwed to timber	36.68	0.39	9.02	–	47.73	m	**56.75**
extra for							
stop end	9.98	0.17	3.93	–	18.62	nr	**22.55**
running outlet	21.67	0.34	7.87	–	20.21	nr	**28.08**
angle	21.62	0.34	7.87	–	24.11	nr	**31.98**
125 mm ogee gutters; pre-painted; on brackets; screwed to timber	38.22	0.43	9.95	–	50.51	m	**60.46**
extra for							
stop end	11.24	0.19	4.40	–	21.07	nr	**25.47**
running outlet	23.36	0.39	9.02	–	23.11	nr	**32.13**
angle	23.09	0.39	9.02	–	29.04	nr	**38.06**
Steel							
3 mm thick magnesium coated galvanized pressed steel gutters and fittings; joggle joints; including bracket and stiffeners							
200 mm × 100 mm (400 mm girth) box gutter; screwed to timber	–	0.60	11.98	–	18.51	m	**30.49**
extra for							
stop end	–	0.32	6.39	–	5.48	nr	**11.87**
running outlet	–	0.65	12.98	–	32.42	nr	**45.40**
stop end outlet	–	0.32	6.39	–	41.81	nr	**48.20**
angle	–	0.65	12.98	–	35.42	nr	**48.40**
381 mm boundary wall gutters (900 mm girth); bent twice; screwed to timber	–	0.60	11.98	–	20.72	m	**32.70**
extra for							
stop end	–	0.37	7.39	–	6.04	nr	**13.43**
running outlet	–	0.65	12.98	–	42.66	nr	**55.64**
stop end outlet	–	0.32	6.39	–	75.14	nr	**81.53**
angle	–	0.65	12.98	–	49.34	nr	**62.32**
457 mm boundary wall gutters (1200 mm girth); bent twice; screwed to timber	–	0.69	13.78	–	86.05	m	**99.83**
extra for							
stop end	–	0.37	7.39	–	29.75	nr	**37.14**
running outlet	–	0.74	14.78	–	60.97	nr	**75.75**
stop end outlet	–	0.37	7.39	–	62.43	nr	**69.82**
angle	–	0.74	14.78	–	65.48	nr	**80.26**

33 DRAINAGE ABOVE GROUND

Item	PC £	Labour hours	Labour £	Plant £	Material £	Unit	Total rate £
750 mm girth; valley gutters; Kalzip Membrane lined composite gutter system	–	–	–	–	–	m	**123.95**
extra for							
stop end	–	–	–	–	–	nr	**92.95**
running outlet	–	–	–	–	–	nr	**117.74**
uPVC							
External rainwater pipes and fittings; slip-in joints							
50 mm pipes; fixing with pipe or socket brackets; plugged and screwed	15.22	0.28	5.59	–	21.04	m	**26.63**
extra for							
shoe	8.95	0.19	3.79	–	11.31	nr	**15.10**
bend	10.46	0.28	5.59	–	12.86	nr	**18.45**
two bends to form offset 229 mm projection	20.93	0.28	5.59	–	20.98	nr	**26.57**
connection to clay pipes; cement and sand (1:2) joint	–	0.12	2.40	–	0.16	nr	**2.56**
68 mm pipes; fixing with pipe or socket brackets; plugged and screwed	11.80	0.31	6.19	–	18.56	m	**24.75**
extra for							
shoe	8.95	0.20	4.00	–	12.29	nr	**16.29**
bend	12.71	0.31	6.19	–	16.16	nr	**22.35**
single branch	25.55	0.41	8.19	–	29.39	nr	**37.58**
two bends to form offset 229 mm projection	25.42	0.31	6.19	–	27.18	nr	**33.37**
loose drain connector; cement and sand (1:2) joint	–	0.14	2.79	–	30.41	nr	**33.20**
110 mm pipes; fixing with pipe or socket brackets; plugged and screwed	24.21	0.33	6.59	–	37.96	m	**44.55**
extra for							
shoe	27.94	0.22	4.40	–	32.30	nr	**36.70**
bend	41.23	0.33	6.59	–	45.99	nr	**52.58**
single branch	62.27	0.44	8.79	–	67.66	nr	**76.45**
two bends to form offset 229 mm projection	82.45	0.33	6.59	–	84.95	nr	**91.54**
loose drain connector; cement and sand (1:2) joint	–	0.32	6.39	–	27.18	nr	**33.57**
65 mm square pipes; fixing with pipe or socket brackets; plugged and screwed	11.80	0.31	6.19	–	18.48	m	**24.67**
extra for							
shoe	8.95	0.20	4.00	–	12.21	nr	**16.21**
bend	12.71	0.31	6.19	–	16.08	nr	**22.27**
single branch	25.55	0.41	8.19	–	29.30	nr	**37.49**
two bends to form offset 229 mm projection	25.42	0.31	6.19	–	27.71	nr	**33.90**
drain connector; square to round; cement and sand (1:2) joint	–	0.32	6.39	–	14.02	nr	**20.41**
Rainwater head; rectangular; for pipes							
50 mm dia.	50.04	0.42	8.38	–	55.78	nr	**64.16**
68 mm dia.	40.65	0.43	8.59	–	48.06	nr	**56.65**
110 mm dia.	79.96	0.51	10.19	–	89.46	nr	**99.65**
65 mm square	40.65	0.43	8.59	–	47.91	nr	**56.50**

33 DRAINAGE ABOVE GROUND

Item	PC £	Labour hours	Labour £	Plant £	Material £	Unit	Total rate £
33.01 RAINWATER INSTALLATIONS – cont							
uPVC – cont							
Gutters and fittings							
76 mm half round gutters; on brackets screwed to timber	11.57	0.28	6.48	–	17.31	m	**23.79**
extra for							
stop end	4.11	0.12	2.78	–	5.42	nr	**8.20**
running outlet	11.59	0.23	5.33	–	10.98	nr	**16.31**
stop end outlet	11.54	0.12	2.78	–	12.12	nr	**14.90**
angle	11.59	0.23	5.33	–	13.33	nr	**18.66**
112 mm half round gutters; on brackets screwed to timber	11.11	0.31	7.17	–	19.28	m	**26.45**
extra for							
stop end	6.45	0.12	2.78	–	8.60	nr	**11.38**
running outlet	12.64	0.26	6.02	–	12.10	nr	**18.12**
stop end outlet	12.64	0.12	2.78	–	14.06	nr	**16.84**
angle	14.13	0.26	6.02	–	17.54	nr	**23.56**
170 mm half round gutters; on brackets; screwed to timber	23.21	0.31	7.17	–	37.21	m	**44.38**
extra for							
stop end	10.14	0.15	3.47	–	14.08	nr	**17.55**
running outlet	22.69	0.29	6.71	–	21.45	nr	**28.16**
stop end outlet	21.57	0.15	3.47	–	23.95	nr	**27.42**
angle	29.53	0.29	6.71	–	35.78	nr	**42.49**
114 mm rectangular gutters; on brackets; screwed to timber	11.43	0.31	7.17	–	21.61	m	**28.78**
extra for							
stop end	6.45	0.12	2.78	–	8.60	nr	**11.38**
running outlet	12.64	0.29	6.71	–	12.08	nr	**18.79**
stop end outlet	12.64	0.12	2.78	–	14.03	nr	**16.81**
angle	14.13	0.26	6.02	–	17.52	nr	**23.54**
Kingspan insulated gutter							
Pre-laminated insulated membrane gutters							
eaves; 1250 mm girth	–	–	–	–	–	m	**130.71**
valley; 1250 mm girth	–	–	–	–	–	m	**130.71**
weir overflows	–	–	–	–	–	nr	**217.86**
stop ends	–	–	–	–	–	nr	**268.69**
T-sections	–	–	–	–	–	nr	**435.71**

33 DRAINAGE ABOVE GROUND

Item	PC £	Labour hours	Labour £	Plant £	Material £	Unit	Total rate £
33.02 FOUL DRAINAGE INSTALLATIONS							
Cast iron							
Cast iron Timesaver pipes and fittings or other equal							
50 mm pipes; primed; 3 m lengths; fixing with							
expanding bolts; to masonry	30.58	0.51	10.19	–	45.00	m	**55.19**
extra for							
fittings with two ends	–	0.51	10.21	–	39.23	nr	**49.44**
fittings with three ends	–	0.69	13.78	–	66.47	nr	**80.25**
bends; short radius	27.23	0.51	10.19	–	39.23	nr	**49.42**
access bends; short radius	67.12	0.51	10.19	–	80.32	nr	**90.51**
boss; 38 BSP	56.39	0.51	10.19	–	68.47	nr	**78.66**
single branch	40.98	0.69	13.78	–	67.88	nr	**81.66**
isolated Timesaver coupling joint	15.45	0.28	5.59	–	15.91	nr	**21.50**
connection to clay pipes; cement and sand							
(1:2) joint	–	0.12	2.40	–	0.15	nr	**2.55**
75 mm pipes; primed; 3 m lengths; fixing with							
standard brackets; plugged and screwed to							
masonry	34.21	0.51	10.19	–	55.75	m	**65.94**
extra for							
bends; short radius	30.81	0.55	10.98	–	44.03	nr	**55.01**
access bends; short radius	72.81	0.51	10.19	–	87.30	nr	**97.49**
boss; 38 BSP	56.39	0.55	10.98	–	70.38	nr	**81.36**
single branch	46.38	0.79	15.78	–	75.73	nr	**91.51**
double branch	68.87	1.02	20.37	–	116.47	nr	**136.84**
offset 115 mm projection	44.20	0.55	10.98	–	53.78	nr	**64.76**
offset 150 mm projection	51.92	0.55	10.98	–	60.67	nr	**71.65**
access pipe, round door	65.53	0.55	10.98	–	75.22	nr	**86.20**
isolated Timesaver coupling joint	17.07	0.32	6.39	–	17.58	nr	**23.97**
connection to clay pipes; cement and sand							
(1:2) joint	–	0.14	2.79	–	0.15	nr	**2.94**
100 mm pipes; primed; 3 m lengths; fixing with							
standard brackets; plugged and screwed to							
masonry	41.35	0.55	10.98	–	77.56	m	**88.54**
extra for							
WC connector	45.55	0.55	10.98	–	54.74	nr	**65.72**
bends; short radius	37.70	0.62	12.38	–	55.39	nr	**67.77**
access bends; short radius	79.73	0.62	12.38	–	98.68	nr	**111.06**
boss; 38 BSP	67.31	0.62	12.38	–	85.89	nr	**98.27**
single branch	58.29	0.93	18.57	–	95.92	nr	**114.49**
double branch	72.08	1.20	23.97	–	133.09	nr	**157.06**
offset 225 mm projection	63.96	0.62	12.38	–	76.91	nr	**89.29**
offset 300 mm projection	61.03	0.62	12.38	–	72.60	nr	**84.98**
access pipe, round door	68.87	0.62	12.38	–	81.11	nr	**93.49**
roof connector; for asphalt	65.11	0.62	12.38	–	82.13	nr	**94.51**
isolated Timesaver coupling joint	22.28	0.39	7.79	–	22.95	nr	**30.74**
transitional clayware socket; cement and sand							
(1:2) joint	44.34	0.37	7.39	–	68.76	nr	**76.15**

33 DRAINAGE ABOVE GROUND

Item	PC £	Labour hours	Labour £	Plant £	Material £	Unit	Total rate £
33.02 FOUL DRAINAGE INSTALLATIONS – cont							
Cast iron – cont							
Cast iron Timesaver pipes and fittings – cont							
150 mm pipes; primed; 3 m lengths; fixing with standard brackets; plugged and screwed to masonry	86.34	0.69	13.78	–	156.99	m	**170.77**
extra for							
bends; short radius	67.31	0.77	15.38	–	101.83	nr	**117.21**
access bends; short radius	113.23	0.77	15.38	–	149.12	nr	**164.50**
boss; 38 BSP	109.88	0.77	15.38	–	143.44	nr	**158.82**
single branch	144.45	1.11	22.17	–	216.43	nr	**238.60**
double branch	202.97	1.48	29.56	–	318.55	nr	**348.11**
access pipe, round door	114.57	0.77	15.38	–	132.35	nr	**147.73**
isolated Timesaver coupling joint	–	0.46	9.19	–	45.83	nr	**55.02**
transitional clayware socket; cement and sand (1:2) joint	77.65	0.48	9.59	–	125.96	nr	**135.55**
Cast iron Ensign lightweight pipes and fittings or other equal							
50 mm pipes; primed; 3 m lengths; fixing with standard brackets; plugged and screwed to masonry	19.87	0.31	6.27	–	105.83	m	**112.10**
extra for							
bends; short radius	15.45	0.27	5.49	–	25.12	nr	**30.61**
single branch	24.79	0.33	6.66	–	43.95	nr	**50.61**
access pipe	41.19	0.27	5.16	–	51.63	nr	**56.79**
70 mm pipes; primed; 3 m lengths; fixing with standard brackets; plugged and screwed to masonry	22.99	0.34	6.92	–	109.36	m	**116.28**
extra for							
bends; short radius	17.39	0.30	6.03	–	28.05	nr	**34.08**
single branch	26.19	0.37	7.46	–	47.26	nr	**54.72**
access pipe	43.57	0.30	6.03	–	55.02	nr	**61.05**
100 mm pipes; primed; 3 m lengths; fixing with standard brackets; plugged and screwed to masonry	27.35	0.37	7.46	–	115.71	m	**123.17**
extra for							
bends; short radius	20.59	0.32	6.53	–	34.39	nr	**40.92**
single branch	33.35	0.39	7.85	–	60.71	nr	**68.56**
double branch	47.97	0.46	9.28	–	88.95	nr	**98.23**
access pipe	47.88	0.32	6.53	–	62.50	nr	**69.03**
connector (transitional)	43.59	0.21	4.19	–	58.08	nr	**62.27**
reducer	34.95	0.32	6.53	–	49.17	nr	**55.70**
150 mm pipes; primed; 3 m lengths; fixing with standard brackets; plugged and screwed to masonry	54.19	0.45	9.15	–	153.23	m	**162.38**
extra for							
bends; short radius	–	0.32	6.53	–	47.02	nr	**53.55**
single branch	79.78	0.39	7.85	–	135.03	nr	**142.88**
double branch	47.97	0.46	9.28	–	128.69	nr	**137.97**

33 DRAINAGE ABOVE GROUND

Item	PC £	Labour hours	Labour £	Plant £	Material £	Unit	Total rate £
access pipe	86.68	0.32	6.53	–	115.71	nr	**122.24**
connector (transitional)	43.59	0.21	4.19	–	71.33	nr	**75.52**
reducer	34.95	0.32	6.53	–	62.42	nr	**68.95**
300 mm pipes; primed; 3 m lengths; fixing with standard brackets; plugged and screwed to masonry	157.56	0.90	18.28	–	298.48	m	**316.76**
extra for							
bends; short radius	–	0.32	6.53	–	118.50	nr	**125.03**
single branch	620.54	0.39	7.85	–	834.97	nr	**842.82**
double branch	47.97	0.46	9.28	–	343.13	nr	**352.41**
access pipe	565.22	0.32	6.53	–	680.09	nr	**686.62**
connector (transitional)	80.18	0.21	4.19	–	180.49	nr	**184.68**
reducer	34.95	0.32	6.53	–	133.90	nr	**140.43**
uPVC							
muPVC waste pipes and fittings; solvent welded joints							
32 mm pipes; fixing with pipe clips; plugged and screwed	3.27	0.23	4.59	–	4.95	m	**9.54**
extra for							
fittings with one end	–	0.16	3.19	–	2.62	nr	**5.81**
fittings with two ends	–	0.23	4.59	–	2.81	nr	**7.40**
fittings with three ends	–	0.31	6.19	–	3.71	nr	**9.90**
access plug	1.89	0.16	3.19	–	2.62	nr	**5.81**
straight coupling	2.04	0.16	3.19	–	2.77	nr	**5.96**
male iron to muPVC coupling	3.17	0.35	6.99	–	3.59	nr	**10.58**
sweep bend	2.08	0.23	4.59	–	2.81	nr	**7.40**
spigot/socket bend	–	0.23	4.59	–	4.17	nr	**8.76**
sweep tee	2.79	0.31	6.19	–	3.71	nr	**9.90**
40 mm pipes; fixing with pipe clips; plugged and screwed	4.04	0.28	5.59	–	5.87	m	**11.46**
extra for							
fittings with one end	–	0.18	3.59	–	2.62	nr	**6.21**
fittings with two ends	–	0.28	5.59	–	3.03	nr	**8.62**
fittings with three ends	–	0.37	7.39	–	4.47	nr	**11.86**
fittings with four ends	8.76	0.49	9.78	–	10.20	nr	**19.98**
access plug	1.89	0.18	3.59	–	2.62	nr	**6.21**
straight coupling	2.03	0.19	3.79	–	2.76	nr	**6.55**
male iron to muPVC coupling	3.71	0.35	6.99	–	4.15	nr	**11.14**
level invert taper	2.61	0.28	5.59	–	3.36	nr	**8.95**
sweep bend	2.29	0.28	5.59	–	3.03	nr	**8.62**
spigot/socket bend	3.89	0.28	5.59	–	4.68	nr	**10.27**
sweep tee	3.52	0.37	7.39	–	4.47	nr	**11.86**
sweep cross	8.76	0.49	9.78	–	10.20	nr	**19.98**
50 mm pipes; fixing with pipe clips; plugged and screwed	6.09	0.32	6.39	–	9.49	m	**15.88**
extra for							
fittings with one end	–	0.19	3.79	–	3.46	nr	**7.25**
fittings with two ends	–	0.32	6.39	–	4.85	nr	**11.24**
fittings with three ends	–	0.43	8.59	–	7.92	nr	**16.51**

Prices for Measured Works

33 DRAINAGE ABOVE GROUND

Item	PC £	Labour hours	Labour £	Plant £	Material £	Unit	Total rate £
33.02 FOUL DRAINAGE INSTALLATIONS – cont							
uPVC – cont							
muPVC waste pipes and fittings – cont							
extra for – cont							
fittings with four ends	–	0.57	11.38	–	10.62	nr	**22.00**
access plug	2.71	0.19	3.79	–	3.46	nr	**7.25**
straight coupling	3.71	0.21	4.19	–	4.49	nr	**8.68**
male iron to muPVC coupling	5.34	0.42	8.38	–	5.84	nr	**14.22**
level invert taper	3.23	0.32	6.39	–	4.00	nr	**10.39**
sweep bend	4.07	0.32	6.39	–	4.85	nr	**11.24**
spigot/socket bend	5.55	0.32	6.39	–	6.39	nr	**12.78**
sweep tee	3.52	0.37	7.39	–	4.47	nr	**11.86**
sweep cross	9.17	0.57	11.38	–	10.62	nr	**22.00**
uPVC overflow pipes and fittings; solvent welded joints							
19 mm pipes; fixing with pipe clips; plugged and							
screwed	2.06	0.20	4.00	–	3.51	m	**7.51**
extra for							
splay cut end	–	0.01	0.20	–	–	nr	**0.20**
fittings with one end	–	0.16	3.19	–	2.54	nr	**5.73**
fittings with two ends	–	0.16	3.19	–	2.95	nr	**6.14**
fittings with three ends	–	0.20	4.00	–	3.34	nr	**7.34**
straight connector	2.15	0.16	3.19	–	2.54	nr	**5.73**
female iron to uPVC coupling	–	0.19	3.79	–	3.55	nr	**7.34**
bend	2.53	0.16	3.19	–	2.95	nr	**6.14**
bent tank connector	4.03	0.19	3.79	–	4.33	nr	**8.12**
uPVC pipes and fittings; with solvent welded joints							
(unless otherwise described)							
82 mm pipes; fixing with holderbats; plugged and							
screwed	15.42	0.37	7.39	–	21.17	m	**28.56**
extra for							
slip coupling; push fit	25.51	0.34	6.79	–	26.28	nr	**33.07**
expansion coupling	12.28	0.37	7.39	–	14.03	nr	**21.42**
sweep bend	20.18	0.37	7.39	–	22.18	nr	**29.57**
boss connector	11.28	0.25	5.00	–	13.00	nr	**18.00**
single branch	28.19	0.49	9.78	–	31.25	nr	**41.03**
access door	27.95	0.56	11.19	–	29.48	nr	**40.67**
110 mm pipes; fixing with holderbats; plugged							
and screwed	14.99	0.41	8.19	–	20.89	m	**29.08**
extra for							
socket plug	13.78	0.20	4.00	–	16.00	nr	**20.00**
slip coupling; push fit	21.46	0.37	7.39	–	22.10	nr	**29.49**
expansion coupling	11.93	0.41	8.19	–	14.08	nr	**22.27**
WC connector	16.59	0.27	5.40	–	18.06	nr	**23.46**
sweep bend	22.95	0.41	8.19	–	25.43	nr	**33.62**
access bend	63.66	0.43	8.59	–	67.36	nr	**75.95**
boss connector	10.96	0.27	5.40	–	13.08	nr	**18.48**
single branch	30.36	0.54	10.78	–	34.03	nr	**44.81**
single branch with access	53.12	0.56	11.19	–	57.47	nr	**68.66**
double branch	49.90	0.68	13.58	–	55.13	nr	**68.71**
WC manifold	22.78	0.27	5.40	–	26.22	nr	**31.62**
access door	–	0.56	11.19	–	28.16	nr	**39.35**

33 DRAINAGE ABOVE GROUND

Item	PC £	Labour hours	Labour £	Plant £	Material £	Unit	Total rate £
access pipe connector	37.64	0.46	9.19	–	40.57	nr	**49.76**
connection to clay pipes; caulking ring and cement and sand (1:2) joint	–	0.39	7.79	–	20.76	nr	**28.55**
160 mm pipes; fixing with holderbats; plugged and screwed	40.05	0.46	9.19	–	55.71	m	**64.90**
extra for							
socket plug	26.51	0.23	4.59	–	31.31	nr	**35.90**
slip coupling; push fit	46.36	0.42	8.38	–	47.75	nr	**56.13**
expansion coupling	28.98	0.46	9.19	–	33.86	nr	**43.05**
sweep bend	60.11	0.46	9.19	–	65.92	nr	**75.11**
boss connector	15.95	0.31	6.19	–	20.44	nr	**26.63**
single branch	49.26	0.61	12.18	–	56.68	nr	**68.86**
double branch	103.60	0.77	15.38	–	114.59	nr	**129.97**
access door	49.05	0.56	11.19	–	51.21	nr	**62.40**
access pipe connector	37.64	0.46	9.19	–	40.57	nr	**49.76**
Weathering apron; for pipe							
82 mm dia.	5.91	0.31	6.19	–	6.78	nr	**12.97**
110 mm dia.	6.49	0.35	6.99	–	7.65	nr	**14.64**
160 mm dia.	20.44	0.39	7.79	–	22.99	nr	**30.78**
Weathering slate; for pipe							
110 mm dia.	68.81	0.83	16.57	–	71.84	nr	**88.41**
Vent cowl; for pipe							
82 mm dia.	5.82	0.31	6.19	–	6.68	nr	**12.87**
110 mm dia.	5.58	0.31	6.19	–	6.72	nr	**12.91**
160 mm dia.	15.66	0.31	6.19	–	18.07	nr	**24.26**
Polypropylene							
Polypropylene (PP) waste pipes and fittings; push fit O – ring joints							
32 mm pipes; fixing with pipe clips; plugged and screwed	2.65	0.20	4.00	–	3.90	m	**7.90**
extra for							
fittings with one end	–	0.15	3.00	–	2.22	nr	**5.22**
fittings with two ends	–	0.20	4.00	–	2.26	nr	**6.26**
fittings with three ends	–	0.28	5.59	–	3.90	nr	**9.49**
access plug	2.16	0.15	3.00	–	2.22	nr	**5.22**
double socket	1.66	0.14	2.79	–	1.71	nr	**4.50**
male iron to PP coupling	2.98	0.26	5.19	–	3.07	nr	**8.26**
sweep bend	2.06	0.20	4.00	–	2.12	nr	**6.12**
spigot bend	2.99	0.23	4.59	–	3.08	nr	**7.67**
40 mm pipes; fixing with pipe clips; plugged and screwed	3.26	0.20	4.00	–	4.55	m	**8.55**
extra for							
fittings with one end	–	0.18	3.59	–	2.34	nr	**5.93**
fittings with two ends	–	0.28	5.59	–	2.68	nr	**8.27**
fittings with three ends	–	0.37	7.39	–	4.12	nr	**11.51**
access plug	2.27	0.18	3.59	–	2.34	nr	**5.93**
double socket	1.66	0.19	3.79	–	1.71	nr	**5.50**
universal connector	3.21	0.23	4.59	–	3.31	nr	**7.90**
sweep bend	2.31	0.28	5.59	–	2.38	nr	**7.97**
spigot bend	2.90	0.28	5.59	–	2.99	nr	**8.58**
reducer 40 mm–32 mm	1.55	0.28	5.59	–	1.60	nr	**7.19**

Prices for Measured Works

33 DRAINAGE ABOVE GROUND

Item	PC £	Labour hours	Labour £	Plant £	Material £	Unit	Total rate £
33.02 FOUL DRAINAGE INSTALLATIONS – cont							
Polypropylene – cont							
Polypropylene (PP) waste pipes and fittings – cont							
50 mm pipes; fixing with pipe clips; plugged and screwed	4.19	0.32	6.39	–	6.68	m	**13.07**
extra for							
fittings with one end	–	0.19	3.79	–	4.13	nr	**7.92**
fittings with two ends	–	0.32	6.39	–	4.43	nr	**10.82**
fittings with three ends	–	0.43	8.59	–	6.13	nr	**14.72**
access plug	4.01	0.19	3.79	–	4.13	nr	**7.92**
double socket	3.39	0.21	4.19	–	3.49	nr	**7.68**
sweep bend	4.41	0.32	6.39	–	4.54	nr	**10.93**
spigot bend	3.70	0.32	6.39	–	3.81	nr	**10.20**
reducer 50 mm–40 mm	2.67	0.32	6.39	–	2.75	nr	**9.14**
Polypropylene ancillaries; screwed joint to waste fitting							
Tubular S trap; bath; low level 38 mm seal							
40 mm dia.	4.48	0.51	10.19	–	4.61	nr	**14.80**
Bottle trap; 75 mm seal							
32 mm dia.	4.29	0.35	6.99	–	4.42	nr	**11.41**
40 mm dia.	5.26	0.42	8.38	–	5.42	nr	**13.80**
Tubular swivel S trap; two piece; 75 mm seal							
32 mm dia.	6.00	0.35	6.99	–	6.18	nr	**13.17**
40 mm dia.	4.48	0.42	8.38	–	4.61	nr	**12.99**
Tubular swivel P trap; 75 mm seal							
32 dia.	3.70	0.35	6.99	–	3.81	nr	**10.80**
40 dia.	4.40	0.42	8.38	–	4.53	nr	**12.91**

34 DRAINAGE BELOW GROUND

Item	PC £	Labour hours	Labour £	Plant £	Material £	Unit	Total rate £
34.01 URBAN AND LANDSCAPE DRAINAGE							
Slot and grate drainage							
ACO RoadDrain one piece channel drainage system for medium to heavy duty highway and distribution yards							
100 F900 units, 500 mm long	–	0.46	6.57	–	217.33	m	**223.90**
200 F900 units, 500 mm long	–	0.48	6.86	–	278.74	m	**285.60**
ACO MultiDrain Monoblock PD100D one piece channel drainage system for pedestrian and medium duty vehicle applications made from high performance recycled materials							
PD100 units, 500 mm long with integral heelguard inlets	–	0.45	6.44	–	155.04	m	**161.48**
ACO MultiDrain M100D polymer concrete channel drainage system; galvanized steel edge trim; nominal bore 100 mm; type of fall constant; bedding and haunching with in situ concrete (not included)							
slotted grating							
galvanized steel grating, load class A15 (pedestrian areas)	–	0.46	6.57	–	71.01	m	**77.58**
galvanized steel grating, load class C250 (cars and light vans)	–	0.46	6.57	–	111.74	m	**118.31**
ductile iron grating, load class D400 (driving lanes of roads)	–	0.46	6.57	–	118.10	m	**124.67**
Heelguard resin composite grating, load class C250 (cars and light vans)	–	0.46	6.57	–	109.80	m	**116.37**
end caps	–	0.09	1.29	–	8.54	nr	**9.83**
sump unit	–	1.39	19.87	–	148.01	nr	**167.88**
ACO universal gully	–	1.50	21.44	–	883.85	nr	**905.29**
ACO MultiDrain M150D polymer concrete channel drainage system; galvanized steel edge trim; nominal bore 150 mm; type of fall constant; bedding and haunching with in situ concrete (not included)							
slotted grating							
galvanized steel grating, load class A15 (pedestrian areas)	–	0.46	6.57	–	128.88	m	**135.45**
galvanized steel grating, load class C250 (cars and light vans)	–	0.46	6.57	–	132.13	m	**138.70**
sump unit	–	1.45	20.73	–	229.32	nr	**250.05**
ACO MultiDrain M200D polymer concrete channel drainage system; galvanized steel edge trim; nominal bore 200 mm; type of fall constant; bedding and haunching with in situ concrete (not included)							
slotted grating							
galvanized steel grating, load class A15 (pedestrian areas)	–	0.46	6.57	–	168.29	m	**174.86**
sump unit	–	1.50	21.44	–	288.32	nr	**309.76**

34 DRAINAGE BELOW GROUND

Item	PC £	Labour hours	Labour £	Plant £	Material £	Unit	Total rate £
34.01 URBAN AND LANDSCAPE DRAINAGE – cont							
Slot and grate drainage – cont							
ACO S Range polymer concrete channel drainage system; bolted ductile iron grating, load class F900 (airfields); bedding and haunching with in situ concrete (not included)							
S100 F900 channel and grate	–	1.00	14.30	–	163.83	m	**178.13**
S150 F900 channel and grate	–	1.00	14.30	–	176.19	m	**190.49**
S200 F900 channel and grate	–	1.10	15.73	–	186.84	m	**202.57**
S300 F900 channel and grate	–	1.20	17.16	–	290.89	m	**308.05**
end caps	–	0.09	1.29	–	23.64	nr	**24.93**
sump unit	–	1.50	21.44	–	299.75	nr	**321.19**
ACO Qmax large capacity slot drainage channel with MDPE body and hot dipped galvanized steel edge rail, up to load class F900; bedding and haunching with in situ concrete (not included)							
ACO Qmax 225	–	0.75	10.72	–	101.23	m	**111.95**
ACO Qmax 350	–	1.00	14.30	–	133.25	m	**147.55**
ACO Qmax 900	–	1.50	21.44	–	254.33	m	**275.77**
shallow access chamber	–	1.50	21.44	–	287.21	nr	**308.65**
deep access chamber	–	2.00	28.59	–	685.25	nr	**713.84**
ACO KerbDrain one-piece polymer concrete combined drainage system, load class D400; bedding and haunching in in situ concrete (not included). Manufactured from recycled and recyclable material							
KerbDrain KD305	–	0.50	7.15	–	97.29	m	**104.44**
KerbDrain KD480	–	0.65	9.29	–	105.64	m	**114.93**
KerbDrain KD305 drop kerb (left drop, one centre stone and right drop) total length 2745 mm	–	2.00	28.59	–	196.83	nr	**225.42**
KerbDrain KD305 mitre unit	–	0.25	3.57	–	106.09	nr	**109.66**
KerbDrain KD end cap	–	0.09	1.29	–	58.28	nr	**59.57**
KerbDrain KD610 shallow gully assembly	–	1.50	21.44	–	836.88	nr	**858.32**
Interconnecting drainage channel; Birco-lite or other equal; Marshalls Plc; galvanized steel grating ref 8041; bedding and haunching in in situ concrete (not included)							
100 mm wide							
laid level or to falls	–	0.46	6.57	–	53.71	m	**60.28**
100 mm dia. trapped outlet unit	–	1.39	19.87	–	118.22	nr	**138.09**
end caps	–	0.09	1.29	–	6.70	nr	**7.99**

34 DRAINAGE BELOW GROUND

Item	PC £	Labour hours	Labour £	Plant £	Material £	Unit	Total rate £
Oil separators (polyethylene single chamber design)							
Supply and install only oil separator complete with lockable cover (excavations, filling etc. measured elsewhere)							
1000 litre ACO Q-ceptor by-pass oil separators NSB3 Class 1 (discharge concentrations of less than 5 mg/litre of oil) for discharges to surface water drains	–	2.00	28.59	63.16	1086.91	nr	**1178.66**
1000 litre ACO Q-ceptor full retention oil separators NS3 Class 1	–	2.25	32.17	72.18	1181.09	nr	**1285.44**
34.02 TRENCHES							
Excavating trenches in firm soil; by machine; grading bottoms; earthwork support; filling with excavated material and compacting; disposal of surplus soil; spreading on site average 50 m from excavations							
Pipes not exceeding 200 mm nominal size							
average depth of trench 0.50 m	–	0.28	4.04	2.03	–	m	**6.07**
average depth of trench 0.75 m	–	0.37	5.34	3.01	–	m	**8.35**
average depth of trench 1.00 m	–	0.79	11.38	4.28	1.79	m	**17.45**
average depth of trench 1.25 m	–	1.16	16.72	4.51	2.46	m	**23.69**
average depth of trench 1.50 m	–	1.48	21.32	5.04	2.91	m	**29.27**
average depth of trench 1.75 m	–	1.85	26.66	5.26	3.58	m	**35.50**
average depth of trench 2.00 m	–	2.13	30.68	6.02	4.03	m	**40.73**
average depth of trench 2.25 m	–	2.64	38.04	7.29	5.38	m	**50.71**
average depth of trench 2.50 m	–	3.10	44.66	8.50	6.27	m	**59.43**
average depth of trench 2.75 m	–	3.42	49.28	9.55	6.94	m	**65.77**
average depth of trench 3.00 m	–	3.75	54.02	10.53	7.61	m	**72.16**
average depth of trench 3.25 m	–	4.07	58.64	11.05	8.29	m	**77.98**
average depth of trench 3.50 m	–	4.35	62.68	11.51	8.96	m	**83.15**
Pipes 225 mm nominal size							
average depth of trench 0.50 m	–	0.28	4.04	2.03	–	m	**6.07**
average depth of trench 0.75 m	–	0.37	5.34	3.01	–	m	**8.35**
average depth of trench 1.00 m	–	0.79	11.38	4.28	1.79	m	**17.45**
average depth of trench 1.25 m	–	1.16	16.72	4.51	2.46	m	**23.69**
average depth of trench 1.50 m	–	1.48	21.32	5.04	2.91	m	**29.27**
average depth of trench 1.75 m	–	1.85	26.66	5.26	3.58	m	**35.50**
average depth of trench 2.00 m	–	2.13	30.68	6.02	4.03	m	**40.73**
average depth of trench 2.25 m	–	2.64	38.04	7.29	5.38	m	**50.71**
average depth of trench 2.50 m	–	3.10	44.66	8.50	6.27	m	**59.43**
average depth of trench 2.75 m	–	3.42	49.28	9.55	6.94	m	**65.77**
average depth of trench 3.00 m	–	3.75	54.02	10.53	7.61	m	**72.16**
average depth of trench 3.25 m	–	4.07	58.64	11.05	8.29	m	**77.98**
average depth of trench 3.50 m	–	4.35	62.68	11.51	8.96	m	**83.15**

34 DRAINAGE BELOW GROUND

Item	PC £	Labour hours	Labour £	Plant £	Material £	Unit	Total rate £
34.02 TRENCHES – cont							
Excavating trenches in firm soil – cont							
Pipes 300 mm nominal size							
average depth of trench 0.75 m	–	0.44	6.33	3.76	–	m	**10.09**
average depth of trench 1.00 m	–	0.93	13.40	4.28	1.79	m	**19.47**
average depth of trench 1.25 m	–	1.25	18.00	4.74	2.46	m	**25.20**
average depth of trench 1.50 m	–	1.62	23.34	5.26	2.91	m	**31.51**
average depth of trench 1.75 m	–	1.85	26.66	5.49	3.58	m	**35.73**
average depth of trench 2.00 m	–	2.13	30.68	6.77	4.03	m	**41.48**
average depth of trench 2.25 m	–	2.64	38.04	7.75	5.38	m	**51.17**
average depth of trench 2.50 m	–	3.10	44.66	8.80	6.27	m	**59.73**
average depth of trench 2.75 m	–	3.42	49.28	9.77	6.94	m	**65.99**
average depth of trench 3.00 m	–	3.75	54.02	10.75	7.61	m	**72.38**
average depth of trench 3.25 m	–	4.07	58.64	11.80	8.29	m	**78.73**
average depth of trench 3.50 m	–	4.35	62.68	12.03	8.96	m	**83.67**
Pipes 375 mm nominal size							
average depth of trench 0.75 m	–	0.46	6.62	4.51	–	m	**11.13**
average depth of trench 1.00 m	–	0.97	13.98	5.04	1.79	m	**20.81**
average depth of trench 1.25 m	–	1.34	19.30	6.02	2.46	m	**27.78**
average depth of trench 1.50 m	–	1.71	24.64	6.24	2.91	m	**33.79**
average depth of trench 1.75 m	–	1.99	28.68	6.77	3.58	m	**39.03**
average depth of trench 2.00 m	–	2.27	32.70	6.99	4.03	m	**43.72**
average depth of trench 2.25 m	–	2.82	40.63	8.50	5.38	m	**54.51**
average depth of trench 2.50 m	–	3.38	48.70	9.77	6.27	m	**64.74**
average depth of trench 2.75 m	–	3.70	53.30	10.53	6.94	m	**70.77**
average depth of trench 3.00 m	–	4.02	57.92	11.28	7.61	m	**76.81**
average depth of trench 3.25 m	–	4.35	62.68	12.26	8.29	m	**83.23**
average depth of trench 3.50 m	–	4.67	67.28	13.01	8.96	m	**89.25**
Pipes 450 mm nominal size							
average depth of trench 0.75 m	–	0.51	7.34	4.51	–	m	**11.85**
average depth of trench 1.00 m	–	1.02	14.70	5.49	1.79	m	**21.98**
average depth of trench 1.25 m	–	1.48	21.32	6.54	2.46	m	**30.32**
average depth of trench 1.50 m	–	1.85	26.66	6.99	2.91	m	**36.56**
average depth of trench 1.75 m	–	2.13	30.68	7.29	3.58	m	**41.55**
average depth of trench 2.00 m	–	2.45	35.30	7.75	4.03	m	**47.08**
average depth of trench 2.25 m	–	3.05	43.94	9.02	5.38	m	**58.34**
average depth of trench 2.50 m	–	3.61	52.01	10.53	6.27	m	**68.81**
average depth of trench 2.75 m	–	3.98	57.34	11.51	6.94	m	**75.79**
average depth of trench 3.00 m	–	4.26	61.38	12.56	7.61	m	**81.55**
average depth of trench 3.25 m	–	4.63	66.70	13.76	8.29	m	**88.75**
average depth of trench 3.50 m	–	5.00	72.04	15.04	8.96	m	**96.04**
Pipes 600 mm nominal size							
average depth of trench 1.00 m	–	1.11	16.00	6.02	1.79	m	**23.81**
average depth of trench 1.25 m	–	1.57	22.62	6.99	2.46	m	**32.07**
average depth of trench 1.50 m	–	2.04	29.39	8.04	2.91	m	**40.34**
average depth of trench 1.75 m	–	2.31	33.28	8.04	3.58	m	**44.90**
average depth of trench 2.00 m	–	2.73	39.34	8.80	4.03	m	**52.17**
average depth of trench 2.25 m	–	3.28	47.26	10.53	5.38	m	**63.17**
average depth of trench 2.50 m	–	3.89	56.04	12.26	6.27	m	**74.57**

34 DRAINAGE BELOW GROUND

Item	PC £	Labour hours	Labour £	Plant £	Material £	Unit	Total rate £
average depth of trench 2.75 m	–	4.30	61.95	13.76	6.94	m	82.65
average depth of trench 3.00 m	–	4.72	68.00	15.04	7.61	m	90.65
average depth of trench 3.25 m	–	5.09	73.34	16.02	8.29	m	97.65
average depth of trench 3.50 m	–	5.46	78.66	16.77	8.96	m	104.39
Pipes 900 mm nominal size							
average depth of trench 1.25 m	–	1.90	27.38	8.50	2.46	m	38.34
average depth of trench 1.50 m	–	2.41	34.72	9.55	2.91	m	47.18
average depth of trench 1.75 m	–	2.78	40.06	9.77	3.58	m	53.41
average depth of trench 2.00 m	–	3.10	44.66	11.28	4.03	m	59.97
average depth of trench 2.25 m	–	3.84	55.32	13.31	5.38	m	74.01
average depth of trench 2.50 m	–	4.53	65.26	15.26	6.27	m	86.79
average depth of trench 2.75 m	–	5.00	72.04	16.77	6.94	m	95.75
average depth of trench 3.00 m	–	5.46	78.66	18.27	7.61	m	104.54
average depth of trench 3.25 m	–	5.92	85.29	19.78	8.29	m	113.36
average depth of trench 3.50 m	–	6.38	91.92	21.05	8.96	m	121.93
Pipes 1200 mm nominal size							
average depth of trench 1.50 m	–	2.73	39.34	10.30	2.91	m	52.55
average depth of trench 1.75 m	–	3.19	45.96	11.80	3.58	m	61.34
average depth of trench 2.00 m	–	3.56	51.29	13.53	4.03	m	68.85
average depth of trench 2.25 m	–	4.35	62.68	16.02	5.38	m	84.08
average depth of trench 2.50 m	–	5.18	74.63	18.27	6.27	m	99.17
average depth of trench 2.75 m	–	5.69	81.98	20.30	6.94	m	109.22
average depth of trench 3.00 m	–	6.20	89.32	22.03	7.61	m	118.96
average depth of trench 3.25 m	–	6.75	97.25	23.83	8.29	m	129.37
average depth of trench 3.50 m	–	7.26	104.60	25.56	8.96	m	139.12
Extra over excavating trenches; irrespective of depth; breaking out existing materials							
brick	–	1.80	25.94	11.14	–	m³	37.08
concrete	–	2.54	36.60	15.38	–	m³	51.98
reinforced concrete	–	3.61	52.01	22.20	–	m³	74.21
Extra over excavating trenches; irrespective of depth; breaking out existing hard pavings; 75 mm thick							
tarmacadam	–	0.19	2.74	1.13	–	m²	3.87
Extra over excavating trenches; irrespective of depth; breaking out existing hard pavings; 150 mm thick							
concrete	–	0.37	5.34	2.49	–	m²	7.83
tarmacadam and hardcore	–	0.28	4.04	1.38	–	m²	5.42
Excavating trenches; by hand; grading bottoms; earthwork support; filling with excavated material and compacting; disposal of surplus soil on site; spreading on site average 25 m from excavations							
Pipes not exceeding 200 mm nominal size							
average depth of trench 0.50 m	–	0.93	13.40	–	–	m	13.40
average depth of trench 0.75 m	–	1.39	20.02	–	–	m	20.02
average depth of trench 1.00 m	–	2.04	29.39	–	1.79	m	31.18
average depth of trench 1.25 m	–	2.87	41.34	–	2.46	m	43.80
average depth of trench 1.50 m	–	3.93	56.62	–	3.00	m	59.62
average depth of trench 1.75 m	–	5.18	74.63	–	3.58	m	78.21
average depth of trench 2.00 m	–	5.92	85.29	–	4.03	m	89.32

34 DRAINAGE BELOW GROUND

Item	PC £	Labour hours	Labour £	Plant £	Material £	Unit	Total rate £
34.02 TRENCHES – cont							
Excavating trenches – cont							
Pipes not exceeding 200 mm nominal size – cont							
average depth of trench 2.25 m	–	7.40	106.62	–	5.38	m	**112.00**
average depth of trench 2.50 m	–	8.88	127.94	–	6.27	m	**134.21**
average depth of trench 2.75 m	–	9.76	140.62	–	6.94	m	**147.56**
average depth of trench 3.00 m	–	10.64	153.29	–	7.61	m	**160.90**
average depth of trench 3.25 m	–	11.52	165.97	–	8.29	m	**174.26**
average depth of trench 3.50 m	–	12.40	178.65	–	8.96	m	**187.61**
Pipes 225 mm nominal size							
average depth of trench 0.50 m	–	0.93	13.40	–	–	m	**13.40**
average depth of trench 0.75 m	–	1.39	20.02	–	–	m	**20.02**
average depth of trench 1.00 m	–	2.04	29.39	–	1.79	m	**31.18**
average depth of trench 1.25 m	–	2.87	41.34	–	2.46	m	**43.80**
average depth of trench 1.50 m	–	3.93	56.62	–	3.00	m	**59.62**
average depth of trench 1.75 m	–	5.18	74.63	–	3.58	m	**78.21**
average depth of trench 2.00 m	–	5.92	85.29	–	4.03	m	**89.32**
average depth of trench 2.25 m	–	7.40	106.62	–	5.38	m	**112.00**
average depth of trench 2.50 m	–	8.88	127.94	–	6.27	m	**134.21**
average depth of trench 2.75 m	–	9.76	140.62	–	6.94	m	**147.56**
average depth of trench 3.00 m	–	10.64	153.29	–	7.61	m	**160.90**
average depth of trench 3.25 m	–	11.52	165.97	–	8.29	m	**174.26**
average depth of trench 3.50 m	–	12.40	178.65	–	8.96	m	**187.61**
Pipes 300 mm nominal size							
average depth of trench 0.75 m	–	1.62	23.34	–	–	m	**23.34**
average depth of trench 1.00 m	–	2.36	34.00	–	1.79	m	**35.79**
average depth of trench 1.25 m	–	3.33	47.98	–	2.46	m	**50.44**
average depth of trench 1.50 m	–	4.44	63.97	–	3.00	m	**66.97**
average depth of trench 1.75 m	–	5.18	74.63	–	3.58	m	**78.21**
average depth of trench 2.00 m	–	5.92	85.29	–	4.03	m	**89.32**
average depth of trench 2.25 m	–	7.40	106.62	–	5.38	m	**112.00**
average depth of trench 2.50 m	–	8.88	127.94	–	6.27	m	**134.21**
average depth of trench 2.75 m	–	9.76	140.62	–	6.94	m	**147.56**
average depth of trench 3.00 m	–	10.64	153.29	–	7.61	m	**160.90**
average depth of trench 3.25 m	–	11.52	165.97	–	8.29	m	**174.26**
average depth of trench 3.50 m	–	12.40	178.65	–	8.96	m	**187.61**
Pipes 375 mm nominal size							
average depth of trench 0.75 m	–	1.80	25.94	–	–	m	**25.94**
average depth of trench 1.00 m	–	2.64	38.04	–	1.79	m	**39.83**
average depth of trench 1.25 m	–	3.70	53.30	–	2.46	m	**55.76**
average depth of trench 1.50 m	–	4.93	71.03	–	3.00	m	**74.03**
average depth of trench 1.75 m	–	5.74	82.70	–	3.58	m	**86.28**
average depth of trench 2.00 m	–	6.57	94.66	–	4.03	m	**98.69**
average depth of trench 2.25 m	–	8.23	118.57	–	5.38	m	**123.95**
average depth of trench 2.50 m	–	9.90	142.63	–	6.27	m	**148.90**
average depth of trench 2.75 m	–	10.87	156.61	–	6.94	m	**163.55**
average depth of trench 3.00 m	–	11.84	170.58	–	7.61	m	**178.19**
average depth of trench 3.25 m	–	12.86	185.28	–	8.29	m	**193.57**
average depth of trench 3.50 m	–	13.88	199.97	–	8.96	m	**208.93**

34 DRAINAGE BELOW GROUND

Item	PC £	Labour hours	Labour £	Plant £	Material £	Unit	Total rate £
Pipes 450 mm nominal size							
average depth of trench 0.75 m	–	2.04	29.39	–	–	m	**29.39**
average depth of trench 1.00 m	–	2.94	42.35	–	1.79	m	**44.14**
average depth of trench 1.25 m	–	4.13	59.50	–	2.46	m	**61.96**
average depth of trench 1.50 m	–	5.41	77.94	–	3.00	m	**80.94**
average depth of trench 1.75 m	–	6.31	90.91	–	3.58	m	**94.49**
average depth of trench 2.00 m	–	7.22	104.02	–	4.03	m	**108.05**
average depth of trench 2.25 m	–	9.05	130.39	–	5.38	m	**135.77**
average depth of trench 2.50 m	–	10.87	156.61	–	6.27	m	**162.88**
average depth of trench 2.75 m	–	11.96	172.31	–	6.94	m	**179.25**
average depth of trench 3.00 m	–	13.04	187.87	–	7.61	m	**195.48**
average depth of trench 3.25 m	–	14.11	203.29	–	8.29	m	**211.58**
average depth of trench 3.50 m	–	15.17	218.56	–	8.96	m	**227.52**
Pipes 600 mm nominal size							
average depth of trench 1.00 m	–	3.24	46.68	–	1.79	m	**48.47**
average depth of trench 1.25 m	–	4.63	66.70	–	2.46	m	**69.16**
average depth of trench 1.50 m	–	6.20	89.32	–	3.00	m	**92.32**
average depth of trench 1.75 m	–	7.17	103.30	–	3.58	m	**106.88**
average depth of trench 2.00 m	–	8.19	118.00	–	4.03	m	**122.03**
average depth of trench 2.25 m	–	9.20	132.55	–	5.38	m	**137.93**
average depth of trench 2.50 m	–	11.56	166.55	–	6.27	m	**172.82**
average depth of trench 2.75 m	–	12.35	177.93	–	6.94	m	**184.87**
average depth of trench 3.00 m	–	14.80	213.23	–	7.61	m	**220.84**
average depth of trench 3.25 m	–	16.03	230.95	–	8.29	m	**239.24**
average depth of trench 3.50 m	–	17.25	248.53	–	8.96	m	**257.49**
Pipes 900 mm nominal size							
average depth of trench 1.25 m	–	5.78	83.28	–	2.46	m	**85.74**
average depth of trench 1.50 m	–	7.63	109.93	–	3.00	m	**112.93**
average depth of trench 1.75 m	–	8.88	127.94	–	3.58	m	**131.52**
average depth of trench 2.00 m	–	10.13	145.95	–	4.03	m	**149.98**
average depth of trench 2.25 m	–	12.72	183.26	–	5.38	m	**188.64**
average depth of trench 2.50 m	–	15.31	220.57	–	6.27	m	**226.84**
average depth of trench 2.75 m	–	16.84	242.62	–	6.94	m	**249.56**
average depth of trench 3.00 m	–	18.32	263.94	–	7.61	m	**271.55**
average depth of trench 3.25 m	–	19.84	285.85	–	8.29	m	**294.14**
average depth of trench 3.50 m	–	21.37	307.89	–	8.96	m	**316.85**
Pipes 1200 mm nominal size							
average depth of trench 1.50 m	–	9.11	131.25	–	3.00	m	**134.25**
average depth of trench 1.75 m	–	10.59	152.57	–	3.58	m	**156.15**
average depth of trench 2.00 m	–	12.12	174.62	–	4.03	m	**178.65**
average depth of trench 2.25 m	–	15.20	218.99	–	5.38	m	**224.37**
average depth of trench 2.50 m	–	18.27	263.23	–	6.27	m	**269.50**
average depth of trench 2.75 m	–	20.07	289.15	–	6.94	m	**296.09**
average depth of trench 3.00 m	–	21.88	315.23	–	7.61	m	**322.84**
average depth of trench 3.25 m	–	23.66	340.88	–	8.29	m	**349.17**
average depth of trench 3.50 m	–	25.44	366.53	–	8.96	m	**375.49**

34 DRAINAGE BELOW GROUND

Item	PC £	Labour hours	Labour £	Plant £	Material £	Unit	Total rate £
34.02 TRENCHES – cont							
Excavating trenches – cont							
Extra over excavating trenches irrespective of depth; breaking out existing materials using compressor and breaker							
brick	–	2.78	40.06	9.13	–	m³	**49.19**
concrete	–	4.16	59.94	15.20	–	m³	**75.14**
reinforced concrete	–	5.55	79.96	21.29	–	m³	**101.25**
150 mm concrete	–	0.65	9.36	2.13	–	m²	**11.49**
75 mm thick tarmacadam	–	0.37	5.34	1.23	–	m²	**6.57**
150 mm thick tarmacadam and hardcore	–	0.46	6.62	1.51	–	m²	**8.13**
34.03 BEDS AND FILLINGS							
Sand filling							
Beds; to receive pitch fibre pipes							
600 mm × 50 mm thick	–	0.07	1.01	0.69	0.93	m	**2.63**
700 mm × 50 mm thick	–	0.09	1.30	0.69	1.08	m	**3.07**
800 mm × 50 mm thick	–	0.11	1.59	0.69	1.24	m	**3.52**
Granular (shingle) filling							
Beds; 100 mm thick; to pipes							
100 mm nominal size	–	0.09	1.30	0.69	3.57	m	**5.56**
150 mm nominal size	–	0.09	1.30	0.69	4.17	m	**6.16**
225 mm nominal size	–	0.11	1.59	0.69	4.77	m	**7.05**
300 mm nominal size	–	0.13	1.87	0.69	5.36	m	**7.92**
375 mm nominal size	–	0.15	2.16	0.69	5.95	m	**8.80**
450 mm nominal size	–	0.17	2.45	0.69	6.55	m	**9.69**
600 mm nominal size	–	0.19	2.74	0.69	7.15	m	**10.58**
Beds; 150 mm thick; to pipes							
100 mm nominal size	–	0.13	1.87	0.69	5.36	m	**7.92**
150 mm nominal size	–	0.15	2.16	0.69	5.95	m	**8.80**
225 mm nominal size	–	0.17	2.45	0.69	6.55	m	**9.69**
300 mm nominal size	–	0.19	2.74	0.69	7.15	m	**10.58**
375 mm nominal size	–	0.22	3.17	0.69	8.94	m	**12.80**
450 mm nominal size	–	0.24	3.46	0.69	9.54	m	**13.69**
600 mm nominal size	–	0.28	4.04	0.69	11.33	m	**16.06**
Beds and benchings; beds 100 mm thick; to pipes							
100 nominal size	–	0.21	3.03	0.69	6.55	m	**10.27**
150 nominal size	–	0.23	3.32	0.69	6.55	m	**10.56**
225 nominal size	–	0.28	4.04	0.69	8.94	m	**13.67**
300 nominal size	–	0.32	4.61	0.69	10.14	m	**15.44**
375 nominal size	–	0.42	6.05	0.69	13.71	m	**20.45**
450 nominal size	–	0.48	6.91	0.69	15.49	m	**23.09**
600 nominal size	–	0.62	8.93	0.69	20.26	m	**29.88**
Beds and benchings; beds 150 mm thick; to pipes							
100 nominal size	–	0.23	3.32	0.69	7.15	m	**11.16**
150 nominal size	–	0.26	3.75	0.69	7.75	m	**12.19**
225 nominal size	–	0.32	4.61	0.69	10.73	m	**16.03**

34 DRAINAGE BELOW GROUND

Item	PC £	Labour hours	Labour £	Plant £	Material £	Unit	Total rate £
300 nominal size	–	0.42	6.05	0.69	13.11	m	**19.85**
375 nominal size	–	0.48	6.91	0.69	15.49	m	**23.09**
450 nominal size	–	0.57	8.21	0.69	18.47	m	**27.37**
600 nominal size	–	0.68	9.80	0.69	23.83	m	**34.32**
Beds and coverings; 100 mm thick; to pipes							
100 nominal size	–	0.33	4.76	0.69	8.94	m	**14.39**
150 nominal size	–	0.42	6.05	0.69	10.73	m	**17.47**
225 nominal size	–	0.56	8.06	0.69	14.89	m	**23.64**
300 nominal size	–	0.67	9.65	0.69	17.88	m	**28.22**
375 nominal size	–	0.80	11.53	0.69	21.44	m	**33.66**
450 nominal size	–	0.94	13.54	0.69	25.62	m	**39.85**
600 nominal size	–	1.22	17.58	0.69	32.77	m	**51.04**
Beds and coverings; 150 mm thick; to pipes							
100 nominal size	–	0.50	7.20	0.69	13.11	m	**21.00**
150 nominal size	–	0.56	8.06	0.69	14.89	m	**23.64**
225 nominal size	–	0.72	10.37	0.69	19.07	m	**30.13**
300 nominal size	–	0.86	12.39	0.69	22.64	m	**35.72**
375 nominal size	–	1.00	14.41	0.69	26.81	m	**41.91**
450 nominal size	–	1.19	17.15	0.69	32.18	m	**50.02**
600 nominal size	–	1.44	20.74	0.69	38.73	m	**60.16**
Plain in situ ready mixed designated concrete; C10–40 mm aggregate							
Beds; 100 mm thick; to pipes							
100 mm nominal size	–	0.17	2.87	0.69	5.58	m	**9.14**
150 mm nominal size	–	0.17	2.87	0.69	5.58	m	**9.14**
225 mm nominal size	–	0.20	3.38	0.69	6.70	m	**10.77**
300 mm nominal size	–	0.23	3.88	0.69	7.82	m	**12.39**
375 mm nominal size	–	0.27	4.56	0.69	8.92	m	**14.17**
450 mm nominal size	–	0.30	5.07	0.69	10.04	m	**15.80**
600 mm nominal size	–	0.33	5.57	0.69	11.15	m	**17.41**
900 mm nominal size	–	0.40	6.76	0.69	13.39	m	**20.84**
1200 mm nominal size	–	0.54	9.13	0.69	17.85	m	**27.67**
Beds; 150 mm thick; to pipes							
100 mm nominal size	–	0.23	3.88	0.69	7.82	m	**12.39**
150 mm nominal size	–	0.27	4.56	0.69	8.92	m	**14.17**
225 mm nominal size	–	0.30	5.07	0.69	10.04	m	**15.80**
300 mm nominal size	–	0.33	5.57	0.69	11.15	m	**17.41**
375 mm nominal size	–	0.40	6.76	0.69	13.39	m	**20.84**
450 mm nominal size	–	0.43	7.26	0.69	14.50	m	**22.45**
600 mm nominal size	–	0.50	8.45	0.69	16.74	m	**25.88**
900 mm nominal size	–	0.63	10.64	0.69	21.20	m	**32.53**
1200 mm nominal size	–	0.77	13.01	0.69	25.66	m	**39.36**
Beds and benchings; beds 100 mm thick; to pipes							
100 mm nominal size	–	0.33	5.57	0.69	10.04	m	**16.30**
150 mm nominal size	–	0.38	6.42	0.69	11.15	m	**18.26**
225 mm nominal size	–	0.45	7.60	0.69	13.39	m	**21.68**
300 mm nominal size	–	0.53	8.95	0.69	15.61	m	**25.25**
375 mm nominal size	–	0.68	11.48	0.69	20.07	m	**32.24**
450 mm nominal size	–	0.80	13.51	0.69	23.42	m	**37.62**
600 mm nominal size	–	1.02	17.23	0.69	30.13	m	**48.05**

Prices for Measured Works

34 DRAINAGE BELOW GROUND

Item	PC £	Labour hours	Labour £	Plant £	Material £	Unit	Total rate £
34.03 BEDS AND FILLINGS – cont							
Plain in situ ready mixed designated concrete – cont							
Beds and benchings – cont							
900 mm nominal size	–	1.65	27.88	0.69	49.08	m	**77.65**
1200 mm nominal size	–	2.44	41.23	0.69	72.50	m	**114.42**
Beds and benchings; beds 150 mm thick; to pipes							
100 mm nominal size	–	0.38	6.42	0.69	11.15	m	**18.26**
150 mm nominal size	–	0.42	7.10	0.69	12.27	m	**20.06**
225 mm nominal size	–	0.53	8.95	0.69	15.61	m	**25.25**
300 mm nominal size	–	0.68	11.48	0.69	20.07	m	**32.24**
375 mm nominal size	–	0.80	13.51	0.69	23.42	m	**37.62**
450 mm nominal size	–	0.94	15.88	0.69	27.88	m	**44.45**
600 mm nominal size	–	1.20	20.28	0.69	35.70	m	**56.67**
900 mm nominal size	–	1.91	32.27	0.69	56.90	m	**89.86**
1200 mm nominal size	–	2.70	45.62	0.69	80.32	m	**126.63**
Beds and coverings; 100 mm thick; to pipes							
100 mm nominal size	–	0.50	8.45	0.69	13.39	m	**22.53**
150 mm nominal size	–	0.58	9.80	0.69	15.61	m	**26.10**
225 mm nominal size	–	0.83	14.03	0.69	22.31	m	**37.03**
300 mm nominal size	–	1.00	16.89	0.69	26.77	m	**44.35**
375 mm nominal size	–	1.21	20.45	0.69	32.34	m	**53.48**
450 mm nominal size	–	1.42	23.99	0.69	37.92	m	**62.60**
600 mm nominal size	–	1.83	30.92	0.69	49.08	m	**80.69**
900 mm nominal size	–	2.79	47.14	0.69	74.75	m	**122.58**
1200 mm nominal size	–	3.83	64.71	0.69	102.63	m	**168.03**
Beds and coverings; 150 mm thick; to pipes							
100 mm nominal size	–	0.75	12.67	0.69	20.07	m	**33.43**
150 mm nominal size	–	0.83	14.03	0.69	22.31	m	**37.03**
225 mm nominal size	–	1.08	18.25	0.69	29.00	m	**47.94**
300 mm nominal size	–	1.30	21.97	0.69	34.58	m	**57.24**
375 mm nominal size	–	1.50	25.35	0.69	40.16	m	**66.20**
450 mm nominal size	–	1.79	30.24	0.69	47.97	m	**78.90**
600 mm nominal size	–	2.16	36.49	0.69	58.01	m	**95.19**
900 mm nominal size	–	3.54	59.81	0.69	94.81	m	**155.31**
1200 mm nominal size	–	5.00	84.48	0.69	133.86	m	**219.03**
Plain in situ ready mixed designated concrete; C20–40 mm aggregate							
Beds; 100 mm thick; to pipes							
100 mm nominal size	–	0.17	2.87	0.69	5.77	m	**9.33**
150 mm nominal size	–	0.17	2.87	0.69	5.77	m	**9.33**
225 mm nominal size	–	0.20	3.38	0.69	6.92	m	**10.99**
300 mm nominal size	–	0.23	3.88	0.69	8.08	m	**12.65**
375 mm nominal size	–	0.27	4.56	0.69	9.23	m	**14.48**
450 mm nominal size	–	0.30	5.07	0.69	10.38	m	**16.14**
600 mm nominal size	–	0.33	5.57	0.69	11.54	m	**17.80**
900 mm nominal size	–	0.40	6.76	0.69	13.84	m	**21.29**
1200 mm nominal size	–	0.54	9.13	0.69	18.46	m	**28.28**

34 DRAINAGE BELOW GROUND

Item	PC £	Labour hours	Labour £	Plant £	Material £	Unit	Total rate £
Beds; 150 mm thick; to pipes							
100 mm nominal size	–	0.23	3.88	0.69	8.08	m	12.65
150 mm nominal size	–	0.27	4.56	0.69	9.23	m	14.48
225 mm nominal size	–	0.30	5.07	0.69	10.38	m	16.14
300 mm nominal size	–	0.33	5.57	0.69	11.54	m	17.80
375 mm nominal size	–	0.40	6.76	0.69	13.84	m	21.29
450 mm nominal size	–	0.43	7.26	0.69	15.00	m	22.95
600 mm nominal size	–	0.50	8.45	0.69	17.30	m	26.44
900 mm nominal size	–	0.63	10.64	0.69	21.92	m	33.25
1200 mm nominal size	–	0.77	13.01	0.69	26.53	m	40.23
Beds and benchings; beds 100 mm thick; to pipes							
100 mm nominal size	–	0.33	5.57	0.69	10.38	m	16.64
150 mm nominal size	–	0.38	6.42	0.69	11.54	m	18.65
225 mm nominal size	–	0.45	7.60	0.69	13.84	m	22.13
300 mm nominal size	–	0.53	8.95	0.69	16.15	m	25.79
375 mm nominal size	–	0.68	11.48	0.69	20.76	m	32.93
450 mm nominal size	–	0.80	13.51	0.69	24.22	m	38.42
600 mm nominal size	–	1.02	17.23	0.69	31.15	m	49.07
900 mm nominal size	–	1.65	27.88	0.69	50.75	m	79.32
1200 mm nominal size	–	2.44	41.23	0.69	74.96	m	116.88
Beds and benchings; beds 150 mm thick; to pipes							
100 mm nominal size	–	0.38	6.42	0.69	11.54	m	18.65
150 mm nominal size	–	0.42	7.10	0.69	12.68	m	20.47
225 mm nominal size	–	0.53	8.95	0.69	16.15	m	25.79
300 mm nominal size	–	0.68	11.48	0.69	20.76	m	32.93
375 mm nominal size	–	0.80	13.51	0.69	24.22	m	38.42
450 mm nominal size	–	0.94	15.88	0.69	28.83	m	45.40
600 mm nominal size	–	1.20	20.28	0.69	36.90	m	57.87
900 mm nominal size	–	1.91	32.27	0.69	58.83	m	91.79
1200 mm nominal size	–	2.70	45.62	0.69	83.05	m	129.36
Beds and coverings; 100 mm thick; to pipes							
100 mm nominal size	–	0.50	8.45	0.69	13.84	m	22.98
150 mm nominal size	–	0.58	9.80	0.69	16.15	m	26.64
225 mm nominal size	–	0.83	14.03	0.69	23.07	m	37.79
300 mm nominal size	–	1.00	16.89	0.69	27.69	m	45.27
375 mm nominal size	–	1.21	20.45	0.69	33.44	m	54.58
450 mm nominal size	–	1.42	23.99	0.69	39.21	m	63.89
600 mm nominal size	–	1.83	30.92	0.69	50.75	m	82.36
900 mm nominal size	–	2.79	47.14	0.69	77.28	m	125.11
1200 mm nominal size	–	3.83	64.71	0.69	106.11	m	171.51
Beds and coverings; 150 mm thick; to pipes							
100 mm nominal size	–	0.75	12.67	0.69	20.76	m	34.12
150 mm nominal size	–	0.83	14.03	0.69	23.07	m	37.79
225 mm nominal size	–	1.08	18.25	0.69	29.98	m	48.92
300 mm nominal size	–	1.30	21.97	0.69	35.75	m	58.41
375 mm nominal size	–	1.50	25.35	0.69	41.52	m	67.56
450 mm nominal size	–	1.79	30.24	0.69	49.59	m	80.52
600 mm nominal size	–	2.16	36.49	0.69	59.98	m	97.16
900 mm nominal size	–	3.54	59.81	0.69	98.04	m	158.54
1200 mm nominal size	–	5.00	84.48	0.69	138.41	m	223.58

34 DRAINAGE BELOW GROUND

Item	PC £	Labour hours	Labour £	Plant £	Material £	Unit	Total rate £
34.04 PIPES, FITTINGS AND ACCESSORIES							
NOTE: The following items unless otherwise described include for all appropriate joints/couplings in the running length. The prices for gullies and rainwater shoes, etc. include for appropriate joints to pipes and for setting on and surrounding accessory with site mixed in situ concrete 10.00 N/mm^2–40 mm aggregate							
Cast Iron							
Timesaver drain pipes and fittings or other equal; coated; with mechanical coupling joints							
100 mm dia. pipes							
laid straight	52.30	0.46	6.57	–	68.59	m	**75.16**
in shorter runs not exceeding 3 m long	52.30	0.63	9.00	–	107.02	m	**116.02**
bend; medium radius	64.25	0.56	8.00	–	93.06	nr	**101.06**
bend; medium radius with access	178.52	0.56	8.00	–	210.76	nr	**218.76**
bend; long radius	106.18	0.56	8.00	–	133.42	nr	**141.42**
rest bend	73.69	0.56	8.00	–	99.96	nr	**107.96**
single branch	85.22	0.69	9.87	–	144.65	nr	**154.52**
single branch; with access	196.62	0.79	11.30	–	259.40	nr	**270.70**
double branch	144.91	0.88	12.58	–	236.97	nr	**249.55**
isolated Timesaver joint	34.34	0.32	4.57	–	35.37	nr	**39.94**
transitional pipe; for WC	45.55	0.46	6.57	–	82.28	nr	**88.85**
150 mm dia. pipes							
laid straight	96.81	0.56	8.00	–	119.27	m	**127.27**
in shorter runs not exceeding 3 m long	96.81	0.76	10.87	–	175.47	m	**186.34**
bend; medium radius	147.83	0.65	9.29	–	179.42	nr	**188.71**
bend; medium radius with access	313.46	0.65	9.29	–	350.00	nr	**359.29**
bend; long radius	201.78	0.65	9.29	–	229.75	nr	**239.04**
rest bend	206.46	0.56	8.00	–	236.40	nr	**244.40**
single branch	184.07	0.79	11.30	–	193.70	nr	**205.00**
single 100 mm branch; with access	384.58	0.79	11.30	–	460.39	nr	**471.69**
taper pipe 150 mm to 100 mm	83.74	0.65	9.29	–	108.16	nr	**117.45**
isolated Timesaver joint	41.60	0.39	5.57	–	42.85	nr	**48.42**
Accessories in Timesaver cast iron or equal; with mechanical coupling joints							
Gully fittings; comprising low invert gully trap and round hopper							
100 mm outlet	83.14	0.88	12.58	–	128.91	nr	**141.49**
150 mm outlet	206.91	1.20	17.16	–	268.21	nr	**285.37**
Gulley inlet for bellmouth; circular plain grate							
100 mm nominal size; 200 mm plain grate	86.60	0.42	6.00	–	138.01	nr	**144.01**
100 mm nominal size; 100 mm horizontal inlet; 200 mm plain grate	116.36	0.42	6.00	–	169.42	nr	**175.42**
100 mm nominal size; 100 mm vertical inlet; 200 mm plain grate	116.36	0.42	6.00	–	169.42	nr	**175.42**
Yard gully (Deans); trapped; galvanized sediment pan; 300 mm round heavy grate							
100 mm outlet	562.11	2.68	38.32	–	678.58	nr	**716.90**

34 DRAINAGE BELOW GROUND

Item	PC £	Labour hours	Labour £	Plant £	Material £	Unit	Total rate £
Yard gully (garage); trapless; galvanized sediment pan; 300 mm round heavy grate							
100 mm outlet	617.95	2.50	35.74	–	694.68	nr	**730.42**
Yard gully (garage); trapped; with rodding eye, galvanized perforated sediment pan; stopper; 405 mm round heavy grate							
100 mm outlet	1165.72	2.50	35.74	–	1399.32	nr	**1435.06**
Grease trap; internal access; galvanized perforated bucket; lid and frame							
100 mm outlet; 20 gallon capacity	1168.21	3.70	52.90	–	1291.30	nr	**1344.20**
Ensign lightweight drain pipes and fittings or other equal; ductile iron couplings							
100 mm dia. pipes							
laid straight	36.61	0.19	3.80	–	43.91	m	**47.71**
bend; long radius	56.65	0.19	3.80	–	76.96	nr	**80.76**
single branch	39.16	0.23	4.70	–	77.57	nr	**82.27**
150 mm dia. pipes							
laid straight	71.16	0.22	4.45	–	85.95	m	**90.40**
bend; medium radius	169.91	0.22	4.45	–	213.01	nr	**217.46**
single branch	92.06	0.28	5.62	–	170.83	nr	**176.45**
3000 mm dia. pipes							
laid straight	220.21	0.45	9.15	–	267.98	m	**277.13**
single branch	815.45	0.56	11.26	–	1086.92	nr	**1098.18**
Clay							
Extra strength vitrified clay pipes and fittings; Hepworth Supersleve or other equal; plain ends with push fit polypropylene flexible couplings							
100 mm dia. pipes (Supersleve House Drain)							
laid straight	13.81	0.19	2.72	–	23.10	m	**25.82**
bend	15.74	0.19	2.72	–	29.68	nr	**32.40**
rest bend	27.89	0.19	2.72	–	42.20	nr	**44.92**
rodding point	89.40	0.19	2.72	–	104.12	nr	**106.84**
socket adaptor	20.57	0.16	2.29	–	28.63	nr	**30.92**
saddle	33.40	0.69	9.87	–	43.27	nr	**53.14**
single junction	34.01	0.23	3.29	–	55.95	nr	**59.24**
single access junction with alloy frame and lid (pedestrian areas only)	119.77	0.25	3.57	–	214.37	nr	**217.94**
150 mm dia. pipes							
laid straight	25.55	0.23	3.29	–	41.85	m	**45.14**
bend	31.29	0.22	3.14	–	55.39	nr	**58.53**
access bend	17.22	0.22	3.14	–	222.09	nr	**225.23**
90° bend	31.29	0.22	3.14	–	55.39	nr	**58.53**
taper pipe	38.72	0.22	3.14	–	56.39	nr	**59.53**
rodding point	112.80	0.22	3.14	–	136.72	nr	**139.86**
socket adaptor	31.73	0.19	2.72	–	45.58	nr	**48.30**
saddle	47.89	0.83	11.87	–	64.86	nr	**76.73**
single junction	41.86	0.28	4.01	–	79.19	nr	**83.20**
single access junction (pedestrian areas only)	171.68	0.28	4.01	–	282.99	nr	**287.00**

34 DRAINAGE BELOW GROUND

Item	PC £	Labour hours	Labour £	Plant £	Material £	Unit	Total rate £
34.04 PIPES, FITTINGS AND ACCESSORIES – cont							
Extra strength vitrified clay pipes and fittings – cont							
225 mm dia. pipes							
laid straight	82.57	0.23	3.29	–	85.05	m	**88.34**
bend	132.04	0.22	3.14	–	170.08	nr	**173.22**
90° bend	132.04	0.22	3.14	–	170.08	nr	**173.22**
taper pipe	120.79	0.22	3.14	–	128.70	nr	**131.84**
socket adaptor	72.34	0.19	2.72	–	95.80	nr	**98.52**
300 mm dia. pipes includiing EPDM coupling							
laid straight	126.40	0.23	3.29	–	191.40	m	**194.69**
bend	360.68	0.24	3.43	–	454.88	nr	**458.31**
90° bend	360.68	0.24	3.43	–	454.88	nr	**458.31**
taper pipe	120.79	0.24	3.43	–	146.57	nr	**150.00**
socket adaptor	108.22	0.20	2.86	–	159.66	nr	**162.52**
Accessories in vitrified clay; set in concrete; with polypropylene coupling joints to pipes							
Rodding point; with square aluminium plate							
100 mm nominal size	76.26	0.46	6.57	–	89.65	nr	**96.22**
Gully fittings; comprising low back trap and square hopper; 150 mm × 150 mm square gully grid							
100 mm nominal size	78.48	0.79	11.30	–	100.80	nr	**112.10**
Gully fittings; comprising low back trap and square hopper with back inlet; 150 mm × 150 mm square gully grid							
100 mm nominal size	107.04	0.85	12.15	–	130.22	nr	**142.37**
Accessories in vitrified clay; set in concrete; with cement and sand (1:2) joints to pipes							
Yard gully; 225 mm dia.; including domestic duty grate and frame (up to 1 tonne) and combined filter and silk bucket							
100 mm outlet	329.79	2.50	35.74	–	340.35	nr	**376.09**
100 mm outlet; 100 mm back inlet	568.38	2.70	38.60	–	586.09	nr	**624.69**
150 mm outlet	329.79	3.50	50.04	–	340.35	nr	**390.39**
150 mm outlet; 150 mm back inlet	577.75	3.70	52.90	–	595.74	nr	**648.64**
Yard gully; 225 mm dia.; including medium duty grate and frame (up to 5 tonnes) and combined filter and silk bucket							
100 mm outlet	429.99	2.50	35.74	–	443.56	nr	**479.30**
100 mm outlet; 100 mm back inlet	459.10	2.70	38.60	–	473.53	nr	**512.13**
150 mm outlet	462.75	3.50	50.04	–	477.29	nr	**527.33**
150 mm outlet; 150 mm back inlet	468.42	3.70	52.90	–	483.14	nr	**536.04**
Road gully; trapped with rodding eye and stopper (grate not included)							
300 mm × 600 mm × 100 mm outlet	232.94	3.05	43.60	–	265.73	nr	**309.33**
300 mm × 600 mm × 150 mm outlet	238.55	3.05	43.60	–	271.51	nr	**315.11**
400 mm × 750 mm × 150 mm outlet	276.67	3.70	52.90	–	324.15	nr	**377.05**
450 mm × 900 mm × 150 mm outlet	374.32	4.65	66.48	–	432.54	nr	**499.02**

34 DRAINAGE BELOW GROUND

Item	PC £	Labour hours	Labour £	Plant £	Material £	Unit	Total rate £
Grease trap; with internal access; galvanized perforated bucket; lid and frame							
600 mm × 450 mm × 600 mm deep; 100 mm outlet	2271.64	3.89	55.61	–	2378.26	nr	**2433.87**
Interceptor; trapped with inspection arm; lever locking stopper; chain and staple; cement and sand (1:2) joints to pipes; building in, and cutting and fitting brickwork around							
100 mm outlet; 100 mm inlet	186.25	3.70	52.90	–	192.42	nr	**245.32**
150 mm outlet; 150 mm inlet	268.83	4.16	59.47	–	277.48	nr	**336.95**
225 mm outlet; 225 mm inlet	838.13	4.63	66.19	–	863.93	nr	**930.12**
Accessories in polypropylene; with coupling joints to pipes							
Mini Access Chamber (MAC); 2 nr branch inlets; ligtweight cover and frame							
300 mm dia. × 600 mm deep	368.72	1.00	14.30	–	427.52	nr	**441.82**
Accessories in polypropylene; cover set in concrete; with coupling joints to pipes							
Inspection chamber; 5 nr 100 mm inlets; cast iron cover and frame							
475 mm dia. × 940 mm deep	674.61	2.13	30.45	–	704.89	nr	**735.34**
475 mm dia. × 595 mm deep	539.89	2.13	30.45	–	566.14	nr	**596.59**
475 mm dia. × 175 mm deep raising piece	73.61	0.25	3.57	–	75.82	nr	**79.39**
Round ductile iron cover + frame	180.59	0.09	1.29	–	186.01	nr	**187.30**
Square ductile iron cover + frame	268.67	0.09	1.29	–	276.73	nr	**278.02**
Accessories; grates and covers							
Aluminium alloy gully grids; set in position							
120 mm × 120 mm	9.36	0.09	1.29	–	9.64	nr	**10.93**
150 mm × 150 mm	8.94	0.09	1.29	–	9.21	nr	**10.50**
225 mm × 225 mm	27.82	0.09	1.29	–	28.65	nr	**29.94**
100 mm dia.	9.36	0.09	1.29	–	9.64	nr	**10.93**
150 mm dia.	14.30	0.09	1.29	–	14.73	nr	**16.02**
225 mm dia.	30.85	0.09	1.29	–	31.78	nr	**33.07**
Polypropylene access covers and frames; supplied by Manhole Covers Ltd or other equal and approved; to suit PPIC inspection chambers; bedding and pointing in frame							
450 mm dia.; class A15	39.11	1.30	18.58	–	43.04	nr	**61.62**
450 mm dia.; class B125; kite-marked	43.36	1.30	18.58	–	47.42	nr	**66.00**
Ductile iron heavy duty road grate and frame; supplied by Manhole Covers Ltd or other equal and approved; bedding and pointing in cement and sand (1:3); one course half brick thick wall in semi-engineering bricks in cement mortar (1:3)							
225 mm × 225 mm × 80 mm hinged and dished road grate and frame; class C250	27.68	2.25	32.17	–	33.52	nr	**65.69**
300 mm × 300 mm × 80 mm hinged and dished road grate and frame; class C250	45.73	2.25	32.17	–	52.12	nr	**84.29**
420 mm × 420 mm × 75 mm hinged road grate and frame; class C250; kite-marked	56.56	2.25	32.17	–	63.27	nr	**95.44**

34 DRAINAGE BELOW GROUND

Item	PC £	Labour hours	Labour £	Plant £	Material £	Unit	Total rate £
34.04 PIPES, FITTINGS AND ACCESSORIES – cont							
Accessories – cont							
Ductile iron heavy duty road grate and frame – cont							
445 mm × 445 mm × 75 mm double triangular road grate and frame; class C250; kite-marked	60.17	2.25	32.17	–	66.98	nr	**99.15**
435 mm × 435 mm × 100 mm pedestrian mesh road grate and frame; class D400	61.38	2.25	32.17	–	68.24	nr	**100.41**
440 mm × 400 mm × 150 mm hinged road grate and frame; class D400; kite-marked	79.44	2.25	32.17	–	86.83	nr	**119.00**
Concrete							
Vibrated concrete pipes and fittings; with flexible joints; BS 5911 Part 1; trench 2.00 m deep							
300 mm dia. pipes							
Class M; laid straight	24.86	0.65	9.29	9.02	25.61	m	**43.92**
bend; <= 45°	–	0.65	9.29	–	256.05	nr	**265.34**
bend; > 45°	–	0.65	9.29	–	384.07	nr	**393.36**
junction; 300 mm × 100 mm	–	0.46	6.57	–	134.43	nr	**141.00**
450 mm dia. pipes							
Class H; laid straight	38.85	1.02	14.58	9.02	40.02	m	**63.62**
bend; <= 45°	–	1.02	14.58	–	400.10	nr	**414.68**
bend; > 45°	–	1.02	14.58	–	600.16	nr	**614.74**
junction; 450 mm × 150 mm	–	0.65	9.29	–	210.06	nr	**219.35**
525 mm dia. pipes							
Class H; laid straight	47.68	1.48	21.16	9.02	49.11	m	**79.29**
bend; <= 45°	–	1.48	21.16	9.02	491.15	nr	**521.33**
bend; > 45°	–	1.48	21.16	9.02	736.72	nr	**766.90**
junction; 600 mm × 150 mm	–	0.83	11.87	9.02	343.80	nr	**364.69**
900 mm dia. pipes							
Class H; laid straight	157.60	2.59	37.03	9.02	162.33	m	**208.38**
bend; <= 45°	–	2.59	37.03	9.02	1623.26	nr	**1669.31**
bend; > 45°	–	2.59	37.03	9.02	2434.88	nr	**2480.93**
junction; 900 mm × 150 mm	–	1.02	14.58	9.02	730.47	nr	**754.07**
1200 mm dia. pipes							
Class H; laid straight	271.55	3.70	52.90	9.02	279.70	m	**341.62**
bend; <= 45°	–	3.70	52.90	9.02	2796.97	nr	**2858.89**
bend; > 45°	–	3.70	52.90	9.02	4195.45	nr	**4257.37**
junction; 1200 mm × 150 mm	–	1.48	21.16	9.02	1258.63	nr	**1288.81**

34 DRAINAGE BELOW GROUND

Item	PC £	Labour hours	Labour £	Plant £	Material £	Unit	Total rate £
Accessories in precast concrete; top set in with rodding eye and stopper; cement and sand (1:2) joint to pipe							
Concrete road gully; trapped with rodding eye and stopper; cement and sand (1:2) joint to pipe							
450 mm dia. × 900 mm deep; 100 mm or 150 mm outlet	69.70	4.39	62.76	–	101.49	nr	**164.25**
450 mm dia. × 1050 mm deep; 100 mm or 150 mm outlet	75.86	4.39	62.76	–	107.84	nr	**170.60**
Gully adapter type 1501	–	–	–	–	6.40	nr	**6.40**
uPVC							
OsmaDrain uPVC pipes and fittings or other equal and approved; with ring seal joints							
110 mm dia. pipes							
laid straight	9.68	0.17	2.43	–	11.92	m	**14.35**
bend; short radius	29.44	0.15	2.14	–	29.73	nr	**31.87**
bend; long radius	55.75	0.15	2.14	–	54.43	nr	**56.57**
spigot/socket bend	24.87	0.15	2.14	–	35.81	nr	**37.95**
socket plug	12.89	0.04	0.58	–	13.28	nr	**13.86**
adjustable double socket bend	35.22	0.15	2.14	–	45.88	nr	**48.02**
adaptor to clay	38.17	0.09	1.29	–	38.85	nr	**40.14**
single junction	3.44	0.21	3.00	–	0.56	nr	**3.56**
sealed access junction	90.80	0.19	2.72	–	90.54	nr	**93.26**
slip coupler	17.03	0.09	1.29	–	17.54	nr	**18.83**
160 mm dia. pipes							
laid straight	22.22	0.21	3.00	–	26.44	m	**29.44**
bend; short radius	70.12	0.18	2.58	–	70.85	nr	**73.43**
spigot/socket bend	63.46	0.18	2.58	–	83.54	nr	**86.12**
socket plug	27.65	0.07	1.00	–	28.48	nr	**29.48**
adaptor to clay	82.93	0.12	1.72	–	84.19	nr	**85.91**
level invert taper	33.94	0.18	2.58	–	51.76	nr	**54.34**
single junction	114.60	0.24	3.43	–	118.04	nr	**121.47**
slip coupler	41.36	0.11	1.58	–	42.60	nr	**44.18**
uPVC Osma Ultra-Rib ribbed pipes and fittings or other equal and approved; WIS approval; with sealed ring push-fit joints							
150 mm dia. pipes							
laid straight	12.57	0.19	2.72	–	12.95	m	**15.67**
bend; short radius	36.31	0.17	2.43	–	36.63	nr	**39.06**
adaptor to 160 mm dia. uPVC	48.32	0.10	1.43	–	48.21	nr	**49.64**
adaptor to clay	59.63	0.10	1.43	–	60.65	nr	**62.08**
level invert taper	17.75	0.18	2.58	–	15.94	nr	**18.52**
single junction	73.08	0.22	3.14	–	71.39	nr	**74.53**
225 mm dia. pipes							
laid straight	32.55	0.22	3.14	–	33.53	m	**36.67**
bend; short radius	170.78	0.20	2.86	–	173.89	nr	**176.75**
level invert taper	29.73	0.20	2.86	–	24.59	nr	**27.45**
single junction	253.51	0.27	3.86	–	251.06	nr	**254.92**

34 DRAINAGE BELOW GROUND

Item	PC £	Labour hours	Labour £	Plant £	Material £	Unit	Total rate £
34.04 PIPES, FITTINGS AND ACCESSORIES – cont							
uPVC Osma Ultra-Rib ribbed pipes and fittings or other equal and approved – cont							
300 mm dia. pipes							
laid straight	46.19	0.32	4.57	–	47.58	m	**52.15**
bend; short radius	269.00	0.29	4.15	–	274.22	nr	**278.37**
level invert taper	96.49	0.29	4.15	–	90.83	nr	**94.98**
single junction	585.78	0.37	5.29	–	589.08	nr	**594.37**
Accessories in uPVC; with ring seal joints to pipes (unless otherwise described)							
Rodding eye							
square top, sealed with coupling	89.60	0.43	6.15	–	97.87	nr	**104.02**
Universal gully fitting; comprising gully trap, plain hopper							
150 mm × 150 mm grate	52.39	0.93	13.30	–	61.78	nr	**75.08**
Bottle gully; square comprising gully with rotating							
217 mm × 217 mm grate	96.76	0.78	11.15	–	107.48	nr	**118.63**
Shallow inspection chamber; 250 mm dia.; 600 mm deep; sealed cover and frame (1.0 tonne loading)							
4 nr 110 mm outlets/inlets	188.86	1.28	18.30	–	222.41	nr	**240.71**
Universal inspection chamber; 450 mm dia.; 4 nr 110 mm branch outlet; single seal cast iron cover and frame (12.5 tonnes loading)							
500 mm deep	384.24	1.35	19.30	–	423.65	nr	**442.95**
730 mm deep	443.11	1.60	22.88	–	489.87	nr	**512.75**
960 mm deep	501.99	1.85	26.45	–	556.09	nr	**582.54**
Equal manhole base; 500 mm dia.							
6 nr 110 mm branch outlets	220.76	1.21	17.29	–	244.12	nr	**261.41**
Unequal manhole base; 500 mm dia.							
2 nr 160 mm, 4 nr 110 mm branch outlets	236.70	1.21	17.29	–	260.54	nr	**277.83**
Kerb to gullies; class B engineering bricks on edge to three sides in cement mortar (1:3) rendering in cement mortar (1:3) to top and two sides and skirting to brickwork 230 mm high; dishing in cement mortar (1:3) to gully; steel trowelled							
230 mm × 230 mm internally	–	1.39	19.87	–	2.97	nr	**22.84**

34 DRAINAGE BELOW GROUND

Item	PC £	Labour hours	Labour £	Plant £	Material £	Unit	Total rate £
34.05 LAND DRAINAGE							
Excavating; by hand; grading bottoms; earthwork support; filling to within 150 mm of surface with gravel rejects; remainder filled with excavated material and compacting; disposal of surplus soil on site; spreading on site average 50 m							
Pipes not exceeding 200 nominal size							
average depth of trench 0.75 m	–	1.57	22.62	–	10.39	m	**33.01**
average depth of trench 1.00 m	–	2.08	29.96	–	16.39	m	**46.35**
average depth of trench 1.25 m	–	2.91	41.92	–	20.53	m	**62.45**
average depth of trench 1.50 m	–	5.00	72.04	–	24.99	m	**97.03**
average depth of trench 1.75 m	–	5.92	85.29	–	29.13	m	**114.42**
average depth of trench 2.00 m	–	6.85	98.69	–	33.59	m	**132.28**
Disposal; load lorry by machine							
Excavated material							
inactive waste off site; to tip not exceeding 13 km (using lorries); including Landfill Tax	–	0.10	1.44	16.51	22.35	m³	**40.30**
active non-hazardous waste off site; to tip not exceeding 13 km (using lorries); including Landfill Tax	–	0.06	1.07	16.51	282.28	m³	**299.86**
Vitrified clay perforated subsoil pipes; BS 65; Hepworth Hepline or other equal and approved							
Pipes; laid straight							
100 mm dia.	17.80	0.20	2.86	–	18.33	m	**21.19**
150 mm dia.	30.60	0.25	3.57	–	31.52	m	**35.09**
225 mm dia.	68.44	0.33	4.72	–	70.49	m	**75.21**
Concrete Canvas cement impregnated cloth lining to form ditch linings; holding ponds; slope protection and similar							
Type CC8, 8 mm thick and sprayed with water	–	0.10	1.69	0.11	89.23	m²	**91.03**
34.06 MANHOLES AND SOAKAWAYS							
Excavating; by machine							
Manholes							
maximum depth not exceeding 1.00 m	–	0.19	2.74	5.26	–	m³	**8.00**
maximum depth not exceeding 2.00 m	–	0.21	3.03	5.79	–	m³	**8.82**
maximum depth not exceeding 4.00 m	–	0.25	3.61	6.77	–	m³	**10.38**
Excavating; by hand							
Manholes							
maximum depth not exceeding 1.00 m	–	3.05	43.94	–	–	m³	**43.94**
maximum depth not exceeding 2.00 m	–	3.61	52.01	–	–	m³	**52.01**
maximum depth not exceeding 4.00 m	–	4.63	66.70	–	–	m³	**66.70**

34 DRAINAGE BELOW GROUND

Item	PC £	Labour hours	Labour £	Plant £	Material £	Unit	Total rate £
34.06 MANHOLES AND SOAKAWAYS – cont							
Earthwork support (average risk prices)							
Maximum depth not exceeding 1.00 m							
distance between opposing faces not exceeding 2.00 m	–	0.14	2.02	–	8.36	m²	**10.38**
Maximum depth not exceeding 2.00 m							
distance between opposing faces not exceeding 2.00 m	–	0.18	2.60	–	16.27	m²	**18.87**
Maximum depth not exceeding 4.00 m							
distance between opposing faces not exceeding 2.00 m	–	0.22	3.17	–	24.18	m²	**27.35**
Disposal; by machine							
Excavated material							
off site; to tip not exceeding 13 km (using lorries) including Landfill Tax based on inactive waste	–	0.10	1.44	16.51	22.35	m³	**40.30**
active non-hazardous waste off site; to tip not exceeding 13 km (using lorries); including Landfill Tax	–	0.06	1.07	16.51	282.28	m³	**299.86**
on site; depositing on site in spoil heaps; average 50 m distance	–	0.14	2.02	4.71	–	m³	**6.73**
Disposal; by hand							
Excavated material							
on site; depositing on site in spoil heaps; average 50 m distance	–	1.20	17.29	–	–	m³	**17.29**
Filling to excavations; by machine							
Average thickness not exceeding 0.25 m							
arising excavations	–	0.14	2.02	2.48	–	m³	**4.50**
Filling to excavations; by hand							
Average thickness not exceeding 0.25 m							
arising from excavations	–	0.93	13.40	–	–	m³	**13.40**
Plain in situ ready mixed designated concrete; C10–40 mm aggregate							
Beds							
thickness not exceeding 150 mm	113.72	2.78	46.97	–	117.13	m³	**164.10**
thickness 150 mm–450 mm	–	2.08	35.14	–	117.13	m³	**152.27**
thickness exceeding 450 mm	–	1.76	29.74	–	117.13	m³	**146.87**
Plain in situ ready mixed designated concrete; C20–20 mm aggregate							
Beds							
thickness not exceeding 150 mm	117.58	2.78	46.97	–	121.11	m³	**168.08**
thickness 150 mm–450 mm	–	2.08	35.14	–	121.11	m³	**156.25**
thickness exceeding 450 mm	–	1.76	29.74	–	121.11	m³	**150.85**

34 DRAINAGE BELOW GROUND

Item	PC £	Labour hours	Labour £	Plant £	Material £	Unit	Total rate £
Plain in situ ready mixed designated concrete; C25–20 mm aggregate; (small quantities)							
Benching in bottoms							
150 mm–450 mm average thickness	118.78	8.33	151.78	–	122.34	m³	**274.12**
Reinforced in situ ready mixed designated concrete; C20–20 mm aggregate; (small quantities)							
Isolated cover slabs							
thickness not exceeding 150 mm	111.98	6.48	109.49	–	115.34	m³	**224.83**
Reinforcement; fabric to BS 4449; lapped; in beds or suspended slabs							
Ref D98 (1.54 kg/m²)							
400 mm minimum laps	1.67	0.11	2.31	–	1.72	m²	**4.03**
Ref A142 (2.22 kg/m²)							
400 mm minimum laps	1.96	0.11	2.31	–	2.02	m²	**4.33**
Ref A193 (3.02 kg/m²)							
400 mm minimum laps	2.66	0.11	2.31	–	2.74	m²	**5.05**
Formwork; basic finish							
Soffits of isolated cover slabs							
horizontal	–	2.64	52.77	–	13.48	m²	**66.25**
Edges of isolated cover slabs							
height not exceeding 250 mm	–	0.78	15.58	–	3.91	m	**19.49**
Precast concrete circular manhole rings; bedding, jointing and pointing in cement mortar (1:3) on prepared bed							
Chamber or shaft rings; plain							
900 mm dia.	77.04	5.09	72.77	–	81.00	m	**153.77**
1050 mm dia.	81.74	6.01	85.92	–	87.49	m	**173.41**
1200 mm dia.	99.60	6.94	99.22	–	107.53	m	**206.75**
Chamber or shaft rings; reinforced							
1350 mm dia.	156.33	7.86	112.37	45.11	167.61	m	**325.09**
1500 mm dia.	183.82	8.79	125.67	45.11	199.22	m	**370.00**
1800 mm dia.	264.33	11.10	158.69	81.21	287.09	m	**526.99**
2100 mm dia.	532.24	13.88	198.43	90.23	567.98	m	**856.64**
step irons built in	8.67	0.14	2.00	–	8.93	nr	**10.93**
integrated ladder system 150 mm projection polypropylene encapusulated steps and rails	10.03	0.50	7.15	–	10.33	step	**17.48**
Reducing slabs							
1200 mm dia.	141.62	5.55	79.34	45.11	149.15	nr	**273.60**
1350 mm dia.	227.06	8.79	125.67	45.11	240.46	nr	**411.24**
1500 mm dia.	258.80	10.18	145.54	45.11	274.80	nr	**465.45**
1800 mm dia.	407.33	12.95	185.14	81.21	432.73	nr	**699.08**
2100 mm dia.	864.33	12.95	185.14	81.21	903.44	nr	**1169.79**

Prices for Measured Works

34 DRAINAGE BELOW GROUND

Item	PC £	Labour hours	Labour £	Plant £	Material £	Unit	Total rate £
34.06 MANHOLES AND SOAKAWAYS – cont							
Precast concrete circular manhole rings – cont							
Heavy duty cover slabs; to suit rings							
900 mm dia.	84.59	2.78	39.75	18.05	88.78	nr	**146.58**
1050 mm dia.	89.73	3.24	46.32	18.05	94.40	nr	**158.77**
1200 mm dia.	108.94	3.70	52.90	20.75	115.50	nr	**189.15**
1350 mm dia.	174.65	4.16	59.47	22.56	184.83	nr	**266.86**
1500 mm dia.	199.08	4.63	66.19	37.89	211.64	nr	**315.72**
1800 mm dia.	313.33	5.55	79.34	43.31	331.88	nr	**454.53**
2100 mm dia.	664.88	6.48	92.64	45.11	697.19	nr	**834.94**
Precast concrete circular manhole; CPM Perfect Manhole or similar; complete with preformed benching and outlets to base; elastomeric seal to joints to rings; single steps as required							
1200 mm with up to four outlets 100 mm or 150 mm outlets; effective internal depth:							
1250 mm deep	–	5.00	74.96	36.09	1530.74	nr	**1641.79**
1500 mm deep	–	5.00	74.96	36.09	1528.72	nr	**1639.77**
1800 mm deep	–	5.00	74.96	36.09	1491.87	nr	**1602.92**
2000 mm deep	–	5.00	74.96	36.09	1504.31	nr	**1615.36**
2500 mm deep	–	4.50	64.33	45.11	1484.63	nr	**1594.07**
1500 mm dia. with up to four outlets; effective internal depth:							
2000 mm deep; upto 450 mm pipe	–	4.50	64.33	45.11	2059.60	nr	**2169.04**
2500 mm deep; up to 450 mm pipe	–	4.75	67.91	50.53	2689.57	nr	**2808.01**
3000 mm deep; up to 450 mm pipe	–	4.75	67.91	50.53	2906.80	nr	**3025.24**
3500 mm deep; up to 450 mm pipe	–	5.00	71.48	54.14	2965.45	nr	**3091.07**
4000 mm deep; up to 450 mm pipe	–	5.00	71.48	54.14	3163.41	nr	**3289.03**
4500 mm deep; up to 450 mm pipe	–	5.00	71.48	54.14	3182.70	nr	**3308.32**
4000 mm deep; 600 mm pipe	–	5.00	71.48	54.14	3612.93	nr	**3738.55**
4500 mm deep; 600 mm pipe	–	5.00	71.48	54.14	3639.18	nr	**3764.80**
Common bricks; in cement mortar (1:3)							
Walls to manholes							
one brick thick	380.00	2.22	49.76	–	70.47	m²	**120.23**
one and a half brick thick	–	3.24	72.62	–	105.72	m²	**178.34**
Projections of footings							
two brick thick	–	4.53	101.53	–	140.96	m²	**242.49**
Class A engineering bricks; in cement mortar (1:3)							
Walls to manholes							
one brick thick (PC £ per 1000)	1144.10	2.50	56.03	–	192.31	m²	**248.34**
one and a half brick thick	–	3.61	80.91	–	197.26	m²	**278.17**
Projections of footings							
two brick thick	–	5.09	114.07	–	384.61	m²	**498.68**

34 DRAINAGE BELOW GROUND

Item	PC £	Labour hours	Labour £	Plant £	Material £	Unit	Total rate £
Class B engineering bricks; in cement mortar (1:3)							
Walls to manholes							
one brick thick (PC £ per 1000)	332.50	2.50	56.03	–	62.90	m²	**118.93**
one and a half brick thick	–	3.61	80.91	–	94.36	m²	**175.27**
Projections of footings							
two brick thick	–	5.09	114.07	–	125.80	m²	**239.87**
Brickwork sundries							
Extra over for fair face; flush smooth pointing							
manhole walls	–	0.19	4.25	–	–	m²	**4.25**
Building ends of pipes into brickwork; making good fair face or rendering							
not exceeding 55 mm nominal size	–	0.09	2.02	–	–	nr	**2.02**
55 mm–110 mm nominal size	–	0.14	3.14	–	–	nr	**3.14**
over 110 mm nominal size	–	0.19	4.25	–	–	nr	**4.25**
Step irons; BS 1247; malleable; galvanized; building into joints							
general purpose pattern	–	0.14	3.14	–	8.93	nr	**12.07**
Cement and sand (1:3) in situ finishings; steel trowelled							
13 mm work to manhole walls; one coat; to							
brickwork base over 300 wide	–	0.65	14.56	–	3.30	m²	**17.86**
Cast iron inspection chambers; with bolted flat covers; BS 437; bedded in cement mortar (1:3); with mechanical coupling joints							
100 mm × 100 mm							
one branch either side	403.41	1.40	20.01	–	417.16	nr	**437.17**
two branches either side	763.04	2.00	28.59	–	787.58	nr	**816.17**
150 mm × 100 mm							
one branch either side	499.49	1.55	22.16	–	516.12	nr	**538.28**
two branches either side	969.40	2.15	30.74	–	1001.78	nr	**1032.52**
150 mm × 150 mm							
one branch either side	618.71	1.80	25.73	–	640.57	nr	**666.30**
two branches either side	1194.23	2.60	37.17	–	1233.35	nr	**1270.52**
Coated cast or ductile iron access covers and frames; to BS EN124; supplied by Manhole Covers Ltd or other equal; bedding frame in cement and sand (1:3); cover in grease and sand							
Light duty; cast iron; rectangular single seal solid top							
450 mm × 450 mm; class A15	46.86	1.50	21.44	–	51.03	nr	**72.47**
600 mm × 450 mm; class A15	46.94	1.50	21.44	–	51.29	nr	**72.73**
600 mm × 600 mm; class A15	62.90	1.50	21.44	–	67.92	nr	**89.36**
750 mm × 600 mm; class A15	105.43	1.50	21.44	–	111.72	nr	**133.16**
900 mm × 600 mm; class A15	120.62	1.50	21.44	–	127.37	nr	**148.81**

34 DRAINAGE BELOW GROUND

Item	PC £	Labour hours	Labour £	Plant £	Material £	Unit	Total rate £
34.06 MANHOLES AND SOAKAWAYS – cont							
Coated cast or ductile iron access covers and frames – cont							
Medium duty; ductile iron; rectangular single seal solid top							
450 mm × 450 mm × 40 mm; class C250; kite-marked	80.72	2.00	28.59	–	86.27	nr	**114.86**
600 mm × 450 mm × 40 mm; slide-out; class C250; kite-marked	80.72	2.00	28.59	–	86.27	nr	**114.86**
600 mm × 600 mm × 40 mm; slide-out; class C250; kite-marked	83.03	2.00	28.59	–	88.65	nr	**117.24**
760 mm × 600 mm × 40 mm; slide-out; class C250; kite-marked	125.37	2.00	28.59	–	132.26	nr	**160.85**
Heavy duty; ductile iron; solid top							
450 mm × 450 mm × 75 mm; single seal; class C250; kite-marked	119.14	2.50	35.74	–	125.85	nr	**161.59**
600 mm × 450 mm × 75 mm; single seal; class C250; kite-marked	132.38	2.50	35.74	–	139.48	nr	**175.22**
600 mm × 600 mm × 75 mm; single seal; class C250; kite-marked	151.63	2.50	35.74	–	159.31	nr	**195.05**
450 mm × 450 mm × 100 mm; double triangular; class D400; kite-marked	108.31	2.50	35.74	–	114.69	nr	**150.43**
600 mm × 450 mm × 100 mm; double triangular; class D400; kite-marked	142.01	2.50	35.74	–	149.40	nr	**185.14**
600 mm × 600 mm × 100 mm; double triangular; class D400; kite-marked	86.66	2.50	35.74	–	92.39	nr	**128.13**
750 mm × 600 mm × 100 mm; double triangular; class D400; kite-marked	221.44	2.50	35.74	–	231.21	nr	**266.95**
1220 mm × 675 mm × 100 mm; double triangular; class D400; kite-marked	246.70	3.50	50.04	–	257.23	nr	**307.27**
British Standard best quality vitrified clay channels; bedding and jointing in cement and sand (1:2)							
Half section straight							
100 mm dia. × 1 m long	15.89	0.74	10.58	–	16.37	nr	**26.95**
150 mm dia. × 1 m long	26.44	0.93	13.30	–	27.23	nr	**40.53**
225 mm dia. × 1 m long	59.38	1.20	17.16	–	61.16	nr	**78.32**
300 mm dia. × 1 m long	121.88	1.48	21.16	–	125.54	nr	**146.70**
Half section bend							
100 mm dia.	17.89	0.56	8.00	–	18.43	nr	**26.43**
150 mm dia.	29.48	0.69	9.87	–	30.36	nr	**40.23**
225 mm dia.	98.32	0.93	13.30	–	101.27	nr	**114.57**
Taper straight							
150 mm–100 mm dia.	74.30	0.65	9.29	–	76.53	nr	**85.82**
300 mm–225 mm dia.	326.16	0.83	11.87	–	335.94	nr	**347.81**
Taper bend							
150 mm–100 mm dia.	113.13	0.83	11.87	–	116.52	nr	**128.39**
225 mm–150 mm dia.	324.16	1.06	15.15	–	333.88	nr	**349.03**

34 DRAINAGE BELOW GROUND

Item	PC £	Labour hours	Labour £	Plant £	Material £	Unit	Total rate £
Three quarter section branch bend							
100 mm dia.	40.29	0.46	6.57	–	41.50	nr	**48.07**
150 mm dia.	67.63	0.69	9.87	–	69.66	nr	**79.53**
Glass fibre septic tank; Klargester or other equal and approved; fixing lockable manhole cover and frame; placing in position (excavations, fill etc. measured elsewhere)							
2800 litre capacity; depth to invert							
1000 mm deep	–	2.45	35.03	72.18	821.09	nr	**928.30**
1500 mm deep	–	2.73	39.03	72.18	870.01	nr	**981.22**
3800 litre capacity; depth to invert							
1000 mm deep	–	2.64	37.74	90.23	1049.10	nr	**1177.07**
1500 mm deep	–	2.91	41.60	99.25	1098.03	nr	**1238.88**
4600 litre capacity; depth to invert							
1000 mm deep	–	2.75	39.32	90.23	1202.74	nr	**1332.29**
1500 mm deep	–	3.20	45.75	99.25	1251.68	nr	**1396.68**
Soakaways							
Soakaway crates, lightweight modular water storage cells for stormwater attenuation with geotextile membrane wrapping							
Polystorm Heavy Duty (60 tonne) crates							
1000 mm × 500 mm × 400 mm deep	–	1.25	17.87	–	102.20	m³	**120.07**
Polystorm Lite Duty (20 tonne) crate kit 1 m³	–	1.25	17.87	–	901.10	nr	**918.97**
Brett Martin Heavy Duty (60 tonne) crates							
1200 mm × 600 mm × 420 mm deep	–	1.25	17.87	–	129.19	m³	**147.06**
Brett Martin Light Duty (20 tonne) crates							
1200 mm × 600 mm × 420 mm deep	–	1.25	17.87	–	91.25	m³	**109.12**
Rainwater harvesting							
FP McCann Ltd Easi-Rain domestic and commercial rainwater harvesting tank system designed to reduce demands on the mains network by providing a sustainable water collection and distribution system ideal for non-potable uses such as toilet flushing, washing and landscape or garden watering. Excluding excavations							
1200 litre tank	2200.00	1.50	21.44	20.86	2266.00	nr	**2308.30**
1500 litre tank	2250.00	1.75	25.02	25.03	2317.50	nr	**2367.55**
3000 litre tank	2680.00	1.80	25.73	29.20	2760.40	nr	**2815.33**
5000 litre tank	2730.00	1.90	27.16	33.37	2811.90	nr	**2872.43**
7500 litre tank	2835.00	2.00	28.59	33.37	2920.05	nr	**2982.01**

35 SITE WORK

Item	PC £	Labour hours	Labour £	Plant £	Material £	Unit	Total rate £
35.01 KERBS, EDGINGS AND CHANNELS							
Excavating; by machine							
Excavating trenches; to receive kerb foundations; average size							
300 mm × 100 mm	–	0.02	0.29	0.37	–	m	**0.66**
450 mm × 150 mm	–	0.02	0.29	0.75	–	m	**1.04**
600 mm × 200 mm	–	0.03	0.46	1.05	–	m	**1.51**
Excavating curved trenches; to receive kerb foundations; average size							
300 mm × 100 mm	–	0.01	0.15	0.60	–	m	**0.75**
450 mm × 150 mm	–	0.03	0.43	0.91	–	m	**1.34**
600 mm × 200 mm	–	0.04	0.58	1.12	–	m	**1.70**
Excavating; by hand							
Excavating trenches; to receive kerb foundations; average size							
150 mm × 50 mm	–	0.02	0.29	–	–	m	**0.29**
200 mm × 75 mm	–	0.06	0.87	–	–	m	**0.87**
250 mm × 100 mm	–	0.10	1.44	–	–	m	**1.44**
300 mm × 100 mm	–	0.13	1.87	–	–	m	**1.87**
Excavating curved trenches; to receive kerb foundations; average size							
150 mm × 50 mm	–	0.03	0.43	–	–	m	**0.43**
200 mm × 75 mm	–	0.07	1.01	–	–	m	**1.01**
250 mm × 100 mm	–	0.11	1.59	–	–	m	**1.59**
300 mm × 100 mm	–	0.14	2.02	–	–	m	**2.02**
Plain in situ ready mixed designated concrete; C7.5–40 mm aggregate; poured on or against earth or unblinded hardcore							
Foundations	109.40	1.16	19.60	–	112.68	m³	**132.28**
Blinding beds							
thickness not exceeding 150 mm	109.40	1.71	28.89	–	112.68	m³	**141.57**
Plain in situ ready mixed designated concrete; C10–40 mm aggregate; poured on or against earth or unblinded hardcore							
Foundations	113.72	1.16	19.60	–	117.13	m³	**136.73**
Blinding beds							
thickness not exceeding 150 mm	113.72	1.71	28.89	–	117.13	m³	**146.02**
Plain in situ ready mixed designated concrete; C20–20 mm aggregate; poured on or against earth or unblinded hardcore							
Foundations	117.58	1.16	19.60	–	121.11	m³	**140.71**
Blinding beds							
thickness not exceeding 150 mm	117.58	1.71	28.89	–	121.11	m³	**150.00**

35 SITE WORK

Item	PC £	Labour hours	Labour £	Plant £	Material £	Unit	Total rate £
Filling to make up levels; by machine							
Average thickness not exceeding 0.25 m							
obtained off site; hardcore	–	0.28	4.04	1.25	42.65	m³	**47.94**
obtained off site; granular fill type one	–	0.28	4.04	1.25	53.46	m³	**58.75**
obtained off site; granular fill type two	–	0.28	4.04	1.25	53.46	m³	**58.75**
Average thickness exceeding 0.25 m							
obtained off site; hardcore	–	0.24	3.46	1.14	36.55	m³	**41.15**
obtained off site; granular fill type one	–	0.24	3.46	1.14	53.46	m³	**58.06**
obtained off site; granular fill type two	–	0.24	3.46	1.14	53.46	m³	**58.06**
Filling to make up levels; by hand							
Average thickness not exceeding 0.25 m							
obtained off site; hardcore	–	0.61	8.79	2.15	42.65	m³	**53.59**
obtained off site; sand	–	0.71	10.23	2.51	86.26	m³	**99.00**
Average thickness exceeding 0.25 m							
obtained off site; hardcore	–	0.51	7.34	1.79	36.55	m³	**45.68**
obtained off site; sand	–	0.60	8.64	2.12	86.26	m³	**97.02**
Surface treatments							
Compacting							
filling; blinding with sand	–	0.04	0.58	0.04	1.97	m²	**2.59**
Precast concrete kerbs, channels, edgings, etc.; BS 340; bedded, jointed and pointed in cement mortar (1:3); including haunching up one side with in situ ready mix designated concrete C10– 40 mm aggregate; to concrete base							
Edgings; straight; square edge							
50 mm × 150 mm	–	0.23	5.74	–	4.92	m	**10.66**
50 mm × 200 mm	–	0.23	5.74	–	7.00	m	**12.74**
50 mm × 255 mm	–	0.23	5.74	–	8.35	m	**14.09**
Kerbs; straight							
125 mm × 255 mm; half battered	–	0.31	7.73	–	9.65	m	**17.38**
125 mm × 255 mm; half battered drop kerb	–	0.31	7.73	–	13.63	m	**21.36**
150 mm × 305 mm; half battered	–	0.31	7.73	–	23.94	m	**31.67**
150 mm × 305 mm; half battered drop kerb	–	0.31	7.73	–	43.12	m	**50.85**
Kerbs; curved							
125 mm × 255 mm; half battered	–	0.46	11.46	–	14.14	m	**25.60**
150 mm × 305 mm; half battered	–	0.46	11.46	–	18.26	m	**29.72**
Channels; 255 × 125 mm							
straight	–	0.31	7.73	–	15.01	m	**22.74**
Quadrants; half battered							
305 mm × 305 mm × 150 mm	–	0.32	7.97	–	21.61	nr	**29.58**
305 mm × 305 mm × 255 mm	–	0.32	7.97	–	25.21	nr	**33.18**
455 mm × 455 mm × 255 mm	–	0.37	9.22	–	27.32	nr	**36.54**

35 SITE WORK

Item	PC £	Labour hours	Labour £	Plant £	Material £	Unit	Total rate £
35.01 KERBS, EDGINGS AND CHANNELS – cont							
Natural stone kerbs; Silver Grey Granite bedded, jointed and pointed in cement mortar (1:3); including haunching up one side with in situ ready mix designated concrete C10–40 mm aggregate; to concrete base							
Kerbs; straight							
Kerbs with granite setts							
100 mm × 100 mm × 200 mm, split all sides	6.62	0.35	8.72	–	15.80	m	**24.52**
Kerbs with granite setts							
300 mm × 50 mm × 600 mm, sawn all sides, flame textured top and front face	53.63	0.35	8.72	–	69.05	m	**77.77**
Kerbs with granite setts							
250 mm × 50 mm × 600 mm, sawn all sides, flame textured top and front face	53.63	0.35	8.72	–	69.05	m	**77.77**
Bullnose kerbs with granite setts							
150 mm × 150 mm × 600 mm, 10 mm rounding to top one long side, flame textured top and front face	21.07	0.50	12.46	–	32.18	m	**44.64**
Bullnose kerbs with granite setts							
300 mm × 150 mm × 600 mm, 10 mm rounding to top one long side, flame textured top and front face	41.53	0.50	12.46	–	55.35	m	**67.81**
Transition Bullnose kerbs with granite setts							
300 mm to 150 mm width change across length × 150 mm × 600 mm, 10 mm rounding to top one long side, flame textured top and front face	11.16	0.03	0.75	–	20.94	nr	**21.69**
Kerbs; curved							
Bullnose kerbs with granite setts							
150 mm × 175 mm × 600 mm, sawn all sides, 10 mm rounding to top one long side, flame textured top and front face, curved on plan 1–12 m	36.12	0.50	12.46	–	49.22	m	**61.68**
Bullnose kerbs with granite setts							
300 mm × 150 mm × 600 mm, 10 mm rounding to top one long side, flame textured top and front face, curved on plan 1–12 m	49.66	0.50	12.46	–	64.56	m	**77.02**

35 SITE WORK

Item	PC £	Labour hours	Labour £	Plant £	Material £	Unit	Total rate £
35.02 IN SITU CONCRETE ROADS AND PAVINGS							
Reinforced in situ ready mixed designated concrete; C10–40 mm aggregate							
Roads; to hardcore base							
thickness not exceeding 150 mm	108.30	1.85	31.26	–	111.55	m³	**142.81**
thickness 150 mm–450 mm	108.30	1.30	21.97	–	111.55	m³	**133.52**
Reinforced in situ ready mixed designated concrete; C20–20 mm aggregate							
Roads; to hardcore base							
thickness not exceeding 150 mm	111.98	1.85	31.26	–	115.34	m³	**146.60**
thickness 150 mm–450 mm	111.98	1.30	21.97	–	115.34	m³	**137.31**
Reinforced in situ ready mixed designated concrete; C25–20 mm aggregate							
Roads; to hardcore base							
thickness ot exceeding 150 mm	118.78	1.85	31.26	–	122.34	m³	**153.60**
thickness 150 mm–450 mm	118.78	1.30	21.97	–	122.34	m³	**144.31**
Formwork; sides of foundations; basic finish							
Plain vertical							
height not exceeding 250 mm	–	0.39	7.80	–	4.14	m	**11.94**
height 250 mm–500 mm	–	0.57	11.39	–	7.25	m	**18.64**
height 500 mm–1.00 m	–	0.83	16.59	–	12.37	m	**28.96**
add to above for curved radius 6 m	–	0.03	0.60	–	0.33	m	**0.93**
Reinforcement; fabric; BS 4449; lapped; in roads, footpaths or pavings							
Ref A142 (2.22 kg/m²)							
400 mm minimum laps	1.96	0.14	2.94	–	2.02	m²	**4.96**
Ref A193 (3.02 kg/m²)							
400 mm minimum laps	–	0.14	2.94	–	2.74	m²	**5.68**
Formed joints; Fosroc Expandite Flexcell impregnated joint filler or other equal and approved							
Width not exceeding 150 mm							
12.50 mm thick	–	0.14	2.80	–	2.91	m	**5.71**
25 mm thick	–	0.19	3.80	–	3.99	m	**7.79**
Width 150–300 mm							
12.50 mm thick	–	0.19	3.80	–	5.30	m	**9.10**
25 mm thick	–	0.19	3.80	–	7.37	m	**11.17**
Width 300–450 mm							
12.50 mm thick	–	0.23	4.59	–	7.95	m	**12.54**
25 mm thick	–	0.23	4.59	–	11.06	m	**15.65**

35 SITE WORK

Item	PC £	Labour hours	Labour £	Plant £	Material £	Unit	Total rate £
35.02 IN SITU CONCRETE ROADS AND PAVINGS – cont							
Sealants; Fosroc Expandite Pliastic N2 hot poured rubberized bituminous compound or other equal and approved							
Width 25 mm							
25 mm depth	–	0.20	4.00	–	2.15	m	**6.15**
Concrete sundries							
Treating surfaces							
unset concrete; grading to cambers; tamping with a 75 mm thick steel shod tamper	–	0.23	3.88	–	–	m²	**3.88**
Sundries							
Line marking							
width not exceeding 300 mm; NB minimum charge usually applies	–	0.04	0.71	–	0.27	m	**0.98**
35.03 COATED MACADAM, ASPHALT ROADS AND PAVINGS							
NOTE: The prices for all bitumen macadam and hot rolled asphalt materials are for individual courses to roads and footpaths and need combining to arrive at complete specifications and costs for full construction. Costs include for work to falls, crossfalls or slopes not exceeding 15° from horizontal; for laying on prepared bases (prices not included) and for rolling with an appropriate roller. The following rates are based on black bitumen macadam. Red bitumen macadam rates are approximately 50% dearer. PSV is Polished Stone Value.							
Dense bitumen macadam base course; bitumen penetration 100/125							
Carriageway, hardshoulder and hardstrip							
100 mm thick; one coat; with 0/32 mm aggregate size	–	–	–	–	–	m²	**22.29**
200 mm thick; one coat; with 0/32 mm aggregate size	–	–	–	–	–	m²	**39.25**
extra for increase thickness in 10 mm increments	–	–	–	–	–	m²	**1.60**
Hot rolled asphalt base course							
Carriageway, hardshoulder and hardstrip							
150 mm thick; one coat; 60% 0/32 mm aggregate size; to column 2/5	–	–	–	–	–	m²	**37.07**
200 mm thick; one coat; 60% 0/32 mm aggregate size; to column 2/5	–	–	–	–	–	m²	**49.30**
extra for increase thickness in 10 mm increments	–	–	–	–	–	m²	**2.04**

35 SITE WORK

Item	PC £	Labour hours	Labour £	Plant £	Material £	Unit	Total rate £
Dense bitumen macadam binder course; bitumen penetration 100/125							
Carriageway, hardsholder and hardstrip							
60 mm thick; one coat; with 0/32 mm aggregate size	–	–	–	–	–	m²	**13.62**
60 mm thick; one coat; with 0/32 mm aggregate size; to clause 6.5	–	–	–	–	–	m²	**13.75**
extra for increase thickness in 10 mm increments	–	–	–	–	–	m²	**1.82**
Hot rolled asphalt binder course							
Carriageway, hardshoulder and hardstrip							
40 mm thick; one coat; 50% 0/14 mm aggregate size; to column 2/2; 55 PSV	–	–	–	–	–	m²	**13.01**
60 mm thick; one coat; 50% 0/14 mm aggregate size; to column 2/2	–	–	–	–	–	m²	**14.98**
60 mm thick; one coat; 50% 0/20 mm aggregate size; to column 2/3	–	–	–	–	–	m²	**14.72**
60 mm thick; one coat; 60% 0/32 mm aggregate size; to column 2/5	–	–	–	–	–	m²	**13.97**
100 mm thick; one coat; 60% 0/32 mm aggregate size; to column 2/5	–	–	–	–	–	m²	**22.55**
extra for increase thickness in 10 mm increments	–	–	–	–	–	m²	**2.70**
Macadam surface course; bitumen penetration 100/125							
Carriageway, hardshoulder and hardstrip							
30 mm thick; one coat; medium graded with 0/6 mm nominal aggregate binder	–	–	–	–	–	m²	**10.52**
40 mm thick; one coat; close graded with 0/14 mm nominal aggregate binder; to clause 7.3	–	–	–	–	–	m²	**9.65**
40 mm thick; one coat; close graded with 0/10 mm nominal aggregate binder; to clause 7.4	–	–	–	–	–	m²	**10.52**
extra over above items for increase/reduction in 10 mm thick increments	–	–	–	–	–	m²	**1.90**
extra over above items for coarse aggregate 60–64 PSV	–	–	–	–	–	m²	**1.91**
extra over above items for coarse aggregate 65–67 PSV	–	–	–	–	–	m²	**2.09**
extra over above items for coarse aggregate 68 PSV	–	–	–	–	–	m²	**2.78**
Hot rolled asphalt surface course; bitumen penetration 40/60							
Carriageway, hardshoulder and hardstrip							
40 mm thick; one coat; 30% mix 0/10 mm aggregate size; to column 3/2; with 20 mm pre-coated chippings 60–64 PSV	–	–	–	–	–	m²	**13.00**
40 mm thick; one coat; 30% mix 0/10 mm aggregate size; to column 3/2; with 14 mm pre-coated chippings 60–64 PSV	–	–	–	–	–	m²	**13.09**
extra over above items for increase/reduction in 10 mm increments	–	–	–	–	–	m²	**2.17**

35 SITE WORK

Item	PC £	Labour hours	Labour £	Plant £	Material £	Unit	Total rate £
35.03 COATED MACADAM, ASPHALT ROADS AND PAVINGS – cont							
Hot rolled asphalt surface course – cont							
Carriageway, hardshoulder and hardstrip – cont							
extra over above items for chippings with 65–67 PSV	–	–	–	–	–	m²	0.13
extra over above items for chippings with 68 PSV	–	–	–	–	–	m²	0.19
extra over above items for 6–10 KN High Traffic Flows	–	–	–	–	–	m²	0.93
Stone mastic asphalt surface course							
Carriageway, hardshoulder and hardstrip							
35 mm thick; one coat; with 0/14 mm nominal aggregate size; 55 PSV	–	–	–	–	–	m²	11.95
35 mm thick; one coat; with 0/10 mm nominal aggregate size; 55 PSV	–	–	–	–	–	m²	11.95
extra for increase thickness in 10 mm increments	–	–	–	–	–	m²	2.68
Thin surface course with 60 PSV							
Carriageway, hardshoulder and hardstrip							
35 mm thick; one coat; with 0/10 mm nominal aggregate size	–	–	–	–	–	m²	11.95
extra for increase thickness in 10 mm increments	–	–	–	–	–	m²	1.35
extra over above items for coarse aggregate 60–64 PSV	–	–	–	–	–	m²	0.32
extra over above items for coarse aggregate 65–67 PSV	–	–	–	–	–	m²	0.32
extra over above items for coarse aggregate 68 PSV	–	–	–	–	–	m²	0.62
Regulating courses							
Carriageway, hardshoulder and hardstrip							
Dense Bitumen Macadam; bitumen penetration 100/125; with 0/20 mm nominal aggregate regulating course	–	–	–	–	–	tonne	102.21
Hot rolled asphalt; 50% 0/20 mm aggregate size	–	–	–	–	–	tonne	113.02
Stone mastic asphalt; 0/6 mm aggregate	–	–	–	–	–	tonne	144.29
Bitumen emulsion tack coats							
Carriageway, hardshoulder and hardstrip							
K1–40; applied 0.35–0.45 l/m²	–	–	–	–	–	m²	0.19
K1–70; applied 0.35–0.45 l/m²	–	–	–	–	–	m²	0.31
Sundries							
Line marking							
width not exceeding 300 mm; NB minimum charge usually applies	–	0.04	0.71	–	0.27	m	0.98

35 SITE WORK

Item	PC £	Labour hours	Labour £	Plant £	Material £	Unit	Total rate £
35.04 GRAVEL PAVINGS							
Two coat gravel paving; level and to falls; first layer course clinker aggregate and wearing layer fine gravel aggregate							
Pavings; over 300 mm wide							
50 mm thick	–	0.07	1.74	0.08	2.76	m²	**4.58**
63 mm thick	–	0.09	2.25	0.10	3.00	m²	**5.35**
Resin bonded gravel paving; level and to falls							
Pavings; over 300 mm wide							
50 mm thick	–	–	–	–	–	m²	**52.36**
35.05 SLAB, BRICK/BLOCK SETTS AND COBBLE PAVINGS							
Artificial stone paving; Charcon's Moordale Textured or other equal and approved; to falls or crossfalls; bedding 25 mm thick in cement mortar (1:3); staggered joints; jointing in coloured cement mortar (1:3), brushed in; to sand base							
Pavings; over 300 mm wide							
600 mm × 600 mm × 50 mm thick; natural	14.88	0.39	9.72	–	21.92	m²	**31.64**
Brick paviors; 215 mm × 103 mm × 65 mm rough stock bricks; to falls or crossfalls; bedding on 50 mm sharp sand; kiln dried sand to joints							
Pavings; over 300 mm wide; straight joints both ways							
bricks laid flat (PC £ per 1000)	450.00	1.50	33.62	–	22.73	m²	**56.35**
bricks laid on edge	–	2.50	56.03	–	32.70	m²	**88.73**
Pavings; over 300 mm wide; laid to herringbone pattern							
bricks laid flat bricks laid flat (PC £ per 1000)	450.00	0.93	20.85	–	22.73	m²	**43.58**
bricks laid on edge	–	1.30	29.14	–	32.70	m²	**61.84**
Add or deduct for variation of £10.00/1000 in PC of brick paviors							
bricks laid flat bricks laid flat (PC £ per 1000)	–	–	–	–	0.46	m²	**0.46**
bricks laid on edge	–	–	–	–	0.70	m²	**0.70**
River washed cobble paving; 50 mm–75 mm; to falls or crossfalls; bedding 13 mm thick in cement mortar (1:3); jointing to a height of two thirds of cobbles in dry mortar (1:3); tightly butted, washed and brushed; to concrete							
Pavings; over 300 mm wide							
regular (PC £ per tonne)	143.00	3.70	92.23	–	38.71	m²	**130.94**
laid to pattern	–	4.63	115.41	–	38.71	m²	**154.12**

Prices for Measured Works

35 SITE WORK

Item	PC £	Labour hours	Labour £	Plant £	Material £	Unit	Total rate £
35.05 SLAB, BRICK/BLOCK SETTS AND COBBLE PAVINGS – cont							
Concrete BSS paving flags; to falls or crossfalls; bedding 25 mm thick in cement and sand mortar (1:4); butt joints straight both ways; jointing in cement and sand (1:3); brushed in; to sand base							
Pavings; over 300 mm wide							
450 mm × 600 mm × 50 mm thick; grey	11.32	0.42	10.46	–	15.54	m²	**26.00**
450 mm × 600 mm × 60 mm thick; coloured	14.98	0.42	10.46	–	19.50	m²	**29.96**
600 mm × 600 mm × 50 mm thick; grey	11.25	0.39	9.72	–	15.49	m²	**25.21**
600 mm × 600 mm × 50 mm thick; coloured	14.62	0.39	9.72	–	19.15	m²	**28.87**
750 mm × 600 mm × 50 mm thick; grey	10.70	0.36	8.97	–	14.87	m²	**23.84**
750 mm × 600 mm × 50 mm thick; coloured	14.00	0.36	8.97	–	18.45	m²	**27.42**
900 mm × 600 mm × 50 mm thick; grey	9.50	0.33	8.23	–	13.58	m²	**21.81**
900 mm × 600 mm × 50 mm thick; coloured	12.42	0.33	8.23	–	16.73	m²	**24.96**
Blister Tactile paving flags; to falls or crossfalls; bedding 25 mm thick in cement and sand mortar (1:4); butt joints straight both ways; jointing in cement and sand (1:3); brushed in; to sand base							
Pavings; over 300 mm wide							
400 mm × 400 mm × 50 mm thick; buff	64.80	0.45	11.22	–	73.38	m²	**84.60**
400 mm × 400 mm × 60 mm thick; buff	58.95	0.45	11.22	–	67.05	m²	**78.27**
450 mm × 450 mm × 50 mm thick; buff	49.50	0.45	11.22	–	56.83	m²	**68.05**
450 mm × 450 mm × 70 mm thick; buff	58.50	0.45	11.22	–	66.57	m²	**77.79**
Concrete rectangular paving blocks; to falls or crossfalls; herringbone pattern; on prepared base (not included here); bedding on 50 mm thick dry sharp sand; filling joints with sharp sand brushed in							
Pavings; Keyblock or other equal and approved; over 300 mm wide; straight joints both ways							
200 mm × 100 mm × 60 mm thick; grey	17.82	1.25	31.16	–	22.24	m²	**53.40**
200 mm × 100 mm × 60 mm thick; coloured	17.82	1.25	31.16	–	22.24	m²	**53.40**
200 mm × 100 mm × 60 mm thick; brindle	17.82	1.25	31.16	–	22.24	m²	**53.40**
200 mm × 100 mm × 80 mm thick; grey	18.00	1.35	33.65	–	22.67	m²	**56.32**
200 mm × 100 mm × 80 mm thick; coloured	18.00	1.35	33.65	–	22.67	m²	**56.32**
200 mm × 100 mm × 80 mm thick; brindle	18.00	1.35	33.65	–	22.67	m²	**56.32**
Pavings; Cemex or other equal; over 300 mm wide; laid to herringbone pattern							
200 mm × 100 mm × 60 mm thick; grey	17.82	1.35	33.65	–	22.24	m²	**55.89**
200 mm × 100 mm × 60 mm thick; coloured	17.82	1.35	33.65	–	22.24	m²	**55.89**
200 mm × 100 mm × 80 mm thick; grey	18.00	1.35	33.65	–	22.67	m²	**56.32**
200 mm × 100 mm × 80 mm thick; coloured	18.00	1.35	33.65	–	22.67	m²	**56.32**
Extra for two row boundary edging to herringbone pavings; 200 mm wide; including a 150 mm high in situ concrete mix C10–40 mm aggregate haunching to one side; blocks laid breaking joint							
200 mm × 100 mm × 60 mm; coloured	–	0.28	6.98	–	3.58	m	**10.56**
200 mm × 100 mm × 80 mm; coloured	–	0.28	6.98	–	3.59	m	**10.57**

35 SITE WORK

Item	PC £	Labour hours	Labour £	Plant £	Material £	Unit	Total rate £
Pavings; Europa or other equal; over 300 mm wide; straight joints both ways							
200 mm × 100 mm × 60 mm thick; grey	17.95	1.25	31.16	–	22.38	m²	**53.54**
200 mm × 100 mm × 60 mm thick; coloured	19.75	1.25	31.16	–	24.33	m²	**55.49**
200 mm × 100 mm × 80 mm thick; grey	21.55	1.35	33.65	–	26.51	m²	**60.16**
200 mm × 100 mm × 80 mm thick; coloured	23.34	1.35	33.65	–	28.45	m²	**62.10**
Pavings; Metropolitan or other equal; over 300 mm wide; straight joints both ways							
200 mm × 100 mm × 80 mm thick; grey	20.34	1.35	33.65	–	25.20	m²	**58.85**
200 mm × 100 mm × 80 mm thick; coloured	20.34	1.35	33.65	–	25.20	m²	**58.85**
Concrete rectangular paving blocks; to falls or crossfalls; 6 mm wide joints; symmetrical layout; bedding in 15 mm semi-dry cement mortar (1:4); jointing and pointing in cement and sand (1:4); on concrete base							
Pavings; Trafica or other equal and approved; over 300 mm wide							
400 mm × 400 mm × 65 mm; Saxon textured; natural	69.59	–	–	–	–	m²	–
400 mm × 400 mm × 65 mm; Saxon textured; buff	57.85	0.44	10.97	–	65.86	m²	**76.83**
400 mm × 400 mm × 65 mm; Perfecta; natural	69.59	0.44	10.97	–	74.97	m²	**85.94**
400 mm × 400 mm × 65 mm; Perfecta; buff	57.85	0.44	10.97	–	65.86	m²	**76.83**
450 mm × 450 mm × 70 mm; Saxon textured; natural	53.51	0.43	10.72	–	61.16	m²	**71.88**
450 mm × 450 mm × 70 mm; Saxon textured; buff	46.71	0.43	10.72	–	53.81	m²	**64.53**
450 mm × 450 mm × 70 mm; Perfecta; natural	53.51	0.43	10.72	–	61.16	m²	**71.88**
450 mm × 450 mm × 70 mm; Perfecta; buff	46.71	0.43	10.72	–	53.81	m²	**64.53**
Blue Grey granite setts tactile pavings (blister or corduroy); 400 mm × 400 mm × 80 mm; standard B dressing; sandblasted top finish, tightly butted to falls or crossfalls; bedding 25 mm thick in cement mortar (1:3); filling joints with dry mortar (1:6); washed and brushed; on concrete base							
Pavings; over 300 mm wide							
straight joints	88.00	1.48	36.89	–	101.20	m²	**138.09**
York stone slab pavings; to falls or crossfalls; bedding 25 mm thick in cement: sand mortar (1:4); 5 mm wide joints; jointing in coloured cement mortar (1:3); brushed in; to sand base							
Pavings; over 300 mm wide							
50 mm thick; random rectangular pattern	101.64	0.69	15.46	–	111.67	m²	**127.13**
600 mm × 600 mm × 50 mm thick	96.80	0.39	8.74	–	106.56	m²	**115.30**
600 mm × 900 mm × 50 mm thick	96.80	0.33	7.40	–	106.56	m²	**113.96**

35 SITE WORK

Item	PC £	Labour hours	Labour £	Plant £	Material £	Unit	Total rate £
35.05 SLAB, BRICK/BLOCK SETTS AND COBBLE PAVINGS – cont							
Granite setts; 200 mm × 100 mm × 100 mm; standard C dressing; tightly butted to falls or crossfalls; bedding 25 mm thick in cement mortar (1:3); filling joints with dry mortar (1:6); washed and brushed; on concrete base							
Pavings; over 300 mm wide							
straight joints (PC £ per m²)	58.58	1.48	36.89	–	70.14	m²	**107.03**
laid to pattern	58.58	1.85	46.11	–	70.14	m²	**116.25**
Boundary edging							
two rows of granite setts as boundary edging; 200 mm wide; including a 150 mm high ready mixed designated concrete C10–40 mm aggregate; haunching to one side; blocks laid breaking joint	–	0.65	16.20	–	15.88	m	**32.08**
Blue Grey granite setts paving bands; 100 mm × 100 mm × 100 mm; standard C dressing; flame textured top, tightly butted to falls or crossfalls; bedding 25 mm thick in cement mortar (1:3); filling joints with dry mortar (1:6); washed and brushed; on concrete base							
Pavings; over 300 mm wide							
straight joints	55.55	1.48	36.89	–	66.94	m²	**103.83**
laid to pattern	55.55	1.85	46.11	–	66.94	m²	**113.05**
Boundary edging							
two rows of granite setts as boundary edging; 200 mm wide; including a 150 mm high ready mixed designated concrete C10–40 mm aggregate; haunching to one side; blocks laid breaking joint	–	0.65	16.20	–	15.53	m	**31.73**
Black Basalt setts paving bands; 100 mm × 100 mm × 100 mm; standard C dressing; flame textured top, tightly butted to falls or crossfalls; bedding 25 mm thick in cement mortar (1:3); filling joints with dry mortar (1:6); washed and brushed; on concrete base							
Pavings; over 300 mm wide							
straight joints	82.35	1.48	36.89	–	95.23	m²	**132.12**
laid to pattern	82.35	1.85	46.11	–	95.23	m²	**141.34**
Boundary edging							
two rows of granite setts as boundary edging; 200 mm wide; including a 150 mm high ready mixed designated concrete C10–40 mm aggregate; haunching to one side; blocks laid breaking joint	–	0.65	16.20	–	21.33	m	**37.53**

35 SITE WORK

Item	PC £	Labour hours	Labour £	Plant £	Material £	Unit	Total rate £
35.06 EXTERNAL UNDERGROUND SERVICE RUNS							
Excavating trenches; by machine; grading bottoms; earthwork support; filling with excavated material and compacting; disposal of surplus soil on site; spreading on site average 50 m from excavations							
Services not exceeding 200 mm nominal size							
average depth of run not exceeding 0.50 m	–	0.28	4.04	1.50	–	m	**5.54**
average depth of run not exceeding 0.75 m	–	0.37	5.34	2.48	–	m	**7.82**
average depth of run not exceeding 1.00 m	–	0.79	11.38	3.23	1.66	m	**16.27**
average depth of run not exceeding 1.25 m	–	1.16	16.72	4.28	2.46	m	**23.46**
average depth of run not exceeding 1.50 m	–	1.48	21.32	5.79	3.00	m	**30.11**
average depth of run not exceeding 1.75 m	–	1.85	26.66	7.52	3.68	m	**37.86**
average depth of run not exceeding 2.00 m	–	2.13	30.68	8.80	4.01	m	**43.49**
Excavating trenches; by hand; grading bottoms; earthwork support; filling with excavated material and compacting; disposal; of surplus soil on site; spreading on site average 50 m from excavations							
Services not exceeding 200 mm nominal size							
average depth of run not exceeding 0.50 m	–	0.93	13.40	–	–	m	**13.40**
average depth of run not exceeding 0.75 m	–	1.39	20.02	–	–	m	**20.02**
average depth of run not exceeding 1.00 m	–	2.04	29.39	–	1.79	m	**31.18**
average depth of run not exceeding 1.25 m	–	2.87	41.34	–	2.46	m	**43.80**
average depth of run not exceeding 1.50 m	–	3.93	56.62	–	3.00	m	**59.62**
average depth of run not exceeding 1.75 m	–	5.18	74.63	–	3.63	m	**78.26**
average depth of run not exceeding 2.00 m	–	5.92	85.29	–	3.99	m	**89.28**
Stop cock pits, valve chambers and the like; excavating; half brick thick walls in common bricks in cement mortar (1:3); on in situ concrete designated mix C20–20 mm aggregate bed; 100 mm thick							
Pits							
100 mm × 100 mm × 750 mm deep; internal holes for one small pipe; polypropylene hinged box cover; bedding in cement mortar (1:3)	–	3.89	87.18	–	75.79	nr	**162.97**

35 SITE WORK

Item	PC £	Labour hours	Labour £	Plant £	Material £	Unit	Total rate £
35.06 EXTERNAL UNDERGROUND SERVICE RUNS – cont							
Communication boxes							
FP McCann Ltd precast reinforced concrete BT approved communication boxes cable and junction protection box. Medium duty cast iron cover. Excavations and bedding excluded							
C2 communication box 1200 mm long × 600 mm wide × 895 mm high	277.82	0.50	7.15	0.83	391.71	nr	**399.69**
DP communication box 910 mm long × 890 mm wide × 1000 mm high	346.43	0.50	7.15	0.83	462.38	nr	**470.36**
J4 communication box 910 mm long × 440 mm wide × 800 mm high	123.72	0.50	7.15	0.83	233.00	nr	**240.98**
MCX communication box 1300 mm long × 850 mm wide × 900 mm high	363.30	0.60	8.58	0.83	558.55	nr	**567.96**
MCX communication box riser piece 1300 mm long × 850 mm wide × 300 mm high	105.72	0.45	6.44	0.42	108.89	nr	**115.75**

36 FENCING

Item	PC £	Labour hours	Labour £	Plant £	Material £	Unit	Total rate £
36.01 FENCING							
NOTE: The prices for all fencing include for setting posts in position, to a depth of 0.60 m for fences not exceeding 1.40 m high and of 0.76 m for fences over 1.40 m high. The prices allow for excavating post holes; filling to within 150 mm of ground level with concrete and all necessary backfilling.							
For more examples please consult *Spon's External Works and Landscape Price Book*.							
Fencing; strained wire							
Strained wire fencing; BS 1722 Part 3; 4 mm dia. galvanized mild steel plain wire threaded through posts and strained with eye bolts							
900 mm high; three line; concrete posts at 2.75 m centres	–	–	–	–	–	m	21.07
end concrete straining post; one strut	–	–	–	–	–	nr	51.23
angle concrete straining post; two struts	–	–	–	–	–	nr	59.50
1.07 m high; six line; concrete posts at 2.75 m centres	–	–	–	–	–	m	21.92
end concrete straining post; one strut	–	–	–	–	–	nr	57.62
angle concrete straining post; two struts	–	–	–	–	–	nr	65.89
1.20 m high; six line; concrete posts at 2.75 m centres	–	–	–	–	–	m	22.05
end concrete straining post; one strut	–	–	–	–	–	nr	59.25
angle concrete straining post; two struts	–	–	–	–	–	nr	67.50
1.40 m high; eight line; concrete posts at 2.75 m centres	–	–	–	–	–	m	22.64
end concrete straining post; one strut	–	–	–	–	–	nr	60.54
angle concrete straining post; two struts	–	–	–	–	–	nr	68.78
Chain link fencing; BS 1722 Part 1; 3 mm dia. galvanized mild steel wire; 50 mm mesh; galvanized mild steel tying and line wire; three line wires threaded through posts and strained with eye bolts and winding brackets							
900 mm high; galvanized mild steel angle posts at 3.00 m centres	–	–	–	–	–	m	29.65
end steel straining post; one strut	–	–	–	–	–	nr	85.40
angle steel straining post; two struts	–	–	–	–	–	nr	98.52
900 mm high; concrete posts at 3.00 m centres	–	–	–	–	–	m	21.48
end concrete straining post; one strut	–	–	–	–	–	nr	45.88
angle concrete straining post; two struts	–	–	–	–	–	nr	54.15
1.20 m high; galvanized mild steel angle posts at 3.00 m centres	–	–	–	–	–	m	21.86
end steel straining post; one strut	–	–	–	–	–	nr	91.14
angle steel straining post; two struts	–	–	–	–	–	nr	116.73
1.20 m high; concrete posts at 3.00 m centres	–	–	–	–	–	m	20.94
end concrete straining post; one strut	–	–	–	–	–	nr	52.52
angle concrete straining post; two struts	–	–	–	–	–	nr	62.05
1.80 m high; galvanized mild steel angle posts at 3.00 m centres	–	–	–	–	–	m	24.56
end steel straining post; one strut	–	–	–	–	–	nr	92.71
angle steel straining post; two struts	–	–	–	–	–	nr	115.32

36 FENCING

Item	PC £	Labour hours	Labour £	Plant £	Material £	Unit	Total rate £
36.01 FENCING – cont							
Fencing – cont							
Chain link fencing – cont							
1.80 m high; concrete posts at 3.00 m centres	–	–	–	–	–	m	**28.37**
end concrete straining post; one strut	–	–	–	–	–	nr	**73.44**
angle concrete straining post; two struts	–	–	–	–	–	nr	**86.64**
Chain link fencing for tennis courts; BS 1722 Part 13; 2.5 dia. galvanized mild wire; 45 mm mesh; line and tying wires threaded through 45 mm × 45 mm × 5 mm galvanized mild steel angle standards, posts and struts; 60 mm × 60 mm × 6 mm straining posts and gate posts							
fencing to tennis court							
36.00 m × 18.00 m × 2.745 m high; standards at 3.00 m centres	–	–	–	–	–	nr	**2878.23**
fencing to tennis court 36.00 m × 18.00 m × 3.66 m high; standards at 2.50 mm centres	–	–	–	–	–	nr	**4409.19**
Fencing; timber							
Cleft chestnut pale fencing; BS 1722 Part 4; pales spaced 51 mm apart; on two lines of galvanized wire; 64 mm dia. posts; 76 mm × 51 mm struts							
900 mm high; posts at 2.50 m centres	–	–	–	–	–	m	**12.25**
straining post; one strut	–	–	–	–	–	nr	**32.45**
corner straining post; two struts	–	–	–	–	–	nr	**32.45**
1.05 m high; posts at 2.50 m centres	–	–	–	–	–	m	**13.74**
straining post; one strut	–	–	–	–	–	nr	**32.70**
corner straining post; two struts	–	–	–	–	–	nr	**32.70**
Closeboarded fencing; BS 1722 Part 5; 76 mm × 38 mm softwood rails; 89 mm × 19 mm softwood pales lapped 13 mm; 152 mm × 25 mm softwood gravel boards; all softwood treated; posts at 3.00 m centres							
Fencing; two rail; concrete posts							
height 1.00 m	–	–	–	–	–	m	**37.23**
height 1.20 m	–	–	–	–	–	m	**37.65**
Fencing; three rail; concrete posts							
height 1.40 m	–	–	–	–	–	m	**41.34**
height 1.60 m	–	–	–	–	–	m	**41.34**
height 1.80 m	–	–	–	–	–	m	**42.86**
Fencing; concrete							
Precast concrete slab fencing; 305 mm × 38 mm × 1753 mm slabs; fitted into twice grooved concrete posts at 1.83 m centres							
height 1.50 m	–	–	–	–	–	m	**73.48**
height 1.80 m	–	–	–	–	–	m	**79.60**

36 FENCING

Item	PC £	Labour hours	Labour £	Plant £	Material £	Unit	Total rate £
Fencing; security							
Mild steel unclimbable fencing; in rivetted panels 2440 mm long; 44 mm × 13 mm flat section top and bottom rails; two 44 mm × 19 mm flat section standards; one with foot plate; and 38 mm × 13 mm raking stay with foot plate; 20 mm dia. pointed verticals at 120 mm centres; two 44 mm × 19 mm supports 760 mm long with ragged ends to bottom rail; the whole bolted together; coated with red oxide primer; setting standards and stays in ground at 2440 mm centres and supports at 815 mm centres							
height 1.67 m	–	–	–	–	–	m	**137.18**
height 2.13 m	–	–	–	–	–	m	**159.22**
Pair of gates and gate posts, to match mild steel unclimbable fencing; with flat section framing, braces, etc., complete with locking bar, lock, handles, drop bolt, gate stop and holding back catches; two 102 mm × 102 mm hollow section gate posts with cap and foot plates							
2.44 m × 1.67 m	–	–	–	–	–	nr	**1181.91**
2.44 m × 2.13 m	–	–	–	–	–	nr	**1371.74**
4.88 m × 1.67 m	–	–	–	–	–	nr	**1837.17**
4.88 m × 2.13 m	–	–	–	–	–	nr	**2314.83**
PVC coated, galvanized mild steel high security fencing; Sentinal Sterling fencing or other equal 50 mm × 50 mm mesh; 3/3.50 mm gauge wire; barbed edge – 1; Sentinal Bi-steel colour coated posts or other equal and approved at 2.44 m centres							
1.80 m	–	0.93	13.40	–	37.14	m	**50.54**
2.10 m	–	1.16	16.72	–	40.17	m	**56.89**
Gates							
Pair of gates and gate posts; gates to match galvanized chain link fencing, with angle framing, braces, etc., complete with hinges, locking bar, lock and bolts; two 100 mm × 100 mm angle section gate posts; each with one strut							
2.44 m × 0.90 m	–	–	–	–	–	nr	**720.28**
2.44 m × 1.20 m	–	–	–	–	–	nr	**743.35**
2.44 m × 1.80 m	–	–	–	–	–	nr	**801.88**

37 SOFT LANDSCAPING

Item	PC £	Labour hours	Labour £	Plant £	Material £	Unit	Total rate £
37.01 SEEDING AND TURFING							
For more examples and depth of information please consult *Spon's External Works and Landscape Price Book.*							
Top soil							
By machine							
Selected from spoil heaps not exceeding 50 m; grading; prepared for turfing or seeding; to general surfaces							
average 100 mm thick	–	–	0.07	1.52	–	m²	**1.59**
average 150 mm thick	–	0.01	0.08	2.03	–	m²	**2.11**
average 200 mm thick	–	0.01	0.11	3.55	–	m²	**3.66**
Selected from spoil heaps; grading; prepared for turfing or seeding; to cuttings or embankments							
average 100 mm thick	–	–	0.07	1.73	–	m²	**1.80**
average 150 mm thick	–	0.01	0.09	2.34	–	m²	**2.43**
average 200 mm thick	–	0.01	0.12	4.07	–	m²	**4.19**
By hand							
Selected from spoil heaps not exceeding 20 m; grading; prepared for turfing or seeding; to general surfaces							
average 100 mm thick	–	0.40	5.77	–	–	m²	**5.77**
average 150 mm thick	–	0.50	7.20	–	–	m²	**7.20**
average 200 mm thick	–	0.60	8.64	–	–	m²	**8.64**
Selected from spoil heaps; grading; prepared for turfing or seeding; to cuttings or embankments							
average 100 mm thick	–	0.51	7.29	–	–	m²	**7.29**
average 150 mm thick	–	0.63	9.12	–	–	m²	**9.12**
average 200 mm thick	–	0.76	10.94	–	–	m²	**10.94**
Imported top soil, planting quality							
By machine							
Grading; prepared for turfing or seeding; to general surfaces							
average 100 mm thick	–	–	0.07	1.52	3.34	m²	**4.93**
average 150 mm thick	–	0.01	0.08	2.03	5.01	m²	**7.12**
average 200 mm thick	–	0.01	0.11	3.55	6.12	m²	**9.78**
Grading; preparing for turfing or seeding; to cuttings or embankments							
average 100 mm thick	–	–	0.07	1.73	4.08	m²	**5.88**
average 150 mm thick	–	0.01	0.09	2.34	6.67	m²	**9.10**
average 200 mm thick	–	0.01	0.12	4.07	8.16	m²	**12.35**
By hand							
Grading; prepared for turfing or seeding; to general surfaces							
average 100 mm thick	–	0.51	7.29	–	4.08	m²	**11.37**
average 150 mm thick	–	0.63	9.12	–	6.67	m²	**15.79**
average 200 mm thick	–	0.76	10.94	–	8.16	m²	**19.10**

37 SOFT LANDSCAPING

Item	PC £	Labour hours	Labour £	Plant £	Material £	Unit	Total rate £
Fertilizer							
Fertilizer 0.07 kg/m²; raking in							
general surfaces (PC £ per 25 kg)	30.36	0.03	0.43	–	0.10	m²	**0.53**
Selected grass seed							
Grass seed; sowing at a rate of 40 g/m² two							
applications; raking in							
general surfaces (PC £ per kg)	3.15	0.17	2.45	–	0.17	m²	**2.62**
cuttings or embankments	–	0.20	2.88	–	0.17	m²	**3.05**
Turfing							
Imported turf; cultivated							
general surfaces	2.83	0.10	1.44	–	2.91	m²	**4.35**
cuttings or embankments; shallow	2.83	0.15	2.16	–	2.91	m²	**5.07**
cuttings or embankments; steep; pegged	2.83	0.28	4.04	–	2.91	m²	**6.95**
Preserved turf from stack on site; lay only							
general surfaces	–	0.19	2.74	–	–	m²	**2.74**
cuttings or embankments; shallow	–	0.20	2.88	–	–	m²	**2.88**
cuttings or embankments; steep; pegged	–	0.28	4.04	–	–	m²	**4.04**
37.02 PLANTING							
Hedge plants							
instant hedge planting; delivered in 2.5 m lengths;							
planting and backfilling with excavted material;							
trench not included; hedge height; hedge 1.0 m							
high × 300 mm wide	–	0.75	10.80	5.86	145.95	nr	**162.61**
Tree planting							
semi-mature or mature tree, 40–45 cm girth	674.73	8.00	115.26	87.97	694.97	nr	**898.20**
semi-mature or mature tree, 80–90 cm girth	5622.75	20.00	288.14	234.59	5791.43	nr	**6314.16**

39 ELECTRICAL SERVICES

Item	PC £	Labour hours	Labour £	Plant £	Material £	Unit	Total rate £
39.01 LIGHTNING PROTECTION							
For more examples and depth of information please consult *Spon's Mechanical and Electrical Services Price Book*.							
Lightning protection equipment							
Copper strip roof or down conductors fixed with bracket or saddle clips							
20 mm × 3 mm flat section	–	–	–	–	–	m	42.05
25 mm × 6 mm flat section	–	–	–	–	–	m	69.05
Joints in tapes	–	–	–	–	–	nr	18.16
Bonding connections to roof and structural metalwork	–	–	–	–	–	nr	86.36
Testing points	–	–	–	–	–	nr	77.72
Earth electrodes							
16 mm dia. driven copper electrodes in 1220 mm long sectional lengths; 2440 mm long overall	–	–	–	–	–	nr	227.20

41 BUILDER'S WORK IN CONNECTION WITH SERVICES

Item	PC £	Labour hours	Labour £	Plant £	Material £	Unit	Total rate £
41.01 HOLES, CHASES FOR SERVICES							
Electrical installations; cutting away for and making good after electrician; including cutting or leaving all holes, notches, mortices, sinkings and chases, in both the structure and its coverings, for the following electrical points							
Exposed installation							
lighting points	–	0.28	5.07	–	–	nr	**5.07**
socket outlet points	–	0.46	8.70	–	–	nr	**8.70**
fitting outlet points	–	0.46	8.70	–	–	nr	**8.70**
equipment points or control gear points	–	0.65	12.48	–	–	nr	**12.48**
Concealed installation							
lighting points	–	0.37	6.88	–	–	nr	**6.88**
socket outlet points	–	0.65	12.48	–	–	nr	**12.48**
fitting outlet points	–	0.65	12.48	–	–	nr	**12.48**
equipment points or control gear points	–	0.93	17.55	–	–	nr	**17.55**
Builders' work for other services installations							
Cutting chases in brickwork							
for one pipe; not exceeding 55 mm nominal size; vertical	–	0.37	5.29	–	–	m	**5.29**
for one pipe; 55 mm–110 mm nominal size; vertical	–	0.65	9.29	–	–	m	**9.29**
Cutting and pinning to brickwork or blockwork; ends of supports							
for pipes not exceeding 55 mm nominal size	–	0.19	4.25	–	–	nr	**4.25**
for cast iron pipes 55 mm–110 mm nominal size	–	0.31	6.95	–	–	nr	**6.95**
Cutting or forming holes for pipes or the like; not exceeding 55 mm nominal size; making good							
reinforced concrete; not exceeding 100 mm deep	–	0.75	12.67	0.84	–	nr	**13.51**
reinforced concrete; 100 mm–200 mm deep	–	1.15	19.44	1.30	–	nr	**20.74**
reinforced concrete; 200 mm–300 mm deep	–	1.50	25.35	1.69	–	nr	**27.04**
half brick thick	–	0.31	5.24	–	–	nr	**5.24**
one brick thick	–	0.51	8.62	–	–	nr	**8.62**
one and a half brick thick	–	0.83	14.03	–	–	nr	**14.03**
100 mm blockwork	–	0.28	4.73	–	–	nr	**4.73**
140 mm blockwork	–	0.37	6.25	–	–	nr	**6.25**
215 mm blockwork	–	0.46	7.78	–	–	nr	**7.78**
plasterboard partition or suspended ceiling	–	0.35	5.91	–	–	nr	**5.91**
Cutting or forming holes for pipes or the like; 55 mm–110 mm nominal size; making good							
reinforced concrete; not exceeding 100 mm deep	–	1.15	19.44	1.30	–	nr	**20.74**
reinforced concrete; 100 mm–200 mm deep	–	1.75	29.57	1.98	–	nr	**31.55**
reinforced concrete; 200 mm–300 mm deep	–	2.25	38.02	2.54	–	nr	**40.56**
half brick thick	–	0.37	6.25	–	–	nr	**6.25**
one brick thick	–	0.65	10.98	–	–	nr	**10.98**
one and a half brick thick	–	1.02	17.23	–	–	nr	**17.23**
100 mm blockwork	–	0.32	5.41	–	–	nr	**5.41**
140 mm blockwork	–	0.46	7.78	–	–	nr	**7.78**
215 mm blockwork	–	0.56	9.47	–	–	nr	**9.47**
plasterboard partition or suspended ceiling	–	0.40	6.76	–	–	nr	**6.76**

41 BUILDER'S WORK IN CONNECTION WITH SERVICES

Item	PC £	Labour hours	Labour £	Plant £	Material £	Unit	Total rate £
41.01 HOLES, CHASES FOR SERVICES – cont							
Builders' work for other services installations – cont							
Cutting or forming holes for pipes or the like; over 110 mm nominal size; making good							
reinforced concrete; not exceeding 100 mm deep	–	1.15	19.44	1.30	–	nr	**20.74**
reinforced concrete; 100 mm–200 mm deep	–	1.75	29.57	1.98	–	nr	**31.55**
reinforced concrete; 200 mm–300 mm deep	–	2.25	38.02	2.54	–	nr	**40.56**
half brick thick	–	0.46	7.78	–	–	nr	**7.78**
one brick thick	–	0.79	13.35	–	–	nr	**13.35**
one and a half brick thick	–	1.25	21.13	–	–	nr	**21.13**
100 mm blockwork	–	0.42	7.10	–	–	nr	**7.10**
140 mm blockwork	–	0.56	9.47	–	–	nr	**9.47**
215 mm blockwork	–	0.69	11.66	–	–	nr	**11.66**
plasterboard partition or suspended ceiling	–	0.45	7.60	–	–	nr	**7.60**
Add for making good fair face or facings one side							
pipe; not exceeding 55 mm nominal size	–	0.07	1.57	–	–	nr	**1.57**
pipe; 55 mm–110 mm nominal size	–	0.09	2.02	–	–	nr	**2.02**
pipe; over 110 mm nominal size	–	0.11	2.46	–	–	nr	**2.46**
Add for fixing sleeve (supply not included)							
for pipe; small	–	0.14	3.14	–	–	nr	**3.14**
for pipe; large	–	0.19	4.25	–	–	nr	**4.25**
for pipe; extra large	–	0.28	6.27	–	–	nr	**6.27**
Add for supplying and fixing two hour intumescent sleeve 300 mm long							
for 55 mm uPVC pipe	–	0.25	4.22	–	18.25	nr	**22.47**
for 110 mm uPVC pipe	–	0.28	4.73	–	22.65	nr	**27.38**
for 200 mm uPVC pipe	–	0.30	5.07	–	54.07	nr	**59.14**
Cutting or forming holes for ducts; girth not exceeding 1.00 m; making good							
half brick thick	–	0.56	9.47	–	–	nr	**9.47**
one brick thick	–	0.93	15.72	–	–	nr	**15.72**
one and a half brick thick	–	1.48	25.01	–	–	nr	**25.01**
100 mm blockwork	–	0.46	7.78	–	–	nr	**7.78**
140 mm blockwork	–	0.65	10.98	–	–	nr	**10.98**
215 mm blockwork	–	0.83	14.03	–	–	nr	**14.03**
plasterboard partition or suspended ceiling	–	0.65	10.98	–	–	nr	**10.98**
Cutting or forming holes for ducts; girth 1.00 m–2.00 m; making good							
half brick thick	–	0.65	10.98	–	–	nr	**10.98**
one brick thick	–	1.11	18.76	–	–	nr	**18.76**
one and a half brick thick	–	1.76	29.74	–	–	nr	**29.74**
100 mm blockwork	–	0.56	9.47	–	–	nr	**9.47**
140 mm blockwork	–	0.74	12.50	–	–	nr	**12.50**
215 mm blockwork	–	0.93	15.72	–	–	nr	**15.72**
plasterboard partition or suspended ceiling	–	0.75	12.67	–	–	nr	**12.67**
Cutting or forming holes for ducts; girth 2.00 m–3.00 m; making good							
half brick thick	–	1.02	17.23	–	–	nr	**17.23**
one brick thick	–	1.76	29.74	–	–	nr	**29.74**
one and a half brick thick	–	2.78	46.97	–	–	nr	**46.97**

41 BUILDER'S WORK IN CONNECTION WITH SERVICES

Item	PC £	Labour hours	Labour £	Plant £	Material £	Unit	Total rate £
100 mm blockwork	–	0.88	14.87	–	–	nr	**14.87**
140 mm blockwork	–	1.20	20.28	–	–	nr	**20.28**
215 mm blockwork	–	1.53	25.85	–	–	nr	**25.85**
plasterboard partition or suspended ceiling	–	1.00	16.89	–	–	nr	**16.89**
Cutting or forming holes for ducts; girth 3.00 m– 4.00 m; making good							
half brick thick	–	1.39	23.48	–	–	nr	**23.48**
one brick thick	–	2.31	39.03	–	–	nr	**39.03**
one and a half brick thick	–	3.70	62.52	–	–	nr	**62.52**
100 mm blockwork	–	1.02	17.23	–	–	nr	**17.23**
140 mm blockwork	–	1.39	23.48	–	–	nr	**23.48**
215 mm blockwork	–	1.76	29.74	–	–	nr	**29.74**
plasterboard partition or suspended ceiling	–	1.25	21.13	–	–	nr	**21.13**
Mortices in brickwork							
for expansion bolt	–	0.19	3.21	–	–	nr	**3.21**
for 20 mm dia. bolt; 75 mm deep	–	0.14	2.37	–	–	nr	**2.37**
for 20 mm dia. bolt; 150 mm deep	–	0.23	3.88	–	–	nr	**3.88**
Mortices in brickwork; grouting with cement mortar (1:1)							
75 mm × 75 mm × 200 mm deep	–	0.28	4.73	–	0.21	nr	**4.94**
75 mm × 75 mm × 300 mm deep	–	0.37	6.25	–	0.31	nr	**6.56**
Holes in softwood for pipes, bars, cables and the like							
12 mm thick	–	0.03	0.69	–	–	nr	**0.69**
25 mm thick	–	0.05	1.15	–	–	nr	**1.15**
50 mm thick	–	0.09	2.08	–	–	nr	**2.08**
100 mm thick	–	0.14	3.23	–	–	nr	**3.23**
Holes in hardwood for pipes, bars, cables and the like							
12 mm thick	–	0.05	1.15	–	–	nr	**1.15**
25 mm thick	–	0.08	1.85	–	–	nr	**1.85**
50 mm thick	–	0.14	3.23	–	–	nr	**3.23**
100 mm thick	–	0.20	4.62	–	–	nr	**4.62**
Diamond drilling for cutting holes and mortices in masonry or concrete							
Cutting holes and mortices in brickwork; per 25 mm depth							
25 mm dia.	–	–	–	–	–	nr	**2.13**
32 mm dia.	–	–	–	–	–	nr	**1.72**
52 mm dia.	–	–	–	–	–	nr	**2.06**
78 mm dia.	–	–	–	–	–	nr	**2.26**
107 mm dia.	–	–	–	–	–	nr	**2.38**
127 mm dia.	–	–	–	–	–	nr	**2.94**
152 mm dia.	–	–	–	–	–	nr	**3.45**
200 mm dia.	–	–	–	–	–	nr	**4.45**
250 mm dia.	–	–	–	–	–	nr	**6.71**
300 mm dia.	–	–	–	–	–	nr	**8.90**

Prices for Measured Works

41 BUILDER'S WORK IN CONNECTION WITH SERVICES

Item	PC £	Labour hours	Labour £	Plant £	Material £	Unit	Total rate £
41.01 HOLES, CHASES FOR SERVICES – cont							
Diamond drilling for cutting holes and mortices in masonry or concrete – cont							
Diamond chasing; per 25 × 25 mm section							
in facing or common brickwork	–	–	–	–	–	m	4.04
in semi-engineering brickwork	–	–	–	–	–	m	8.10
in engineering brickwork	–	–	–	–	–	m	11.29
in lightweight blockwork	–	–	–	–	–	m	3.17
in heavyweight blockwork	–	–	–	–	–	m	6.37
in render/screed	–	–	–	–	–	m	12.52
Forming boxes; 100 × 100 mm; per 25 mm depth							
in facing or common brickwork	–	–	–	–	–	nr	1.62
in semi-engineering brickwork	–	–	–	–	–	nr	3.24
in engineering brickwork	–	–	–	–	–	nr	4.51
in lightweight blockwork	–	–	–	–	–	nr	1.29
in heavyweight blockwork	–	–	–	–	–	nr	2.55
in render/screed	–	–	–	–	–	nr	5.02
Other items							
diamond track mount or ring sawing brickwork	–	–	–	–	–	m	7.97
diamond floor sawing asphalte	–	–	–	–	–	m	1.34
stitch drilling 107 mm dia. hole in brickwork	–	–	–	–	–	nr	1.72
41.02 INTERNAL FLOOR DUCTS FOR SERVICES							
Screed floor ducting; with side flanges; laid within floor screed; galvanized mild steel							
Floor ducting							
100 mm wide × 50 mm deep	13.07	0.19	4.40	–	13.46	m	17.86
bend	–	0.09	2.08	–	20.84	nr	22.92
tee section	–	0.09	2.08	–	20.84	nr	22.92
connector/stop end	–	0.09	2.08	–	2.40	nr	4.48
ply cover 15 mm/16 mm thick WBP exterior grade	–	0.09	2.08	–	2.99	m	5.07
100 mm wide × 70 mm deep	14.35	0.20	4.62	–	14.78	m	19.40
bend	–	0.09	2.08	–	20.84	nr	22.92
tee section	–	0.09	2.08	–	20.84	nr	22.92
connector/stop end	–	0.09	2.08	–	2.40	nr	4.48
ply cover 15 mm/16 mm thick WBP exterior grade	–	0.09	2.08	–	2.99	m	5.07
200 mm wide × 50 mm deep	18.06	0.19	4.40	–	18.60	m	23.00
bend	–	0.09	2.08	–	23.20	nr	25.28
tee section	–	0.09	2.08	–	23.20	nr	25.28
connector/stop end	–	0.09	2.08	–	2.40	nr	4.48
ply cover 15 mm/16 mm thick WBP exterior grade	–	0.09	2.08	–	5.42	m	7.50

41 BUILDER'S WORK IN CONNECTION WITH SERVICES

Item	PC £	Labour hours	Labour £	Plant £	Material £	Unit	Total rate £
41.03 EXTERNAL SERVICES							
Service runs							
Water main; all laid in trenches including excavation and backfill with type 1 hardcore. Surface finish not included							
MDPE pipe; 20 mm dia.	–	2.85	41.06	10.25	57.31	m	**108.62**
MDPE pipe; 25 mm dia.	–	2.85	41.06	10.25	57.30	m	**108.61**
MDPE pipe; 32 mm dia.	–	2.85	41.06	10.25	57.71	m	**109.02**
MDPE pipe; 50 mm dia.	–	2.85	41.06	10.25	59.51	m	**110.82**
Electric mains 25 mm 2 core armoured cable laid in service duct, including excavation and backfill with type 1 hardcore. Surface finish not included							
10 mm 2 core armoured cable	–	3.00	43.22	10.25	88.35	m	**141.82**
16 mm 3 core armoured cable	–	3.00	43.22	10.25	93.21	m	**146.68**
16 mm 5 core armoured cable	–	3.00	43.22	10.25	98.45	m	**151.92**

Leadership in the Construction Industry

George Ofori and Shamas-ur-Rehman Toor

This book presents a new framework for leadership in the construction industry which draws from the authentic leadership construct. The framework has three major themes: self-leadership, self-transcendent leadership, and sustainable leadership.

Despite its significance, leadership has not been given due importance in the construction industry as focus is placed on managerial functionalism. At the project level, even with the technological advances in the industry in recent years, construction is realized in the form of people undertaking distinct interdependent activities which require effective leadership. The industry faces many challenges including: demanding client requirements and project parameters; more stringent regulations, codes and systems; intense competition in the industry; and threats from disruptive enterprise. In such a complex environment, technology-driven and tool-based project and corporate management is insufficient. It must be complemented by a strategic, genuine, stakeholder-focused and ethical leadership.

Leadership in the Construction Industry is based on a study on authentic leadership and its development in Singapore. Leadership theories and concepts are reviewed; the importance of leadership in the construction industry is discussed; and the grounded theory approach which was applied in the study is explained. Many eminent construction professionals in Singapore were interviewed in the field study. Emerging from the experiences of the leaders documented in this book are three major themes: (1) self-leadership: how leaders engage in various self-related processes such as self-awareness, self-regulation, and role modelling. (2) self-transcendent leadership: how leaders go beyond leading themselves to leading others through servant leadership, shared leadership, spiritual leadership, and socially-responsible leadership; and, finally, (3) sustainable leadership or the strategies leaders employ to make the impact of their leadership lasting. A synthesis of these themes and their implications for leadership development is presented before the book concludes with some recommendations for current and aspiring leaders about how they can engage with them. This book is essential reading for all construction practitioners from all backgrounds; and researchers on leadership and management in construction.

March 2021: 364 pp
ISBN: 9780367482152

To Order
Tel:+44 (0) 1235 400524
Email: tandf@bookpoint.co.uk

For a complete listing of all our titles visit:
www.tandf.co.uk

Taylor & Francis
Taylor & Francis Group

PART 5

Fees for Professional Services

This part contains the following sections:

Metric Handbook: Planning and Design Data, 7th edition

Edited by Pamela Buxton

- Consistently updated since 2015 by expert authors in the field
- Significantly revised in reference to changing building types and construction standards
- New chapters added on data centres and logistics facilities
- Sustainable design integrated into chapters throughout
- Over 100,000 copies sold to successive generations of architects and designers
- This book belongs in every design office.

The *Metric Handbook* is the major handbook of planning and design data for architects and architecture students. Covering basic design data for all the major building types it is the ideal starting point for any project. For each building type, the book gives the basic design requirements and all the principal dimensional data, and succinct guidance on how to use the information and what regulations the designer needs to be aware of.

As well as buildings, the *Metric Handbook* deals with broader aspects of design such as materials, acoustics and lighting, and general design data on human dimensions and space requirements. The *Metric Handbook* is the unique reference for solving everyday planning problems.

November 2021: 880 pp
ISBN: 9780367511395

To Order
Tel:+44 (0) 1235 400524
Email: tandf@bookpoint.co.uk

For a complete listing of all our titles visit:
www.tandf.co.uk

Taylor & Francis
Taylor & Francis Group

QUANTITY SURVEYORS' FEES

Guidance on basic quantity surveying services is set out in the RICS Standard Form of Consultant's Appointment for Quantity Surveyors. Services are separated into core services for a variety of contracts and supplementary services and it is advisable to refer to this guidance. Copies can be obtained from the Royal Institution of Chartered Surveyors (RICS) at www.ricsbooks.com.

Quantity Surveying Services

Preparation:

- liaising with clients and the professional team
- advice on cost
- preparation of initial budget/cost plan/cash flow forecasts

Design:

- prepare and maintain cost plan
- advise design team on impact of design development on cost

Pre-construction:

- liaise with professional team
- advise on procurement strategy
- liaise with client's legal advisors on contract matters
- prepare tender documents
- define prospective tenderers
- obtain tenders/check tenders/prepare recommendation for client
- maintain and develop cost plan

Construction:

- visit the site
- prepare interim valuations
- advise on the cost of variations
- agree the cost of claims
- advise on contractual matters

Supplementary services may include:

- preparation of mechanical and electrical tender documentation
- preparation of cost analyses
- advice on insurance claims
- facilitate value management exercises
- prepare life cycle calculations/sustainability
- capital allowance advice/VAT
- attend adjudication/mediation proceedings

Fee Guide

These do not represent minimum or maximum values.

Quantity Surveying Services Benchmark	Mean	Lower Quartile (Not complicated projects)	Upper Quartile (Complicated/Sophisticated project)
	2.2%	1.9%	3.4%

The level of fees above, are expressed as a percentage of the contract value of £4,000,000 for a new build project and do not include VAT.

Fee levels vary depending on many factors, including, but not limited to the following; type of project; complexity; procurement route.

ARCHITECTS' FEES

RIBA Agreements are designed to be:

- in line with current working practices, legislative changes and procurement methods
- attractive to clients, architects and other consultants, with robust but fair terms
- a flexible system of components that can be assembled and customized to create tailored and bespoke contracts
- suitable for a wide range of projects and services
- based upon the updated RIBA Plan of Work
- available in electronic and printed formats

RIBA Agreements	Suitable for
Standard Agreement	• a commission where detailed contract terms are necessary for a wide range of projects using most procurement methods • where the client is acting for business or commercial purposes • where the commission is for work to the client's home where the size or value of the Project merits use of the JCT Standard or Intermediate forms of building contract or similar and the terms have been negotiated with the client as a consumer
Concise Agreement	• for a commission where the concise contract terms are compatible with the complexity of the Project and the risks to each party • where the Client is acting for business or commercial purposes • where the commission is for work to the Client's home and the terms have been negotiated with the Client as a 'consumer'. A consumer is 'a natural person acting for purposes outside his trade, business or profession • where the building works, including extensions and alterations, will be carried out using forms of building contract, such as JCT Agreement for Minor Works or JCT Intermediate Form of Building Contract
Domestic Project Agreement	• the commission relates to work to the client's home, provided that they have elected to use these conditions in their own name, i.e. not as a limited company or other legal entity • the contract terms are compatible with the complexity of the project and the risks to each party and have been negotiated with the client as a consumer • the building works, including extensions and alterations, will be carried out using forms of building contract, such as the JCT Building Contract for a homeowner/occupier, JCT Agreement for Minor Works or JCT Intermediate Form of Building Contract
Sub-consultant's Agreement	• a consultant wishes or perhaps is required by the client to appoint another consultant (thus, a sub-consultant) to perform part of the consultant's services • the contract terms are compatible with the (head) agreement between the consultant and the client, with the complexity of the project and the risks to each party
Electronic and print formats	• all the RIBA Agreements 2013 and their components are available as electronic files. A limited number of the conditions and core components are published in print. The electronic and printed versions may be used in combination
Electronic components	• conditions, notes and guides are available as locked PDFs and all other components, e.g. schedules and model letters, are available in Rich Text Format (RTF), which can be customized using most commonly used word processing software to meet project requirements or modified to match the house style of the practice

Each agreement comprises the selected Conditions of Appointment (i.e. Standard, Concise or Domestic), related components, and a schedule or schedules of Services.

Notes on use and completion and model letters for business clients and domestic clients are included with each pack.

For further information, readers are advised to log onto the RIBA Publications website at *www.ribabookshops.com/agreements*

ARCHITECTS' FEES

RIBA Plan of Work 2020

The RIBA Plan of Work is still the definitive design and process management tool for the UK construction industry and is gaining traction internationally too. This 2020 update brings into focus the trends and innovations that are changing the construction industry and provides space for these to thrive on our projects while ensuring a simple and robust framework remains in place.

The table below shows the new Plan of Work stages

RIBA Plan of Work

0	Strategic definition
1	Preparation and briefing
2	Concept design
3	Spatial coordination
4	Technical design
5	Manufacturing and construction
6	Handover
7	Use

More details can be found at the RIBA website; www.ribaplanofwork.com.

Appointment Guidance

A guide, 'A Client's Guide to Engaging an Architect', is available from RIBA Bookshops at www.ribabookshops.com.

This guide includes an introduction to the services an Architect can be expected to provide, advice on the forms to use, linking the RIBA Plan of Work Stages with fees (which are a matter of negotiation) and classifying buildings according to three levels of complexity.

Generally, the more complex the building the higher the level of fee.

Example categories include:

- Simple: for buildings such as car parks, warehouses, factories and speculative retail schemes.
- Average: for buildings such as offices, most retail outlets, general housing, schools etc.
- Complex: for multi-purpose developments, specialist buildings e.g. hospitals, research laboratories etc.

Procurement option can also influence the architects' fees.

ARCHITECTS' FEES

Fee Guide

These do not represent minimum or maximum values.

Architectural Services Benchmark	Mean	Lower Quartile (Not complicated projects)	Upper Quartile (Complicated/Sophisticated project)
	4.5%	3.6%	7.0%

The level of fees above, are expressed as a percentage of the contract value of £4,000,000 for a new build project and do not include VAT.

Projects carried out under the traditional form of contracts tend to attract higher fees than for design and build contract led jobs.

Sectors project have less of an impact on fee levels, however health and education projects attract slightly higher fees than private housing and industrial.

CONSULTING ENGINEERS' FEES

CONDITIONS OF APPOINTMENT

A scale of professional charges for consulting engineering services is published by the Association for Consultancy and Engineering (ACE).

Copies of the document can be obtained direct from:

Association for Consultancy and Engineering
Alliance House
12 Caxton Street
London SW1H OQL
Tel 020 7222 6557
https://www.acenet.co.uk/

Fee Guide

These do not represent minimum and maximum values. Based on value of service not project cost.

Structural Engineering Services Benchmark	Mean	Lower Quartile	Upper Quartile
	2.5%	1.7%	2.9%

Services Engineering Benchmark	Mean	Lower Quartile	Upper Quartile
	2.0%	1.5%	2.6%

THE TOWN AND COUNTRY PLANNING FEES AND BUILDING REGULATION FEES

THE TOWN AND COUNTRY PLANNING APPLICATION FEES

Author's Note

There is a very useful online calculator which will give prices for planning applications at

http://www.planningportal.gov.uk

The fee is paid at the time of application. You should always check your fees with your local planning authority

THE BUILDING (LOCAL AUTHORITY CHARGES) REGULATIONS

CHARGE SCHEDULES

Fees vary from one authority to another, so always check with your Local Authority if you decide to use their officers to provide building control services.

The Construction Industry Council (CIC) has been designated by the government as a body for approving inspectors (AI). Individual and Corporate Approved Inspectors registered with CIC are qualified to undertake building control work in accordance with section 49 of the Building Act 1984 and regulation 4 of the Building (Approved Inspectors etc.) Regulations 1985, and the Building (Approved Inspectors etc.) Regulations 2010.

Approved Inspectors provide building control services on all types of construction projects. The Construction Industry Council (CIC) maintains a list of approved inspectors, see www.cic.org.uk.

Fees vary. Each Local Authority publishes a schedule of rates applicable to project values commonly up to £250,000. The fee for a project of £250,000 in value is approximately £1,500

For higher value projects the fees are individually determined but are typically approximately 0.1% of the estimated project cost.

PART 6

Daywork and Prime Cost

NEC4: 100 Questions and Answers

Kelvin Hughes

This book details some of the most important and interesting questions raised about the NEC4 family of contracts and provides clear, comprehensive answers to those questions.

Written by an NEC expert with over 20 years' experience using, advising and training others, the book has several distinctive features:

- It covers the *whole* NEC4 family
- It is written by a very experienced NEC author who explains sometimes complex issues in a simple and accessible style
- The questions and answers range from beginner level up to a masterclass level
- The questions are real life questions asked by actual NEC practitioners on real projects.

The book includes questions and answers relating to tendering, early warnings, programme issues, quality management, payment provisions, compensation events, liabilities, insurances, adjudication, termination and much more. It is essential reading for anyone working with the NEC4 family of contracts, whether professionals or students in construction, architecture, project management and engineering.

May 2019: 246pp
ISBN: 9781138365254

To Order
Tel: +44 (0) 1235 400524
Email: book.orders@tandf.co.uk

For a complete listing of all our titles visit:
www.tandf.co.uk

BUILDING INDUSTRY

When work is carried out which cannot be valued in any other way it is customary to assess the value on a cost basis with an allowance to cover overheads and profit. The basis of costing is a matter for agreement between the parties concerned, but definitions of prime cost for the building industry have been prepared and published jointly by the Royal Institution of Chartered Surveyors and the National Federation of Building Trades Employers (now the Construction Confederation) for the convenience of those who wish to use them. These documents are reproduced with the permission of the Royal Institution of Chartered Surveyors, which owns the copyright.

The daywork schedule published by the Civil Engineering Contractors Association is included in the *Architects' & Builders'* companion title, *Spon's Civil Engineering and Highway Works Price Book*.

For larger Prime Cost contracts the reader is referred to the form of contract issued by the Royal Institute of British Architects.

DEFINITION OF PRIME COST OF BUILDING WORKS OF A JOBBING OR MAINTENANCE CHARACTER (2010 EDITION)

This definition of Prime Cost is published by the Royal Institution of Chartered Surveyors and the National Federation of Building Trades Employers, for convenience and for use by people who choose to use it. Members of the National Federation of Building Trades Employers are not in any way debarred from defining Prime Cost and rendering their accounts for work carried out on that basis in any way they choose. Building owners are advised to reach agreement with contractors on the Definition of Prime Cost to be used prior to issuing instructions.

SECTION 1 – APPLICATION
1.1.   This definition provides a basis for the valuation of work of a jobbing or maintenance character executed under such building contracts as provide for its use.
1.2.   It is not applicable in any other circumstances, such as daywork executed under or incidental to a building contract.

SECTION 2 – COMPOSITION OF TOTAL CHARGES
2.1.   The prime cost of jobbing work comprises the sum of the following costs:
   (a)   Labour as defined in Section 3
   (b)   Materials and goods as defined in Section 4.
   (c)   Plant, consumable stores and services as defined in Section 5.
   (d)   Subcontracts as defined in Section 6.

2.2.   Incidental costs, overhead and profit as defined in Section 7 and expressed as percentage adjustments are applicable to each of 2.1 (a)–(d).

SECTION 3 – LABOUR
3.1.   Labour costs comprise all payments made to or in respect of all persons directly engaged upon the work, whether on or off the site, except those included in Section 7.
3.2.   Such payments are based upon the standard wage rates, emoluments and expenses as laid down for the time being in the rules or decisions of the National Joint Council for the Building Industry and the terms of the Building and Civil Engineering Annual and Public Holiday Agreements applying to the works, or the rules of decisions or agreements of such other body as may relate to the class of labour concerned, at the time when and in the area where the work is executed, together with the Contractor's statutory obligations, including:
   (a)   Guaranteed minimum weekly earnings (e.g. Standard Basic Rate of Wages and Guaranteed Minimum Bonus Payment in the case of NJCBI rules).
   (b)   All other guaranteed minimum payments (unless included in Section 7).
   (c)   Payments in respect of incentive schemes or productivity agreements applicable to the works.
   (d)   Payments in respect of overtime normally worked; or necessitated by the particular circumstances of the work; or as otherwise agreed between the parties.
   (e)   Differential or extra payments in respect of skill, responsibility, discomfort or inconvenience.
   (f)   Tool allowance.
   (g)   Subsistence and periodic allowances.
   (h)   Fares, travelling and lodging allowances.
   (i)   Employer's contributions to annual holiday credits.
   (j)   Employer's contributions to death benefit schemes.

BUILDING INDUSTRY

(k)   Any amounts which may become payable by the Contractor to or in respect of operatives arising from the operation of the rules referred to in 3.2 which are not provided for in 3.2 (a)–(k) or in Section 7.

(l)   Employer's National Insurance contributions and any contribution, levy or tax imposed by statute, payable by the Contractor in his capacity as employer.

Note:

Any payments normally made by the Contractor which are of a similar character to those described in 3.2 (a)–(c) but which are not within the terms of the rules and decisions referred to above are applicable subject to the prior agreement of the parties, as an alternative to 3.2 (a)–(c).

3.3.   The wages or salaries of supervisory staff, timekeepers, storekeepers, and the like, employed on or regularly visiting site, where the standard wage rates, etc., are not applicable, are those normally paid by the Contractor together with any incidental payments of a similar character to 3.2 (c)–(k).

3.4.   Where principals are working manually their time is chargeable, in respect of the trades practised, in accordance with 3.2.

SECTION 4 – MATERIALS AND GOODS

4.1.   The prime cost of materials and goods obtained by the Contractor from stockists or manufacturers is the invoice cost after deduction of all trade discounts but including cash discounts not exceeding 5 per cent, and includes the cost of delivery to site.

4.2.   The prime cost of materials and goods supplied from the Contractor's stock is based upon the current market prices plus any appropriate handling charges.

4.3.   The prime cost under 4.1 and 4.2 also includes any costs of:
(a)   non-returnable crates or other packaging.
(b)   returning crates and other packaging less any credit obtainable.

4.4.   Any value added tax which is treated, or is capable of being treated, as input tax (as defined in the Finance Act, 1972 or any re-enactment thereof) by the Contractor is excluded.

SECTION 5 – PLANT, CONSUMABLE STORES AND SERVICES

5.1.   The prime cost of plant and consumable stores as listed below is the cost at hire rates agreed between the parties or in the absence of prior agreement at rates not exceeding those normally applied in the locality at the time when the works are carried out, or on a use and waste basis where applicable:
(a)   Machinery in workshops.
(b)   Mechanical plant and power-operated tools.
(c)   Scaffolding and scaffold boards.
(d)   Non-mechanical plant excluding hand tools.
(e)   Transport including collection and disposal of rubbish.
(f)   Tarpaulins and dust sheets.
(g)   Temporary roadways, shoring, planking and strutting, hoarding, centring, formwork, temporary fans, partitions or the like.
(h)   Fuel and consumable stores for plant and power-operated tools unless included in 5.1 (a), (b), (d) or (e) above.
(i)   Fuel and equipment for drying out the works and fuel for testing mechanical services.

5.2.   The prime cost also includes the net cost incurred by the Contractor of the following services, excluding any such cost included under Sections 3, 4 or 7:
(a)   Charges for temporary water supply including the use of temporary plumbing and storage.
(b)   Charges for temporary electricity or other power and lighting including the use of temporary installations.
(c)   Charges arising from work carried out by local authorities or public undertakings.
(d)   Fees, royalties and similar charges.
(e)   Testing of materials.
(f)   The use of temporary buildings including rates and telephone and including heating and lighting not charged under (b) above.

BUILDING INDUSTRY

(g)   The use of canteens, sanitary accommodation, protective clothing and other provision for the welfare of persons engaged in the work in accordance with the current Working Rule Agreement and any Act of Parliament, statutory instrument, rule, order, regulation or bye-law.

(h)   The provision of safety measures necessary to comply with any Act of Parliament.

(i)   Premiums or charges for any performance bonds or insurances which are required by the Building Owner and which are not referred to elsewhere in this Definition.

SECTION 6 – SUBCONTRACTS

6.1.   The prime cost of work executed by subcontractors, whether nominated by the Building Owner or appointed by the Contractor, is the amount which is due from the Contractor to the subcontractors in accordance with the terms of the subcontracts after deduction of all discounts except any cash discount offered by any subcontractor to the Contractor not exceeding 2.5%.

SECTION 7 – INCIDENTAL COSTS, OVERHEADS AND PROFIT

7.1.   The percentage adjustments provided in the building contract, which are applicable to each of the totals of Sections 3–6, provide for the following:

(a)   Head Office charges.

(b)   Off-site staff including supervisory and other administrative staff in the Contractor's workshops and yard.

(c)   Payments in respect of public holidays.

(d)   Payments in respect of apprentices' study time.

(e)   Sick pay or insurance in respect thereof.

(f)   Third party employer's liability insurance.

(g)   Liability in respect of redundancy payments made to employees.

(h)   Use, repair and sharpening of non-mechanical hand tools.

(i)   Any variations to basic rates required by the Contractor in cases where the building contract provides for the use of a specified schedule of basic plant charges (to the extent that no other provision is made for such variation).

(j)   All other liabilities and obligations whatsoever not specifically referred to in this Section nor chargeable under any other section.

(k)   Profit.

BUILDING INDUSTRY

SPECIMEN ACCOUNT FORMAT

If this Definition of Prime Cost is followed the Contractor's account could be in the following format:

	£
Labour (as defined in Section 3)	
Add _ % (see Section 7)	
Materials and goods (as defined in Section 4)	
Add _ % (see Section 7)	£
Plant, consumable stores and services (as defined in Section 5)	
Add _ % (see Section 7)	£
Subcontracts (as defined in Section 6)	
Add _ % (see Section 7)	£
Total	£

VAT to be added if applicable.

Example Calculations of Prime Cost of Labour in Daywork

Example of calculation of typical standard hourly base rate (as defined in Section 3) for CIJC Building Craft operative and General Operative based upon assumed rates applicable June 2021.

		Craft Operative		General Operative	
		Rate £	Annual Cost £	Rate £	Annual Cost £
Basic Wages	52 weeks	507.72	26,401.37	381.89	19,858.33
CITB levy (0.5% of payroll)	0.5%		132.01		99.29
Pension and welfare benefit	52 weeks	12.86	668.72	12.86	668.72
Employer's National Insurance contribution *(13.8% after the first £184 per week)*	46.2 weeks		2,470.28		1,567.34
Annual labour cost:			**29,672.37**		**22,193.68**
Hourly base rate:			**16.54**		**12.37**

This is the prime cost of employment per person, which the employer has to meet even if there is no work for the employee to do

Note:

(1)   Calculated following Definition of Prime Cost of Daywork carried out under a Building Contract, published by the Royal Institution of Chartered Surveyors and the Construction Confederation.

(2)   Standard basic assumed rates effective from June 2021.

(3)   Standard working hours per annum calculated as follows:

BUILDING INDUSTRY

52 weeks @ 39 hours =	2028
Less	
22 days holiday @ 39 hours =	171.6
8 days public holidays @ 7.8 hours =	62.4
	= −234
Standard working hours per year =	1794

(4) All labour costs incurred by the contractor in his capacity as an employer, other than those contained in the hourly base rate, are to be considered under Section 6.

(5) The above example is for guidance only and does not form part of the Definition; all the basic costs are subject to re-examination according to the time when and in the area where the daywork is executed.

(6) NI payments are at not contracted out rates applicable from April 2021.

(7) Basic rate and GMP number of weeks =
52 weeks − 4.4 weeks annual holiday − 1.6 weeks public holiday = 46 weeks

BUILDING INDUSTRY

SCHEDULE OF BASIC PLANT CHARGES (1 JULY 2010 ISSUE)

This Schedule is published by the Royal Institution of Chartered Surveyors, Parliament Square, London, and is for use in connection with Dayworks under a Building Contract.

EXPLANATORY NOTES

1. The rates in the Schedule are intended to apply solely to daywork carried out under and incidental to a Building Contract. They are NOT intended to apply to:
   (i)   Jobbing or any other work carried out as a main or separate contract; or
   (ii)   Work carried out after the date of commencement of the Defects Liability Period.
2. The rates apply only to plant and machinery already on site, whether hired or owned by the Contractor.
3. The rates, unless otherwise stated, include the cost of fuel and power of every description, lubricating oils, grease, maintenance, sharpening of tools, replacement of spare parts, all consumable stores and for licences and insurances applicable to items of plant.
4. The rates, unless otherwise stated, do not include the costs of drivers and attendants.
5. The rates do not include for any possible discounts which may be given to the Contractor if the plant or machinery was hired.
6. The rates are base costs and may be subject to the overall adjustment for price movement, overheads and profit, quoted by the Contractor prior to the placing of the Contract.
7. The rates should be applied to the time during which the plant is actually engaged in daywork.
8. Whether or not plant is chargeable on daywork depends on the daywork agreement in use and the inclusion of an item of plant in this schedule does not necessarily indicate that the item is chargeable.
9. Rates for plant not included in the Schedule or which is not already on site and is specifically provided or hired for daywork shall be settled at prices which are reasonably related to the rates in the Schedule having regard to any overall adjustment quoted by the Contractor in the Conditions of Contract.

Item of plant	Size/Rating	Unit	Rate per Hour (£)
PUMPS			
Mobile Pumps			
Including pump hoses, values and strainers, etc.			
Diaphragm	50 mm dia.	Each	1.17
Diaphragm	76 mm dia.	Each	1.89
Diaphragm	102 mm dia.	Each	3.54
Submersible	50 mm dia.	Each	0.76
Submersible	76 mm dia.	Each	0.86
Submersible	102 mm dia.	Each	1.03
Induced flow	50 mm dia.	Each	0.77
Induced flow	76 mm dia.	Each	1.67
Centrifugal, self-priming	25 mm dia.	Each	1.30
Centrifugal, self-priming	50 mm dia.	Each	1.92
Centrifugal, self-priming	75 mm dia.	Each	2.74
Centrifugal, self-priming	102 mm dia.	Each	3.35
Centrifugal, self-priming	152 mm dia.	Each	4.27

BUILDING INDUSTRY

Item of plant	Size/Rating	Unit	Rate per Hour (£)
SCAFFOLDING, SHORING, FENCING			
Complete Scaffolding			
Mobile working towers, single width	2.0 m × 0.72 m base × 7.45 m high	Each	3.36
Mobile working towers, single width	2.0 m × 0.72 m base × 8.84 m high	Each	3.79
Mobile working towers, double width	2.0 m × 1.35 m × 7.45 m high	Each	3.79
Mobile working towers, double width	2.0 m × 1.35 m × 15.8 m high	Each	7.13
Chimney scaffold, single unit		Each	1.92
Chimney scaffold, twin unit		Each	3.59
Push along access platform	1.63–3.1 m	Each	5.00
Push along access platform	1.80 m × 0.70 m	Each	1.79
Trestles			
Trestle, adjustable	any height	Pair	0.41
Trestle, painters	1.8 m high	Pair	0.31
Trestle, painters	2.4 m high	Pair	0.36
Shoring, Planking and Strutting			
'Acrow' adjustable prop	sizes up to 4.9 m (open)	Each	0.06
'Strong boy' support attachment		Each	0.22
Adjustable trench strut	sizes up to 1.67 m (open)	Each	0.16
Trench sheet		Metre	0.03
Backhoe trench box	base unit	Each	1.23
Backhoe trench box	top unit	Each	0.87
Temporary Fencing			
Including block and coupler			
Site fencing steel grid panel	3.5 m × 2.0 m	Each	0.05
Anti-climb site steel grid fence panel	3.5 m × 2.0 m	Each	0.08
Solid panel Heras	2.0 m × 2.0 m	Each	0.09
Pedestrian gate		Each	0.36
Roadway gate		Each	0.60
LIFTING APPLIANCES AND CONVEYORS			
Cranes			
<u>Mobile Cranes</u>			
Rates are inclusive of drivers			
Lorry mounted, telescopic jib			
Two-wheel drive	5 tonnes	Each	19.00
Two-wheel drive	8 tonnes	Each	42.00
Two-wheel drive	10 tonnes	Each	50.00

BUILDING INDUSTRY

Item of plant	Size/Rating		Unit	Rate per Hour (£)
LIFTING APPLIANCES AND CONVEYORS				
Two-wheel drive	12 tonnes		Each	77.00
Two-wheel drive	20 tonnes		Each	89.69
Four-wheel drive	18 tonnes		Each	46.51
Four-wheel drive	25 tonnes		Each	35.90
Four-wheel drive	30 tonnes		Each	38.46
Four-wheel drive	45 tonnes		Each	46.15
Four-wheel drive	50 tonnes		Each	53.85
Four-wheel drive	60 tonnes		Each	61.54
Four-wheel drive	70 tonnes		Each	71.79

Static Tower Crane

Rates inclusive of driver

Note: Capacity equals maximum lift in tonnes times maximum radius at which it can be lifted

	Capacity (Metre/tonnes) Up to	Height under hook above ground (m) Up to	Unit	Rate per Hour (£)
Tower crane	30	22	Each	22.23
Tower crane	40	22	Each	26.62
Tower crane	40	30	Each	33.33
Tower crane	50	22	Each	29.16
Tower crane	60	22	Each	35.90
Tower crane	60	36	Each	35.90
Tower crane	70	22	Each	41.03
Tower crane	80	22	Each	39.12
Tower crane	90	42	Each	37.18
Tower crane	110	36	Each	47.62
Tower crane	140	36	Each	55.77
Tower crane	170	36	Each	64.11
Tower crane	200	36	Each	71.95
Tower crane	250	36	Each	84.77
Tower crane with luffing jig	30	25	Each	22.23
Tower crane with luffing jig	40	30	Each	26.62
Tower crane with luffing jig	50	30	Each	29.16
Tower crane with luffing jig	60	36	Each	41.03
Tower crane with luffing jig	65	30	Each	33.13
Tower crane with luffing jig	80	22	Each	48.72
Tower crane with luffing jig	100	45	Each	48.72
Tower crane with luffing jig	125	30	Each	53.85

BUILDING INDUSTRY

Item of plant	Size/Rating		Unit	Rate per Hour (£)
LIFTING APPLIANCES AND CONVEYORS				
Tower crane with luffing jig	160	50	Each	53.85
Tower crane with luffing jig	200	50	Each	74.36
Tower crane with luffing jig	300	60	Each	100.00
Crane Equipment				
Muck tipping skip	up to 200 litres		Each	0.67
Muck tipping skip	500 litres		Each	0.82
Muck tipping skip	750 litres		Each	1.08
Muck tipping skip	1000 litres		Each	1.28
Muck tipping skip	1500 litres		Each	1.41
Muck tipping skip	2000 litres		Each	1.67
Mortar skip	250 litres, plastic		Each	0.41
Mortar skip	350 litres steel		Each	0.77
Boat skip	250 litres		Each	0.92
Boat skip	500 litres		Each	1.08
Boat skip	750 litres		Each	1.23
Boat skip	1000 litres		Each	1.38
Boat skip	1500 litres		Each	1.64
Boat skip	2000 litres		Each	1.90
Boat skip	3000 litres		Each	2.82
Boat skip	4000 litres		Each	3.23
Master flow skip	250 litres		Each	0.77
Master flow skip	500 litres		Each	1.03
Master flow skip	750 litres		Each	1.28
Master flow skip	1000 litres		Each	1.44
Master flow skip	1500 litres		Each	1.69
Master flow skip	2000 litres		Each	1.85
Grand master flow skip	500 litres		Each	1.28
Grand master flow skip	750 litres		Each	1.64
Grand master flow skip	1000 litres		Each	1.69
Grand master flow skip	1500 litres		Each	1.95
Grand master flow skip	2000 litres		Each	2.21
Cone flow skip	500 litres		Each	1.33
Cone flow skip	1000 litres		Each	1.69
Geared rollover skip	500 litres		Each	1.28
Geared rollover skip	750 litres		Each	1.64
Geared rollover skip	1000 litres		Each	1.69

Daywork and Prime Cost

BUILDING INDUSTRY

Item of plant	Size/Rating	Unit	Rate per Hour (£)
LIFTING APPLIANCES AND CONVEYORS			
Geared rollover skip	1500 litres	Each	1.95
Geared rollover skip	2000 litres	Each	2.21
Multi-skip, rope operated	200 mm outlet size, 500 litres	Each	1.49
Multi-skip, rope operated	200 mm outlet size, 750 litres	Each	1.64
Multi-skip, rope operated	200 mm outlet size, 1000 litres	Each	1.74
Multi-skip, rope operated	200 mm outlet size, 1500 litres	Each	2.00
Multi-skip, rope operated	200 mm outlet size, 2000 litres	Each	2.26
Multi-skip, man riding	200 mm outlet size, 1000 litres	Each	2.00
Multi-skip	4 point lifting frame	Each	0.90
Multi-skip	Chain brothers	Set	0.87
Crane Accessories			
Multi-purpose crane forks	1.5 and 2 tonnes S.W.L.	Each	1.13
Self-levelling crane forks		Each	1.28
Man cage	1 man, 230 kg S.W.L.	Each	1.90
Man cage	2 man, 500 kg S.W.L.	Each	1.95
Man cage	4 man, 750 kg S.W.L.	Each	2.15
Man cage	8 man, 1000 kg S.W.L.	Each	3.33
Stretcher cage	500 kg, S.W.L.	Each	2.69
Goods carrying cage	1500 kg, S.W.L.	Each	1.33
Goods carrying cage	3000 kg, S.W.L.	Each	1.85
Builders' skip lifting cradle	12 tonnes, S.W.L.	Each	2.31
Board/pallet fork	1600 kg, S.W.L.	Each	1.90
Gas bottle carrier	500 kg, S.W.L.	Each	0.92
Hoists			
Scaffold hoist	200 kg	Each	2.46
Rack and pinion (goods only)	500 kg	Each	4.56
Rack and pinion (goods only)	1100 kg	Each	5.90
Rack and pinion (goods and passenger)	8-person, 80 kg	Each	7.44
Rack and pinion (goods and passenger)	14-person, 1400 kg	Each	8.72
Wheelbarrow chain sling		Each	1.67

BUILDING INDUSTRY

Item of plant	Size/Rating		Unit	Rate per Hour (£)
LIFTING APPLIANCES AND CONVEYORS				
Conveyors				
<u>Belt Conveyors</u>				
Conveyor	8 m long × 450 mm wide		Each	5.90
Miniveyor, control box and loading hopper	3 m unit		Each	4.49
<u>Other Conveying Equipment</u>				
Wheelbarrow			Each	0.62
Hydraulic superlift			Each	4.56
Pavac slab lifter (tile hoist)			Each	4.49
High lift pallet truck			Each	3.08
Lifting Trucks	Payload	Maximum Lift		
Fork lift, two-wheel drive	1100 kg	up to 3.0 m	Each	5.64
Fork lift, two-wheel drive	2540 kg	up to 3.7 m	Each	5.64
Fork lift, four-wheel drive	1524 kg	up to 6.0 m	Each	5.64
Fork lift, four-wheel drive	2600 kg	up to 5.4 m	Each	7.44
Fork life, four-wheel drive	4000 kg	up to 17 m	Each	10.77
Lifting Platforms				
Hydraulic platform (Cherry picker)	9 m		Each	4.62
Hydraulic platform (Cherry picker)	12 m		Each	7.56
Hydraulic platform (Cherry picker)	15 m		Each	10.13
Hydraulic platform (Cherry picker)	17 m		Each	15.63
Hydraulic platform (Cherry picker)	20 m		Each	18.13
Hydraulic platform (Cherry picker)	25.6 m		Each	32.38
Scissor lift	7.6 m, electric		Each	3.85
Scissor lift	7.8 m, electric		Each	5.13
Scissor lift	9.7 m, electric		Each	4.23
Scissor lift	10 m, diesel		Each	6.41
Telescopic handler	7 m, 2 tonnes		Each	5.13
Telescopic handler	13 m, 3 tonnes		Each	7.18
Lifting and Jacking Gear				
Pipe winch including gantry	1 tonne		Set	1.92
Pipe winch including gantry	3 tonnes		Set	3.21
Chain block	1 tonne		Each	0.35
Chain block	2 tonnes		Each	0.58
Chain block	5 tonnes		Each	1.14
Pull lift (Tirfor winch)	1 tonne		Each	0.64
Pull lift (Tirfor winch)	1.6 tonnes		Each	0.90

Daywork and Prime Cost

BUILDING INDUSTRY

Item of plant	Size/Rating	Unit	Rate per Hour (£)
LIFTING APPLIANCES AND CONVEYORS			
Pull lift (Tirfor winch)	3.2 tonnes	Each	1.15
Brother or chain slings, two legs	not exceeding 3.1 tonnes	Set	0.21
Brother or chain slings, two legs	not exceeding 4.25 tonnes	Set	0.31
Brother or chain slings, four legs	not exceeding 11.2 tonnes	Set	1.09
CONSTRUCTION VEHICLES			
Lorries			
Plated lorries (Rates are inclusive of driver)			
Platform lorry	7.5 tonnes	Each	16.21
Platform lorry	17 tonnes	Each	22.90
Platform lorry	24 tonnes	Each	30.68
Extra for lorry with crane attachment	up to 2.5 tonnes	Each	3.25
Extra for lorry with crane attachment	up to 5 tonnes	Each	6.00
Extra for lorry with crane attachment	up to 7.5 tonnes	Each	9.10
Tipper Lorries			
(Rates are inclusive of driver)			
Tipper lorry	up to 11 tonnes	Each	15.78
Tipper lorry	up to 17 tonnes	Each	23.95
Tipper lorry	up to 25 tonnes	Each	31.35
Tipper lorry	up to 31 tonnes	Each	37.79
Dumpers			
Site use only (excl. tax, insurance			
and extra cost of DERV etc. when			
operating on highway)	Makers Capacity		
Two-wheel drive	1 tonne	Each	1.71
Four-wheel drive	2 tonnes	Each	2.43
Four-wheel drive	3 tonnes	Each	2.44
Four-wheel drive	5 tonnes	Each	3.08
Four-wheel drive	6 tonnes	Each	3.85
Four-wheel drive	9 tonnes	Each	5.65
Tracked	0.5 tonnes	Each	3.33
Tracked	1.5 tonnes	Each	4.23
Tracked	3.0 tonnes	Each	8.33
Tracked	6.0 tonnes	Each	16.03
Dumper Trucks *(Rates are inclusive of drivers)*			

BUILDING INDUSTRY

Item of plant	Size/Rating	Unit	Rate per Hour (£)
CONSTRUCTION VEHICLES			
Dumper truck	up to 15 tonnes	Each	28.56
Dumper truck	up to 17 tonnes	Each	32.82
Dumper truck	up to 23 tonnes	Each	54.64
Dumper truck	up to 30 tonnes	Each	63.50
Dumper truck	up to 35 tonnes	Each	73.02
Dumper truck	up to 40 tonnes	Each	87.84
Dumper truck	up to 50 tonnes	Each	133.44
Tractors			
Agricultural Type			
Wheeled, rubber-clad tyred	up to 40 kW	Each	8.63
Wheeled, rubber-clad tyred	up to 90 kW	Each	25.31
Wheeled, rubber-clad tyred	up to 140 kW	Each	36.49
Crawler Tractors			
With bull or angle dozer	up to 70 kW	Each	29.38
With bull or angle dozer	up to 85 kW	Each	38.63
With bull or angle dozer	up to 100 kW	Each	52.59
With bull or angle dozer	up to 115 kW	Each	55.85
With bull or angle dozer	up to 135 kW	Each	60.43
With bull or angle dozer	up to 185 kW	Each	76.44
With bull or angle dozer	up to 200 kW	Each	96.43
With bull or angle dozer	up to 250 kW	Each	117.68
With bull or angle dozer	up to 350 kW	Each	160.03
With bull or angle dozer	up to 450 kW	Each	219.86
With loading shovel	0.8 m³	Each	26.92
With loading shovel	1.0 m³	Each	32.59
With loading shovel	1.2 m³	Each	37.53
With loading shovel	1.4 m³	Each	42.89
With loading shovel	1.8 m³	Each	52.22
With loading shovel	2.0 m³	Each	57.22
With loading shovel	2.1 m³	Each	60.12
With loading shovel	3.5 m³	Each	87.26
Light vans			
VW Caddivan or the like		Each	5.26
VW Transport transit or the like	1.0 tonne	Each	6.03
Luton Box Van or the like	1.8 tonnes	Each	9.87

BUILDING INDUSTRY

Item of plant	Size/Rating	Unit	Rate per Hour (£)
CONSTRUCTION VEHICLES			
Water/Fuel Storage			
Mobile water container	110 litres	Each	0.62
Water bowser	1100 litres	Each	0.72
Water bowser	3000 litres	Each	0.87
Mobile fuel container	110 litres	Each	0.62
Fuel bowser	1100 litres	Each	1.23
Fuel bowser	3000 litres	Each	1.87
EXCAVATIONS AND LOADERS			
Excavators			
Wheeled, hydraulic	up to 11 tonnes	Each	25.86
Wheeled, hydraulic	up to 14 tonnes	Each	30.82
Wheeled, hydraulic	up to 16 tonnes	Each	34.50
Wheeled, hydraulic	up to 21 tonnes	Each	39.10
Wheeled, hydraulic	up to 25 tonnes	Each	43.81
Wheeled, hydraulic	up to 30 tonnes	Each	55.30
Crawler, hydraulic	up to 11 tonnes	Each	25.86
Crawler, hydraulic	up to 14 tonnes	Each	30.82
Crawler, hydraulic	up to 17 tonnes	Each	34.50
Crawler, hydraulic	up to 23 tonnes	Each	39.10
Crawler, hydraulic	up to 30 tonnes	Each	43.81
Crawler, hydraulic	up to 35 tonnes	Each	55.30
Crawler, hydraulic	up to 38 tonnes	Each	71.73
Crawler, hydraulic	up to 55 tonnes	Each	95.63
Mini excavator	1000/1500 kg	Each	4.87
Mini excavator	2150/2400 kg	Each	6.67
Mini excavator	2700/3500 kg	Each	7.31
Mini excavator	3500/4500 kg	Each	8.21
Mini excavator	4500/6000 kg	Each	9.23
Mini excavator	7000 kg	Each	14.10
Micro excavator	725 mm wide	Each	5.13
Loaders			
Shovel loader	0.4 m³	Each	7.69
Shovel loader	1.57 m³	Each	8.97
Shovel loader, four-wheel drive	1.7 m³	Each	4.83
Shovel loader, four-wheel drive	2.3 m³	Each	4.38

BUILDING INDUSTRY

Item of plant	Size/Rating	Unit	Rate per Hour (£)
EXCAVATIONS AND LOADERS			
Shovel loader, four-wheel drive	3.3 m³	Each	5.06
Skid steer loader wheeled	300/400 kg payload	Each	7.31
Skid steer loader wheeled	625 kg payload	Each	7.67
Tracked skip loader	650 kg	Each	4.42
Excavator Loaders			
Wheeled tractor type with black-hoe excavator			
Four-wheel drive			
Four-wheel drive, 2-wheel steer	6 tonnes	Each	6.41
Four-wheel drive, 2-wheel steer	8 tonnes	Each	8.59
Attachments			
Breakers for excavator		Each	8.72
Breakers for mini excavator		Each	1.75
Breakers for back-hoe excavator/loader		Each	5.13
COMPACTION EQUIPMENT			
Rollers			
Vibrating roller	368–420 kg	Each	1.43
Single roller	533 kg	Each	1.94
Single roller	750 kg	Each	3.43
Twin roller	up to 650 kg	Each	6.03
Twin roller	up to 950 kg	Each	6.62
Twin roller with seat end steering wheel	up to 1400 kg	Each	7.68
Twin roller with seat end steering wheel	up to 2500 kg	Each	10.61
Pavement roller	3–4 tonnes dead weight	Each	6.00
Pavement roller	4–6 tonnes	Each	6.86
Pavement roller	6–10 tonnes	Each	7.17
Pavement roller	10–13 tonnes	Each	19.86
Rammers			
Tamper rammer 2 stroke-petrol	225 mm–275 mm	Each	1.52
Soil Compactors			
Plate compactor	75 mm–400 mm	Each	1.53
Plate compactor rubber pad	375 mm–1400 mm	Each	1.53
Plate compactor reversible plate-petrol	400 mm	Each	2.44

BUILDING INDUSTRY

Item of plant	Size/Rating	Unit	Rate per Hour (£)
CONCRETE EQUIPMENT			
Concrete/Mortar Mixers			
Open drum without hopper	0.09/0.06 m³	Each	0.61
Open drum without hopper	0.12/0.09 m³	Each	1.22
Open drum without hopper	0.15/0.10 m³	Each	0.72
Concrete/Mortar Transport Equipment			
Concrete pump incl. hose, valve and couplers			
Lorry mounted concrete pump	24 m max. distance	Each	50.00
Lorry mounted concrete pump	34 m max. distance	Each	66.00
Lorry mounted concrete pump	42 m max. distance	Each	91.50
Concrete Equipment			
Vibrator, poker, petrol type	up to 75 mm dia.	Each	0.69
Air vibrator (*excluding compressor and hose*)	up to 75 mm dia.	Each	0.64
Extra poker heads	25/36/60 mm dia.	Each	0.76
Vibrating screed unit with beam	5.00 m	Each	2.48
Vibrating screed unit with adjustable beam	3.00–5.00 m	Each	3.54
Power float	725 mm–900 mm	Each	2.56
Power float finishing pan		Each	0.62
Floor grinder	660 × 1016 mm, 110 V electric	Each	4.31
Floor plane	450 × 1100 mm	Each	4.31
TESTING EQUIPMENT			
Pipe Testing Equipment			
Pressure testing pump, electric		Set	2.19
Pressure test pump		Set	0.80
SITE ACCOMODATION AND TEMPORARY SERVICES			
Heating equipment			
Space heater – propane	80,000 Btu/hr	Each	1.03
Space heater – propane/electric	125,000 Btu/hr	Each	2.09
Space heater – propane/electric	250,000 Btu/hr	Each	2.33
Space heater – propane	125,000 Btu/hr	Each	1.54
Space heater – propane	260,000 Btu/hr	Each	1.88
Cabinet heater		Each	0.82
Cabinet heater, catalytic		Each	0.57

BUILDING INDUSTRY

Item of plant	Size/Rating	Unit	Rate per Hour (£)
SITE ACCOMODATION AND TEMPORARY SERVICES			
Electric halogen heater		Each	1.27
Ceramic heater	3 kW	Each	0.99
Fan heater	3 kW	Each	0.66
Cooling fan		Each	1.92
Mobile cooling unit, small		Each	3.60
Mobile cooling unit, large		Each	4.98
Air conditioning unit		Each	2.81
Site Lighting and Equipment			
Tripod floodlight	500 W	Each	0.48
Tripod floodlight	1000 W	Each	0.62
Towable floodlight	4 × 100 W	Each	3.85
Hand-held floodlight	500 W	Each	0.51
Rechargeable light		Each	0.41
Inspection light		Each	0.37
Plasterer's light		Each	0.65
Lighting mast		Each	2.87
Festoon light string	25 m	Each	0.55
Site Electrical Equipment			
Extension leads	240 V/14 m	Each	0.26
Extension leads	110 V/14 m	Each	0.36
Cable reel	25 m 110 V/240 V	Each	0.46
Cable reel	50 m 110 V/240 V	Each	0.88
4-way junction box	110 V	Each	0.56
Power Generating Units			
Generator – petrol	2 kVA	Each	1.23
Generator – silenced petrol	2 kVA	Each	2.87
Generator – petrol	3 kVA	Each	1.47
Generator – diesel	5 kVA	Each	2.44
Generator – silenced diesel	10 kVA	Each	1.90
Generator – silenced diesel	15 kVA	Each	2.26
Generator – silenced diesel	30 kVA	Each	3.33
Generator – silenced diesel	50 kVA	Each	4.10
Generator – silenced diesel	75 kVA	Each	4.62
Generator – silenced diesel	100 kVA	Each	5.64
Generator – silenced diesel	150 kVA	Each	7.18
Generator – silenced diesel	200 kVA	Each	9.74

Daywork and Prime Cost

BUILDING INDUSTRY

Item of plant	Size/Rating	Unit	Rate per Hour (£)
SITE ACCOMODATION AND TEMPORARY SERVICES			
Generator – silenced diesel	250 kVA	Each	11.28
Generator – silenced diesel	350 kVA	Each	14.36
Generator – silenced diesel	500 kVA	Each	15.38
Tail adaptor	240 V	Each	0.10
Transformers			
Transformer	3 kVA	Each	0.32
Transformer	5 kVA	Each	1.23
Transformer	7.5 kVA	Each	0.59
Transformer	10 kVA	Each	2.00
Rubbish Collection and Disposal			
Equipment			
Rubbish Chutes			
Standard plastic module	1 m section	Each	0.15
Steel liner insert		Each	0.30
Steel top hopper		Each	0.22
Plastic side entry hopper		Each	0.22
Plastic side entry hopper liner		Each	0.22
Dust Extraction Plant			
Dust extraction unit, light duty		Each	2.97
Dust extraction unit, heavy duty		Each	2.97
SITE EQUIPMENT – Welding Equipment			
Arc-(Electric) Complete With Leads			
Welder generator – petrol	200 amp	Each	3.53
Welder generator – diesel	300/350 amp	Each	3.78
Welder generator – diesel	4000 amp	Each	7.92
Extra welding lead sets		Each	0.69
Gas-Oxy Welder			
Welding and cutting set (including oxygen and acetylene, excluding underwater equipment and thermic boring)			
Small		Each	2.24
Large		Each	3.75
Lead burning gun		Each	0.50
Mig welder		Each	1.38
Fume extractor		Each	2.46

BUILDING INDUSTRY

Item of plant	Size/Rating	Unit	Rate per Hour (£)
SITE EQUIPMENT – Welding Equipment			
Road Works Equipment			
Traffic lights, mains/generator	2-way	Set	10.94
Traffic lights, mains/generator	3-way	Set	11.56
Traffic lights, mains/generator	4-way	Set	12.19
Flashing light		Each	0.10
Road safety cone	450 mm	Each	0.08
Safety cone	750 mm	Each	0.10
Safety barrier plank	1.25 m	Each	0.13
Safety barrier plank	2 m	Each	0.15
Safety barrier plank post		Each	0.13
Safety barrier plank post base		Each	0.10
Safety four gate barrier	1 m each gate	Set	0.77
Guard barrier	2 m	Each	0.19
Road sign	750 mm	Each	0.23
Road sign	900 mm	Each	0.31
Road sign	1200 mm	Each	0.42
Speed ramp/cable protection	500 mm section	Each	0.14
Hose ramp open top	3 m section	Each	0.07
DPC Equipment			
Damp-proofing injection machine		Each	2.56
Cleaning Equipment			
Vacuum cleaner (industrial wet) single motor		Each	1.08
Vacuum cleaner (industrial wet) twin motor	30 litre capacity	Each	1.79
Vacuum cleaner (industrial wet) twin motor	70 litre capacity	Each	2.21
Steam cleaner	diesel/electric 1 phase	Each	3.33
Steam cleaner	diesel/electric 3 phase	Each	3.85
Pressure washer, light duty electric	1450 PSI	Each	0.72
Pressure washer, heavy duty, diesel	2500 PSI	Each	1.33
Pressure washer, heavy duty, diesel	4000 PSI	Each	2.18
Cold pressure washer, electric		Each	2.39
Hot pressure washer, petrol		Each	4.19
Hot pressure washer, electric		Each	5.13
Cold pressure washer, petrol		Each	2.92

Daywork and Prime Cost

BUILDING INDUSTRY

Item of plant	Size/Rating	Unit	Rate per Hour (£)
SITE EQUIPMENT – Welding Equipment			
Sandblast attachment to last washer		Each	1.23
Drain cleaning attachment to last washer		Each	1.03
Surface Preparation Equipment			
Rotavator	5 h.p.	Each	2.46
Rotavator	9 h.p.	Each	5.00
Scabbler, up to three heads		Each	1.53
Scabbler, pole		Each	2.68
Scrabbler, multi-headed floor		Each	3.89
Floor preparation machine		Each	1.05
Compressors and Equipment			
Portable Compressors			
Compressor – electric	4 cfm	Each	1.36
Compressor – electric	8 cfm lightweight	Each	1.31
Compressor – electric	8 cfm	Each	1.36
Compressor – electric	14 cfm	Each	1.56
Compressor – petrol	24 cfm	Each	2.15
Compressor – electric	25 cfm	Each	2.10
Compressor – electric	30 cfm	Each	2.36
Compressor – diesel	100 cfm	Each	2.56
Compressor – diesel	250 cfm	Each	5.54
Compressor – diesel	400 cfm	Each	8.72
Mobile Compressors			
Lorry mounted compressor			
(machine plus lorry only)	up to 3 m³	Each	41.47
(machine plus lorry only)	up to 5 m³	Each	48.94
Tractor mounted compressor			
(machine plus rubber tyred tractor)	up to 4 m³	Each	21.03
Accessories (Pneumatic Tools) *(with and including up to 15 m of air hose)*			
Demolition pick, medium duty		Each	0.90
Demolition pick, heavy duty		Each	1.03
Breakers (with six steels) light	up to 150 kg	Each	1.19
Breakers (with six steels) medium	295 kg	Each	1.24
Breakers (with six steels) heavy	386 kg	Each	1.44
Rock drill (for use with compressor) hand held		Each	1.18

BUILDING INDUSTRY

Item of plant	Size/Rating	Unit	Rate per Hour (£)
SITE EQUIPMENT – Welding Equipment			
Additional hoses	15 m	Each	0.09
Breakers			
Demolition hammer drill, heavy duty, electric		Each	1.54
Road breaker, electric		Each	2.41
Road breaker, 2 stroke, petrol		Each	4.06
Hydraulic breaker unit, light duty, petrol		Each	3.06
Hydraulic breaker unit, heavy duty, petrol		Each	3.46
Hydraulic breaker unit, heavy duty, diesel		Each	4.62
Quarrying and Tooling Equipment			
Block and stone splitter, hydraulic	600 mm × 600 mm	Each	1.90
Block and stone splitter, manual		Each	1.64
Steel Reinforcement Equipment			
Bar bending machine – manual	up to 13 mm dia. rods	Each	1.03
Bar bending machine – manual	up to 20 mm dia. rods	Each	1.41
Bar shearing machine – electric	up to 38 mm dia. rods	Each	3.08
Bar shearing machine – electric	up to 40 mm dia. rods	Each	4.62
Bar cropper machine – electric	up to 13 mm dia. rods	Each	2.05
Bar cropper machine – electric	up to 20 mm dia. rods	Each	2.56
Bar cropper machine – electric	up to 40 mm dia. rods	Each	4.62
Bar cropper machine – 3 phase	up to 40 mm dia. rods	Each	4.62
Dehumidifiers			
110/240 V Water	68 litres extraction per 24 hours	Each	2.46
110/240 V Water	90 litres extraction per 24 hours	Each	3.38
SMALL TOOLS			
Saws			
Masonry bench saw	350 mm–500 mm dia.	Each	1.13
Floor saw	125 mm max. cut	Each	1.15
Floor saw	150 mm max. cut	Each	3.83
Floor saw, reversible	350 mm max. cut	Each	3.32
Wall saw, electric		Each	2.05
Chop/cut off saw, electric	350 mm dia.	Each	1.79
Circular saw, electric	230 mm dia.	Each	0.72
Tyrannosaw		Each	1.74
Reciprocating saw		Each	0.79

BUILDING INDUSTRY

Item of plant	Size/Rating	Unit	Rate per Hour (£)
SMALL TOOLS			
Door trimmer		Each	1.17
Stone saw	300 mm	Each	1.44
Chainsaw, petrol	500 mm	Each	3.92
Full chainsaw safety kit		Each	0.41
Worktop jig		Each	1.08
Pipe Work Equipment			
Pipe bender	15 mm–22 mm	Each	0.92
Pipe bender, hydraulic	50 mm	Each	1.76
Pipe bender, electric	50 mm–150 mm dia.	Each	2.19
Pipe cutter, hydraulic		Each	0.46
Tripod pipe vice		Set	0.75
Ratchet threader	12 mm–32 mm	Each	0.93
Pipe threading machine, electric	12 mm–75 mm	Each	3.07
Pipe threading machine, electric	12 mm–100 mm	Each	4.93
Impact wrench, electric		Each	1.33
Hand-held Drills and equipment			
Impact or hammer drill	up to 25 mm dia.	Each	1.03
Impact or hammer drill	35 mm dia.	Each	1.29
Dry diamond core cutter		Each	0.99
Angle head drill		Each	0.90
Stirrer, mixer drill		Each	1.13
Paint, Insulation Application Equipment			
Airless spray unit		Each	4.13
Portaspray unit		Each	1.16
HPVL turbine spray unit		Each	2.23
Compressor and spray gun		Each	1.91
Other Handtools			
Staple gun		Each	0.96
Air nail gun	110 V	Each	1.01
Cartridge hammer		Each	1.08
Tongue and groove nailer complete with mallet		Each	1.59
Diamond wall chasing machine		Each	2.63
Masonry chain saw	300 mm	Each	5.49
Floor grinder		Each	3.99
Floor plane		Each	1.79

BUILDING INDUSTRY

Item of plant	Size/Rating	Unit	Rate per Hour (£)
SMALL TOOLS			
Diamond concrete planer		Each	1.93
Autofeed screwdriver, electric		Each	1.38
Laminate trimmer		Each	0.91
Biscuit jointer		Each	1.49
Random orbital sander		Each	0.97
Floor sander		Each	1.54
Palm, delta, flap or belt sander		Each	0.75
Disc cutter, electric	300 mm	Each	1.49
Disc cutter, 2 stroke petrol	300 mm	Each	1.24
Dust suppressor for petrol disc cutter		Each	0.51
Cutter cart for petrol disc cutter		Each	1.21
Grinder, angle or cutter	up to 225 mm	Each	0.50
Grinder, angle or cutter	300 mm	Each	1.41
Mortar raking tool attachment		Each	0.19
Floor/polisher scrubber	325 mm	Each	1.76
Floor tile stripper		Each	2.44
Wallpaper stripper, electric		Each	0.81
Hot air paint stripper		Each	0.50
Electric diamond tile cutter	all sizes	Each	2.42
Hand tile cutter		Each	0.82
Electric needle gun		Each	1.29
Needle chipping gun		Each	1.85
Pedestrian floor sweeper	250 mm dia.	Each	0.82
Pedestrian floor sweeper	petrol	Each	2.20
Diamond tile saw		Each	1.84
Blow lamp equipment and glass		Set	0.50

Net Zero Energy Building

Ming Hu

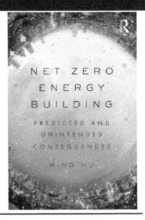

What do we mean by net zero energy? Zero operating energy? Zero energy costs? Zero emissions? There is no one answer: approaches to net zero building vary widely across the globe and are influenced by different environmental and cultural contexts.

Net Zero Energy Building: Predicted and Unintended Consequences presents a comprehensive overview of variations in 'net zero' building practices. Drawing on examples from countries such as the United States, United Kingdom, Germany, Japan, Hong Kong, and China, Ming Hu examines diverse approaches to net zero and reveals their intended and unintended consequences.

Existing approaches often focus on operating energy: how to make buildings more efficient by reducing the energy consumed by climate control, lighting, and appliances. Hu goes beyond this by analyzing overall energy consumption and environmental impact across the entire life cycle of a building—ranging from the manufacture of building materials to transportation, renovation, and demolition. Is net zero building still achievable once we look at these factors?

With clear implications for future practice, this is key reading for professionals in building design, architecture, and construction, as well as students on sustainable and green architecture courses.

April 2019: 162 pp
ISBN: 9780815367802

To Order
Tel: +44 (0) 1235 400524
Email: book.orders@tandf.co.uk

For a complete listing of all our titles visit:
www.tandf.co.uk

Taylor & Francis
Taylor & Francis Group

PART 7

Useful Addresses for Further Information

ACOUSTICAL INVESTIGATION & RESEARCH
ORGANISATION LTD (AIRO)
Tel: 01442 247 146
Website: www.airo.co.uk

Aluminium Federation Ltd (ALFED)
Tel: 0121 601 6361
Website: www.alfed.org.uk

AMERICAN HARDWOOD EXPORT COUNCIL (AHEC)
Tel: 020 7626 4111
Website: www.ahec-europe.org

ANCIENT MONUMENTS SOCIETY (AMS)
Tel: 020 7236 3934
Website: www.ancientmonumentssociety.org.uk

APA – THE ENGINEERED WOOD ASSOCIATION
Tel: 0845 123 3721
Website: http://www.apa-europe.org/

ARBORICULTURAL ASSOCIATION
Tel: 01242 522152
Website: www.trees.org.uk

ARCHITECTURAL ADVISORY SERVICE CENTRE
(POWDER/ANODIC METAL FINISHES)
Tel: 01844 342 425
Website: http://www.aasc.org.uk/

ARCHITECTURAL ASSOCIATION (AA)
Tel: 020 7887 4000
Website: http://www.aaschool.ac.uk/

ARCHITECTURAL CLADDING ASSOCIATION (ACA)
Tel: 0116 253 6161
Website: http://www.architectural-cladding-association.org.uk/

ASBESTOS INFORMATION CENTRE (AIC)
Tel: 01283 531 126
Website: http://www.aic.org.uk

ASSOCIATION FOR CONSULTANCY AND ENGINEERING
Tel: 020 7222 6557
Website: www.acenet.co.uk

ASSOCIATION OF INTERIOR SPECIALISTS
Tel: 0121 707 0077
Website: www.ais-interiors.org.uk

ASSOCIATION OF LOADING AND ELEVATING
EQUIPMENT MANUFACTURERS
Tel: 020 8253 4501
Website: www.alem.org.uk

ASSOCIATION OF PROJECT MANAGEMENT
Tel: 0845 458 1944
Website: www.apm.org.uk

BOX CULVERT ASSOCIATION (BCA)
Tel: 0116 253 6161
Website: www.boxculvert.org.uk

BRITISH ADHESIVES AND SEALANTS ASSOCIATION
Tel: 03302 233290
Website: www.basa.uk.com

BRITISH APPROVALS FOR FIRE EQUIPMENT (BAFE)
Tel: 0844 335 0897
Website: www.bafe.org.uk

BRITISH APPROVALS SERVICE FOR CABLES (BASEC)
Tel: 01908 267 300
Website: www.basec.org.uk

BRITISH ARCHITECTURAL LIBRARY (BAL)
Tel: 020 7580 5533
Website: www.architecture.com

BRITISH ASSOCIATION OF LANDSCAPE INDUSTRIES (BALI)
Tel: 024 7669 0333
Website: www.bali.co.uk

BRITISH ASSOCIATION OF REINFORCEMENT
Tel: 07802 747031
Website: www.uk-bar.org

BRITISH BATHROOM COUNCIL
(BATHROOM MANUFACTURERS ASSOCIATION)
Tel: 01782 747 123
Website: www.bathroom-assciation.org

BRITISH BOARD OF AGREMENT (BBA)
Tel: 01923 665 300
Website: www.bbacerts.co.uk

BRITISH CABLES ASSOCIATION (BCA)
Tel: 020 8941 4079
Website: www.bcauk.org

BRITISH CARPET MANUFACTURERS ASSOCIATION LTD (BCMA)
Tel: 01562 755 568
Website: www.carpetfoundation.com

BRITISH CEMENT ASSOCIATION (BCA)
CENTRE FOR CONCRETE INFORMATION
Tel: 01344 466 007
Website: www.cementindustry.co.uk

BRITISH CERAMIC CONFEDERATION (BCC)
Tel: 01782 744 631
Website: www.ceramfed.co.uk

BRITISH CERAMIC RESEARCH LTD (BCR)
Tel: 01782 764 444
Website: www.ceram.co.uk

BRITISH CERAMIC TILE COUNCIL (BCTC TILE ASSOCIATION)
Tel: 01782 747 147
Website: www.tpb.org.uk/

BRITISH COMBUSTION EQUIPMENT MANUFACTURERS ASSOCIATION (BCEMA)
Tel: 0116 275 7111
Website: bcema.co.uk

BRITISH CONSTRUCTIONAL STEELWORK ASSOCIATION LTD (BCSA)
Tel: 020 7839 8566
Website: www.steelconstruction.org

BRITISH CONTRACT FURNISHING ASSOCIATION (BCFA)
Tel: 020 7226 6641
Website: thebcfa.com

BRITISH ELECTROTECHNICAL APPROVALS BOARD (BEAB)
Tel: 01483 455 466
Website: www.beab.co.uk

BRITISH FURNITURE MANUFACTURERS FEDERATION LTD (BFM Ltd)
Tel: 020 7724 0851
Website: www.bfm.org.uk

BRITISH GEOLOGICAL SURVEY (BGS)
Tel: 0115 936 3100
Website: www.thebgs.co.uk

BRITISH INSTITUTE OF ARCHITECTURAL TECHNOLOGISTS (BIAT)
Tel: 020 7278 2206
Website: www.biat.org.uk

BRITISH LAMINATE FABRICATORS ASSOCIATION
Tel: 0845 056 8496
Website: www.blfa.co.uk

BRITISH LIBRARY BIBLIOGRAPHIC SERVICE AND
DOCUMENT SUPPLY
Tel: 01937 546 548
Website: www.bl.uk

BRITISH LIBRARY ENVIRONMENTAL INFORMATION
SERVICE
Tel: 020 7412 7000
Website: www.bl.uk/environment

BRITISH PLASTICS FEDERATION (BPF)
Tel: 020 7457 5000
Website: www.bpf.co.uk

BRITISH PRECAST CONCRETE FEDERATION LTD
Tel: 0116 232 5170
Website: www.britishprecast.org

BRITISH PROPERTY FEDERATION (BPF)
Tel: 020 7828 0111
Website: www.bpf.org.uk

BRITISH RUBBER MANUFACTURERS' ASSOCIATION
LTD (BRMA)
Tel: 020 7457 5040
Website: www.brma.co.uk

BRITISH STAINLESS STEEL ASSOCIATION
Tel: 0114 267 1260
Website: www.bssa.org.uk

BRITISH STANDARDS INSTITUTION (BSI)
Tel: 020 8996 9001
Website: www.bsigroup.com

BRITISH WATER
Tel: 020 7957 4554
Website: www.britishwater.co.uk

BRITISH WOOD PRESERVING & DAMP-PROOFING
ASSOCIATION (BWPDA)
Tel: 020 8519 2588
Website: www.bwpda.co.uk

BRITISH WOODWORKING FEDERATION
Tel: 0870 458 6939
Website: www.bwf.org.uk

BUILDERS MERCHANTS FEDERATION
Tel: 020 7439 1753
Website: www.bmf.org.uk

BUILDING & ENGINEERING SERVICES ASSOCIATION
Tel: 020 7313 4900
Website: www.hvca.org.uk

BUILDING CENTRE
Tel: 020 7692 4000
Website: www.buildingcentre.co.uk

BUILDING COST INFORMATION SERVICE LTD (BCIS)
Tel: 020 7695 1500
Website: www.bcis.co.uk

BUILDING EMPLOYERS CONFEDERATION (BEC)
Tel: 0870 898 9090
Website: www.thecc.org.uk

BUILDING MAINTENANCE INFORMATION (BMI)
Tel: 020 7695 1500
Website: www.bcis.co.uk

BUILDING RESEARCH ESTABLISHMENT (BRE)
Tel: 01923 664 000
Website: www.bre.co.uk

BUILDING RESEARCH ESTABLISHMENT: SCOTLAND
(BRE)
Tel: 01355 576 200
Website: www.bre.co.uk

BUILDING SERVICES RESEARCH AND INFORMATION
ASSOCIATION LTD
Tel: 01344 465 600
Website: www.bsria.co.uk

CASTINGS TECHNOLOGY INTERNATIONAL
Tel: 0114 272 8647
Website: www.castingsdev.com

CATERING EQUIPMENT SUPPLIERS ASSOCIATION
(CEMA)
Tel: 020 7793 3030
Website: www.cesa.org.uk

CEMENT ADMIXTURES ASSOCIATION
Tel: 01564 776 362
Website: www.admixtures.org.uk

CHARTERED INSTITUTE OF ARBITRATORS
Tel: 0207 421 7444
Website: www.ciarb.org

CHARTERED INSTITUTE OF ARCHITECTURAL
TECHNOLOGISTS
Tel: 0207 278 2206
Website: www.ciat.org.uk

CHARTERED INSTITUTE OF BUILDING (CIOB)
Tel: 01344 630 700
Website: www.ciob.org.uk

CHARTERED INSTITUTE OF WASTES MANAGEMENT
Tel: 01604 620 426
Website: www.iwm.co.uk

CHARTERED INSTITUTION OF BUILDING SERVICES
ENGINEERS (CIBSE)
Tel: 020 8675 5211
Website: www.cibse.org

CIVIL ENGINEERING CONTRACTORS ASSOCIATION
Tel: 020 7340 0450
Website: www.ceca.co.uk

CLAY PIPE DEVELOPMENT ASSOCIATION (CPDA)
Tel: 01494 791 456
Website: www.cpda.co.uk

CLAY ROOF TILE COUNCIL
Tel: 01782 744 631
Website: www.clayroof.co.uk

COLD ROLLED SECTIONS ASSOCIATION (CRSA)
Tel: 0121 601 6350
Website: www.crsauk.com

COMMONWEALTH ASSOCIATION OF ARCHITECTS
Tel: 020 8951 0550
Website: www.comarchitect.org

CONCRETE BRIDGE DEVELOPMENT GROUP
Tel: 01276 33777
Website: www.concrete.org.uk

CONCRETE PIPE ASSOCIATION (CPA)
Tel: 0116 232 5170
Website: www.concretepipes.co.uk

CONCRETE REPAIR ASSOCIATION (CRA)
Tel: 01252 321 302
Website: www.concreterepair.org.uk

CONCRETE SOCIETY ADVISORY SERVICE
Tel: 01276 607 140
Website: www.concrete.org.uk

CONFEDERATION OF BRITISH INDUSTRY (CBI)
Tel: 020 7379 7400
Website: www.cbi.org.uk

CONSTRUCT – CONCRETE STRUCTURES GROUP LTD
Tel: 01276 38444
Website: www.construct.org.uk

CONSTRUCTION EMPLOYERS FEDERATION LTD (CEF)
Tel: 028 9087 7143
Website: www.cefni.co.uk

CONSTRUCTION INDUSTRY JOINT COUNCIL (CIJC)
Tel: 0870 898 9090
Website: http: //www.nscc.org.uk/Website:

CONSTRUCTION INDUSTRY RESEARCH AND
INFORMATION ASSOCIATION
Tel: 020 7549 3300
Website: www.ciria.org.uk

CONSTRUCTION PLANT-HIRE ASSOCIATION (CPA)
Tel: 020 7796 3366
Website: www.cpa.uk.net

CONTRACT FLOORING ASSOCIATION (CFA)
Tel: 0115 941 1126
Website: www.cfa.org.uk

CONTRACTORS MECHANICAL PLANT ENGINEERS
(CMPE)
Tel: 023 925 70011
Website: www.cmpe.co.uk/

COPPER DEVELOPMENT ASSOCIATION
Tel: 01727 731 200
Website: www.cda.org.uk

COUNCIL FOR ALUMINIUM IN BUILDING (CAB)
Tel: 01453 828851
Website: www.c-a-b.org.uk

DEPARTMENT FOR BUSINESS, ENERGY & INDUSTRIAL
STRATEGY
Website: https://www.gov.uk/government/organizations/
department-for-business-energy-and-industrial-
strategy#content

DEPARTMENT FOR TRANSPORT
Tel: 0300 330 3000
Website: www.dft.gov.uk

DOORS & HARDWARE FEDERATION
Tel: 01827 52337
Website: www.abhm.org.uk

DRY STONE WALLING ASSOCIATION OF GREAT
BRITAIN (DSWA)
Tel: 01539 567 953
Website: www.dswa.org.uk

ELECTRICAL CONTRACTORS ASSOCIATION (ECA)
Tel: 020 7313 4800
Website: www.eca.co.uk

ELECTRICAL CONTRACTORS ASSOCIATION OF
SCOTLAND (SELECT)
Tel: 0131 445 5577
Website: www.select.org.uk

ELECTRICAL INSTALLATION EQUIPMENT
MANUFACTURERS ASSOCIATION LTD (EIEMA)
Tel: 020 7793 3000
Website: www.eiema.org.uk

EUROPEAN LIQUID ROOFING ASSOCIATION (ELRA)
Tel: 01444 417 458
Website: www.elra.org.uk

FEDERATION OF ENVIRONMENTAL TRADE
ASSOCIATIONS
Tel: 0118 940 3416
Website: www.feta.co.uk

FEDERATION OF MANUFACTURERS OF
CONSTRUCTION EQUIPMENT & CRANES
Tel: 020 8665 5727
Website: www.coneq.org.uk

FEDERATION OF MASTER BUILDERS
Tel: 020 7242 7583
Website: www.fmb.org.uk/

FEDERATION OF PILING SPECIALISTS
Tel: 020 8663 0947
Website: www.fps.org.uk

FEDERATION OF PLASTERING & DRYWALL
CONTRACTORS
Tel: 020 7608 5092
Website: www.fpdc.org

FENCING CONTRACTORS ASSOCIATION
Tel: 07000 560 722
Website: www.fencingcontractors.org

FINNISH PLYWOOD INTERNATIONAL
Tel: 01438 798 746

FLAT ROOFING ALLIANCE
Tel: 01444 440 027
Website: www.fra.org.uk

FURNITURE INDUSTRY RESEARCH ASSOCIATION
(FIRA INTERNATIONAL LTD)
Tel: 01438 777 700
Website: www.fira.co.uk

GLASS & GLAZING FEDERATION (GGF)
Tel: 020 7403 7177
Website: www.ggf.org.uk

HOUSING CORPORATION HEADQUARTERS
Tel: 0845 230 7000
Website: www.housingcorp.gov.uk

ICOM Energy Association
Tel: 01926 513748
Website: www.icome.org.uk

INSTITUTE OF ACOUSTICS
Tel: 01727 848 195
Website: www.ioa.org.uk

INSTITUTE OF ASPHALT TECHNOLOGY
Tel: 01306 742 792
Website: www.instofasphalt.org

INSTITUTE OF MAINTENANCE AND BUILDING
MANAGEMENT
Tel: 01252 710 994
Website: www.imbm.org.uk

INSTITUTE OF MATERIALS
Tel: 020 7451 7300
Website: www.materials.org.uk

INSTITUTE OF PLUMBING
Tel: 01708 472 791
Website: www.plumbers.org.uk

INSTITUTE OF WOOD SCIENCE
Tel: 01494 565 374
Website: www.iwsc.org.uk

INSTITUTION OF CIVIL ENGINEERS
Tel: 020 7222 7722
Website: www.ice.org.uk

INSTITUTION OF CIVIL ENGINEERS (ICE)
Tel: 020 7222 7722
Website: www.ice.org.uk

INSTITUTION OF ENGINEERING AND TECHNOLOGY
Tel: 01438 313 311
Website: www.theiet.org

INSTITUTION OF MECHANICAL ENGINEERS
Tel: 0207 222 7899
Website: www.imeche.org

INSTITUTION OF STRUCTURAL ENGINEERS (ISE)
Tel: 020 7235 4535
Website: www.istructe.org

INTERNATIONAL LEAD ASSOCIATION
Tel: 020 7499 8422
Website: www.ldaint.org

INTERPAVE (THE PRECAST CONCRETE PAVING
& KERB ASSOCIATION)
Tel: 0116 253 6161
Website: www.paving.org.uk/

JOINT CONTRACTS TRIBUNAL LTD
Tel: 020 7630 8650
Web site: www.jctltd.co.uk

KITCHEN SPECIALISTS ASSOCIATION
Tel: 01905 621 787
Website: www.kbsa.co.uk

LIGHTING ASSOCIATION LTD
Tel: 01952 290 905
Website: www.lightingassociation.com/

MASTIC ASPHALT COUNCIL LTD
Tel: 01424 814 400
Website: www.masticasphaltcouncil.co.uk

METAL CLADDING & ROOFING MANUFACTURERS
ASSOCIATION
Tel: 0151 652 3846
Website: www.mcrma.co.uk

METAL GUTTER MANUFACTURERS ASSOCIATION
Tel: 01633 891584
Website: www.mgma.co.uk

MET OFFICE
Tel: 0870 900 0100
Website: www.metoffice.gov.uk

NATIONAL ACCESS AND SCAFFOLDING
CONFEDERATION
Tel: 0207 822 7400
Website: www.nasc.org.uk

NATIONAL ASSOCIATION OF STEEL STOCKHOLDERS
Tel: 0121 200 2288
Website: www.nass.org.uk

NATIONAL HOUSE-BUILDING COUNCIL (NHBC)
Tel: 0844 633 1000
Website: www.nhbc.co.uk

NATURAL SLATE QUARRIES ASSOCIATION
Tel: 020 7323 3770
Website: http://slateassociation.org

NHS ESTATES
Tel: 0113 254 7000
Website: http://www.buyingsolutions.gov.uk/healthcms/

ORDNANCE SURVEY
Tel: 08456 050 504
Website: www.ordnancesurvey.co.uk

PAINTING AND DECORATING ASSOCIATION
Tel: 01203 353 776
Website: www.paintingdecoratingassociation.co.uk

PIPELINE INDUSTRIES GUILD
Tel: 020 7235 7938
Website: www.pipeguild.co.uk

PLASTIC PIPES GROUP
Tel: 020 7457 5024
Website: www.plasticpipesgroup.com

PRECAST FLOORING FEDERATION
Tel: 0116 253 6161
Website: www.pff.org.uk

PRESTRESSED CONCRETE ASSOCIATION
Tel: 0116 253 6161
Website: www.britishprecast.org

PROPERTY CONSULTANTS SOCIETY LTD
Tel: 01903 883 787
Website: www.p-c-s.org.uk

QUARRY PRODUCTS ASSOCIATION
Tel: 020 7963 8000
Website: www.qpa.org

READY-MIXED CONCRETE BUREAU
Tel: 01344 725 732
Website: www.rcb.org.uk

REINFORCED CONCRETE COUNCIL
Tel: 01276 607140
Website: www.rcc-info.org.uk

ROYAL INCORPORATION OF ARCHITECTS IN
SCOTLAND (RIAS)
Website: www.rias.org.uk

ROYAL INSTITUTE OF BRITISH ARCHITECTS (RIBA)
Tel: 020 7580 5533
Website: www.architecture.com

ROYAL INSTITUTION OF CHARTERED SURVEYORS
(RICS)
Tel: 020 7222 7000
Website: www.rics.org

ROYAL TOWN PLANNING INSTITUTE (RTPI)
Tel: 020 7929 9494
Website: www.rtpi.org.uk/

RURAL DESIGN AND BUILDING ASSOCIATION
Tel: 01449 676 049
Website: www.rdba.org.uk

SCOTTISH BUILDING EMPLOYERS FEDERATION
Tel: 01324 555 550
Website: www.scottish-building.co.uk

SCOTTISH HOMES – Community Scotland
Tel: 0131 313 0044
Website: www.scotland.gov.uk

SCOTTISH NATURAL HERITAGE
Tel: 0131 447 4784
Website: www.snh.org.uk

SINGLE PLY ROOFING ASSOCIATION
Tel: 0115 914 4445
Website: www.spra.co.uk/

SMOKE CONTROL ASSOCIATION
Tel: 0118 940 3416
Website: www.feta.co.uk

SOCIETY FOR THE PROTECTION OF ANCIENT
BUILDINGS (SPAB)
Tel: 020 7377 1644
Website: www.spab.org.uk

SOCIETY OF GLASS TECHNOLOGY
Tel: 0114 263 4455
Website: www.sgt.org

SOIL SURVEY AND LAND RESEARCH INSTITUTE
Tel: 01525 863 000
Website: www.cranfield.ac.uk/sslrc

SOLAR ENERGY SOCIETY
Tel: 01865 741 111
Website: www.uk-ises.org/

SPON'S PRICE BOOK AUTHORS
AECOM LTD
Website: www.aecom.com

SPORT ENGLAND
Tel: 0845 850 8508
Website: www.sportengland.org

SPORT SCOTLAND
Tel: 0131 317 7200
Website: www.sportscotland.org.uk

SPORTS COUNCIL FOR WALES
Tel: 0845 045 0904
Website: www.sports-council-wales.co.uk

SPORTS TURF RESEARCH INSTITUTE (STRI)
Tel: 01274 565 131
Website: www.stri.co.uk

SPRAYED CONCRETE ASSOCIATION
Tel: 01252 321 302
Website: www.sca.org.uk

STEEL CONSTRUCTION INSTITUTE
Tel: 01344 636 505
Website: www.steel-sci.org

STEEL WINDOW ASSOCIATION
Tel: 020 7637 3571
Website: www.steel-window-association.co.uk

STONE FEDERATION GREAT BRITAIN
Tel: 01303 856123
Website: www.stone-federationgb.org.uk

SUSPENDED ACCESS EQUIPMENT
MANUFACTURERS ASSOCIATION
Tel: 020 7608 5098
Website: www.saema.org

SWIMMING POOL & ALLIED TRADES ASSOCIATION
(SPATA)
Tel: 01264 356210
Website: www.spata.co.uk

THE PIPELINE INDUSTRIES GUILD
Tel: 020 7235 7938
Website: www.pipeline.com

THERMAL INSULATION CONTRACTORS ASSOCIATION
Tel: 01325 466 704
Website: www.tica-acad.co.uk

TIMBER RESEARCH & DEVELOPMENT ASSOCIATION
(TRADA)
Tel: 01494 569 600
Website www.trada.co.uk

TIMBER TRADE FEDERATION
Tel: 020 7839 1891
Website: www.ttf.co.uk

TOWN & COUNTRY PLANNING ASSOCIATION (TCPA)
Tel: 020 7930 8903
Website: tcpa.org.uk

TREE COUNCIL
Tel: 020 7407 9992
Website: www.treecouncil.org.uk

TRUSSED RAFTER ASSOCIATION
Tel: 01777 869 281
Website: www.tra.org.uk

TWI
Tel: 01223 899 000
Website: www.twi.co.uk

UNDERFLOOR HEATING MANUFACTURERS'
ASSOCIATION
Tel: 020 8941 7177
Website: www.uhma.org.uk

VERMICULITE INFORMATION SERVICE
Tel: 01483 242 100

WALLCOVERING MANUFACTURERS ASSOCIATION
Tel: 01372 360 660

WATERHEATER MANUFACTURERS ASSOCIATION
Tel: 07775 754456
Website: www.waterheating.fsnet.co.uk/wma.htm

WATER RESEARCH CENTRE
Tel: 01491 636 500
Website: www.wrcplc.co.uk

WATER UK
Tel: 020 7344 1844
Website: www.water.org.uk/

WELDING MANUFACTURERS' ASSOCIATION
Tel: 020 7793 3041
Website: www.wma.uk.com

WOOD PANEL INDUSTRIES FEDERATION
Tel: 01476 563 707
Website: www.wpif.org.uk

WRAP
Tel: 01295 819 900
Website: www.wrap.org.uk

ZINC DEVELOPMENT ASSOCIATION
Tel: 020 7499 6636
Website: www.zincinfocentre.org

New Aspects of Quantity Surveying Practice, 4th edition

Duncan Cartlidge

In this fourth edition of *New Aspects of Quantity Surveying Practice,* renowned quantity surveying author Duncan Cartlidge reviews the history of the quantity surveyor, examines and reflects on the state of current practice with a concentration on new and innovative practice, and attempts to predict the future direction of quantity surveying practice in the UK and worldwide.

The book champions the adaptability and flexibility of the quantity surveyor, whilst covering the hot topics which have emerged since the previous edition's publication, including:

- the RICS 'Futures' publication;
- Building Information Modelling (BIM);
- mergers and acquisitions;
- a more informed and critical evaluation of the NRM;
- greater discussion of ethics to reflect on the renewed industry interest;
- and a new chapter on Dispute Resolution.

As these issues create waves throughout the industry whilst it continues its global growth in emerging markets, such reflections on QS practice are now more important than ever. The book is essential reading for all Quantity Surveying students, teachers and professionals. It is particularly suited to undergraduate professional skills courses and non-cognate postgraduate students looking for an up to date understanding of the industry and the role.

December 2017: 282 pp
ISBN: 9781138673762

To Order
Tel:+44 (0) 1235 400524
Email: tandf@bookpoint.co.uk

For a complete listing of all our titles visit:
www.tandf.co.uk

PART 8

Tables and Memoranda

This part contains the following sections:

Professional Ethics in Construction and Surveying

Greg Watts, Jason Challender, Anthony Higham and Peter McDermott

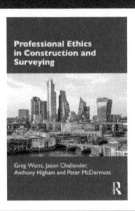

This textbook responds to the increasing demand for practical, industry aligned, ethical practices in quantity surveying, construction management and related AEC professions.

Professional Ethics for Construction and Surveying addresses how existing ethical standards can be pragmatically applied to both private and contracting practice, with case studies aligned with the ethical requirements of the main professional bodies. After an introduction to ethics, the authors present real-world situations where the minimum legal and contractual requirements necessitate the combination of professional judgement and ethical decision-making. They outline how such situations arise, then address how decisions can and should be made that are in keeping with the moral, contractual and CSR requirements, with cases covering the building lifecycle from procurement to handover. Consequently, the book brings together ethical theory, existing worldwide ethical standards and the requirements of the RICS, the CIOB and the ICES, with the authors' experiences of examining candidates for entry into the professional bodies.

The result is a professionally focused textbook aimed at vocational learners (at both undergraduate and postgraduate taught levels) and practitioners in construction, engineering, architecture and the wider built environment.

May 2021: 188 pp
ISBN: 9780367354190

To Order
Tel:+44 (0) 1235 400524
Email: tandf@bookpoint.co.uk

For a complete listing of all our titles visit:
www.tandf.co.uk

Taylor & Francis
Taylor & Francis Group

CONVERSION TABLES

CONVERSION TABLES

Length	Unit	Conversion factors			
Millimetre	mm	1 in	= 25.4 mm	1 mm	= 0.0394 in
Centimetre	cm	1 in	= 2.54 cm	1 cm	= 0.3937 in
Metre	m	1 ft	= 0.3048 m	1 m	= 3.2808 ft
		1 yd	= 0.9144 m		= 1.0936 yd
Kilometre	km	1 mile	= 1.6093 km	1 km	= 0.6214 mile

Note:

1 cm	= 10 mm	1 ft	= 12 in
1 m	= 1 000 mm	1 yd	= 3 ft
1 km	= 1 000 m	1 mile	= 1 760 yd

Area	Unit	Conversion factors			
Square Millimetre	mm^2	$1\ in^2$	$= 645.2\ mm^2$	$1\ mm^2$	$= 0.0016\ in^2$
Square Centimetre	cm^2	$1\ in^2$	$= 6.4516\ cm^2$	$1\ cm^2$	$= 1.1550\ in^2$
Square Metre	m^2	$1\ ft^2$	$= 0.0929\ m^2$	$1\ m^2$	$= 10.764\ ft^2$
		$1\ yd^2$	$= 0.8361\ m^2$	$1\ m^2$	$= 1.1960\ yd^2$
Square Kilometre	km^2	$1\ mile^2$	$= 2.590\ km^2$	$1\ km^2$	$= 0.3861\ mile^2$

Note:

$1\ cm^2$	$= 100\ mm^2$	$1\ ft^2$	$= 144\ in^2$
$1\ m^2$	$= 10\ 000\ cm^2$	$1\ yd^2$	$= 9\ ft^2$
$1\ km^2$	$= 100$ hectares	1 acre	$= 4\ 840\ yd^2$
		$1\ mile^2$	$= 640$ acres

Volume	Unit	Conversion factors			
Cubic Centimetre	cm^3	$1\ cm^3$	$= 0.0610\ in^3$	$1\ in^3$	$= 16.387\ cm^3$
Cubic Decimetre	dm^3	$1\ dm^3$	$= 0.0353\ ft^3$	$1\ ft^3$	$= 28.329\ dm^3$
Cubic Metre	m^3	$1\ m^3$	$= 35.3147\ ft^3$	$1\ ft^3$	$= 0.0283\ m^3$
		$1\ m^3$	$= 1.3080\ yd^3$	$1\ yd^3$	$= 0.7646\ m^3$
Litre	l	1 l	= 1.76 pint	1 pint	= 0.5683 l
			= 2.113 US pt		= 0.4733 US l

Note:

$1\ dm^3$	$= 1\ 000\ cm^3$	$1\ ft^3$	$= 1\ 728\ in^3$	1 pint	= 20 fl oz
$1\ m^3$	$= 1\ 000\ dm^3$	$1\ yd^3$	$= 27\ ft^3$	1 gal	= 8 pints
1 l	$= 1\ dm^3$				

Neither the Centimetre nor Decimetre are SI units, and as such their use, particularly that of the Decimetre, is not widespread outside educational circles.

Mass	Unit	Conversion factors			
Milligram	mg	1 mg	= 0.0154 grain	1 grain	= 64.935 mg
Gram	g	1 g	= 0.0353 oz	1 oz	= 28.35 g
Kilogram	kg	1 kg	= 2.2046 lb	1 lb	= 0.4536 kg
Tonne	t	1 t	= 0.9842 ton	1 ton	= 1.016 t

Note:

1 g	= 1000 mg	1 oz	= 437.5 grains	1 cwt	= 112 lb
1 kg	= 1000 g	1 lb	= 16 oz	1 ton	= 20 cwt
1 t	= 1000 kg	1 stone	= 14 lb		

Force	Unit	Conversion factors			
Newton	N	1 lbf	= 4.448 N	1 kgf	= 9.807 N
Kilonewton	kN	1 lbf	= 0.004448 kN	1 ton f	= 9.964 kN
Meganewton	MN	100 tonf	= 0.9964 MN		

CONVERSION TABLES

Pressure and stress	Unit	Conversion factors
Kilonewton per square metre	kN/m^2	1 lbf/in^2 = 6.895 kN/m^2
		1 bar = 100 kN/m^2
Meganewton per square metre	MN/m^2	1 tonf/ft^2 = 107.3 kN/m^2 = 0.1073 MN/m^2
		1 kgf/cm^2 = 98.07 kN/m^2
		1 lbf/ft^2 = 0.04788 kN/m^2

Coefficient of consolidation (Cv) or swelling	Unit	Conversion factors
Square metre per year	$m^2/year$	1 cm^2/s = 3 154 m^2/year
		1 ft^2/year = 0.0929 m^2/year

Coefficient of permeability	Unit	Conversion factors
Metre per second	m/s	1 cm/s = 0.01 m/s
Metre per year	m/year	1 ft/year = 0.3048 m/year
		= 0.9651 × (10)^8 m/s

Temperature	Unit	Conversion factors	
Degree Celsius	°C	°C = 5/9 × (°F − 32)	°F = (9 × °C)/ 5 + 32

CONVERSION TABLES

SPEED CONVERSION

km/h	m/min	mph	fpm
1	16.7	0.6	54.7
2	33.3	1.2	109.4
3	50.0	1.9	164.0
4	66.7	2.5	218.7
5	83.3	3.1	273.4
6	100.0	3.7	328.1
7	116.7	4.3	382.8
8	133.3	5.0	437.4
9	150.0	5.6	492.1
10	166.7	6.2	546.8
11	183.3	6.8	601.5
12	200.0	7.5	656.2
13	216.7	8.1	710.8
14	233.3	8.7	765.5
15	250.0	9.3	820.2
16	266.7	9.9	874.9
17	283.3	10.6	929.6
18	300.0	11.2	984.3
19	316.7	11.8	1038.9
20	333.3	12.4	1093.6
21	350.0	13.0	1148.3
22	366.7	13.7	1203.0
23	383.3	14.3	1257.7
24	400.0	14.9	1312.3
25	416.7	15.5	1367.0
26	433.3	16.2	1421.7
27	450.0	16.8	1476.4
28	466.7	17.4	1531.1
29	483.3	18.0	1585.7
30	500.0	18.6	1640.4
31	516.7	19.3	1695.1
32	533.3	19.9	1749.8
33	550.0	20.5	1804.5
34	566.7	21.1	1859.1
35	583.3	21.7	1913.8
36	600.0	22.4	1968.5
37	616.7	23.0	2023.2
38	633.3	23.6	2077.9
39	650.0	24.2	2132.5
40	666.7	24.9	2187.2

CONVERSION TABLES

km/h	m/min	mph	fpm
41	683.3	25.5	2241.9
42	700.0	26.1	2296.6
43	716.7	26.7	2351.3
44	733.3	27.3	2405.9
45	750.0	28.0	2460.6
46	766.7	28.6	2515.3
47	783.3	29.2	2570.0
48	800.0	29.8	2624.7
49	816.7	30.4	2679.4
50	833.3	31.1	2734.0

GEOMETRY

Two dimensional figures

Figure	Diagram of figure	Surface area	Perimeter
Square		a^2	$4a$
Rectangle		ab	$2(a+b)$
Triangle		$\frac{1}{2}ch$	$a+b+c$
Circle		πr^2 $\frac{1}{4}\pi d^2$ where $2r = d$	$2\pi r$ πd
Parallelogram		ah	$2(a+b)$
Trapezium		$\frac{1}{2}h(a+b)$	$a+b+c+d$
Ellipse		Approximately πab	$\pi(a+b)$
Hexagon		$2.6 \times a^2$	

GEOMETRY

Figure	Diagram of figure	Surface area	Perimeter
Octagon		$4.83 \times a^2$	$6a$
Sector of a circle		$\frac{1}{2}rb$ or $\frac{q}{360}\pi r^2$ note $b = $ angle $\frac{q}{360} \times \pi 2r$	
Segment of a circle		$S - T$ where S = area of sector, T = area of triangle	
Bellmouth		$\frac{3}{14} \times r^2$	

GEOMETRY

Three dimensional figures

Figure	Diagram of figure	Surface area	Volume
Cube		$6a^2$	a^3
Cuboid/ rectangular block		$2(ab + ac + bc)$	abc
Prism/ triangular block		$bd + hc + dc + ad$	$\frac{1}{2}hcd$
Cylinder		$2\pi r^2 + 2\pi h$	$\pi r^2 h$
Sphere		$4\pi r^2$	$\frac{4}{3}\pi r^3$
Segment of sphere		$2\pi Rh$	$\frac{1}{6}\pi h(3r^2 + h^2)$ $\frac{1}{3}\pi h^2(3R - H)$
Pyramid		$(a + b)l + ab$	$\frac{1}{3}abh$

GEOMETRY

Figure	Diagram of figure	Surface area	Volume
Frustum of a pyramid		$l(a + b + c + d) + \sqrt{(ab + cd)}$ [rectangular figure only]	$\frac{h}{3}(ab + cd + \sqrt{abcd})$
Cone		πrl (excluding base) $\pi rl + \pi r^2$ (including base)	$\frac{1}{3}\pi r^2 h$ $\frac{1}{12}\pi d^2 h$
Frustum of a cone		$\pi r^2 + \pi R^2 + \pi l(R + r)$	$\frac{1}{3}\pi(R^2 + Rr + r^2)$

FORMULAE

Formulae

Formula	Description
Pythagoras Theorem	$A^2 = B^2 + C^2$ where A is the hypotenuse of a right-angled triangle and B and C are the two adjacent sides
Simpsons Rule	The Area is divided into an even number of strips of equal width, and therefore has an odd number of ordinates at the division points $$\text{area} = \frac{S(A + 2B + 4C)}{3}$$ where S = common interval (strip width) A = sum of first and last ordinates B = sum of remaining odd ordinates C = sum of the even ordinates The Volume can be calculated by the same formula, but by substituting the area of each coordinate rather than its length
Trapezoidal Rule	A given trench is divided into two equal sections, giving three ordinates, the first, the middle and the last $$\text{volume} = \frac{S \times (A + B + 2C)}{2}$$ where S = width of the strips A = area of the first section B = area of the last section C = area of the rest of the sections
Prismoidal Rule	A given trench is divided into two equal sections, giving three ordinates, the first, the middle and the last $$\text{volume} = \frac{L \times (A + 4B + C)}{6}$$ where L = total length of trench A = area of the first section B = area of the middle section C = area of the last section

TYPICAL THERMAL CONDUCTIVITY OF BUILDING MATERIALS

(Always check manufacturer's details – variation will occur depending on product and nature of materials)

	Thermal conductivity (W/mK)		Thermal conductivity (W/mK)
Acoustic plasterboard	0.25	Oriented strand board	0.13
Aerated concrete slab ($500 \, kg/m^3$)	0.16	Outer leaf brick	0.77
Aluminium	237	Plasterboard	0.22
Asphalt ($1700 \, kg/m^3$)	0.5	Plaster dense ($1300 \, kg/m^3$)	0.5
Bitumen-impregnated fibreboard	0.05	Plaster lightweight ($600 \, kg/m^3$)	0.16
Blocks (standard grade $600 \, kg/m^3$)	0.15	Plywood ($950 \, kg/m^3$)	0.16
Blocks (solar grade $460 \, kg/m^3$)	0.11	Prefabricated timber wall panels (check manufacturer)	0.12
Brickwork (outer leaf $1700 \, kg/m^3$)	0.84	Screed ($1200 \, kg/m^3$)	0.41
Brickwork (inner leaf $1700 \, kg/m^3$)	0.62	Stone chippings ($1800 \, kg/m^3$)	0.96
Dense aggregate concrete block $1800 \, kg/m^3$ (exposed)	1.21	Tile hanging ($1900 \, kg/m^3$)	0.84
Dense aggregate concrete block $1800 \, kg/m^3$ (protected)	1.13	Timber ($650 \, kg/m^3$)	0.14
Calcium silicate board ($600 \, kg/m^3$)	0.17	Timber flooring ($650 \, kg/m^3$)	0.14
Concrete general	1.28	Timber rafters	0.13
Concrete (heavyweight $2300 \, kg/m^3$)	1.63	Timber roof or floor joists	0.13
Concrete (dense $2100 \, kg/m^3$ typical floor)	1.4	Roof tile ($1900 \, kg/m^3$)	0.84
Concrete (dense $2000 \, kg/m^3$ typical floor)	1.13	Timber blocks ($650 \, kg/m^3$)	0.14
Concrete (medium $1400 \, kg/m^3$)	0.51	Cellular glass	0.045
Concrete (lightweight $1200 \, kg/m^3$)	0.38	Expanded polystyrene	0.034
Concrete (lightweight $600 \, kg/m^3$)	0.19	Expanded polystyrene slab ($25 \, kg/m^3$)	0.035
Concrete slab (aerated $500 \, kg/m^3$)	0.16	Extruded polystyrene	0.035
Copper	390	Glass mineral wool	0.04
External render sand/cement finish	1	Mineral quilt ($12 \, kg/m^3$)	0.04
External render ($1300 \, kg/m^3$)	0.5	Mineral wool slab ($25 \, kg/m^3$)	0.035
Felt – Bitumen layers ($1700 \, kg/m^3$)	0.5	Phenolic foam	0.022
Fibreboard ($300 \, kg/m^3$)	0.06	Polyisocyanurate	0.025
Glass	0.93	Polyurethane	0.025
Marble	3	Rigid polyurethane	0.025
Metal tray used in wriggly tin concrete floors ($7800 \, kg/m^3$)	50	Rock mineral wool	0.038
Mortar ($1750 \, kg/m^3$)	0.8		

EARTHWORK

Weights of Typical Materials Handled by Excavators

The weight of the material is that of the state in its natural bed and includes moisture.
Adjustments should be made to allow for loose or compacted states

Material	Mass (kg/m³)	Mass (lb/cu yd)
Ashes, dry	610	1028
Ashes, wet	810	1365
Basalt, broken	1954	3293
Basalt, solid	2933	4943
Bauxite, crushed	1281	2159
Borax, fine	849	1431
Caliche	1440	2427
Cement, clinker	1415	2385
Chalk, fine	1221	2058
Chalk, solid	2406	4055
Cinders, coal, ash	641	1080
Cinders, furnace	913	1538
Clay, compacted	1746	2942
Clay, dry	1073	1808
Clay, wet	1602	2700
Coal, anthracite, solid	1506	2538
Coal, bituminous	1351	2277
Coke	610	1028
Dolomite, lumpy	1522	2565
Dolomite, solid	2886	4864
Earth, dense	2002	3374
Earth, dry, loam	1249	2105
Earth, Fullers, raw	673	1134
Earth, moist	1442	2430
Earth, wet	1602	2700
Felsite	2495	4205
Fieldspar, solid	2613	4404
Fluorite	3093	5213
Gabbro	3093	5213
Gneiss	2696	4544
Granite	2690	4534
Gravel, dry ¼ to 2 inch	1682	2835
Gravel, dry, loose	1522	2565
Gravel, wet ¼ to 2 inch	2002	3374
Gypsum, broken	1450	2444
Gypsum, solid	2787	4697
Hardcore (consolidated)	1928	3249
Lignite, dry	801	1350
Limestone, broken	1554	2619
Limestone, solid	2596	4375
Magnesite, magnesium ore	2993	5044
Marble	2679	4515
Marl, wet	2216	3735
Mica, broken	1602	2700
Mica, solid	2883	4859
Peat, dry	400	674
Peat, moist	700	1179

EARTHWORK

Material	Mass (kg/m³)	Mass (lb/cu yd)
Peat, wet	1121	1889
Potash	1281	2159
Pumice, stone	640	1078
Quarry waste	1438	2423
Quartz sand	1201	2024
Quartz, solid	2584	4355
Rhyolite	2400	4045
Sand and gravel, dry	1650	2781
Sand and gravel, wet	2020	3404
Sand, dry	1602	2700
Sand, wet	1831	3086
Sandstone, solid	2412	4065
Shale, solid	2637	4444
Slag, broken	2114	3563
Slag, furnace granulated	961	1619
Slate, broken	1370	2309
Slate, solid	2667	4495
Snow, compacted	481	810
Snow, freshly fallen	160	269
Taconite	2803	4724
Trachyte	2400	4045
Trap rock, solid	2791	4704
Turf	400	674
Water	1000	1685

Transport Capacities

Type of vehicle	Capacity of vehicle	
	Payload	Heaped capacity
Wheelbarrow	150	0.10
1 tonne dumper	1250	1.00
2.5 tonne dumper	4000	2.50
Articulated dump truck (Volvo A20 6 × 4)	18500	11.00
Articulated dump truck (Volvo A35 6 × 6)	32000	19.00
Large capacity rear dumper (Euclid R35)	35000	22.00
Large capacity rear dumper (Euclid R85)	85000	50.00

EARTHWORK

Machine Volumes for Excavating and Filling

Machine type	Cycles per minute	Volume per minute (m³)
1.5 tonne excavator	1	0.04
	2	0.08
	3	0.12
3 tonne excavator	1	0.13
	2	0.26
	3	0.39
5 tonne excavator	1	0.28
	2	0.56
	3	0.84
7 tonne excavator	1	0.28
	2	0.56
	3	0.84
21 tonne excavator	1	1.21
	2	2.42
	3	3.63
Backhoe loader JCB3CX excavator Rear bucket capacity 0.28 m³	1	0.28
	2	0.56
	3	0.84
Backhoe loader JCB3CX loading Front bucket capacity 1.00 m³	1	1.00
	2	2.00

Machine Volumes for Excavating and Filling

Machine type	Loads per hour	Volume per hour (m³)
1 tonne high tip skip loader Volume 0.485 m³	5	2.43
	7	3.40
	10	4.85
3 tonne dumper Max volume 2.40 m³ Available volume 1.9 m³	4	7.60
	5	9.50
	7	13.30
	10	19.00
6 tonne dumper Max volume 3.40 m³ Available volume 3.77 m³	4	15.08
	5	18.85
	7	26.39
	10	37.70

EARTHWORK

Bulkage of Soils (after excavation)

Type of soil	Approximate bulking of 1 m³ after excavation
Vegetable soil and loam	25–30%
Soft clay	30–40%
Stiff clay	10–20%
Gravel	20–25%
Sand	40–50%
Chalk	40–50%
Rock, weathered	30–40%
Rock, unweathered	50–60%

Shrinkage of Materials (on being deposited)

Type of soil	Approximate bulking of 1 m³ after excavation
Clay	10%
Gravel	8%
Gravel and sand	9%
Loam and light sandy soils	12%
Loose vegetable soils	15%

Voids in Material Used as Subbases or Beddings

Material	m³ of voids/m³
Alluvium	0.37
River grit	0.29
Quarry sand	0.24
Shingle	0.37
Gravel	0.39
Broken stone	0.45
Broken bricks	0.42

Angles of Repose

Type of soil		Degrees
Clay	– dry	30
	– damp, well drained	45
	– wet	15–20
Earth	– dry	30
	– damp	45
Gravel	– moist	48
Sand	– dry or moist	35
	– wet	25
Loam		40

EARTHWORK

Slopes and Angles

Ratio of base to height	Angle in degrees
5:1	11
4:1	14
3:1	18
2:1	27
1½:1	34
1:1	45
1:1½	56
1:2	63
1:3	72
1:4	76
1:5	79

Grades (in Degrees and Percents)

Degrees	Percent	Degrees	Percent
1	1.8	24	44.5
2	3.5	25	46.6
3	5.2	26	48.8
4	7.0	27	51.0
5	8.8	28	53.2
6	10.5	29	55.4
7	12.3	30	57.7
8	14.0	31	60.0
9	15.8	32	62.5
10	17.6	33	64.9
11	19.4	34	67.4
12	21.3	35	70.0
13	23.1	36	72.7
14	24.9	37	75.4
15	26.8	38	78.1
16	28.7	39	81.0
17	30.6	40	83.9
18	32.5	41	86.9
19	34.4	42	90.0
20	36.4	43	93.3
21	38.4	44	96.6
22	40.4	45	100.0
23	42.4		

EARTHWORK

Bearing Powers

Ground conditions		Bearing power		
		kg/m^2	lb/in^2	Metric t/m^2
Rock,	broken	483	70	50
	solid	2415	350	240
Clay,	dry or hard	380	55	40
	medium dry	190	27	20
	soft or wet	100	14	10
Gravel,	cemented	760	110	80
Sand,	compacted	380	55	40
	clean dry	190	27	20
Swamp and alluvial soils		48	7	5

Earthwork Support

Maximum depth of excavation in various soils without the use of earthwork support

Ground conditions	Feet (ft)	Metres (m)
Compact soil	12	3.66
Drained loam	6	1.83
Dry sand	1	0.3
Gravelly earth	2	0.61
Ordinary earth	3	0.91
Stiff clay	10	3.05

It is important to note that the above table should only be used as a guide. Each case must be taken on its merits and, as the limited distances given above are approached, careful watch must be kept for the slightest signs of caving in

CONCRETE WORK

Weights of Concrete and Concrete Elements

Type of material		kg/m³	lb/cu ft
Ordinary concrete (dense aggregates)			
Non-reinforced plain or mass concrete			
Nominal weight		2305	144
Aggregate	– limestone	2162 to 2407	135 to 150
	– gravel	2244 to 2407	140 to 150
	– broken brick	2000 (av)	125 (av)
	– other crushed stone	2326 to 2489	145 to 155
Reinforced concrete			
Nominal weight		2407	150
Reinforcement	– 1%	2305 to 2468	144 to 154
	– 2%	2356 to 2519	147 to 157
	– 4%	2448 to 2703	153 to 163
Special concretes			
Heavy concrete			
Aggregates	– barytes, magnetite	3210 (min)	200 (min)
	– steel shot, punchings	5280	330
Lean mixes			
Dry-lean (gravel aggregate)		2244	140
Soil-cement (normal mix)		1601	100

CONCRETE WORK

Type of material		kg/m² per mm thick	lb/sq ft per inch thick
Ordinary concrete (dense aggregates)			
Solid slabs (floors, walls etc.)			
Thickness:	75 mm or 3 in	184	37.5
	100 mm or 4 in	245	50
	150 mm or 6 in	378	75
	250 mm or 10 in	612	125
	300 mm or 12 in	734	150
Ribbed slabs			
Thickness:	125 mm or 5 in	204	42
	150 mm or 6 in	219	45
	225 mm or 9 in	281	57
	300 mm or 12 in	342	70
Special concretes			
Finishes etc.			
	Rendering, screed etc. Granolithic, terrazzo	1928 to 2401	10 to 12.5
	Glass-block (hollow) concrete	1734 (approx)	9 (approx)
Prestressed concrete		Weights as for reinforced concrete (upper limits)	
Air-entrained concrete		Weights as for plain or reinforced concrete	

CONCRETE WORK

Average Weight of Aggregates

Materials	Voids %	Weight kg/m³
Sand	39	1660
Gravel 10–20 mm	45	1440
Gravel 35–75 mm	42	1555
Crushed stone	50	1330
Crushed granite (over 15 mm)	50	1345
(n.e. 15 mm)	47	1440
'All-in' ballast	32	1800–2000

Material	kg/m³	lb/cu yd
Vermiculite (aggregate)	64–80	108–135
All-in aggregate	1999	125

Applications and Mix Design

Site mixed concrete

Recommended mix	Class of work suitable for	Cement (kg)	Sand (kg)	Coarse aggregate (kg)	Nr 25 kg bags cement per m³ of combined aggregate
1:3:6	Roughest type of mass concrete such as footings, road haunching over 300 mm thick	208	905	1509	8.30
1:2.5:5	Mass concrete of better class than 1:3:6 such as bases for machinery, walls below ground etc.	249	881	1474	10.00
1:2:4	Most ordinary uses of concrete, such as mass walls above ground, road slabs etc. and general reinforced concrete work	304	889	1431	12.20
1:1.5:3	Watertight floors, pavements and walls, tanks, pits, steps, paths, surface of 2 course roads, reinforced concrete where extra strength is required	371	801	1336	14.90
1:1:2	Works of thin section such as fence posts and small precast work	511	720	1206	20.40

CONCRETE WORK

Ready mixed concrete

Application	Designated concrete	Standardized prescribed concrete	Recommended consistence (nominal slump class)
Foundations			
Mass concrete fill or blinding	GEN 1	ST2	S3
Strip footings	GEN 1	ST2	S3
Mass concrete foundations			
Single storey buildings	GEN 1	ST2	S3
Double storey buildings	GEN 3	ST4	S3
Trench fill foundations			
Single storey buildings	GEN 1	ST2	S4
Double storey buildings	GEN 3	ST4	S4
General applications			
Kerb bedding and haunching	GEN 0	ST1	S1
Drainage works – immediate support	GEN 1	ST2	S1
Other drainage works	GEN 1	ST2	S3
Oversite below suspended slabs	GEN 1	ST2	S3
Floors			
Garage and house floors with no embedded steel	GEN 3	ST4	S2
Wearing surface: Light foot and trolley traffic	RC30	ST4	S2
Wearing surface: General industrial	RC40	N/A	S2
Wearing surface: Heavy industrial	RC50	N/A	S2
Paving			
House drives, domestic parking and external parking	PAV 1	N/A	S2
Heavy-duty external paving	PAV 2	N/A	S2

CONCRETE WORK

Prescribed Mixes for Ordinary Structural Concrete

Weights of cement and total dry aggregates in kg to produce approximately one cubic metre of fully compacted concrete together with the percentages by weight of fine aggregate in total dry aggregates

Conc. grade	Nominal max size of aggregate (mm)	40		20		14		10	
	Workability	Med.	High	Med.	High	Med.	High	Med.	High
	Limits to slump that may be expected (mm)	50–100	100–150	25–75	75–125	10–50	50–100	10–25	25–50
7	Cement (kg)	180	200	210	230	–	–	–	–
	Total aggregate (kg)	1950	1850	1900	1800	–	–	–	–
	Fine aggregate (%)	30–45	30–45	35–50	35–50	–	–	–	–
10	Cement (kg)	210	230	240	260	–	–	–	–
	Total aggregate (kg)	1900	1850	1850	1800	–	–	–	–
	Fine aggregate (%)	30–45	30–45	35–50	35–50	–	–	–	–
15	Cement (kg)	250	270	280	310	–	–	–	–
	Total aggregate (kg)	1850	1800	1800	1750	–	–	–	–
	Fine aggregate (%)	30–45	30–45	35–50	35–50	–	–	–	–
20	Cement (kg)	300	320	320	350	340	380	360	410
	Total aggregate (kg)	1850	1750	1800	1750	1750	1700	1750	1650
	Sand								
	Zone 1 (%)	35	40	40	45	45	50	50	55
	Zone 2 (%)	30	35	35	40	40	45	45	50
	Zone 3 (%)	30	30	30	35	35	40	40	45
25	Cement (kg)	340	360	360	390	380	420	400	450
	Total aggregate (kg)	1800	1750	1750	1700	1700	1650	1700	1600
	Sand								
	Zone 1 (%)	35	40	40	45	45	50	50	55
	Zone 2 (%)	30	35	35	40	40	45	45	50
	Zone 3 (%)	30	30	30	35	35	40	40	45
30	Cement (kg)	370	390	400	430	430	470	460	510
	Total aggregate (kg)	1750	1700	1700	1650	1700	1600	1650	1550
	Sand								
	Zone 1 (%)	35	40	40	45	45	50	50	55
	Zone 2 (%)	30	35	35	40	40	45	45	50
	Zone 3 (%)	30	30	30	35	35	40	40	45

REINFORCEMENT

Weights of Bar Reinforcement

Nominal sizes (mm)	Cross-sectional area (mm²)	Mass (kg/m)	Length of bar (m/tonne)
6	28.27	0.222	4505
8	50.27	0.395	2534
10	78.54	0.617	1622
12	113.10	0.888	1126
16	201.06	1.578	634
20	314.16	2.466	405
25	490.87	3.853	260
32	804.25	6.313	158
40	1265.64	9.865	101
50	1963.50	15.413	65

Weights of Bars (at specific spacings)

Weights of metric bars in kilogrammes per square metre

Size (mm)	Spacing of bars in millimetres									
	75	100	125	150	175	200	225	250	275	300
6	2.96	2.220	1.776	1.480	1.27	1.110	0.99	0.89	0.81	0.74
8	5.26	3.95	3.16	2.63	2.26	1.97	1.75	1.58	1.44	1.32
10	8.22	6.17	4.93	4.11	3.52	3.08	2.74	2.47	2.24	2.06
12	11.84	8.88	7.10	5.92	5.07	4.44	3.95	3.55	3.23	2.96
16	21.04	15.78	12.63	10.52	9.02	7.89	7.02	6.31	5.74	5.26
20	32.88	24.66	19.73	16.44	14.09	12.33	10.96	9.87	8.97	8.22
25	51.38	38.53	30.83	25.69	22.02	19.27	17.13	15.41	14.01	12.84
32	84.18	63.13	50.51	42.09	36.08	31.57	28.06	25.25	22.96	21.04
40	131.53	98.65	78.92	65.76	56.37	49.32	43.84	39.46	35.87	32.88
50	205.51	154.13	123.31	102.76	88.08	77.07	68.50	61.65	56.05	51.38

Basic weight of steelwork taken as 7850 kg/m³
Basic weight of bar reinforcement per metre run = 0.00785 kg/mm²
The value of π has been taken as 3.141592654

REINFORCEMENT

Fabric Reinforcement

Preferred range of designated fabric types and stock sheet sizes

Fabric reference	Longitudinal wires			Cross wires			
	Nominal wire size (mm)	Pitch (mm)	Area (mm²/m)	Nominal wire size (mm)	Pitch (mm)	Area (mm²/m)	Mass (kg/m²)
Square mesh							
A393	10	200	393	10	200	393	6.16
A252	8	200	252	8	200	252	3.95
A193	7	200	193	7	200	193	3.02
A142	6	200	142	6	200	142	2.22
A98	5	200	98	5	200	98	1.54
Structural mesh							
B1131	12	100	1131	8	200	252	10.90
B785	10	100	785	8	200	252	8.14
B503	8	100	503	8	200	252	5.93
B385	7	100	385	7	200	193	4.53
B283	6	100	283	7	200	193	3.73
B196	5	100	196	7	200	193	3.05
Long mesh							
C785	10	100	785	6	400	70.8	6.72
C636	9	100	636	6	400	70.8	5.55
C503	8	100	503	5	400	49.0	4.34
C385	7	100	385	5	400	49.0	3.41
C283	6	100	283	5	400	49.0	2.61
Wrapping mesh							
D98	5	200	98	5	200	98	1.54
D49	2.5	100	49	2.5	100	49	0.77

Stock sheet size 4.8 m × 2.4 m, Area 11.52 m²

Average weight kg/m³ of steelwork reinforcement in concrete for various building elements

Substructure	kg/m³ concrete	Substructure	kg/m³ concrete
Pile caps	110–150	Plate slab	150–220
Tie beams	130–170	Cant slab	145–210
Ground beams	230–330	Ribbed floors	130–200
Bases	125–180	Topping to block floor	30–40
Footings	100–150	Columns	210–310
Retaining walls	150–210	Beams	250–350
Raft	60–70	Stairs	130–170
Slabs – one way	120–200	Walls – normal	40–100
Slabs – two way	110–220	Walls – wind	70–125

Note: For exposed elements add the following %:
Walls 50%, Beams 100%, Columns 15%

FORMWORK

Formwork Stripping Times – Normal Curing Periods

Conditions under which concrete is maturing	Minimum periods of protection for different types of cement					
	Number of days (where the average surface temperature of the concrete exceeds 10°C during the whole period)			Equivalent maturity (degree hours) calculated as the age of the concrete in hours multiplied by the number of degrees Celsius by which the average surface temperature of the concrete exceeds 10°C		
	Other	SRPC	OPC or RHPC	Other	SRPC	OPC or RHPC
1. Hot weather or drying winds	7	4	3	3500	2000	1500
2. Conditions not covered by 1	4	3	2	2000	1500	1000

KEY
OPC – Ordinary Portland Cement
RHPC – Rapid-hardening Portland Cement
SRPC – Sulphate-resisting Portland Cement

Minimum Period before Striking Formwork

	Minimum period before striking		
	Surface temperature of concrete		
	16°C	17°C	t°C (0–25)
Vertical formwork to columns, walls and large beams	12 hours	18 hours	300 hours t+10
Soffit formwork to slabs	4 days	6 days	100 days t+10
Props to slabs	10 days	15 days	250 days t+10
Soffit formwork to beams	9 days	14 days	230 days t+10
Props to beams	14 days	21 days	360 days t+10

MASONRY

Number of Bricks Required for Various Types of Work per m² of Walling

Description	Brick size	
	215 × 102.5 × 50 mm	215 × 102.5 × 65 mm
Half brick thick		
Stretcher bond	74	59
English bond	108	86
English garden wall bond	90	72
Flemish bond	96	79
Flemish garden wall bond	83	66
One brick thick and cavity wall of two half brick skins		
Stretcher bond	148	119

Quantities of Bricks and Mortar Required per m² of Walling

	Unit	No of bricks required	Mortar required (cubic metres)		
Standard bricks			**No frogs**	**Single frogs**	**Double frogs**
Brick size 215 × 102.5 × 50 mm					
half brick wall (103 mm)	m²	72	0.022	0.027	0.032
2 × half brick cavity wall (270 mm)	m²	144	0.044	0.054	0.064
one brick wall (215 mm)	m²	144	0.052	0.064	0.076
one and a half brick wall (322 mm)	m²	216	0.073	0.091	0.108
Mass brickwork	m³	576	0.347	0.413	0.480
Brick size 215 × 102.5 × 65 mm					
half brick wall (103 mm)	m²	58	0.019	0.022	0.026
2 × half brick cavity wall (270 mm)	m²	116	0.038	0.045	0.055
one brick wall (215 mm)	m²	116	0.046	0.055	0.064
one and a half brick wall (322 mm)	m²	174	0.063	0.074	0.088
Mass brickwork	m³	464	0.307	0.360	0.413
Metric modular bricks			**Perforated**		
Brick size 200 × 100 × 75 mm					
90 mm thick	m²	67	0.016	0.019	
190 mm thick	m²	133	0.042	0.048	
290 mm thick	m²	200	0.068	0.078	
Brick size 200 × 100 × 100 mm					
90 mm thick	m²	50	0.013	0.016	
190 mm thick	m²	100	0.036	0.041	
290 mm thick	m²	150	0.059	0.067	
Brick size 300 × 100 × 75 mm					
90 mm thick	m²	33	–	0.015	
Brick size 300 × 100 × 100 mm					
90 mm thick	m²	44	0.015	0.018	

Note: Assuming 10 mm thick joints

MASONRY

Mortar Required per m² Blockwork (9.88 blocks/m²)

Wall thickness	75	90	100	125	140	190	215
Mortar m³/m²	0.005	0.006	0.007	0.008	0.009	0.013	0.014

Mortar Group	Cement: lime: sand	Masonry cement: sand	Cement: sand with plasticizer
1	1:0–0.25:3		
2	1:0.5:4–4.5	1:2.5-3.5	1:3–4
3	1:1:5–6	1:4–5	1:5–6
4	1:2:8–9	1:5.5–6.5	1:7–8
5	1:3:10–12	1:6.5–7	1:8

Group 1: strong inflexible mortar
Group 5: weak but flexible

All mixes within a group are of approximately similar strength
Frost resistance increases with the use of plasticizers
Cement: lime: sand mixes give the strongest bond and greatest resistance to rain penetration
Masonry cement equals ordinary Portland cement plus a fine neutral mineral filler and an air entraining agent

Calcium Silicate Bricks

Type	Strength	Location
Class 2 crushing strength	14.0 N/mm²	not suitable for walls
Class 3	20.5 N/mm²	walls above dpc
Class 4	27.5 N/mm²	cappings and copings
Class 5	34.5 N/mm²	retaining walls
Class 6	41.5 N/mm²	walls below ground
Class 7	48.5 N/mm²	walls below ground

The Class 7 calcium silicate bricks are therefore equal in strength to Class B bricks
Calcium silicate bricks are not suitable for DPCs

Durability of Bricks	
FL	Frost resistant with low salt content
FN	Frost resistant with normal salt content
ML	Moderately frost resistant with low salt content
MN	Moderately frost resistant with normal salt content

MASONRY

Brickwork Dimensions

No. of horizontal bricks	Dimensions (mm)	No. of vertical courses	Height of vertical courses (mm)
½	112.5	1	75
1	225.0	2	150
1½	337.5	3	225
2	450.0	4	300
2½	562.5	5	375
3	675.0	6	450
3½	787.5	7	525
4	900.0	8	600
4½	1012.5	9	675
5	1125.0	10	750
5½	1237.5	11	825
6	1350.0	12	900
6½	1462.5	13	975
7	1575.0	14	1050
7½	1687.5	15	1125
8	1800.0	16	1200
8½	1912.5	17	1275
9	2025.0	18	1350
9½	2137.5	19	1425
10	2250.0	20	1500
20	4500.0	24	1575
40	9000.0	28	2100
50	11250.0	32	2400
60	13500.0	36	2700
75	16875.0	40	3000

TIMBER

TIMBER

Weights of Timber

Material	kg/m³	lb/cu ft
General	806 (avg)	50 (avg)
Douglas fir	479	30
Yellow pine, spruce	479	30
Pitch pine	673	42
Larch, elm	561	35
Oak (English)	724 to 959	45 to 60
Teak	643 to 877	40 to 55
Jarrah	959	60
Greenheart	1040 to 1204	65 to 75
Quebracho	1285	80
Material	**kg/m² per mm thickness**	**lb/sq ft per inch thickness**
Wooden boarding and blocks		
Softwood	0.48	2.5
Hardwood	0.76	4
Hardboard	1.06	5.5
Chipboard	0.76	4
Plywood	0.62	3.25
Blockboard	0.48	2.5
Fibreboard	0.29	1.5
Wood-wool	0.58	3
Plasterboard	0.96	5
Weather boarding	0.35	1.8

TIMBER

Conversion Tables (for timber only)

Inches	Millimetres	Feet	Metres
1	25	1	0.300
2	50	2	0.600
3	75	3	0.900
4	100	4	1.200
5	125	5	1.500
6	150	6	1.800
7	175	7	2.100
8	200	8	2.400
9	225	9	2.700
10	250	10	3.000
11	275	11	3.300
12	300	12	3.600
13	325	13	3.900
14	350	14	4.200
15	375	15	4.500
16	400	16	4.800
17	425	17	5.100
18	450	18	5.400
19	475	19	5.700
20	500	20	6.000
21	525	21	6.300
22	550	22	6.600
23	575	23	6.900
24	600	24	7.200

Planed Softwood

The finished end section size of planed timber is usually 3/16" less than the original size from which it is produced. This however varies slightly depending upon availability of material and origin of the species used.

Standards (timber) to cubic metres and cubic metres to standards (timber)

Cubic metres	Cubic metres standards	Standards
4.672	1	0.214
9.344	2	0.428
14.017	3	0.642
18.689	4	0.856
23.361	5	1.070
28.033	6	1.284
32.706	7	1.498
37.378	8	1.712
42.050	9	1.926
46.722	10	2.140
93.445	20	4.281
140.167	30	6.421
186.890	40	8.561
233.612	50	10.702
280.335	60	12.842
327.057	70	14.982
373.779	80	17.122

TIMBER

1 cu metre = 35.3148 cu ft = 0.21403 std

1 cu ft = 0.028317 cu metres

1 std = 4.67227 cu metres

Basic sizes of sawn softwood available (cross-sectional areas)

Thickness (mm)	Width (mm)								
	75	100	125	150	175	200	225	250	300
16	X	X	X	X					
19	X	X	X	X					
22	X	X	X	X					
25	X	X	X	X	X	X	X	X	X
32	X	X	X	X	X	X	X	X	X
36	X	X	X	X					
38	X	X	X	X	X	X	X		
44	X	X	X	X	X	X	X	X	X
47*	X	X	X	X	X	X	X	X	X
50	X	X	X	X	X	X	X	X	X
63	X	X	X	X	X	X	X		
75	X	X	X	X	X	X	X	X	
100		X		X		X		X	X
150				X		X			X
200						X			
250								X	
300									X

* This range of widths for 47 mm thickness will usually be found to be available in construction quality only

Note: The smaller sizes below 100 mm thick and 250 mm width are normally but not exclusively of European origin. Sizes beyond this are usually of North and South American origin

Basic lengths of sawn softwood available (metres)

1.80	2.10	3.00	4.20	5.10	6.00	7.20
	2.40	3.30	4.50	5.40	6.30	
	2.70	3.60	4.80	5.70	6.60	
		3.90			6.90	

Note: Lengths of 6.00 m and over will generally only be available from North American species and may have to be recut from larger sizes

TIMBER

Reductions from basic size to finished size by planning of two opposed faces

Purpose	Reductions from basic sizes for timber			
	15–35 mm	36–100 mm	101–150 mm	over 150 mm
a) Constructional timber	3 mm	3 mm	5 mm	6 mm
b) Matching interlocking boards	4 mm	4 mm	6 mm	6 mm
c) Wood trim not specified in BS 584	5 mm	7 mm	7 mm	9 mm
d) Joinery and cabinet work	7 mm	9 mm	11 mm	13 mm

Note: The reduction of width or depth is overall the extreme size and is exclusive of any reduction of the face by the machining of a tongue or lap joints

Maximum Spans for Various Roof Trusses

Maximum permissible spans for rafters for Fink trussed rafters

Basic size (mm)	Actual size (mm)	Pitch (degrees)								
		15 (m)	17.5 (m)	20 (m)	22.5 (m)	25 (m)	27.5 (m)	30 (m)	32.5 (m)	35 (m)
38 × 75	35 × 72	6.03	6.16	6.29	6.41	6.51	6.60	6.70	6.80	6.90
38 × 100	35 × 97	7.48	7.67	7.83	7.97	8.10	8.22	8.34	8.47	8.61
38 × 125	35 × 120	8.80	9.00	9.20	9.37	9.54	9.68	9.82	9.98	10.16
44 × 75	41 × 72	6.45	6.59	6.71	6.83	6.93	7.03	7.14	7.24	7.35
44 × 100	41 × 97	8.05	8.23	8.40	8.55	8.68	8.81	8.93	9.09	9.22
44 × 125	41 × 120	9.38	9.60	9.81	9.99	10.15	10.31	10.45	10.64	10.81
50 × 75	47 × 72	6.87	7.01	7.13	7.25	7.35	7.45	7.53	7.67	7.78
50 × 100	47 × 97	8.62	8.80	8.97	9.12	9.25	9.38	9.50	9.66	9.80
50 × 125	47 × 120	10.01	10.24	10.44	10.62	10.77	10.94	11.00	11.00	11.00

TIMBER

Sizes of Internal and External Doorsets

Description	Internal size (mm)	Permissible deviation	External size (mm)	Permissible deviation
Coordinating dimension: height of door leaf height sets	2100		2100	
Coordinating dimension: height of ceiling height set	2300 2350 2400 2700 3000		2300 2350 2400 2700 3000	
Coordinating dimension: width of all doorsets S = Single leaf set D = Double leaf set	600 S 700 S 800 S&D 900 S&D 1000 S&D 1200 D 1500 D 1800 D 2100 D		900 S 1000 S 1200 D 1800 D 2100 D	
Work size: height of door leaf height set	2090	± 2.0	2095	± 2.0
Work size: height of ceiling height set	2285 2335 2385 2685 2985	± 2.0	2295 2345 2395 2695 2995	± 2.0
Work size: width of all doorsets S = Single leaf set D = Double leaf set	590 S 690 S 790 S&D 890 S&D 990 S&D 1190 D 1490 D 1790 D 2090 D	± 2.0	895 S 995 S 1195 D 1495 D 1795 D 2095 D	± 2.0
Width of door leaf in single leaf sets F = Flush leaf P = Panel leaf	526 F 626 F 726 F&P 826 F&P 926 F&P	± 1.5	806 F&P 906 F&P	± 1.5
Width of door leaf in double leaf sets F = Flush leaf P = Panel leaf	362 F 412 F 426 F 562 F&P 712 F&P 826 F&P 1012 F&P	± 1.5	552 F&P 702 F&P 852 F&P 1002 F&P	± 1.5
Door leaf height for all doorsets	2040	± 1.5	1994	± 1.5

ROOFING

Total Roof Loadings for Various Types of Tiles/Slates

	Roof load (slope) kg/m²		
	Slate/Tile	Roofing underlay and battens²	Total dead load kg/m
Asbestos cement slate (600 × 300)	21.50	3.14	24.64
Clay tile interlocking	67.00	5.50	72.50
plain	43.50	2.87	46.37
Concrete tile interlocking	47.20	2.69	49.89
plain	78.20	5.50	83.70
Natural slate (18" × 10")	35.40	3.40	38.80
	Roof load (plan) kg/m²		
Asbestos cement slate (600 × 300)	28.45	76.50	104.95
Clay tile interlocking	53.54	76.50	130.04
plain	83.71	76.50	60.21
Concrete tile interlocking	57.60	76.50	134.10
plain	96.64	76.50	173.14

ROOFING

Tiling Data

Product		Lap (mm)	Gauge of battens	No. slates per m²	Battens (m/m²)	Weight as laid (kg/m²)
CEMENT SLATES						
Eternit slates	600 × 300 mm	100	250	13.4	4.00	19.50
(Duracem)		90	255	13.1	3.92	19.20
		80	260	12.9	3.85	19.00
		70	265	12.7	3.77	18.60
	600 × 350 mm	100	250	11.5	4.00	19.50
		90	255	11.2	3.92	19.20
	500 × 250 mm	100	200	20.0	5.00	20.00
		90	205	19.5	4.88	19.50
		80	210	19.1	4.76	19.00
		70	215	18.6	4.65	18.60
	400 × 200 mm	90	155	32.3	6.45	20.80
		80	160	31.3	6.25	20.20
		70	165	30.3	6.06	19.60
CONCRETE TILES/SLATES						
Redland Roofing						
Stonewold slate	430 × 380 mm	75	355	8.2	2.82	51.20
Double Roman tile	418 × 330 mm	75	355	8.2	2.91	45.50
Grovebury pantile	418 × 332 mm	75	343	9.7	2.91	47.90
Norfolk pantile	381 × 227 mm	75	306	16.3	3.26	44.01
		100	281	17.8	3.56	48.06
Renown interlocking tile	418 × 330 mm	75	343	9.7	2.91	46.40
'49' tile	381 × 227 mm	75	306	16.3	3.26	44.80
		100	281	17.8	3.56	48.95
Plain, vertical tiling	265 × 165 mm	35	115	52.7	8.70	62.20
Marley Roofing						
Bold roll tile	420 × 330 mm	75	344	9.7	2.90	47.00
		100	–	10.5	3.20	51.00
Modern roof tile	420 × 330 mm	75	338	10.2	3.00	54.00
		100	–	11.0	3.20	58.00
Ludlow major	420 × 330 mm	75	338	10.2	3.00	45.00
		100	–	11.0	3.20	49.00
Ludlow plus	387 × 229 mm	75	305	16.1	3.30	47.00
		100	–	17.5	3.60	51.00
Mendip tile	420 × 330 mm	75	338	10.2	3.00	47.00
		100	–	11.0	3.20	51.00
Wessex	413 × 330 mm	75	338	10.2	3.00	54.00
		100	–	11.0	3.20	58.00
Plain tile	267 × 165 mm	65	100	60.0	10.00	76.00
		75	95	64.0	10.50	81.00
		85	90	68.0	11.30	86.00
Plain vertical tiles (feature)	267 × 165 mm	35	110	53.0	8.70	67.00
		34	115	56.0	9.10	71.00

ROOFING

Slate Nails, Quantity per Kilogram

Length	Type			
	Plain wire	Galvanized wire	Copper nail	Zinc nail
28.5 mm	325	305	325	415
34.4 mm	286	256	254	292
50.8 mm	242	224	194	200

Metal Sheet Coverings

Thicknesses and weights of sheet metal coverings								
Lead to BS 1178								
BS Code No	3	4	5	6	7	8		
Colour code	Green	Blue	Red	Black	White	Orange		
Thickness (mm)	1.25	1.80	2.24	2.50	3.15	3.55		
Density kg/m^2	14.18	20.41	25.40	30.05	35.72	40.26		
Copper to BS 2870								
Thickness (mm)		0.60	0.70					
Bay width								
Roll (mm)		500	650					
Seam (mm)		525	600					
Standard width to form bay	600	750						
Normal length of sheet	1.80	1.80						
Zinc to BS 849								
Zinc Gauge (Nr)	9	10	11	12	13	14	15	16
Thickness (mm)	0.43	0.48	0.56	0.64	0.71	0.79	0.91	1.04
Density (kg/m^2)	3.1	3.2	3.8	4.3	4.8	5.3	6.2	7.0
Aluminium to BS 4868								
Thickness (mm)	0.5	0.6	0.7	0.8	0.9	1.0	1.2	
Density (kg/m^2)	12.8	15.4	17.9	20.5	23.0	25.6	30.7	

ROOFING

Type of felt	Nominal mass per unit area (kg/10 m)	Nominal mass per unit area of fibre base (g/m²)	Nominal length of roll (m)
Class 1			
1B fine granule	14	220	10 or 20
surfaced bitumen	18	330	10 or 20
	25	470	10
1E mineral surfaced bitumen	38	470	10
1F reinforced bitumen	15	160 (fibre) 110 (hessian)	15
1F reinforced bitumen, aluminium faced	13	160 (fibre) 110 (hessian)	15
Class 2			
2B fine granule surfaced bitumen asbestos	18	500	10 or 20
2E mineral surfaced bitumen asbestos	38	600	10
Class 3			
3B fine granule surfaced bitumen glass fibre	18	60	20
3E mineral surfaced bitumen glass fibre	28	60	10
3E venting base layer bitumen glass fibre	32	60*	10
3H venting base layer bitumen glass fibre	17	60*	20

* Excluding effect of perforations

GLAZING

GLAZING

Nominal thickness (mm)	Tolerance on thickness (mm)	Approximate weight (kg/m²)	Normal maximum size (mm)
Float and polished plate glass			
3	+ 0.2	7.50	2140 × 1220
4	+ 0.2	10.00	2760 × 1220
5	+ 0.2	12.50	3180 × 2100
6	+ 0.2	15.00	4600 × 3180
10	+ 0.3	25.00)	6000 × 3300
12	+ 0.3	30.00)	
15	+ 0.5	37.50	3050 × 3000
19	+ 1.0	47.50)	3000 × 2900
25	+ 1.0	63.50)	
Clear sheet glass			
2 *	+ 0.2	5.00	1920 × 1220
3	+ 0.3	7.50	2130 × 1320
4	+ 0.3	10.00	2760 × 1220
5 *	+ 0.3	12.50)	2130 × 2400
6 *	+ 0.3	15.00)	
Cast glass			
3	+ 0.4 / − 0.2	6.00)	2140 × 1280
4	+ 0.5	7.50)	
5	+ 0.5	9.50	2140 × 1320
6	+ 0.5	11.50)	3700 × 1280
10	+ 0.8	21.50)	
Wired glass (Cast wired glass)			
6	+ 0.3 / − 0.7	−))	3700 × 1840
7	+ 0.7	−)	
(Polished wire glass)			
6	+ 1.0	−	330 × 1830

* The 5 mm and 6 mm thickness are known as *thick drawn sheet*. Although 2 mm sheet glass is available it is not recommended for general glazing purposes

METAL

Weights of Metals

Material	kg/m³	lb/cu ft
Metals, steel construction, etc.		
Iron		
– cast	7207	450
– wrought	7687	480
– ore – general	2407	150
– (crushed) Swedish	3682	230
Steel	7854	490
Copper		
– cast	8731	545
– wrought	8945	558
Brass	8497	530
Bronze	8945	558
Aluminium	2774	173
Lead	11322	707
Zinc (rolled)	7140	446

	g/mm² per metre	lb/sq ft per foot
Steel bars	7.85	3.4

Structural steelwork	Net weight of member @ 7854 kg/m³	
riveted	+ 10% for cleats, rivets, bolts, etc.	
welded	+ 1.25% to 2.5% for welds, etc.	
Rolled sections		
beams	+ 2.5%	
stanchions	+ 5% (extra for caps and bases)	
Plate		
web girders	+ 10% for rivets or welds, stiffeners, etc.	

	kg/m	lb/ft
Steel stairs: industrial type		
1 m or 3 ft wide	84	56
Steel tubes		
50 mm or 2 in bore	5 to 6	3 to 4
Gas piping		
20 mm or ¾ in	2	1¼

METAL

Universal Beams BS 4: Part 1: 2005

Designation	Mass (kg/m)	Depth of section (mm)	Width of section (mm)	Thickness		Surface area (m²/m)
				Web (mm)	Flange (mm)	
1016 × 305 × 487	487.0	1036.1	308.5	30.0	54.1	3.20
1016 × 305 × 438	438.0	1025.9	305.4	26.9	49.0	3.17
1016 × 305 × 393	393.0	1016.0	303.0	24.4	43.9	3.15
1016 × 305 × 349	349.0	1008.1	302.0	21.1	40.0	3.13
1016 × 305 × 314	314.0	1000.0	300.0	19.1	35.9	3.11
1016 × 305 × 272	272.0	990.1	300.0	16.5	31.0	3.10
1016 × 305 × 249	249.0	980.2	300.0	16.5	26.0	3.08
1016 × 305 × 222	222.0	970.3	300.0	16.0	21.1	3.06
914 × 419 × 388	388.0	921.0	420.5	21.4	36.6	3.44
914 × 419 × 343	343.3	911.8	418.5	19.4	32.0	3.42
914 × 305 × 289	289.1	926.6	307.7	19.5	32.0	3.01
914 × 305 × 253	253.4	918.4	305.5	17.3	27.9	2.99
914 × 305 × 224	224.2	910.4	304.1	15.9	23.9	2.97
914 × 305 × 201	200.9	903.0	303.3	15.1	20.2	2.96
838 × 292 × 226	226.5	850.9	293.8	16.1	26.8	2.81
838 × 292 × 194	193.8	840.7	292.4	14.7	21.7	2.79
838 × 292 × 176	175.9	834.9	291.7	14.0	18.8	2.78
762 × 267 × 197	196.8	769.8	268.0	15.6	25.4	2.55
762 × 267 × 173	173.0	762.2	266.7	14.3	21.6	2.53
762 × 267 × 147	146.9	754.0	265.2	12.8	17.5	2.51
762 × 267 × 134	133.9	750.0	264.4	12.0	15.5	2.51
686 × 254 × 170	170.2	692.9	255.8	14.5	23.7	2.35
686 × 254 × 152	152.4	687.5	254.5	13.2	21.0	2.34
686 × 254 × 140	140.1	383.5	253.7	12.4	19.0	2.33
686 × 254 × 125	125.2	677.9	253.0	11.7	16.2	2.32
610 × 305 × 238	238.1	635.8	311.4	18.4	31.4	2.45
610 × 305 × 179	179.0	620.2	307.1	14.1	23.6	2.41
610 × 305 × 149	149.1	612.4	304.8	11.8	19.7	2.39
610 × 229 × 140	139.9	617.2	230.2	13.1	22.1	2.11
610 × 229 × 125	125.1	612.2	229.0	11.9	19.6	2.09
610 × 229 × 113	113.0	607.6	228.2	11.1	17.3	2.08
610 × 229 × 101	101.2	602.6	227.6	10.5	14.8	2.07
533 × 210 × 122	122.0	544.5	211.9	12.7	21.3	1.89
533 × 210 × 109	109.0	539.5	210.8	11.6	18.8	1.88
533 × 210 × 101	101.0	536.7	210.0	10.8	17.4	1.87
533 × 210 × 92	92.1	533.1	209.3	10.1	15.6	1.86
533 × 210 × 82	82.2	528.3	208.8	9.6	13.2	1.85
457 × 191 × 98	98.3	467.2	192.8	11.4	19.6	1.67
457 × 191 × 89	89.3	463.4	191.9	10.5	17.7	1.66
457 × 191 × 82	82.0	460.0	191.3	9.9	16.0	1.65
457 × 191 × 74	74.3	457.0	190.4	9.0	14.5	1.64
457 × 191 × 67	67.1	453.4	189.9	8.5	12.7	1.63
457 × 152 × 82	82.1	465.8	155.3	10.5	18.9	1.51
457 × 152 × 74	74.2	462.0	154.4	9.6	17.0	1.50
457 × 152 × 67	67.2	458.0	153.8	9.0	15.0	1.50
457 × 152 × 60	59.8	454.6	152.9	8.1	13.3	1.50
457 × 152 × 52	52.3	449.8	152.4	7.6	10.9	1.48
406 × 178 × 74	74.2	412.8	179.5	9.5	16.0	1.51
406 × 178 × 67	67.1	409.4	178.8	8.8	14.3	1.50
406 × 178 × 60	60.1	406.4	177.9	7.9	12.8	1.49

METAL

Designation	Mass (kg/m)	Depth of section (mm)	Width of section (mm)	Thickness Web (mm)	Thickness Flange (mm)	Surface area (m²/m)
406 × 178 × 50	54.1	402.6	177.7	7.7	10.9	1.48
406 × 140 × 46	46.0	403.2	142.2	6.8	11.2	1.34
406 × 140 × 39	39.0	398.0	141.8	6.4	8.6	1.33
356 × 171 × 67	67.1	363.4	173.2	9.1	15.7	1.38
356 × 171 × 57	57.0	358.0	172.2	8.1	13.0	1.37
356 × 171 × 51	51.0	355.0	171.5	7.4	11.5	1.36
356 × 171 × 45	45.0	351.4	171.1	7.0	9.7	1.36
356 × 127 × 39	39.1	353.4	126.0	6.6	10.7	1.18
356 × 127 × 33	33.1	349.0	125.4	6.0	8.5	1.17
305 × 165 × 54	54.0	310.4	166.9	7.9	13.7	1.26
305 × 165 × 46	46.1	306.6	165.7	6.7	11.8	1.25
305 × 165 × 40	40.3	303.4	165.0	6.0	10.2	1.24
305 × 127 × 48	48.1	311.0	125.3	9.0	14.0	1.09
305 × 127 × 42	41.9	307.2	124.3	8.0	12.1	1.08
305 × 127 × 37	37.0	304.4	123.3	7.1	10.7	1.07
305 × 102 × 33	32.8	312.7	102.4	6.6	10.8	1.01
305 × 102 × 28	28.2	308.7	101.8	6.0	8.8	1.00
305 × 102 × 25	24.8	305.1	101.6	5.8	7.0	0.992
254 × 146 × 43	43.0	259.6	147.3	7.2	12.7	1.08
254 × 146 × 37	37.0	256.0	146.4	6.3	10.9	1.07
254 × 146 × 31	31.1	251.4	146.1	6.0	8.6	1.06
254 × 102 × 28	28.3	260.4	102.2	6.3	10.0	0.904
254 × 102 × 25	25.2	257.2	101.9	6.0	8.4	0.897
254 × 102 × 22	22.0	254.0	101.6	5.7	6.8	0.890
203 × 133 × 30	30.0	206.8	133.9	6.4	9.6	0.923
203 × 133 × 25	25.1	203.2	133.2	5.7	7.8	0.915
203 × 102 × 23	23.1	203.2	101.8	5.4	9.3	0.790
178 × 102 × 19	19.0	177.8	101.2	4.8	7.9	0.738
152 × 89 × 16	16.0	152.4	88.7	4.5	7.7	0.638
127 × 76 × 13	13.0	127.0	76.0	4.0	7.6	0.537

METAL

Universal Columns BS 4: Part 1: 2005

Designation	Mass (kg/m)	Depth of section (mm)	Width of section (mm)	Thickness		Surface area (m²/m)
				Web (mm)	Flange (mm)	
356 × 406 × 634	633.9	474.7	424.0	47.6	77.0	2.52
356 × 406 × 551	551.0	455.6	418.5	42.1	67.5	2.47
356 × 406 × 467	467.0	436.6	412.2	35.8	58.0	2.42
356 × 406 × 393	393.0	419.0	407.0	30.6	49.2	2.38
356 × 406 × 340	339.9	406.4	403.0	26.6	42.9	2.35
356 × 406 × 287	287.1	393.6	399.0	22.6	36.5	2.31
356 × 406 × 235	235.1	381.0	384.8	18.4	30.2	2.28
356 × 368 × 202	201.9	374.6	374.7	16.5	27.0	2.19
356 × 368 × 177	177.0	368.2	372.6	14.4	23.8	2.17
356 × 368 × 153	152.9	362.0	370.5	12.3	20.7	2.16
356 × 368 × 129	129.0	355.6	368.6	10.4	17.5	2.14
305 × 305 × 283	282.9	365.3	322.2	26.8	44.1	1.94
305 × 305 × 240	240.0	352.5	318.4	23.0	37.7	1.91
305 × 305 × 198	198.1	339.9	314.5	19.1	31.4	1.87
305 × 305 × 158	158.1	327.1	311.2	15.8	25.0	1.84
305 × 305 × 137	136.9	320.5	309.2	13.8	21.7	1.82
305 × 305 × 118	117.9	314.5	307.4	12.0	18.7	1.81
305 × 305 × 97	96.9	307.9	305.3	9.9	15.4	1.79
254 × 254 × 167	167.1	289.1	265.2	19.2	31.7	1.58
254 × 254 × 132	132.0	276.3	261.3	15.3	25.3	1.55
254 × 254 × 107	107.1	266.7	258.8	12.8	20.5	1.52
254 × 254 × 89	88.9	260.3	256.3	10.3	17.3	1.50
254 × 254 × 73	73.1	254.1	254.6	8.6	14.2	1.49
203 × 203 × 86	86.1	222.2	209.1	12.7	20.5	1.24
203 × 203 × 71	71.0	215.8	206.4	10.0	17.3	1.22
203 × 203 × 60	60.0	209.6	205.8	9.4	14.2	1.21
203 × 203 × 52	52.0	206.2	204.3	7.9	12.5	1.20
203 × 203 × 46	46.1	203.2	203.6	7.2	11.0	1.19
152 × 152 × 37	37.0	161.8	154.4	8.0	11.5	0.912
152 × 152 × 30	30.0	157.6	152.9	6.5	9.4	0.901
152 × 152 × 23	23.0	152.4	152.2	5.8	6.8	0.889

Tables and Memoranda

METAL

Joists BS 4: Part 1: 2005 (retained for reference, Corus have ceased manufacture in UK)

Designation	Mass (kg/m)	Depth of section (mm)	Width of section (mm)	Thickness Web (mm)	Flange (mm)	Surface area (m²/m)
254 × 203 × 82	82.0	254.0	203.2	10.2	19.9	1.210
203 × 152 × 52	52.3	203.2	152.4	8.9	16.5	0.932
152 × 127 × 37	37.3	152.4	127.0	10.4	13.2	0.737
127 × 114 × 29	29.3	127.0	114.3	10.2	11.5	0.646
127 × 114 × 27	26.9	127.0	114.3	7.4	11.4	0.650
102 × 102 × 23	23.0	101.6	101.6	9.5	10.3	0.549
102 × 44 × 7	7.5	101.6	44.5	4.3	6.1	0.350
89 × 89 × 19	19.5	88.9	88.9	9.5	9.9	0.476
76 × 76 × 13	12.8	76.2	76.2	5.1	8.4	0.411

Parallel Flange Channels

Designation	Mass (kg/m)	Depth of section (mm)	Width of section (mm)	Thickness Web (mm)	Flange (mm)	Surface area (m²/m)
430 × 100 × 64	64.4	430	100	11.0	19.0	1.23
380 × 100 × 54	54.0	380	100	9.5	17.5	1.13
300 × 100 × 46	45.5	300	100	9.0	16.5	0.969
300 × 90 × 41	41.4	300	90	9.0	15.5	0.932
260 × 90 × 35	34.8	260	90	8.0	14.0	0.854
260 × 75 × 28	27.6	260	75	7.0	12.0	0.79
230 × 90 × 32	32.2	230	90	7.5	14.0	0.795
230 × 75 × 26	25.7	230	75	6.5	12.5	0.737
200 × 90 × 30	29.7	200	90	7.0	14.0	0.736
200 × 75 × 23	23.4	200	75	6.0	12.5	0.678
180 × 90 × 26	26.1	180	90	6.5	12.5	0.697
180 × 75 × 20	20.3	180	75	6.0	10.5	0.638
150 × 90 × 24	23.9	150	90	6.5	12.0	0.637
150 × 75 × 18	17.9	150	75	5.5	10.0	0.579
125 × 65 × 15	14.8	125	65	5.5	9.5	0.489
100 × 50 × 10	10.2	100	50	5.0	8.5	0.382

METAL

Equal Angles BS EN 10056-1

Designation	Mass (kg/m)	Surface area (m²/m)
200 × 200 × 24	71.1	0.790
200 × 200 × 20	59.9	0.790
200 × 200 × 18	54.2	0.790
200 × 200 × 16	48.5	0.790
150 × 150 × 18	40.1	0.59
150 × 150 × 15	33.8	0.59
150 × 150 × 12	27.3	0.59
150 × 150 × 10	23.0	0.59
120 × 120 × 15	26.6	0.47
120 × 120 × 12	21.6	0.47
120 × 120 × 10	18.2	0.47
120 × 120 × 8	14.7	0.47
100 × 100 × 15	21.9	0.39
100 × 100 × 12	17.8	0.39
100 × 100 × 10	15.0	0.39
100 × 100 × 8	12.2	0.39
90 × 90 × 12	15.9	0.35
90 × 90 × 10	13.4	0.35
90 × 90 × 8	10.9	0.35
90 × 90 × 7	9.61	0.35
90 × 90 × 6	8.30	0.35

Unequal Angles BS EN 10056-1

Designation	Mass (kg/m)	Surface area (m²/m)
200 × 150 × 18	47.1	0.69
200 × 150 × 15	39.6	0.69
200 × 150 × 12	32.0	0.69
200 × 100 × 15	33.7	0.59
200 × 100 × 12	27.3	0.59
200 × 100 × 10	23.0	0.59
150 × 90 × 15	26.6	0.47
150 × 90 × 12	21.6	0.47
150 × 90 × 10	18.2	0.47
150 × 75 × 15	24.8	0.44
150 × 75 × 12	20.2	0.44
150 × 75 × 10	17.0	0.44
125 × 75 × 12	17.8	0.40
125 × 75 × 10	15.0	0.40
125 × 75 × 8	12.2	0.40
100 × 75 × 12	15.4	0.34
100 × 75 × 10	13.0	0.34
100 × 75 × 8	10.6	0.34
100 × 65 × 10	12.3	0.32
100 × 65 × 8	9.94	0.32
100 × 65 × 7	8.77	0.32

METAL

Structural Tees Split from Universal Beams BS 4: Part 1: 2005

Designation	Mass (kg/m)	Surface area (m²/m)
305 × 305 × 90	89.5	1.22
305 × 305 × 75	74.6	1.22
254 × 343 × 63	62.6	1.19
229 × 305 × 70	69.9	1.07
229 × 305 × 63	62.5	1.07
229 × 305 × 57	56.5	1.07
229 × 305 × 51	50.6	1.07
210 × 267 × 61	61.0	0.95
210 × 267 × 55	54.5	0.95
210 × 267 × 51	50.5	0.95
210 × 267 × 46	46.1	0.95
210 × 267 × 41	41.1	0.95
191 × 229 × 49	49.2	0.84
191 × 229 × 45	44.6	0.84
191 × 229 × 41	41.0	0.84
191 × 229 × 37	37.1	0.84
191 × 229 × 34	33.6	0.84
152 × 229 × 41	41.0	0.76
152 × 229 × 37	37.1	0.76
152 × 229 × 34	33.6	0.76
152 × 229 × 30	29.9	0.76
152 × 229 × 26	26.2	0.76

Universal Bearing Piles BS 4: Part 1: 2005

Designation	Mass (kg/m)	Depth of Section (mm)	Width of section (mm)	Thickness Web (mm)	Thickness Flange (mm)
356 × 368 × 174	173.9	361.4	378.5	20.3	20.4
356 × 368 × 152	152.0	356.4	376.0	17.8	17.9
356 × 368 × 133	133.0	352.0	373.8	15.6	15.7
356 × 368 × 109	108.9	346.4	371.0	12.8	12.9
305 × 305 × 223	222.9	337.9	325.7	30.3	30.4
305 × 305 × 186	186.0	328.3	320.9	25.5	25.6
305 × 305 × 149	149.1	318.5	316.0	20.6	20.7
305 × 305 × 126	126.1	312.3	312.9	17.5	17.6
305 × 305 × 110	110.0	307.9	310.7	15.3	15.4
305 × 305 × 95	94.9	303.7	308.7	13.3	13.3
305 × 305 × 88	88.0	301.7	307.8	12.4	12.3
305 × 305 × 79	78.9	299.3	306.4	11.0	11.1
254 × 254 × 85	85.1	254.3	260.4	14.4	14.3
254 × 254 × 71	71.0	249.7	258.0	12.0	12.0
254 × 254 × 63	63.0	247.1	256.6	10.6	10.7
203 × 203 × 54	53.9	204.0	207.7	11.3	11.4
203 × 203 × 45	44.9	200.2	205.9	9.5	9.5

METAL

Hot Formed Square Hollow Sections EN 10210 S275J2H & S355J2H

Size (mm)	Wall thickness (mm)	Mass (kg/m)	Superficial area (m²/m)
40 × 40	2.5	2.89	0.154
	3.0	3.41	0.152
	3.2	3.61	0.152
	3.6	4.01	0.151
	4.0	4.39	0.150
	5.0	5.28	0.147
50 × 50	2.5	3.68	0.194
	3.0	4.35	0.192
	3.2	4.62	0.192
	3.6	5.14	0.191
	4.0	5.64	0.190
	5.0	6.85	0.187
	6.0	7.99	0.185
	6.3	8.31	0.184
60 × 60	3.0	5.29	0.232
	3.2	5.62	0.232
	3.6	6.27	0.231
	4.0	6.90	0.230
	5.0	8.42	0.227
	6.0	9.87	0.225
	6.3	10.30	0.224
	8.0	12.50	0.219
70 × 70	3.0	6.24	0.272
	3.2	6.63	0.272
	3.6	7.40	0.271
	4.0	8.15	0.270
	5.0	9.99	0.267
	6.0	11.80	0.265
	6.3	12.30	0.264
	8.0	15.00	0.259
80 × 80	3.2	7.63	0.312
	3.6	8.53	0.311
	4.0	9.41	0.310
	5.0	11.60	0.307
	6.0	13.60	0.305
	6.3	14.20	0.304
	8.0	17.50	0.299
90 × 90	3.6	9.66	0.351
	4.0	10.70	0.350
	5.0	13.10	0.347
	6.0	15.50	0.345
	6.3	16.20	0.344
	8.0	20.10	0.339
100 × 100	3.6	10.80	0.391
	4.0	11.90	0.390
	5.0	14.70	0.387
	6.0	17.40	0.385
	6.3	18.20	0.384
	8.0	22.60	0.379
	10.0	27.40	0.374
120 × 120	4.0	14.40	0.470
	5.0	17.80	0.467
	6.0	21.20	0.465

Tables and Memoranda

METAL

Size (mm)	Wall thickness (mm)	Mass (kg/m)	Superficial area (m²/m)
	6.3	22.20	0.464
	8.0	27.60	0.459
	10.0	33.70	0.454
	12.0	39.50	0.449
	12.5	40.90	0.448
140 × 140	5.0	21.00	0.547
	6.0	24.90	0.545
	6.3	26.10	0.544
	8.0	32.60	0.539
	10.0	40.00	0.534
	12.0	47.00	0.529
	12.5	48.70	0.528
150 × 150	5.0	22.60	0.587
	6.0	26.80	0.585
	6.3	28.10	0.584
	8.0	35.10	0.579
	10.0	43.10	0.574
	12.0	50.80	0.569
	12.5	52.70	0.568
Hot formed from seamless hollow	16.0	65.2	0.559
160 × 160	5.0	24.10	0.627
	6.0	28.70	0.625
	6.3	30.10	0.624
	8.0	37.60	0.619
	10.0	46.30	0.614
	12.0	54.60	0.609
	12.5	56.60	0.608
	16.0	70.20	0.599
180 × 180	5.0	27.30	0.707
	6.0	32.50	0.705
	6.3	34.00	0.704
	8.0	42.70	0.699
	10.0	52.50	0.694
	12.0	62.10	0.689
	12.5	64.40	0.688
	16.0	80.20	0.679
200 × 200	5.0	30.40	0.787
	6.0	36.20	0.785
	6.3	38.00	0.784
	8.0	47.70	0.779
	10.0	58.80	0.774
	12.0	69.60	0.769
	12.5	72.30	0.768
	16.0	90.30	0.759
250 × 250	5.0	38.30	0.987
	6.0	45.70	0.985
	6.3	47.90	0.984
	8.0	60.30	0.979
	10.0	74.50	0.974
	12.0	88.50	0.969
	12.5	91.90	0.968
	16.0	115.00	0.959

METAL

Size (mm)	Wall thickness (mm)	Mass (kg/m)	Superficial area (m²/m)
300 × 300	6.0	55.10	1.18
	6.3	57.80	1.18
	8.0	72.80	1.18
	10.0	90.20	1.17
	12.0	107.00	1.17
	12.5	112.00	1.17
	16.0	141.00	1.16
350 × 350	8.0	85.40	1.38
	10.0	106.00	1.37
	12.0	126.00	1.37
	12.5	131.00	1.37
	16.0	166.00	1.36
400 × 400	8.0	97.90	1.58
	10.0	122.00	1.57
	12.0	145.00	1.57
	12.5	151.00	1.57
	16.0	191.00	1.56
(Grade S355J2H only)	20.00*	235.00	1.55

Note: * SAW process

METAL

Hot Formed Square Hollow Sections JUMBO RHS: JIS G3136

Size (mm)	Wall thickness (mm)	Mass (kg/m)	Superficial area (m²/m)
350 × 350	19.0	190.00	1.33
	22.0	217.00	1.32
	25.0	242.00	1.31
400 × 400	22.0	251.00	1.52
	25.0	282.00	1.51
450 × 450	12.0	162.00	1.76
	16.0	213.00	1.75
	19.0	250.00	1.73
	22.0	286.00	1.72
	25.0	321.00	1.71
	28.0 *	355.00	1.70
	32.0 *	399.00	1.69
500 × 500	12.0	181.00	1.96
	16.0	238.00	1.95
	19.0	280.00	1.93
	22.0	320.00	1.92
	25.0	360.00	1.91
	28.0 *	399.00	1.90
	32.0 *	450.00	1.89
	36.0 *	498.00	1.88
550 × 550	16.0	263.00	2.15
	19.0	309.00	2.13
	22.0	355.00	2.12
	25.0	399.00	2.11
	28.0 *	443.00	2.10
	32.0 *	500.00	2.09
	36.0 *	555.00	2.08
	40.0 *	608.00	2.06
600 × 600	25.0 *	439.00	2.31
	28.0 *	487.00	2.30
	32.0 *	550.00	2.29
	36.0 *	611.00	2.28
	40.0 *	671.00	2.26
700 × 700	25.0 *	517.00	2.71
	28.0 *	575.00	2.70
	32.0 *	651.00	2.69
	36.0 *	724.00	2.68
	40.0 *	797.00	2.68

Note: * SAW process

METAL

Hot Formed Rectangular Hollow Sections: EN10210 S275J2h & S355J2H

Size (mm)	Wall thickness (mm)	Mass (kg/m)	Superficial area (m²/m)
50 × 30	2.5	2.89	0.154
	3.0	3.41	0.152
	3.2	3.61	0.152
	3.6	4.01	0.151
	4.0	4.39	0.150
	5.0	5.28	0.147
60 × 40	2.5	3.68	0.194
	3.0	4.35	0.192
	3.2	4.62	0.192
	3.6	5.14	0.191
	4.0	5.64	0.190
	5.0	6.85	0.187
	6.0	7.99	0.185
	6.3	8.31	0.184
80 × 40	3.0	5.29	0.232
	3.2	5.62	0.232
	3.6	6.27	0.231
	4.0	6.90	0.230
	5.0	8.42	0.227
	6.0	9.87	0.225
	6.3	10.30	0.224
	8.0	12.50	0.219
76.2 × 50.8	3.0	5.62	0.246
	3.2	5.97	0.246
	3.6	6.66	0.245
	4.0	7.34	0.244
	5.0	8.97	0.241
	6.0	10.50	0.239
	6.3	11.00	0.238
	8.0	13.40	0.233
90 × 50	3.0	6.24	0.272
	3.2	6.63	0.272
	3.6	7.40	0.271
	4.0	8.15	0.270
	5.0	9.99	0.267
	6.0	11.80	0.265
	6.3	12.30	0.264
	8.0	15.00	0.259
100 × 50	3.0	6.71	0.292
	3.2	7.13	0.292
	3.6	7.96	0.291
	4.0	8.78	0.290
	5.0	10.80	0.287
	6.0	12.70	0.285
	6.3	13.30	0.284
	8.0	16.30	0.279

METAL

Size (mm)	Wall thickness (mm)	Mass (kg/m)	Superficial area (m²/m)
100 × 60	3.0	7.18	0.312
	3.2	7.63	0.312
	3.6	8.53	0.311
	4.0	9.41	0.310
	5.0	11.60	0.307
	6.0	13.60	0.305
	6.3	14.20	0.304
	8.0	17.50	0.299
120 × 60	3.6	9.70	0.351
	4.0	10.70	0.350
	5.0	13.10	0.347
	6.0	15.50	0.345
	6.3	16.20	0.344
	8.0	20.10	0.339
120 × 80	3.6	10.80	0.391
	4.0	11.90	0.390
	5.0	14.70	0.387
	6.0	17.40	0.385
	6.3	18.20	0.384
	8.0	22.60	0.379
	10.0	27.40	0.374
150 × 100	4.0	15.10	0.490
	5.0	18.60	0.487
	6.0	22.10	0.485
	6.3	23.10	0.484
	8.0	28.90	0.479
	10.0	35.30	0.474
	12.0	41.40	0.469
	12.5	42.80	0.468
160 × 80	4.0	14.40	0.470
	5.0	17.80	0.467
	6.0	21.20	0.465
	6.3	22.20	0.464
	8.0	27.60	0.459
	10.0	33.70	0.454
	12.0	39.50	0.449
	12.5	40.90	0.448
200 × 100	5.0	22.60	0.587
	6.0	26.80	0.585
	6.3	28.10	0.584
	8.0	35.10	0.579
	10.0	43.10	0.574
	12.0	50.80	0.569
	12.5	52.70	0.568
	16.0	65.20	0.559
250 × 150	5.0	30.40	0.787
	6.0	36.20	0.785
	6.3	38.00	0.784
	8.0	47.70	0.779
	10.0	58.80	0.774
	12.0	69.60	0.769
	12.5	72.30	0.768
	16.0	90.30	0.759

METAL

Size (mm)	Wall thickness (mm)	Mass (kg/m)	Superficial area (m²/m)
300 × 200	5.0	38.30	0.987
	6.0	45.70	0.985
	6.3	47.90	0.984
	8.0	60.30	0.979
	10.0	74.50	0.974
	12.0	88.50	0.969
	12.5	91.90	0.968
	16.0	115.00	0.959
400 × 200	6.0	55.10	1.18
	6.3	57.80	1.18
	8.0	72.80	1.18
	10.0	90.20	1.17
	12.0	107.00	1.17
	12.5	112.00	1.17
	16.0	141.00	1.16
450 × 250	8.0	85.40	1.38
	10.0	106.00	1.37
	12.0	126.00	1.37
	12.5	131.00	1.37
	16.0	166.00	1.36
500 × 300	8.0	98.00	1.58
	10.0	122.00	1.57
	12.0	145.00	1.57
	12.5	151.00	1.57
	16.0	191.00	1.56
	20.0	235.00	1.55

METAL

Hot Formed Circular Hollow Sections EN 10210 S275J2H & S355J2H

Outside diameter (mm)	Wall thickness (mm)	Mass (kg/m)	Superficial area (m²/m)
21.3	3.2	1.43	0.067
26.9	3.2	1.87	0.085
33.7	3.0	2.27	0.106
	3.2	2.41	0.106
	3.6	2.67	0.106
	4.0	2.93	0.106
42.4	3.0	2.91	0.133
	3.2	3.09	0.133
	3.6	3.44	0.133
	4.0	3.79	0.133
48.3	2.5	2.82	0.152
	3.0	3.35	0.152
	3.2	3.56	0.152
	3.6	3.97	0.152
	4.0	4.37	0.152
	5.0	5.34	0.152
60.3	2.5	3.56	0.189
	3.0	4.24	0.189
	3.2	4.51	0.189
	3.6	5.03	0.189
	4.0	5.55	0.189
	5.0	6.82	0.189
76.1	2.5	4.54	0.239
	3.0	5.41	0.239
	3.2	5.75	0.239
	3.6	6.44	0.239
	4.0	7.11	0.239
	5.0	8.77	0.239
	6.0	10.40	0.239
	6.3	10.80	0.239
88.9	2.5	5.33	0.279
	3.0	6.36	0.279
	3.2	6.76	0.27
	3.6	7.57	0.279
	4.0	8.38	0.279
	5.0	10.30	0.279
	6.0	12.30	0.279
	6.3	12.80	0.279
114.3	3.0	8.23	0.359
	3.2	8.77	0.359
	3.6	9.83	0.359
	4.0	10.09	0.359
	5.0	13.50	0.359
	6.0	16.00	0.359
	6.3	16.80	0.359

METAL

Outside diameter (mm)	Wall thickness (mm)	Mass (kg/m)	Superficial area (m²/m)
139.7	3.2	10.80	0.439
	3.6	12.10	0.439
	4.0	13.40	0.439
	5.0	16.60	0.439
	6.0	19.80	0.439
	6.3	20.70	0.439
	8.0	26.00	0.439
	10.0	32.00	0.439
168.3	3.2	13.00	0.529
	3.6	14.60	0.529
	4.0	16.20	0.529
	5.0	20.10	0.529
	6.0	24.00	0.529
	6.3	25.20	0.529
	8.0	31.60	0.529
	10.0	39.00	0.529
	12.0	46.30	0.529
	12.5	48.00	0.529
193.7	5.0	23.30	0.609
	6.0	27.80	0.609
	6.3	29.10	0.609
	8.0	36.60	0.609
	10.0	45.30	0.609
	12.0	53.80	0.609
	12.5	55.90	0.609
219.1	5.0	26.40	0.688
	6.0	31.50	0.688
	6.3	33.10	0.688
	8.0	41.60	0.688
	10.0	51.60	0.688
	12.0	61.30	0.688
	12.5	63.70	0.688
	16.0	80.10	0.688
244.5	5.0	29.50	0.768
	6.0	35.30	0.768
	6.3	37.00	0.768
	8.0	46.70	0.768
	10.0	57.80	0.768
	12.0	68.80	0.768
	12.5	71.50	0.768
	16.0	90.20	0.768
273.0	5.0	33.00	0.858
	6.0	39.50	0.858
	6.3	41.40	0.858
	8.0	52.30	0.858
	10.0	64.90	0.858
	12.0	77.20	0.858
	12.5	80.30	0.858
	16.0	101.00	0.858

METAL

Outside diameter (mm)	Wall thickness (mm)	Mass (kg/m)	Superficial area (m²/m)
323.9	5.0	39.30	1.02
	6.0	47.00	1.02
	6.3	49.30	1.02
	8.0	62.30	1.02
	10.0	77.40	1.02
	12.0	92.30	1.02
	12.5	96.00	1.02
	16.0	121.00	1.02
355.6	6.3	54.30	1.12
	8.0	68.60	1.12
	10.0	85.30	1.12
	12.0	102.00	1.12
	12.5	106.00	1.12
	16.0	134.00	1.12
406.4	6.3	62.20	1.28
	8.0	79.60	1.28
	10.0	97.80	1.28
	12.0	117.00	1.28
	12.5	121.00	1.28
	16.0	154.00	1.28
457.0	6.3	70.00	1.44
	8.0	88.60	1.44
	10.0	110.00	1.44
	12.0	132.00	1.44
	12.5	137.00	1.44
	16.0	174.00	1.44
508.0	6.3	77.90	1.60
	8.0	98.60	1.60
	10.0	123.00	1.60
	12.0	147.00	1.60
	12.5	153.00	1.60
	16.0	194.00	1.60

METAL

Spacing of Holes in Angles

Nominal leg length (mm)	Spacing of holes						Maximum diameter of bolt or rivet		
	A	B	C	D	E	F	A	B and C	D, E and F
200		75	75	55	55	55		30	20
150		55	55					20	
125		45	60					20	
120									
100	55						24		
90	50						24		
80	45						20		
75	45						20		
70	40						20		
65	35						20		
60	35						16		
50	28						12		
45	25								
40	23								
30	20								
25	15								

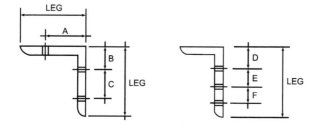

KERBS, PAVING, ETC.

KERBS/EDGINGS/CHANNELS

Precast Concrete Kerbs to BS 7263

Straight kerb units: length from 450 to 915 mm

150 mm high × 125 mm thick		
bullnosed	type BN	
half battered	type HB3	
255 mm high × 125 mm thick		
45° splayed	type SP	
half battered	type HB2	
305 mm high × 150 mm thick		
half battered	type HB1	
Quadrant kerb units		
150 mm high × 305 and 455 mm radius to match	type BN	type QBN
150 mm high × 305 and 455 mm radius to match	type HB2, HB3	type QHB
150 mm high × 305 and 455 mm radius to match	type SP	type QSP
255 mm high × 305 and 455 mm radius to match	type BN	type QBN
255 mm high × 305 and 455 mm radius to match	type HB2, HB3	type QHB
225 mm high × 305 and 455 mm radius to match	type SP	type QSP
Angle kerb units		
305 × 305 × 225 mm high × 125 mm thick		
bullnosed external angle	type XA	
splayed external angle to match type SP	type XA	
bullnosed internal angle	type IA	
splayed internal angle to match type SP	type IA	
Channels		
255 mm wide × 125 mm high flat	type CS1	
150 mm wide × 125 mm high flat type	CS2	
255 mm wide × 125 mm high dished	type CD	

KERBS, PAVING, ETC.

Transition kerb units			
from kerb type SP to HB	left handed	type TL	
	right handed	type TR	
from kerb type BN to HB	left handed	type DL1	
	right handed	type DR1	
from kerb type BN to SP	left handed	type DL2	
	right handed	type DR2	

Number of kerbs required per quarter circle (780 mm kerb lengths)

Radius (m)	Number in quarter circle
12	24
10	20
8	16
6	12
5	10
4	8
3	6
2	4
1	2

Precast Concrete Edgings

Round top type ER	Flat top type EF	Bullnosed top type EBN
150 × 50 mm	150 × 50 mm	150 × 50 mm
200 × 50 mm	200 × 50 mm	200 × 50 mm
250 × 50 mm	250 × 50 mm	250 × 50 mm

KERBS, PAVING, ETC.

BASES

Cement Bound Material for Bases and Subbases

CBM1:	very carefully graded aggregate from 37.5–75 mm, with a 7-day strength of 4.5 N/mm^2
CBM2:	same range of aggregate as CBM1 but with more tolerance in each size of aggregate with a 7-day strength of 7.0 N/mm^2
CBM3:	crushed natural aggregate or blast furnace slag, graded from 37.5–150 mm for 40 mm aggregate, and from 20–75 mm for 20 mm aggregate, with a 7-day strength of 10 N/mm^2
CBM4:	crushed natural aggregate or blast furnace slag, graded from 37.5–150 mm for 40 mm aggregate, and from 20–75 mm for 20 mm aggregate, with a 7-day strength of 15 N/mm^2

INTERLOCKING BRICK/BLOCK ROADS/PAVINGS

Sizes of Precast Concrete Paving Blocks

Type R blocks
200 × 100 × 60 mm
200 × 100 × 65 mm
200 × 100 × 80 mm
200 × 100 × 100 mm

Type S
Any shape within a 295 mm space

Sizes of clay brick pavers
200 × 100 × 50 mm
200 × 100 × 65 mm
210 × 105 × 50 mm
210 × 105 × 65 mm
215 × 102.5 × 50 mm
215 × 102.5 × 65 mm

Type PA: 3 kN
Footpaths and pedestrian areas, private driveways, car parks, light vehicle traffic and over-run

Type PB: 7 kN
Residential roads, lorry parks, factory yards, docks, petrol station forecourts, hardstandings, bus stations

KERBS, PAVING, ETC.

PAVING AND SURFACING

Weights and Sizes of Paving and Surfacing

Description of item	Size	Quantity per tonne
Paving 50 mm thick	900 × 600 mm	15
Paving 50 mm thick	750 × 600 mm	18
Paving 50 mm thick	600 × 600 mm	23
Paving 50 mm thick	450 × 600 mm	30
Paving 38 mm thick	600 × 600 mm	30
Path edging	914 × 50 × 150 mm	60
Kerb (including radius and tapers)	125 × 254 × 914 mm	15
Kerb (including radius and tapers)	125 × 150 × 914 mm	25
Square channel	125 × 254 × 914 mm	15
Dished channel	125 × 254 × 914 mm	15
Quadrants	300 × 300 × 254 mm	19
Quadrants	450 × 450 × 254 mm	12
Quadrants	300 × 300 × 150 mm	30
Internal angles	300 × 300 × 254 mm	30
Fluted pavement channel	255 × 75 × 914 mm	25
Corner stones	300 × 300 mm	80
Corner stones	360 × 360 mm	60
Cable covers	914 × 175 mm	55
Gulley kerbs	220 × 220 × 150 mm	60
Gulley kerbs	220 × 200 × 75 mm	120

KERBS, PAVING, ETC.

Weights and Sizes of Paving and Surfacing

Material	kg/m³	lb/cu yd
Tarmacadam	2306	3891
Macadam (waterbound)	2563	4325
Vermiculite (aggregate)	64–80	108–135
Terracotta	2114	3568
Cork – compressed	388	24
	kg/m²	lb/sq ft
Clay floor tiles, 12.7 mm	27.3	5.6
Pavement lights	122	25
Damp-proof course	5	1
	kg/m² per mm thickness	lb/sq ft per inch thickness
Paving slabs (stone)	2.3	12
Granite setts	2.88	15
Asphalt	2.30	12
Rubber flooring	1.68	9
Polyvinyl chloride	1.94 (avg)	10 (avg)

Coverage (m²) Per Cubic Metre of Materials Used as Subbases or Capping Layers

Consolidated thickness laid in (mm)	Square metre coverage		
	Gravel	Sand	Hardcore
50	15.80	16.50	–
75	10.50	11.00	–
100	7.92	8.20	7.42
125	6.34	6.60	5.90
150	5.28	5.50	4.95
175	–	–	4.23
200	–	–	3.71
225	–	–	3.30
300	–	–	2.47

KERBS, PAVING, ETC.

Approximate Rate of Spreads

Average thickness of course (mm)	Description	Approximate rate of spread			
		Open Textured		Dense, Medium & Fine Textured	
		(kg/m²)	(m²/t)	(kg/m²)	(m²/t)
35	14 mm open textured or dense wearing course	60–75	13–17	70–85	12–14
40	20 mm open textured or dense base course	70–85	12–14	80–100	10–12
45	20 mm open textured or dense base course	80–100	10–12	95–100	9–10
50	20 mm open textured or dense, or 28 mm dense base course	85–110	9–12	110–120	8–9
60	28 mm dense base course, 40 mm open textured of dense base course or 40 mm single course as base course		8–10	130–150	7–8
65	28 mm dense base course, 40 mm open textured or dense base course or 40 mm single course	100–135	7–10	140–160	6–7
75	40 mm single course, 40 mm open textured or dense base course, 40 mm dense roadbase	120–150	7–8	165–185	5–6
100	40 mm dense base course or roadbase	–	–	220–240	4–4.5

KERBS, PAVING, ETC.

Surface Dressing Roads: Coverage (m²) per Tonne of Material

Size in mm	Sand	Granite chips	Gravel	Limestone chips
Sand	168	–	–	–
3	–	148	152	165
6	–	130	133	144
9	–	111	114	123
13	–	85	87	95
19	–	68	71	78

Sizes of Flags

Reference	Nominal size (mm)	Thickness (mm)
A	600 × 450	50 and 63
B	600 × 600	50 and 63
C	600 × 750	50 and 63
D	600 × 900	50 and 63
E	450 × 450	50 and 70 chamfered top surface
F	400 × 400	50 and 65 chamfered top surface
G	300 × 300	50 and 60 chamfered top surface

Sizes of Natural Stone Setts

Width (mm)		Length (mm)		Depth (mm)
100	×	100	×	100
75	×	150 to 250	×	125
75	×	150 to 250	×	150
100	×	150 to 250	×	100
100	×	150 to 250	×	150

SEEDING/TURFING AND PLANTING

Topsoil Quality

Topsoil grade	Properties
Premium	Natural topsoil, high fertility, loamy texture, good soil structure, suitable for intensive cultivation.
General purpose	Natural or manufactured topsoil of lesser quality than Premium, suitable for agriculture or amenity landscape, may need fertilizer or soil structure improvement.
Economy	Selected subsoil, natural mineral deposit such as river silt or greensand. The grade comprises two subgrades; 'Low clay' and 'High clay' which is more liable to compaction in handling. This grade is suitable for low-production agricultural land and amenity woodland or conservation planting areas.

Forms of Trees

Standards:	Shall be clear with substantially straight stems. Grafted and budded trees shall have no more than a slight bend at the union. Standards shall be designated as Half, Extra light, Light, Standard, Selected standard, Heavy, and Extra heavy.
Sizes of Standards	
Heavy standard	12–14 cm girth × 3.50 to 5.00 m high
Extra Heavy standard	14–16 cm girth × 4.25 to 5.00 m high
Extra Heavy standard	16–18 cm girth × 4.25 to 6.00 m high
Extra Heavy standard	18–20 cm girth × 5.00 to 6.00 m high
Semi-mature trees:	Between 6.0 m and 12.0 m tall with a girth of 20 to 75 cm at 1.0 m above ground.
Feathered trees:	Shall have a defined upright central leader, with stem furnished with evenly spread and balanced lateral shoots down to or near the ground.
Whips:	Shall be without significant feather growth as determined by visual inspection.
Multi-stemmed trees:	Shall have two or more main stems at, near, above or below ground.

Seedlings grown from seed and not transplanted shall be specified when ordered for sale as:

1+0	one year old seedling
2+0	two year old seedling
1+1	one year seed bed, one year transplanted = two year old seedling
1+2	one year seed bed, two years transplanted = three year old seedling
2+1	two years seed bed, one year transplanted = three year old seedling
1u1	two years seed bed, undercut after 1 year = two year old seedling
2u2	four years seed bed, undercut after 2 years = four year old seedling

Cuttings

The age of cuttings (plants grown from shoots, stems, or roots of the mother plant) shall be specified when ordered for sale. The height of transplants and undercut seedlings/cuttings (which have been transplanted or undercut at least once) shall be stated in centimetres. The number of growing seasons before and after transplanting or undercutting shall be stated.

0 + 1	one year cutting
0 + 2	two year cutting
0 + 1 + 1	one year cutting bed, one year transplanted = two year old seedling
0 + 1 + 2	one year cutting bed, two years transplanted = three year old seedling

Grass Cutting Capacities in m² per hour

Speed mph	Width of cut in metres												
	0.5	0.7	1.0	1.2	1.5	1.7	2.0	2.0	2.1	2.5	2.8	3.0	3.4
1.0	724	1127	1529	1931	2334	2736	3138	3219	3380	4023	4506	4828	5472
1.5	1086	1690	2293	2897	3500	4104	4707	4828	5069	6035	6759	7242	8208
2.0	1448	2253	3058	3862	4667	5472	6276	6437	6759	8047	9012	9656	10944
2.5	1811	2816	3822	4828	5834	6840	7846	8047	8449	10058	11265	12070	13679
3.0	2173	3380	4587	5794	7001	8208	9415	9656	10139	12070	13518	14484	16415
3.5	2535	3943	5351	6759	8167	9576	10984	11265	11829	14082	15772	16898	19151
4.0	2897	4506	6115	7725	9334	10944	12553	12875	13518	16093	18025	19312	21887
4.5	3259	5069	6880	8690	10501	12311	14122	14484	15208	18105	20278	21726	24623
5.0	3621	5633	7644	9656	11668	13679	15691	16093	16898	20117	22531	24140	27359
5.5	3983	6196	8409	10622	12834	15047	17260	17703	18588	22128	24784	26554	30095
6.0	4345	6759	9173	11587	14001	16415	18829	19312	20278	24140	27037	28968	32831
6.5	4707	7322	9938	12553	15168	17783	20398	20921	21967	26152	29290	31382	35566
7.0	5069	7886	10702	13518	16335	19151	21967	22531	23657	28163	31543	33796	38302

Number of Plants per m² at the following offset spacings

(All plants equidistant horizontally and vertically)

Distance mm	Nr of Plants
100	114.43
200	28.22
250	18.17
300	12.35
400	6.72
500	4.29
600	3.04
700	2.16
750	1.88
900	1.26
1000	1.05
1200	0.68
1500	0.42
2000	0.23

SEEDING/TURFING AND PLANTING

Grass Clippings Wet: Based on 3.5 m³/tonne

Annual kg/100 m²		Average 20 cuts kg/100 m²		m²/tonne	m²/m³
32.0		1.6		61162.1	214067.3

Nr of cuts	22	20	18	16	12	4
kg/cut	1.45	1.60	1.78	2.00	2.67	8.00
Area capacity of 3 tonne vehicle per load						
m²	206250	187500	168750	150000	112500	37500
Load m³	100 m² units/m³ of vehicle space					
1	196.4	178.6	160.7	142.9	107.1	35.7
2	392.9	357.1	321.4	285.7	214.3	71.4
3	589.3	535.7	482.1	428.6	321.4	107.1
4	785.7	714.3	642.9	571.4	428.6	142.9
5	982.1	892.9	803.6	714.3	535.7	178.6

Transportation of Trees

To unload large trees a machine with the necessary lifting strength is required. The weight of the trees must therefore be known in advance. The following table gives a rough overview. The additional columns with root ball dimensions and the number of plants per trailer provide additional information, for example about preparing planting holes and calculating unloading times.

Girth in cm	Rootball diameter in cm	Ball height in cm	Weight in kg	Numbers of trees per trailer
16–18	50–60	40	150	100–120
18–20	60–70	40–50	200	80–100
20–25	60–70	40–50	270	50–70
25–30	80	50–60	350	50
30–35	90–100	60–70	500	12–18
35–40	100–110	60–70	650	10–15
40–45	110–120	60–70	850	8–12
45–50	110–120	60–70	1100	5–7
50–60	130–140	60–70	1600	1–3
60–70	150–160	60–70	2500	1
70–80	180–200	70	4000	1
80–90	200–220	70–80	5500	1
90–100	230–250	80–90	7500	1
100–120	250–270	80–90	9500	1

Data supplied by Lorenz von Ehren GmbH
The information in the table is approximate; deviations depend on soil type, genus and weather

FENCING AND GATES

Types of Preservative

Creosote (tar oil) can be 'factory' applied	by pressure to BS 144: pts 1&2 by immersion to BS 144: pt 1 by hot and cold open tank to BS 144: pts 1&2
Copper/chromium/arsenic (CCA)	by full cell process to BS 4072 pts 1&2
Organic solvent (OS)	by double vacuum (vacvac) to BS 5707 pts 1&3 by immersion to BS 5057 pts 1&3
Pentachlorophenol (PCP)	by heavy oil double vacuum to BS 5705 pts 2&3

Boron diffusion process (treated with disodium octaborate to BWPA Manual 1986)

Note: Boron is used on green timber at source and the timber is supplied dry

Cleft Chestnut Pale Fences

Pales	Pale spacing	Wire lines	
900 mm	75 mm	2	temporary protection
1050 mm	75 or 100 mm	2	light protective fences
1200 mm	75 mm	3	perimeter fences
1350 mm	75 mm	3	perimeter fences
1500 mm	50 mm	3	narrow perimeter fences
1800 mm	50 mm	3	light security fences

Close-Boarded Fences

Close-boarded fences 1.05 to 1.8 m high
Type BCR (recessed) or BCM (morticed) with concrete posts 140 × 115 mm tapered and Type BW with timber posts

Palisade Fences

Wooden palisade fences
Type WPC with concrete posts 140 × 115 mm tapered and Type WPW with timber posts

For both types of fence:
Height of fence 1050 mm: two rails
Height of fence 1200 mm: two rails
Height of fence 1500 mm: three rails
Height of fence 1650 mm: three rails
Height of fence 1800 mm: three rails

FENCING AND GATES

Post and Rail Fences

Wooden post and rail fences
Type MPR 11/3 morticed rails and Type SPR 11/3 nailed rails
Height to top of rail 1100 mm
Rails: three rails 87 mm, 38 mm

Type MPR 11/4 morticed rails and Type SPR 11/4 nailed rails
Height to top of rail 1100 mm
Rails: four rails 87 mm, 38 mm

Type MPR 13/4 morticed rails and Type SPR 13/4 nailed rails
Height to top of rail 1300 mm
Rail spacing 250 mm, 250 mm, and 225 mm from top
Rails: four rails 87 mm, 38 mm

Steel Posts

Rolled steel angle iron posts for chain link fencing

Posts	Fence height	Strut	Straining post
1500 × 40 × 40 × 5 mm	900 mm	1500 × 40 × 40 × 5 mm	1500 × 50 × 50 × 6 mm
1800 × 40 × 40 × 5 mm	1200 mm	1800 × 40 × 40 × 5 mm	1800 × 50 × 50 × 6 mm
2000 × 45 × 45 × 5 mm	1400 mm	2000 × 45 × 45 × 5 mm	2000 × 60 × 60 × 6 mm
2600 × 45 × 45 × 5 mm	1800 mm	2600 × 45 × 45 × 5 mm	2600 × 60 × 60 × 6 mm
3000 × 50 × 50 × 6 mm with arms	1800 mm	2600 × 45 × 45 × 5 mm	3000 × 60 × 60 × 6 mm

Concrete Posts

Concrete posts for chain link fencing

Posts and straining posts	Fence height	Strut
1570 mm 100 × 100 mm	900 mm	1500 mm × 75 × 75 mm
1870 mm 125 × 125 mm	1200 mm	1830 mm × 100 × 75 mm
2070 mm 125 × 125 mm	1400 mm	1980 mm × 100 × 75 mm
2620 mm 125 × 125 mm	1800 mm	2590 mm × 100 × 85 mm
3040 mm 125 × 125 mm	1800 mm	2590 mm × 100 × 85 mm (with arms)

FENCING AND GATES

Rolled Steel Angle Posts

Rolled steel angle posts for rectangular wire mesh (field) fencing

Posts	Fence height	Strut	Straining post
1200 × 40 × 40 × 5 mm	600 mm	1200 × 75 × 75 mm	1350 × 100 × 100 mm
1400 × 40 × 40 × 5 mm	800 mm	1400 × 75 × 75 mm	1550 × 100 × 100 mm
1500 × 40 × 40 × 5 mm	900 mm	1500 × 75 × 75 mm	1650 × 100 × 100 mm
1600 × 40 × 40 × 5 mm	1000 mm	1600 × 75 × 75 mm	1750 × 100 × 100 mm
1750 × 40 × 40 × 5 mm	1150 mm	1750 × 75 × 100 mm	1900 × 125 × 125 mm

Concrete Posts

Concrete posts for rectangular wire mesh (field) fencing

Posts	Fence height	Strut	Straining post
1270 × 100 × 100 mm	600 mm	1200 × 75 × 75 mm	1420 × 100 × 100 mm
1470 × 100 × 100 mm	800 mm	1350 × 75 × 75 mm	1620 × 100 × 100 mm
1570 × 100 × 100 mm	900 mm	1500 × 75 × 75 mm	1720 × 100 × 100 mm
1670 × 100 × 100 mm	600 mm	1650 × 75 × 75 mm	1820 × 100 × 100 mm
1820 × 125 × 125 mm	1150 mm	1830 × 75 × 100 mm	1970 × 125 × 125 mm

Cleft Chestnut Pale Fences

Timber Posts

Timber posts for wire mesh and hexagonal wire netting fences

Round timber for general fences

Posts	Fence height	Strut	Straining post
1300 × 65 mm dia.	600 mm	1200 × 80 mm dia.	1450 × 100 mm dia.
1500 × 65 mm dia.	800 mm	1400 × 80 mm dia.	1650 × 100 mm dia.
1600 × 65 mm dia.	900 mm	1500 × 80 mm dia.	1750 × 100 mm dia.
1700 × 65 mm dia.	1050 mm	1600 × 80 mm dia.	1850 × 100 mm dia.
1800 × 65 mm dia.	1150 mm	1750 × 80 mm dia.	2000 × 120 mm dia.

Squared timber for general fences

Posts	Fence height	Strut	Straining post
1300 × 75 × 75 mm	600 mm	1200 × 75 × 75 mm	1450 × 100 × 100 mm
1500 × 75 × 75 mm	800 mm	1400 × 75 × 75 mm	1650 × 100 × 100 mm
1600 × 75 × 75 mm	900 mm	1500 × 75 × 75 mm	1750 × 100 × 100 mm
1700 × 75 × 75 mm	1050 mm	1600 × 75 × 75 mm	1850 × 100 × 100 mm
1800 × 75 × 75 mm	1150 mm	1750 × 75 × 75 mm	2000 × 125 × 100 mm

FENCING AND GATES

Steel Fences to BS 1722: Part 9: 1992

	Fence height	Top/bottom rails and flat posts	Vertical bars
Light	1000 mm	40 × 10 mm 450 mm in ground	12 mm dia. at 115 mm cs
	1200 mm	40 × 10 mm 550 mm in ground	12 mm dia. at 115 mm cs
	1400 mm	40 × 10 mm 550 mm in ground	12 mm dia. at 115 mm cs
Light	1000 mm	40 × 10 mm 450 mm in ground	16 mm dia. at 120 mm cs
	1200 mm	40 × 10 mm 550 mm in ground	16 mm dia. at 120 mm cs
	1400 mm	40 × 10 mm 550 mm in ground	16 mm dia. at 120 mm cs
Medium	1200 mm	50 × 10 mm 550 mm in ground	20 mm dia. at 125 mm cs
	1400 mm	50 × 10 mm 550 mm in ground	20 mm dia. at 125 mm cs
	1600 mm	50 × 10 mm 600 mm in ground	22 mm dia. at 145 mm cs
	1800 mm	50 × 10 mm 600 mm in ground	22 mm dia. at 145 mm cs
Heavy	1600 mm	50 × 10 mm 600 mm in ground	22 mm dia. at 145 mm cs
	1800 mm	50 × 10 mm 600 mm in ground	22 mm dia. at 145 mm cs
	2000 mm	50 × 10 mm 600 mm in ground	22 mm dia. at 145 mm cs
	2200 mm	50 × 10 mm 600 mm in ground	22 mm dia. at 145 mm cs

Notes: Mild steel fences: round or square verticals; flat standards and horizontals. Tops of vertical bars may be bow-top, blunt, or pointed. Round or square bar railings

Timber Field Gates to BS 3470: 1975

Gates made to this standard are designed to open one way only
All timber gates are 1100 mm high
Width over stiles 2400, 2700, 3000, 3300, 3600, and 4200 mm
Gates over 4200 mm should be made in two leaves

Steel Field Gates to BS 3470: 1975

All steel gates are 1100 mm high
Heavy duty: width over stiles 2400, 3000, 3600 and 4500 mm
Light duty: width over stiles 2400, 3000, and 3600 mm

FENCING AND GATES

Domestic Front Entrance Gates to BS 4092: Part 1: 1966

Metal gates:	Single gates are 900 mm high minimum, 900 mm, 1000 mm and 1100 mm wide

Domestic Front Entrance Gates to BS 4092: Part 2: 1966

Wooden gates:	All rails shall be tenoned into the stiles
	Single gates are 840 mm high minimum, 801 mm and 1020 mm wide
	Double gates are 840 mm high minimum, 2130, 2340 and 2640 mm wide

Timber Bridle Gates to BS 5709:1979 (Horse or Hunting Gates)

Gates open one way only	
Minimum width between posts	1525 mm
Minimum height	1100 mm

Timber Kissing Gates to BS 5709:1979

Minimum width	700 mm
Minimum height	1000 mm
Minimum distance between shutting posts	600 mm
Minimum clearance at mid-point	600 mm

Metal Kissing Gates to BS 5709:1979

Sizes are the same as those for timber kissing gates	
Maximum gaps between rails 120 mm	

Categories of Pedestrian Guard Rail to BS 3049:1976

Class A for normal use
Class B where vandalism is expected
Class C where crowd pressure is likely

DRAINAGE

Width Required for Trenches for Various Diameters of Pipes

Pipe diameter (mm)	Trench n.e. 1.50 m deep	Trench over 1.50 m deep
n.e. 100 mm	450 mm	600 mm
100–150 mm	500 mm	650 mm
150–225 mm	600 mm	750 mm
225–300 mm	650 mm	800 mm
300–400 mm	750 mm	900 mm
400–450 mm	900 mm	1050 mm
450–600 mm	1100 mm	1300 mm

Weights and Dimensions – Vitrified Clay Pipes

Product	Nominal diameter (mm)	Effective length (mm)	BS 65 limits of tolerance min (mm)	max (mm)	Crushing strength (kN/m)	Weight (kg/pipe)	(kg/m)
Supersleve	100	1600	96	105	35.00	14.71	9.19
	150	1750	146	158	35.00	29.24	16.71
Hepsleve	225	1850	221	236	28.00	84.03	45.42
	300	2500	295	313	34.00	193.05	77.22
	150	1500	146	158	22.00	37.04	24.69
Hepseal	225	1750	221	236	28.00	85.47	48.84
	300	2500	295	313	34.00	204.08	81.63
	400	2500	394	414	44.00	357.14	142.86
	450	2500	444	464	44.00	454.55	181.63
	500	2500	494	514	48.00	555.56	222.22
	600	2500	591	615	57.00	796.23	307.69
	700	3000	689	719	67.00	1111.11	370.45
	800	3000	788	822	72.00	1351.35	450.45
Hepline	100	1600	95	107	22.00	14.71	9.19
	150	1750	145	160	22.00	29.24	16.71
	225	1850	219	239	28.00	84.03	45.42
	300	1850	292	317	34.00	142.86	77.22
Hepduct (conduit)	90	1500	–	–	28.00	12.05	8.03
	100	1600	–	–	28.00	14.71	9.19
	125	1750	–	–	28.00	20.73	11.84
	150	1750	–	–	28.00	29.24	16.71
	225	1850	–	–	28.00	84.03	45.42
	300	1850	–	–	34.00	142.86	77.22

DRAINAGE

Weights and Dimensions – Vitrified Clay Pipes

Nominal internal diameter (mm)	Nominal wall thickness (mm)	Approximate weight (kg/m)
150	25	45
225	29	71
300	32	122
375	35	162
450	38	191
600	48	317
750	54	454
900	60	616
1200	76	912
1500	89	1458
1800	102	1884
2100	127	2619

Wall thickness, weights and pipe lengths vary, depending on type of pipe required

The particulars shown above represent a selection of available diameters and are applicable to strength class 1 pipes with flexible rubber ring joints

Tubes with Ogee joints are also available

DRAINAGE

Weights and Dimensions – PVC-u Pipes

	Nominal size	Mean outside diameter (mm)		Wall thickness	Weight
		min	max	(mm)	(kg/m)
Standard pipes	82.4	82.4	82.7	3.2	1.2
	110.0	110.0	110.4	3.2	1.6
	160.0	160.0	160.6	4.1	3.0
	200.0	200.0	200.6	4.9	4.6
	250.0	250.0	250.7	6.1	7.2
Perforated pipes heavy grade	As above	As above	As above	As above	As above
thin wall	82.4	82.4	82.7	1.7	–
	110.0	110.0	110.4	2.2	–
	160.0	160.0	160.6	3.2	–

Width of Trenches Required for Various Diameters of Pipes

Pipe diameter (mm)	Trench n.e. 1.5 m deep (mm)	Trench over 1.5 m deep (mm)
n.e. 100	450	600
100–150	500	650
150–225	600	750
225–300	650	800
300–400	750	900
400–450	900	1050
450–600	1100	1300

DRAINAGE

DRAINAGE BELOW GROUND AND LAND DRAINAGE

Flow of Water Which Can Be Carried by Various Sizes of Pipe

Clay or concrete pipes

	Gradient of pipeline							
	1:10	1:20	1:30	1:40	1:50	1:60	1:80	1:100
Pipe size	Flow in litres per second							
DN 100 15.0	8.5	6.8	5.8	5.2	4.7	4.0	3.5	
DN 150 28.0	19.0	16.0	14.0	12.0	11.0	9.1	8.0	
DN 225 140.0	95.0	76.0	66.0	58.0	53.0	46.0	40.0	

Plastic pipes

	Gradient of pipeline							
	1:10	1:20	1:30	1:40	1:50	1:60	1:80	1:100
Pipe size	Flow in litres per second							
82.4 mm i/dia.	12.0	8.5	6.8	5.8	5.2	4.7	4.0	3.5
110 mm i/dia.	28.0	19.0	16.0	14.0	12.0	11.0	9.1	8.0
160 mm i/dia.	76.0	53.0	43.0	37.0	33.0	29.0	25.0	22.0
200 mm i/dia.	140.0	95.0	76.0	66.0	58.0	53.0	46.0	40.0

Vitrified (Perforated) Clay Pipes and Fittings to BS En 295-5 1994

Length not specified		
75 mm bore	250 mm bore	600 mm bore
100	300	700
125	350	800
150	400	1000
200	450	1200
225	500	

Precast Concrete Pipes: Prestressed Non-pressure Pipes and Fittings: Flexible Joints to BS 5911: Pt. 103: 1994

Rationalized metric nominal sizes: 450, 500	
Length:	500–1000 by 100 increments
	1000–2200 by 200 increments
	2200–2800 by 300 increments
Angles: length:	450–600 angles 45, 22.5,11.25°
	600 or more angles 22.5, 11.25°

DRAINAGE

Precast Concrete Pipes: Unreinforced and Circular Manholes and Soakaways to BS 5911: Pt. 200: 1994

Nominal sizes:	
Shafts:	675, 900 mm
Chambers:	900, 1050, 1200, 1350, 1500, 1800, 2100, 2400, 2700, 3000 mm
Large chambers:	To have either tapered reducing rings or a flat reducing slab in order to accept the standard cover
Ring depths:	1. 300–1200 mm by 300 mm increments except for bottom slab and rings below cover slab, these are by 150 mm increments
	2. 250–1000 mm by 250 mm increments except for bottom slab and rings below cover slab, these are by 125 mm increments
Access hole:	750 × 750 mm for DN 1050 chamber 1200 × 675 mm for DN 1350 chamber

Calculation of Soakaway Depth

The following formula determines the depth of concrete ring soakaway that would be required for draining given amounts of water.

$$h = \frac{4ar}{3\pi D^2}$$

h = depth of the chamber below the invert pipe
a = the area to be drained
r = the hourly rate of rainfall (50 mm per hour)
π = pi
D = internal diameter of the soakaway

This table shows the depth of chambers in each ring size which would be required to contain the volume of water specified. These allow a recommended storage capacity of ⅓ (one third of the hourly rainfall figure).

Table Showing Required Depth of Concrete Ring Chambers in Metres

Area m²	50	100	150	200	300	400	500
Ring size							
0.9	1.31	2.62	3.93	5.24	7.86	10.48	13.10
1.1	0.96	1.92	2.89	3.85	5.77	7.70	9.62
1.2	0.74	1.47	2.21	2.95	4.42	5.89	7.37
1.4	0.58	1.16	1.75	2.33	3.49	4.66	5.82
1.5	0.47	0.94	1.41	1.89	2.83	3.77	4.72
1.8	0.33	0.65	0.98	1.31	1.96	2.62	3.27
2.1	0.24	0.48	0.72	0.96	1.44	1.92	2.41
2.4	0.18	0.37	0.55	0.74	1.11	1.47	1.84
2.7	0.15	0.29	0.44	0.58	0.87	1.16	1.46
3.0	0.12	0.24	0.35	0.47	0.71	0.94	1.18

DRAINAGE

Precast Concrete Inspection Chambers and Gullies to BS 5911: Part 230: 1994

Nominal sizes:	375 diameter, 750, 900 mm deep
	450 diameter, 750, 900, 1050, 1200 mm deep
Depths:	from the top for trapped or untrapped units:
	centre of outlet 300 mm
	invert (bottom) of the outlet pipe 400 mm
Depth of water seal for trapped gullies:	
	85 mm, rodding eye int. dia. 100 mm
Cover slab:	65 mm min

Bedding Flexible Pipes: PVC-u Or Ductile Iron

Type 1 =	100 mm fill below pipe, 300 mm above pipe: single size material
Type 2 =	100 mm fill below pipe, 300 mm above pipe: single size or graded material
Type 3 =	100 mm fill below pipe, 75 mm above pipe with concrete protective slab over
Type 4 =	100 mm fill below pipe, fill laid level with top of pipe
Type 5 =	200 mm fill below pipe, fill laid level with top of pipe
Concrete =	25 mm sand blinding to bottom of trench, pipe supported on chocks, 100 mm concrete under the pipe, 150 mm concrete over the pipe

DRAINAGE

Bedding Rigid Pipes: Clay or Concrete
(for vitrified clay pipes the manufacturer should be consulted)

Class D:	Pipe laid on natural ground with cut-outs for joints, soil screened to remove stones over 40 mm and returned over pipe to 150 m min depth. Suitable for firm ground with trenches trimmed by hand.
Class N:	Pipe laid on 50 mm granular material of graded aggregate to Table 4 of BS 882, or 10 mm aggregate to Table 6 of BS 882, or as dug light soil (not clay) screened to remove stones over 10 mm. Suitable for machine dug trenches.
Class B:	As Class N, but with granular bedding extending half way up the pipe diameter.
Class F:	Pipe laid on 100 mm granular fill to BS 882 below pipe, minimum 150 mm granular fill above pipe: single size material. Suitable for machine dug trenches.
Class A:	Concrete 100 mm thick under the pipe extending half way up the pipe, backfilled with the appropriate class of fill. Used where there is only a very shallow fall to the drain. Class A bedding allows the pipes to be laid to an exact gradient.
Concrete surround:	25 mm sand blinding to bottom of trench, pipe supported on chocks, 100 mm concrete under the pipe, 150 mm concrete over the pipe. It is preferable to bed pipes under slabs or wall in granular material.

PIPED SUPPLY SYSTEMS

Identification of Service Tubes From Utility to Dwellings

Utility	Colour	Size	Depth
British Telecom	grey	54 mm od	450 mm
Electricity	black	38 mm od	450 mm
Gas	yellow	42 mm od rigid 60 mm od convoluted	450 mm
Water	may be blue	(normally untubed)	750 mm

ELECTRICAL SUPPLY/POWER/LIGHTING SYSTEMS

Electrical Insulation Class En 60.598 BS 4533

Class 1:   luminaires comply with class 1 (I) earthed electrical requirements
Class 2:   luminaires comply with class 2 (II) double insulated electrical requirements
Class 3:   luminaires comply with class 3 (III) electrical requirements

Protection to Light Fittings

BS EN 60529:1992 Classification for degrees of protection provided by enclosures.
(IP Code – International or ingress Protection)

1st characteristic: against ingress of solid foreign objects

The figure	2	indicates that fingers cannot enter
	3	that a 2.5 mm diameter probe cannot enter
	4	that a 1.0 mm diameter probe cannot enter
	5	the fitting is dust proof (no dust around live parts)
	6	the fitting is dust tight (no dust entry)

2nd characteristic: ingress of water with harmful effects

The figure	0	indicates unprotected
	1	vertically dripping water cannot enter
	2	water dripping 15° (tilt) cannot enter
	3	spraying water cannot enter
	4	splashing water cannot enter
	5	jetting water cannot enter
	6	powerful jetting water cannot enter
	7	proof against temporary immersion
	8	proof against continuous immersion

Optional additional codes:		A–D protects against access to hazardous parts
	H	high voltage apparatus
	M	fitting was in motion during water test
	S	fitting was static during water test
	W	protects against weather

Marking code arrangement:	(example) IPX5S = IP (International or Ingress Protection)
	X (denotes omission of first characteristic)
	5 = jetting
	S = static during water test

RAIL TRACKS

	kg/m of track	lb/ft of track
Standard gauge		
Bull-head rails, chairs, transverse timber (softwood) sleepers etc.	245	165
Main lines		
Flat-bottom rails, transverse prestressed concrete sleepers, etc.	418	280
Add for electric third rail	51	35
Add for crushed stone ballast	2600	1750
	kg/m²	**lb/sq ft**
Overall average weight – rails connections, sleepers, ballast, etc.	733	150
	kg/m of track	**lb/ft of track**
Bridge rails, longitudinal timber sleepers, etc.	112	75

RAIL TRACKS

Heavy Rails

British Standard Section No.	Rail height (mm)	Foot width (mm)	Head width (mm)	Min web thickness (mm)	Section weight (kg/m)
Flat Bottom Rails					
60 A	114.30	109.54	57.15	11.11	30.62
70 A	123.82	111.12	60.32	12.30	34.81
75 A	128.59	114.30	61.91	12.70	37.45
80 A	133.35	117.47	63.50	13.10	39.76
90 A	142.88	127.00	66.67	13.89	45.10
95 A	147.64	130.17	69.85	14.68	47.31
100 A	152.40	133.35	69.85	15.08	50.18
110 A	158.75	139.70	69.85	15.87	54.52
113 A	158.75	139.70	69.85	20.00	56.22
50 'O'	100.01	100.01	52.39	10.32	24.82
80 'O'	127.00	127.00	63.50	13.89	39.74
60R	114.30	109.54	57.15	11.11	29.85
75R	128.59	122.24	61.91	13.10	37.09
80R	133.35	127.00	63.50	13.49	39.72
90R	142.88	136.53	66.67	13.89	44.58
95R	147.64	141.29	68.26	14.29	47.21
100R	152.40	146.05	69.85	14.29	49.60
95N	147.64	139.70	69.85	13.89	47.27
Bull Head Rails					
95R BH	145.26	69.85	69.85	19.05	47.07

Light Rails

British Standard Section No.	Rail height (mm)	Foot width (mm)	Head width (mm)	Min web thickness (mm)	Section weight (kg/m)
Flat Bottom Rails					
20M	65.09	55.56	30.96	6.75	9.88
30M	75.41	69.85	38.10	9.13	14.79
35M	80.96	76.20	42.86	9.13	17.39
35R	85.73	82.55	44.45	8.33	17.40
40	88.11	80.57	45.64	12.3	19.89
Bridge Rails					
13	48.00	92	36.00	18.0	13.31
16	54.00	108	44.50	16.0	16.06
20	55.50	127	50.00	20.5	19.86
28	67.00	152	50.00	31.0	28.62
35	76.00	160	58.00	34.5	35.38
50	76.00	165	58.50	–	50.18
Crane Rails					
A65	75.00	175.00	65.00	38.0	43.10
A75	85.00	200.00	75.00	45.0	56.20
A100	95.00	200.00	100.00	60.0	74.30
A120	105.00	220.00	120.00	72.0	100.00
175CR	152.40	152.40	107.95	38.1	86.92

RAIL TRACKS

Fish Plates

British Standard Section No.	Overall plate length		Hole diameter	Finished weight per pair	
	4 Hole (mm)	6 Hole (mm)	(mm)	4 Hole (kg/pair)	6 Hole (kg/pair)
For British Standard Heavy Rails: Flat Bottom Rails					
60 A	406.40	609.60	20.64	9.87	14.76
70 A	406.40	609.60	22.22	11.15	16.65
75 A	406.40	–	23.81	11.82	17.73
80 A	406.40	609.60	23.81	13.15	19.72
90 A	457.20	685.80	25.40	17.49	26.23
100 A	508.00	–	pear	25.02	–
110 A (shallow)	507.00	–	27.00	30.11	54.64
113 A (heavy)	507.00	–	27.00	30.11	54.64
50 'O' (shallow)	406.40	–	–	6.68	10.14
80 'O' (shallow)	495.30	–	23.81	14.72	22.69
60R (shallow)	406.40	609.60	20.64	8.76	13.13
60R (angled)	406.40	609.60	20.64	11.27	16.90
75R (shallow)	406.40	–	23.81	10.94	16.42
75R (angled)	406.40	–	23.81	13.67	–
80R (shallow)	406.40	609.60	23.81	11.93	17.89
80R (angled)	406.40	609.60	23.81	14.90	22.33
For British Standard Heavy Rails: Bull head rails					
95R BH (shallow)	–	457.20	27.00	14.59	14.61
For British Standard Light Rails: Flat Bottom Rails					
30M	355.6	–	–	–	2.72
35M	355.6	–	–	–	2.83
40	355.6	–	–	3.76	–

FRACTIONS, DECIMALS AND MILLIMETRE EQUIVALENTS

Fractions	Decimals	(mm)	Fractions	Decimals	(mm)
1/64	0.015625	0.396875	33/64	0.515625	13.096875
1/32	0.03125	0.79375	17/32	0.53125	13.49375
3/64	0.046875	1.190625	35/64	0.546875	13.890625
1/16	0.0625	1.5875	9/16	0.5625	14.2875
5/64	0.078125	1.984375	37/64	0.578125	14.684375
3/32	0.09375	2.38125	19/32	0.59375	15.08125
7/64	0.109375	2.778125	39/64	0.609375	15.478125
1/8	0.125	3.175	5/8	0.625	15.875
9/64	0.140625	3.571875	41/64	0.640625	16.271875
5/32	0.15625	3.96875	21/32	0.65625	16.66875
11/64	0.171875	4.365625	43/64	0.671875	17.065625
3/16	0.1875	4.7625	11/16	0.6875	17.4625
13/64	0.203125	5.159375	45/64	0.703125	17.859375
7/32	0.21875	5.55625	23/32	0.71875	18.25625
15/64	0.234375	5.953125	47/64	0.734375	18.653125
1/4	0.25	6.35	3/4	0.75	19.05
17/64	0.265625	6.746875	49/64	0.765625	19.446875
9/32	0.28125	7.14375	25/32	0.78125	19.84375
19/64	0.296875	7.540625	51/64	0.796875	20.240625
5/16	0.3125	7.9375	13/16	0.8125	20.6375
21/64	0.328125	8.334375	53/64	0.828125	21.034375
11/32	0.34375	8.73125	27/32	0.84375	21.43125
23/64	0.359375	9.128125	55/64	0.859375	21.828125
3/8	0.375	9.525	7/8	0.875	22.225
25/64	0.390625	9.921875	57/64	0.890625	22.621875
13/32	0.40625	10.31875	29/32	0.90625	23.01875
27/64	0.421875	10.71563	59/64	0.921875	23.415625
7/16	0.4375	11.1125	15/16	0.9375	23.8125
29/64	0.453125	11.50938	61/64	0.953125	24.209375
15/32	0.46875	11.90625	31/32	0.96875	24.60625
31/64	0.484375	12.30313	63/64	0.984375	25.003125
1/2	0.5	12.7	1.0	1	25.4

IMPERIAL STANDARD WIRE GAUGE (SWG)

SWG No.	Diameter (inches)	Diameter (mm)	SWG No.	Diameter (inches)	Diameter (mm)
7/0	0.5	12.7	23	0.024	0.61
6/0	0.464	11.79	24	0.022	0.559
5/0	0.432	10.97	25	0.02	0.508
4/0	0.4	10.16	26	0.018	0.457
3/0	0.372	9.45	27	0.0164	0.417
2/0	0.348	8.84	28	0.0148	0.376
1/0	0.324	8.23	29	0.0136	0.345
1	0.3	7.62	30	0.0124	0.315
2	0.276	7.01	31	0.0116	0.295
3	0.252	6.4	32	0.0108	0.274
4	0.232	5.89	33	0.01	0.254
5	0.212	5.38	34	0.009	0.234
6	0.192	4.88	35	0.008	0.213
7	0.176	4.47	36	0.008	0.193
8	0.16	4.06	37	0.007	0.173
9	0.144	3.66	38	0.006	0.152
10	0.128	3.25	39	0.005	0.132
11	0.116	2.95	40	0.005	0.122
12	0.104	2.64	41	0.004	0.112
13	0.092	2.34	42	0.004	0.102
14	0.08	2.03	43	0.004	0.091
15	0.072	1.83	44	0.003	0.081
16	0.064	1.63	45	0.003	0.071
17	0.056	1.42	46	0.002	0.061
18	0.048	1.22	47	0.002	0.051
19	0.04	1.016	48	0.002	0.041
20	0.036	0.914	49	0.001	0.031
21	0.032	0.813	50	0.001	0.025
22	0.028	0.711			

PIPES, WATER, STORAGE, INSULATION

WATER PRESSURE DUE TO HEIGHT

Imperial

Head (Feet)	Pressure (lb/in^2)		Head (Feet)	Pressure (lb/in^2)
1	0.43		70	30.35
5	2.17		75	32.51
10	4.34		80	34.68
15	6.5		85	36.85
20	8.67		90	39.02
25	10.84		95	41.18
30	13.01		100	43.35
35	15.17		105	45.52
40	17.34		110	47.69
45	19.51		120	52.02
50	21.68		130	56.36
55	23.84		140	60.69
60	26.01		150	65.03
65	28.18			

Metric

Head (m)	Pressure (bar)		Head (m)	Pressure (bar)
0.5	0.049		18.0	1.766
1.0	0.098		19.0	1.864
1.5	0.147		20.0	1.962
2.0	0.196		21.0	2.06
3.0	0.294		22.0	2.158
4.0	0.392		23.0	2.256
5.0	0.491		24.0	2.354
6.0	0.589		25.0	2.453
7.0	0.687		26.0	2.551
8.0	0.785		27.0	2.649
9.0	0.883		28.0	2.747
10.0	0.981		29.0	2.845
11.0	1.079		30.0	2.943
12.0	1.177		32.5	3.188
13.0	1.275		35.0	3.434
14.0	1.373		37.5	3.679
15.0	1.472		40.0	3.924
16.0	1.57		42.5	4.169
17.0	1.668		45.0	4.415

1 bar	=	14.5038 lbf/in^2
1 lbf/in^2	=	0.06895 bar
1 metre	=	3.2808 ft or 39.3701 in
1 foot	=	0.3048 metres
1 in wg	=	2.5 mbar (249.1 N/m^2)

PIPES, WATER, STORAGE, INSULATION

Dimensions and Weights of Copper Pipes to BSEN 1057, BSEN 12499, BSEN 14251

Outside diameter (mm)	Internal diameter (mm)	Weight per metre (kg)	Internal diameter (mm)	Weight per metre (kg)	Internal siameter (mm)	Weight per metre (kg)
	Formerly Table X		Formerly Table Y		Formerly Table Z	
6	4.80	0.0911	4.40	0.1170	5.00	0.0774
8	6.80	0.1246	6.40	0.1617	7.00	0.1054
10	8.80	0.1580	8.40	0.2064	9.00	0.1334
12	10.80	0.1914	10.40	0.2511	11.00	0.1612
15	13.60	0.2796	13.00	0.3923	14.00	0.2031
18	16.40	0.3852	16.00	0.4760	16.80	0.2918
22	20.22	0.5308	19.62	0.6974	20.82	0.3589
28	26.22	0.6814	25.62	0.8985	26.82	0.4594
35	32.63	1.1334	32.03	1.4085	33.63	0.6701
42	39.63	1.3675	39.03	1.6996	40.43	0.9216
54	51.63	1.7691	50.03	2.9052	52.23	1.3343
76.1	73.22	3.1287	72.22	4.1437	73.82	2.5131
108	105.12	4.4666	103.12	7.3745	105.72	3.5834
133	130.38	5.5151	–	–	130.38	5.5151
159	155.38	8.7795	–	–	156.38	6.6056

Dimensions of Stainless Steel Pipes to BS 4127

Outside siameter (mm)	Maximum outside siameter (mm)	Minimum outside diameter (mm)	Wall thickness (mm)	Working pressure (bar)
6	6.045	5.940	0.6	330
8	8.045	7.940	0.6	260
10	10.045	9.940	0.6	210
12	12.045	11.940	0.6	170
15	15.045	14.940	0.6	140
18	18.045	17.940	0.7	135
22	22.055	21.950	0.7	110
28	28.055	27.950	0.8	121
35	35.070	34.965	1.0	100
42	42.070	41.965	1.1	91
54	54.090	53.940	1.2	77

PIPES, WATER, STORAGE, INSULATION

Dimensions of Steel Pipes to BS 1387

Nominal Size	Approx. Outside Diameter	Outside diameter				Thickness		
		Light		Medium & Heavy		Light	Medium	Heavy
		Max	Min	Max	Min			
(mm)	(mm)	(mm)	(mm)	(mm)	(mm)	(mm)	(mm)	(mm)
6	10.20	10.10	9.70	10.40	9.80	1.80	2.00	2.65
8	13.50	13.60	13.20	13.90	13.30	1.80	2.35	2.90
10	17.20	17.10	16.70	17.40	16.80	1.80	2.35	2.90
15	21.30	21.40	21.00	21.70	21.10	2.00	2.65	3.25
20	26.90	26.90	26.40	27.20	26.60	2.35	2.65	3.25
25	33.70	33.80	33.20	34.20	33.40	2.65	3.25	4.05
32	42.40	42.50	41.90	42.90	42.10	2.65	3.25	4.05
40	48.30	48.40	47.80	48.80	48.00	2.90	3.25	4.05
50	60.30	60.20	59.60	60.80	59.80	2.90	3.65	4.50
65	76.10	76.00	75.20	76.60	75.40	3.25	3.65	4.50
80	88.90	88.70	87.90	89.50	88.10	3.25	4.05	4.85
100	114.30	113.90	113.00	114.90	113.30	3.65	4.50	5.40
125	139.70	–	–	140.60	138.70	–	4.85	5.40
150	165.1*	–	–	166.10	164.10	–	4.85	5.40

* 165.1 mm (6.5in) outside diameter is not generally recommended except where screwing to BS 21 is necessary
All dimensions are in accordance with ISO R65 except approximate outside diameters which are in accordance with ISO R64
Light quality is equivalent to ISO R65 Light Series II

Approximate Metres Per Tonne of Tubes to BS 1387

Nom. size	BLACK						GALVANIZED					
	Plain/screwed ends			Screwed & socketed			Plain/screwed ends			Screwed & socketed		
	L	M	H	L	M	H	L	M	H	L	M	H
(mm)	(m)	(m)	(m)	(m)	(m)	(m)	(m)	(m)	(m)	(m)	(m)	(m)
6	2765	2461	2030	2743	2443	2018	2604	2333	1948	2584	2317	1937
8	1936	1538	1300	1920	1527	1292	1826	1467	1254	1811	1458	1247
10	1483	1173	979	1471	1165	974	1400	1120	944	1386	1113	939
15	1050	817	688	1040	811	684	996	785	665	987	779	661
20	712	634	529	704	628	525	679	609	512	673	603	508
25	498	410	336	494	407	334	478	396	327	474	394	325
32	388	319	260	384	316	259	373	308	254	369	305	252
40	307	277	226	303	273	223	296	268	220	292	264	217
50	244	196	162	239	194	160	235	191	158	231	188	157
65	172	153	127	169	151	125	167	149	124	163	146	122
80	147	118	99	143	116	98	142	115	97	139	113	96
100	101	82	69	98	81	68	98	81	68	95	79	67
125	–	62	56	–	60	55	–	60	55	–	59	54
150	–	52	47	–	50	46	–	51	46	–	49	45

The figures for 'plain or screwed ends' apply also to tubes to BS 1775 of equivalent size and thickness
Key:
L – Light
M – Medium
H – Heavy

PIPES, WATER, STORAGE, INSULATION

Flange Dimension Chart to BS 4504 & BS 10

Normal Pressure Rating (PN 6) 6 Bar

Nom. size	Flange outside dia.	Table 6/2 Forged Welding Neck	Table 6/3 Plate Slip on	Table 6/4 Forged Bossed Screwed	Table 6/5 Forged Bossed Slip on	Table 6/8 Plate Blank	Raised face		Nr. bolt hole	Size of bolt
							Dia.	T'ness		
15	80	12	12	12	12	12	40	2	4	M10 × 40
20	90	14	14	14	14	14	50	2	4	M10 × 45
25	100	14	14	14	14	14	60	2	4	M10 × 45
32	120	14	16	14	14	14	70	2	4	M12 × 45
40	130	14	16	14	14	14	80	3	4	M12 × 45
50	140	14	16	14	14	14	90	3	4	M12 × 45
65	160	14	16	14	14	14	110	3	4	M12 × 45
80	190	16	18	16	16	16	128	3	4	M16 × 55
100	210	16	18	16	16	16	148	3	4	M16 × 55
125	240	18	20	18	18	18	178	3	8	M16 × 60
150	265	18	20	18	18	18	202	3	8	M16 × 60
200	320	20	22	–	20	20	258	3	8	M16 × 60
250	375	22	24	–	22	22	312	3	12	M16 × 65
300	440	22	24	–	22	22	365	4	12	M20 × 70

Normal Pressure Rating (PN 16) 16 Bar

Nom. size	Flange outside dia.	Table 6/2 Forged Welding Neck	Table 6/3 Plate Slip on	Table 6/4 Forged Bossed Screwed	Table 6/5 Forged Bossed Slip on	Table 6/8 Plate Blank	Raised face		Nr. bolt hole	Size of bolt
							Dia.	T'ness		
15	95	14	14	14	14	14	45	2	4	M12 × 45
20	105	16	16	16	16	16	58	2	4	M12 × 50
25	115	16	16	16	16	16	68	2	4	M12 × 50
32	140	16	16	16	16	16	78	2	4	M16 × 55
40	150	16	16	16	16	16	88	3	4	M16 × 55
50	165	18	18	18	18	18	102	3	4	M16 × 60
65	185	18	18	18	18	18	122	3	4	M16 × 60
80	200	20	20	20	20	20	138	3	8	M16 × 60
100	220	20	20	20	20	20	158	3	8	M16 × 65
125	250	22	22	22	22	22	188	3	8	M16 × 70
150	285	22	22	22	22	22	212	3	8	M20 × 70
200	340	24	24	–	24	24	268	3	12	M20 × 75
250	405	26	26	–	26	26	320	3	12	M24 × 90
300	460	28	28	–	28	28	378	4	12	M24 × 90

PIPES, WATER, STORAGE, INSULATION

Minimum Distances Between Supports/Fixings

Material	BS Nominal pipe size		Pipes – Vertical	Pipes – Horizontal on to low gradients
	(inch)	(mm)	Support distance in metres	Support distance in metres
Copper	0.50	15.00	1.90	1.30
	0.75	22.00	2.50	1.90
	1.00	28.00	2.50	1.90
	1.25	35.00	2.80	2.50
	1.50	42.00	2.80	2.50
	2.00	54.00	3.90	2.50
	2.50	67.00	3.90	2.80
	3.00	76.10	3.90	2.80
	4.00	108.00	3.90	2.80
	5.00	133.00	3.90	2.80
	6.00	159.00	3.90	2.80
muPVC	1.25	32.00	1.20	0.50
	1.50	40.00	1.20	0.50
	2.00	50.00	1.20	0.60
Polypropylene	1.25	32.00	1.20	0.50
	1.50	40.00	1.20	0.50
uPVC	–	82.40	1.20	0.50
	–	110.00	1.80	0.90
	–	160.00	1.80	1.20
Steel	0.50	15.00	2.40	1.80
	0.75	20.00	3.00	2.40
	1.00	25.00	3.00	2.40
	1.25	32.00	3.00	2.40
	1.50	40.00	3.70	2.40
	2.00	50.00	3.70	2.40
	2.50	65.00	4.60	3.00
	3.00	80.40	4.60	3.00
	4.00	100.00	4.60	3.00
	5.00	125.00	5.50	3.70
	6.00	150.00	5.50	4.50
	8.00	200.00	8.50	6.00
	10.00	250.00	9.00	6.50
	12.00	300.00	10.00	7.00
	16.00	400.00	10.00	8.25

PIPES, WATER, STORAGE, INSULATION

Litres of Water Storage Required Per Person Per Building Type

Type of building	Storage (litres)
Houses and flats (up to 4 bedrooms)	120/bedroom
Houses and flats (more than 4 bedrooms)	100/bedroom
Hostels	90/bed
Hotels	200/bed
Nurses homes and medical quarters	120/bed
Offices with canteen	45/person
Offices without canteen	40/person
Restaurants	7/meal
Boarding schools	90/person
Day schools – Primary	15/person
Day schools – Secondary	20/person

Recommended Air Conditioning Design Loads

Building type	Design loading
Computer rooms	500 W/m² of floor area
Restaurants	150 W/m² of floor area
Banks (main area)	100 W/m² of floor area
Supermarkets	25 W/m² of floor area
Large office block (exterior zone)	100 W/m² of floor area
Large office block (interior zone)	80 W/m² of floor area
Small office block (interior zone)	80 W/m² of floor area

PIPES, WATER, STORAGE, INSULATION

Capacity and Dimensions of Galvanized Mild Steel Cisterns – BS 417

Capacity (litres)	BS type (SCM)	Dimensions		
		Length (mm)	Width (mm)	Depth (mm)
18	45	457	305	305
36	70	610	305	371
54	90	610	406	371
68	110	610	432	432
86	135	610	457	482
114	180	686	508	508
159	230	736	559	559
191	270	762	584	610
227	320	914	610	584
264	360	914	660	610
327	450/1	1220	610	610
336	450/2	965	686	686
423	570	965	762	787
491	680	1090	864	736
709	910	1070	889	889

Capacity of Cold Water Polypropylene Storage Cisterns – BS 4213

Capacity (litres)	BS type (PC)	Maximum height (mm)
18	4	310
36	8	380
68	15	430
91	20	510
114	25	530
182	40	610
227	50	660
273	60	660
318	70	660
455	100	760

PIPES, WATER, STORAGE, INSULATION

Minimum Insulation Thickness to Protect Against Freezing for Domestic Cold Water Systems (8 Hour Evaluation Period)

Pipe size (mm)	Insulation thickness (mm)					
	Condition 1			Condition 2		
	λ = 0.020	λ = 0.030	λ = 0.040	λ = 0.020	λ = 0.030	λ = 0.040
Copper pipes						
15	11	20	34	12	23	41
22	6	9	13	6	10	15
28	4	6	9	4	7	10
35	3	5	7	4	5	7
42	3	4	5	8	4	6
54	2	3	4	2	3	4
76	2	2	3	2	2	3
Steel pipes						
15	9	15	24	10	18	29
20	6	9	13	6	10	15
25	4	7	9	5	7	10
32	3	5	6	3	5	7
40	3	4	5	3	4	6
50	2	3	4	2	3	4
65	2	2	3	2	3	3

Condition 1: water temperature 7°C; ambient temperature –6°C; evaluation period 8 h; permitted ice formation 50%; normal installation, i.e. inside the building and inside the envelope of the structural insulation
Condition 2: water temperature 2°C; ambient temperature –6°C; evaluation period 8 h; permitted ice formation 50%; extreme installation, i.e. inside the building but outside the envelope of the structural insulation
λ = thermal conductivity [W/(mK)]

Insulation Thickness for Chilled And Cold Water Supplies to Prevent Condensation

On a Low Emissivity Outer Surface (0.05, i.e. Bright Reinforced Aluminium Foil) with an Ambient Temperature of +25°C and a Relative Humidity of 80%

Steel pipe size (mm)	t = +10			t = +5			t = 0		
	Insulation thickness (mm)			Insulation thickness (mm)			Insulation thickness (mm)		
	λ = 0.030	λ = 0.040	λ = 0.050	λ = 0.030	λ = 0.040	λ = 0.050	λ = 0.030	λ = 0.040	λ = 0.050
15	16	20	25	22	28	34	28	36	43
25	18	24	29	25	32	39	32	41	50
50	22	28	34	30	39	47	38	49	60
100	26	34	41	36	47	57	46	60	73
150	29	38	46	40	52	64	51	67	82
250	33	43	53	46	60	74	59	77	94
Flat surfaces	39	52	65	56	75	93	73	97	122

t = temperature of contents (°C)
λ = thermal conductivity at mean temperature of insulation [W/(mK)]

Insulation Thickness for Non-domestic Heating Installations to Control Heat Loss

Steel pipe size (mm)	t = 75			t = 100			t = 150		
	Insulation thickness (mm)			Insulation thickness (mm)			Insulation thickness (mm)		
	$\lambda = 0.030$	$\lambda = 0.040$	$\lambda = 0.050$	$\lambda = 0.030$	$\lambda = 0.040$	$\lambda = 0.050$	$\lambda = 0.030$	$\lambda = 0.040$	$\lambda = 0.050$
10	18	32	55	20	36	62	23	44	77
15	19	34	56	21	38	64	26	47	80
20	21	36	57	23	40	65	28	50	83
25	23	38	58	26	43	68	31	53	85
32	24	39	59	28	45	69	33	55	87
40	25	40	60	29	47	70	35	57	88
50	27	42	61	31	49	72	37	59	90
65	29	43	62	33	51	74	40	63	92
80	30	44	62	35	52	75	42	65	94
100	31	46	63	37	54	76	45	68	96
150	33	48	64	40	57	77	50	73	100
200	35	49	65	42	59	79	53	76	103
250	36	50	66	43	61	80	55	78	105

t = hot face temperature (°C)
λ = thermal conductivity at mean temperature of insulation [W/(mK)]

Index

Ebook Single-User Licence Agreement

We welcome you as a user of this Spon Price Book ebook and hope that you find it a useful and valuable tool. Please read this document carefully. **This is a legal agreement** between you (hereinafter referred to as the "Licensee") and Taylor and Francis Books Ltd. (the "Publisher"), which defines the terms under which you may use the Product. **By accessing and retrieving the access code on the label inside the front cover of this book you agree to these terms and conditions outlined herein. If you do not agree to these terms you must return the Product to your supplier intact, with the seal on the label unbroken and with the access code not accessed.**

1. **Definition of the Product**
   The product which is the subject of this Agreement, (the "Product") consists of online and offline access to the VitalSource ebook edition of *Spon's Architects' & Builders' Price Book 2022*.

2. **Commencement and licence**
   2.1   This Agreement commences upon the breaking open of the document containing the access code by the Licensee (the "Commencement Date").
   2.2   This is a licence agreement (the "Agreement") for the use of the Product by the Licensee, and not an agreement for sale.
   2.3   The Publisher licenses the Licensee on a non-exclusive and non-transferable basis to use the Product on condition that the Licensee complies with this Agreement. The Licensee acknowledges that it is only permitted to use the Product in accordance with this Agreement.

3. **Multiple use**
   Use of the Product is not provided or allowed for more than one user or for a wide area network or consortium.

4. **Installation and Use**
   4.1   The Licensee may provide access to the Product for individual study in the following manner: The Licensee may install the Product on a secure local area network on a single site for use by one user.
   4.2   The Licensee shall be responsible for installing the Product and for the effectiveness of such installation.
   4.3   Text from the Product may be incorporated in a coursepack. Such use is only permissible with the express permission of the Publisher in writing and requires the payment of the appropriate fee as specified by the Publisher and signature of a separate licence agreement.
   4.4   The Product is a free addition to the book and the Publisher is under no obligation to provide any technical support.

5. **Permitted Activities**
   5.1   The Licensee shall be entitled to use the Product for its own internal purposes;
   5.2   The Licensee acknowledges that its rights to use the Product are strictly set out in this Agreement, and all other uses (whether expressly mentioned in Clause 6 below or not) are prohibited.

6. **Prohibited Activities**
   The following are prohibited without the express permission of the Publisher:
   6.1   The commercial exploitation of any part of the Product.
   6.2   The rental, loan, (free or for money or money's worth) or hire purchase of this product, save with the express consent of the Publisher.
   6.3   Any activity which raises the reasonable prospect of impeding the Publisher's ability or opportunities to market the Product.
   6.4   Any networking, physical or electronic distribution or dissemination of the product save as expressly permitted by this Agreement.
   6.5   Any reverse engineering, decompilation, disassembly or other alteration of the Product save in accordance with applicable national laws.
   6.6   The right to create any derivative product or service from the Product save as expressly provided for in this Agreement.
   6.7   Any alteration, amendment, modification or deletion from the Product, whether for the purposes of error correction or otherwise.

7. General Responsibilities of the License

7.1 The Licensee will take all reasonable steps to ensure that the Product is used in accordance with the terms and conditions of this Agreement.

7.2 The Licensee acknowledges that damages may not be a sufficient remedy for the Publisher in the event of breach of this Agreement by the Licensee, and that an injunction may be appropriate.

7.3 The Licensee undertakes to keep the Product safe and to use its best endeavours to ensure that the product does not fall into the hands of third parties, whether as a result of theft or otherwise.

7.4 Where information of a confidential nature relating to the product of the business affairs of the Publisher comes into the possession of the Licensee pursuant to this Agreement (or otherwise), the Licensee agrees to use such information solely for the purposes of this Agreement, and under no circumstances to disclose any element of the information to any third party save strictly as permitted under this Agreement. For the avoidance of doubt, the Licensee's obligations under this sub-clause 7.4 shall survive the termination of this Agreement.

8. Warrant and Liability

8.1 The Publisher warrants that it has the authority to enter into this agreement and that it has secured all rights and permissions necessary to enable the Licensee to use the Product in accordance with this Agreement.

8.2 The Publisher warrants that the Product as supplied on the Commencement Date shall be free of defects in materials and workmanship, and undertakes to replace any defective Product within 28 days of notice of such defect being received provided such notice is received within 30 days of such supply. As an alternative to replacement, the Publisher agrees fully to refund the Licensee in such circumstances, if the Licensee so requests, provided that the Licensee returns this copy of *Spon's Architects & Builders' Price Book 2022* to the Publisher. The provisions of this sub-clause 8.2 do not apply where the defect results from an accident or from misuse of the product by the Licensee.

8.3 Sub-clause 8.2 sets out the sole and exclusive remedy of the Licensee in relation to defects in the Product.

8.4 The Publisher and the Licensee acknowledge that the Publisher supplies the Product on an "as is" basis. The Publisher gives no warranties:

8.4.1 that the Product satisfies the individual requirements of the Licensee; or

8.4.2 that the Product is otherwise fit for the Licensee's purpose; or

8.4.3 that the Product is compatible with the Licensee's hardware equipment and software operating environment.

8.5 The Publisher hereby disclaims all warranties and conditions, express or implied, which are not stated above.

8.6 Nothing in this Clause 8 limits the Publisher's liability to the Licensee in the event of death or personal injury resulting from the Publisher's negligence.

8.7 The Publisher hereby excludes liability for loss of revenue, reputation, business, profits, or for indirect or consequential losses, irrespective of whether the Publisher was advised by the Licensee of the potential of such losses.

8.8 The Licensee acknowledges the merit of independently verifying the price book data prior to taking any decisions of material significance (commercial or otherwise) based on such data. It is agreed that the Publisher shall not be liable for any losses which result from the Licensee placing reliance on the data under any circumstances.

8.9 Subject to sub-clause 8.6 above, the Publisher's liability under this Agreement shall be limited to the purchase price.

9. Intellectual Property Rights

9.1 Nothing in this Agreement affects the ownership of copyright or other intellectual property rights in the Product.

9.2 The Licensee agrees to display the Publishers' copyright notice in the manner described in the Product.

9.3 The Licensee hereby agrees to abide by copyright and similar notice requirements required by the Publisher, details of which are as follows:

"© 2022 Taylor & Francis. All rights reserved. All materials in *Spon's Architects' & Builders' Price Book 2022* are copyright protected. All rights reserved. No such materials may be used, displayed, modified, adapted, distributed, transmitted, transferred, published or otherwise reproduced in any form or by any means now or hereafter developed other than strictly in accordance with the terms of the licence agreement enclosed with *Spon's Architects' & Builders' Price Book 2022*. However, text and images may be printed and copied for research and private study within the preset program limitations. Please note the copyright notice above, and that any text or images printed or copied must credit the source."

9.4 This Product contains material proprietary to and copyedited by the Publisher and others. Except for the licence granted herein, all rights, title and interest in the Product, in all languages, formats and media throughout the world, including copyrights therein, are and remain the property of the Publisher or other copyright holders identified in the Product.

10. Non-assignment

This Agreement and the licence contained within it may not be assigned to any other person or entity without the written consent of the Publisher.

11. Termination and Consequences of Termination.

11.1 The Publisher shall have the right to terminate this Agreement if:

11.1.1 the Licensee is in material breach of this Agreement and fails to remedy such breach (where capable of remedy) within 14 days of a written notice from the Publisher requiring it to do so; or

11.1.2 the Licensee becomes insolvent, becomes subject to receivership, liquidation or similar external administration; or

11.1.3 the Licensee ceases to operate in business.

11.2 The Licensee shall have the right to terminate this Agreement for any reason upon two month's written notice. The Licensee shall not be entitled to any refund for payments made under this Agreement prior to termination under this sub-clause 11.2.

11.3 Termination by either of the parties is without prejudice to any other rights or remedies under the general law to which they may be entitled, or which survive such termination (including rights of the Publisher under sub-clause 7.4 above).

11.4 Upon termination of this Agreement, or expiry of its terms, the Licensee must destroy all copies and any back up copies of the product or part thereof.

12. General

12.1 *Compliance with export provisions*

The Publisher hereby agrees to comply fully with all relevant export laws and regulations of the United Kingdom to ensure that the Product is not exported, directly or indirectly, in violation of English law.

12.2 *Force majeure*

The parties accept no responsibility for breaches of this Agreement occurring as a result of circumstances beyond their control.

12.3 *No waiver*

Any failure or delay by either party to exercise or enforce any right conferred by this Agreement shall not be deemed to be a waiver of such right.

12.4 *Entire agreement*

This Agreement represents the entire agreement between the Publisher and the Licensee concerning the Product. The terms of this Agreement supersede all prior purchase orders, written terms and conditions, written or verbal representations, advertising or statements relating in any way to the Product.

12.5 *Severability*

If any provision of this Agreement is found to be invalid or unenforceable by a court of law of competent jurisdiction, such a finding shall not affect the other provisions of this Agreement and all provisions of this Agreement unaffected by such a finding shall remain in full force and effect.

12.6 *Variations*

This agreement may only be varied in writing by means of variation signed in writing by both parties.

12.7 *Notices*

All notices to be delivered to: Spon's Price Books, Taylor & Francis Books Ltd., 4 Park Square, Milton Park, Abingdon, Oxfordshire, OX14 4RN, UK.

12.8 *Governing law*

This Agreement is governed by English law and the parties hereby agree that any dispute arising under this Agreement shall be subject to the jurisdiction of the English courts.

If you have any queries about the terms of this licence, please contact:

Spon's Price Books
Taylor & Francis Books Ltd.
4 Park Square, Milton Park, Abingdon, Oxfordshire, OX14 4RN
Tel: +44 (0) 20 7017 6000
www.sponpress.com

Spon Press
an imprint of Taylor & Francis

Ebook options

Your print copy comes with a free ebook on the VitalSource® Bookshelf platform. Further copies of the ebook are available from https://www.crcpress.com/search/results?kw=spon+2022
Pick your price book and select the ebook under the 'Select Format' option.

To buy Spon's ebooks for five or more users of in your organization please contact:

Spon's Price Books
eBooks & Online Sales
Taylor & Francis Books Ltd.
4 Park Square, Milton Park, Oxfordshire, OX14 4RN
Tel: (020) 7017 6000
onlinesales@informa.com